Natural Gas

edited by
Primož Potočnik

Natural Gas

Edited by Primož Potočnik

CBS, Edition 2016

Published by InTech

Janeza Trdine 9, 51000 Rijeka, Croatia

Publishing Process Manager Iva Lipovic

Technical Editor Goran Bajac

Cover Designer Martina Sirotic

Image Copyright Kenneth V. Pilon, 2010. Used under license from Shutterstock.com

Additional hard copies can be obtained from orders@intechopen.com

Natural Gas, Edited by Primož Potočnik
p. cm.
ISBN 978-953-307-112-1

Contents

Preface IX

Chapter 1 **Natural gas** 1
Wan Azelee Wan Abu Bakar and Rusmidah Ali

Chapter 2 **Natural gas: physical properties and combustion features** 39
Le Corre Olivier and Loubar Khaled

Chapter 3 **The importance of natural gas reforming** 71
Laédna Souto Neiva and Lucianna Gama

Chapter 4 **Natural gas odorization** 87
Daniel Tenkrat, Tomas Hlincik and Ondrej Prokes

Chapter 5 **Synthetic Natural Gas (SNG) from coal and biomass: a survey of existing process technologies, open issues and perspectives** 105
Maria Sudiro and Alberto Bertucco

Chapter 6 **Environmental technology assessment of natural gas compared to biogas** 127
Ola Eriksson

Chapter 7 **Natural Gas Hydrates** 147
Geir Ersland and Arne Graue

Chapter 8 **The effect of H_2S on hydrogen and carbon black production from sour natural gas** 163
M. Javadi, M. Moghiman and Seyyed Iman Pishbin

Chapter 9 **Soil gas geochemistry: significance and application in geological prospectings** 183
Nunzia Voltattorni and Salvatore Lombardi

Chapter 10 **Adsorption of methane in porous materials as the basis for the storage of natural gas** 205
Cecilia Solar, Andrés García Blanco, Andrea Vallone and Karim Sapag

Chapter 11 **Industrial application of natural gas** 245
Alejandro Sáez

Chapter 12 **Combined operational planning of natural gas and electric power systems: state of the art** 271
Ricardo Rubio-Barros, Diego Ojeda-Esteybar, Osvaldo Añó and Alberto Vargas

Chapter 13 **Compressed natural gas direct injection (spark plug fuel injector)** 289
Taib Iskandar Mohamad

Chapter 14 **Hydrogen-enriched compressed natural gas as a fuel for engines** 307
Fanhua Ma, Nashay Naeve, Mingyue Wang, Long Jiang, Renzhe Chen and Shuli Zhao

Chapter 15 **Looking for clean energy considering LNG assessment to provide energy security in Brazil and GTL from Bolivia natural gas reserves** 333
Miguel Edgar Morales Udaeta, Jonathas Luiz de Oliveira Bernal, Geraldo Francisco Burani and José Aquiles Baesso Grimoni

Chapter 16 **Practical results of forecasting for the natural gas market** 371
Primož Potočnik and Edvard Govekar

Chapter 17 **Statistical model of segment-specific relationship between natural gas consumption and temperature in daily and hourly resolution** 393
Marek Brabec, Marek Malý, Emil Pelikán and Ondřej Konár

Chapter 18 **Molecular dynamics simulations of volumetric thermophysical properties of natural gases** 417
Santiago Aparicio and Mert Atilhan

Chapter 19 **Static behaviour of natural gas and its flow in pipes** 435
Ohirhian, P. U.

Chapter 20 **Steady State Compressible Fluid Flow in Porous Media** 467
Peter Ohirhian

Chapter 21 **Natural gas properties and flow computation** 501
Ivan Marić and Ivan Ivek

Chapter 22 **Rarefied natural gas transport** 531
Huei Chu Weng

Chapter 23 **Consequence analysis of large-scale liquefied natural gas spills on water** 549
Hideyuki Oka

Chapter 24 **Risk assessment of marine LNG operations** **571**
Tarek Elsayed

Chapter 25 **Reliability measures for liquefied natural gas receiving terminal based on the failure information of emergency shutdown system** **591**
Bi-Min Hsu, Ming-Hung Shu, and Min Tsao

Preface

The contributions in this book present an overview of utting edge research on natural gas which is a vital component of world's supply of energy. Natural gas is combustible mixture of hydrocarbon gases, primarily methane but also heavier gaseous hydrocarbons such as ethane, propane and butane. Unlike other fossil fuels, natural gas is clean burning and emits lower levels of potentially harmful by-products into the air. Therefore, it is considered as one of the cleanest, safest, and most useful of all energy sources applied in variety of residential, commercial and industrial fields.

The book is organised in 25 chapters that are collected into groups related to technology, applications, forecasting, numerical simulations, transport and risk assessment of natural gas.

The introductory chapter 1 provides a soft introduction about the background and history of natural gas, what exactly it is and where it can be found in nature, and presents an overview of applications and technologies related to natural gas. The introduction is extended in the second chapter providing the physical properties and combustion features of natural gas.

The next group of chapters 3-10 is related to various technological aspects of natural gas and describes the importance of natural gas reforming, its odorisation, synthetic natural gas, comparison with biogas, natural gas hydrates, thermal decomposition of sour natural gas, soil-gas geochemistry, and storage of natural gas in porous materials.

Chapter 11 provides an introduction to industrial applications of natural gas and the application topics are further discussed in chapters 12-15. The relation of natural gas to electric power systems is discussed in chapter 12. The next two chapters consider engine applications, namely compressed natural gas direct injection engine and hydrogen-enriched compressed natural gas as a fuel for engines. Chapter 15 discusses clean energy considerations provided by the liquid natural gas.

Two chapters are dedicated to short-term forecasting of natural gas consumption. Daily and hourly forecasting models for natural gas distributors are presented in chapter 16, and statistical daily forecasting models for households and small and medium size commercial customers are discussed in chapter 17.

Chapters 18-20 are concerned with numerical simulations in the field of natural gas and discuss molecular dynamics simulations of volumetric thermophysical properties of natural gases, static behaviour of natural gas and its flow in pipes, and simulations of steady state compressible flow in porous media.

Chapters 21 and 22 are related to transport of natural gas and discuss natural gas properties and methods of precise flow computation, and transportation of rarefied natural gas in pipelines.

The last three chapters 23-25 are concerned with risk estimation in various natural gas operations and discuss analysis of large-scale liquefied natural gas spills on water, risk assessment of marine liquid natural gas operations, and reliability measures for liquefied natural gas receiving terminal.

Editor

Primož Potočnik
University of Ljubljana
Ljubljana, Slovenia

1

Natural gas

Wan Azelee Wan Abu Bakar and Rusmidah Ali
Department of Chemistry, Universiti Teknologi Malaysia, Skudai, Johor, Malaysia

1. Introduction

This chapter contains a description of background of natural gas: what exactly natural gas is?, how it is formed and how it is found in nature; history of natural gas: a brief history and development of modern natural gas; resources: how much abundance, where to find and what is the composition of natural gas; Uses: application and the important of energy source; natural gas versus environment: emission from the combustion of natural gas; natural gas technology: role of technology in the evolution of the natural gas industry; Purification of crude natural gas: various technologies used to convert sour to sweet natural gas; synthesis of artificial natural gas: methanation reaction.

2. Background of Natural Gas

A mixture of gaseous hydrocarbons occurring in reservoirs of porous rock (commonly sand or sandstone) capped by impervious strata. It is often associated with petroleum, with which it has a common origin in the decomposition of organic matter in sedimentary deposits. Natural gas consists largely of methane (CH_4) and ethane (C_2H_6), with also propane (C_3H_8) and butane (C_4H_{10})(separated for bottled gas), some higher alkanes (C_5H_{12} and above) (used for gasoline), nitrogen (N_2) , oxygen (O_2), carbon dioxide (CO_2), hydrogen sulfide (H_2S), and sometimes valuable helium (He). It is used as an industrial and domestic fuel, and also to make carbon-black and chemical synthesis. Natural gas is transported by large pipelines or (as a liquid) in refrigerated tankers. Natural gas is combustible mixture of hydrocarbon gases, and when burned it gives off a great deal of energy. We require energy constantly, to heat our homes, cook our food, and generate our electricity . Unlike other fossil fuels, however, natural gas is clean burning and emits lower levels of potentially harmful byproducts into the air. It is this need for energy that has elevated natural gas to such a level of importance in our society, and in our lives.

Natural Gas is a vital component of the world's supply of energy. It is one of the cleanest, safest, and most useful of all energy sources. Despite its importance, however, there are many misconceptions about natural gas. For instance, the word 'gas' itself has a variety of different uses, and meanings. When we fuel our car, we put 'gas' in it. However, the gasoline that goes into your vehicle, while a fossil fuel itself, is very different from natural gas. The 'gas' in the common barbecue is actually propane, which, while closely associated and commonly found in natural gas, is not really natural gas itself. While commonly grouped in with other fossil fuels and sources of energy, there are many characteristics of natural gas that make it unique. Below is a bit of background information about natural gas, what exactly it is, how it is formed, and how it is found in nature

2.1 History of Natural Gas

Naturally occurring natural gas was discovered and identified in America as early as 1626, when French explorers discovered natives igniting gases that were seeping into and around Lake Erie. The American natural gas industry got its beginnings in this area. In 1859, Colonel Edwin Drake (a former railroad conductor who adopted the title 'Colonel' to impress the townspeople) dug the first well. Drake hit oil and natural gas at 69 feet below the surface of the earth.

Fig. 1. A Reconstruction of 'Colonel' Drake's First Well in Titusville, Pa (Source: API)

Most in the industry characterize this well (Fig.1) as the beginning of the natural gas industry in America. A two-inch diameter pipeline was built, running 5 and ½ miles from the well to the village of Titusville, Pennsylvania. The construction of this pipeline proved that natural gas could be brought safely and relatively easy from its underground source to be used for practical purposes.

In 1821, the first well specifically intended to obtain natural gas was dug in Fredonia, New York, by William Hart. After noticing gas bubbles rising to the surface of a creek, Hart dug a 27 foot well to try and obtain a larger flow of gas to the surface. Hart is regarded by many as the 'father of natural gas' in America. Expanding on Hart's work, the Fredonia Gas Light Company was eventually formed, becoming the first American natural gas company.

In 1885, Robert Bunsen invented what is now known as the Bunsen burner (Fig.2). He managed to create a device that mixed natural gas with air in the right proportions, creating a flame that could be safely used for cooking and heating. The invention of the Bunsen burner opened up new opportunities for the use of natural gas in America, and throughout the world. The invention of temperature-regulating thermostatic devices allowed for better use of the heating potential of natural gas, allowing the temperature of the flame to be adjusted and monitored.

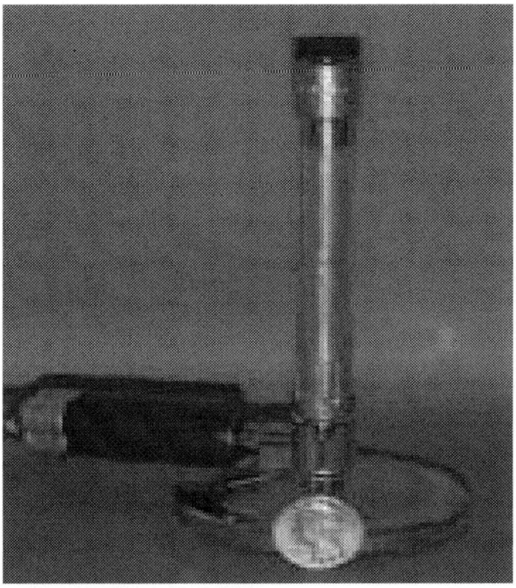

Fig. 2. A Typical Bunsen Burner (Source:DOE)

Without any way to transport it effectively, natural gas discovered pre-world war II was usually just allowed to vent into the atmosphere, or burnt, when found alongside coal and oil, or simply left in the ground when found alone.

One of the first lengthy pipelines was constructed in 1891. This pipeline was 120 miles long, and carried natural gas from wells in central Indiana to the city of Chicago. However, this early pipeline was very rudimentary, and did not transport natural gas efficiently. It wasn't until the 1920's that any significant effort was put into building a pipeline infrastructure. After World War II welding techniques, pipe rolling, and metallurgical advances allowed for the construction of reliable pipelines. This led to a post-war pipeline construction boom lasting well into the 60's, creating thousands of miles of pipeline in America.

Once the transportation of natural gas was possible, new uses for natural gas were discovered. These included using natural gas to heat homes and operate appliances such as water heaters and oven ranges. Industry began to use natural gas in manufacturing and processing plants. Also, natural gas was used to heat boilers used to generate electricity. The transportation infrastructure made natural gas easier to obtain, and as a result expanded its uses.

2.2 How Natural Gas is Formed

Millions of years ago, the remains of plants and animals decayed and built up in thick layers. This decayed matter from plants and animals is called organic material -a compound that capable of decay or sometime refers as a compound consists mainly carbon. Over time, the mud and soil changed to rock, covered the organic material and trapped it beneath the rock. Pressure and heat changed some of this organic material into coal, some into oil (petroleum), and some into natural gas - tiny bubbles of odorless gas. The main ingredient in natural gas is methane, a gas (or compound) composed of one carbon atom and four hydrogen atoms, CH_4. It is colorless, shapeless, and odorless in its pure form.

In some places, gas escapes from small gaps in the microscopic plants and animals living in the ocean rocks into the air; then, if there is enough activation energy from lightning or a fire, it burns. When people first saw the flames, they experimented with them and learned they could use them for heat and light. The formation of natural gas can be explained starting with microscopic plants and animals living in the ocean.

The process began in amillions of years ago, when microscopic plants and animals living in the ocean absorbed energy from the sun, which was stored as carbon molecules in their bodies. When they died, they sank to the bottom of the sea. Over millions of years, layer after layer of sediment and other plants and bacteria were formed.

As they became buried ever deeper, heat and pressure began to rise. The amount of pressure and the degree of heat, along with the type of biomass (biological materials derived from living organisms), determined if the material became oil or natural gas. More heat produced lighter oil. At higher heat or biomass made predominantly of plant material produced natural gas.

After oil and natural gas were formed, they tended to migrate through tiny pores in the surrounding rock. Some oil and natural gas migrated all the way to the surface and escaped. Other oil and natural gas deposits migrated until they were caught under impermeable layers of rock or clay where they were trapped. These trapped deposits are where we find oil and natural gas wells today where drilling process was conducted to obtain the gas.

In a modern technology, machines called "digesters" is used to turn today's organic material (plants, animal wastes, etc.) into synthetic natural gas (SNG). This replaces waiting for thousands of years for the gas to form naturally and could overcome the depletion of natural resources. The conventional route for SNG production is based on gasification of biomass to produce synthesis gas and then the subsequent methanation of the synthesis gas turn it to synthesis natural gas. Woody biomass contain 49.0% carbon and 5.7% hydrogen that can be converted to 76.8% methane, CH_4.

2.3 How Natural Gas is Obtained

Now imagine how to obtain the invisible treasure? That's the challenge face by geologist when exploring for natural gas. Sometimes there are clues on the earth's surface. An oil seeps is a possible sign of natural gas below, since oil and gas are sometimes found together. Geologists also have sensitive machines that can "sniff" surface soil and air for small amounts of natural gas that may have leaked from below ground. The search for natural gas begins with geologists who locate the types of rock that are known to contain gas and oil deposits. Today their tools include seismic surveys that are used to find the right places to drill wells. Seismic surveys use echoes from a vibration source at the Earth's surface (usually a vibrating pad under a truck built for this purpose) to collect information about the rocks beneath. They send sound waves into the ground and measure how fast the waves bounce back. This tells them how hard and how thick the different rock layers are underground. The data is fed into a computer, which draws a picture of the rock layers. This picture is called a seismogram. Sometimes, it is necessary to use small amounts of dynamite to provide the vibration that is needed.

The next task are taken by scientists and engineers who explore a chosen area by studying rock samples from the earth and taking measurements. If the site seems promising, drilling begins. Some of these areas are on land but many are offshore, deep in the ocean. Once the gas is found, it flows up through the well to the surface of the ground and into large pipelines. Some of the gases that are produced along with methane, such as butane and propane, are separated and the other sour gases such as carbon dioxide and hydrogen sulfide are cleaned at a gas processing plant (normally called as sweetening process). The by-products, once removed, are used in a number of ways. For example, propane and butane can be used for cooking gas.

Because natural gas is colorless, odorless and tasteless, mercaptan (a sulfur-containing organic compound with the general formula RSH where R is any radical, especially ethyl mercaptan, C_2H_5SH) is added before distribution, to give it a distinct unpleasant odor (like that of rotten eggs). This serves as a safety device by allowing it to be detected in the atmosphere, in cases where leaks occur.

Most of the natural gas consumed in the United States is produced in the United States. Some is imported from Canada and shipped to the United States in pipelines. Increasingly natural gas is also being shipped to the United States as liquefied natural gas (LNG).

2.4 How Natural Gas is Stored and Delivered

Natural gas is normally produced far away from the consumption regions, therefore they requires an extensive and elaborate transportation system to reach its point of use. The transportation system for natural gas consists of a complex network of pipeline, designed to quickly and efficiently transport natural gas from the origin to areas of high natural gas demand. Transportation of natural gas is closely linked with its storage since the demand of the gas is depend on the season.

Since natural gas demand is greater in the winter, gas is stored along the way in large underground storage systems, such as old oil and gas wells or caverns formed in old salt beds in western country. The gas remains there until it is added back into the pipeline when people begin to use more gas, such as in the winter to heat homes. In Malaysia, and other tropical country, gas is supplied throughout the year, therefore it was storage in a large tank in the processing plant, either in Bintulu, Sarawak, or at Kertih, Terengganu.

Three major types of pipeline available along the transportation route, the gathering system, the interstate pipeline and the distribution system. The gathering system consists of low pressure, low diameter pipelines that transport raw natural gas from the wellhead to the processing plant. In Malaysia, the natural gas is transported from oil rig offshore to the processing plant at Petronas Gas Berhad at Kertih, Terengganu, and Bintulu LNG Tanker, Sarawak. Since Malaysia natural gas and other producing country contain high sulfur and carbon dioxide (sour gaseous) it must used specialized sour gas gathering pipe. Natural wet gas from the wellhead contain high percentage of water therefore it will react with sour gaseous to form acids, which are extremely corrosive and dangerous, thus its transportation from the wellhead to the sweetening plant must be done carefully. The topic will be discussed in depth in the treatment and processing of natural gas.

Pipeline can be classified as interstate or intrastate either it carries natural gas across the state boundary (interstate) or within a particular state (intrastate). Natural gas pipelines are

subject to regulatory oversight, which in many ways determines the manner in which pipeline companies must operate. When the gas gets to the communities where it will be used (usually through large pipelines), the gas is measured as it flows into smaller pipelines called mains. Very small lines, called services, connect to the mains and go directly to homes or buildings where it will be used. This method is used by rich country such as in the United State, Canada or European country, such as United Kingdom, France etc.
The used of pipeline for natural gas delivery is costly, therefore some countries prefer to use trucks for inland delivery. Using this method the natural gas should be liquefied to minimize the size of the tanker truck. In certain country, the natural gas is transported by trucks tankers to the end users. For example in Malaysia the natural gas was transported as Liquefied Natural Gas (LNG) using tanker trucks to different state in peninsular of Malaysia and in East Malaysia. The gas was supplied by Petronas Gas Berhad, at Kertih, Terengganu while in east Malaysia, Sabah and Sarawak, the gas was supplied by Bintulu Plant. The natural is exported by large ships equipped with several domed tanks.
When chilled to very cold temperatures, approximately -260°F, natural gas changes into a liquid and can be stored in this form. Because it takes up only 1/600th of the space that it would in its gaseous state, Liquefied natural gas (LNG) can be loaded onto tankers (large ships with several domed tanks) and moved across the ocean to deliver gas to other countries. When this LNG is received in the United States, it can be shipped by truck to be held in large chilled tanks close to users or turned back into gas to add to pipelines. The whole process to obtain the natural gas to the end user can be simplified by the diagram shown in Fig. 3.

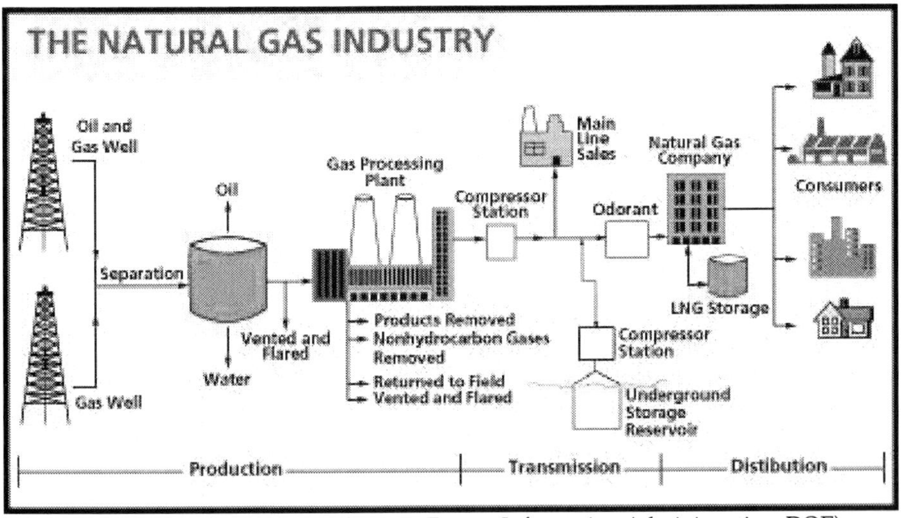

Fig. 3. Natural gas industry. Image (source: Energy Information Administration, DOE)

2.5 What is the Composition of Natural Gas

Natural gas, in itself, might be considered a very uninteresting gas - it is colorless, shapeless, and odorless in its pure form. Quite uninteresting - except that natural gas is combustible, and when burned it gives off a great deal of energy. Unlike other fossil fuels, however, natural gas is clean burning and emits lower levels of potentially harmful byproducts into the air. We require energy constantly, to heat our homes, cook our food, and generate our electricity. It is this need for energy that has elevated natural gas to such a level of importance in our society, and in our lives.

Natural gas is a combustible mixture of hydrocarbon gases. While natural gas is formed primarily of methane, it can also include ethane, propane, butane and pentane. The composition of natural gas can vary widely, but below is a chart outlining the typical makeup of natural gas before it is refined.

Chemical Name	Chemical Formula	Percentage (%)
Methane	CH_4	70-90%
Ethane	C_2H_6	
Propane	C_3H_8	0-20%
Butane	C_4H_{10}	
Carbon Dioxide	CO_2	0-8%
Oxygen	O_2	0-0.2%
Nitrogen	N_2	0-5%
Hydrogen sulphide	H_2S	0-5%
Rare gases	A, He, Ne, Xe	trace

Table 1. Typical composition of Natural Gas

In its purest form, such as the natural gas that is delivered to your home, it is almost pure methane. Methane is a molecule made up of one carbon atom and four hydrogen atoms, and is referred to as CH_4. Malaysia producing sour natural gas. Before purification process, Malaysia's natural gas is consists of several gaseous and impurities. The chemical composition of Malaysia natural gas before it is being refined is shown in Table 2.

Chemical Name	Chemical Formula	Percentage (%)
Methane	CH_4	40-50%
Ethane	C_2H_6	5-10%
Propane	C_3H_8	1-5%
Carbon Dioxide	CO_2	20-3-%
Hydrogen sulphide	H_2S	0-1%

Table 2. Chemical composition in crude natural gas provided by Bergading Platform offshore of Terengganu, Malaysia.

2.6 How Much Natural Gas is there

There is an abundance of natural gas in North America, but it is a non-renewable resource, the formation of which takes thousands and possibly millions of years. Therefore, understanding the availability of our supply of natural gas is important as we increase our use of this fossil fuel. This section will provide a framework for understanding just how much natural gas there is in the ground available for our use, as well as links to the most recent statistics concerning the available supply of natural gas.

As natural gas is essentially irreplaceable (at least with current technology), it is important to have an idea of how much natural gas is left in the ground for us to use. However, this becomes complicated by the fact that no one really knows exactly how much natural gas exists until it is extracted. Measuring natural gas in the ground is no easy job, and it involves a great deal of inference and estimation. With new technologies, these estimates are becoming more and more reliable; however, they are still subject to revision.

Natural Gas Resource Category	As of January 1, 2007(Trillion Cubic Feet)
Nonassociated Gas	
Undiscovered	**373.20**
Onshore	113.61
Offshore	259.59
Inferred Reserves	**220.14**
Onshore	171.05
Offshore	49.09
Unconventional Gas Recovery	**644.92**
Tight Gas	309.58
Shale Gas	267.26
Coalbed Methane	68.09
Associated-Dissolved Gas	**128.69**
Total Lower 48 Unproved	**1366.96**
Alaska	**169.43**
Total U.S. Unproved	**1536.38**
Proved Reserves	**211.09**
TOTAL NATURAL GAS	**1747.47**

Table 3. Natural Gas Technically Recoverable Resources (Source: Energy Information Administration - Annual Energy Outlook 2009)

A common misconception about natural gas is that we are running out, and quickly. However, this couldn't be further from the truth. Many people believe that price spikes, seen in the 1970's, and more recently in the winter of 2000, indicate that we are running out of natural gas. The two aforementioned periods of high prices were not caused by waning natural gas resources - rather, there were other forces at work in the marketplace. In fact, there is a vast amount of natural gas estimated to still be in the ground. In order to understand exactly what these estimates mean, and their importance, it is useful first to learn a bit of industry terminology for the different types of estimates.

The EIA provides classification system for natural gas resources. Unconventional natural gas reservoirs are also extremely important to the nation's supply of natural gas.

Below are three estimates of natural gas reserves in the United States. The first (Table 3), compiled by the Energy Information Administration (EIA), estimates that there are 1,747.47 Tcf of technically recoverable natural gas in the United States. This includes undiscovered, unproved, and unconventional natural gas. As seen from the table, proved reserves make up a very small proportion of the total recoverable natural gas resources in the U.S.

The following table includes an estimate of natural gas resources compiled by the National Petroleum Council (NPC) in 1999 in its report Natural Gas - Meeting the Challenges of the Nation's Growing Natural Gas Demand. This estimate places U.S. natural gas resources higher than the EIA, at 1,779 Tcf remaining. It is important to note that different methodologies and systems of classification are used in various estimates that are completed. There is no single way that every industry player quantifies estimates of natural gas. Therefore, it is important to delve into the assumptions and methodology behind each study to gain a complete understanding of the estimate itself.

	1992 NPC Study	**1999 NPC Study**
	As of Jan 1, 1991	As of Jan 1, 1998
Lower 48 Resources		
Proved Reserves	160	157
Assessed Additional Resources	1135	1309
Old Fields (Reserve Appreciation)	236	305
New Fields	493	633
Nonconventional	406	371
Total Remaining Resources	**1295**	**1466**
Alaskan Resources		
Proved Reserves	9	10
Assessed Additonal Resources	171	303
Old Fields (Reserve Appreciation)	30	32
New Fields	84	214
Nonconventional	57	57
Total Remaining Resources	**180**	**313**
Total U.S. Remaining Resources	**1475**	**1779**

Table 4. U.S. Natural Gas Resources (Trillion Cubic Feet) (Source: National Petroleum Council - Meeting the Challenges of the Nation's Growing Natural Gas Demand, 2007)

Below (Table 5) is a third estimate completed by the Potential Gas Committee. This estimate places total U.S. natural gas resources at just over 1,836 Tcf. This estimate classifies natural gas resources into three categories: probable resources, possible resources, and speculative resources, which are added together to reach a total potential resource estimate. Only this total is shown below.

	Total Potential Resource
Traditional Resources	
Lower 48 States	
Total Lower 48	**1479.6**
Alaska	
Onshore	94.432
Offshore	99.366
Total Alaska	**193.831**
Total Traditional	**1,673.4**
Coalbed Methane	**163.0**
Total United States	**1,836.4**

Table 5. Potential Natural Gas Resources of the U.S. (Trillion Cubic Feet) (Source: Potential Gas Committee - Potential Supply of Natural Gas in the United States, 2009)

There are a myriad of different industry participants that formulate their own estimates regarding natural gas supplies, such as production companies, independent geologists, the government, and environmental groups, to name a few. While this leads to a wealth of information, it also leads to a number of difficulties. Each estimate is based on a different set of assumptions, completed with different tools, and even referred to with different language. It is thus difficult to get a definitive answer to the question of how much natural gas exists. In addition, since these are all essentially educated guesses as to the amount of natural gas in the earth, there are constant revisions being made. New technology, combined with increased knowledge of particular areas and reservoirs mean that these estimates are in a constant state of flux. Further complicating the scenario is the fact that there are no universally accepted definitions for the terms that are used differently by geologists, engineers, accountants, and others.

Natural gas has been discovered on all continents except Antarctica. World natural gas reserves total approximately 150 trillion cu m (5.3 quadrillion cu ft). The world's largest natural gas reserves, totaling, 50 trillion cu m (1.9 quadrillion cu ft) are located in Russia. The second-largest reserves, 48 trillion cu m (1.7 quadrillion cu ft), are found in the Middle East. Vast deposits are also located in other parts of Asia, in Africa, and in

Australia. Natural gas reserves in the United States total 5 trillion cu m (177 trillion cu ft). In Asia-Oceania, natural gas reserves total 12.6 trillion cu m (Table 6). Malaysia has the 14th largest gas reserves as at January 2008. As at January 2008, Malaysia's gas reserves stood at 88.0 trillion standard cubic feet (tscf) or 14.67 billion barrels of oil equivalent, approximately three times the size of crude oil reserves of 5.46 billion barrels.

	Proven reserves (Tm^3)	Annual production (Gm^3)	Reserve to product (years)
Australia	2.5	34.5	72.5
China	1.5	32.6	46.0
India	0.8	28.4	28.2
Indonesia	2.6	70.6	36.8
Malaysia	2.1	50.3	41.8
Others	3.1	85.3	36.3
Total	**12.6**	**301.7**	**41.8**

Table 6. Proven reserves and Annual production, Asia-Oceania. (Taken from BP Statistical Review, 2003)

Most of this gas reserves are located at offshore Peninsular Malaysia, Sarawak and Sabah. The Malaysian natural gas reserves are as shown in Figure 4 [4].

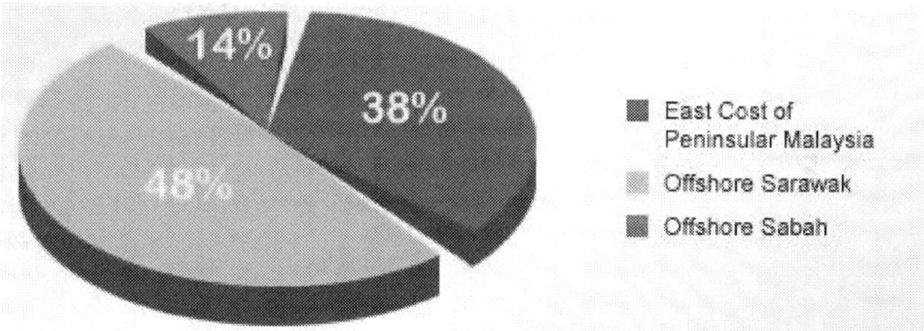

Fig. 4. Malaysian Natural Gas Reserve (Taken from Oil and Gas Exploration and Production-Reserves, Costs, Contract, 2004)

Currently, Malaysia is a net exporter of natural gas and is the third largest exporter after Algeria and Indonesia. In 2001, the country exported 49.7% of its natural gas production to the Republic of Korea and Taiwan under long-term contracts. The other 50.3% of Malaysia natural gas was delivered to the gas processing plants.

2.7 Uses of Natural Gas

For hundreds of years, natural gas has been known as a very useful substance. The Chinese discovered a very long time ago that the energy in natural gas could be harnessed, and used to heat water. In the early days of the natural gas industry, the gas was mainly used to light streetlamps, and the occasional house. However, with much improved distribution channels and technological advancements, natural gas is being used in ways never thought possible.
There are so many different applications for this fossil fuel that it is hard to provide an exhaustive list of everything it is used for. And no doubt, new uses are being discovered all the time. Natural gas has many applications, commercially, in your home, in industry, and even in the transportation sector! While the uses described here are not exhaustive, they may help to show just how many things natural gas can do.
According to the Energy Information Administration, total energy (Fig. 5) from natural gas accounts for 23% of total energy consumed in the developing countries, making it a vital component of the nation's energy supply.

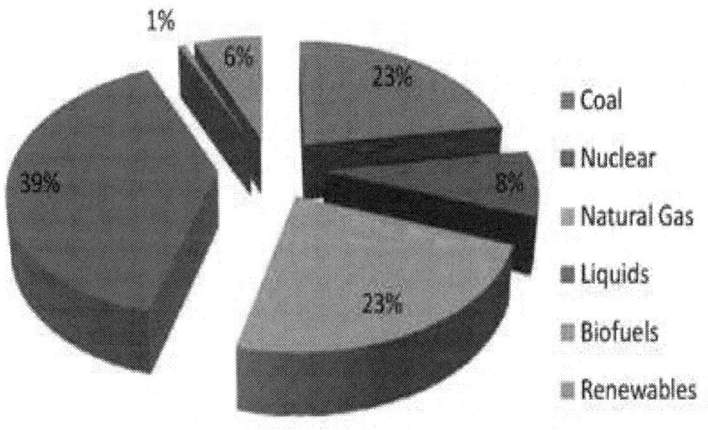

Fig. 5. Total Energy Consumed in the U.S. - 2007 (Source: EIA - Annual Energy Outlook 2009)

Natural gas is used across all sectors, in varying amounts. The pie chart below (Fig. 6) gives an idea of the proportion of natural gas use per sector. The residential sector accounts for the greatest proportion of natural gas use in the most of the developing countries, with the residential sector consuming the greatest quantity of natural gas.

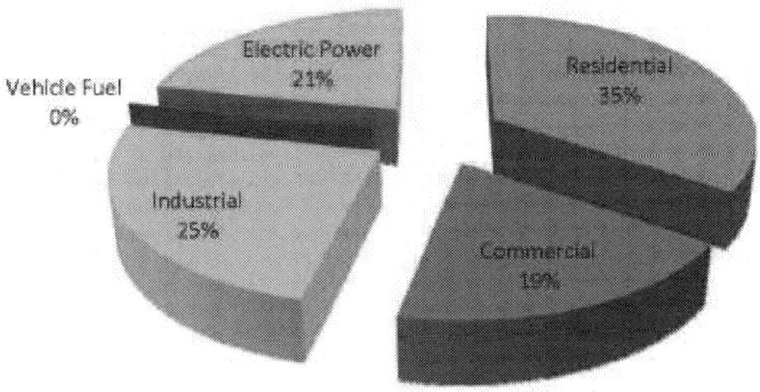

Fig. 6. Natural Gas Use By Sector (Source: EIA - Annual Energy Outlook 2009)

Commercial uses of natural gas are very similar to electric power uses. The commercial sector includes public and private enterprises, like office buildings, schools, churches, hotels, restaurants, and government buildings. The main uses of natural gas in this sector include space heating, water heating, and cooling. For restaurants and other establishments that require cooking facilities, natural gas is a popular choice to fulfill these needs.

According to the Energy Information Administration (EIA), as of the year 2003, the commercial sector consumes about 6,523 trillion Btu's of energy a year (aside from electrical system losses), most of which is required for space heating, lighting, and cooling. Of this 6,523 trillion Btu, about 2,100 trillion Btu (or 32.2%) are supplied by natural gas.

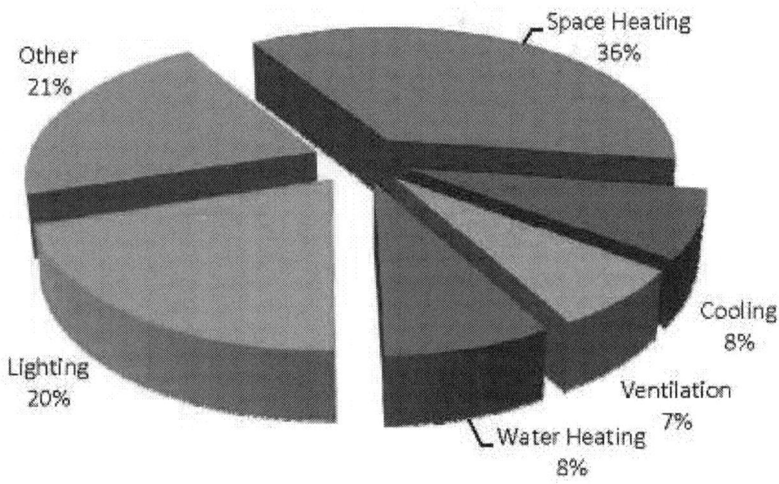

Fig. 7. Commercial Energy Use (Source: EIA Major Fuel Consumption by End Use, 2003.)

Natural gas space and water heating for commercial buildings is very similar to that found in residential houses. Natural gas is an extremely efficient, economical fuel for heating in all types of commercial buildings. Although space and water heating account for a great deal of natural gas use in commercial settings, non-space heating applications are expected to account for the majority of growth in natural gas use in the commercial sector. Cooling and cooking represent two major growth areas for the use of natural gas in commercial settings. Natural gas currently accounts for 13 percent of energy used in commercial cooling, but this percentage is expected to increase due to technological innovations in commercial natural gas cooling techniques. There are three types of natural gas driven cooling processes. Engine driven chillers use a natural gas engine, instead of an electric motor, to drive a compressor. With these systems, waste heat from the gas engine can be used for heating applications, increasing energy efficiency. The second category of natural gas cooling devices consist of what are called absorption chillers, which provide cool air by evaporating a refrigerant like water or ammonia. These absorption chillers are best suited to cool large commercial buildings, like office towers and shopping malls. The third type of commercial cooling system consists of gas-based desiccant systems (Fig. 8). These systems cool by reducing humidity in the air. Cooling this dry air requires much less energy than it would to cool humid air.

Fig. 8. A Desiccant Unit Atop the Park Hyatt Hotel, Washington D.C. (Source: National Renewable Energy Laboratory, DOE)

Another area of growth in commercial natural gas use is in the food service industry. Natural gas is an excellent choice for commercial cooking requirements, as it is a flexible energy source in being able to supply the food service industry with appliances that can cook food in many different ways. Natural gas is also an economical, efficient choice for large commercial food preparation establishments. New developments such as Nontraditional Restaurant Systems, which provide compact, multifunctional natural gas appliances for smaller sized food outlets such as those found in shopping malls and airports, are expanding the commercial use of natural gas. These types of systems can integrate a gas-fired fryer, griddle, oven, hot and cold storage areas, and multiple venting options in a relatively small space - providing the ease and efficiency of natural gas cooking while being compact enough to serve small kiosk type establishments.

In addition to traditional uses of natural gas for space heating, cooling, cooking and water heating, a number of technological advancements have allowed natural gas to be used to increase energy efficiency in commercial settings. Many buildings, because of their high electricity needs, have on-site generators that produce their own electricity. Natural gas powered reciprocating engines, turbines, and fuel cells are all used in commercial settings to generate electricity. These types of 'distributed generation' units offer commercial environments more independence from power disruption, high-quality consistent electricity, and control over their own energy supply.

Another technological innovation brought about is combined heating and power and combined cooling, heating and power systems, which are used in commercial settings to increase energy efficiency. These are integrated systems that are able to use energy that is normally lost as heat. For example, heat that is released from natural gas powered electricity generators can be harnessed to run space or water heaters, or commercial boilers. Using this normally wasted energy can dramatically improve energy efficiency.

Natural gas fired electric generation, and natural gas powered industrial applications, offer a variety of environmental benefits and environmentally friendly uses, including:

- Fewer Emissions - combustion of natural gas, used in the generation of electricity, industrial boilers, and other applications, emits lower levels of NO_x, CO_2, and particulate emissions, and virtually no SO_2 and mercury emissions. Fig. 9 shows a picture of emissions from Industrial Smokestacks (Source: EPA). Natural gas can be used in place of, or in addition to, other fossil fuels, including coal, oil, or petroleum coke, which emit significantly higher levels of these pollutants.
- Reduced Sludge - coal fired power plants and industrial boilers that use scrubbers to reduce SO_2 emissions levels generate thousands of tons of harmful sludge. Combustion of natural gas emits extremely low levels of SO_2, eliminating the need for scrubbers, and reducing the amounts of sludge associated with power plants and industrial processes.
- Reburning - This process involves injecting natural gas into coal or oil fired boilers. The addition of natural gas to the fuel mix can result in NO_x emission reductions of 50 to 70 percent, and SO_2 emission reductions of 20 to 25 percent.
- Cogeneration - the production and use of both heat and electricity can increase the energy efficiency of electric generation systems and industrial boilers, which translates to requiring the combustion of less fuel and the emission of fewer pollutants. Natural gas is the preferred choice for new cogeneration applications.
- Combined Cycle Generation - Combined cycle generation units generate electricity and capture normally wasted heat energy, using it to generate more electricity. Like cogeneration applications, this increases energy efficiency, uses less fuel, and thus produces fewer emissions. Natural gas fired combined cycle generation units can be up to 60 percent energy efficient, whereas coal and oil generation units are typically only 30 to 35 percent efficient.

- Fuel Cells - Natural gas fuel cell technologies are in development for the generation of electricity. Fuel cells are sophisticated devices that use hydrogen to generate electricity, much like a battery. No emissions are involved in the generation of electricity from fuel cells, and natural gas, being a hydrogen rich source of fuel, can be used. Although still under development, widespread use of fuel cells could in the future significantly reduce the emissions associated with the generation of electricity.
- Essentially, electric generation and industrial applications that require energy, particularly for heating, use the combustion of fossil fuels for that energy. Because of its clean burning nature, the use of natural gas wherever possible, either in conjunction with other fossil fuels, or instead of them, can help to reduce the emission of harmful pollutants.

Fig. 9. Emissions from Industrial Smokestacks (Source: EPA)

3. Purification of Natural Gas

Gas processing of acidic crude natural gas is necessary to ensure that the natural gas intended for use is clean-burning and environmentally acceptable. Natural gas used by consumers is composed almost entirely of methane but natural gas that emerges from the reservoir at the wellhead contains many components that need to be extracted. Although, the processing of natural gas is less complicated rather than the processing and refining of crude oil, it is equal and necessary before it can be used by end user.

One of the most important parts of gas processing is the removal of carbon dioxide and hydrogen sulfide. The removal of acid gases (CO_2, H_2S and other sulfur components) from natural gas is often referred to as gas sweetening process. There are many acid gas treating processes available for removal of CO_2 and H_2S from natural gas. These processes include Chemical solvents, Physical solvents, Adsorption Processes Hybrid solvents and Physical separation (Membrane) (Kohl and Nielsen, 1997).

3.1 Various Technologies Used to Convert Sour to Sweet Natural Gas

According to previous research done by Hao *et al.* (2002), there are ways to upgrading the low quality natural gas with selective polymer membranes. The membrane processes were designed to reduce the concentrations of CO_2 and H_2S in the natural gas pipeline specifications. However, this technique incurs high cost and low selectivity towards toxic gas separation. This technique also needs further development because the performance of membrane depends upon the specific characteristics of flue gas composition, and the specific features of the separation (i.e. large volumetric flow rate, low pressure source, high temperature, and the relative low commodity value of H_2S and CO_2) (Rangwala, 1996).
Another method of H_2S removal and one that leaves the CO_2 in the natural gas is called the Iron Sponge process. The disadvantage of this is that it is called a batch-type function and is not easily adapted to continuous operating cycle. The Iron Sponge is simply the process of passing the sour gas through a bed of wood chips that have been impregnated with a special hydrated form of iron oxide that has a high affinity for H_2S. Regeneration of the bed incurs excessive maintenance and operating costs, making this method inconsistent with an efficient operating program. If there are any real advantages in using this process, it is fact that CO_2 remains in the gas, thereby reducing the shrinkage factor which could be significant for very large volumes with an otherwise high CO_2 content (Curry, 1981).
Chemical absorption processes with aqueous alkanolamine solutions are used for treating gas streams containing CO_2. They offer good reactivity at low cost and good flexibility in design and operation. However, depending on the composition and operating conditions of the feed gas, different amines can be selected to meet the product gas specification (Mokhatab et al., 2006). Some of the commonly used alkanolamine for absorption desulfurization are monoethanolamine (MEA), diethanolamine (DEA), triethanolamine (TEA), diglycolamine (DGA), di-isopropanolamine (DIPA) and methyldiethanolamine (MDEA). MDEA allows the selective absorption of H_2S in the presence of CO_2 but can be use effectively to remove CO_2 from natural gas in the present of additives (Salako and Gudmundsson, 2005).
In the other hand, CO_2 can be removed from natural gas via chemical conversion techniques. Catalysts for CO_2 methanation have been extensively studied because of their application in the conversion of CO_2 gas to produce methane, which is the major component in natural gas (Wan Abu Bakar *et al.*, 2008a). Usually, the catalysts are prepared from the metal oxide because of the expensiveness of pure metal. This process can increase the purity and quality of the natural gas without wasting the undesired components but fully used them to produce high concentration of methane (Ching Kuan Yong, 2008).

3.2 Synthesis of Artificial Natural Gas: Methanation Reaction

Methane (CH_4) gas was formed from the reaction of hydrogen gas and carbon dioxide gas through methanation process by reduction reaction as in Equation 1.1 below:-

$$CO_{2\,(g)} + 4H_{2\,(g)} \rightarrow CH_{4\,(g)} + 2H_2O_{\,(l)} \qquad (1.1)$$

This reaction is moderately exothermic, H^o = -165 kJ/mol. In order for this method to be effective, a suitable catalyst must be applied to promote selectively CO_2 methanation because of the main side product under this reaction also will be form (Eq 1.2), which obviously should be avoided. Thus, high selectivity of the catalyst in promoting CO_2

methanation is paramount importance. In Equation 1.2, carbon monoxide produced by this reaction can also be used to form methane by reaction with hydrogen.

$$CO_{2\,(g)} + H_{2\,(g)} \rightarrow CO_{(g)} + H_2O_{(l)} \quad (1.2)$$

$$CO_{(g)} + 3H_{2\,(g)} \rightarrow CH_{4\,(g)} + H_2O_{(l)} \quad (1.3)$$

3.2.1 Mechanism of Methanation Reaction

Mechanism of methanation reaction has been studied a long time ago. A lot of researcher agreed that methanation process involves Langmuir-Hinshelwood (LH) mechanism to support the reaction process between active species and catalyst surface.

For the simplest possible reaction, methanation process can be described as follows:

$$CO_2 + S \rightleftharpoons CO_{2(ads)} \quad (1.4)$$

$$H_2 + S \rightleftharpoons H_{2(ads)} \quad (1.5)$$

$$CO_{2(ads)} + H_{2(ads)} \rightleftharpoons CH_{4(ads)} + H_2O_{(ads)} \quad (1.6)$$

$$CH_{4(ads)} \rightleftharpoons CH_{4(desorp)} + S \quad (1.7)$$

$$H_2O_{(ads)} \rightleftharpoons H_2O_{(desorp)} + S \quad (1.8)$$

Where S = Catalyst surface; ads = adsorbed species on the catalyst surface; desorp = desorbed species from catalyst surface.

According to Equation 1.4, carbon dioxide is reacting with the catalyst surface, (S) by chemisorptions and creates an active species that adsorbed onto catalyst surface. This is followed by hydrogen compound that also react with catalyst surface by chemisorptions and adsorbed onto catalyst surface as an active species. Both active species than react each other to produce products that is methane and water. Finally, (Equation 1.7 & 1.8) both products were dissociated from the catalyst surface.

4. Catalysts Used in Methanation Reaction

Metal oxide supported catalysts have been widely used in research for investigating the CO and CO_2 methanation reaction. Depending on the metal used and the reaction conditions, a variety of products may be formed including methane. However, fewer researches on the catalyst for in-situ reactions of CO_2 methanation and H_2S desulfurization have been carried out. In fact, there is also presence of H_2S in real natural gas. Therefore, H_2S should be considered in invention of methanation catalyst, since it could cause poisoning of the nickel catalyst (Wan Abu Bakar *et al.*, 2008b). As been said by Xu *et al.* (2003), a good methanation catalyst is physically durable and reducible at temperature not more than 300°C with high performance ability and these properties should retained in the catalyst while in use with a life span up to 10 years.

4.1 Nickel Oxide Based used in Methanation Catalysts

The methanation of carbon dioxide on Ni catalysts was studied in detail by fewer researchers because of the theoretical significance and possible practical application of this reaction. The methanation activity of Ni/Al_2O_3 catalyst depended intimately on the surface chemical state of Ni and different active phases formed from the reduction of different nickel species in the oxidated states. Nickel oxides appeared in Ni/Al_2O_3 in two forms prior to reduction as "free" and "fixed oxide", and formed large and small crystallites, respectively, when reduced (Zielinski, 1982). Studied done by Rodriguez *et al.* (2001) showed that NiO catalyst has ability to gives higher catalytic activity with higher methane formation due to the malformation sites which converted to active sites on the surface of nickel oxide. This property is important as reference to construct excellent catalysts for CO_2 conversion

Previously, it was shown that nickel particles change their morphology during catalytic reactions by cluster growth processes and that part of the active clusters are lifted from the support due to carbon deposition and carbon whisker formation (Czekaj *et al.*, 2007). Early study by Douglas *et al.*, (2001) found that Ni catalysts are promising catalysts since they are active and more resistant to sulfur poisoning thus high dispersion of Ni and is expected to be used in catalytic reaction that proceeds at relatively low temperature (Takahashi *et al.*, 2007). Moreover, Inui (1996) claimed that NiO has a bimodal pore structure, which will enhance the higher activity for CO_2 methanation. A bimodal pore structure was found to be beneficial to catalyst preparation and methanation rate (Inui, 1979) which will serve as an optimum pore size for the adsorption of both the reactants. Therefore, Ni based catalyst are commonly used as catalysts in hydrogenation and hydrogenolysis reaction.

Aksoylu and Onsan (1997) reported 5.5×10^{-5} % of CH_4 was produced at 250°C over the Ni/Al_2O_3 catalyst prepared by conventional impregnation method at 350°C for 3 hours under reduction environment. They also investigated the 15%-Ni/Al_2O_3 prepared by coprecipitation method for methanation of carbon dioxide. The result achieved 30% of conversion with 99.7% selectivity towards methane at 510 K (Aksoylu *et al.*, 1996). Some previous research was only focused on conversion of CO_2 without mentioned the yielded of CH_4. Similarly to Chang *et al.* (2003) who had investigated CO_2 methanation over NiO supported on rice husk ash-Al_2O_3 and SiO_2-Al_2O_3 which had been synthesized by impregnation method and calcined at 500°C. At reaction temperature of 400°C, there were 30% conversion of CO_2 over the rice husk ash-Al_2O_3 supported catalyst, while only 5% conversion of CO_2 over the SiO_2-Al_2O_3 catalyst.

Moreover, Ni/SiO catalyst prepared by conventional impregnation method was also studied by Shi and Liu, (2009). The sample was treated by glow discharge plasma for 1 hour and followed by calcinations thermally at 500°C for 4 hours. Such prepared catalyst presents smaller metal particles (17.5 and 7.9 nm) and higher conversion of CO at 400°C around 90% for methantion reaction. However, Ni/SiO_2 catalyst prepared by a sol gel process showed better quality when compared to the Ni/SiO_2 catalyst prepared by conventional impregnation (Tomiyama *et al.*, 2003). Thus, Takahashi *et al.* (2007) investigated the bimodal pore structure of Ni/SiO_2 prepared by the sol-gel method of silicon tetraethoxide and nickel nitrate in the presence of poly(ethylene oxide) (PEO) and urea.

They found that the catalyst shows steady activity which around 30-40% without decay within the reaction period until 240 min with total flow rate of 360 cm^3/min. The performance of the catalyst influenced strongly by Ni surface area rather than the presence of macropores. As been shown that, nickel oxide can be prepared through various methods such as wetness impregnation, co-precipitation, sol gel method, ion-exchange, adsorption, deposition-precipitation and else. These preparation methods are, however very complicated and difficult to control except for wetness impregnation method. Therefore, most of the work published has focused on the use of impregnation technique for their catalyst preparation.

Research done by Liu *et al.* (2008) on the removal of CO contained in hydrogen-rich reformed gases was conducted by selective methanation over Ni/ZrO_2 catalysts prepared by conventional wetness impregnation method. The catalyst achieved CO conversion of more than 96% and held a conversion of CO_2 under 7% at temperature range 260°C-280°C. The results showed that only methane was observed as a hydrogenated product. Furthermore, the maximum of CO_2 conversion was found by Perkas *et al.* (2009) which achieved about 80% at 350°C on the Ni/meso-ZrO_2 catalyst. Around 100% selectivity to CH_4 formation was obtained at the same reaction temperature. This catalyst was prepared by an ultrasound-assisted method and testing with gas hourly space velocity (GHSV) of 5400 h-1 at all temperatures. They also reported that none modified mesoporous Ni/ZrO_2 catalyst and with the Ni/ZrO_2 modified with Ce and Sm did not effect the conversion of CO_2. Previous work by Sominski *et al.* (2003), a Ni catalyst supported on a mesoporous yttria-stabilized-zirconia composite was successfully prepared by a sonochemical method using templating agent ofvsodium dodecyl sulfate (SDS). However, the result is not as good as the catalyst that had been obtained by Perkas *et al.*

In a research done by Rostrup-Nielsen *et al.* (2007), supported nickel catalyst containing 22 wt% Ni on a stabilized support was exposed to a synthesis gas equilibrated at 600°C and 3000kPa for more than 8000h. The CO_2 conversion is 57.87% while methane formed is 42.76%. The research showed that at 600°C, loss of active surface area proceeds via the atom migration sintering mechanism. The methanation reaction is structure sensitive and it was suggested that atomic step sites play the important role as the active sites of the reaction. High temperature methanation may play a role in manufacture of substitute natural gas (SNG). The key problem is resistance to sintering, which results in a decrease of both the metal surface area and the specific activity.

Modification of the catalyst by some appropriate additives may effect the conversion of CO_2 which then methane production. Ni catalysts were modified by alkali metal, alkaline earth metals, transition metal, noble metal or rare earth metal just to select which promoters could increase the conversion of CO_2 as well as the methane formation. The effect of cerium oxide as a promoter in supported Ni catalysts was studied by Xavier et al. (1999). They claimed that the highest activity of CeO_2 promoter for Ni/Al_2O_3 catalysts could be attributed to the electronic interactions imparted by the dopant on the active sites under reducing conditions. The testing was evaluated in a high pressure catalytic reactor consists of a stainless steel reactor of 25 mm diameter and 180 mm length which is mounted vertically inside a furnace. Methanation activity and metal dispersion was found to decrease with increasing of metal loading. It is observed that the catalyst doped with 1.5 wt% CeO_2 exhibited highest conversion of CO and CO_2 with percentage of conversion increase 3.674 moles/second, which is 86.34%. The presence of CeO_2 in impregnated $Ni/\gamma\text{-}Al_2O_3$ catalysts was associated with easier reduction of chemical interaction between nickel and alumina support hence increase its reducibility and higher nickel dispersion Zhuang *et al.* (1991). It showed a beneficial effect by not only decreasing the carbon deposition rate but also increasing and maintaining the catalytic activity.

The study of Yoshida *et al.* (1997) in a bench scale test at ambient temperature and 350°C for carbon recycling system using Ni ferrite process was carried out in LNG power plant. The feed gas was passed at a flow rate of 10 mL/min. They found that the amount of methane formed after CO_2 decomposition was 0.22 g (conversion CO_2 of to CH_4: 77%) in the latter and 0.49 g (conversion of CO_2 to CH_4: 35%) in the former. According to their study, the methanation and carbon recycling system could also be applied to other CO_z sources such as IGCC power plant and depleted natural gas plant. Hence, pure CH_4 gas can be theoretically synthesized from CO_2 with low concentration in flue gas and H_2 gas with the minimum process energy loss, while conventional catalytic processes need an additional separation process of CH_4 gas formed.

Hashimoto *et al.* (2002) who revealed that the catalysts obtained by oxidation-reduction treatment of amorphous Ni-Zr alloys exhibited high catalytic activity with 100% selectivity formation of CH_4 at 1 atm. Around 80% of CO_2 was converted at 573 K. They found the number of surface nickel atom decreases with nickel content of catalyst, because of coagulation of surface nickel atoms leading to a decrease in dispersion of nickel atoms in the catalysts. Moreover, Habazaki *et al.* (1998) reported that over the catalysts prepared from amorphous Ni-Zr (-Sm) and Co-Zr, nickel-containing catalysts show higher activity than the Co-Zr catalyst. CO reacted preferentially with H_2 and was almost completely converted into CH_4 at or above 473 K in the CO-CO_2-H_2. The maximum conversion of carbon dioxide under the present reactant gas composition is about 35% at 575 K.

Most of the previous work used rare earth oxide as a dopant over Ni/Al_2O_3 catalysts for hydrogenation reaction. Su and Guo (1999) also reported an improvement in catalytic activity and resistance to Ni sintering of doped with rare earth oxides. The growth of Ni particles and the formation of inactive NiO and $NiAl_2O_4$ phases were suppressed by addition of rare earth oxides. The combinations of two oxides lead to creation of new systems with new physicochemical properties which may exhibit high catalytic performance as compared to a single component system (Luo *et al.*, 1997). However, the catalytic and physicochemical properties of different oxide catalysts are dependent mainly on the chemical composition, method of preparation and calcination temperatures (Selim and El-Aihsy, 1994).

Ando and Co-workers (1995) had studied on intermetallic compounds synthesized by arc-melting metal constituent in a copper crucible under 66.7 kPa argon atmosphere. The hydrogenation of carbon dioxide took place under 5 Mpa at a reaction temperature at 250°C over $LaNi_4X$. They found that the conversion of CO_2 was 93% over $LaNi_5$ and the selectivities to methane and ethane in the product were 98% and 2%, respectively. The source of activity can be attributed to the new active sites generated by decomposition of the intermetallic compounds. However, even under atmospheric pressure, 56% of CO_2 converted to CH_4 and CO with selectivities of 98% and 2%, respectively.

The promotion of lanthanide to the nickel oxide based catalyst gives positive effects which are easier reduction of oxide based, smaller particles size and larger surface area of active nickel (Zhang *et al.*, 2001). Moreover, the highly dispersed nickel crystallites is obtained over nickel catalyst containing of lanthanide promoter (Rivas *et al.*, 2008). Furthermore, the methanation of carbon dioxide over Ni-incorporated MCM-41 catalyst was carried out by Du *et al.* (2007). At 873 K, 1 wt% of Ni-MCM-41 with space velocity of 115001 $kg^{-1}h^{-1}$ showed only 46.5% CO_2 conversion and a selectivity of 39.6% towards CH_4. Almost no catalytic activity was detected at 373-473 K and only negligible amounts of products were detected at 573 K. However, this catalyst structure did not change much after CO_2 methanation for several hours, producing the high physical stability of this catalytic system.

In addition, nickel based catalysts that used more than one dopants had been studied by Liu *et al.* (2009). Ni-Ru-B/ZrO_2 catalyst was prepared by means of chemical reduction and dried at 80°C for 18 h in air with total gas flow rate of 100 cm^3/min. They found that CO_2 methanation occurred only when temperature was higher than 210°C. At reaction temperature of 230°C, the CO conversion reached 99.93% but CO_2 conversion only 1.55%. Meanwhile, Ni-Fe-Al oxide nano-composites catalyst prepared by the solution-spray plasma technique for the high temperature water-gas shift reaction was investigated by Watanabe *et al.* (2009). The CO conversion over 39 atom% Ni-34 atom% Fe-27 atom% Al catalyst achieved around 58% and yielded about 6% of methane at 673 K.

On the other hand, Kodama *et al.* (1997) had synthesized ultrafine $Ni_xFe_{3-x}O_4$ with a high reactivity for CO_2 methanation by the hydrolysis of Ni^{2+}, Fe^{2+} and Fe^{3+} ions at 60-90°C followed by heating of the co-precipitates to 300°C. At reaction temperature of 300°C, the maximum yield (40%) and selectivity (95%) for CH_4 were obtained. Moreover, the conversion of CO_2 over NiO-YSZ-CeO_2 catalyst prepared by impregnation method was 100% at temperature above 800°C. This catalyst was investigated by Kang *et al.* (2007). No NiC phase was detected on the surface of NiO-YSZ-CeO_2 catalyst. Yamasaki *et al.* (1999) reported that amorphous alloy of Ni-25Zr-5Sm catalyzed the methanation reaction with 90% conversion of CO_2 and 100% selectivity towards CH_4 at 300°C.

Furthermore, Ocampo *et al.* (2009) had investigated the methanation of carbon dioxide over 5 wt% nickel based $Ce_{0.72}Zr_{0.28}O_2$ catalyst which prepared by pseudo sol-gel method. The catalyst exhibited high catalytic activity with 71.5% CO_2 conversion and achieved 98.5% selectivity towards methane gas at 350°C. However, it never stabilized and slowly deactivated with a constant slope and ended up with 41.1% CO_2 conversion and its CH_4 selectivity dropped to 94.7% after 150 h on stream. Catalytic testing was performed under operating conditions at pressure of 1 atm and a $CO_2/H_2/N_2$ ratio is 36/9/10 with a total gas flow of 55 mL/min.

Meanwhile, Kramer *et al.* (2009) also synthesize $Re_2Zr_{10}Ni_{88}O_x$ catalyst by modified sol gel method based on the molar ratio metal then dried for 5 days at room temperature followed by 2 days at 40°C and lastly calcined at 350°C for 5 h. The catalytic performance was carried out by the reactant gas mixture of $CO/CO_2/N_2/H_2$ = 2/14.9/19.8/63.3 enriched with water at room temperature under pressure of 1 bar and total flow rate of 125 mL/min. At reaction temperature of 230°C, almost 95% conversion of CO was occurred and less than 5% for conversion of CO_2 over this catalyst.

The novel catalyst development to achieve both low temperature and high conversion of sour gasses of H_2S and CO_2 present in the natural gas was investigated by Wan Abu Bakar *et al.* (2008c). It was claimed that conversion of H_2S to elemental sulfur achieved 100% and methanation of CO_2 in the presence of H_2S yielded 2.9% of CH_4 over Fe/Co/Ni-Al_2O_3 catalyst at maximum studied temperature of 300°C. This exothermic reaction will generate a significant amount of heat which caused sintering effect towards the catalysts (Hwang and Smith, 2009). Moreover, exothermic reaction is unfavorable at low temperature due to its low energy content. Thus, the improvement of catalysts is needed for the in-situ reactions of methanation and desulfurization to be occurred at lower reaction temperature.

4.2 Manganese Based used in Methanation Catalysts

Manganese has been widely used as a catalyst for many types of reactions including solid state chemistry, biotechnology, organic reactions and environmental management. Due to its properties, numerous field of research has been investigated whereby manganese is employed as the reaction catalyst. Although nickel also reported to be applicable in many process as a good and cheap catalysts, it seems that using nickel as based catalysts will deactivated the active site by deposition of carbon (Luna *et al.,* 2008). Hence, it is essential to use other metal to improve the activity and selectivity as well as to reduce formation of carbon. A proof that manganese improves the stability of catalysts can be shown in researched done by Seok. H. S *et al* (2001). He have proven that manganese improve the stability of the catalyst in CO_2 reforming methane. Added Mn to Ni/Al_2O_3 will promotes adsorption CO_2 by forming carbonate species and it was responsible for suppression of carbon deposition over Ni/MnO-Ni/Al_2O_3. Other research done by Li. J *et al.* (2009) prove that when manganese doped in appropriate amount, it will cause disorder in the spinal structure of metal surface and can enhance the catalytic activity of the reactive ion.

According to Ouaguenouni *et al,* [33], in the development of manganese oxide doped with nickel catalyst, they found that the spinel $NiMn_2O_4$ was active in the reaction of the partial oxidation of methane. The catalysts show higher methane conversion when calcined at 900°C. This is because the stability of the structure which led to good dispersion of nickel species. Indeed, the presence of the oxidized nickel limited the growth of the particles probably by the formation of interaction between metallic nickel out of the structure and the nickel oxide of the structure.

Ching [34] in his studies found that, 5% of manganese that had been introduced into cobalt containing nickel oxide supported alumina catalyst will converted only 17.71% of CO_2 at reaction temperature of 300°C. While when Mn was introduced into iron containing nickel oxide supported alumina catalyst, the percentage of CO_2 conversion does not differ much as in the Co:Ni catalyst. This may be because manganese is not a good dopant for nickel based catalyst. This is in agreement by Wachs *et al.* [35], where some active basic metal oxide components such as MnO and CeO did not interact strongly with the different oxide functionalities present on oxide support and consequently, did not disperse very well to form crystalline phases. Therefore, in research done by Wachs *et al.* [36] stated that Ru could be assigned as a good dopant towards MnO based catalyst. They are active basic metal oxides that usually anchor to the oxide substrate by preferentially titrating the surface Lewis acid sites, such as surface M-oxide vacancies, of the oxide support.

In addition, the hydrogenation of carbon oxides was also performed over promoted iron-manganese catalysts. Herranz *et al.* [37] in their research found that manganese containing catalyst showed higher activity towards formation of hydrocarbons. When these catalysts were promoted with copper, sodium and potassium, carbon dioxide conversion was favoured by alkaline addition, especially by potassium, due to the promotion of the water-gas shift reaction.

When Najwa Sulaiman (2010) incorporated ruthenium into the manganese oxide based catalyst system with the ratio of 30:70 that was Ru/Mn (30:70)/Al_2O_3, it gave a positive effect on the methanation reaction. The percentage conversion keeps on increasing at 200°C with a percentage of CO_2 conversion of 17.18% until it reaches its maximum point at 400°C, whereby the percentage of CO_2 conversion is at the highest which is 89.01%. At reaction temperature of 200°C and 400°C, it showed Mn and Ru enhances the catalytic activity because H_2 and CO_2 are easily chemisorbed and activated on these surfaces. Murata *et al.* (2009) suggested that the high CO_2 conversion was probably due to the manganese species which causes the removal of chlorine atoms from $RuCl_3$ precursor and increases the density of active ruthenium oxide species on the catalyst which resulted in high catalytic activity.

Furthermore, it is very important to used stable and effective metal oxide catalyst with improved resistance to deactivation caused by coking and poisoning. Baylet. A *et al.* (2008) studies on effect of Pd on the reducibility of Mn based material. They found that in H_2-TPR and XPS test, only Mn^{3+}/Mn^{2+} is proportional to the total Mn content in the solid support that leads to the stable catalyst to avoid cooking and poisoning effect. Additionally, Hu. J *et al.* (2008) in their research on Mn/Al_2O_3 calcined at 500°C, shows that manganese oxide proved to have a good performance for catalytic oxidation reaction and also show better catalytic performance compare using support SiO_2 and TiO_2. It is shows that not only doped material are important in producing good catalyst, based catalyst also play major role in giving high catalytic activity in catalysts. El-Shobaky *et al.* (2003) studied, the doping process did not change the activation energy of the catalyzed reaction but much increased the concentration of the catalytically reactive constituents without changing their energetic nature.

Other research made by Chen. H. Y *et al* (1998) revealed the important of promoting manganese in catalyst. The studies shows that when $Cu/ZnO/Al_2O_3$ promoted Mn as based catalyst, it shows increasing in catalytic activity, larger surface area of Cu concentration and elevated Cu reduction temperature compare catalyst without Mn. In XPS studies also revealed that reaction between Mn and Cu resulting reduction of Mn^{4+} to Mn^{3+} as well as oxidation of Cu^0 and Cu^+ to higher oxidation state. The most important result is, added Mn enhanced methanation yield up to 5-10%. This is an agreement with Wojciechowska. M *et al.* (2007) where in they found that when using manganese as based in copper catalyst increase methane yield and activated the catalyst more compare to copper-cooper catalyst.

Wachs *et al.* (2005) found that some active basic metal oxide components such as MnO and CeO did not interact strongly with the different oxide functionalities present on oxide support and consequently, did not disperse very well to form crystalline phases. Therefore, in research done by Wachs *et al.* (1996) stated that Ru could be assigned as a good dopant towards MnO based catalyst. They are active basic metal oxides that usually anchor to the oxide substrate by preferentially titrating the surface Lewis acid sites, such as surface M-oxide vacancies, of the oxide support.

4.3 Noble Metals used in Methanation Catalysts

Nickel oxide will lose its catalytic ability after a few hours when it undergoes carbon formation process. The carbon formation can be avoided by adding dopants towards the Ni catalyst. Therefore, incorporating of noble metals will overcome this problem. Noble metals such as rhodium, ruthenium, platinum and iridium exhibit promising CO_2/H_2 methanation performance, high stability and less sensitive to coke deposition. However, from a practical point of view, noble metals are expensive and little available. In this way, the addition of dopants and support is good alternative to avoid the high cost of this precious metal. For the same metal loading, activity is mainly governed by the type of metal but also depends on precursor selection (Yaccato *et al.*, 2005). While, the reaction selectivity depends on support type and addition of modifier (Kusmierz, 2008).

Methane production rates for noble metals based catalysts were found to decrease in order Ru > Rh > Pt > Ir ~ Pd. It may be suggested that the high selectivities to CH_4 of Ru and Rh are attributed to the rapid hydrogenation of the intermediate CO, resulting in higher CO_2 methanation activities. Panagiotopoulou *et al.* (2008) had claimed the selectivity towards methane which typically higher than 70%, increases with increasing temperature and approaches 100% when CO_2 conversion initiated at above 250°C. A different ranking of noble metals is observed with respect to their activity for CO_2 hydrogenation, where at 350°C decreases by about one order of magnitude in the order of Pt > Ru > Pd ~ Rh. From the research of Ali *et al.* (2000), the rate of hydrogenation can be increased by loading noble metals such as palladium, ruthenium and rhodium. The results showed that all of them perform excellently in the process of selective oxidation of CO, achieving more than 90% conversion in most of the temperature region tested between 200°C to 300°C.

Finch and Ripley (1976) claimed that the noble metal promoters may enhance the activity of the cobalt supported catalysts to increase the conversion to methane. In addition, the noble metals promoted catalysts maintained greater activity for methane conversion than the non-promoted catalysts in the presence of sulfur poison. The addition of small contents of noble metals on cobalt oxides has been proposed in order to increase the reduction degree on the catalytic activity of Co catalysts (Profeti *et al.*, 2007). Research done by Miyata *et al.* (2006) revealed that the addition of Rh, Pd and Pt noble metals drastically improved the behavior of Ni/Mg(Al)O catalysts. The addition of noble metals on Ni resulted in a decrease in the reduction temperature of Ni and an increase in the amount of H_2 uptake on Ni on the catalyst.

It well known that ruthenium is the most active methanation catalyst and highly selective towards methane where the main products of the reaction were CH_4 and water. However, the trace amount of CO was present among the products and methanol was completely absent (Kusmierz, 2008). Takeishi and Aika (1995) who had studied on Raney Ru catalysts found a small amount of methanol was produced on supported Ru catalyst but the methane gas was produced thousands of times more than the amount of methanol from CO_2 hydrogenation. The selectivity to methane was 96-97% from CO_2. Methane production rate from CO_2 and H_2 at 500 K on their Raney Ru was estimated to be 0.25 mol g^{-1} h^{-1}. The activity for methane production from $CO_2 \pm H_2$ at 433 K under 1.1 MPa was much higher than that under atmospheric pressure. The rate of methane synthesis was 3.0 mmol g^{-1} h^{-1} and the selectivity for methane formation was 98% at 353 K, suggesting the practical use of this catalyst (Takeish *et al.*, 1998).

Particularly suitable for the methanation of carbon dioxide are Ru/TiO_2 catalysts. Such catalysts display their maximum activity at relatively low temperatures which is favorable with respect to the equilibrium conversion of the strongly exothermic reaction and form small amount of methane even at room temperature (Traa and Weitkamp, 1999). It can be prove by VanderWiel *et al.* (2000) who had studied on the production of methane from CO_2 via Sabatier reaction. The conversion reaches nearly 85% over 3 wt% Ru/TiO_2 catalyst at 250°C and the selectivity towards methane for this catalyst was 100%.

Meanwhile, a microchannel reactor has been designed and demonstrated by Brooks *et al.* (2007) to implement the Sabatier process for CO_2 reduction of H_2, producing H_2O and CH_4. From the catalyst prepared, the powder form of Ru/TiO_2 catalyst is found to provide good performance and stability which is in agreement with Abe *et al.* (2008). They claimed that the CO_2 methanation reaction on Ru/TiO_2 prepared by barrel-sputtering method produced a

100% yield of CH_4 at 160°C which was significantly higher than that required in the case of Ru/TiO_2 synthesize by wetness impregnation method and Gratzel method. Barrel-sputtering method gives highly dispersed Ru nano particles deposited on the TiO_2 support which then strongly increase its methanation activity.

Another research regarding CO-selective methanation over Ru-based catalyst was done by Galletti *et al.* (2009). The γ-Al_2O_3 to be used as Ru carrier was on purpose prepared through the solution combustion synthesis (SCS) method. The active element Ru was added via the incipient wetness impregnation (IWI) technique by using $RuCl_3$ as precursor. Three Ru loads were prepared: 3%, 4% and 5% by weight. All of the catalysts reached complete CO conversion in different temperature ranges where simultaneously both the CO_2 methanation was kept at a low level and the reverse water gas shift reaction was negligible. The best results were obtained with 4% Ru/γ-Al_2O_3 in the range of 300–340°C, which is 97.40% of CO conversion.

For further understanding about methanation over Ru-based catalysts, Dangle *et al.* (2007) conducted a research of selective CO methanation catalysts prepared by a conventional impregnation method for fuel processing applications. It well known that metal loading and crystallite size have an affect towards the catalyst activity and selectivity. Therefore, they was studied the crystallite size by altering metal loading, catalyst preparation method, and catalyst pretreatment conditions to suppress CO_2 methanation. These carefully controlled conditions result in a highly active and selective CO methanation catalyst that can achieve very low CO concentrations while keeping hydrogen consumption relatively low. Even operating at a gas hourly space velocity as high as 13500 h^{-1}, a 3% Ru/Al_2O_3 catalyst with a 34.2 nm crystallite was shown to be capable of converting 25–78% of CO_2 to CH_4 over a wide temperature range from 240 to 280°C, while keeping hydrogen consumption below 10%.

In addition, Gorke and Co-workers (2005) had carried out research on the microchannel reactor which coated with a Ru/SiO_2 and a Ru/Al_2O_3 catalyst. They found that the Ru/SiO_2 catalyst exhibits its highest CH_4 selectivity of only 82% with 90% CO_2 conversion at a temperature of 305°C, whereas a selectivity of 99% is obtained by the Ru/Al_2O_3 catalyst at 340°C with CO_2 conversion of 78%. However, Weatherbee and Bartholomew (1984) achieved a CH_4 selectivity of 99.8% with CO_2 conversion of only 5.7% at reaction temperature of 230°C using Ru/SiO_2 catalyst.

Mori *et al.* (1996) investigated the effect of reaction temperature on CH_4 yield using Ru-MgO under mixing and miling conditions at initial pressures of 100 Torr CO_2 and 500 Torr H_2. No CH_4 formation was observed at the temperatures below 80°C under mixing conditions over Ru-MgO catalyst. It reached 31% at 130°C but leveled off at 180°C. CH_4 formation over this catalyst under milling condition increased from 11% at 80°C to 96% at 180°C. They found that incorporating of MgO, a basic oxide to the Ru, promotes the catalytic activity by strongly adsorbing an acidic gas of CO_2. According to Chen *et al.* (2007), Ru impregnated on alumina and modified with metal oxide (K_2O and La_2O_3) showed that the activity temperature was lowered approximately 30°C compared with pure Ru supported on alumina. The conversion of CO on Ru-K_2O/Al_2O_3 and Ru-La_2O_3/Al_2O_3 was above 99% at 140–160°C, suitable to remove CO in a hydrogen-rich gas and the selectivity of Ru-La_2O_3/Al_2O_3 was higher than that of Ru-K_2O/Al_2O_3 in the active temperature range. While methanation reaction was observed at temperature above 200°C.

Other than that, Szailer *et al.* (2007) had studied the methanation of CO_2 on noble metal supported on TiO_2 and CeO_2 catalysts in the presence of H_2S at temperature 548 K. It was observed that in the reaction gas mixture containing 22 ppm H_2S, the reaction rate increased on TiO_2 and on CeO_2 supported metals (Ru, Rh, Pd) but when the H_2S content up to 116 ppm, the all supported catalysts was poisoned. In the absence of H_2S, the result showed that 27% conversion of CO_2 and 39% conversion of CO_2 to methane with the presence of 22 ppm H_2S after 4 hours of the reaction.

Moreover, the addition of Rh strongly improves the activity and stability of the catalysts (Wu and Chou, 2009), resistance to deactivation and carbon formation can be significantly reduced (Jozwiak *et al.*, 2005). Erdohelyi *et al.* (2004) studied the hydrogenation of CO_2 on Rh/TiO_2. The rate of methane formation was unexpectedly higher in the CO_2 + H_2 reaction on Rh/TiO_2 in the presence of H_2S. At higher temperature of 673 K, around 75% of selectivity for CH_4 formation and CO was also formed from the reaction. Choudhury *et al.* (2006) presented the result of an Rh-modified Ni-La_2O_3-Ru catalyst for the selective methanation of CO. However, the performance of the prepared catalysts was reported to be that CO_2 conversion appeared to be less than 30% when CO converted completely.

It had been reported that the addition of Pd had a positive effect for hydrogenation of CO or CO_2 because of its higher electronegativity with greater stability of Pd^0 species compared to those of Ni^0 under on stream conditions (Castaño *et al.*, 2007). In contrast, Pd/SiO_2 and Pt/SiO_2 catalysts showed poor activities at temperature lower than 700 K with the CO conversion was not greater than 22% at temperature 823 K over these catalysts (Takenaka *et al.*, 2004). A Pd–Mg/SiO_2 catalyst synthesized from a reverse microemulsion has been found to be active and selective for CO_2 methanation (Park *et al.*, 2009). At 450°C, the Pd-Mg/SiO_2 catalyst had greater than 95% selectivity to CH_4 at a carbon dioxide conversion of 59%. They claimed that the similar catalyst without Mg has an activity only for CO_2 reduction to CO. these results support a synergistic effect between the Pd and Mg/Si oxide.

Furthermore, platinum-based catalysts present an activity and a selectivity that are almost satisfactory. Finch and Ripley (1976) claimed that the tungsten-nickel-platinum catalyst was substantially more active as well as sulfur resistant than the catalyst in the absence of platinum. It was capable to show a conversion of 84% of CO after on stream for 30 minutes in the presence of less than 0.03% CS_2. No catalytic activity was observed under the poison of 0.03% CS_2 without the addition of Pt. The platinum group promoters enabled the catalysts to maintain good activity until the critical concentration of poison was reached. Pt catalysts were most well known as effective desulfurizing catalysts. Panagiotopolou and Kodarides (2007) found that the platinum catalyst is inactive in the temperature range of 200°C-400°C, since temperatures higher than 450°C are required in order to achieve conversion above 20%.

Moreover, Nishida *et al.* (2008) found that the addition of 0.5 wt% Pt towards *cp*-Cu/Zn/Al (45/45/10) catalyst was the most effective for improving both activity and sustainability of the catalyst. At 250°C, the conversion of CO was achieved around 77.1% under gas mixture of $CO/H_2O/H_2/CO_2/N_2 = 0.77/2.2/4.46/0.57/30$ mL/min. Pierre *et al* (2007) found that the conversion of CO over 5.3% PtCeOx catalyst prepared by urea gelation co-precipitation (UGC) which was calcined at 400°C reached about 92%. This catalyst is more active and shows excellent activity and stability with time on stream at 300°C under water gas shift reaction.

Recently, Bi *et al.* (2009) found that Pt/$Ce_{0.6}Zr_{0.4}O_2$ catalyst exhibited a markedly higher activity with 90.4% CO conversion at 623 K for the water gas shift (WGS) reaction. The methane selectivity was only 0.9% over this catalyst which has been prepared by wetness impregnation method. Meanwhile, Utaka *et al.* (2003) examined the reaction of a simulated reforming gas over Pt-catalysts. At temperatures from 100°C to 250°C, high CO conversions of more than 90% were obtained but most of the conversion was caused by water gas shift reaction. The use of platinum catalyst in conversion of cyclohexane was conducted by Songrui et al. (2006). They found that the cyclohexane conversion over Pt/Ni catalyst prepared by impregnation method (55%-53%) was obviously higher than that over Pt/Al_2O_3 catalyst (30%-20%).

4.4 Supports for the Methanation Catalysts

The presence of the support was recognized to play an important role since it may influence both the activity and selectivity of the reaction as well as control the particle morphology. Insulating oxides such as SiO_2, y-Al_2O_3, V_2O_5, TiO_2 and various zeolites usually used as material supports. These supports are used to support the fine dispersion of metal crystallites, therefore preparing them to be available for the reactions. These oxides will possess large surface area, numerous acidic/basic sites and metal-support interaction that offer particular catalytic activity for many reactions (Wu and Chou, 2009).

Alumina is often used as support for nickel catalyst due to its high resistance to attrition in the continuously stirred tank reactor or slurry bubble column reactor and its favorable ability to stabilize a small cluster size (Xu *et al.*, 2005). Alumina does not exhibit methanation activity, it was found to be active for CO_2 adsorption and the reverse spillover from alumina to nickel increases the methane production especially for co-precipitated catalyst with low nickel loading (Chen and Ren, 1997). Happel and Hnatow (1981) also said that alumina could increase the methanation activity although there was presence of low concentration of H_2S. Additionally, Chang *et al.* (2003) believed that Al_2O_3 is a good support to promote the nickel catalyst activity for CO_2 methanation by modifying the surface properties. Supported Ni based catalyst in the powder form usually used by many researchers and only few researchers used a solid support for their study.

Mori *et al.* (1998) had studied the effect of nickel oxide catalyst on the various supports material prepared using impregnation technique. They revealed that the reactivity of the catalysts depended on the type of supports used which follow the order of $Al_2O_3 > SiO_2 > TiO_2 > SiO_2{\cdot}Al_2O_3$. The reason for the higher activity of Ni/Al_2O_3 catalyst with 70% methanation of CO_2 at 500°C was attributed to the basic properties of the Al_2O_3 support on which CO_2 could be strongly adsorbed and kept on the catalyst even at higher temperatures. However, Takenaka *et al.* (2004) found that the conversion of CO at 523 K were higher in the order of Ni/MgO (0%) < Ni/Al_2O_3 (7.9%) < Ni/SiO_2 (30.0%) < Ni/TiO_2 (42.0%) < Ni/ZrO_2 (71.0%). These results implied that Al_2O_3 support was not appropriate for the CO conversion but suitable support for CO_2 conversion.

In addition, Seok *et al.* (2002) also prepared various Ni-based catalysts for the carbon dioxide reforming of methane in order to examine the effects of supports (Al_2O_3, ZrO_2, CeO_2, La_2O_3 and MnO) and preparation methods (co-precipitated and impregnated) on the catalytic activity and stability. Catalytic activity and stability were tested at 923 K with a feed gas ratio CH_4/CO_2 of 1 without a diluent gas. Co-precipitated Ni/Al_2O_3, Ni/ZrO_2, and Ni/CeO_2 showed high initial activities but reactor plugging occurred due to the formation of large amounts of coke. The gradual decrease in the activity was observed for Ni/La_2O_3 and Ni/MnO, in which smaller amounts of coke were formed than in Ni/Al_2O_3, Ni/ZrO_2, and Ni/CeO_2. The catalyst deactivation due to coke formation also occurred for impregnated 5 wt% Ni/γ-Al_2O_3 catalysts. Addition of MnO onto this Ni/γ -Al_2O_3 catalyst decreased the amount of deposited coke drastically and 90% of initial CO_2 conversion was maintained after 25 h.

Zhou *et al.* (2005) had synthesized the Co-Ni catalyst support with activated carbon for CO selective catalytic oxidation. The conversion of Co-Ni/AC catalyst always keeps at above 97% in a wide reaction temperature of 120-160°C. However, the CO conversion dramatically decreases with the increasing temperature when the reaction temperature is beyond 160°C. Activated carbons are used efficiently in many environmental remediation processes due to their high adsorption capacity, which makes their use possible in the removal of great variety of pollutants present in air aqueous medium. This is because, besides their high surface area, they possess several functional surface groups with an affinity for several adsorbates, justifying the extreme relevance of this adsorbent for the treatment of the pollutant (Avelar *et al.*, 2010).

Kowalczky *et al.* (2008) had revealed that the reactions of trace CO_2 amount with hydrogen (low CO_x /H_2 ratios) are dependent on the Ru dispersion and the kind of support for the metal. Among the supports used in the present study (low and high surface area graphitized carbons, magnesia, alumina and a magnesium-aluminum spinel), alumina was found to be the most advantageous material. For similar Ru dispersions, CO methanation over Ru/Al_2O_3 at 220°C was about 25 times and CO_2 methanation was about 8 times as high as ruthenium deposited on carbon B (Ru3/CB). For high metal dispersion, the following of sequence was obtained: Ru/Al_2O_3 > Ru/$MgAl_2O_3$ > Ru/MgO > Ru/C, both for CO and CO_2 methanation. It is suggested that the catalytic properties of very small ruthenium particles are strongly affected by metal-support interactions. In the case of Ru/C systems, the carbon support partly covers the metal surface, thus lowering the number of active sites (site blocking).

Takenaka *et al.* (2004) also found that at 473 K, Ru/TiO_2 catalyst showed the highest activity among all the catalysts but when the reaction temperature increased to 523 K, the CO conversion was follows the order of Ru/MgO (0%) < Ru/Al_2O_3 (62.0%) < Ru/SiO_2 (85.0%) < Ru/ZrO_2 (100.0%) = Ru/TiO_2 (100.0%). Similarly to Görke *et al.* (2005) who found that Ru/SiO_2 catalysts exhibits higher CO conversion and selectivity, compared to Ru/Al_2O_3. Meanwhile, Panagiotopolou and Kodarides (2007) demonstrated that Pt/TiO_2 is the most active catalyst at low temperatures exhibiting measurable CO conversion at temperature as low as 150°C. Conversion of CO over this catalyst increases with increasing temperature and reach 100% at temperature of 380°C. While, platinum catalyst supported on Nd_2O_3, La_2O_3 and CeO_2 become active at temperature higher than 200°C and reach 100% above 400°C. MgO and SiO_2 supported platinum catalysts are practically inactive in the temperature range of interest.

The detailed studied on the SiO_2 and Al_2O_3 support was investigated by Nurunnabi *et al.* (2008) had investigated the performance of γ-Al_2O_3, α-Al_2O_3 and SiO_2 supported Ru catalysts prepared using conventional impregnation method. They found that γ-Al_2O_3 support is more effective than catalysts α-Al_2O_3 and SiO_2 supports for Fischer-Tropsch synthesis under reaction condition of P=20 bar, H_2/CO=2 and GHSV=1800/h. γ-Al_2O_3 showed a moderate pore and particle size around 8 nm which achieved higher catalytic activity about 82.6% with 3% methane selectivity than those of α-Al_2O_3 and SiO_2 catalysts. Pentasil-type zeolite also could be used as a support for the catalysts which exhibited high

activity and stability due to not only the basicity of alkaline promoters but also the incorporation with zeolite support (Park *et al.*, 1995).
Furthermore, Solymosi *et al.* (1981) reported a sequence of activity of supported rhodium catalysts of Rh/TiO_2 > Rh/Al_2O_3 > Rh/SiO_2. This order of CO_2 methanation activity and selectivity was the same as observed for Ni on the same support by Vance and Bartheolomew *et al.* (1983). These phenomena can be attributed to the different metal-support electronic interactions which affects the bonding and the reactivity of the chemisorbed species.

5. Conclusion

Natural gas fuel is a green fuel and becoming very demanding because it is environmental safe and clean. Furthermore, this fuel emits lower levels of potentially harmful by-products into the atmosphere. Most of the explored crude natural gas is of sour gas and yet, very viable and cost effective technology is still need to be developed. Above all, methanation technology is considered a future potential treatment method for converting the sour natural gas to sweet natural gas.

6. References

Annual Energy Outlook.(2009). U.S Department of Energy

Abe, T.; Tanizawa, M.; Watanabe, K. & Taguchi, A. (2008). CO_2 methanation property of Ru nanoparticle-loaed TiO_2 prepared by a polygonal barrel-sputtering method. *Energy & Environmental Science*, Vol. 2, 315-321. Royal Society of Chemistry.

Ali, A. M.; Suzuki, Y.; Inui, T.; Kimura, T.; Hamid, H. & Al-Yami, M. A. (2000), Hydrocracking Activity of Noble Metal Modified Clay-Based Catalysts Compared with a Commercial Catalyst. *Journal of Power Source,* Vol. 142, 70–74.

Aksoylu E. A. & Önsan, İ. Z. (1997). Hydrogenation of carbon dioxides using coprecipitated and impregnated Ni/Al_2O_3 catalysts. *Applied Catalysis A: General,* Vol. 164, 1-11. Elsevier.

Aksoylu, A. E.; Akin, A.N.; Onsan, Z.I. & Trimm, D.L. (1996). Structure./ activity relationships in coprecipitated nickel-alumina catalysts using CO_2 dsorption and methanation. *Applied Catalysis A: Genera,* Vol. 145, 185-193. Elsevier.

Ando, H.; Fujiwara, M.; Matsumura, Y.; Miyamura, H.; Tanaka, H. & Souma, Y. (1995). Methanation of carbon dioxide over $LaNi_4X$-type intermetallic compounds as catalyst precursor. *Journal of Alloys and Compounds,* Vol. 223, 139-141.

Avelar, F.F.; Bianchi M. L.; Gonçalves, M. & Mota, E.G.D. (2010). The use of piassava fibers (Attalea funifera) in the preparation of activated carbon. *Bioresource Technology,* Vol. 101, 4639–4645. Elsevier.

Bi, Y.; Xu, H.; Li, W. & Goldbach, A. (2009). Water-gas shift reaction in a Pd membrane reactor over $Pt/Ce_{0.6}Zr_{0.4}O_2$ catalyst. *International Journal of Hydrogen Energy,* Vol. 34, 2965-2971. Elsevier.

Brooks, K.P.; Hu, J.; Zhu, H. & Kee, R.J. (2007). Methanation of carbon dioxide by hydrogen reduction using the Sabatier process in microchannel reactors. *Chemical Engineering Science,* Vol. 62. 1161-1170.

Bianchi (2001). TPR and XPS Investigations of Co/Al_2O_3 Catalysts Promoted With Ru, Ir and Pt. *Catalysis Letters*. 76.

Brooks, K.P.; Hu, J.; Zhu, H. & Kee, R.J. (2007). Methanation of Carbon Dioxide by Hydrogen Reduction Using the Sabatier Process in Microchannel Reactors. *Chemical Engineering Science*. 62. 1161-1170.

Baylet, A.; Royer, S.; Labrugere, C.; Valencia, H.; Marecot, P.; Tatibouet, M. J & Duprez, D. (2008). Effect of Palladium On The Reducibility of Mn Based Materials: Correlation With Methane Oxidation Activity. *Physical Chemistry Chemical Physics*. 10 .5983-5992.

Castaño, P.; Pawelec, B.; Fierro, J.L.G.; Arandes, J.M. & Bilbao, J. (2007). Enhancement of pyrolysis gasoline hydrogenation over Pd-promoted Ni/ SiO_2-Al_2O_3 catalysts. *Fuel*, Vol. 86, 2262-2274. Elsevier.

Chang, F.W.; Kuo, M.S.; Tsay, M.T. & Hsieh, M.C. (2003). Hydrogenation of CO_2 over nickel catalysts on rice husk ash-alumina prepared by incipient wetness impregnation. *Applied catalysis A: General*, Vol. 247, 309-320. Elsevier.

Chen, X.; Zou, H.; Chen, S.; Dong, X. & Lin, W. (2007). Selective oxidation of CO in excess H_2 over Ru/Al_2O_3 catalyst modified with metal oxide. *Journal of Natural Gas Chemistry*, Vol. 16. 409-414.

Chen, Y., & Ren, S. (1997). Effect of Addition on Ni/Al_2O_3 Catalysts over CO_2 Methanation. *Applied Catalysis A: General*, Vol. 164, 127-140. Elsevier.

Choudhury, M.; Shakeel, A.; Shalabi, M. & Inui, T. (2006). Preferential methanation of CO in a syngas involving CO_2 at lower temperature range. *Applied Catalysis A: General*, Vol. 314, 47-53.

Czekaj, I.; Loviat, F.; Raimondi, F.; Wambach, J.; Biollaz, S. & Wokaun, A. (2007). Characterization of surface processes at the Ni-based catalyst during the metantaion of biomass-derived synthesis gas: X-ray Photoelectron Spectroscopy (XPS). *Applied Catalysis A: General*, Vol. 329, 68-78. Elsevier.

Chen, H.; Lin, Y.; Tan, L. K & Li, J (1998). Comparative Studies of Manganese-doped Coppr-based Catalysts: The Promoter Effect of Mn on Methanol Synthesis. *Applied Surface Scince*. 126. 323-331

Chen, C.; Lin, C.; Tsai, M.; Tsay, C.; Lee, C. & Chen, G. (2008). Characterization of Nanocrystalline Manganese Oxide Powder Prepared by Inert Gas Condensation. *Ceramics International*. 34. 1661-1666.

Dangle, R. A.; Wang, Y.; Xia, G-G.; Strohm, J. J.; Holladay, J. & Palo, D. R. (2007). Selective CO_2 methanation catalysts for fuel processing applications. *Applied Catalysis A: General*, Vol. 326. 213–218.

Du, G.; Lim, S.; Yang, Y.; Wang, C.; Pfefferle, L. & Haller, G.L. (2007). Methanation of carbon dioxide on Ni-incorporated MCM-41 catalyts: The influence of catalyst pretreatment and study of steady - state reaction. *Journal of catalysis*, Vol. 249, 370-379.

Erdohelyi, A.; Fodor, K. & Szailer, T. (2004). Effect of H_2S on the reaction of methane with carbon dioxide over supported Rh catalysts. *Applied Catalysis B: Environmental*, Vol. 53, 153-160. Elsevier.

El-Shobaky, G.A.; El-Molla, S.A. & Ali, A.M.I. (2003). Catalytic Promotion of NiO/MgO System by Doping With Some Transition Metal Cations. *Applied Catalysis A: General*. 253, 417-425.

Finch, J.N. & Ripley, D.L. (1976). United States Patent 3988334. Retrieved on October 26, 1976 from http://www.freepatentsonline.com

Gorke, O.; Pfeifer, P. & Schubert, K. (2005). Highly selective methanation by the use of a microchannel reactor. *Catalysis Today*, Vol. 110, 132-139.

Galetti, C.; Speechia, S.; Saracco, G & Speechia, V. (2010). CO- Selective Methanation Over Ru-γ- Al_2O_3 Catalyst in H_2 Rich Gas for PEM FC applications. *Chemical Engineering Science*.65. 590-596.

Habazaki, H.; Yamasaki, M.; Zhang, B.; Kawashima, A.; Kohno, S.; Takai, T. & Hashimoto, K. (1998). Co-Methanation of Carbon Monoxide and Carbon Dioxide on Supported Nickel and Cobalt Catalysts Prepared from Amorphous Alloy. *Applied Catalysis A: General*, Vol. 172, 131-140. Elsevier.

Happel, J. & Hnatow, M. A. (1981). United States Patent 4260553. Retrieved on April 7, 1981 from http://patft.uspto.gov/

Happel, J. & Hnatow, M. A. (1976) Resolution of Kinetic Moles by Steady State Racing. *Journal of Catalysis*. 42. 54-59

Hashimoto, K.; Yamasaki, M.; Meguro, S.; Sasaki, T.; Katagiri, H.; Izumiya, K.; Kumagai, N.; Habazaki, H.; Akiyama, E. & Asami, K. (2002). Materials for global carbon dioxide recycling. *Corrosion Science*, Vol. 44, 371-386. Elsevier.

Hwang, S. & Smith, R. (2009). Optimum reactor design in methanation processes with nonuniform catalysts. *Chemical Engineering Comunications*, Vol. 196, No. 5, 616-642.

Hu, J.; Chu, W & Shi, L. (2008). Effect of Carrier and Mn Loading On Supported Manganese Oxide Catalysts for Catalytic Combustion of Methane. *Journal of Natural gas Chemistry*. 17. 159-164.

Inui, T. (1996). Highly effective conversion of carbon dioxide to valuable compounds on composite catalysts. *Catalysis Today*, Vol. 29, 329-337. Elsevier.

Inui, T.; Funabiki, M.; Suehiro, M. & Sezume, T. (1979). Methanation of CO_2 and CO on supported nickel-based composite catalysts. *Journal of the Chemical Society, Faraday Transaction*, Vol. 75, 787-802.

Ishihara, A.; Qian, W. E.; Finahari, N. I.; Sutrisma, P. I & Kabe, T. (2005). Addition Effect of Ruthenium in Nickel Steam Reforming Catalysts. *Fuel*. 84. 1462-1468.

Jóźwiak, W.K.; Nowosielska, M. & Rynkowski, J. (2005). Reforming of methane with carbon dioxide over supported bimetallic catalysts containing Ni and noble metal I. Characterization and activity of SiO_2 supported Ni-Rh catalysts. *Applied Catalysis A: General*, Vol. 280, No. 2, 233-244. Elsevier.

Jose, A. R.; Jonathan, C. H.; Anatoly, I. F.; Jae, Y. K. & Manuel, P. (2001). Experimental and Theoretical Studies on The Reaction of H_2 With NiO. Role of O Vacancies and Mechanism for Oxide Reduction. *Journal of the American Chemical Society*. 124, 346-354

Kang, J.S.; Kim, D.H.; Lee, S.D.; Hong, S.I. & Moon, D.J. (2007). Nickel based tri-reforming catalyst for production of synthesis gas. *Applied Catalysis A: General*, Vol. 332, 153-158. Elsevier.

Kiennemann, A.; Kieffer, R. & Chornet, E. (1981). CO/ H_2 and CO_2/ H_2 reactions with amorphous carbon-metal catalysts. *Reaction Kinetics and Catalysis Letters*, Vol. 16, No. 4, 371-376. Springer.

Kodama, T.; Kitayama, Y.; Tsuji, M. & Tamaura, Y. (1997). Methanation of CO_2 using ultrafine $Ni_xFe_{3-x}O_4$. *Energy*, Vol. 22, No. 2-3, 183-187. Elsevier.

Kowalczyk, Z.; Stolecki, K.; Rarog-Pilecka, W. & Miskiewicz, E. (2008). Supported Ruthenium Catalysts for Selective Methanation of Carbon Oxides at very Low CO_x/H_2 Ratios. *Applied Catalysis A: General,* Vol. 342, 35-39. Elsevier.

Kowalczky, Z.; Jodzis, S.; Rarog, W.; Zielinski, J & Pielaszek, J. (1998). Effect of Potassium and Barium on the Stability of a Carbon-Supported Ruthenium Catalyst for the Synthesis of Ammonia. *Applied Catalyst A: General.* 173. 153-160.

Kramer, M.; Stowe, K.; Duisberg, M.; Muller, F.; Reiser, M.; Sticher, S. & Maier, W.F. (2009). The impact of dopants on the activity and selectivity of a Ni-based methanation catalyst. *Applied Catalysis A: General,* Vol. 369, 42-52. Elsevier.

Kusmierz, M. (2008). Kinetic Study on Carbon Dioxide Hydrogenation over $Ru/\gamma\text{-}Al_2O_3$ Catalysts. *Catalysis Today,* Vol. 137, 429-432.

Liu, Q.; Dong, X.; Mo, X. & Lin, W. (2008). Selective Catalytic Methanation of CO in Hydrogen Gases over Ni/ZrO_2 Catalyst. *Journal of Natural Gas Chemistry,* Vol. 17, 268-272.

Liu, Q.H.; Dong, X.F. & Lin, W.M. (2009). Highly selective CO methanation over amorphous $Ni\text{–}Ru\text{–}B/ZrO_2$ catalyst. *Chinese Chemical Letters,* Vol. 20, No. 8, 889-892. Elsevier.

Li, J., Liang, X., Xu, S and Hao, J. (2009). Catalytic Performance of Manganese Cobalt Oxides on Methane Combustion at Low Temperature. *Applied Catalysis B: Environmental.* 90.

Luo, M.F.; Zhong, Y.J.; Yuan, X.X. & Zheng, X.M. (1997). TPR and TPD studies of CuO/CeO_2 catalysts for low temperature CO oxidation. *Applied Catalysis A: General,* Vol. 162, 121-131. Elsevier.

Luna, A. E. C and Iriate, M. E. (2008). Carbon Dioxide Reforming of Methane over a Metal Modified Ni- Al_2O_3 Catalyst. *Applied Catalysts A*: General. 343. 10-15.

Miyata, T.; Li, D.; Shiraga, M.; Shishido, T.; Oumi, Y.; Sano, T. & Takehira, K. (2006). Promoting Effect of Rh, Pd and Pt Noble Metals to the Ni/Mg(Al)O catalysts for the DSS-like Operation in CH_4 Steam Reforming. *Applied Catalysis A: General,* Vol. 310, 97-104. Elsevier.

Mills, G. A and Steffgen, F. W. (1973). Catalytic Methanation. *Catalysis Review 8.* 2 159-210.

Mori, S., Xu, W.C., Ishidzuki, T., Ogasawara, N., Imai, J. & Kobayashi, K. (1996). Mechanochemical activation of catalysts for CO_2 methanation. *Applied Catalysis A: General,* Vol. 137, 255-268. Elsevier.

Murata, K.; Okabe, K.; Inaba, M.; Takahara, I. & Liu, Y. (2009). Mn-Modified Ru Catalysts Supported on Carbon Nanotubes for Fischer-Tropsch Synthesis. *Journal of the Japan Petroleum Institute.* 52. 16-20.

Najwa Binti Sulaiman. (2009). Manganese Oxide Doped Nobel Metals Supported Catalyst for Carbon Dioxide Methanation Reaction. Universiti Teknologi Malaysia, Skudai.

Neal, M. L.; Hernandez, D & Weaver, H.E.H (2009). Effects of Nanoparticles and Porous Metal Oxide Supports on the Activity of Palladium Catalysts in the Oxidative Coupling of 4-Methylpyridine. *Journal of Molecule. Catalyst A: Chemical.* 307. 29-26.

Nishida, K.; Atake, I.; Li, D.; Shishido, T.; Oumi, Y.; Sano, T. & Takehira, K. (2008). Effects of noble metal-doping on Cu/ZnO/Al2O3 catalysts for water-gas shift reaction catalyst preparation by adopting "memory effect" of hydrotalcite. *Applied Catalysis A: General,* Vol. 337, 48-57. Elsevier.

Nurunnabi, M.; Muruta, K.; Okabe, K.; Inaba, M. & Takahara, I. (2008). Performance and Characterization of Ru/Al_2O_3 and Ru/SiO_2 Catalysts Modified with Mn for Fisher-Tropsch Synthesis. *Applied Catalysis A: General,* Vol. 340, 203-211. Elsevier.

Ocampo, F.; Louis, B. & Roger, A.C. (2009). Methanation of carbon dioxide over nickel-based $Ce_{0.72}Zr_{0.28}O_2$ mixed oxide catalysts prepared by sol-gel method. *Applied Catalysis A: General,* Vol. 369, 90-96. Elsevier.

Panagiotopolou, P. & Kondarides, D.I. (2007). Acomparative study of the water-gas shift activity of Pt catalysts supported on single (MO_x) and composite (MO_x/Al_2O_3, MO_x/TiO_2) metal oxide carriers. *Catalysis Today.* 127. 319-329

Panagiotopoulou, P.; Kondarides, D.I. & Verykios, X. (2008). Selective Methanation of CO over Supported Noble Metal Catalysts: Effects of the Nature of the Metallic Phase on Catalytic Performance. *Applied Catalysis A: General,* Vol. 344, 45-54. Elsevier.

Panagiotopoulou. ; Dimitris I. Kondarides, Xenophon E & Verykios (2009). Selective Methanation of CO over Supported Ru Catalysts. *Applied Catalysis B: Environmental.* 88. 470–478.

Park, S.E.; Nam, S.S.; Choi, M.J. & Lee, K.W. (1995). Catalytic Reduction of CO_2: The Effects of Catalysts and Reductants. *Energy Conversion Management,* Vol. 26, 6-9. Peragon.

Park, J-N. & McFarland, E. W. (2009). A highly dispersed Pd–Mg/SiO_2 catalyst active for methanation of CO_2. *Journal of Catalysis,* Vol. 266. 92–97.

Park, S. E; Chang, S.J & Chon, H. (2003). Catalytic Activity and Coke Resistence in the Carbon Dioxide Reforming of Methane to Synthesis gas over zeolite-supported Ni Catalysts. *Applied Catalysis A: General.* 145. 114-124

Perkas, N.; Amirian, G.; Zhong, Z.; Teo, J.; Gofer, Y. & Gedanken, A. (2009). Methanation of carbon dioxide on ni catalysts on mesoporous ZrO_2 doped with rare earth oxides. *Catalysis Letters,* Vol. 130, No. 3-4, 455-462. Elsevier.

Pierre, D.; Deng, W. & Flytzani-Stephanopoulos, M. (2007). The importance of strongly bound Pt-CeO_x species for the water-gas shift reaction: catalyst activity and stability evaluation. *Topic Catalysis,* Vol. 46, 363-373. Elsevier.

Profeti, L.P.R.; Ticianelli, E.A. & Assaf, E.M. (2008). Co/Al_2O_3 catalysts promoted with noble metals for production of hydrogen by methane steam reforming. *Fuel,* Vol. 87, 2076-2081.

Radler M. (2003). Worldwide Look at Reserves and Production. *Oil & Gas Journal.* 49. 46-47

Riedel, T. & Schaub, G. (2003). Low-temperature Fischer-Tropsch synthesis on cobalt catalysts - effects of CO_2. *Topics in Catalysis,* Vol. 26, 145-156. Springer.

Rivas, M.E.; Fierro, J.L.G.; Guil-Lopez, R.; Pena, M.A.; La Parola, V. & Goldwasser, M.R. (2008). Preparation and characterization of nickel-based mixed-oxides and their performance for catalytic methane decomposition. *Catalysis Today,* Vol. 133-135, 367-373.

Rodriguez, J.A.; Hanson, J.C.; Frenkel, A.I.; Kim, J.Y. & Pérez, M. (2001). Experimental and theoretical studies on the reaction of H_2 with NiO. Role of O vacancies and mechanism for oxide reduction. *Journal of the American Chemical Society,* Vol. 124, 346-354. America Chemical Society.

Rostrup-Nielsen, J. R.; Pedersen, K. & Sehested, J. (2007), High temperature methanation-Sintering and structure sensitivity, *Applied Catalysis A: General,* Vol. 330. 134–138.

Selim, M.M. & El-Aishsy, M.K. (1994). Solid-solid interaction between manganese carbonate and molybdic acid and the stability of the formed thermal products. *Materials Letters*, Vol. 21, No. 3-4, 265-270. Elsevier.\

Seok, S.H.; Choi, S.H.; Park, E.D.; Han, S.H. & Lee, J.S. (2002). Mn-Promoted Ni/Al_2O_3 Catalysts for Stable Carbon Dioxide Reforming of Methane. *Journal of Catalysis*, Vol. 209, 6-15.

Seok, H. S., Han, H. S and Lee, S. J. (2001). The Role of MnO in Ni/MnO-Al_2O_3 Catalysts for Carbon Dioxide Reforming of Methane. *Applied Catalysis A: General*. 215. 31-38.

Shi, P. & Liu, C.J. (2009). Characterization of silica supported nickel catalyst for methanation with improved activity by room temperature plasma treatment. *Catalysis Letters*, Vol. 133, No. 1-2, 112-118.

Solymosi, F.; Erdehelyi, A. & Bansagi, T. (1981). Methanation of CO_2 on supported rhodium catalyst. *Journal of Catalysis*, Vol. 68, 371-382.

Solymosi, F. & Erdehelyi, A. (1981). Methanation of CO_2 on supported rhodium catalyst. *Studies in Surface Science and Catalysis*. 7. 1448-1449

Sominski, E.; Gedanken, A.; Perkas, N.; Buchkremer, H.P.; Menzler, N.H.; Zhang, L.Z. & Yu, J.C. (2003). The sonochemical preparation of a mesoporous NiO/yttria stabilized zirconia composite. *Microporous and Mesoporous Materials*, Vol. 60, No. 1-3, 91-97. Elsevier.

Songrui, W.; Wei, L.; Yuexiang, Z.; Youchang, X. & Chen, J.G. (2006). Preparation and catalytic activity of monolayer-dispersed Pt/Ni bimetallic catalyst for C=C and C=O hydrogenation. *Chinese Journal of Catalysis*, Vol. 27, 301-303.

Stoop, F.; Verbiest, A.M.G. & Van Der Wiele, K. (1986). The influence of the support on the catalytic properties of Ru catalysts in the CO hydrogenation. *Applied Catalysis*, Vol. 25, 51-57.

Stoop, F., Verbiest, A.M.G. and Van Der Wiele, K. (1986). The Influence of The Support on The Catalytic Properties of Ru Catalysts in the CO Hydrogenation. *Applied Catalysis*. 25, 51-57.

Su, B.L. & Guo, S.D. (1999). Effects of rare earth oxides on stability of Ni/α-Al_2O_3 catalysts for steam reforming of methane. *Studies in Surface Science and Catalysis*, Vol. 126, 325-332.

Suh, D. J.; Kwak, C.; Kim, J-H.; Kwon, S. M. & Park, T-J. (2004). Removal of carbon monoxide from hydrogen-rich fuels by selective low-temperature oxidation over base metal added platinum catalysts. *Journal of Power Sources*, Vol. 142, 70–74.

Szailer, E.N.; Albert, O. & Andra, E. (2007). Effect of H_2S on the Hydrogenation of Carbon Dioxide over supported Rh Catalysts. *Topics in Catalysis*, Vol. 46, No. 1-2, 79-86.

Szailer, Eva Novaka, Albert Oszko and Andra Erdohelyia (2007). Effect of H_2S on the Hydrogenation of Carbon Dioxide over Supported Rh Catalysts. *Topics in Catalysis*. 46.

Takahashi, R.; Sato, S.; Tomiyama, S.; Ohashi, T. & Nakamura, N. (2007). Pore structure control in Ni/SiO_2 catalysts with both macropores and mesopores. *Microporous and Mesoporous Materials*, Vol. 98, No. 1-3, 107-114. Elsevier.

Takeishi, K. & Aika, K.I. (1995). Comparison of Carbon Dioxide and Carbon Monoxide with Respect to Hydrogenation on Raney Ruthenium Catalysts. *Applied Catalysis A: General*, Vol. 133, 31-45. Elsevier.

Takeishi, K.; Yamashita, Y. & Aika, K.I. (1998). Comparison of carbon dioxide and carbon monoxide with respects to hydrogenation on Raney ruthenium catalysts under 1.1 and 2.1 MPa. *Applied Catalysis A: General,* Vol. 168, 345-351. Elsevier.

Takenaka, S.; Shimizu, T. & Otsuka, K. (2004). Complete removal of carbon monoxide in hydrogen-rich gas stream through methanation over supported metal catalysts. *International Journal of Hydrogen Energy,* Vol. 29, 1065-1073. Elsevier.

Tomiyama, S.; Takahashi, R.; Sato, S.; Sodesawa, T. & Yoshida, S. (2003). Preparation of Ni/SiO_2 catalyst with high thermal stability for CO_2 reforming of CH_4. *Applied Catalysis A: General,* Vol. 241, 349-361. Elsevier.

Traa, Y. & Weitkamp, J. (1999). Kinetics of the methanation of carbon dioxide over ruthenium on titania. *Chemistry Engineering Technology,* Vol. 21, 291-293.

Trimm, D.L. (1980). *Design Industrial Catalysts. Netherland,* USA: Elsevier Science Publisher. 11.

Utaka, T.; Takeguchi, T.; Kikuchi, R. & Eguchi, K. (2003). CO removal from reformed fuels over Cu and precious metal catalysts. *Applied Catalysis A: General,* Vol. 246, 117-124. Elsevier.

Vance, C.K. & Bartholomew, C.H. (1983). Hydrogenation of CO_2 on Group VIII metals III, effects of support on activity/selectivity and adsorption properties of nickel. *Applied Catalysis,* Vol. 7, 169-173.

Van Rossum, G. J. (1986). Gas Quality. Netherleand, USA: Elsevier Science Publisher

Vanderwiel, D.P.; Zilka-Marco, J.L.; Wang, Y.; Tonkovich, A.Y. & Wegeng, R.S. (2000). Carbon dioxide conversions in microreactors. *Pasific Northwest National Laboratory.*

Wachs, I.E. (1996). Raman and IR Studies of Surface Metal Oxide Species on Oxide Supports: Supported Metal Oxide Catalysts. *Catalysis Today.* 27. 437-455.

Wan Abu Bakar, W.A.; Othman,M.Y. & Ching, K.Y. (2008c). Cobalt Nickel and Manganese-Nickel Oxide Based Catalysts for the In-situ Reactions of Methanation and Desulfurization in the Removal of Sour Gases from Simulated Natural gas. *International Conference on Environmental Research and Technology (ICERT).* Universiti Teknologi Malaysia, Skudai.

Wan Abu Bakar, W.A. (2006). *Personnel Communications.* Universiti Teknologi Malaysia, Skudai.

Wan Abu Bakar, W.A., Othman,M.Y., Ali, R. and Ching, K.Y (2008b). Nickel Oxide Based Supported Catalysts for the In-situ Reactions of Methanation and Desulfurization in the Removal of Sour Gases from Simulated Natural. *Catalyst Letter,* Vol. 128, No. 1-2, 127-136. Springer.

Watanabe, K.; Miyao, T.; Higashiyama, K.; Yamashita, H. & Watanabe, M. (2009). High temperature water-gas shift reaction over hollow Ni-Fe-Al oxide nano-composite catalysts prepared by the solution-spray plasma technique. *Catalysis Cominications.* Vol. 10, 1952-1955. Elsevier.

Weatherbee, G.D. & Bartholomew, C.H. (1984). Hydrogenation of CO_2 on Group VIII metals IV. Specific activities and selectivities of silica-supported Co, Fe, and Ru. *Journal of Catalysis,* Vol. 87. 352-362.

Wojciechowska, M., Przystajko, W and Zielinski, M. (2007). CO Oxidation Catalysts Based on Copper and Manganese or Cobalt Oxides Supported on MgF_2 and Al_2O_3. *Catalysis Today.* 119. 338-348.

Wu, J.C.S. & Chou, H.C. (2009). Bimetallic Rh-Ni/BN catalyst for methane reforming with CO_2. *Chemical Engineering Journal*, Vol. 148, 539-545. Elsevier.

Xavier, K. O.; Sreekala, R.; Rashid, K. K. A.; Yusuff, K. K. M. & Sen, B. (1999). Doping effects of cerium oxide on Ni/Al_2O_3 catalysts for methanation. *Catalysis Today*, Vol. 49, 17-21.

Xu, W.L.; Duan, H.; Ge, Q. & Xu, H. (2005). Reaction Performance and Characterization of Co/Al2O3 Fisher-tropsch Catalysts Promoted with Pt, Pd and Ru. *Catal Letter*, Vol. 102.

Xu, B.; Wei, J.; Yu, Y.; Li, J. & Zhu, Q. (2003). Size Limit of Support Particles in an Oxide-Supported Metal Catalyst: Nanocomposite Ni/ZrO_2 for Utilization of Natural Gas. *J. Phys. Chem. B*, Vol. 107, 5203-5207.

Yaccato, K.; Carhart, R.; Hagemeyer, A.; Lesik, A.; Strasser, P.; Jr, A.F.V.; Turner, H.; Weinberg, H.; Grasselli, R.K. & Brooks, C. (2005). Competitive CO and CO_2 Methanation over Supported Noble Metal Catalysts in High Throughout Scanning Mass Spectrometer. *Applied Catalysis A: General*, Vol. 296, 30-48. Elsevier.

Yamasaki, M.; Komori, M.; Akiyama, E.; Habazaki, H.; Kawashima, A.; Asami, K. & Hashimoto, K. (1999). CO_2 methanation catalysts prepared from amorphous Ni-Zr-Sm and Ni-Zr-misch metal alloy precursors. *Materials Science and Engineering A*, Vol. 267, 220-226. Elsevier.

Yoshida, T.; Tsuji, M.; Tamaura, Y.; Hurue, T.; Hayashida, T. & Ogawa, K. (1997). Carbon recycling system through methanation of CO_2 in flue gas in LNG power plant. *Energy Convers. Mgmt*, Vol. 38. 44 -448.

Zhang, R.; Li, F.; Shi, Q. & Luo, L. (2001). Effects of rare earths on supported amorphous NiB/Al_2O_3 catalysts. *Applied Catalysis A: General*, Vol. 205, 279-284. Elsevier.

Zhou, G.; Jiang, Y.; Xie, H. & Qiu, F. (2005). Non-noble metal catalyst for carbon monoxide selective oxidation in excess hydrogen. *Chemical Engineering Journal*, Vol. 109, 141-145. Elsevier.

Zhou, H. J.; Sui, J. Z.; Li, P.; Chen, D.; Dai, C. Y & Yuan, K. W.(2006). Structural Characterization of Carbon Nanofibers Formed from Different Carbon-Containing Gas. *Carbon* .44.3255-3262

Zhuang, Q.; Qin, Y. & Chang, L. (1991). Promoting effect of cerium oxide in supported nickel catalyst for hydrocarbon steam-reforming. *Applied Catalyst*, Vol. 70, No. 1, 1-8.

Zielinski, J. (1982). Morphology of nickel / alumina catalyst. *Journal of catalysis*, Vol. 76, No. 1, 157-163. Elsevier.

2

Natural gas: physical properties and combustion features

Le Corre Olivier and Loubar Khaled
GEPEA, Ecole des Mines de Nantes, CNRS, UMR 6144
Ecole des Mines de Nantes, NATech, GEM, PRES UNAM
La Chantrerie, 4, rue Alfred Kastler, B.P. 20722,
F-44307, Nantes, Cedex 3, France

1. Introduction

One calls combustible natural gas or simply natural gas, any combustible gas fluid coming from the basement. The concept of a unique "natural gas" is incorrect. It is more exact to speak about natural gases. In fact, the chemical composition of available natural gas (at the final customer) depends on its geographic origin and various mixtures carried out by networks operators.

The majority of natural gases are mixtures of saturated hydrocarbons where methane prevails; they come from underground accumulations of gases alone or gases associated with oil. There are thus as many compositions of natural gases as exploited hydrocarbon layers. Apart from the methane which is the prevailing element, the crude natural gas usually contains decreasing volumetric percentages of ethane, propane, butane, pentane, etc. The ultimate analysis of a natural gas thus includes/understands the molar fraction of hydrocarbons in CH_4, C_2H_6, C_3H_8, C_4H_{10} and the remainder of heavier hydrocarbons is generally indicated under the term C_{5+}. Table 1 gives typical compositions. Apart from these hydrocarbons, one often finds one or more minor elements, or impurities, quoted hereafter:

- nitrogen N_2: it has as a disadvantage its inert character which decreases the commercial value of gas,
- carbon dioxide CO_2: it is harmful by its corrosive properties,
- hydrogen sulfide H_2S: it is harmful by its corrosive properties,
- helium He: it can be developed commercially,
- water H_2O: the natural gas of a layer is generally saturated with steam. To be exploited, it undergoes a partial dehydration.

In this chapter, the characteristics of natural gas in term of composition and physical properties and combustion features are presented. The physical models for the calculation of the physical properties are developed and a synthesis of the models selected is carried out.

Fuel	CH_4	C_2H_6	C_3H_8	C_4H_{10}	C_5H_{12}	N_2	CO_2	MN
No.1	87.1	8.8	2.5	0.8	0	0.8	0	70.7
No.2	97.3	2.1	0.2	0.1	0	0.3	0	90.6
No. 3	87.0	9.4	2.6	0.6	0	0.4	0	70.9
No.4	91.2	6.5	1.1	0.2	0	1.0	0	79.3
No.5	88.6	4.6	1.1	0.3	0.1	3.9	1.4	82.2
No.6	82.9	3.2	0.6	0.2	0.1	12	1	87.9
No.7	92.3	3.2	0.6	0.2	0.1	3	0.4	85.7
No.8	89.5	3.1	3.6	0.2	0.1	2.9	0.4	76.3
No.9	87.7	3.0	5.6	0.2	0.1	2.9	0.4	71.8
No10	84.9	2.9	8.5	0.2	0.1	2.7	0.3	66.5

Table 1. Sample group of fuel gases (Saikaly et al., 2008).

Various techniques of determination of combustion features such as equivalence ratio, the low heating value and Wobbe index are exposed. These techniques are based on direct or indirect methods. The section *"Physical Properties"* is a toolbox to calculate transport properties (dynamic viscosity and thermal conductivity) and other important properties such as speed of sound, refractive index and density. Regards time, the ultimate consumer burns a fuel whose chemical composition varies, see Figure 1. These variations bring problems for plant operation, whatever is the prime mover (Internal Combustion engine, gas turbine or boiler).

The section *"Combustion features"* details:

- Air-fuel ratio is the ratio of air to fuel in stoichiometric conditions.
- Network operator sells natural gas volume but final customer needs heat. Low heating value *LHV* is the link and is very important. By contract, network operator takes obligations on the *LHV* minimum value.
- Wobbe index (*W*) is an important criterion of inter-changeability of gases in the industrial applications (engines, boilers, burners, etc). Gas composition variation does not involve any notable change of the factor of air and the velocity burning when the index of Wobbe remains almost constant.
- Methane number (*MN*) characterizes gaseous fuel tendency to auto-ignition. By convention, this index has a value 100 for methane and 0 for hydrogen (Leiker et al., 1972). The gaseous fuels are thus compared with a methane-hydrogen binary mixture. Two gases with same value *MN* have the same resistance against the spontaneous combustion.

2. Physical Properties

2.1 Introduction

Physical models of transport properties relating to the gases (viscosity, conductivity) result from the kinetic theory of gases, see (Hirschfelder et al., 1954) and (Chapman & Cowling, 1970).

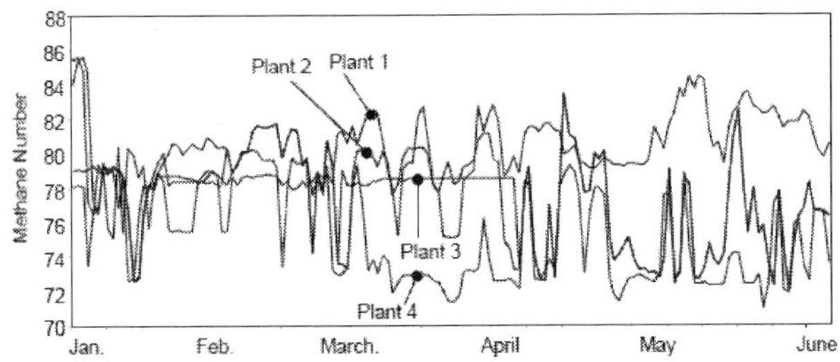

Fig. 1. Methane Number during 5 consecutive months (Saikaly et al., 2008)

The assumptions with regards to the kinetic theory of gases are:

1. The average distance between the molecules is sufficiently important so that the molecular interactions (other than shocks) are negligible,
2. The number of molecules per unit volume is large and constant (gas homogeneity on a macroscopic scale).

The following assumptions are relating to kinematics:

1. Between two shocks, presumed elastic, the movement of each molecule is rectilinear and uniform,
2. The direction of the Speed Vectors of the various molecules obeys a uniform space distribution,
3. The module of the Speed Vectors varies according to a law of distribution which does not depend on time when the macroscopic variables of state are fixed.

Natural gases are a mixture of ν components. Their physical properties such as dynamic viscosity and thermal conductivity, evaluated on the basis of kinetics of gases, are obtained starting from the properties of pure gases and corrective factors (related on the mixtures, the polar moments, etc).

2.2 Dynamic viscosity

Natural gas viscosity is required to carry out flow calculations at the various stages of the production and in particular to determine pressure network losses. Natural gas generally behaves as a Newtonian fluid, see (Rojey et al., 2000) and, in this case, dynamic viscosity η in unit [Pa.s] is defined by Equation (1):

$$\tau = \eta \frac{du}{dy} \tag{1}$$

With τ the shear stress and $\frac{du}{dy}$ the shear rate.

2.2.1 Pure gases

Considering brownian motion of the molecules regards to the intermolecular forces, Chapman and Enskog theory can be applied. This approach considers in detail the interactions between molecules which enter in collision and is based on equation of Maxwell-Boltzmann function distribution, see (Chapman & Cowling, 1970).
For mono-atomic gases, analytic solution of this equation gives the viscosity depending of a two double integrals $\Omega^{(2,2)}$, corresponding to molecules binary collisions, often called "collision integral for viscosity". However, this theoretical approach is only applicable to mono-atomic gases under low pressures and high temperatures. To apply this model to polyatomic gases, a correction for energy storage and transfer are required, see (Le Neindre, 1998). In general terms, the solution obtained for the dynamic viscosity of the mono-atomic gases which do not have degree of freedom of rotation or vibration is written:

$$\eta = 2.6693 \cdot 10^{-6} \frac{\sqrt{M\,T}}{\sigma^2\, \Omega^{(2,2)^*}} \tag{2}$$

With M the molar mass in [g mol^{-1}], T the absolute temperature in [K], σ a characteristic diameter of the molecules, often called "the collision diameter" in [1 $A°$], $\Omega^{(2,2)^*}$ the collision integral depending on the reduced temperature T^* defined as $T^* = kT/\varepsilon$, where k is the Boltzmann constant and ε is the maximum energy of attraction. Correlations exist to approximate the collision integral.

- For nonpolar gases, Neufeld et al. (1972) have proposed the expression:

$$\Omega^{(2,2)^*} = A\left(T^*\right)^{-B} + C\, e^{\left(-D\,T^*\right)} + E\, e^{\left(-F\,T^*\right)} \tag{3}$$

Where A=1.16145, B=0.14874, C=0.52487, D=0.77320, E=2.16178 and F=2.43787. Equation (3) is valuable in the range $0.3 \le T_r \le 100$, where $T_r = T/T_c$, T_c being the critical temperature, with a standard deviation of 0.064%.

Chung et al. (1984) and (1988) have experimentally obtained:

$$\frac{\varepsilon}{k} = \frac{T_c}{1.2593} \tag{4}$$

$$\sigma = 0.809\, V_c^{1/3} \tag{5}$$

To take into account molecule shapes, Chung et al. have introduced a corrective factor F_c:

$$\eta = 4.0785 \cdot 10^{-6}\, F_c \frac{\sqrt{M\,T}}{V_c^{2/3}\, \Omega^{(2,2)^*}} \tag{6}$$

With $F_c = 1 - 0.2756\,\varpi + 0.059035\,\zeta_r^4 + \kappa$; ϖ is the acentric factor, κ is a correction for gases being strongly polar; the dimensionless dipole moment ζ_r being given by $\zeta_r = 131.3\,\zeta/(V_c\,T_c)^{1/2}$

- Reichenberg (1974) have chosen a linear dependence:

$$\Omega^{(2,2)^*} = a\left(T^*\right)^{-n} \tag{7}$$

Dynamic viscosity is then expressed by:

$$\eta = 4.0785 \cdot 10^{-6}\, M^{1/2}\, T^{1/2-n}\, a^{-1}\, \frac{(T_c/1.2593)^n}{V_c^{2/3}} \tag{8}$$

2.2.2 Gaseous blends

At low pressure, dynamic viscosity of gases blend, noted η_m, can be estimated from the viscosity of pure gases. For a mixture of υ components, gaseous blend viscosity is given by the expression:

$$\eta_m = \sum_{i=1}^{\upsilon} K_i \left(1 + 2\sum_{j=1}^{i-1} H_{ij}\, K_i + \sum_{j=1,\neq i}^{\upsilon}\ \sum_{k=1,\neq i}^{\upsilon} H_{ij}\, H_{ik}\, K_j\, K_k\right) \tag{9}$$

Where

$$K_i = \frac{x_i\,\eta_i}{x_i + \eta_i \sum_{k=1,\neq i}^{\upsilon} x_k\, H_{ik}\,[3 + 2\, M_k / M_i]} \tag{10}$$

Where η_i is the dynamic viscosity of ith pure gas, M_i its molar mass, x_i its molar fraction and coefficients H_{ij} are obtained by

$$H_{ij} = \left[\frac{M_i\, M_j}{32(M_i + M_j)^3}\right]^{1/2} (C_i + C_j)^2\, \frac{\left[1 + 0.36\, T_{r,ij}\,(T_{r,ij} - 1)\right]^{1/6} F_{R,ij}}{T_{r,ij}^{1/2}} \tag{11}$$

Reduced temperature $T_{r,ij}$ is based on critical temperature of pure gases *i* and *j*:

$$T_{r,ij} = \frac{T}{(T_{c,i}\, T_{c,j})^{1/2}} \tag{12}$$

Correction coefficients $F_{R,ij}$ is given by:

$$F_{R,ij} = \frac{T_{r,ij}^{7/2} + \left[10\left(\zeta_{r,i}\,\zeta_{r,j}\right)^{1/2}\right]^7}{T_{r,ij}^{7/2}\left[1 + \left(10\left(\zeta_{r,i}\,\zeta_{r,j}\right)^{1/2}\right)^7\right]} \tag{13}$$

Coefficients C_i is obtained by:

$$C_i = \frac{M_i^{1/4}}{\left(\eta_i\,U_i\right)^{1/2}} \tag{14}$$

With:

$$U_i = \frac{F_{R,ii}\left[1 + 0.36\,T_{r,ii}\,(T_{r,ii} - 1)\right]^{1/6}}{T_{r,ii}^{1/2}} \tag{15}$$

Wilke (1950) have introduced simplifications into equation (9) by neglecting the term of the second order. The expression of dynamic viscosity obtained makes easier the application:

$$\eta_m = \sum_{i=1}^{\upsilon} \frac{x_i\,\eta_i}{\sum_{j=1}^{\upsilon} x_j\,\phi_{ij}} \tag{16}$$

With $\phi_{ij} = \dfrac{\left[1 + \left(\eta_i / \eta_j\right)^{1/2}\left(M_i / M_j\right)^{1/4}\right]^2}{\left[8\left(1 + M_i / M_j\right)\right]^{1/2}}$

In the literature, specific correlations were established to calculate the viscosity of gas hydrocarbons. In particular, to calculate the viscosity of methane, an equation of the following general form was proposed by Hanley et al (1975) and included by Vogel et al. (2000):

$$\eta_m = \eta_0(T) + \eta_1(T)\,\rho + \Delta\eta(\rho, T) \tag{17}$$

where $\eta_0(T)$ represents dynamic viscosity in extreme cases of $\rho = 0$. The sum $\eta_1(T)\,\rho + \Delta\eta(\rho, T)$ is the residual dynamic viscosity which takes account of the increase in viscosity from $\eta_0(T)$.

```
function viscosity = func_viscosity(compo)
% compo is a vector in volume fraction
% [CH4 C2H6 C3H8 i-C4H10 n-C4H10 C5H12 CO2 N2 O2 H2 H2S CO]
P = 101325; % current gas pressure in Pa
T = 273.15; % current gas temperature in K
M = [16.043 30.069 44.096 58.123 58.123 72.151 44.01 28.013 32 2.016 34 28.01]; % molar
mass in g mol-1
Tc = [190.58 305.42 369.82 408.14 425.18 469.65 304.19 126.1 154.58 33.18 373.53
132.92];% Critical temperature
Vc = [99.2 148.3 203 263 255 304 93.9 89.8 73.4 64.3 98.6 93.2];%Critical Volume
cm3/mol
Dip = [0 0 0 0.1 0 0 0 0 0 0 0.9 0.1];% Dipolar Moment
omega = [0.011 0.099 0.1518 0.1770 0.1993 0.2486 0.2276 0.0403 0.0218 -0.215 0.0827
0.0663];
T_et = 1.2593*T/Tc; %
omegaV = 1.16145*T_et^(-0.14874)+0.52487*(exp(-0.77320*T_et))+ 2.16178*(exp(-2.43787*T_et));
mu_r = 131.3*Dip./sqrt(Vc.*Tc);
Fc = ones(1,12)-0.2756*omega+0.05903*mu_r.^4;
eta = 40.785*(Fc.*sqrt(T.*M))./(Vc.^(2/3).*omegaV)/10000000;
for i = 1:12
    for j = 1:12
        A(i,j) = (1 + sqrt(eta(i)/eta(j))*(M(i)/M(j))^(1/4))^2/sqrt(8*(1+M(i)/M(j))); %
    end
end
p1 = compo.*eta;
for i = 1:12
    p2(i) = p1(i)/sum(compo.*A(i,:)); %
end
viscosity = sum(p2); %Pa s-1
```

Sandia National Laboratory (www.sandia.gov) has developed CHEMKIN, a reference tool for chemical. The Gas Research Group (www.me.berkeley.edu/gri_mech/overview.html), carried out by the University of California at Berkeley, Stanford University, The University of Texas at Austin, and SRI International, has set up the description of methane and its co-products. The hand-made **Matlab©** function is compared to this reference code. Error is defined as:

$$\forall T \in [300 \ldots 500]\ \max(\varepsilon_\eta) = \max\left|\frac{\eta_{hm}(T) - \eta_{CHEM}(T)}{\eta_{CHEM}(T)}\right| \tag{19}$$

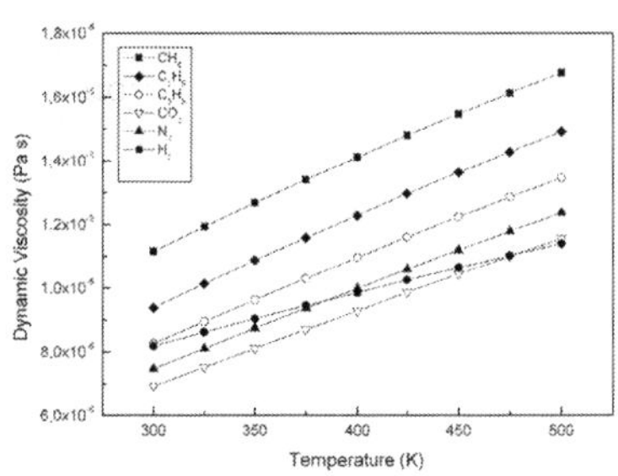

Fig. 2. Dynamic viscosity for main constituents of natural gases

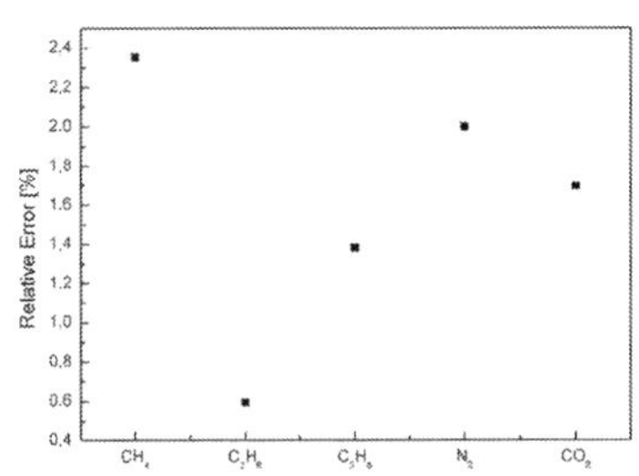

Fig. 3. Relative error between hand-made function and CHEMKIN for dynamic viscosity

The variation of the viscosity of the various components of natural gas according to the temperature is presented on Figure 2 at atmospheric pressure. Good agreement is obtained for the 5 major gases constituting a natural gas, see Figure 3.

2.2.3 Viscometer

Various methods exist to measure the dynamic viscosity of a gas (Guérin, 1981):

- U-tubes of Fagelson (1929) are an extension of Rankine apparatus (1910)
- Double-Helmholtz resonator is first conceived (Greenspan and Wimenitz, 1953). The precision have been extended (Wilhem et al, 2000).
- Rotational viscometers are available products.

2.3 Thermal conductivity

Fourier law characterizes heat conduction: the heat conduction flux φ crossing surface S in a given direction is proportional to the gradient of temperature $\partial T/\partial y$. This factor of proportionality is called thermal conductivity λ.

$$\varphi = \lambda\, S \frac{\partial T}{\partial y} \tag{20}$$

2.3.1 Pure gases

Thermal conductivity of a mono-atomic gas, for which only the energy of translation acts, is given by the traditional expression (Reid et al., 1987):

$$\lambda = 2.63\, 10^{-23} \frac{\sqrt{T/M}}{\sigma^2\, \Omega^{(2,2)^*}} \tag{21}$$

Where λ is in $[Wm^{-1}K^{-1}]$

Using Equation (2), thermal conductivity is expressed from dynamic viscosity by:

$$\lambda = \frac{15}{4} \frac{R}{M} \eta \tag{22}$$

For polyatomic gases (constituents of natural gases), Euken number Eu is introduced:

$$Eu = \frac{\lambda\, M}{\eta\, C_v} \tag{23}$$

Where C_v is the heat capacity at constant volume.

For mono-atomic gases, Euken Number is close to 5/2. For polyatomic gases, Euken Number is modified by separating the contributions due to translation energy from those due to internal energy (Reid et al., 1987):

$$Eu = \frac{\lambda\, M}{\eta\, C_v} = f_{tr}\left(\frac{C_{tr}}{C_v}\right) + f_{in}\left(\frac{C_{in}}{C_v}\right) \tag{24}$$

With C_{tr} in $[J\ mol^{-1}\ K^{-1}]$ the part of the heat capacity due to translation modes $C_{tr} = 3/2\ R$ and C_{in}, related to internal modes, is defined as: $C_{in} = C_v - C_{tr}$, see (Reid et al., 1987) .

$$Eu = \frac{\lambda\, M}{\eta\, C_v} = 1 + \frac{\frac{9}{4}}{\frac{C_p}{R} - 1} \tag{25}$$

Where C_p is the heat capacity at constant pressure.

A modified Euken relation was proposed for which f_{in} is related to a coefficient of molecular diffusion too. This new relation is written as, see (Reid et al., 1987):

$$Eu = \frac{\lambda\, M}{\eta\, C_v} = 1.32 + \frac{1.77}{\frac{C_p}{R} - 1} \tag{26}$$

Mason and Monchick (1962) worked out a theory based on a dynamic formalism to calculate the conductivity of polyatomic gases. They obtained for non-polar gases, by supposing the contributions of the negligible modes of vibrations, the following expression:

$$Eu = \frac{\lambda\, M}{\eta\, C_v} = 1.32 + \frac{1.77}{\frac{C_p}{R} - 1} - 0.886\,\frac{(C_{rot}/C_v)}{Z_{rot}} \tag{27}$$

With C_{rot} in $[J\ mol^{-1}\ K^{-1}]$ the part of the heat capacity due to rotation modes and Z_{rot} the number of collisions necessary to change a quantum of rotation energy into translation energy.

Equation (27) was applied to hydrogen, nitrogen and carbon dioxide, but the main problem for their use remains the precise determination of the number of collisions of rotation Z_{rot} which is function of the temperature.

Chung and al. (1984) used similar method to Mason and Monchick (1962) and obtained the relation of thermal conductivity. Indeed, Euken number is expressed in this case according to a coefficient of correction v as follows:

$$Eu = \frac{\lambda\, M}{\eta\, C_v} = 3.75\frac{v}{C_p/R - 1} \tag{28}$$

Coefficient ν is given by the following formula:

$$\nu = 1 + \alpha \frac{0.215 + 0.28288\,\alpha - 1.061\,\beta + 0.26665\,\Gamma}{0.6366 + \beta\,\Gamma + 1.061\,\alpha\,\beta} \quad (29)$$

With $\alpha = (C_v/R) - 3/2$; $\beta = 0.7862 - 0.7109\varpi + 1.3168\varpi^2$ and $\Gamma = 2.0 + 10.50\,T_r^2$.

Term β is given by an empirical correlation for the contribution of translation energy of the molecules to thermal conductivity for polyatomic gases and applies for the non-polar molecules. As the two main components of the natural gas (methane and ethane) are non-polar and that the other components have weak dipole moment, this correlation represents well the behaviour of natural gases. In the case of the polar molecules, a default value of 0,758 should be used. Term α corresponds to the heat-storage capacity due to the internal degrees of freedom. Thus, term ν can be included/understood as being a shape factor pointing out the deviations of the polyatomic molecules with respect to the model of the rigid sphere.

2.3.2 Gaseous blends

Thermal conductivity of blends is estimated in the same manner as for viscosity. The thermal conductivity of a gas mixture λ_m can be thus calculated starting from a standard formula in the same way than Equation (16), see (Reid et al., 1987):

$$\lambda_m = \sum_{i=1}^{\upsilon} \frac{x_i\,\lambda_i}{\sum_{j=1}^{\upsilon} x_j\,A_{ij}} \quad (30)$$

Mason and Saxena (1958) proposed the following expression for coefficient A_{ij}:

$$A_{ij} = \frac{\varepsilon\left[1 + \left(\lambda_{tr,i}/\lambda_{tr,j}\right)^{1/2}\left(M_i/M_j\right)^{1/4}\right]^2}{\left[8\left(1 + M_i/M_j\right)\right]^{1/2}} \quad (31)$$

Where λ_{tr} represents thermal conductivity of monoatomic gas and ε is a constant close to 1.0; Mason and Saxena (1958) proposed $\varepsilon = 1.065$. Heat conductivities ratio due to the energy of translation of the molecules can be obtained in a purely empirical way:

$$\lambda_{tr,i}/\lambda_{tr,j} = \frac{\Gamma_j\left[e^{0.0464\,T_{r,i}} - e^{0.2412\,T_{r,i}}\right]}{\Gamma_i\left[e^{0.0464\,T_{r,j}} - e^{0.2412\,T_{r,j}}\right]} \quad (32)$$

With $\Gamma_i = 210.0\left(\frac{T_{c,i}\,M_i^3}{P_{c,i}^4}\right)^{1/6}$; $P_{c,i}$ is the critical pressure of the ith component.

```
function thermal_conductivity = func_conductivity(compo)

P = 101325; % current gas pressure in Pa
T = 273.15; % current gas temperature in K
R = 8.314; %ideal gas constant J/K/mol
M = [16.043 30.069 44.096 58.123 58.123 72.151 44.01 28.013 32 2.016 34 28.01]; % molar
mass in g mol-1
Tc = [190.58 305.42 369.82 408.14 425.18 469.65 304.19 126.1 154.58 33.18 373.53
132.92];% Critical temperature
Vc = [99.2 148.3 203 263 255 304 93.9 89.8 73.4 64.3 98.6 93.2];%Critical Volume
cm3/mol
Pc = [4.604 4.88 4.249 3.648 3.797 3.369 7.382 3.394 5.043 1.313 8.963 3.499];%
Critical pressure
Dip = [0 0 0 0.1 0 0 0 0 0 0 0.9 0.1];% Dipolar Moment
omega = [0.011 0.099 0.1518 0.1770 0.1993 0.2486 0.2276 0.0403 0.0218 -0.215 0.0827
0.0663];

methane = -672.87+439.74*(T/100)^0.25-24.875*(T/100)^0.75+323.88*(T/100)^(-0.5);
ethane = 6.895+17.26*(T/100)-0.6402*(T/100)^2+0.00728*(T/100)^3;
propane = -4.092+30.46*(T/100)-1.571*(T/100)^2+0.03171*(T/100)^3;
ibutane = 3.954+37.12*(T/100)-1.833*(T/100)^2+0.03498*(T/100)^3;
nbutane = 3.954+37.12*(T/100)-1.833*(T/100)^2+0.03498*(T/100)^3;
pentane = R*(1.878+4.1216*(T/100)+0.12532*(T/100)^2-0.037*(T/100)^3+0.001525*(T/100)^4);
diocarbone = -3.7357+30.529*(T/100)^0.5-4.1034*(T/100)+0.024198*(T/100)^2;
azote = 39.060-512.79*(T/100)^(-1.5)+1072.7*(T/100)^(-2)-820.4*(T/100)^(-3);
oxygene = 37.432+0.020102*(T/100)^1.5-178.57*(T/100)^(-1.5)+236.88*(T/100)^(-2);
hydrogene = 56.505-702.74*(T/100)^(-0.75)+1165*(T/100)^(-1)-560.7*(T/100)^(-1.5);
hydrosulf = R*(3.071029+0.5578*(T/100)-0.1031*(T/100)^2+0.01202*(T/100)^3-0.0004838*(T/100)^4);
monocarbone = 69.145-0.70463*(T/100)^0.75-200.77*(T/100)^(-0.5)+176.76*(T/100)^(-0.75);

Cpmol = [methane ethane propane ibutane nbutane pentane diocarbone azote oxygene
hydrogene hydrosulf monocarbone];
cp=Cpmol.*1000./M;

alpha = (cp.*M*1e-3-R*ones(1,12))./(R*ones(1,12))-1.5*ones(1,12);
beta = 0.7862*ones(1,12)-0.7109*omega+1.3168*omega.^2;
zed = 2*ones(1,12)+10.5*(T./Tc).^2;
psi = ones(1,12)+alpha.*(0.215*ones(1,12)+0.28288*alpha-1.061*beta+0.26665*zed)./ ...
    (0.6366*ones(1,12)+beta.*zed+1.061*alpha.*beta);
%**************Dynamic Viscosity*********************************
T_et = 1.2593*T./Tc;
omegaV = 1.16145.*T_et.^(-0.14874)+0.52487*(exp(-0.77320*T_et))+ ...
    2.16178*(exp(-2.43787*T_et));
mu_r = 131.3*Dip./sqrt(Vc.*Tc);
Fc = ones(1,12)-0.2756*omega+0.05903*mu_r.^4;
eta = 40.785*(Fc.*sqrt(T.*M))./(Vc.^(2/3).*omegaV)/10000000;
%*****************************************************************
lambda = 3.75*R*eta.*psi./M*1000;

%for mixture
temp = 210*(Tc.*M.^3./(Pc*10).^4).^(1/6);
lambda_tr = temp.*(exp(0.0464.*(T./Tc))-exp(-0.2412.*(T./Tc)));

for i = 1:12
    for j = 1:12
        A(i,j) = (1 + sqrt(lambda_tr(i)/lambda_tr(j))*(M(i)/M(j))^(1/4))^2/ ...
            sqrt(8*(1+M(i)/M(j)));
    end
end
p1 = lambda.*compo;

for i = 1:12
    p2(i) = p1(i)/sum(compo.*A(i,:));
end
thermal_conductivity = sum(p2);
```

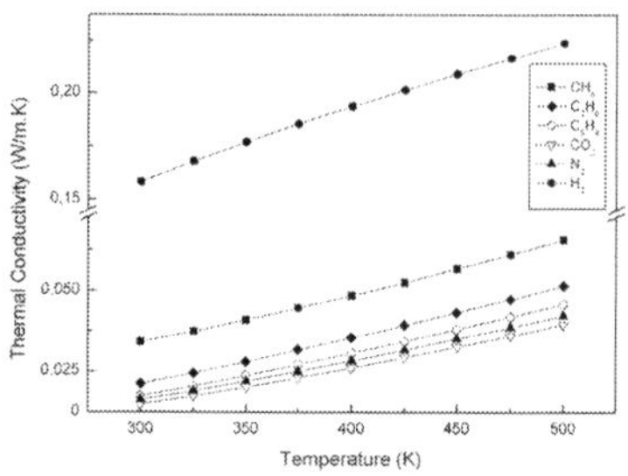

Fig. 4. Thermal conductivity for main constituents of natural gases

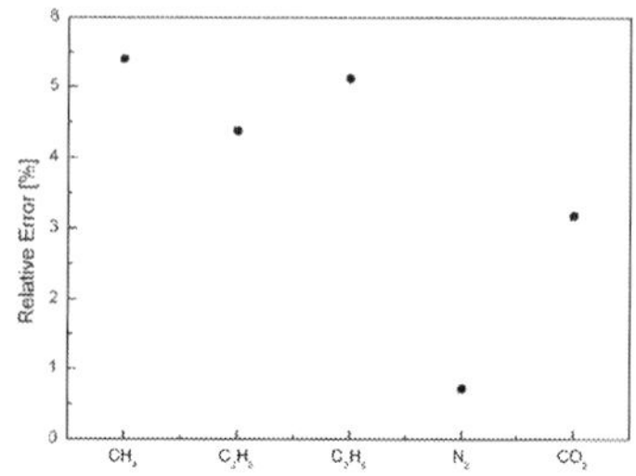

Fig. 5. Relative error between hand-made function and CHEMKIN for thermal conductivity

The variation of the thermal conductivity of the various components of natural gas according to the temperature is presented on Figure 4 at atmospheric pressure. Good agreement is obtained for the 5 major gases constituting a natural gas, see Figure 5.

$$\forall T \in [300\ldots500]\ \max(\varepsilon_\lambda) = \max\left|\frac{\lambda_{hm}(T) - \lambda_{CHEM}(T)}{\lambda_{CHEM}(T)}\right| \quad (33)$$

2.3.3 Thermal conductivity measurement

Different techniques can be used to measure the thermal conductivity:

Katharometer: Thermal conductivity determination of a gas is commonly based on the method of hot wires (Guérin, 1981). A wire is tended in the axis of a metal cylindrical room whose walls are maintained at constant temperature and traversed by a gas, constituting a cell. If one applies a constant electromotive force at the ends of this wire, its temperature rises until the energy spent by Joule effect is, at each time, compensated by the energy dissipated by radiation, convection and thermal conduction. By choosing conditions such as the losses other than the last are negligible (temperature of the wire lower than 400°C, diameter maximum of the tube of 1 cm, rather slow gas flow: 6 to 12 l/h), the temperature of the wire depends primarily on the nature of the gas which surrounds it. If the wire has a resistivity whose temperature coefficient is raised, resistance is function of the thermal conductivity of this gas.

Guarded Hot Plate Method: Guarded hot plate is a widely used and versatile method for measuring the thermal conductivity. A flat, electrically heated metering section surrounded on all lateral sides by a guard heater section controlled through differential thermocouples, supplies the planar heat source introduced over the hot face of the specimens (gas). The most common measurement configuration is the conventional, symmetrically arranged guarded hot plate where the heater assembly is sandwiched between two specimens, see Figure 6. It is an absolute method of measurement and its applicability requires: (a) the establishment of steady-state conditions, and (b) the measurement of the unidirectional heat flux in the metered region, the temperatures of the hot and cold surfaces, the thickness of the specimens and other parameters which may affect the unidirectional heat flux through the metered area of the specimen.

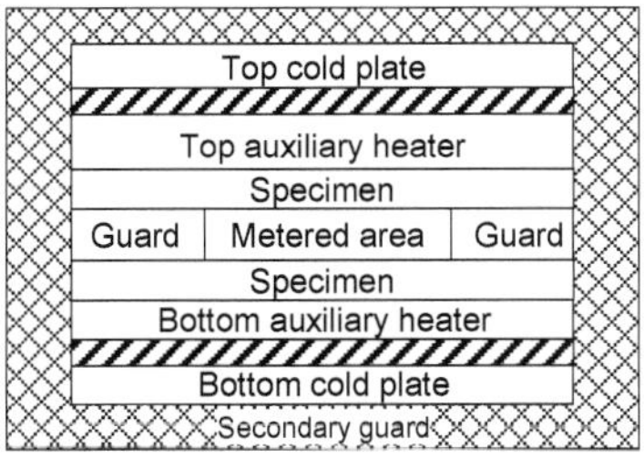

Fig. 6. Guarded hot plate method configuration.

2.4 Speed of sound

Speed of sound is connected to thermodynamic scale of the fluid by the relation:

$$c = \left(\frac{\partial P}{\partial \rho} \right)_S \tag{34}$$

where P and ρ represent the pressure and the density respectively, and S the entropy. The previous relation shows the direct link between the speed of sound and state equation of gas.

2.4.1 Speed of sound for ideal gas

For ideal gas, speed of sound is:

$$c = \sqrt{\gamma \, R \, T / M} \tag{35}$$

For a mixture of ideal gases, speed of sound is:

$$c_m^2 = \frac{\gamma_m \, R \, T}{M_m} = \left(\frac{\sum_{i=1}^{\upsilon} x_i \, C_{p,i}}{\sum_{i=1}^{\upsilon} x_i \, C_{v,i}} \right) \frac{R \, T}{\sum_{i=1}^{\upsilon} x_i \, M_i} \tag{36}$$

Ideal gas law is a good approximation for low pressure. However, in order to take into account the real behavior of gases, several state laws were proposed. Van Der Waals equation thus introduces two corrective terms:

$$P = \frac{R \, T}{(V - b)} + \frac{a}{V^2} \tag{37}$$

Then, in this case, speed of sound is:

$$c^2 = \gamma_r \left(\frac{R\,T}{\left(1-\frac{b}{V}\right)^2} + \frac{2\,a}{V} \right) \tag{38}$$

Thermodynamic properties models based on state equation provide value of compressibility factor. It is useful, in the field of gas industry, to have specific methods of calculation for natural gas of commercial type. The equation derived from virial equation, established by Groups European of Gas Research - GEGR (Jaescheke et al., 2003), gives calculation for the compressibility factor of commercial gas with an average error of about 0.06% for a pressure up to 12 MPa. However, one of the methods most used in this field is based on the model AGA8-DC92 developed by American Gas Association (Starling & Savidge, 1992). This model makes it possible to estimate the density with an average absolute deviation (*AAD*) of 0.04% and the speed of sound with *AAD* of 0.08%. In addition, Estela-Uribe et al. (2003, 2005) used another formulation for natural gas in the range $270 \le T[K] \le 330$ and $P \le 12\ MPa$. This model presents compressibility factor according to the density by:

$$Z = 1 + B_m\,\rho + C_m\,\rho^2 \tag{39}$$

Coefficients B_m and C_m respectively represents the second and the third coefficient of the virial development of the gas mixture. They are given according to temperature and composition of natural gas by the relations:

$$B_m = \sum_i \sum_j x_i\,x_j\,B_{ij} \tag{40}$$

$$C_m = \sum_i \sum_j \sum_k x_i\,x_j\,x_k\,C_{ijk} \tag{41}$$

Where B_{ij} and C_{ijk} are given by:

$$B_{ij} = b_{ij,0} + \frac{b_{ij,1}}{T} + \frac{b_{ij,2}}{T^2} \tag{42}$$

$$C_{ij} = c_{ijk,0} + \frac{c_{ijk,1}}{T} + \frac{c_{ijk,2}}{T^2} \tag{43}$$

Reader is referred to Estela-Uribe et al. (2003, 2005) for coefficients b_{ij} and c_{ijk}.

Speed of sound is written:

$$c_m^2 = \left(\frac{RT}{M_m}\right)\left[Z + \rho\left(\frac{\partial Z}{\partial \rho}\right)_T + \frac{R}{C_{v,m}}\left(Z + T\left(\frac{\partial Z}{\partial T}\right)_\rho\right)^2\right] \tag{44}$$

Where $C_{v,m}$ is heat capacity at constant volume of the mixture calculated by:

$$C_{v,m} = C_{v,m}^{IGL} + C_{v,m}^{res} \tag{45}$$

$C_{v,m}^{IGL}$ is heat capacity calculated by ideal gas law, see (Jaeschke & Schley, 1995), and $C_{v,m}^{res}$ is residual correction, calculated by:

$$C_{v,m}^{res} = -\left[2T\frac{dB_m}{dT} + T^2\frac{d^2B_m}{dT^2}\right]\rho - \left[T\frac{dC_m}{dT} + \frac{T^2}{2}\frac{d^2C_m}{dT^2}\right]\rho^2 \tag{46}$$

```
function speedofsound = func_speedofsound(compo)
P = 101325; % current gas pressure in Pa
T = 273.15; % current gas temperature in K
R = 8.314; %ideal gas constant J/K/mol
M = [16.043 30.069 44.096 58.123 58.123 72.151 44.01 28.013 32 2.016 34 28.01];
methane = -672.87+439.74*(T/100)^0.25-24.875*(T/100)^0.75+323.88*(T/100)^(-0.5);
ethane = 6.895+17.26*(T/100)-0.6402*(T/100)^2+0.00728*(T/100)^3;
propane = -4.092+30.46*(T/100)-1.571*(T/100)^2+0.03171*(T/100)^3;
ibutane = 3.954+37.12*(T/100)-1.833*(T/100)^2+0.03498*(T/100)^3;
nbutane = 3.954+37.12*(T/100)-1.833*(T/100)^2+0.03498*(T/100)^3;
pentane = R*(1.878+4.1216*(T/100)+0.12532*(T/100)^2-0.037*(T/100)^3+0.001525*(T/100)^4);
diocarbone = -3.7357+30.529*(T/100)^0.5-4.1034*(T/100)+0.024198*(T/100)^2;
azote = 39.060-512.79*(T/100)^(-1.5)+1072.7*(T/100)^(-2)-820.4*(T/100)^(-3);
oxygene = 37.432+0.020102*(T/100)^1.5-178.57*(T/100)^(-1.5)+236.88*(T/100)^(-2);
hydrogene = 56.505-702.74*(T/100)^(-0.75)+1165*(T/100)^(-1)-560.7*(T/100)^(-1.5);
hydrosulf = R*(3.071029+0.5578*(T/100)-0.1031*(T/100)^2+0.01202*(T/100)^3-0.0004838*(T/100)^4);
monocarbone = 69.145-0.70463*(T/100)^0.75-200.77*(T/100)^(-0.5)+176.76*(T/100)^(-0.75);
Cpmol = [methane ethane propane ibutane nbutane pentane diocarbone azote oxygene
hydrogene hydrosulf monocarbone];
MassMol =1/100*sum(M.*compo);
HeatCapacity = 1/100*sum(Cpmol.*compo)*1000./MassMol;
speedofsound = sqrt(HeatCapacity /( HeatCapacity -1000*R/MassMol)*R*T/MassMol*1000
```

The variation of the speed of sound of the various components of natural gas according to the temperature is presented on Figure 7 at atmospheric pressure. Good agreement is obtained for the 5 major gases constituting a natural gas, see figure 8.

$$\forall T \in [300\ldots500]\ \max(\varepsilon_c) = \max\left|\frac{c_{hm}(T) - c_{CHEM}(T)}{c_{CHEM}(T)}\right| \tag{47}$$

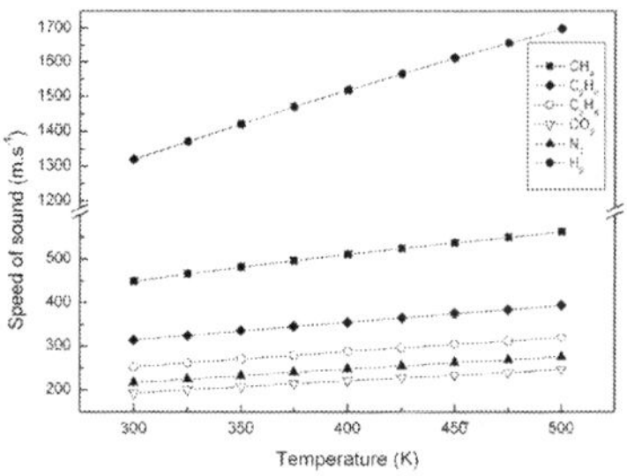

Fig. 7. Speed of sound for main constituents of natural gases

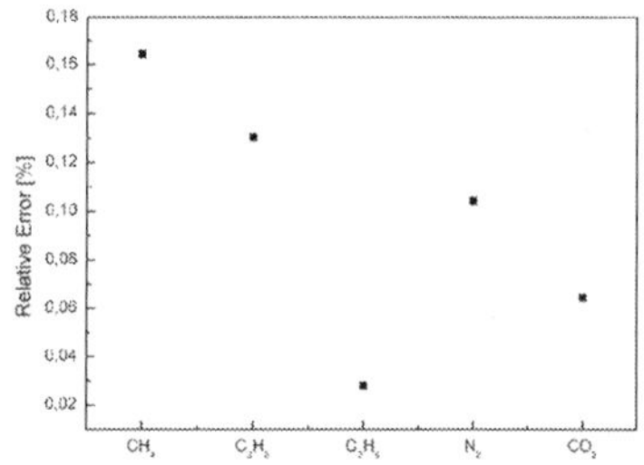

Fig. 8. Relative error between hand-made function and CHEMKIN for speed of sound

2.4.2 Sound velocity sensor

Acoustic wave propagation is characterized by the speed of sound *c* in the propagation medium. Several techniques allow the measurement of this characteristic in gases. Three methods of measurement can be distinguished such as:

- the acoustic waves dephasing,
- the acoustic resonator,
- the time of transit.

The last method is largely used in industrial applications such as level measurement, flow metering, etc… (Hauptmann et al., 2002). It involves measurement of the transit time of an ultrasonic pulse travelling over a known propagation distance in the gas. This technique typically employs one or more piezoelectric transducers to generate and detect sound waves in the frequency range of about 20 kHz to 1 MHz and higher. A particular technique known as a "pulse echo" technique uses a single transducer as both the transmitter and the receiver see Figure 9. The generated sound wave is reflected back to the source transducer from a target located at a known distance from the transducer, and is received by the same transducer. If the distance between the transducer and the reflecting target is *D*, and the measured two-way travel time is *t*, then the speed of sound is represented by:

$$c = \frac{2D}{t} \tag{48}$$

This method is advantageous because it uses only one transducer. However, in applications requiring high precision speed of sound measurements, the method has the disadvantage of introducing time delay errors associated with imperfectly defined and variable distance, *D*, and an imperfect ability to determine the exact time delay with respect to the time of the transmitted pulse and the time instant when the reflected sound wave is received at the transducer.

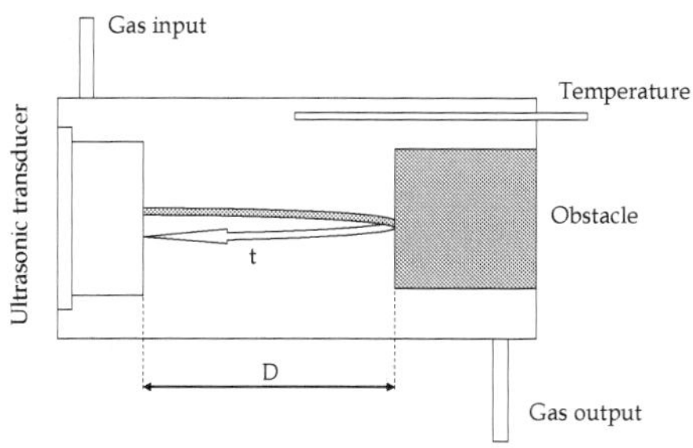

Fig. 9. "Pulse echo" technique

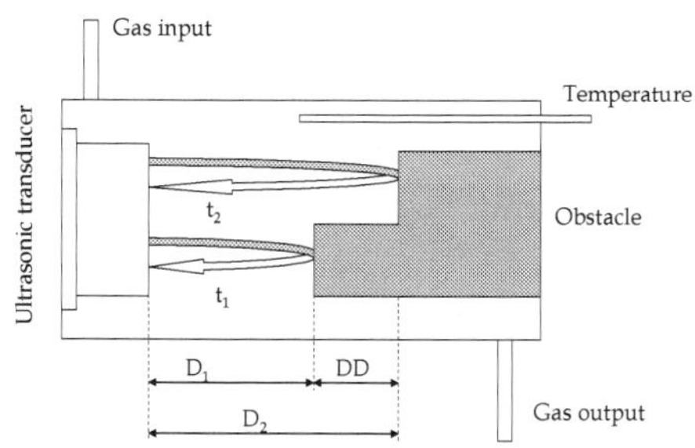

Fig. 10. Modified "pulse echo" technique

To reduce the time delay error, the pulse echo method may be modified to measure a time difference between two received signals (Kelner et al., 2004). A transmitted wave is reflected from two different targets rather than a single target, see Figure 10. The distance, *DD*, between the two targets is known. Using this method, the speed of sound is represented by:

$$c_{gas} = \frac{2\,DD}{\Delta t} \tag{49}$$

where Δt is the time difference between the two received signals.

2.5 Refractive index

Guérin (1981) expressed refractive index n_g of a gas, for radiation of wavelength μ, according to the density:

$$n_g^{\mu^0} - 1 = q\,\rho_g \overset{def}{=} IR \tag{50}$$

Where q is a constant.

Noting RI_0 the value of RI corresponding to the normal conditions (273,15 K, 1 atm) and assuming that the gases follow ideal gas law, the value of RI (called co-index of refraction, but named improperly refractive index too) relates to temperature T (in Kelvin) and pressure P (in atmosphere) is given by:

$$RI = RI_0 \frac{T_0}{T} \frac{P}{P_0} \tag{51}$$

Co-index of refraction has an additive property:

$$RI = \sum_{i=1}^{\upsilon} x_i\, RI_i \tag{52}$$

Equations (51-52) are enough to calculate with precision the co-index of refraction of natural gases.

2.6 Density and specific density

In the case of a gas mixture, the expression of the specific density d_m is written:

$$d_m = d_m^{IGL} \frac{Z_{air}(T,P)}{Z_m(T,P)} \tag{53}$$

with

$$Z_m = 1 - \left(\sum_{i=1}^{\upsilon} x_i \sqrt{1 - Z_i} \right)^2 + 0.0005 \left(2\, x_H - x_H^2 \right) \tag{54}$$

With Z_i compressibility factor of component i, x_H molar fraction of hydrogen.

Specific density d_m^{IGL} is independent of any state of reference and is calculated starting from the equation:

$$d_m^{IGL} = \sum_{i=1}^{\upsilon} x_i \frac{M_i}{M_{air}} \tag{55}$$

In the same way, the density is obtained by:

$$\rho(T,P) = \frac{\rho_m^{IGL}(T,P)}{Z_m(T,P)} \tag{56}$$

$$\rho_m^{IGL} = \left(\frac{P}{RT} \right) \sum_{i=1}^{\upsilon} x_i\, M_i \tag{57}$$

2.7 Synthesis

Quality of natural gas, mainly composed of methane, varies according to the various sources of supply (layers). Consequently, physical properties and energy content are subject to variations. As a result, one of the important information required for natural gas exploitation relates to its physical properties. Besides the properties of transport (viscosity, thermal conductivity), various models of determination speed of sound, index of refraction and density were presented.

3. Combustion features

Combustion features of a gas such as the low heating value, Wobbe index and air-fuel equivalence ratio are of a great industrial interest. These properties interest both engine manufacturers and business activities of CHP installations and boilers. The commercial transactions on natural gas are generally based on the energy content of gas, obtained by multiplying the volumes measured by the higher heating value.

3.1 Air Fuel Ratio

Air Fuel ratio ε is defined as the ratio of air volume (or mass) V_a (at normal conditions of temperature and pressure) required to the theoretical complete combustion per fuel volume unit (or mass). Complete combustion of generic fuel $C_xH_yO_zN_u$ under stoichiometric conditions gives equivalence ratio [Nm3/Nm3]:

$$C_xH_yO_zN_u + \varepsilon\left(79\%\,N_2 + 21\%\,O_2\right) \xrightarrow{stoich} \nu_{N_2} N_2 + \nu_{CO_2}\,CO_2 + \nu_{H_2O} H_2O \tag{58}$$

$$\begin{cases} x = \dfrac{x_{CH_4} + 2\,x_{C_2H_6} + 3\,x_{C_3H_8} + 4\,x_{C_4H_{10}} + 5\,x_{C_5H_{12}} + x_{CO_2}}{x_{CH_4} + x_{C_2H_6} + x_{C_3H_8} + x_{C_4H_{10}} + x_{C_5H_{12}} + x_{CO_2} + x_{O_2} + x_{N_2}} \\ y = \dfrac{4\,x_{CH_4} + 6\,x_{C_2H_6} + 8\,x_{C_3H_8} + 10\,x_{C_4H_{10}} + 12\,x_{C_5H_{12}} + x_{CO_2}}{x_{CH_4} + x_{C_2H_6} + x_{C_3H_8} + x_{C_4H_{10}} + x_{C_5H_{12}} + x_{CO_2} + x_{O_2} + x_{N_2}} \\ z = \dfrac{2\,x_{CO_2}}{x_{CH_4} + x_{C_2H_6} + x_{C_3H_8} + x_{C_4H_{10}} + x_{C_5H_{12}} + x_{CO_2} + x_{O_2} + x_{N_2}} \\ u = \dfrac{2\,x_{N_2}}{x_{CH_4} + x_{C_2H_6} + x_{C_3H_8} + x_{C_4H_{10}} + x_{C_5H_{12}} + x_{CO_2} + x_{O_2} + x_{N_2}} \end{cases} \tag{59}$$

$$\varepsilon = \frac{1}{21\%}\left(x + \frac{y}{4} - \frac{z}{2}\right) \tag{60}$$

$$\varepsilon = \sum_{i=1}^{\upsilon} x_i\,\varepsilon_i \tag{61}$$

Industrial combustion is never complete, dissociations/recombinations occurred.

$$\begin{aligned} C_xH_yO_zN_u + \lambda\,\varepsilon\left(79\%\,N_2 + 21\%\,O_2\right) \longrightarrow\ & \nu_{N_2} N_2 + \nu_{CO_2}\,CO_2 + \nu_{H_2O} H_2O + \\ & \nu_{O_2} O_2 + \nu_{CO}\,CO + \nu_{H_2} H_2 + \\ & \nu_{OH} OH + \nu_{NO}\,NO + \nu_{NO_2} NO_2 + \ldots \end{aligned} \tag{62}$$

Where λ is the relative air fuel ratio.

3.2 Heating value

Low heating value is the energy released during fuel combustion (of unit of mass or volume) under stoichiometric condition and thermodynamic conditions (*P*, T) giving CO_2 and H_2O products. Through the world, different thermodynamic reference conditions are considered as reference conditions.

$$LHV = \sum_{i=1}^{\upsilon} x_i \; LHV_i \tag{63}$$

High heating value *HHV* is deduced from low heating value *LHV* and is defined as the heat that can be obtained by condensing the water vapor produced by combustion.

3.3 Wobbe index

Wobbe index (*W*) is an important criterion of inter-changeability of gases in the industrial applications (engines, boilers, burners, etc). Gas composition variation does not involve any notable change of air factor and of flame speed when Wobbe index remains almost constant. Wobbe index can be calculated starting from the high heating value (*HHV*) and specific gas density (*d*) by:

$$W = \frac{HHV}{\sqrt{d}} \tag{64}$$

This parameter is usually used to characterize gas quality. Indeed, two gases with the same Wobbe index deliver the same quantity of heat for the same supply pressure. Thus, for an industrial burner for example, one maintains heat flow with a constant value by the output control of gas according to the index of Wobbe.

In DOE report (2007), a modified Wobbe index is used in real applications:

$$W_r = \frac{LHV}{\sqrt{d\,T}} \tag{65}$$

This modified Wobbe index takes account for heating of the fuel and the uncovered heat from water vapour formed during combustion.

3.4 Methane number

Methane number (*MN*) characterizes gaseous fuel tendency to auto-ignition. By convention, this index has the value 100 for methane and 0 for hydrogen (Leiker et al., 1972). The gaseous fuels are thus compared with a methane-hydrogen binary mixture. Two gases with same value of *MN* have the same resistance against the spontaneous combustion

4. Measuring instruments

Combustion features can be determined according to two types of methods: direct or indirect. Direct methods are based on calorimetric measures where the energy released by the combustion of a gas sample is measured. Indirect methods are issued of either calculation from gas composition, or of measurements of gas physical properties.

4.1 Calorimeter

This direct method is based on calorimetric measures. Ulbig & Hoburg (2002) synthesized measurement of heat value by:

- combustion of a gas sample inside a calorimetric bomb (isochoric combustion),
- combustion of a gas with a gas-burner (isobar combustion),
- catalytic combustion (isobar combustion without flame) by oxidation of a gas on a catalyst.

Combustion technique with a gas-burner is largely used. Various types of calorimeters, based on this technique, are employed: Junkers, Reinke, Thomas--Cambridge and Culter--Hammer. Operation principle, presented on Figure 11, is identical. Specific quantity of gas is measured then burned completely. In a heat exchanger, energy released by combustion heats a coolant (water or air). Consequently, coolant temperature increase makes it possible to measure gas heating value. Apparatus calibration is done using reference gas which its specific heating value is known (in general pure methane).

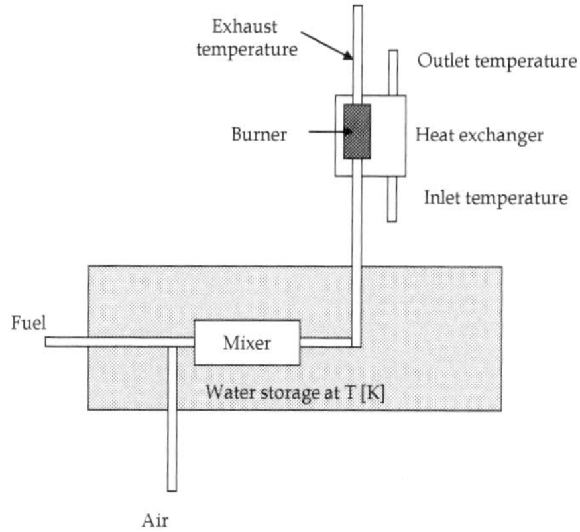

Fig. 11. Calorimeter operation principle

Catalytic combustion is safe way (flameless) to measure high heating value of gases (Hornemann, 1995), (Heyden & Berg, 1998). This batch method is based on the following principle: gas mixture and air are introduced on a noble metal (platinum). Air quantity introduced is sufficient for gas mixture oxidation. Hydrocarbons are oxidized over noble metal being a catalyst. The procedure is renewed thereafter with an unknown gas mixture. Heat released can be measured either starting from temperature changes related to the catalytic reaction, or starting from electric output changes required to keep catalyst at constant temperature. This method can however be subject at two errors: incomplete gas oxidation or catalyst poisoning.

4.2 Stoichiometric combustion

For saturated linear hydrocarbons (alkanes), there exists a linear relation between air fuel ratio and low heating value of gas mixtures, see Figure 12. This measurement principle is thus issued on air volume knowledge per unit of gas volume under stoichiometric combustion. Consequently, that makes it possible to reach the calorific value of gas starting from the following relation see (Ingrain, 1990):

$$LHV = K\left(\frac{V_a}{V_m}\right) \tag{66}$$

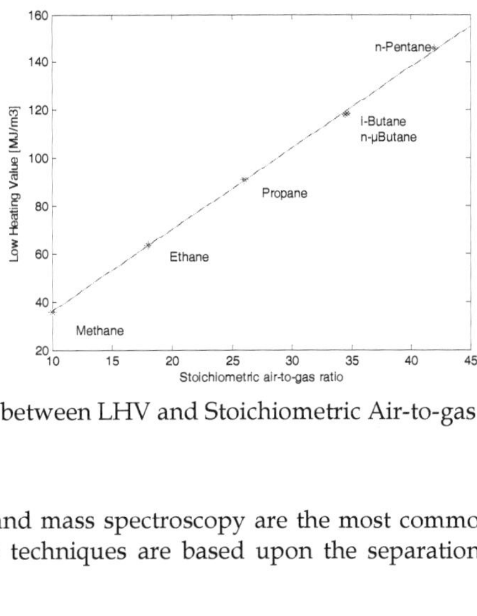

Fig. 12. Linear relation between LHV and Stoichiometric Air-to-gas ratio

4.3 Gas composition

Gas chromatography and mass spectroscopy are the most commonly employed laboratory techniques. These two techniques are based upon the separation of gas species followed detection.

4.3.1 Gas chromatography

Gas chromatography is a partition method. It is based on components distribution of a sample between mobile phase (the gas) and stationary phase (liquid or solid), see Figure 13 upon a column. The column provides a pathway, which aims to separate the species based upon molecular size, charge, polarizability, and other physical parameters which limit interactions between the gas species and the column materials. If the components of the sample have different partition coefficients between the two phases, they migrate with different speeds. An inert carrier gas (e.g. nitrogen or helium) is used to transport the gas sample through the columns.

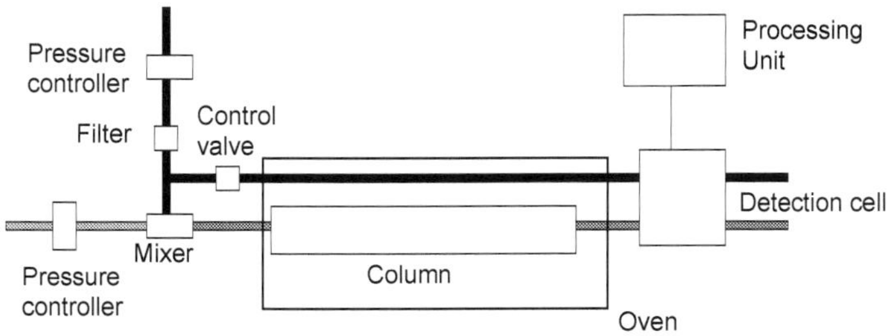

Fig. 13. Gas chromatography principle

Gas chromatography is considered accurate and reliable. However, it is usually slow and the columns require maintenance. It is not considered practical for a continuous on-line application.

Calculation of heating value and Wobbe index of gas are obtained regards to the standard procedure ISO 6976. For a natural gas, the higher heating value is written see (Ingrain, 1990):

$$HHV = \frac{\sum_{i=1}^{\upsilon} x_i \; HHV_i}{Z_m} \tag{67}$$

Where Z_m is the compressibility factor, see Equation (54).

4.3.2 Mass Spectrometer

Mass spectrometer is based mainly upon the mass-to-charge ratio of ionized species. The mass spectrum is generated by first ionizing natural gas and the accelerating it with an electric field. The ions are separated by their momentum distribution.

Most mass spectrometers have software that determines gas concentrations from peak intensities and allows real time calculations

4.4 Gas composition by Infrared spectroscopy

Infrared spectroscopy method exploits the property that natural gas components absorb light in a given wavelengh of the infra-red spectrum. Only the hydrogen, which do not absorb the infrared radiation, and the carbon monoxide, which absorbs in another area of the spectrum, do not take part in this phenomenon. The general diagram is represented on the figure 14.

From calibration, using the absorption band of the methane and the band of hydrocarbons higher than methane, the measurement of radiation absorption lead to the determination of the heating value of natural gas. However, the components higher than the C_4, as well as hydrogen, do not absorb the infra-red radiation. Consequently, those components are not taken into account for the calculation of the calorific value

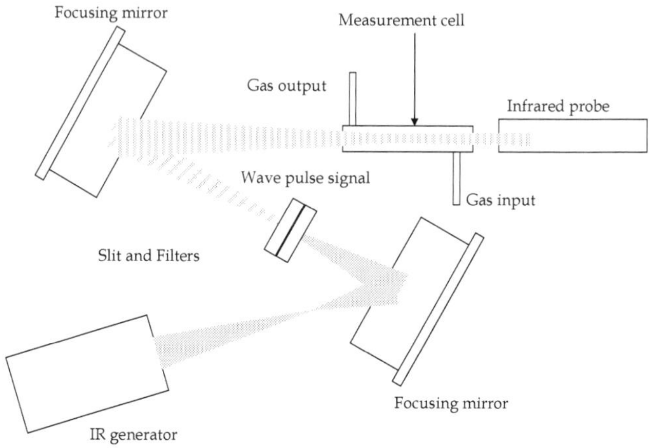

Fig. 14. Infrared spectroscopy principle

5. Physical properties methods

Besides the combustion or analysis of fuel gases, it is also possible to correlate and, respectively, calculate the calorific value by measuring the physical properties of the gas mixture. Correlations have to be determined, allowing the calorific value to be reliably calculated as a function of one or several different physical properties.

- For gases distributed in the United Kingdom, British Gas (Thurston et al., 2002) established correlation to estimate low heating value *LHV* regards to speed of sound measurement and thermal conductivity of natural gas. Low heating value is thus given by the expression :

$$LHV = a_1\ \lambda_{T_H} + a_2\ \lambda_{T_L} + a_3\ c + a_4\ T_a + a_5\ T_a^2 + a_6 \tag{68}$$

  Where λ_{T_H} is the thermal conductivity of natural gas at temperature T_H, respectively λ_{T_L} the thermal conductivity at temperature T_L, c is the speed of sound, T_a the ambient temperature, a_1 to a_6 are the fitted coefficients.

  Thurston et al. (2002) proposed : $a_1 = 36.25649$, $a_2 = -45.5768$, $a_3 = 0.047029$, $a_4 = 0.091067$, $a_5 = 0.00074$ and $a_6 = 24.18731$ under conditions $T_H - T_a = 70\ K$ and $T_L - T_a = 50\ K$.

- In partnership with the department of energy of the United States, Morrow and Behring (1999), of Southwest Research Institute, have developed a correlation based on speed of sound, nitrogen and carbon dioxide contents. *HHV* is evaluated by:

$$HHV = \frac{54343.048 + B\, M_{CH}}{M} \tag{69}$$

$$B = 20442.406 + \left(0.04552871 - 0.02523803\, x_{N_2} - 0.02568212\, x_{CO_2}\right)c \tag{70}$$

Where x_{N_2} is the volume fraction of N_2, respectively x_{CO_2} the volume fraction of CO_2, and c is the speed of sound, M is the mass molar of the mixture, evaluated by:

$$M = \sum_{i=0}^{2}\left(a_0^i + a_1^i\, x_{N_2} + a_2^i\, x_{CO_2}\right)c^i \tag{71}$$

- Bonne (1996) proposed a general expression :

$$HHV = 3643.53 + 1050.71\left(\frac{102}{\eta}\right)^3 - 7.60221\, \lambda_{T_H} - 2294.2\left(\frac{\lambda_{T_L}}{\lambda_{T_H}}\right) \tag{72}$$

Where λ is the thermal conductivity (cal/s m °C) and T_L, T_H are the low and high temperatures respectively (°C). η is the gas viscosity (μpoise). This HHV estimation has a maximal error of 0.067 $MJ.m^{-3}$ with a standard error of 0.01831 $MJ.m^{-3}$.

$$HHV = -1287.7 + 808700\, C_p^{0.73846} - 1048800\, \lambda^{-1.742} - 9.0189\, 10^{-4}\left(\eta\, M\right)^{1.7514} \tag{73}$$

- Pinvidic et al. (2000), of Gaz de France, developed a correlation of high heating value regards to light beam optical absorption by gas components. Light beam defines three bands of wave length measure, located to the near infrared, with a width of 10 to 20 10^{-9} *m*. *HHV* is deduced from the measurement of temperature *T*, pressure *P* and transmission coefficients $\tau_i, i \in \{1, 2, 3\}$ of natural gas in three bands. *HHV* is expressed thus by the following relation:

$$HHV = \sum_{i=1}^{3}\left(a_{0,i} + a_{1,i}\,\frac{T}{P}\right)\ln(\tau_i) + \left(a_0 + a_1\,\frac{T}{P}\right) \tag{74}$$

Where $a_{0,i}$, $a_{1,i}$, a_0 and a_1 are obtained from reference gases. This fitting has a maximum error of 1%.

- Tacke and Kastner (2003) developed measure device of *HHV*, in particular, for natural gas. The apparatus set up uses infrared radiation dispersion regards to wavelengths getting a spectrum. This spectrum is detected by a system to be analyzed thereafter. *HHV* is deduced through the correlation:

$$HHV = \sum_i \alpha_{\mu_i} \; HHV_{\mu_i} \tag{75}$$

Where α_{μ_i} is the absorption coefficient for wavelength μ_i and HHV_{μ_i} is the high heating value of the component which has its wavelength μ_i.

- Florisson and Burrie (1989) determined Wobbe index based on density estimation and nitrogen and carbon dioxide contents. The correlation is valid only for gases whose Wobbe index is in the range 43.4 and 44.4 $MJ\ m^{-3}$.

$$W = \frac{5.671 + 61.36\ d - 98.97\ x_{CO_2} - 64.57\ x_{N_2}}{\sqrt{d}} \tag{76}$$

- Pickenäcker et al. (2000) showed that Wobbe index can also be given by a technique based on measurements of dynamic viscosity. They obtained the following correlation:

$$W = -28.486\ \ln\left[\eta\left(\frac{T_{ref}}{T}\right)^{0.8405}\right] + 311.10 \tag{77}$$

- Rahmouni et al. (2003-2004)

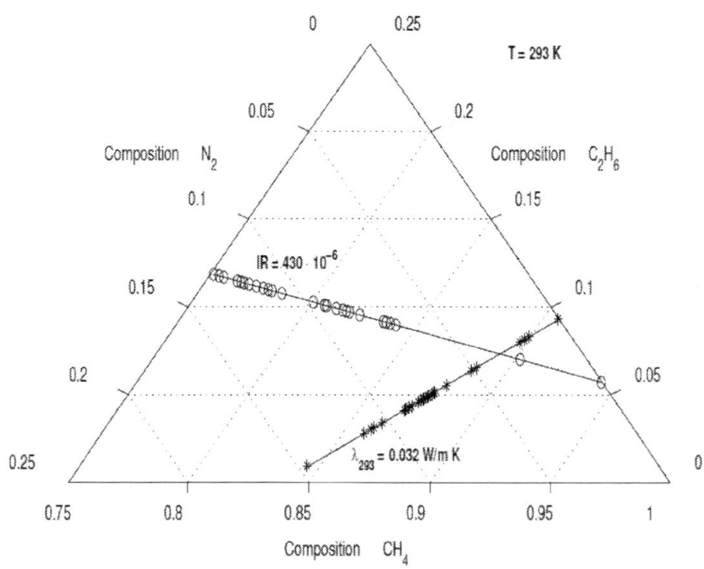

Fig. 15. Determination of the ternary composition by using a mixture diagram.

A non-correlative method was proposed by Rahmouni et al. (2004). Combustion properties of natural gas were calculated from a ternary composition that has the same physical properties as the tested gas. As illustrated on Figure 15, when measured physical properties of natural gas are represented in a ternary diagram mixture, an 'equivalent' ternary composition is determined. That pseudo-composition has no connection to the composition of the real gas (natural gas of five or more constituents). The term 'equivalent' means that the three compounds – gas, has the same two physical properties than the real gas (same refraction index and same thermal conductivity for instance). Then, the combustion properties are evaluated using the 'equivalent' ternary composition (or pseudo-composition) and match the real gas combustion properties.

The choice of these properties was based on a statistical analysis (numerical experiment plan, principal component analysis) of a natural gas base distributed in Europe. The basic assumption is that if the measurement of two physical properties corresponds to the description of a ternary gas, then the properties of combustion of this ternary gas will be also those of the real gas which models the ternary gas, with the help of a reasonable error.

The statistical analysis concluded that methane and ethane, then comes propane and nitrogen, are the most influent component of natural gas. Thus, two ternary mixtures can be considered: CH_4-C_2H_6-C_3H_8 or CH_4-C_2H_6-N_2.

Three combinations Ternary diagram / physical properties were used:

- The measurement of thermal conductivity and co-index of refraction for T = 293 K and using CH_4-C_2H_6-C_3H_8 diagram,
- Combination of the precedent measurements with the CH_4-C_2H_6-N_2 diagram,
- The measurement of thermal conductivity for two temperature levels 293 K and 343 K.

The relations below describe the intersection of two physical properties in a ternary diagram which give the ternary pseudo-composition:

$$x_1 = x_{10} + \left(1 - \frac{a_1(\phi_1)}{\sqrt{3}}\right) \cdot \left(\frac{b_1(\phi_1) - b_2(\phi_2)}{a_2(\phi_2) - a_1(\phi_1)}\right) + \frac{b_1(\phi_1)}{\sqrt{3}} \tag{78}$$

$$x_2 = x_{20} + \frac{2\,a_1(\phi_1)}{\sqrt{3}} \cdot \left(\frac{b_2(\phi_2) - b_1(\phi_1)}{a_1(\phi_1) - a_2(\phi_2)}\right) + \frac{2\,b_1(\phi_1)}{\sqrt{3}} \tag{79}$$

$$x_3 = 1 - x_1 - x_2 \tag{80}$$

Where ϕ_i : Physical property *i*, x_1, x_2 and x_3 : Ternary gas composition, x_{10} : Lower limit of x_1 axis, x_{20} Lower limit of x_2 axis, a_1, b_1, a_2 and b_2 : Coefficients depending on physical properties and temperature. Once the composition of the ternary mixture is determined, the combustion properties can be calculated using Equations 63, 64 and 65.

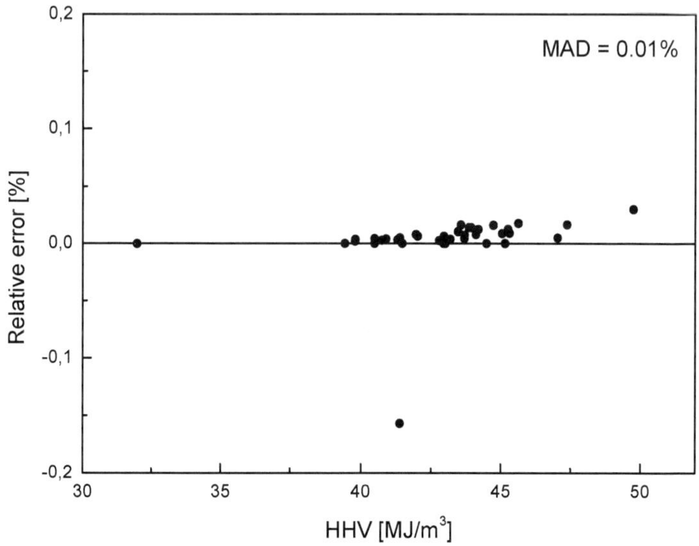

Fig. 16. Relative error between pseudo-composition and direct GC analysis for high heating value

Based on the work of Rahmouni et al. (2003-a and 2003-b), Loubar et al. (2007) have used a quinternary pseudo-composition in order to improve the accuracy of combustion properties determination. However, with a pseudo-composition of five components, it is impossible to use the previous graphical method involving mixture diagrams. So, they determined the compositions by solving a nonlinear system of equations expressing, for quinternary gases, the thermal conductivity (Equation 30), at two fixed temperatures (T_1 = 333 K and T_2 = 383 K), the speed of sound determined at T_3 = 303 K (Equation 36) and the carbon dioxide content. Figure 16 shows the percentage of the relative error of high heating value, calculated from pseudo-composition and the GC analysis. The mean absolute deviation (MAD) is about 0.01%.

6. Summary

The actual composition of natural gas depends primarily on the production field from which it is extracted and limited variations in composition must therefore be accepted. Moreover, at a local distribution level, seasonal adjustments by the local gas distributor may cause significant variations in the gas composition. Consequently, physical properties and energy content are subject to variations and their calculation / estimation is of great importance for technical and economical aspects.

In this chapter, physical models for the calculation of physical properties of natural gas are presented. Physical models of transport properties (viscosity, conductivity) result from the kinetic theory of gases. Viscosity and thermal conductivity of gases mixture (like natural gas) can be estimated from the properties of pure gases and requiring some correction factors. These physicals models offer a good tool for industrial calculations and applications. Speed of sound can be easily determined from the composition of natural gas. Besides, different methods and devices are developed to measure this property with a good accuracy. A particular technique known as a "pulse echo" was presented. This technique uses a single ultrasonic transducer as both the transmitter and the receiver and two obstacles separated by a known distance. The speed of sound of natural gas is then determined from the transit time and the distance between the targets.

In addition, properties such as refractive index and the specific density were presented. These properties can be used in correlations in order to estimate other properties that are difficult to measure or need expensive instruments.

Various techniques of determination of combustion features such as air-fuel ratio, the low heating value and Wobbe index are exposed. These techniques are based on direct or indirect methods. Besides the combustion or analysis of fuel gases, it is possible to correlate and, respectively, calculate the combustion features by measuring the physical properties of the gas mixture. Some correlations can be limited to a region, a quality of natural gas or a specified range of combustion properties variation. So, it is important to consider those correlations according to their specified conditions. Researchers and industrials continue their efforts in developing methods and devices to reach rapid and accurate measurement / determination of natural gas properties regarding the stakes.

7. References

Bonne U. (1996). Sensing fuel properties with thermal microsensors. Proc. SPIE Vol. 2722, p. 165-176, Smart Structures and Materials 1996: Smart Electronics and MEMS, Vijay K. Varadan; Paul J. McWhorter; Eds.

Chapman C. & Cowling T.G. (1970). *The mathematical theory of non-uniform gases*, Press Syndicate of the University of Cambridge, ISBN 0 521 40844.

Chung T.H., Lee L.L. & Starling K.E. (1984). Applications of Kinetic Gas Theories and Multiparameter Correlation for Prediction of Dilute Gas Viscosity and Thermal Conductivity, *Industrial & engineering chemistry fundamentals*, Vol. 23, No. 1, p 8-13, ISSN 0196-4313

Chung T.H., Ajlan M., Lee L.L. & Starling K.E. (1988). Generalized Multiparameter Correlation for Nonpolar and Polar Fluid Transport Properties, *Industrial & engineering chemistry research*, Vol. 27, No. 4, p 671-679, ISSN 0888-5885.

DOE (2007) LNG Interchangeability/Gas Quality: Results of the National Energy Technology Laboratory's Research for the FERC on Natural Gas Quality and Interchangeability, *DOE/NETL-207/1290*, prepared by U.S. Department of Energy, National Energy Technology Laboratory

Estela-Uribe J.F., Jaramillo J., Salazar M.A. & Trusler J.P.M. (2003). Viriel equation of state for natural gas systems. *Fluid Phase Equilibria*, Vol. 204, No. 2, pp. 169--182. ISSN 0378-3812

Estela-Uribe J.F. and Jaramillo J. (2005). Generalised viriel equation of state for natural gas systems. *Fluid Phase Equilibria* Vol. 231, No.1, pp. 84--98, ISSN 0378-3812.

Florisson O. & Burrie P.H. (1989). Rapid determination of the Wobbe index of natural gas. *Journal of physics. Part E. Scientific instruments* ISSN 0022-3735, Vol. 22, No.2, p. 123-128.

Guérin H. (1981) *Traité de manipulation et d'analyse des gaz.* Ed. Masson 2nd édition.

Greenspan M. and Wimenitz F.N. (1953). *An acoustic viscometer for gases- Part I.* National Bureau of Standard Report No. 2658. pages 25. OSTI ID: 4379711.

Hanley H.J., McCarty R.D. & Haynes W.N. (1975). Equation for the viscosity and thermal conductivity coefficients of methane, *Cryogenics,* Vol. 15, No. 7, p413-417

Hauptmann P., Hoppe N. and Püttmer A. (2002). Application of ultrasonic sensors in the process industry. *Meas. Sci. Technol.* 13 R73 - R83, ISSN 0957-0233

Heyden W.H.V. & Berg R.A. (1998) Measuring heating value using catalytic combustion. *United States Patent,* US 0 575 986 2A, pp. 1--14, 1998.

Hirschfelder J.O., Curtiss C.F. & Bird R.B. (1954). *Molecular theory of gases and liquids,* Ed. John Wiley, ISBN 978-0-471-40065-3.

Hornemann J.A.T. (1995) *Method for determining the calorific value of a gas and/or the Wobbe index of natural gas.* European Patent, EP 0 665 953 B1, pp. 1--9.

Ingrain D. (1990). Caloric value of gases (in french). *Technique de l'ingénieur,* ISSN 0399-4147 Vol. RC3, No. R 2980, pp. 1--10.

Jaescheke M., Benito A., Fredheim A., Henault J. M., Viglietti B., van Wesenbeeck P., Klimeck R., Kunz O. & Wagner W. (2003), GERG Project : Wide-range reference equation of state for natural gases. *Gas- und Wasserfach. Gas Erdgas,* Vol. 144, No. 7, p. 430 – 435, ISSN 0016-4909.

Jaeschke M. and Schley P. (1995). Ideal gas thermodynamic properties for natural gas applications. *International Journal of Thermophysics,* Vol. 16, No. 6, p. 1381 – 1392, ISSN 0195-928X, DOI 10.1007/BF02083547.

Kelner E., Minachi A., Owen T. E, Burzynski JR. M. and Petullo S. P. (2004). Device for precision measurement of speed of sound in a gas. *United States Patent,* US 2004/0093948 A1.

Leiker M., Christoph K., Rankl M., Cartellieri W. & Pleifer U. (1972). Evaluation of anti-knocking property of gaseous fuels by means of methane number and its practical application to gas engines, *ASME paperNo. 72-DGP-4, Vol. 94, Issue 7, p55*

Le Neindre B. (1998). *Heat conductivity of gases and liquids* (in french), Ed. Technique de l'ingénieur, K 427, p1-43.

Loubar K., Rahmouni C., Le Corre O. and Tazerout M. (2007). A combustionless determination method for combustion properties of natural gases, *Fuel,* Vol. 86, No. 16, p 2535-2544, ISSN 0016-2361, DOI: 10.1016/j.fuel.2007.02.024.

Morrow T.B. and Behring K.A. (1999). Energy flow measurement technology, and the promise of reduced operating costs. *4th International Symposium on Fluid Flow Measurement, Denver,* CO. June 28--30, pp. 1--12, 1999.

Mason E.A. and Monchick I. (1962). Heat Conductivity of Polyatomic and Polar Gases, *The journal of Chemical Physics,* Vol. 36, p. 1622- 1962. ISSN 0021-9606 doi:10.1063/1.1732790

Mason E.A. and Saxena S.C. Approximate formula for the thermal conductivity of gas mixtures, *The Physics of Fluids,* Vol.1, No. 5, p. 361-369, ISSN 0031- 9171 doi:10.1063/1.1724352.

Neufeld P.D., Jansen A.R. & Aziz R.A. (1972). Empirical equations to calculate 16 of the transport collision integrals $\Omega^{(l,s)^*}$ for the (12-6) potential, *Journal of Chemical Physics*, Vol. 57, p 1100-1102, ISSN 0021-9606

Pickenäcker K., Trimis D. & Wawrzinek K. (2000). Excess air controlled operation of boilers and furnaces by means of Wobbe number measurement. *5th European Conference on Industrial Furnaces and Boilers (INFUB5)*, ISBN 972-8034-04-0, Espinho-Porto, Portugal (11/04/2000)

Pinvidic J.J., Juen G.L. & Pelous G.P. (2000). Method and apparatus for determining the calorific value of a natural gas optically and in real time. *United States Patent*, US S6157455A.

Saikaly K., Rousseau S., Rahmouni C., Le Corre O. & Truffet L. (2008). Safe operating conditions determination for stationary SI gas engines, *Fuel Processing Technology*, Vol. 89, No. 11, 1p169-1179, ISSN 0378-3820

Rahmouni C., Le Corre O. and Tazerout M. (2003-a). Online determination of natural gas properties, *Comptes Rendus Mecanique*, Vol. 331, No. 8, Pages 545-550, ISSN 1631-0721, DOI: 10.1016/S1631-0721(03)00126-8.

Rahmouni C., Tazerout M. and O. Le Corre (2003-b)., Determination of the combustion properties of natural gases by pseudo-constituents*, Fuel, Vol. 82, No. 11, Pages 1399-1409, ISSN 0016-2361, DOI: 10.1016/S0016-2361(03)00029-2.

Rahmouni C., Tazerout M. and Le Corre O. (2004). Method for determining at least one energetic property of a gas fuel mixture by measuring physical properties of the gas mixture, Patent US2004195531, 2004-10-07, Also published as US7091509, FR2827961, RU2004102387, WO03012435, EP1412742

Reid R.C., Prausnitz J.M. and Poling B.E. (1987). *The properties of gases & liquids*. Ed. McGraw Hill Book Company, Fourth Edition, ISBN 978-0-07-118971-2.

Rojey A., Durand B., Jaffret C., Jullian S. & Valais M. (2000). *Natural gas : Production - processing - transport (translate by N. Marshall)*, Eds Technip, ISBN-13 978-2710806936

Starling K. E. and Savidge J. L. (1992), *Compressibility factors of natural gas and other related hydrocarbon gases. AGA Transmission Measurement Committee*. Report 8, American Gas Association.

Tacke M. & Kastner J. (2003). Photometric device and photometric method for determining the gross calorific value of a test gas. *United States Patent*, US 6555820B1.

Thurston R.R., Hammond P.S. and Price B.L. (2002). Method and apparatus for measuring the calorific value of a gas. *United States Patent*, US 6442996B1.

Ulbig P. & Hoburg D. (2002). Determination of the calorific value of natural gas by different methods. *Thermochimica acta*. Vol. 382, No.1-2, pp. 27—35, ISSN 0040-6031.

Vogel E., Wilhelm J., Küchenmeister C. & Jaeschke M. (2000). High-precision viscosity measurements on methane, *High temperatures-high pressures*, Vol. 32, No. 1, p. 73-81, doi:10.1068/htwu359.

Wilhem J., Gillis K.A., Mehl J.B. and Moldover M.R. (2000). An improved Greenspan acoustic viscometer. *International Journal of Thermophysics*, Vol .21, No. 5, p. 983-997, ISSN 0195-928X, DOI: 10.1023/A:1026471901657

Wilke C.R. (1950). A viscosity for gas mixtures, *Journal of Chemical Physics*, Vol. 18, p517-519, ISSN 0021-9606

3

The importance of natural gas reforming

Laédna Souto Neiva and Lucianna Gama
Federal University of Campina Grande, Paraíba, Brazil
National Agency of Petroleum, Natural Gas and Biofuels (ANP/PRH-25)

1. Introduction

Natural gas is a fossil fuel found in nature reserves, associated or not with petroleum. Its composition is a mixture of light hydrocarbons, generally alkanes, which are normally gaseous at room temperature. Methane is the most abundant gas, accounting for more than 85% of the natural gas, and the other constituents are light alkanes such as ethane, propane, butane, among others. The percentages of each constituent of natural gas vary depending on factors such as geological formation of the reservoir rock, as well as the type of organic matter that gave rise to the natural gas found. Recently, natural gas has attracted the interest of many researchers and the large amount of methane contained in natural gas has been considered an input in the production of other high-value products such as syngas and high purity hydrogen. Considering the global trend toward environmental preservation, which emphasizes clean and sustainable energy generation, it can be said that the interest of researchers for natural gas will increase significantly from now on (Odell & Rosing, 1983).

The interest in natural gas is directly related to the search for alternatives to replace petroleum-based fuels and for generating energy from sources less aggressive to the environment. This behavior resulted in the intensification of research and exploration, particularly among developing countries. The result was not only the increase in proven oil reserves but also in its geographic expansion (the existence of reserves and the possibility of their exploitation must be proven by tests). Until 1970, these reserves were concentrated in a few regions of the world, like North America and the former Soviet Union (ANEEL, 2008).

Awareness of the imminent scarcity of oil in the next decades is stimulating the search for a fuel that can partially replace petroleum-based fuels. Worldwide reserves of natural gas are under-exploited because they are not as valuable as the petroleum reserves. In some cases, when natural gas is associated with oil reserves, while the valuable petroleum is fully exploited, the associated gas of the same reserves is considered undesirable, volatilized into the atmosphere or burnt in the platform's flare. Fortunately, though, this situation is gradually changing, and natural gas is getting more attention, due to the growing need to produce hydrogen from hydrocarbons. Among the fossil fuels, natural gas is the most suitable for this application (Fishtik et al., 2000).

Energy generation is fundamental to the socioeconomic development of a country or region. Somehow, it is present in the entire chain of production, distribution and comsumption of goods and services. Equally important is the role of technology in the balanced and sustainable development of various economic sectors, especially power generation. The more it brings new knowledge and technology to a product or service, the higher its market value and its benefits to society, such as generating skilled jobs, improving the distribution of income and quality of life, impelling the economy and increasing the country's sovereignty (Pompermayer, 2009).

Meeting the energy demands has been a constant challenge for many countries, especially the least developed. Aware of this, Brazil has invested considerable resources in infrastructure and power supply, and has developed important technologies in specific segments such as hydroelectric power generation, transmission over long distances and integration of new electrical systems. This leadership has proved to be essential and will remain important to Brazil, but we must go further. In this new business context, we must be able to provide quality, safe, environmentally sustainable and low-cost energy services that require more leadership in specific segments. We need a broad technology-based supply chain of the energy sector, which includes electronics and nanostructured materials, among other items that involve technologies which are a privilege that few countries have afforded (Pompermayer, 2009).

In order to use natural gas to produce a clean fuel like high purity hydrogen to fuel cells for electric energy generation it is first necessary to bring natural gas to a catalytic process called natural gas reforming. This catalytic process is also known as reforming of methane. Natural gas reforming is based on a catalytic chemical reaction that aims to convert methane, the main constituent of natural gas, in a mixture of hydrogen and carbon monoxide. This mixture of gases (H_2 + CO), the product of natural gas reforming, is called syngas. Syngas is commonly used in the synthesis of important products of the petrochemical industry such as methanol and ammonia (Rostrup-Nielsen, 1984; Armor, 1999).

In this chapter, we set out the general approach we adopted concerning the importance of natural gas in the worldwide energy matrix, and also on the basic principles that guide the reforming of natural gas catalytic processes.

2. History of the Use of Natural Gas as Fuel

The use of natural gas by ancient civilizations (1000 BC) to make fire to light candles in religious temples or to fire kilns to bake ceramics is widely reported in the literature.

At the end of the XIX century, natural gas was already used in North America as a fuel to generate thermal energy for heating homes and other applications such as cooking. Since then, the use of natural gas has increased and was present in several areas, such as welding processes and other processes in the metallurgical industry, water heaters, illuminator place, clothes dryers, in addition to the applications already mentioned. Thus, natural gas has spent decades, throughout of the XIX and XX centuries, being used as fuel for generating thermal energy of various forms (Olah et al., 2006).

The use of natural gas as fuel has become even more widespread when its transport and storage processes were mastered and became more reliable. Large quantities of natural gas have already been lost during the processes of petroleum and gas extraction, and this is still happening today. In many cases, when the unique purpose of a platform is to extract petroleum from a reserve, the associated gas found in the same reserve is considered as a byproduct of the petroleum extraction process. This natural gas considered an undesirable byproduct is often released into the atmosphere or burnt in the platform of extraction.

Natural gas has been growing on the worldwide scenery after the discovery of its great potential for generating electricity. Thereafter, this fuel began to attract the attention of researchers, industry and environmentalists (Hoffmann, 2002). As a consequence, some developed countries began to recognize natural gas as a highly valuable raw material to be used in energy generation.

Since environmental preservation has become a major global concern, alternative sources of energy generation must be sought, so that the growing worldwide energy demand is met without damage to the environment, particularly with respect to the minimization of the major factors of global warming.

Currently, water and petroleum are considered the main fuels for power generation worldwide. However, these fuels are natural resources that are getting scarce and because they are so valuable and non-renewables it is of is vital and urgent studies related to the development of alternative forms of energy generation.

Within this context, natural gas is believed to be the most appropriate fossil fuel to generate electricity in an alternative and sustainable form, that may help preserve the natural reserves of water for more noble and humanitarian applications.

3. The growing need for extraction of hydrogen from hydrocarbons

Hydrocarbons are formed by molecules made up of carbon and hydrogen atoms. Methane, the simplest hydrocarbon molecule (CH_4) is the main constituent in natural gas. In the methane molecule, a single carbon atom is surrounded by four hydrogen atoms. Besides methane, the gas composition contains other light hydrocarbons such as ethane, propane, butane, and so on. Figure 1 shows two examples of constituent molecules of the hydrocarbons in natural gas. Hydrocarbons may have direct or branched-chain molecules. Carbon can also form multiple bonds with other carbon atoms, resulting in unsaturated hydrocarbons with double or triple bonds between carbon atoms (Olah et al., 2006).

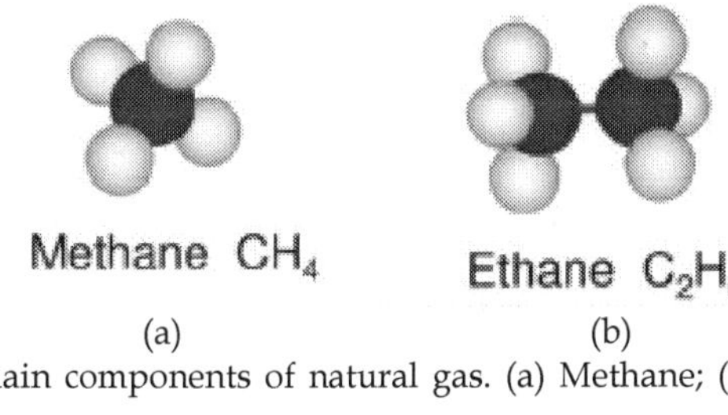

Fig. 1. Examples of main components of natural gas. (a) Methane; (b) Ethane (Olah et al., 2006).

All fossil fuels, natural gas, oil and coal, are basically composed of hydrocarbons, but they differ significantly regarding the number of hydrogen atoms and carbon atoms in their molecules. The main constituent of natural gas is methane (typically at concentrations above 80-90%) but are also found in varying proportions ethane, propane, butane, carbon dioxide, nitrogen, water, hydrochloric acid, methanol, and others. The proportion of each constituent in the final composition depends on a number of natural variables such as the formation and accumulation conditions in the reservoir (Odell & Rosing, 1983; ANEEL, 2008).

Hydrogen can be produced from hydrocarbons by their reforming or partial oxidation. Compared to other fossil fuels, natural gas is the most appropriate input for hydrogen production because of its availability for this purpose compared to oil, and also because it has the highest ratio hydrogen to carbon ratio, which minimizes the amount of CO_2 produced as a byproduct. Methane can be converted into hydrogen by steam reforming or dry, or by means of partial oxidation, or by both processes performed in sequence (Autothermic reforming). Steam reforming is the preferred method, which represents 50% of the global processes of conversion of natural gas for hydrogen production. This percentage reaches 90% in the U.S. In this process, natural gas (methane) reacts with water in vapor form in the presence of a metal catalyst in a reactor under high temperature and pressure conditions to form a mixture of carbon monoxide (CO) and hydrogen as reaction product, this product mixture being called synthesis gas. In a subsequent catalytic process for the reform process, the flow of hydrogen contaminated with CO will be oxidized to produce CO_2 and hydrogen as products. In this purification process, the hydrogen is recovered, while the byproduct CO_2 is generally volatilized to the atmosphere. In the future, however, the CO_2 shall be captured and isolated, Obec the environmental protection measures that support the control or combat global warming. The concept of producing hydrogen from oil, although established, is not attractive, since it is not expected to meet the global demand for energy in the long run, due to the scarcity of oil reserves. Coal, with the largest reserves among all other fossil fuels, may provide significant amounts of hydrogen, and the current technology to achieve this goal is called integrated gasification combined cycle (IGCC). As it occurs in the reforming of methane, coal is gasified by partial oxidation at high temperature and pressure. The synthesis gas generated in a mixture containing mainly CO and H_2 (also CO_2) must be subsequently subjected to catalytic processes to treat CO and, thus, purify the hydrogen stream. However, as coal has a low ratio of hydrogen / carbon, the process of obtaining hydrogen from coal would lead to a greater production of CO_2 by methane or even oil. A great amount of energy is required for the processes of capture and sequestration of CO_2, which makes it very expensive, and consequently, avoided by the industries of this area (Romm, 2004).

4. The Reforming of the Natural Gas

In order to insert natural gas into the energy worldwide matrix as an input for power generation, this gas must be subjected to some chemical catalysts for the removal of excess carbon in its composition. Thus, three catalytic chemical processes are used in the conversion of natural gas, composed of hydrocarbons, in a gas hydrogen flow of high purity. These three catalytic chemical processes are used sequentially and are as follows, respectively: 1. Natural gas reforming; 2. WGSR process (Water Gas Shift Reaction) and 3. PROX or SELOX reaction (Preferential Reaction Oxidation of the CO).

This chapter will discuss only the first catalytic chemical process, that is, the chemical process called natural gas reforming.

Natural gas reforming also known as reforming of methane can be accomplished by means of an exothermic or endothermic reaction depending on the chemical process selected to perform catalytic reforming of methane.

There are basically four different types of processes that can be used to carry out the reforming of methane. They are: 1. Steam reforming; 2. Dry reforming; 3. Autothermal

reforming and 4. Partial oxidation. All these four types of reforming of methane processes have the same purpose and lead to same final product. The purpose of the reforming of methane process, whatever it is, is to convert natural gas, mainly composed of methane molecules, in syngas. The product of the reforming of methane, the syngas, is a mixture of hydrogen and carbon monoxide.
In order to obtain a gas hydrogen flow of high purity from natural gas it is necessary that the syngas (H_2 + CO) obtained as a product of the reforming of the natural gas process be subjected to the two previously mentioned catalytic chemical processes: WGSR process and PROX or SELOX reaction, in this sequential order.
A brief approach on the four types of catalytic chemical processes that can be used to carry out the reforming of methane follows.

4.1. Steam Reforming

The process of steam reforming of methane produces syngas (H_2 + CO) with a ratio $H_2/CO = 3$. In this catalytic process, methane reacts with water steam in the presence of a catalyst. The product of this reaction is the syngas (Rostrup-Nielsen, 1984). The scheme of the reaction of steam reforming of methane is shown below.

$$CH_4 + H_2O \longrightarrow CO + 3H_2$$

Because the process of steam reforming of methane is the reforming process that leads to the obtaining of syngas with the major H_2/CO ratio, this type of reforming process is considered ideal to obtain a gas hydrogen flow of high purity from syngas.
The steam reforming of methane is an endothermic process and, therefore, requires very high temperatures, which makes his process very expensive. Therefore, research on alternative processes to reforming of methane to ensure the economic viability according to the destination of the syngas obtained would be interesting. The concern with the economic viability issue led to the development of alternative processes to reforming of methane, such as dry reforming, autothermal reforming and partial oxidation, which are being considered in scientific research for conversin of methane to syngas (Rostrup-Nielsen, 1984; Armor, 1999).

4.2. Dry Reforming

The dry reforming of natural gas is a process where methane reacts with carbon dioxide in the presence of a catalyst, and syngas at a $H_2/CO = 1$ ratio (Rostrup-Nielsen, 1984; Lercher et al., 1999) is obtained as a product of this reaction. The scheme of the dry reforming of methane reaction is shown below.

$$CH_4 + CO_2 \longrightarrow 2CO + 2H_2$$

Due to the value of the H_2/CO ratio shown by the syngas obtained in the dry reforming of methane, this process is considered the ideal type of reforming process when it comes to use the syngas produced as a raw material for the synthesis of important fuel liquids which require H_2 and CO as raw materials. On the other hand, this type of reforming process is considered very expensive because, being an endothermic process, it consumes a great amount of energy. The main disadvantage of dry reforming of methane is the significant formation of structures (coke) that are subsequently deposited on the surface of the catalyst that is active in the reaction. The deposition of coke on the surface of the catalyst contributes to the reduction of its useful life. The large formation of coke occurred in this process is explained by the presence of the CO_2 reagent introduced in the catalytic process input, the share of CO_2 reagent increasing the production of coke. Thus, dry reforming is the unique process for reforming of methane powered by two reagents that contain carbon (CH_4 and CO_2) (Rostrup-Nielsen, 1984; Cheng et al., 2008; Lercher et al., 1999).

4.3. Partial Oxidation

The partial oxidation of methane is a catalytic process in which methane reacts directly with oxygen in the presence of a catalyst, and the product of this reaction is the syngas which is shown with a H_2/CO good ratio (Fathi et al., 2000). The scheme of the partial oxidation of methane is shown below.

$$CH_4 + 1/2O_2 \longrightarrow CO + 2H_2$$

The partial oxidation of methane is an exothermic process and, thus, considered more economic than the processes of steam reforming or dry reforming, because it requires a smaller amount of thermal energy. On the other hand, the partial oxidation is considered an expensive process because it requires a flow of pure oxygen. Thus, there is a warning of danger inherent in the process of partial oxidation of methane, since the two reagents (CH_4 and O_2) can cause an explosion if the reaction is not conducted with the necessary care (Peña et al., 1996).

4.4. Autothermal Reforming

The autothermal reforming of methane is a combination of both procedures: steam reforming and partial oxidation. Thus, in the steam reforming there is contact with a gas oxygen flow, in the presence of a catalyst (Armor, 1999). Hence, this process of catalytic reforming of methane involves three reagents (CH_4, H_2O and O_2).

The autothermal reforming of methane process was designed to save energy, because the thermal energy required is generated in the partial oxidation of methane. As this process consumes the thermal energy that it produces, it is called autothermal (Ayabe et al., 2003; Wilhem et al., 2001).

Like other reforming processes of methane, the purpose of the autothermal reforming is the production of syngas. The value of the H_2/CO ratio of the syngas obtained in the autothermal reforming is a function of the gaseous reactant fractions introduced in the process input. Thus, the value of H_2/CO ratio can be 1 or 2 (Palm, 2002).

4.5. Comparison between the types of reforming of methane

Overall, regardless the type of process, the reforming of methane is an important chemical operation in the energy worldwide matrix, because this chemical process is the first catalytic step of the natural gas conversion to make way for the subsequent chemical catalytic processes necessary to obtain the valuable gas hydrogen flow of high purity.

According to the definitions presented in this chapter for the four types of reforming processes of methane, it was found that the main type of reforming is the process called steam reforming, because it has the greatest value for H_2/CO ratio, ie, the product of the reforming process is a gas flow considered ideal for the development of the catalytic process of obtaining a gas hydrogen flow of high purity. However, as the process of steam reforming is considered too expensive, the other three types of catalytic chemical processes are considered as alternative processes for carrying out the reforming of methane and they were developed with the aim of making savings in thermal energy consumption required for catalytic process to occur. The choice of the catalytic chemical process type to reforming of methane must take into consideration the economic viability of the process related to the destination to be given to the syngas produced as a product, ie, in general the ultimate purpose is to obtain a gas hydrogen flow of high purity. The types of catalytic processes of reforming of methane called partial oxidation and autothermal reforming are good choices to produce syngas when the value of H_2/CO ratio is adequate and specially when it comes to reduce the consumption of thermal energy, a most important factor. In short, it can be said that the selection of the type of catalytic chemical process of reforming of methane depends on the type of application of the syngas produced.

5. Catalysts commonly used in the reforming of methane

Reports on the development of scientific research involving the use of catalysts on noble metal supported in metal oxides to carry out reforming of the methane are widely reported in the literature.
The main noble metals used in catalytic processes of reforming of methane are Pt, Rh, Pd and Ru, according to scientific publications. Each noble metal considered individually has characteristics and peculiarities when submitted to the reaction conditions of the reforming of methane processes (Seo et al., 2002; Wang et al., 2005; Bulushev & Froment., 1999). Therefore, scientific research is essential to define the catalytic action of each active species individually analyzed, showing the strength points in their catalytic performance as well as stressing their limitations, such as restrictions on activity, selectivity limits, low thermal stability, among others. Thus, in general, the published scientific studies are unanimous in stating that the noble metals, particularly Pt and Rh metals, are excellent for use as active species in catalytic reforming processes of the methane. These are ideal for this application because they have the exact catalytic characteristics that are necessary to reaction conditions of the reforming of the methane process. The characteristics of the catalytic performances of noble metals that make them so valued for this application are: high activity, ie, the great high capacity of methane to convert in syngas, good thermal stability, good selectivity and high resistance to deposition of coke on its surface, this latter characteristic helps increase the lifetime of the catalyst. The use of noble metals, particularly Pt and Rh as active species for catalytic reforming of the methane processes attract much interest because they lead to excellent results. However, they are very expensive (Hickman & Shimidt, 1992; Monnet et al., 2000; Fathi et al., 2000).
Through scientific research was discovered that Ni when tested under reaction conditions of reforming of methane process, the catalytic performance and the quality of the product output are equivalent to the final results obtained by noble metals such as Pt and Rh. Thus, the Ni has attracted much interest from researchers, because this metal exhibits the catalytic

performance of a noble metal combined with the advantage of low cost. Thus, regardless of the type of reforming of the methane process, Ni is considered the main active catalytic species to convert methane in syngas. The Ni can be considered a classic catalyst for the reforming of methane processes (Seo et al., 2002; Torniainen et al., 1994; Eriksson et al., 2005).

However, the catalytic system that operates in the reaction is not solely formed by the active catalytic species. In order to incorporate the active catalytic species, the catalyst system needs a catalytic support for the active species. Thus, the catalytic system consists of two components of equal importance: the active catalytic species also known as active catalytic phase and catalytic support.

The active catalytic species consists of a noble or non-noble metal in the reforming of the methane process, usually Ni, and the catalytic support consists of a metal oxide. The function of the catalytic support is to assist the active species so that their catalytic action is undertaken, ie, the active species can not perform its catalytic action alone. The catalytic support acts as a material substrate where the catalytically active species must be physically supported to act.

The catalytic systems are generally composed of active catalytic species + catalytic support. They are usually represented as follows: metal/metal oxide. Example: Ni/Al_2O_3.

5.1. Importance of the structural characteristics of the catalytic system

The process of steam reforming, the main type of catalytic process of reforming of methane involves a highly endothermic reaction that reachs very high temperatures, in most cases varying between 700 and 1000°C (Rostrup-Nielsen, 1984). Thus, the catalytic system (active species + catalytic support) of this process must be refractory to ensure the thermal stability of the catalytic system. In this case, the aluminum oxide (Al_2O_3) is a good option to be used as catalytic support, because this oxide is highly refractory, supports inert form values of temperatures above 1000°C. Therefore, there are many scientific publications on its use as catalytic support for reforming of methane processes. However, other oxides can also be used as supports for catalysts for reforming of methane. The scientific publications on the issue generally report the use of different oxides such as Al_2O_3, TiO_2, SiO_2, Fe_2O_3, CeO_2, ZnO and others as catalyst supports, though the use of Al_2O_3 is the most common, certainly due to its ability to promote the thermal stability of catalytic chemical processes.

Since the catalytic chemical process of reforming of methane involves tough operating conditions, special attention must be paid to the characteristics of the catalytic refractory support to avoid or minimize the sintering of active species. The sintering of active species is one of the factors that lead to the deactivation of the catalyst, so it must be fought with the use of catalyst refractory supports. Nevertheless, the selection of the type of catalytic support material must be made according to the catalytic process in which this support will act. For example, if the catalytic process requires a large amount of oxygen to occur, it is preferable to to use a metal oxide capable of storing oxygen in its atomic structure, e.g. CeO_2, as catalytic support.

Most times, the catalytic support may consist of a mixture of metal oxides. In general, two oxides are mixed in a doping process, where an oxide is used as host matrix for the incorporation of the second oxide that will be used as the dopant substance of the support. In these cases, the selection of the metal oxides to form the mixture is based on their individual characteristics. The purpose of mixing a metal oxide with another one to compose a catalytic support is to optimize the performance of the catalytic system as a whole. It has been exhaustively proven in scientific publications that certain compounds or substances (metal oxides) incorporated with other types of oxides, have a positive influence on the outcome of the catalytic process. Thus, optimizations such as increases in activity, selectivity and resistance to coke deposit are observed (Carreño et al, 2002; Neiva, 2007; Neiva et al., 2009).

Some metal oxides are more suitable to the optimization of the catalytic system. As demonstrated by Neiva (2007) in a study involving the addition of Fe_2O_3, ZnO and CeO_2, as doping substances in the catalytic system Ni/Al_2O_3, the oxide that most favored the optimization of the catalytic activity was ZnO added in a concentration of 0.01 mol in the structure of the catalytic system Ni/Al_2O_3. The present study stresses the importance of the concentration value of the doping substance added to a catalytic system, which must be within a given range of values. If the concentration of the doping substance exceeds this limit the catalytic activity of the system may be harmed. Also, according to the research carried out by Neiva (2007), a comparison of the catalytic activity of the system (1.5%) Ni/Al_2O_3 doped with the following oxides Fe_2O_3, ZnO and CeO_2 is shown in figure 2. According to the graphs of Figure 2, the catalytic system (1.5%) Ni/Al_2O_3 doped ZnO showed higher catalytic activity, ie, higher peaks of methane conversion. These catalytic systems with the performances shown in Figure 2 were synthesized by the combustion method.

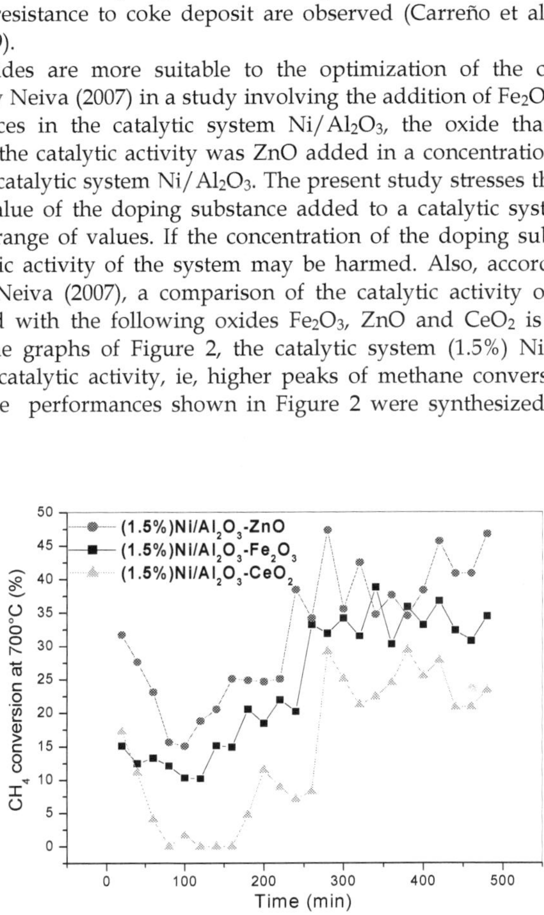

Fig. 2. Comparison between the performances of the catalytic system Ni/Al_2O_3 doped with the oxides Fe_2O_3, ZnO and CeO_2 in the reforming of the methane reaction held at 700°C (Neiva, 2007).

In general, these catalytic supports consisting of more than one oxide are called doped or modified catalytic supports. Is called of dopant substance or dopant element of the catalytic system the metal oxide added in small quantities in the atomic structure of metal oxide which is most of the support structure, ie, inside of the hospitable matrix structure. The

functions of the two metal oxides, dopant substance and host matrix are defined in the reforming of the methane catalytic process.
The stages formed by the dopant substance are active phases that optimize the catalytic activity of the system as a whole, by helping the catalytic action of the main phase that was deposited on the support doped or modified **(Neiva et al., 2008)**.
The atomic structure of the doped or non-doped catalytic support must have a porosity suitable to the deposition of the active catalytic species on the support and also should allow that active species have a satisfactory performance in the catalytic process. The active species shouls be deposed on the porous structure of the catalytic structure as smoothly as possible, so that the catalytic activity is carried out all along the catalytic system and not merely in isolated points. Catalyst supports that have highly crystalline atomic structures favor the occurrence of deposition with very homogeneous dispersion of the active catalytic species (Figueiredo & Ribeiro, 1987; Neiva, 2007).

5.2. Synthesis of catalytic systems for reforming of the methane

Currently, it is possible to develop catalytic supports with controllable physical and structural characteristics. Thus, we can affirm that physical characteristics such as type of porosity, degree of crystallinity and particle size are a function of the type of synthesis method employed in the process of obtaining the metal oxide. Also, these structural characteristics are strongly dependent on the preparation conditions used in the synthesis process, such as the type of precursor chemical used and the possible heat treatments (Neiva, 2007).
The catalytic supports formed by a unique metal oxide or a mixture of oxides usually occur in the form of a ceramic powder made by smaller particles. In some cases, the referred powders are composed of nano size particles. Thus, in general, the synthesis methods used to prepare the catalytic supports are the same methods used in the synthesis of ceramic powders. The synthesis methods commonly used for the development of catalytic supports are called combustion reaction, Pechini method and co-coprecipitation method. Of these, the most versatile is the method of combustion reaction, because it is faster, more efficient and can be performed from any heat source, such as a hot plate, conventional oven, microwave oven, among others. The great advantage of this synthesis method is its fastness, because the synthesis of a ceramic powder sample obtained by using the combustion reaction method lasts in average 5 minutes. Consequently, the ceramic powder final product has small-sized particles that can reach the nano scale, which represents an advantage in catalysis. Since the catalytic chemical processes involve adsorption of gases, the use of small particles such as nano is recommended, because small-sized particles have a greater contact area between the adsorbent (particle) and adsorbate (gaseous reactants of the catalytic reaction). On the other hand, the combustion reaction synthesis method is not the method of synthesis of ceramic powder most suitable for the development of catalysts for the reforming of the methane process, because since it does not include a thermal treatment such as calcination to remove undesirable elements aggregates, the ceramic powder obtained as final product of this synthesis method contains highly contaminated waste arising from the carbon precursor used as fuel in the combustion reaction. Such waste carbon will interact with the reagents of reforming of the methane process and, as a consequence, will significantly increase the formation of coke, strongly contributing to the deactivation of the catalyst. The utilization of chemical methods for nanosize particles preparation, with physical chemical

properties and wished structural has been being the main focus of several researchers in different areas of the science and technology, due to the molecular stability and good chemical homogeneity that can be reached. These methods, also enable a good control in the particle size form and distribution and/or agglomerates. Among lots of existing chemical methods, the synthesis for combustion reaction has been being used with success for obtainment of several ceramic systems. It is an easy technique, it holds and fast to produce nanosize particles, with excellent control of the purity, chemical homogeneity and with good reproduction possibility of the post in pilot scale (Costa et al., 2007). The use of synthesis methods of ceramic powders that include calcination steps in their synthesis procedure are more appropriate for the development of catalysts for reforming of the methane process. The use of calcination as a heat treatment is very important to remove the carbon waste of the synthesized catalysts. The synthesis method of ceramic powders called polymeric precursor method or Pechini method has proven to be very suitable for the development of catalysts for reforming of the methane process, as the Pechini method includes three steps in its synthesis procedure, the last step being calcination that can be performed at temperatures sufficiently high to cause the volatilization of residual carbon-based substances. Generally, depending on the type of synthesized metal oxide, temperatures values ranging between 500 and 1000° C can be used in calcination. Another advantage of the heat treatment of the Pechini method is that it favors the formation of an atomic structure with high percentage of crystallinity and the formation of size controlled particles. The co-precipitation method is also widely used for the synthesis of catalytic supports for reforming of the methane. Also, this method can synthesize pure or mixed metal oxide. The co-precipitation method is capable of producing metal oxide consisting of particles with controlled sizes, including particles with nanometer dimensions, which play a significant role in various catalytic process. The disadvantage of this method is the existence of multiple steps in the synthesis procedure. However, the metal oxides in the form of ceramic powders can be synthesized by less disseminated methods such as spray dry, freeze dry, sol-gel, hydrothermal method, among others. Regardless the synthesis method used for obtaining a catalytic support formed of metal oxide pure or mixed, in many cases, the active catalytic species is deposed on the support at a later stage of the catalyst synthesis procedure. The active catalytic species can be deposed on the catalytic support by means of different methods. In most cases, in the catalysts for reforming of the methane, the active species are deposed on the catalytic support by the impregnation method also known as humid impregnation method with incipient humidity. In this impregnation method, specific quantities of the catalytic support and of the precursor source of ions of the active catalytic species (usually a metal nitrate) are immersed in aqueous solution. Impregnation is performed by means of rotation followed by drying and calcining to ensure the elimination of the humidity adsorbed in the structure of the catalytic developed material (Figueiredo and Ribeiro, 1987). However, the classic method of preparation of catalysts (humid impregnation) induces carbon condensation (derived from reagent CH_4) on the exposed crystal of Ni impregnated on the catalyst surface, reducing the catalytic stability and accelerating catalyst deactivation (Leite at al., 2002).

During the impregnation process, after the calcination stage, which is usually performed at a temperature range of 300 - 500°C , this step is concluded and s the catalytic system developed is then ready to be forwarded to the catalytic reaction. The temperature value used in the calcination stage of the impregnation process should be selected according to the

material that forms the catalyst system. Factors such as temperature limit before the phase transformations of material structure, type of porosity of atomic structure are assessed here. In order to minimize the coke deposit on the surfaces of catalysts, some alternative methods to suppress the poisoning of the active site metal and the formation of carbon nanotubes are available. Through consecutive reactivation of the catalyst (Ni supported on alumina) with CO_2-rich atmosphere, it was possible to eliminate the remaining carbon from the catalytic oxidation of the same, with formation of CO (Ito et al., 1999).
The future prospects for the reforming of methane indicate the need for further improvement of this process to optimize its implementation and results. Catalytic systems that are more resistant to coke formation and increasingly appropriate operating conditions of this chemical process will always be the focus of researchers in this area. Certainly, new materials will be produced and tested in the process of reforming of methane, always aiming to reduce the factors leading to deactivation of the catalyst.
The technological innovation in this area may also focus on the discovery or development of new methods for obtaining catalytic systems, in order to ensure the complete mastering of the process of obtaining materials with increasingly controllable physical and structural characteristics. This would lead to the development of catalytic systems more suitable for the reforming of the methane process.

5.3. The action of the catalytic system in the reforming of the methane process

The calcination stage of the impregnation process does not make sure that the active catalytic species of the developed catalytic system have an effective action on the catalytic reaction. So, the active species of the developed catalytic system can only be effectively activated by a heat treatment called Temperature Program Reduction - TPR. This heat treatment is aimed to reducing the active catalytic species deposed in the oxide form on the catalytic support, e.g NiO, in a catalytically active metallic phase, e.g. Ni. This reduction is essential for the occurrence of the catalytic action of the developed material. If the TPR process does not occur, the material developed will be catalytically inert. The treatment of TPR is performed in situ inside of reactor, a few minutes before the catalytic reaction, for example a few minutes before the occurrence of the reforming of methane. In the case of catalytic reactions other than reforming of methane also uses the TPR process to reduce the metal oxide in effectively active species, ie, in metallic phase catalytically active.
In general, at the end of the process of catalytic reforming of methane, the catalyst is recovered and sent for analysis to help characterize this catalyst. The main characteristic to be assessed is the amount of coke deposited on the surface of the catalyst until deactivation. The amount of coke detected on the catalyst system is a function of the type of reforming of methane process accomplished, other factors that influence the formation of coke are the fractions of gaseous reactants injected in the catalytic process input and conditions, e.g. temperature, and especially the type of material that constitutes the catalytic system. Thus, it can be said that the amount of coke detected on the catalyst system characterizes the resistance of this material to the formation of coke. Smaller amounts of coke on the catalyst indicate high resistance to the formation of carbon-based substances.
Finally, it is important to monitor the reforming of the methane process as a whole in order to determine the main factors in this catalytic process, such as the lifetime of the catalyst and the peak values of the temperatures recorded throughout the process. All these combined aspects help define and clarify the success or failure of methane in syngas conversion.

6. Conclusion

At the end of this chapter, we believe that we have clarified the importance of a catalytic system (catalyst) in the reforming of the methane process. In fact, the catalyst is a indispensable element in the reforming of the methane process, as well as the subsequent chemical processes that are aimed to obtain high purity hydrogen from syngas, the product of reforming of methane. In the absence of the catalyst, there is no or insufficient interaction between methane and the other reactant (water steam, CO_2 or O_2). Therefore, we can affirm that the catalyst is an element of the chemical process of reforming of methane which has basically the following functions during the performance of the catalytic process: activate, accelerate, optimize, direct interactions or block interactions. The occurrence of each of these functions depends on the type of reforming of the methane process performed and also on the type of material that constitutes the catalytic system in operation.

Within the operating conditions of reforming of the methane processes, the reagents of these processes interact in gaseous state and in the presence of a catalyst in solid state. Thus, according to the classical definitions of catalysis, the reforming of the methane process is defined as a heterogeneous catalytic process because the reagents and the catalyst interact with distinct phases.

Also, in conclusion of this chapter, we hope that the importance of natural gas in the worldwide energy matrix has become clear. The trend is that natural gas will become even more space in the energy generation area from now on, keeping in view the scarcity of natural resources used as energy generators currently.

Water will remain the most important source of electricity generation worldwide, in the long-term. Nevertheless, it will be hard to construct new hydroelectric dams and reservoirs due to the current policies of environmental preservation. Consequently, alternative forms of energy generation shall be given greater consideration throughout the world.

7. References

ANEEL - Agência Nacional de Energia Elétrica (2008). *Atlas de energia elétrica do Brasil*, Editora Brasília, 3rd edition.

Armor, J. N. (1999). The Multiple Roles for Catalysis in the Production of H_2. *Applied Catalysis A: General*. No. 21, pp. 159-176. ISSN: 0926-860X.

Ayabe, S.; Omoto, H.; Utaka T.; Kikuchi R.; Sasaki K.; Teraoka Y. & Eguchi, K. (2003). Catalytic autothermal reforming of methane and propane over supported metal catalysts. *Applied Catalysis A: General*. No. 241, pp. 261-269. ISSN: 0926-860X.

Bulushev, D. A. & Froment, G. F. (1999). A drifts study of the stability and reactivity of adsorbed CO species on a Rh/γ-AlB$_2$BOB$_3$B catalyst with a very low metal content. *Journal of Molecular Catalysis A: Chemical*. No. 139, pp. 63-72. ISSN: 1381-1169.

Carreño, N. L. V., Valentini, A., Maciel, A. P., Weber, I. T., Leite, E. R., Probst, L. F. D. & Longo, E. (2002). Nanopartículas catalisadoras suportadas por materiais cerâmicos, *Journal Materials Research*, Vol. 48, pp. 1-17. ISSN: 0884-2914.

Cheng, Z. X.; Zhao, J. L.; Li, J. L. & Zhu, Q. M. (2001). Role of support in CO_2 reforming of CH_4 over a Ni/α-Al_2O_3 catalyst. *Applied Catalysis A: General*. No . 205, pp. 31-36. ISSN: 0926-860X.

Costa, A. C. F. M.; Kiminami, R. H. G. A.; Moreli,.M. R. (2007). Microstructure and magnetic properties of $Ni_{1-x}Zn_xFe_2O_4$ synthesized by combustion reaction, Journal of Materials Science, Vol. 42, pp. 779-783.

Rostrup-Nielsen, J. R. (1984). *Catalysis, Science and Technology* (Anderson, J.R. & Boudart, M., eds.). Springer Ed., Berlin Heidelberg New York, Vol. 5, pp. 1-117.

Eriksson, S.; Nilsson, M.; Boutonnet, M. & Jaras, S. (2005). Partial oxidation of methane over rhodium catalysts for power generation applications. *Catalysis Today*, No. 100, pp. 447-451. ISSN 0920-5861.

Fathi, M.; Bjorgum, E.; Viig, T. & Rokstad, O. A. (2000). Partial oxidation of methane to synthesis gas: elimination of gas phase oxygen. *Catalysis Today*. No. 63, pp. 489-497. ISSN 0920-5861.

Figueiredo, J. L. & Ribeiro, F. R. (1987). *Catálise Heterogênea*; Editora Fundação Calouste Gulbenkian, Lisboa, Potugal.

Fishtik, I.; Alexander, A.; Datta, R. & Geanna, D. A. (2000). Thermodynamic analysis of hydrogen production by stem reforming of ethanol via response reactions; *International Journal of Hydrogen Energy*, Vol. 25, pp. 31-45. ISSN: 0360-3199.

Hickman, D. A. & Schmidt, L. D. (1992). Synthesis gas-formation by direct oxidation of methane over Pt monoliths, *Journal Catalysis*, Vol. 138, pp. 267-282. ISSN: 0021-9517.

Hoffmann, P. (2002). *Tomorrow's Energy, Hydrogen, Fuel Cells and the Prospects for a Cleaner Planet*, 2nd Edition, The MIT Press, Cambridge, Massachussets, USA.

Ito, S., Fujimori, T., Nagashima, K., Yuzaki, K. & Kunimori, K. (1999). Strong rhodium-niobia interaction in Rh/Nb2O5, Nb2O5-Rh/SiO2 and RhNbO4/SiO2 catalysts - Application to selective CO oxidation and CO hydrogenation, *Catalysis Today*, 57, pp. 247-254. ISSN 0920-5861.

Leite, E. R.; Carreño, N. L. V.; Longo, E.; Valentini, A. & Probst, L. F. D. (2002). Development of metal - SiO_2 nanocomposites in a single-step process by the polymerizable complex method, *Chemistry of Materials*, Vol. 14, No. 9, pp. 3722-3729. ISSN: 1520-5002

Lercher, J. A.; Bitter, J. H.; Steghuis, A. G.; Van Ommen, J. G. & Seshan, K. (1999). Methane Utilization via Synthesis Gas Generation - Catalytic Chemistry and Technology. *Environmental Catalysis, Catalytic Science Series*, Vol. 1. pp. 12-19. ISSN: 1793-1398.

Monnet, F.; Schuurman, Y.; Aires, F. J. C. S.; Bertolini, J-C. & Mirodatos, C. (2000). Toward new Pt- and Rh-based catalysts for methane partial oxidation at high temperatures and short contact times, *Surface chemistry and catalysis, Comptes Rendus de l'Académie des Sciences - Series IIC - Chemistry*, Vol. 3, Issue 7, pp. 577-581.

Neiva, L. S., Andrade, H. M. C., Costa, A. C. F. M. & Gama, L. (2009). Synthesis gas (syngas) production over Ni/Al_2O_3 catalysts modified with Fe_2O_3, *Brazilian Journal of Petroleum and Gas*, Vol. 3, No. 3, pp. 085-093. ISSN 1982-0593.

Neiva, L. S. (2007). Preparação de catalisadores de Ni/Al_2O_3 dopados com Fe, Zn e Ce para aplicação em processos de reforma do gás natural, Master Dissertation, Engineering of Materials, Federal University of Campina Grande, Brazil.

Neiva, L. S.; Gama, L.; Freitas, N. L.; Andrade, H. M. C.; Mascarenhas, A. J. S. & Costa, A. C. F. M. (2008). Ni/α-Al_2O_3 catalysts modified with ZnO and Fe_2O_3 for steam reforming of the natural gas, *Materials Science Forum*, Vol. 591, pp. 729-733. ISSN: 0255-5476.

Odell, P. R. & Rosing, K. E. (1983). *The Future of Oil; world Oil Resources and Use*, Editor Kogan Page Ltd., 2nd Edition, London, UK.

Olah, G. A., Goeppert, A. & Prakash, G. K. S. (2006). *Beyond Oil and Gas: The Metanol Economy*, Wiley-VCH Editor, Weinheim, Germany.

Palm, C.; Cremer, P.; Peters, R. & Stolten, D. (2002). Small-scale testing of a precious metal catalyst in the autothermal reforming of various hydrocarbon feeds, *Journal of Power Sources*, No. 106, pp. 231-237. ISSN: 0378-7753.

Peña, M. A.; Gómez, J. P. & Fierro, J. L. G. (1996). New Catalytic Routes for Syngas and Hydrogen Production, *Applied Catalysis A: General*, No. 144, pp. 7-57. ISSN: 0926-860X.

Pompermayer, M. L. (2009). Desafios e perspectivas para a inovação tecnológica no setor de energia elétrica, *Revista Pesquisa e Desenvolvimento da Agência Nacional de Energia Elétrica - ANEEL*, No. 3, pp. 11.

Seo, Y.-S., Shirley, S. T. & Kolaczkowski, S. T. (2002). Evaluation of thermodynamically favourable operating conditions for production of hydrogen in three different reforming Technologies, *Journal of Power Sources*, Vol. 108; pp. 213-225. ISSN: 0378-7753.

Torniainen, P. M.; Chu, X. & Schmidt, L. D. (1994). Comparison of monolith-supported metals for the direct oxidation of methane to syngas, *Journal of Catalysis*, No.146, pp. 1-10. ISSN: 0021-9517.

Wang, J. A.; Lopes, T.; Bokhimi, X. & Novaro, O. (2005). Phase composition, reducibility and catalytic activity of Rh/zirconia and Rh/zirconia-ceria catalysts, *Journal of Molecular Catalysis*, Vol. 239, No 1-2, pp. 249-256. ISSN 1381-1169.

Wilhelm, D. J.; Simbeck, D. R.; Karp, A. D.; Dickenson, R. L. (2001). Syngas production for gas-to-liquids applications: technologies, issues and Outlook, *Fuel Processing Technology*, No. 71, pp. 139-148. ISSN: 0378-3820.

4

Natural gas odorization

Daniel Tenkrat, Tomas Hlincik and Ondrej Prokes
Institute of Chemical Technology Prague
Czech Republic

1. Introduction

Natural gas is an odorless and colorless flammable gas. Natural gas odorization means operations involving addition of an odorant to gas to ensure characteristic odor of natural gas in order for people the odor to be distinctive and unpleasant so that the presence of gas in air in concentrations below the lower explosive limit (LEL) is readily detectable. By the odorant addition any physical or chemical property (except the smell) of natural gas cannot be changed. Generally speaking, in the process of natural gas delivering for both public and industrial use, odorization provides safety for those who use it.

Starting with the year 1807 when Pall-Mall in London was experimentally illuminated, the beginnings of gas industry in the European countries were exclusively associated with town gas. This gas, produced by carbonization of coal, contained mainly hydrogen and carbon monoxide. Besides other components, gas produced from coal contained a wide range of sulfur compounds which made it easily detectable in case of leaks and lent it the typical "gassy odor". With the development of the use of natural gas or gas produced by cracking of hydrocarbons or coal pressure gasification the need to odorize these gases became ever more evident.

Historically, first gas odorization was carried out in Germany in 1880's by Von Quaglio who used ethyl mercaptan for detecting gas leakages of blue water gas. However, the real begging of widespread odorization started in US in 1930's as a consequence of the New London's disaster.
Early in 1937, the New London school board cancelled their natural gas contract in order to save money. Instead, plumbers installed a tap into a residual gas line associated with oil production. This practice, while not explicitly authorized by local oil companies, was widespread in the area. The natural gas extracted with the oil was seen as a waste product and thus was flared off. Odorless and therefore undetectable natural gas had been leaking from the connection to the residual line and had built up inside an enclosed crawlspace which ran the entire length of the building. A spark is believed to have ignited the accumulated gas-air mixture leaving behind totally collapsed building and approximately 319 casualties (P&GJ, 2006).

As a consequence of this accident the use of odorants in USA and Canada was enacted. The currently applicable Federal Regulation, 49CFR, 192.625, "Odorization of Gas", requires a combustible gas which is transmitted interstate or distributed to be odorized either with natural odorant which is present in that gas or by odorant addition so that at a concentration in air of one-fifth of the lower explosive limit, the gas is readily detectable by a person with a normal sense of smell. It means that the presence of natural gas at 1.26% in air must be detectable by smell.

Regulations in force in most European countries are similar (e.g. DVGW G280 in Germany), differing only in that there is a requirement for detectability of gas when 1/5 of lower flammable limit (LFL) is achieved. In practice, this represents 1% concentration of natural gas in the air. Used as an example may be Japan where natural gas used as CNG (compressed natural gas) must be detectable by smell whenever concentration in the air reaches 1.000ppm. In practice this represents the value of 0.1%.

2. Gas Odorants

As high quality natural gas replaced manufactured gas the need for odorization of this gas with little (if any) detectable smell arose. In beginnings, the "gassy odor" was supplied by cheap refinery and coke industry by products. However, these products varied in quality and were quite unreliable. After the World War II these by-products are being replaced by low molecular weight synthetic chemicals (such as mercaptans and sulfides) so that in 60's nearly all odorization of natural gas was performed either with pure or blended synthetic chemicals.

Modern gas odorants can be divided into two basic groups. The "classic" sulfur-based odorants which are further subdivided to alkyl mercaptans, alkyl sulfides and cyclic sulfides and new types of sulfur-free odorants based on acrylates which are being introduced to the market in recent years and have their special potential especially in environmental issues due to the zero sulfur dioxide emissions after gas combustion.

Basic requirements for odorants apply both to their physiological effects and on their physiochemical properties. Ideally odorants should have a characteristic "gassy odor". As for physiological properties these inlcude in particular:

- Piercing, strong and unmistakable odor
- Odor must remain perceptible as long as the fault of technical equipment is detected and removed
- Odorant combustion must not produce toxic and irritating products

The most important physiochemical properties include:

- Odorants must be chemically stable, must not react with gas components, piping material, rust, etc.
- Must have high enough vapor pressure in order to avoid condensation at operating pressure
- Must not have a corrosive effect on gas equipment in concentrations used
- Must have a minimum tendency to soil adsorption during gas leaks from pipes
- Odorant smell must not be masked by the presence of higher hydrocarbons
- Odorants must not contain water and must not be diluted with water due to possible subsequent corrosion of the equipment.

The selection of the suitable odorant to be injected into natural gas grid is the key aspect of properly operated odorization system. Selecting the specific odorant involves knowledge of the chemical and physical characteristic of available odorants, properties of the gas to be odorized, the layout of the pipeline (e.g. soil properties, constructing material and pipeline condition), ambient conditions and also the recognition of smell of the local population.

2.1 Types of odorants

Tetrahydrothiophene (THT)

THT is the sole representative of cyclic sulfides used in odorization of gas and is the archetype of "stand alone" odorants; due to poor soil permeability it is nevertheless used in blends with e.g. TBM. THT is most resistant to pipeline oxidation a due to its low odor impact it is difficult to over-odorize with this type of odorant. THT is slightly skin irritant and has a moderate narcotic effect.

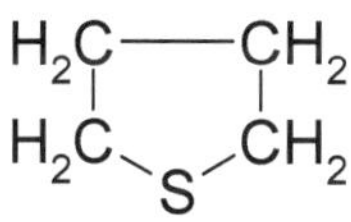

Fig. 1. Tetrahydrothiophene

Formula: C_4H_8S
Molecular weight: 88.172
CAS reg. number: 110-01-0
Specific gravity: 1.000
Boiling point: 115 - 124 °C
Freezing point: -96°C
Flash point: -7 °C
Total sulfur content: 36.37 (Wt. %)

NFPA Ratings:

Dimethyl sulfide (DMS)

DMS is characterized by good oxidation stability and good soil permeability. It is mainly used in blends with TBM, but thanks to its relatively high pressure of vapor in blends with DMS it is not quite suitable for vaporization type odorizers. DMS is a "garlic stinking" compound that causes nausea in higher concentrations. With its effect it first stimulates and then frustrates the nervous system.

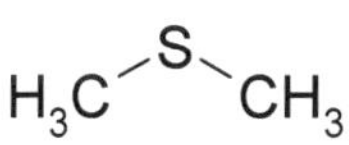

Fig. 2. Dimethyl Sulfide

Formula: C_2H_6S
Molecular weight: 62.135
CAS reg. number: 75-18-3
Specific gravity: 0.8
Boiling point: 37 °C
Freezing point: -98°C
Flash point: -38 °C
Total sulfur content: 51.61 (Wt. %)

NFPA Ratings:

Diethyl sulfide (DES)

DES has good oxidation stability, low odor threshold but its high boiling point is limiting for using in odorant blends.

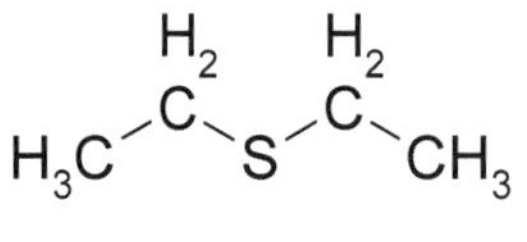

Fig. 3. Diethyl Sulfide

Formula: $C_4H_{10}S$
Molecular weight: 90.188
CAS reg. number: 352-93-2
Specific gravity: 0.837
Boiling point: 90 °C
Freezing point: -100°C
Flash point: -9 °C
Total sulfur content: 35.55 (Wt. %)

NFPA Ratings:

Methylethyl sulfide (MES)

MES has a good oxidation stability in pipelines and a vapor pressure similar with TBM and thus blends of TBM/MES are suitable for both vaporization and injection type odorizers. From the toxicological point of view MES has similar properties with NPM.

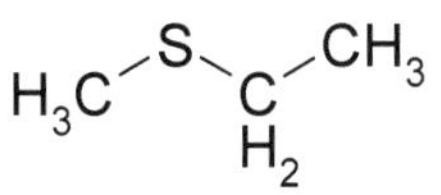

Fig. 4. Methylethyl Sulfide

Formula: C_3H_8S
Molecular weight: 76.162
CAS reg. number: 624-89-5
Specific gravity: 0.8422
Boiling point: 65 - 67 °C
Freezing point: -106°C
Flash point: -15 °C
Total sulfur content: 42.10 (Wt. %)

NFPA Ratings:

Ethyl mercaptan (EM)

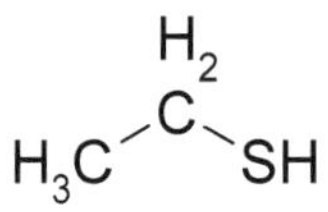

Fig. 5. Ethyl Mercaptan

Formula: C_2H_6S
Molecular weight: 62.135
CAS reg. number: 75-08-1
Specific gravity: 0.839
Boiling point: 34 - 37 °C
Freezing point: -148 - -121°C
Flash point: -48 °C
Total sulfur content: 51.61 (Wt. %)

NFPA Ratings:

Sec-butyl mercaptan (SBM)

SBM is one of the least used components in odorant blends. Originates as a by product or impurity in TBM manufacturing and is seldom used and only in low concentrations. This branched chain mercaptan has good oxidation stability but a relatively high boiling point.

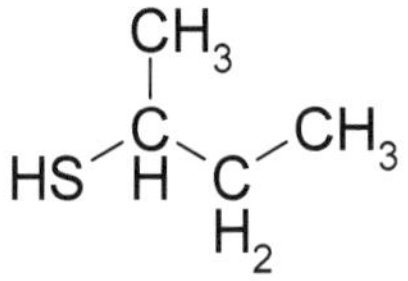

Fig. 6. Sec-butyl mercaptan

Formula: C_3H_8S
Molecular weight: 90.188
CAS reg. number: 513-53-1
Specific gravity: 0.8299
Boiling point: 84 - 85 °C
Freezing point: -165°C
Flash point: -23 °C
Total sulfur content: 35.55 (Wt. %)

NFPA Ratings:

Tert-butyl mercaptan (TBM)

Typical "gassy odor", low odor threshold, high oxidation resistance (highest among mercaptans) and good soil penetration is what make TBM the most used component of gas odorants. The main disadvantage is its high freezing point which disables using TBM as a stand alone odorant and thus TBM has to be blended with other types of odorant.

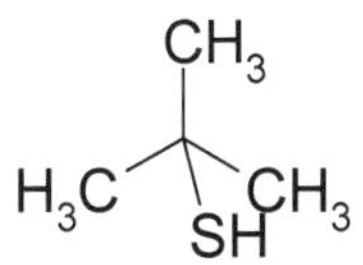

Fig. 7. Tert-Butyl Mercaptan

Formula: $C_4H_{10}S$
Molecular weight: 90.188
CAS reg. number: 75-66-1
Specific gravity: 0.8002
Boiling point: 64 °C
Freezing point: 1°C
Flash point: <-29 °C
Total sulfur content: 35.55 (Wt. %)

NFPA Ratings:

N-Propyl mercaptan (NPM)

NPM has a low freezing point and a strong odor but is not used in high concentrations (typically 3-6%) due to its low oxidation stability. From the toxicological point of view it has a depressive effect on central nervous system.

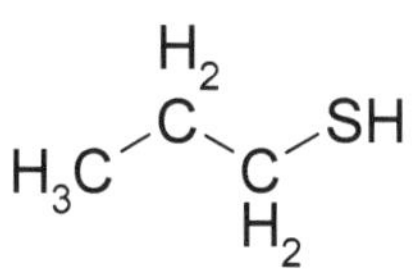

Fig. 8. N-Propyl mercaptan

Formula: C_3H_8S
Molecular weight: 76.162
CAS reg. number: 107-03-9
Specific gravity: 0.8411
Boiling point: 67 – 68 °C
Freezing point: -113°C
Flash point: -21 °C
Total sulfur content: 42.10 (Wt. %)

NFPA Ratings:

Isopropyl mercaptan (IPM)

IPM is the second most resistant to oxidation from mercaptans, has a strong "gassy odor" and low freezing point. IPM is commonly used in blends with TBM in order to decrease the freezing point. In some cases IPM should be used as a stand alone odorant. IPM has similar toxicological effects with NPM.

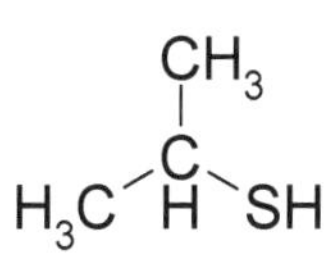

Fig. 9. Isopropyl mercaptan

Formula: C_3H_8S
Molecular weight: 76.162
CAS reg. number: 75-33-2
Specific gravity: 0.8143
Boiling point: 53 °C
Freezing point: -113°C
Flash point: -34 °C
Total sulfur content: 42.10 (Wt. %)

NFPA Ratings:

Methyl acrylate (MA)

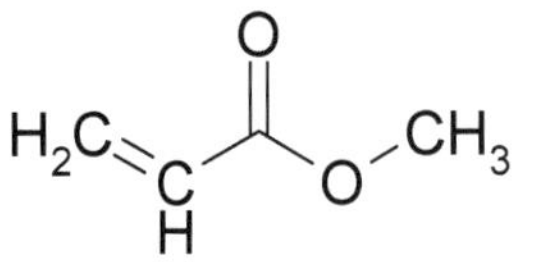

Fig. 10. Methyl acrylate

Formula: C_4H_6O
Molecular weight: 86.0892
CAS reg. number: 96-33-3
Specific gravity: 0.9535 – 0.9574
Boiling point: 78 - 81 °C
Freezing point: -75°C
Flash point: -3 °C
Total sulfur content: - (Wt. %)

NFPA Ratings:

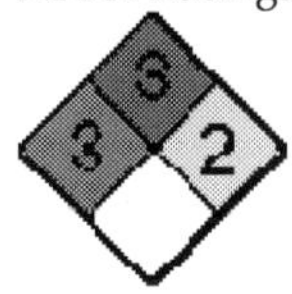

MA and EA are the main components (together with Methylethyl Pyrazine) of the sulfur-free odorant. They perform good permeability through soil (which is slightly lower in case of dry soil) and low odor threshold. Under certain circumstances they can be "washed out" from the gas stream particularly if hydrocarbon condensate occurs within the pipeline.

Ethyl acrylate (EA)

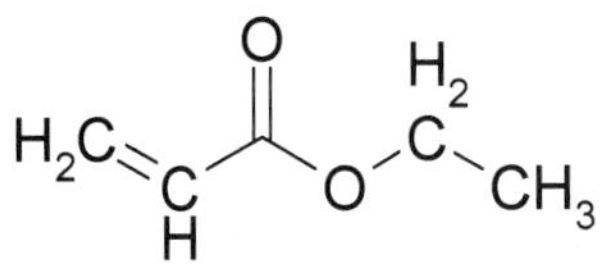

Fig. 11. Ethyl acrylate

Formula: $C_5H_8O_2$
Molecular weight: 100.1158
CAS reg. number: 140-88-5
Specific gravity: 0.9
Boiling point: 99 - 100 °C
Freezing point: -72°C
Flash point: 8.3 °C
Total sulfur content: - (Wt. %)

NFPA Ratings:

2.2 Odorant blends

The odorants used today are typically a blend made and they fall into four main categories, which are:

- All mercaptan blends
- Mercaptan/ Alkyl sulfide blends
- Tetrahydrothiophene/mercaptan blends
- Acrylates blends (sulfur free).

The main reason for odorant blending is to reach specific properties of an odorant for use under different conditions or to improve some of its characteristic. A list of some common blends is given in table 1, other widespread odorants are e.g. Scentinel® odorants by Chevron Philips.

Blend Type	Composition	Specific density (20°C)	Boiling point [°C]	Flash point [°C]	Viscosity (20 °C) [cP]	Odor threshold
Alerton 88 Spotleak 1013	THT 100 %	1.000 (20°C)	115	<13	1.04	1 ppb
Alerton 452 Spotleak 1001	TBM 80 % DMS 20 %	0.816 (20°C)	50	<-32	0.52	0.1 ppb
Alerton 541	TBM 50 % DMS 50 %	0.830 (20°C)	36	<-34	0.41	N/A
Alerton 841 Penndorant 1005	THT 70% TBM 30 %	0.930 (20°C)	60	<-18	0.93	N/A
Alerton 841 P	THT 65 % TBM 35 %	0.931 (20°C)	65	<-20	0.92	N/A
Alerton 842	THT 95 % TBM 5%	0.991 (20°C)	65	<-4.4	0.98	N/A
Alerton 843	THT 85 % TBM 15 %	0.969 (20°C)	65	<-6.8	0.96	N/A
Alerton 1440	IPM 80 % NPM 10 % TBM 10 %	0.820 (20°C)	50	<-17	N/A	N/A
Spotleak 1007	TBM 80 % MES 20 %	0.815 (15.5°C)	63	<-18	0.55	0.1 ppb
Spotleak 1009	TBM 79 % IPM 15 % NPM 6 %	0.812 (15.5°C)	62	<-18	0.570	0.1 ppb
Spotleak 1039	THT 50 % TBM 50 %	0.904 (15.5°C)	67	<-12	N/A	N/A
Spotleak 1420	TBM 75 % DMS 25 %	0.825 (15.5°C)	54	<-18	0.49	0.1 ppb
Spotleak 1450	IPM 70% TBM 10 % DMS 10 % NPM 10 %	0.825 (15.5°C)	53	<-18	0.570	0.1 ppb
Spotleak 2323	TBM 50 % NPM 50 %	0.826 (15.5°C)	62	<-18	N/A	0.1 ppb
Gasodor S-free	Methyl acrylate 37.4 % Ethyl acrylate 60 % Methylethyl pyrazine 2.5 %	0.930 (20°C)	<130	<5	N/A	N/A

Table 1. Basic properties of common odorant blends (Sources: Arkema; Symrise)

3. Odorizing systems

In the odorization process an essential step is to select the right tool in this case a suitable odorizing system. From the technical point of view odorizers should be divided into two basic groups according to the system in which odorants are introduced into the gas stream which are:

- Chemical vaporization
- Chemical injection.

Vaporization based system rely on diffusion of odorant into a flowing natural gas stream. Examples of vaporization systems are wick odorizers and bypass type systems. The main advantage of these odorizers is their simplicity however they are generally suitable for low and stable gas flows.
The injection type systems rely on direct injection of an odorant which is stored away from the pipeline directly into the flowing stream. These systems are generally used for wide range of flow rates.

3.1 Wick odorizers

Odorization by means of wick odorizers is one of the oldest and simplest methods. It is based on free evaporation of the odorant from the wick into the gas stream. It was and is still used for odorization of small amounts of gas. The device consists of a storage tank with odorant into which the wick extends through a hole. The other end of the wick is placed directly in the stream of fuel gas. Dosage was controlled only by setting the size of the wick. The disadvantage of the original device was that during low gas flow gas could be over odorized and vice versa the intensity of odorization could be insufficient during high gas flow.

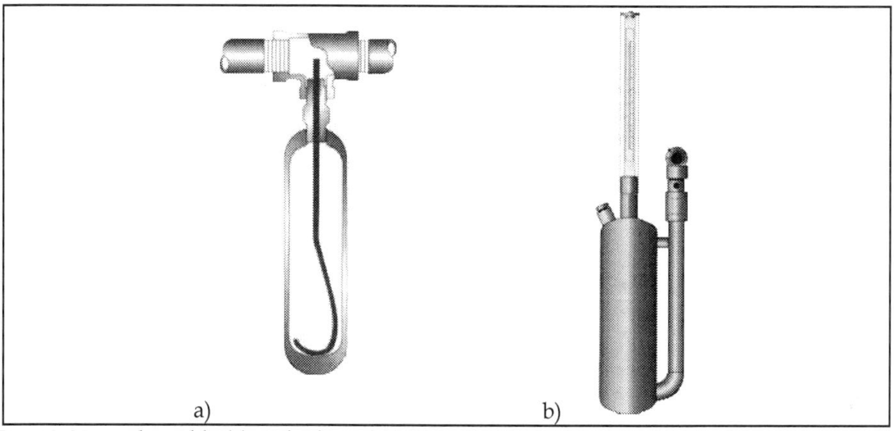

Fig. 12. Non-adjustable (a) and adjustable (b) wick odorizer [Source: King tool company]

3.2 By-pass systems

Due to its simplicity this method of odorization was widely used. By strangling the mainstream of natural gas in the pipeline (by means of an orifice, Venturi tube, slide or ram pipe with sideway cant embedded into gas stream) difference of pressures is reached so that partial flow of fuel gas proportional to the mainstream of fuel gas passes through the tank with odorant above its surface, saturates with odorant vapors and returns to main gas stream. Odorant dosage can be changed by changing the strangling of fuel gas mainstream. The device was used for fuel gas odorization up to the flow of 10 000 m^3/h.. These devices are suitable for both local odorization and additional odorization of fuel gas in central odorization.

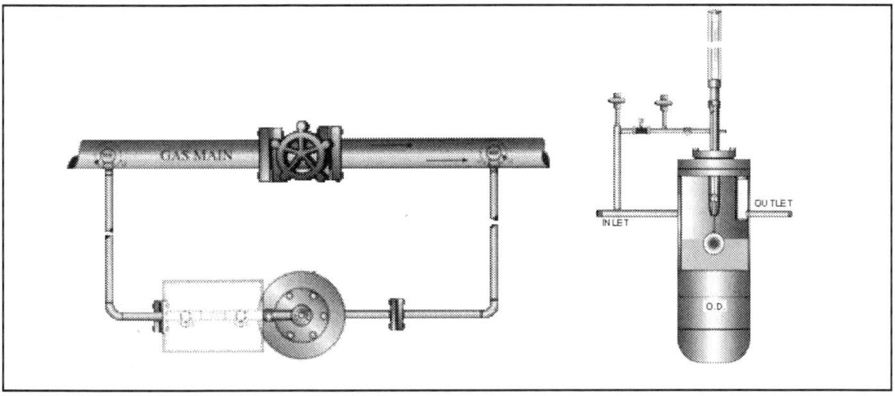

Fig. 13. Bypass odorizer [Source: King tool company]

3.3 Pulse Bypass

The operating principal of Pulse Bypass Odorization is to use higher pressure gas supply from the transmission line to introduce odorant vapors into a lower pressure feeder or distribution line. This is accomplished by diverting or bypassing un-odorized natural gas through an odorant filled tank to mix with odorant vapors. Odorization occurs when the odorant saturated bypass gas is returned to the down stream line. A signal from a meter interface switch is received to actuate the pulse bottle solenoid valve.

3.4 Bourdon Tube

In these rarely used odorizers the amount of odorant injected is controlled by a bourdon tube activated by a differential-pressure transmitter which senses the gas flow across an orifice plate in the pipeline.

3.5 Drip systems

This system was used for the odorization of high amounts of low-varying stream of fuel gas with stable temperature and pressure. Odorant dripping into fuel gas stream was controlled by a needle valve and monitored through a peep-hole. This type of odorization device required regular supervision because of frequent clogging of the needle valve due to variation of viscosity, density or odorant deposits.

In recent years Smart Drip systems appeared on the market. It is an odorization system composed of age old proven drip technology combined with modern measurement, computational processing, and feedback control electronics. The result is a precision dispensing system capable of supplying odorant over a wide range of natural gas flow rates.

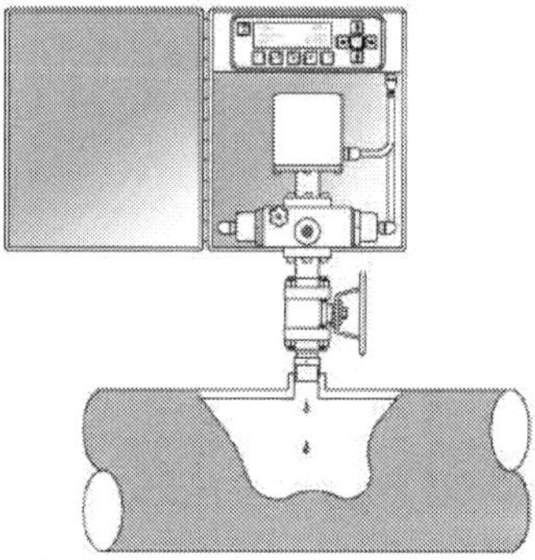

Fig. 14. Smart drip system (Source: Z systems, Midland, TX)

A conventional gravity feed drip odorization system was modified by adding electrically operated valves. The valves are pulsed on and off with a duty cycle sufficient to permit the required mass of odorant flow though a drip tube. The duty cycle and rate of valve operation is controlled to follow changes in gas flow and changes in head pressure resulting from varying levels in the run tank. The drip tubes are contained within a stainless steel measurement chamber equipped with optical interrupters sending electrical pulse signals with each drop of odorant dispensed. Drop size is dependent on the mass (weight) of the drop, surface tension of the fluid, and surface area of the drip tube tip. Surface tension is weakly dependent on fluid temperature, requiring a simple linear adjustment. Therefore the only variable required to calibrate drop mass is odorant temperature, which may be assumed to be the ambient temperature of the drip chamber. This methodology of odorant metering is more direct than measuring volume and converting to mass.

3.6 Electrically or pneumatically driven pump

Odorant is brought into the pipeline with flowing fuel gas by means of a dosage pump. The pump is controlled by an electronic system on the basis of gas flow data. Devices of this design are suitable for gas flow rates above 5000 m^3/h and allow for accurate dosage. In simpler devices with built-in gas meter the energy required for dosage pump drive was discharged from fuel gas mainstream and the dosage pump was driven by gas meter rotational movement. Thus the appropriate odorant dose was also controlled.

Another system uses a diaphragm proportioning pump. Depending on a real flow of gas, impulses from gas meters, or counters actuate a pump by way of electronics of the equipment. A diaphragm proportioning pump which is controlled by a microprocessor and powered by a magnet injects the adjusted quantity of odorizing liquid by injection apparatus to the gas flow. Through a primary tank the pump sucks in the odorizing liquid from the exchangeable tank.

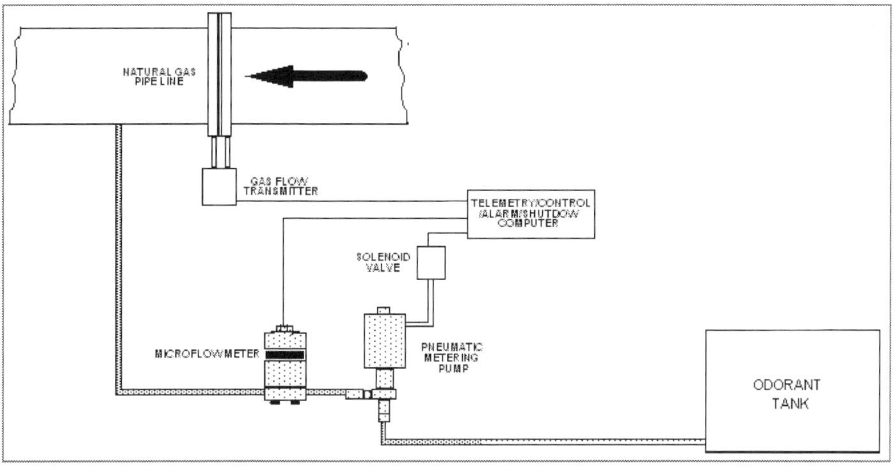

Fig. 15. Scheme of an odorant injection system (Source: DEA Engineering)

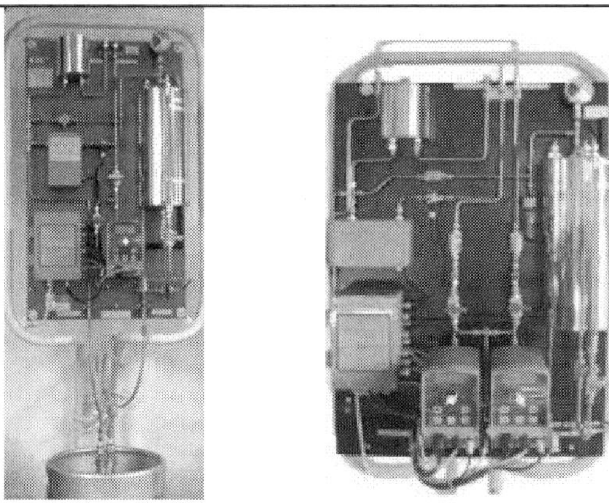

Fig. 16. Odorization device with a diaphragm proportioning pump (Source: Gascontrol)

4. Odorization monitoring

The main task of natural gas odorization is to ensure such operating condition when natural gas in every part of the distribution grid fulfils the requirement of a „warning odor level".

In case of a gas leakage the warning odor level (see table 2.) must be reached until the 20% of lower explosive/flammable limit (LEL/LFL) is reached. Odorisation level can be verified by:

1. The odorization level control - which can be done by olfactometry in selected points on distribution grid or by means of questionnaires at selected representative sample of customers. In both cases indirect indicators are taken into account so that both forms are considered to be subjective methods.

2. Odorant concentration measurement - in natural gas can be estimated continuously or discontinuously in selected points on grid. In this case particular concentration of odorant in NG is measured. This is so called objective method.

The most important task is to estimate optimal odorant concentration. For calculating the safety-relevant, minimum odorant concentration necessary to reach the warning odor level (grade 3, see table 2), an experimentally determined *K*-value is used.

Minimal odorant concentration represents the odorant content in natural gas ($mg.m^{-3}$) which fulfills the requirement for creating warning odor level - grade 3.
Estimation of the minimal odorant concentration is determined by:

- *K*-value ($mg.m^{-3}$) which represents the minimal concentration of an odorant in natural gas-air mixture which reliably ensures the warning odor level,
- lower explosive/flammable limit (LEL/LFL) –expressed by vol. % of natural gas in air,
- and from the requirement to evoke the warning odor level before one fifth (i.e. 20 %) of LEL/LFL of natural gas in air is reached.

Minimal odorant concentration c_n can be estimated according to the following formula:

$$c_n = \frac{100 \cdot K}{0.2 \cdot LEL} \text{ (mg. m}^3\text{).} \tag{1}$$

K-values are obtained by olfactometry measurements using defined sample of population. Typical *K*-values of commonly used odorants are 0.08 for tetrahydrothiophene, 0.03 for mercaptans and 0.07 $mg.m^{-3}$ for the GASODOR S-free odorant.

Odor intensity is the extent of odor perception which is by the odor evoked. Commonly the odor intensity is evaluated as an odorization level. List of odorization levels can be found in the table 2.

Odorization Level (grade)	Olfactory perception	Comment
0	Odor not detected	-
1	Very low intensity	Odor threshold
2	Weak odor	-
3	Mean odor	Warning odor level
4	Strong odor	-
5	Very strong odor	-
6	Extremely strong odor	Upper limit of intensity

Table 2. Odorization levels

4.1 Subjective odorization control

By subjective odorization control odorant concentration is tested primarily with the use of electronic instruments. These instruments all employ the use of the human nose to determine the gas in air mixture at which an individual can detect the smell of odorant. These quantitative olfactory tests are commonly called "sniff test".

There are only several manufacturers of such units. Some of them are e.g. the DTEX made by YZ Systems, the Odorometer made by Bacharach, and the Heath Odorator (see Fig. 17.) All three units are designed to mix gas and air and move them to a sniffing chamber. The air is drawn in through each unit, and mixed with gas. The technician smells the gas and air mixture, gradually raising the level of gas in the mixture until he or she detects an odor of gas.

The Bacharach Odorometer was the first device designed to monitor odor levels and is still available today. The Odorometer uses a rotameter (balls floating up and down on the air stream created by opening the gas stream). The results of a test are read off of the bottom of the balls and compared to a chart on the unit door prepared for each Odorometer.

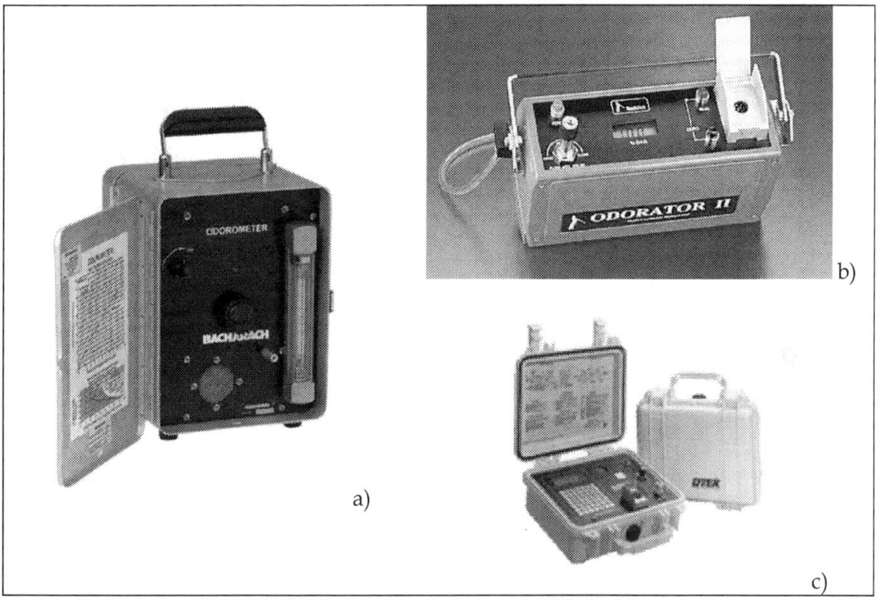

Fig. 17. Bacharach ODOROMETR (a) YZ Industries DTEX (b) and Heath Consultants ODORATOR

The Heath Odorator is another unit designed to test for odor intensity. First step with this device is to zero the unit following the instructions printed on the side of the box. Next opening the gas valve while positioning your nose above the sniff chamber until the odor intensity reaches the threshold level. Push the display button and copy down the reading.

Again with your nose above the sniff chamber, open the valve until the odor intensity reaches a readily detectable level. After the readily detectable level is reached, you push the display button and read the display. Then compare the two display readings to the chart for correction on the side of the unit to get your test results.

To take a test with the DTEX the operator turns on the power and the unit puts itself through a series of self-diagnostic checks. After the operator logs on with a private password, he or she can choose to do a test at a pre-entered test location, or a new location can be entered via the keypad on the unit.

4.2 Objective odorization control

The use of titrators, analyzers and chromatographs are several methods employed for quantitative sulfur analysis. A variety of detectors are used including lead acetate tapes, chemiluminescence, flame photometric and technologies with electro-chemical detectors. These instruments can be configured either for laboratory use or placed directly on the pipeline for real-time calculations. These instruments provide for real-time determinations of total sulfur and in many case individual mercaptan and sulfide component levels.

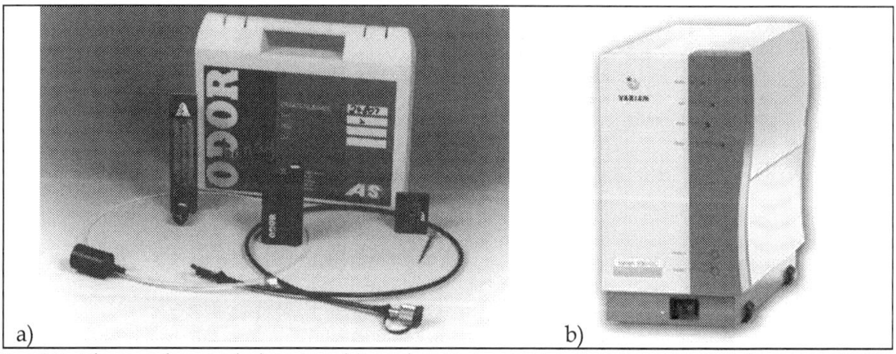

a) b)

Fig. 18. Electro-chemical detector (a) and micro gas chromatography (b) for quantitative estimation of odorant concentrations

Although these "quantitative" methods of determining actual odorant concentrations in the gas stream does not meet the Federal requirement for odorant reporting (not only) under DOT 192.625 it does, however, provide another piece of information in terms of evaluating the overall effectiveness of the odorization program.

To determine the concentrations of Gasodor S-Free odorants a number of analyzers based on different principles may be used. Table 3 provides an overview of suitable analyzers.

Type	(Micro-)GC	IMS-Odor	µIMS-Odor	CMS-Analyzer
		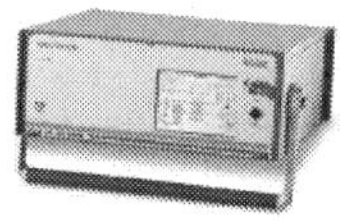		
Measuring principle	Chromatography	Ion Mobility Spectrometr	Ion Mobility Spectrometr	Colorimetric Chemical
Application	Stationary, mobile	Stationary	Mobile	mobile
Measuring range [mg/m³]	>1.5	4 - 23	0 - 23	3 -30

Table 3. Equipment suitable for sulfur free odorant monitoring (Source: Graf 2007)

Impact odorization

Impact (temporarily increased) odorization which is sometimes performed is a targeted, one, two or threefold increase in the dosage of odorant into fuel gas compared to standard operating condition. Its aim is to verify the technical condition of gas distribution and gas supply facilities, usually before the winter season. It is advisable that public in the area where impact odorization is to be carried out be alerted.

5. Preodorization and odor fade

When a new pipeline is constructed preodorization must be carried out. When gas with odorant is injected into the new pipeline absorption and reaction between the pipeline inner wall and the odorant occurs during the passage of this gas mixture through the pipeline. Gas at the exit from the pipeline is then odorless and may pose a serious security risk

If a new steel pipe is ready, the porous inner wall of the pipe contains metal oxides (rust) which react with the odorant; in reaction with TBM disulfides may for example form which are less odorous than TBM proper. Therefore the steel pipe must be clean and free of oxides, otherwise it could happen that the exiting gas is odorless and may pose a potential risk. This effect occurs even when plastic pipes are used and this phenomenon must be given increased attention when putting the pipeline into operation. In order to ensure sufficient security to end users the new line must be saturated with odorant prior to its commissioning. This is done by overodorizing the gas entering the new line. The process of pipeline preodorization and saturation with odorant is often referred to as "pickling".

Used for gas pipeline preodorization and pickling are three basic methods:

1. Injection of highly odorized gas (more than 40 ppm of odorant).
2. Slugging, i.e. pouring a bulk amount of liquid odorant directly into the pipeline and letting the pipe "pickle" for a prolonged period without any flow through the pipe.
3. Continuous injection of controlled dosage of liquid odorant into the gas stream flowing through the pipe.

Odor Fade

Even though odorant is added to natural gas, contractors should not rely solely on the sense of smell to determine if natural gas is present in the ambient air of a work space. This is because it may be possible that:

- Some individuals suffer an impaired sense of smell (chronic or transient) and cannot detect the odorant;
- The odorant can at times be disguised by other odors present on the job site or naturally occurring in the environment;
- Some individuals who have worked around natural gas odorant for an extended period of time may suffer from odor fatigue and may be unable to recognize the presence or change in odor levels; and
- In some rare cases, odor fade (loss of odorant) may occur making it difficult to detect the presence of natural gas in the air. In general, odor fade occurs when physical and/or chemical processes cause the level of odorant in the gas to be reduced. Odor fade can occur in both existing pipe and new installations but is most likely to occur in new steel pipe of larger diameters and longer lengths. Odorant fade can also occur in plastic pipe and in smaller and/or shorter pipe installations.

6. Odorants decontamination

Although odorization systems are designed as leakage-free systems odorant may still leak in current operation e.g. due to an accident or improper handling. Recommended methods of odorant disposal may be divided into three areas according to the quantity and nature of odorant to be disposed.

When disposing smaller amounts of odorants and in order to remove the repulsive smell of spilled odorant the smell is eliminated by means of absorption of spilled odorant in chips impregnated with special oil and masked by a suitable pleasantly smelling substance such as ALAMASK THT-X, ALAMASK VET, ALDOR 1052, PLANAROME 877, or by using PENNCOVER.

For cleaning barrels and tanks from odorants a solution of 2 kg of calcium hypochlorite in 50 l of water with addition of hydrogen peroxide is used. This solution must be left to work 2 – 3 days and is sufficient to clean one 200 l barrel. The residue from cleaning barrels and the barrels proper must be disposed of separately as hazardous waste.

When disposing larger odorant leakages the odorant must first be drained by appropriate binding substances (peat, diatomaceous earth, sawdust, cleaning wool) and the soil soaked with odorant together with soaked binding substance must be stored in closable containers. These containers must be disposed of separately as hazardous waste. This means disposal in special equipment intended for this purpose. The location where odorant spilled may be decontaminated by oxidizing means, either by 1.5% solution of sodium hypochlorite (bleaching process), or by 5% solution of potassium permanganate

7. References

Graf F., Kröger K. & Reimert R. (2007). Sulfur-Free Odorization with Gasodor S-Free—A Review of the Accompanying Research and Development Activities. *Energy Fuels,* 21, 6, (2007) 3322-3333, ISSN: 0887-0624.

Ivanov I., Strmen J., Jones L.(2009). Pre-odorization or "Pickling" of new natural gas pipe. *Pipeline & Gas Journal,* 236, 11, (2009) 48-54, ISSN: 32-0188.

Stefl P. (1997). In *Plynarenska prirucka,* Praha: Gas s.r.o., ISBN 80-902339-6-1, p. 309-320.

Silliman S. R. (2002). *Natural Gas Testing Instruments and Applications,* International School of Hydrocarbon Measurement

Hannah S. (2002). *Odorants and Odorization, A Manufacturers Perspective,* Appalachian Gas Measurement Short Course

Wehnert P.D. (2003). *Determination of Proper Odorization of Natural Gas,* International School of Hydrocarbon Measurement

Code of Federal Regulations, Title 49 Part 192.625.

Usher M. J. (1999). *Odor fade -Possible Causes and Remedies,* CGA Gas measurement School London, Elf Atochem North America, Inc., Philadelphia

Tenkrat D., Prokes O., Beranek J.(2008). Possibility of Objective Control of Natural Gas Odorization. *Chemicke listy,* 2008, S, p. 487-488.

Zhidkov M.A., Shaikhutdinov R.M. (2004). *Odorization Instalation.* Springer New York, ISSN 1573-8329, New York

https://www.yankeegas.com/HomeCustomer/pdfs/OdorFade.pdf

http://help.intellisitesuite.com/ASGMT%20White%20Papers/papers/031.pdf

5

Synthetic Natural Gas (SNG) from coal and biomass: a survey of existing process technologies, open issues and perspectives

Maria Sudiro and Alberto Bertucco
Department of Chemical Engineering (DIPIC), University of Padova
Italy

1. Introduction

Natural gas is a well known energy carrier. It is often used for producing heat and power, but can also be applied as a fuel in the transport sector. The production of synthetic natural gas (SNG) from coal or biomass is an interesting opportunity for both exploiting coal and biomass, and for replacing oil products for transportation and other uses.

SNG has many important advantages with respect to other synfuels: it can be transported efficiently and cheaply using existing natural gas pipelines and distributing networks, it is an easily convertible feedstock, both in natural-gas combined-cycle power plants and in petrochemical facilities, it can count on a high social acceptance with respect to coal, and it can be stored underground, enabling efficient operation throughout the year independently of a fluctuating demand.

Unfortunately, the commercial deployment of technologies for the production of SNG is currently constrained by technical barriers, so that more research is required before extensive applications on the industrial scale can be achieved. An important issue to be addressed is the strong exothermicity of the methane formation reactions, so that conventional fixed-bed catalytic reactors cannot be safely used (Sudiro et al., 2009).

Following the 1970s energy crisis much work has been initiated in the US on coal-to-SNG: SNG process technologies and catalysts were developed and tested extensively. But most have been cancelled in the 1980 because of the changing energy picture. One industrial plant has actually been realized in North Dakota, which began operating in July 1984 and today produces more than 54 billion standard cubic feet of synthetic natural gas annually (1.53 billion Nm^3/yr). Coal consumption is about 6 million tons each year. The heart of this plant is a building containing 14 gasifiers (www.dakotagas.com).

Nowadays, the rise of natural gas prices have created a strong interest in producing SNG from the cheaper and much more abundant coal. A renewed interest in basing more energy consumption on coal and petcoke has resulted in a revival of several older technologies that have been enhanced to improve efficiency and to lower investment cost.

Methanation is used as the final syngas purification step in the production of ammonia, but methanation for SNG production is more complex. The main industrial application of methanation has been the removal of CO from H_2-rich feed gases in ammonia plants. With the aim of producing SNG the methanation reaction changes from a gas cleaning step to the main synthesis process. Methanation for SNG production is more complex because it involves much higher concentrations of CO and CO_2. The high reaction heat results in large potential adiabatic temperature increase that may cause catalyst sintering and possibly leads to carbon particles formation.

The methanation catalysts should be active and stable at both high and low temperatures for the methane production processes because this is the key problem in optimizing the coal-based SNG methanation process. The optimal recovery of the reaction heat from the methanation reaction is also critical (www.syngasrefiner.com/SNG/agenda.asp).

2. The methanation process

2.1 Principles of methanation

The principle of catalytic synthetic production of methane from carbon monoxide and hydrogen was discovered in 1902 by Sabatier and Senderens (Ulmann, 1989). It is described by the CO methanation reaction:

$$CO + 3H_2 \leftrightarrow CH_4 + H_2O \qquad (1)$$

Carbon dioxide can also be converted to methane according to the following reaction

$$CO_2 + 4H_2 \leftrightarrow CH_4 + 2H_2O \qquad (2)$$

Both reactions are linked by the water gas shift conversion, which is always observed simultaneously whenever active catalysts are used:

$$CO + H_2O \leftrightarrow CO_2 + H_2 \qquad (3)$$

A number of observations, reported in Ulmann (1989), indicate that the transformation of carbon dioxide to methane is initiated by a reverse shift conversion reaction with hydrogen to yield carbon monoxide and steam. The carbon monoxide formed then reacts to yield methane.

Both reactions (1) and (2) are strongly exothermic: -206 kJ/mol and -165 kJ/mol, respectively. Also low temperature and high pressure are required to achieve high methane yield.

Interest in SNG production is concentrated on the gasification step, that may yield high methane content in the raw gas. This is possible, for instance, with Lurgi pressure gasification of coal, especially when gasification pressures of 80-100 bar are applied, as has been successfully tested in recent years (Ulmann, 1989).

Methanation processes with little methane in the raw syngas suffer principally from:

1. high exothermic heat release during methanation;
2. need of handling very large quantities of synthesis gas (four to five volumes of dry synthesis gas yield one volume of methane);
3. high proportion of steam formed during methane synthesis, which limits the directly achievable SNG quality in wet methanation steps.

2.1 Earlier methane synthesis plants

The oil crisis in the 1970s intensified the development of lignite and coal gasification processes coupled with SNG production; United States, Germany and Great Britain were involved in these activities (Kopyscinski et al., 2010). A few demonstration and pilot plants were constructed during this period but only one commercial plant was built (in North Dakota, as reported above).

Four types of methane synthesis process have been developed for commercial operation (Ulmann, 1989). They limit the temperature increase by recycling of reacted gas or steam dilution, or by special techniques such as isothermal reactors or fluidized beds, each with indirect cooling by evaporating water.

A first configuration is the "methane synthesis plant with hot recycle", in which a two-stage recycle system with a final countercurrent-cooled reactor is used, as shown in Figure 1. In this process the syngas, after being preheated, is routed to the first methanation reactor with a part of the stream coming from the exit of the second reactor. Then the outlet of the first methanator is cooled and sent to the second reactor step with a part of fresh synthesis gas; these two reactors are run adiabatically. The outlet of the second reactor is cooled and the part not recycled to the first methanator is sent, after a cooling section, to the final reactor which is a countercurrent cooling reactor. The two streams from the two adiabatic reactors are cooled using boiler feed water (BFW) in order to produce saturated steam at 10 bar. The SNG produced is then cooled, dried and recovered.

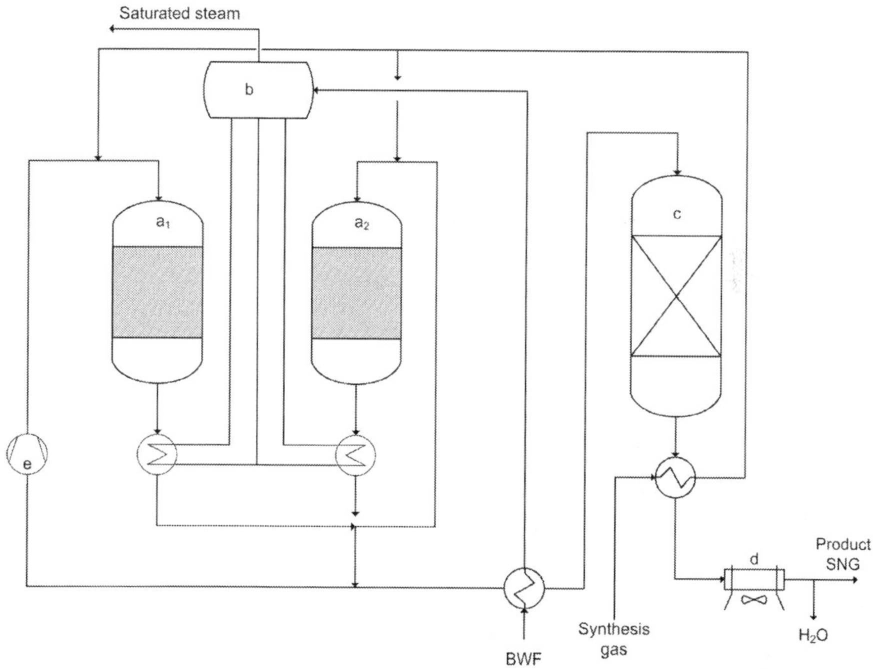

Fig. 1. Recycle methane synthesis: a_1, a_2) recycle reactors; b) steam drum; c) countercurrent cooling reactor; d) air-cooled cooler; e) hot recycle compressor, adapted from Ulmann, 1989

This process was originally designed with a recycle quantity such that 500°C is not exceeded at the outlet of the adiabatic reactors. Current systems are available working at maximum temperatures up to 650°C. Inlet temperatures are around 300°C, which is also the operating range of the recycle compressor. An advantage of the hot recycle is that water vapour formed during the methanation reaction is not condensed.

A second example of methane synthesis process is the "steam quenching methane synthesis" and is sketched in Figure 2. For synthesis gases with low H_2/CO molar ratio, steam must be added to prevent carbon particle formation. For optimum results, the required quantity of steam is mixed with a portion of the feed gas in such a way that the temperature in the first methanation step is kept within allowable limits and, after heat removal, a further portion of the feed gas is admitted to the first intermediate product gas stream as quenching stream. The quenching procedure after cooling is repeated until all the feed gas is consumed. Thus, a multistage concept results, depending on the inlet gas composition, as shown in Figure 2. With the addition of steam, the equilibrium composition of any intermediate product gas is not favourable enough to enable direct production of specification-grade SNG in a wet methanation step. The removal of carbon dioxide, therefore, is followed by a final, dry methanation stage.

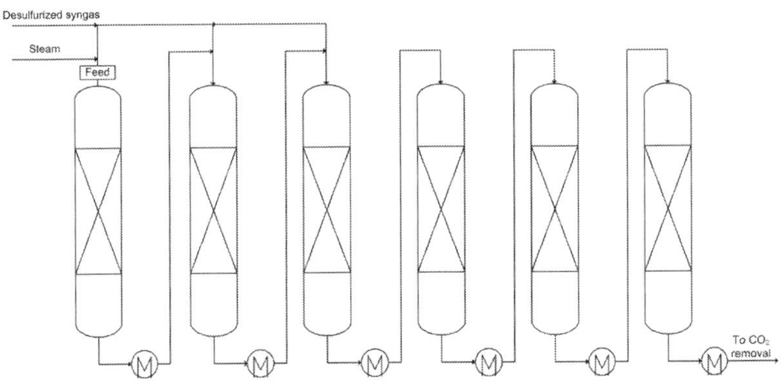

Fig. 2. Scheme of an example of multi stage quenching system, adapted from Ulmann, 1989, where a fixed bed reactor followed by a cooling section is repeated until the feed is consumed

A third example of methane synthesis is the "fluidized-bed methanation". The Comflux® process is a plant of this type, in which feed gases of unfavorable composition are sent, with little or no addition of steam, into a cooled fluidized-bed reactor in one step. The dangers of carbon formation, catalyst deactivation by high carbon monoxide partial pressure, and excessive catalyst consumption by the fluidized-bed process are said to have been overcome, and operation up to 60 bar has been demonstrated successfully in the pilot plant (Ulmann, 1989).

A final example of methane synthesis is the "isothermal operation". By increasing the allowable temperatures for methanation catalysts, methane synthesis can be performed by a once-through method in quasi-isothermal reactors cooled by evaporating water which generates saturated steam. Under favorable conditions, such systems produce specification-grade SNG in only one catalytic step (Ulmann, 1989). Any traces of sulfur compounds must be eliminated carefully in order to avoid catalyst poisoning.

3. SNG: state of the art

3.1 Existing process technologies

Methanation reactor technologies can be classified into three categories: fixed bed, fluidised bed and other types of reactor.

About fixed bed methanation the following processes have been developed: Lurgi, Tremp™, Conoco/BGC, HICOM, Linde, RMP and ICI/Koppers.

In the 1930s Lurgi developed a methanation unit including two adiabatic fixed bed reactors with internal recycle. One pilot plant was designed and erected by Lurgi and Sasol in Sasolburg (South Africa) and another pilot plant in Austria, erected by Lurgi and El Paso Natural gas Corporation. The methanation unit of these two pilot plants consisted of two adiabatic fixed bed reactors with internal gas recycle. Based on the results obtained by Lurgi and Sasol, the first and only commercial SNG from coal plant (Great Plains Synfuels Plant) was commissioned in North Dakota USA (Kopyscinski et al., 2010).

The Tremp™ process (Topsoe's Recycle Energy-efficient Methanation Process) was developed by Haldor Topsoe's laboratory. This technology addresses the essential question of heat recovery in the most efficient manner by recovering the heat as high pressure steam. In order to apply such a heat recovery concept, it is essential to recover the reaction heat at high temperature. Due to the unique MCR-2X methanation catalyst, Tremp™ can operate at high temperatures, up to 700°C (Topsoe, 2009). This catalyst has the following advantages:

- recovery of reaction heat in the form of high pressure superheated steam to be used directly for turbine drive;
- low recycle ratio to ensure energy savings;
- reduced gas flow resulting in lower equipment cost.

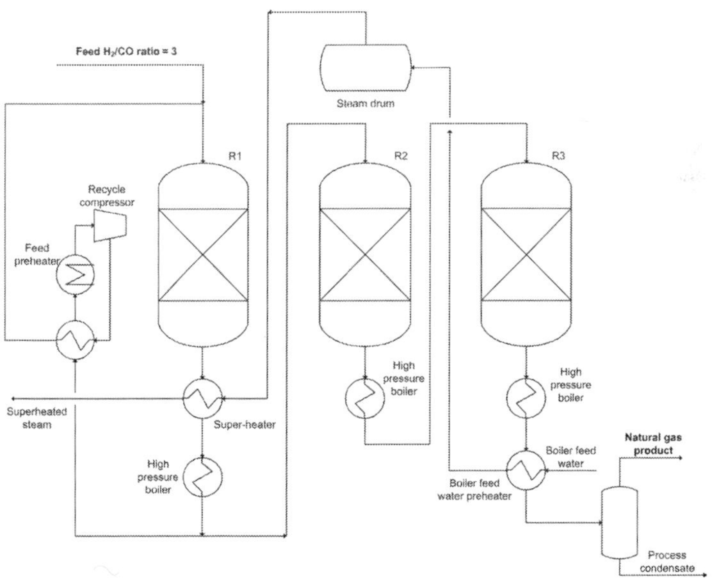

Fig. 3. Scheme of the Tremp™ technology, adapted from Topsoe, 2009

With reference to Figure 3 the process steps upstream the methanation unit are designed to provide a near stoichiometric ratio of hydrogen to carbon monoxide in the gas according to the methanation reaction (eq. 1). The exit gas from the first reactor is cooled in two steps where the first step serves to superheat the high pressure steam generated in the second step. After cooling, the gas enters the following methanation stages.
The CO methanation takes place in adiabatic reactors. The heat of the reaction results in a high temperature increase, and recycle is used to control this temperature rise in the first methanation reactor.
This technology is characterized by (Topsoe, 2009):

- low energy consumption for recycle;
- production of high pressure superheated steam;
- low investments;
- producing a natural gas compatible with pipeline specification, ensuring an easy access to distribution of the product.

Any recycle involves a loss of energy but the MCR-2X catalyst is a good choice to minimize the amount of recycle gas. This catalyst is stable and operable at low as well as high temperature (from 250 to 700°C).
The experience with operation of this technology dates back to 1978 and a substantial process demonstration has taken place ensuring that the technology can be applied. A semi-commercial process has been demonstrated in a plant producing 2000 Nm^3/h of natural gas. However no industrial plants have been constructed until now. The project was closed down in 1981 for political change and lower energy prices (Undergaard, 2008). Presently Topsoe's TREMP™ technology has been approved for a US plant. This methanation technology was selected for use in Power Holding's coal gasification plant in Jefferson County, Illinois (USA). The plant will convert about 4 million tons per year of coal into pipeline-quality natural gas. Along with GE Energy and Lurgi, Haldor Topsoe has been selected as technology provider. It is expected the coal-to-gas plant will startup in 2010 (www.zeuslibrary.com).
In 1972 in Scotland (Westfield Coal Gasification plant) the first worldwide demonstration plant producing SNG from coal has been accomplished by ConocoPhilips and the British Gas Corporation (BGC), with a production of 59 Million Nm^3/day. The methanation unit, consisting of a fixed bed adiabatic reactor with gas recycle, was connected to an existing Lurgi fixed bed gasifier and gas cleaning section was a Rectisol unit. Unfortunately, no plant data can be found (Kopyscinski et al., 2010).
A further development of the British Gas Corporation was the HICOM process in which shift and methanation are combined. In this type of process (see Figure 4) the syngas, after purification, is heated and saturated with hot water in a countercurrent flow packed bed. After that, the syngas is passed through a series of fixed bed reactors. The temperature is controlled by recycling the cooled product gas. Excess steam is added to the first methanation reactor to avoid carbon particle deposition. A part of the product gas from the main methanation reactors is recycled and the other part is passed through one or more low temperature fixed bed methanation reactors. In the last reactor the remaining CO and H_2 are converted to CH_4 and CO_2. Almost all the heat released from these reactions is used to generate high pressure steam except the one of the last reactor, which is applied to warm the saturated boiler feed water. With this type of configuration a bench-scale reactor for screening of catalysts and process conditions was erected, also a pilot plant was built where tests for about 2000h were

done. Finally, a semi-commercial scale plant was constructed at the Westfield Development Center but no data about these plants were found (Kopyscinski et al., 2010).
In the 1970s, Linde AG (Germany) developed an isothermal fixed bed reactor with indirect heat exchange. In this reactor the cooling tube bundles are embedded in the catalyst bed, so the reactor is able to produce steam from the exothermic reactions and a part of this steam can be added to the syngas at the inlet of the methanator in order to minimize the risk of carbon deposition. No information are available about the use of this type of reactor in SNG production (Kopyscinski et al., 2010).
A high temperature methanation without gas recycle was proposed by the Ralph M. Parsons Company (USA). This process, called RMP, consists of 4-6 adiabatic fixed bed reactors in series with intermediate gas cooling. The syngas is distributed in different ratios in the first four reactors (Figure 2 shows the idea of this process). Working pressures are between 4.5 and 77 bar and temperatures are in the range between 315°C (inlet) and 538°C (outlet). Data about gas composition of different tests were available whereas no data about the catalyst and reactor dimensions were published. After 1977 no more information about this project are available (Kopyscinski et al., 2010.

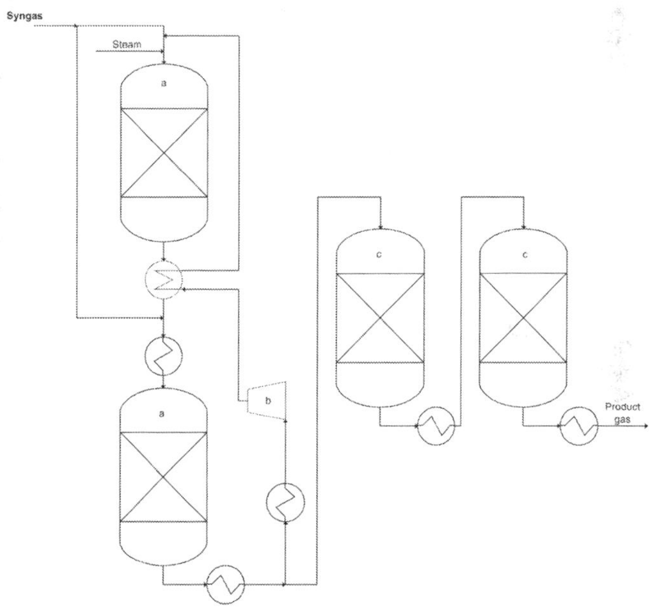

Fig. 4. Scheme of the HICOM process: a) main methanation stages, b) recycle compressor, c) non-recycle methanation stages, adapted from Kopyscinski et al., 2010

Similar to the RMP process, the Imperial Chemical Industries ICI (Great Britain) developed a catalyst and a high temperature once-through methanation process, using a catalyst with a high nickel content (up to 60%). This process consists of three adiabatic fixed bed reactors in series with intermediate gas cooling (see Figure 5), where it is possible to see that steam is

added to the second reactor in order to maintain the temperature below 750°C. On the basis of this scheme no large scale plant have been built (Kopyscinski et al., 2010).
About fluidised bed methanation several projects have been set forth. In the first project, started in 1952 by the United States Department of the Interior, one fixed bed and two different fluidised bed methanation reactors were developed, which were operated in total for more than 1000h. About this project no data are available after 1956 (Kopyscinski et al., 2010).
A second project, the Bi-Gas project, was initiated in 1963 by Bituminous Coal Research Inc. (USA) for producing SNG from coal, via gasification in a entrained flow gasifier. The methanation reactor developed within this project is a gas-solid fluidised bed reactor including a second feed inlet in the middle of the reactor and two in-tube heat exchanger. Experimental tests for about 2200h were done obtaining conversion of CO between 70 and 95%. After the last publication in 1979, no more reports on the Bi-Gas project have been found in the literature (Kopyscinski et al., 2010).

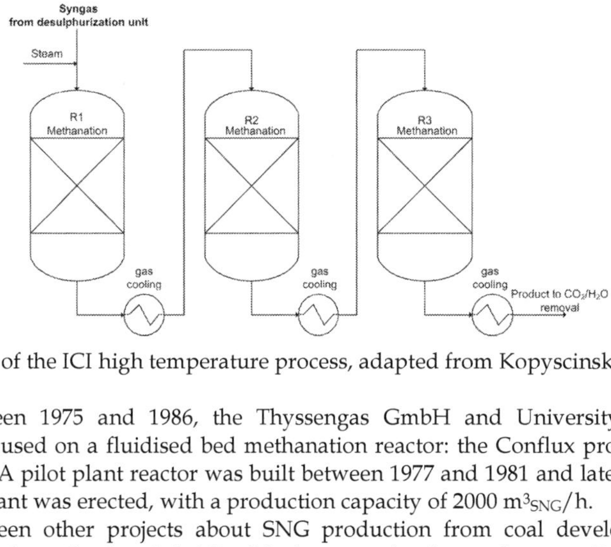

Fig. 5. Scheme of the ICI high temperature process, adapted from Kopyscinski et al., 2010

Finally, between 1975 and 1986, the Thyssengas GmbH and University of Karlsruhe (Germany) focused on a fluidised bed methanation reactor: the Conflux process, described in section 2.1. A pilot plant reactor was built between 1977 and 1981 and later in 1981 a pre-commercial plant was erected, with a production capacity of 2000 m^3_{SNG}/h.
There have been other projects about SNG production from coal developing different configurations from fixed and fluidised bed reactor for the methanator. For example, the Synthane project, developed by the Pittsburg Energy Technology Center (USA), the catalytic coal gasification by the Exxon Research and Engineering Company (USA) and the liquid phase methanation proposed by Chem System Inc. (USA). The first and the third project were terminated in 1980-1981.

3.2 Patents

About patents dealing with SNG production process, some are recent and described below, other are older, for instance before 1976 (Müller et al., 1976 and Schultz & Hemsath, 1976).
In the patent by Jahnke & Parab (2007) an invention related to methanation of synthesis gas is reported and, in particular, to a methanation assembly using multiple reactors for controlled methanation. Object of the invention is to produce a gas having a desired

temperature control and methane composition. Also direct water injection is used as a cooling medium to control the temperature in the methanation reactors as well as to avoid deposition of soot on the methanation catalyst.

The process is realized in a methanation assembly for use with a water supply and a gas supply containing gas to be methanated. The reactor assembly has a plurality of methanation reactors each for methanating gas input to the assembly and a gas delivery and cooling assembly adapted to deliver gas from the gas supply to each of the methanation reactors. The system is also to combine water from the water supply with the output of each methanation reactor being conveyed to a next methanation reactor and to carry the mixture to such next methanation reactor.

Three methanation reactors are employed and the gas delivery and cooling assembly includes one or more water injection units, gas dividing units, one or more water routing units and lines connecting these units.

In another recent patent (Ravikumar & Sabbadini, 2007) the invention includes one or more methanation reactors producing a primary methanation product that is cooled to a temperature sufficient to condense water, which is removed in a separator. The dry methanation product is then split to provide a reflux stream to the methanation reactors and a feed stream to an adiabatic trim reactor. The plant comprises at least two methanation reactors that are operated in series, wherein the first reactor receives the recycle steam and the second one a portion of the first methanation reactor effluent and a portion of the first methanation reactor feed. Most preferably a recycle conduit is coupled to the separator and the first primary reactor such that a first portion of the dried effluent is fed to the first primary reactor.

Another patent (Mozaffarian, 2000) reports an invention related to a process for producing methane-rich product gas from biomass or fossil fuels. This patent is focused on the synthesis gas production system, which is a hydrogasification reactor using biomass or fossil fuels as feedstock together with hydrogen from an external source.

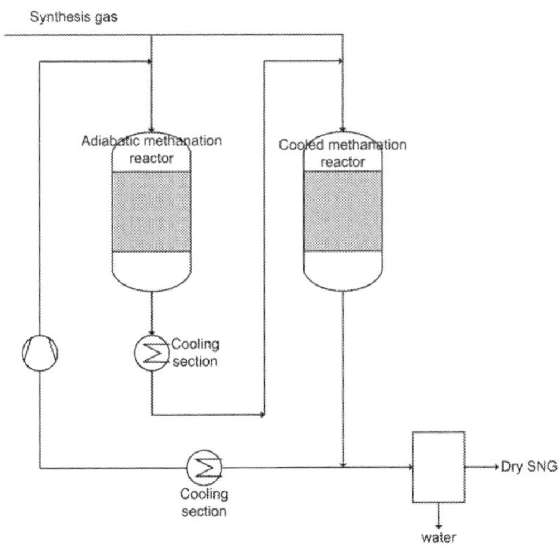

Fig. 6. Simplified Block Flow Diagram of the process patented by Haldor Topsoe A/S, Lyngby, Denmark, adapted from Skov, 1981

A patent, not recent but very interesting, with Haldor Topsoe A/S, Lyngby, Denmark as assignee, is the one by Skov (1981). The scheme of this process is a quite interesting modification of TREMP™ and is reported in Figure 6.

The invention relates to an improved catalytic methanation process, where a feed gas containing predominantly hydrogen and being rich in carbon oxides (CO and/or CO_2) is divided into two part streams of which the first is methanated partially in an adiabatic methanation reactor by a methanation catalyst. After that, the effluent from the adiabatic methanation reactors is mixed, after cooling, with the second feed gas part stream and the thus-combined stream is methanated in a cooled methanation reactor by a catalyst, preferably the same as that used in the adiabatic methanation reactor. It is possible, but not always necessary, to recycle part of the produced gas to the adiabatic methanation reactor to keep the temperature in a moderate level. It is advantageous because it can be operated to produce superheated steam for producing electricity from the cooling sections at the end of the adiabatic reactors. This process has the great advantages that practically all of the heat of reaction can be utilized for producing superheated steam, and that the superheated steam may be produced within the ranges of pressure and temperature which are convenient for the production of electricity. Superheated steam for the production of power has normally a pressure of 90-160 atm and a temperature of 500-550°C.

By the methanation of gases having high content of carbon oxides the amount of heat generated in accordance with the reaction equations 1 and 2 will be so considerable and the temperature so high that the catalyst in an adiabatic reactor may be destroyed, and possibly even the reactor may be damaged (Skov, 1981). One way of solving this problem involves the cooling and recycling a part of the methanated gas from the outlet of the reactor. It is a drawback of this process that considerable amounts of energy is used for the recycling, whereby the total useful effect of the process is reduced.

In summary, this new process consists of these steps:

- dividing the feed gas into two streams, a first feed gas part stream comprising 30-70% by volume of the total feed gas stream and a second feed gas part stream comprising the remainder of the feed gas;
- subjecting the first feed gas part stream to a catalytic methanation in at least one adiabatic methanation reactor containing a bed of catalyst;
- cooling the outlet gas stream from the adiabatic methanation reactor to 250-400°C;
- mixing the cooled outlet stream of the previous step with the second feed gas part stream to form a combined stream;
- subjecting the combined stream from the previous step to a catalytic methanation in at least one cooled methanation reactor containing a bed of catalyst; and finally
- recovering the outlet gas from the cooled methanation reactor totally or partially as a product gas for use or further treatment.

About older patents we quote one by Müller et al. (1976) about the design of the methanator reactor, and a second one (Schultz & Hemsath, 1976) which studies an apparatus and a method for heat removal in a methanation plant.

3.3 Research studies

Among others, Moeller et al. were involved in research projects concerning methanation. They demonstrated the feasibility of methanation of syngas from coal. In a first work (Moeller et al., 1974), tests in a semi-technical pilot plant prove that CO-rich syngas can be methanated without carbon formation to yield specification grade SNG with a residual hydrogen of less than 1% (vol.) and residual CO less than 0.1% (vol.). Also, it has been demonstrated that trace components left in the synthesis gas after coal gasification and Rectisol wash have little influence on catalyst activity and life. The catalyst used is a special methanation catalyst developed by BASF with a high nickel content supported on Al_2O_3 and activated by reduction with hydrogen. The configuration of the plant consists of two adiabatic methanators. Effluent gas from the first reactor is cooled and a part of this effluent gas is recycled, while the rest is reheated and fed to the final methanation reactor. In fact syngas with an H_2/CO molar ratio equal to 8 are mixed with recycle gas and then the total feed is heated and sent to the first methanation reactor with addition of steam (as inert agent). Effluent gas from the first reactor is cooled, condensing the steam. Part of the reactor effluent gas is recycled, while the rest is reheated and fed to the final methanation reactor. At the inlet of the first reactor methane content is about 51.6% vol. whereas at the exit is 55.6% vol. At the exit of the second methanator the methane content is about 75.1% vol. and the rest is mainly carbon dioxide (21.1% vol.) and inerts, i.e. N_2 and Ar (2.0% vol.) (Moeller et al., 1974).

Tucci and Thomson (Tucci & Thomson, 1979) carried out a comparative study of methanation over ruthenium catalyst both in pellet and in honeycomb form. In addition to pressure drops lower by two orders of magnitude they found also significantly higher selectivities (97% versus 83%) over the monolith catalyst.

Recent studies on SNG production have been performed by Duret et al. (Duret et al., 2005), by Zwart and Boerritger (Zwart & Boerrigter, 2005), by Waldner and Vogel (Waldner & Vogel, 2005) and more recently by Sudiro et al. (Sudiro et al., 2009), Juraščik et al. (Juraščik et al., 2009) and Gassner and Maréchal (Gassner & Maréchal, 2009).

Objective of the work of Duret et al. (Duret et al., 2005) was to perform a study of the process in order to find its optimal operating parameters. The methodology used combines process modelling and process integration techniques. It passes through two steps: a thermodynamic model of the process and a process integration to identify the energy saving opportunities. The process design of a 10-20 MWth Synthetic Natural Gas (SNG) production process from wood has been performed.

Methanation reactor is based on the Comflux® process, in which the reactor is a pressurized fluidized bed reactor with an internal cooling system which allows performing an isothermal once through methanation of coal gas. Note that methanation reactor has been modelled by using a simplified model (thermodynamic equilibrium, pressure of 60 bar and outlet temperature of 400°C) without considering heat transfer problem.

This work demonstrated that the process can transform wood into pipeline quality methane with a thermal efficiency of 57.9% based on the Lower Heating Value (LHV). The process integration study shows that the heat surplus of the process can be used to almost satisfy the mechanical work required by the process; only 7% of the mechanical needs should come from an external source, for example by converting the excess of heat produced in the system.

Objective of the study of Zwart and Boerritger (Zwart & Boerrigter, 2005) was to determine the technical and economic feasibility of large-scale systems for the co-generation of "green" Fischer-Tropsch (FT) transportation fuels and "green" SNG from biomass. The systems were assessed assuming a targeted annual production of 50 PJ (1 PJ = 10^{15} J) of FT transportation fuels and 150 PJ of SNG. The evaluated overall system is composed of the entire chain of biomass collection, transport, syngas production via gasification, gas cleaning, and FT and SNG synthesis. In case of co-production, some of the thermal biomass input is converted to liquid fuels by FT synthesis and the off-gas is methanated to produce SNG. In the integrated co-production concepts, some of the product gas is used for FT synthesis and the other portion is used for SNG synthesis, whereas in the parallel co-production concepts, two different gasification processes are used.

For all the systems evaluated, an Aspen Plus™ model was constructed, to determine the mass, heat, and work balances of the processes. Six combinations of gasifier type, operating pressures, and pressurization gas were considered.

The major conclusions, with respect to the technical feasibility of producing synthetic natural gas (SNG) as co-product of FT liquids are (Zwart & Boerrigter, 2005):

- there is no incentive to produce either SNG or FT liquids, because the conversion efficiencies to both products are essentially equal;
- the overall efficiencies (FT liquids plus SNG) are higher for circulating fluidized bed and indirect gasification concepts, compared to gasification with oxygen, because a significant amount of CH_4 and C_2 compounds is already present in the product gas;
- additional SNG can be produced either by "integrated co-production", in which a side-stream of the product gas of the gasifier is used for dedicated methanation, or by "parallel co-production", in which some of the biomass is fed to a second gasifier that is coupled to a dedicated stand-alone methanation reactor.

Another research work is that by Waldner and Vogel (Waldner & Vogel, 2005). Here, the production of SNG from wood by a catalytic hydrothermal process was studied in a laboratory batch reactor suitable for high feed concentrations (10-30 wt %) at 300-410°C and 12-34 MPa with Raney nickel as the catalyst. A maximum methane yield of 0.33 (g of CH_4)/(g of wood) was obtained, corresponding to the thermodynamic equilibrium yield.

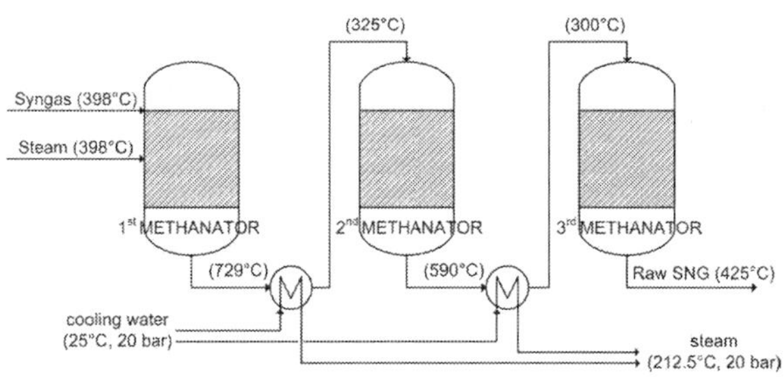

Fig. 7. Scheme of the ICI methanation process, adapted from Juraščík et al., 2009

Another recent work by Juraščík (Juraščík et al., 2009) performed a detailed exergy analysis for the SNG process based on woody biomass gasification: an overall energy efficiency of 72.6% was found. To simulate the methane synthesis the steam-moderated ICI high-temperature once-through methanation process was chosen. This process, which is shown in Figure 7, consists of three methanation reactors and two heat exchangers placed between them in order to control the temperature of gas entering the 2nd and 3rd methanation reactor. The indicated temperatures of the streams entering and leaving the reactors are the original temperatures of the ICI technology.

Gassner and Maréchal (Gassner & Maréchal, 2009) developed a detailed thermo-economic model considering different technological alternatives for thermochemical production of SNG from lignocellulosic biomass (wood) investigating the energetic performances of the processes. Gasification and methanation reactors have been represented by using simplified models (i.e. thermodynamic equilibrium ones) which is a reasonable assumption for methanation when the amount of catalytic material is suitable, as in this case product's composition obtained is very similar to that at equilibrium (Duret et al., 2005). In the work by (Gassner & Maréchal, 2009) there is no particular attention to methanation reactor but authors report only that common industrial installations use product gas recycle loops or multiple intercooled reactors with prior steam addition to obtain a suitable temperature control. The model they proposed is based on data from existing plants and pilot installations; it was shown that the conversion of woody biomass to SNG is a viable option with respect to both energetic and economic aspects, and the overall energy efficiency of the process is in the range 69-76%.

Sudiro et al. (Sudiro et al., 2009) developed and simulated a process to produce SNG from petcoke via gasification, facing the main issue of this process: the temperature control of the methanator. For the methanation section the problem of temperature control has been resolved with a proper suitable use of recycle streams. The process consists in three main sections: petcoke gasification, syngas purification system and methanation reactor. The attention is focused on the syngas generation, obtained with a dual bed petcoke gasification system, and the methanation reactor. For the first section a detailed model including kinetics and mass transfer was investigated, for the methanation section three different possible configurations (A, B, and C) of the plant was developed. Figure 8 shows configuration A, where cooled and purified syngas is sent to methanation, after being split into three streams: the first one is sent to the first methanator together with part of the outlet stream from this reactor, which is recycled by a compressor. The part not recycled is sent to a second methanator with fresh syngas and then, in a similar way, the outlet from this second reactor is sent to the third methanator with part of the fresh syngas. The outlet from the third methanation reactor is sent to a cooling section, then to a unit to remove carbon dioxide, and finally the gas is dried and the SNG product is recovered.

The system has two main disadvantages. Firstly, it requires many Acid Gas Removal (AGR) units: one unit at the output of gasifier in order to remove CO_2 but especially H_2S, which is a poison for the methanation catalyst, a second one at the output of shift reactor and a final one to separate the product, i.e. SNG, from carbon dioxide. The second disadvantage is the use of a compressor, which complicates the plant, and represents a relevant additional energy consumption.

Performances of the global process to produce SNG from petcoke were simulated with Aspen Plus™ and evaluated with respect to product yield, CO_2 emissions and overall energy efficiency. They are shown in Table 1. The value of product yield was found to be 39.7%, CO_2 emissions amount to 2.2 kg per kg of SNG produced and the overall energy efficiency is 67.7%, similar to that of a conventional Gas-to-Liquid (GTL) process (Sudiro & Bertucco, 2007).

The second configuration (B) proposed is similar to the first one with the difference that the water condensed and recovered from the product (SNG), after being pumped, is partly sent to the second methanator, and partly to the third methanator, while another portion is purged out of the system. In this way the inert content in the stream sent to reactors is higher, facilitating temperature control inside the reactors. The third configuration proposed (C) is also similar to the second one, except for the second recycle, which is now part of the SNG produced, sent to the compressor together with the outlet stream of the first reactor. In this way the two streams are mixed and then divided into four parts: one to the first methanator, one to the second and one to the third methanator and one to the product.

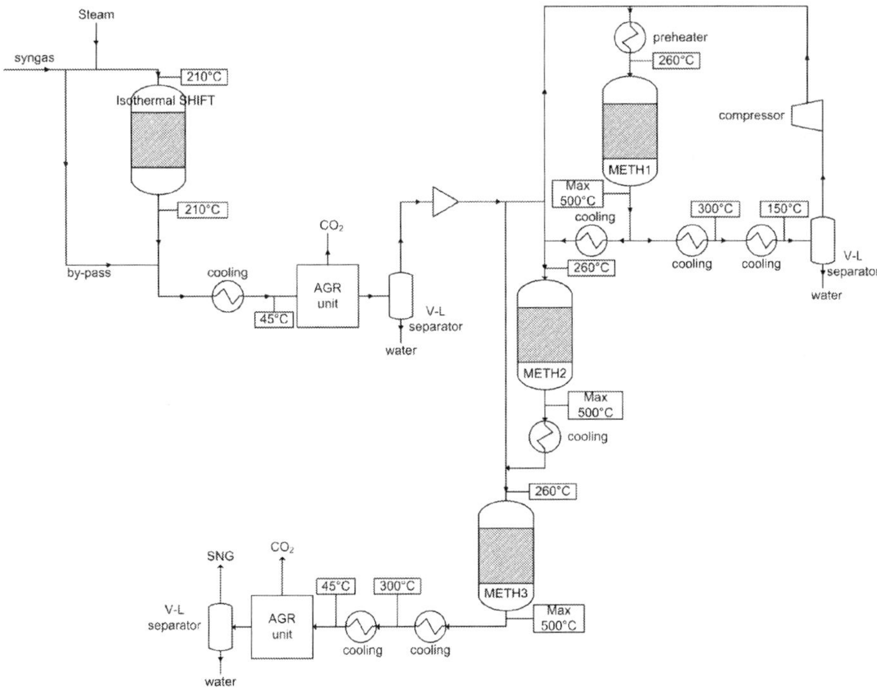

Fig. 8. Scheme of the methanation plant (configuration A)

Overall energy efficiency (*)	67.7%
kg CO_2/kg SNG	2.2
kg CO_2/MJ SNG	0.044
Mass yield % (kg SNG/kg petcoke)	39.7

(*) defined as the ratio between the energy content in the product (SNG) and in the feedstock (petcoke), based on lower heating value.

Table 1. Performances for the configuration A simulated for the methanation section

It was concluded that one method to control the temperature in SNG processes is operating with a lower H_2/CO molar ratio than stoichiometric, using the recycle, in order to control the temperature with the inerts. However, several reactors in series are needed to obtain acceptable conversion of CO and CO_2. The best solution would be to have a process that works without the use of the compressor, thereby reducing both the plant complexity and the operating costs.

4. Coal-to-SNG projects in the world

The only commercial-scale coal-to-SNG plant is located in Beulah, North Dakota USA, owned by Dakota Gasification company. This plant began operating in 1984 and uses 6 million tons of coal per year with an average yearly production of approximately 54 billion standard cubic feet (scf). Synthetic natural gas leaves the plant through a 2-foot-diameter pipeline, travelling 34 miles south.

In addition to natural gas, this synfuel plant produces fertilizers, solvents, phenol, carbon dioxide and other chemicals. Carbon dioxide is now part of an international venture for enhanced oil recovery in Canada (www.dakotagas.com).

The plant had a cost of $2.1 billion and a work force of more than 700 people (www.gasification.org/Docs/Conferences/2007/45FAGE.pdf).

The heart of the Dakota plant is a building containing 14 gasifier, which are cylindrical pressure vessels 40 feet high with an inside diameter of 13 feet. Each day 16000 tons of lignite are fed into the top of the gasifiers. Steam and oxygen are fed into the bottom of the coal beds causing intense combustion (2200°F (~1094°C)). Ash is discharged from the bottom of the gasifiers. The raw gas goes to the gas cooling area where the tar, oils, phenols, ammonia and water are condensed from the gas stream. These byproducts are sent on for purification and transportation. Other byproducts are stored for later use as boiler fuel for steam generation. The gas is moved to a cleaning area where further impurities are removed. Methanation is the next step, which takes place by passing the cleaned gas over a nickel catalyst causing carbon monoxide and most remaining carbon dioxide to react with free hydrogen to form methane. Final cleanup removes traces of carbon monoxide. The gas is then cooled, dried and compressed and enters the pipeline (www.dakotagas.com).

Today in the United States many SNG plants are planned and some of them are expected to be operational in the decade 2010-2020 (Petrucci, 2009). Table 2 reports coal-to-SNG projects in the United States.

Coal-to-SNG plants are becoming the new focus in China's coal chemical industry. Currently there are about 15 coal-to-SNG projects proposed in China. It is expected that China will have around 20 billion Nm^3/a SNG capacity in 2015 (www.chemconsulting.com.cn/info_detail01.asp?id=7677). Shenhua Group has different projects for SNG plant in China: in Yijinhuoluo County, Ordos City and Inner Mongolia (Petrucci, 2009).

For biomass, the only commercial project is in Sweden. In the Gothenburg Biomass Gasification Project (GoBiGas), started in 2008, SNG will be produced from forest residues. A 20 MW_{SNG} plant is scheduled to be commissioned in 2012 and a further 80 MW_{SNG} plant is scheduled to be in operation by 2016 (Kopyscinski et al., 2010). These plants will use PSI technology for methanation process and the FICFB gasifier similar of that of Güssing.

Project Owner	Project Name	Location	Feedstock	Status	SNG Capacity
Secure Energy Systems, Siemens SFG	Secure Energy Systems SNG	Decatur, Illinois	Bituminous coal	Commissioning 2010	20 Billion scf/yr
Peabody Energy, Conoco-Phillips	Kentucky NewGas Energy Center	Central City, Kentucky	Coal	Planning-Development	60-70 Billion scf/yr
TransGas Development Systems	Scriba Coal Gasification Plant	Scriba, New York	Coal	Fully operational in late 2010	-
Great Northern power Development/ Allied Syngas	South Heart Gasification Project	South Heart, North Dakota	Lignite	Construction to begin 2010. To be complete in 2012	100 Million scf/day
Lackawanna Clean Energy	Lackawanna Clean Energy	Lackawanna, NY	Petcoke	In operation by 2012	85 Million scf/day
C Change Investments, NC12	-	Louisiana	Coal-Petcoke	Commissioning 2012 (estimated)	300 Billion scf/yr
Cash Creek Generation LLC	-	Henderson County, Kentucky	Coal	Construction to be completed in 2012	720 MW natural gas combined-cycle power plant
Indiana Gasification LLC	-	Spencer County, Illinois	Coal	Expected to be operational in late 2012 or 2013	-
Christian County Generation, LLC	Taylorville Energy Center (TEC)	Taylorville, Illinois	Bituminous Coal	Construction to begin in 2010. Commercial operation in 2014.	500 MW IGCC and SNG production
Hunton Energy (US)	Freeport plant (HE)	Freeport (Texas)	Petcoke-coal-biomass	Completion in 2015	180 Million scf/day
Power Holdings, LLC	Southern Illinois Coal-to-SNG	Jefferson County, Illinois	Coal	Planning	65 Billion scf/yr

Table 2. USA Coal-to-SNG projects (Petrucci, 2009)

5. Research and recent developments about SNG processes from coal and biomass

The technical assessment of different technological systems for SNG production is currently an important research topic. Some new ideas are briefly reviewed in the following. Three processes have been recently developed in the USA (Kopyscinski et al., 2010):

1. Bluegas™ process by Great Point Energy;
2. fluid-bed methanation process by Research Triangle Institute (RTI);
3. hydro-gasification process by Arizona Public Service (APS).

The first one is proposed by Great Point Energy and is a hydro-methanation process, called Bluegas™, where gasification and methanation reactions occur in the same catalytic reactor working at temperatures between 600 and 700°C. The Bluegas™ gasification system is an optimized catalytic process for combining coal, steam and a catalyst in a pressurized reactor vessel to produce pipeline-grade methane (about 99% CH_4) instead of the low quality syngas obtained by conventional coal gasification. This technology employs a novel catalyst to "crack" the carbon bonds and transforms coal into clean burning methane (www.greatpoint). The first step is to feed the coal or biomass and the catalyst into the methanation reactor. Inside the reactor, pressurized steam is injected to "fluidize" the mixture and ensure constant contact between the catalyst and the carbon particles. In this environment, unlike the conventional gasification, the catalysts facilitates multiple chemical reactions between the carbon and the steam on the surface of the coal (or biomass).

$$C + H_2O \leftrightarrow CO + H_2 \tag{4}$$

$$CO + H_2O \leftrightarrow CO_2 + H_2 \tag{5}$$

$$C + 2H_2 \leftrightarrow CH_4 \tag{6}$$

The overall reaction is the following:

$$2C + 2H_2O \leftrightarrow CH_4 + CO_2 \tag{7}$$

Accordingly, in a single reactor a mixture predominantly composed of a mixture of methane and CO_2 is generated.

The proprietary catalyst formulation is made up of abundant, low cost metal materials specifically designed to promote gasification at the low temperatures where water gas shift and methanation take place. The catalyst is continuously recycled and reused within the process. Unlike many conventional gasifiers, the Bluegas™ process is ideally suited for lowest cost feedstocks such as petroleum coke from the Canadian oil sands (a waste product produced in the upgrading process) as well as a number of biomass feedstocks. The result is a technology with improved economics and an environmental footprint equivalent to that of natural gas, the most environmentally-friendly fossil fuel.

The Bluegas™ technology has several advantages:

- it produces methane in a single step and in a single reactor, called catalytic coal methanation (with no need for external water gas shift reactor and for external methanation reactor);

- it produces CO_2 as a valuable sequestration-ready byproduct;
- it significantly reduces operating temperatures so there are lower costs for reactor components, lower maintenance costs. Costs for high temperature cooling are eliminated;
- costs for air separation plant are eliminated because it utilizes steam methanation;
- it has an high efficiency (65% of overall efficiency).

The catalyst is able to "crack" the carbon bonds and transforms coal into clean burning methane (Lesemann, 2004). Tests on a pilot plant in Illinois has been done for more than 1200h but no data are available (Kopyscinski et al., 2010).

As a second process, we quote the one by the Research Triangle Institute (RTI), which has developed a system for producing SNG and electric power from coal. Coal is sent to a pyrolyzer where products are char and a gaseous mixture; char is used to generate electricity and the gas is sent to a methanation fluidized-bed reactor in order to produce a syngas rich in methane. About this process no experimental data are available. At this time RTI has a fluid-bed methanation bench scale system (Lesemann, 2004). Also Peabody Energy and GasPoint Energy are working towards this project (www.trib.com/news/state-and-regional/article_03676d79-d722-525e-98d9-946be031fcd2.html).

Finally, Arizona Public Service (APS) are developing a hydro-gasification process where coal is gasified with hydrogen at moderate temperatures (870°C) and high pressures (70 bar). Methane is directly produced in the gasifier without using a catalyst. In this process electric power is produced by burning the unconverted char, and a part of the SNG obtained must be use to generate the hydrogen required for hydro-gasification.

Among other projects in progress we are studying an innovative solution where the problem of temperature control, typical of fixed-bed methanation reactors, can be overcome by using monolith catalyst supports (Sudiro et al., 2010). The use of monolith catalyst supports offers at least two advantages with respect to conventional packed-bed reactors: pressure drops are greatly reduced (to less than 1%) and the radial heat transport can be more favorable.

The possibility of using monolithic reactors carrying out exothermic methanation reactions from syngas was investigated by process simulation. The reactor is an externally cooled fixed-bed reactor, loaded with honeycomb catalysts. It was shown that synthetic natural gas can be produced in a single pass monolithic catalyst reactor, with acceptable CO conversion values (around 80%) and temperature hot spots compatible with the catalyst stability. This system improves the presently adopted process configurations (Sudiro et al., 2009), as it overcomes the problem of temperature control typical of fixed-bed methanation reactors. The use of monolith reactors is also useful in view of process intensification: we have verified that the GHSV can be increased up to 20000 h^{-1}, with minimal pressure drops, increasing the cooling temperature correspondingly (Sudiro et al., 2010).

For what concerns SNG from biomass a number of centers in Europe are addressing this problem (Kopyscinski et al., 2010).

For example the Energy Research Centre of the Netherlands (ECN) began in 2002 a preliminary study to investigate the feasibility of SNG production from biomass (wood, sewage sludge and lignite) via indirectly heated gasification (MILENA) (www.ecn.nl). In 2003 ECN used a fixed bed catalytic reactor where tests about methanation reactions from gas produced by a wood gasifier were carried out for about 150h. The ongoing activity focuses on the construction of an 800 kWth pilot plant.

In Germany the Center for Solar Energy and Hydrogen Research (ZSW) has developed the Absorption Enhanced Gasification/Reforming (AER) process where a gas rich in H_2 is produced from biomass in a dual fluidized bed and recent activities are about the production of SNG from this gas in a molten salt cooled multi-tubular reactor. Preliminary results about these tests are available.

Finally, at the Paul Scherrer Institute (PSI) in Switzerland a research about the conversion of dry biomass to SNG has been carried out for about ten years. This project started from an idea promoted by Gazobois SA since the early 1990s. PSI started this project in 1999 and at the end of 2002 a preliminary study was successfully finished, in which theoretical and experimental investigations about gasification and methanation technologies were done. The Fast Internally Circulating Fluidized Bed (FICFB) gasification process was selected as a gasification technology, so that a plant was built in Güssing (Austria); for methanation the selected technology was the Comflux® fluidized bed process. A bench scale reactor was designed and connected to the FICFB gasifier in Güssing in 2003 and tested for 120h; in addition, before the end of 2004 a 10 kW_{SNG} reactor (in term of chemical energy content of the SNG) was designed and built at PSI. Different tests were performed until 2007; after that, based on the results obtained, a 1 MW_{SNG} Process Development Unit (PDU) has been erected with the aim to demonstrate the complete process chain from wood to SNG including gasification, gas cleaning, methanation and gas purification in a semi commercial scale. In December 2008 the FICFB produced gas was converted to methane rich gas in the PDU and in June 2009 the PDU was operated during 250h at up to 1 MW_{SNG}, producing 100 m^3/h of high quality synthetic natural gas.

At the Paul Scherrer Institute, a process converting microalgae to a methane-rich gas is under study (Haiduc et al., 2009). This new technology, called SunCHem, produces bio-methane via hydrothermal catalytic gasification of microalgae, where nutrients, water and the CO_2 produced are recycled. The two main parts of this process: growing of microalgae and hydrothermal gasification biomass, have been previously studied independently; in this work the fact of coupling these two parts into a sustainable process is a novel concept. The experimental work at PSI is, for example, about the investigation of the supercritical catalytic gasification of different species of microalgae (for example *Phaeodactylum tricornutum* and *Spirulina platensis*), and about the influence on the growth of algae of nickel, which is a trace contaminant that may be present in the effluent recycled from the gasification-methanation step.

A simplified scheme of this process is presented in Figure 9. The process consists of five steps. In the first one biomass is produced in a photobioreactor, after the excess water is removed mechanically from the biomass to approx. 15–20% wt. dry mass. The separated water, which contains a part of the nutrients, is recycled to the algae growth system. As a third step, the biomass slurry is liquefied hydrothermally by heating it up to a temperature of 400–450°C at 30 MPa, and the remaining nutrients are separated from the liquefied slurry for reuse as nutrients. The stream containing the organic fraction and the water is catalytically gasified under hydrothermal conditions to methane by using a catalyst such as ruthenium or nickel. Finally, CO_2 is separated from the product gas and recycled to the photobioreactor.

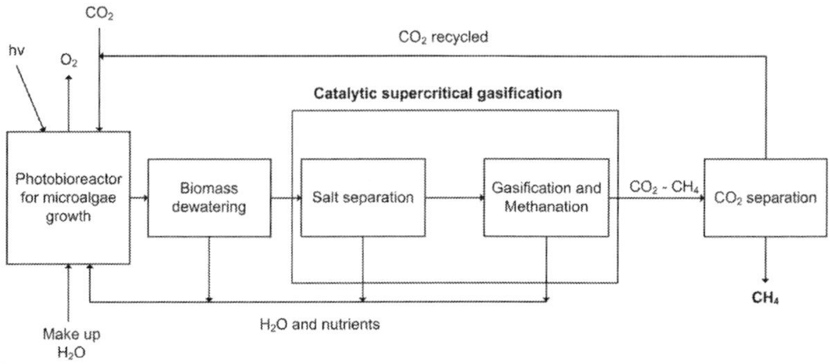

Fig. 9. Simplified scheme of Sun CHem process for the production of methane using microalgae, adapted from Haiduc et al., 2009

With regard to conventional gasification, which requires a dry biomass feed (moisture content lower than 15%), the hydrothermal process is suitable to convert/gasify wet biomass (moisture higher than 60%) into a fuel gas with a high heating value. This process operates under pressure, but at temperatures much lower (ca. 400°C) than typical gasification temperatures (800-900°C).
Also, in another work (Stucky et al., 2009) experimental tests showed that catalytic hydrothermal gasification of microalgae (in this particular case *Spirulina platensis*) can be a viable option for sustainable production of fuel with carbon capture.

6. Conclusions and Outlook

SNG production from coal or biomass is currently under strong investigation due to rising prices of natural gas and the wish for less dependency from natural gas imports. The interest is high especially in USA and China. So, the technical assessment of different technological alternatives for SNG production is an important research topic, even though, at present, only one industrial plant is in operation (North Dakota gasification plant).
Starting from the 1970s, as a consequence of the energy crisis, a number of methanation processes were developed comprising both fixed beds and fluidised beds. The large amount of heat of reaction in fixed bed reactors is usually removed by a combination of gas recycle cooling and steam adding, resulting in a high number of reactors, heat exchangers and compressors. Nowadays many projects are in progress especially for the conversion of wood biomass and wet biomass (microalgae) into SNG. At the ECN in the Netherlands there is a 800 kWth pilot plant for studying the production of methane from wood, sewage sludge and lignite. At the Paul Scherrer Institute (PSI) in Switzerland a research about the conversion of dry biomass to SNG has been carried out for about ten years with the construction in June 2009 of a semi commercial plant (1 MW_{SNG}) including gasification, gas cleaning, methanation and gas purification, producing 100 m^3/h of high quality synthetic natural gas. At PSI also an innovative technology is under investigation: the Sun CHem process for the production of methane using microalgae via catalytic supercritical gasification. Also in our research group different process configurations for methane synthesis plant from syngas

were developed to overcome the problem of temperature control typical of fixed-bed methanation; the best scheme proposed consists in the use of a monolithic reactor with the advantage of avoiding the recycle and reducing at minimum the numbers of reactors required to convert all the carbon monoxide in the feed.

7. References

Duret, A.; Friedli C. & Maréchal, F. (2005). Process design of synthetic natural gas (SNG) production using wood gasification. *Journal of Cleaner Production*, 13, 1434-1446.

Gassner, M. & Maréchal, F. (2009). Thermo-economic process model fro thermochemical production of Synthetic Natural Gas (SNG) from lignocellulosic biomass. *Biomass & Bioenergy*, 33, 1587-1604.

Haiduc, A.G.; Brandenberger, M.; Suquet., S; Vogel, F.; Bernier-Latmani, R. & Ludwig, C. (2009). SunCHem: an integrated process for the hydrothermal production of methane from microalgae and CO_2 mitigation. *Journal of Applied Phycology*, 21, 529-541.

Jahnke, F. C. & Parab, S. C. (2007). Methanation Assembly using Multiple Reactors. U.S. Patent 7247281 B2 (Jul. 24, 2007). Assignee: FuelCell Energy, Inc., Danbury, CT (US).

Juraščik, M.; Sues, A. & Ptasinski, K.J. (2009). Exergetic evaluation and improvement of biomass-to-synthetic natural gas conversion. *Energy & Environmental Science*, 2, 791-801.

Kopyscinski, J.; Schildhauer, T.J. & Biollaz, S.M.A. (2010). Production of synthetic natural gas (SNG) from coal and dry biomass - A technology review from 1950 to 2009. *Fuel*, in Press.

Lesemann, M. (2004). An overview of energy and power generation RD&D at RTI International. Available at the website:
https://www.rtec.rtp.org/Userfiles/File/Presentations_Nov14-Nov16/04-RTEC_Lesemann.pdf

Moeller, F. W.; Roberts, H. & Britz, B. (1974). Methanation of coal gas for SNG. *Hydrocarbon Processing*, 53, 69-74.

Mozaffarian, M. (2000). Process for Converting Hydrogen into Substitute Natural Gas. World Intellectual Property Organization, International Publication Number WO 00/21911 (Apr. 20, 2000).

Müller, W.; Möller, F.W. & Pirl, K. (1976). Methanation Reactor. U.S. Patent 3996014 (Dec. 7, 1976). Assignee: Metallgesellschaft Aktiengesellschaft, Frankfurt am Maim, Germany.

Petrucci, D. (2009). Summary of activities: Coal-to-SNG and Coal-to-Chemicals, US Department of Energy - Office of Sequestration, Hydrogen, and Clean Coal Fuels. Available at the website:
http://www.energy.psu.edu/osd/reports/DPettruci_2009.pdf

Ravikumar, R. & Sabbadini, G. (2007). Configurations and Methods of SNG Production. World Intellectual Property Organization, International Publication Number WO 2007/117587 A2 (Oct. 18, 2007).

Schultz, T. J. & Hemsath, K. H. (1976). Apparatus and Method for Methanation. U.S. Patent 3970435 (Jul. 20, 1976). Assignee: Midland-Ross Corporation, Cleveland, Ohio.

Stucky, S; Vogel, F.; Ludwig, C; Haiduc, A.G. & Brandenberger, M. (2009). Catalytic gasification of algae in supercritical water for biofuel production and carbon capture. *Energy & Environmental Science*, 2, 535-541.

Sudiro, M. & Bertucco, A. (2007). Synthetic fuels by a limited CO_2 emissions process which uses both fossil and solar energy. *Energy & Fuels*, 21, 3668-3675.

Sudiro, M.; Zanella, C.; Bressan, L.; Fontana, M. & Bertucco, A. Synthetic natural gas (SNG) from petcoke: model development and simulation. *Proceedings of ICheaP-9, The ninth International Conference on Chemical & Process Engineering*, pp. 1251-1256, Rome-Italy, May 10-13, 2009.

Sudiro, M.; Bertucco, A.; Groppi, G. & Tronconi, E. Simulation of a structured catalytic reactor for exothermic methanation reactions producing synthetic natural gas. Proceedings of the 20th European Symposium on Computer Aided Process Engineering - ESCAPE20, pp. 691-696, Ischia-Italy, June 6-9, 2010.

Skov, A. (1981). Process and a Plant for Preparing a Gas Rich in Methane. U.S. Patent 4298694 (Nov. 3, 1981). Assignee: Haldor Topsoe A/S, Lyngby, Denmark.

Topsoe, H. (2009). From Coal to Substitute Natural Gas using Tremp™. Available at the website: www.topsoe.com

Tucci, E.R. & Thomson, W.J. (1979). Monolith catalyst favoured for methanation. *Hydrocarbon Processing*, 58, 123-126.

Ulmann's Encyclopedia of Industrial Chemistry, 5th completely revised edition (1989). VHC Verlagsgesell schaft mbH, D-6940 Weinheim, Federal Republic of Germany.

Waldner, M. H. & Vogel, F. (2005). Renewable Production of Methane from Woody Biomass by Catalytic Hydrothermal Gasification. *Industrial & Engineering Chemistry Research*, 44, 4543-4551.

Zwart, R. W. R. & Boerritger, H. (2005). High efficiency co-production of synthetic natural gas (SNG) and Fischer-Trospch (FT) transportation fuels from biomass. *Energy & Fuels*, 19, 591-597.

Web sites:

http://www.chemconsulting.com.cn/info_detail01.asp?id=7677

http://www.dakotagas.com

http://www.greatpointenergy.com

http://www.trib.com/news/state-and-regional/article_03676d79-d722-525e-98d9-946be031fcd2.html 318359

http://www.syngasrefiner.com/SNG/agenda.asp

http://www.zeuslibrary.com

http://www.gasification.org/Docs/Conferences/2007/45FAGE.pdf

6

Environmental technology assessment of natural gas compared to biogas

Ola Eriksson
University of Gävle
Sweden

1. Introduction

The aim of this chapter is to bring about information on how the renewable competitor to natural gas - biogas - is produced, and to make a comparison of natural gas and biogas from primarily an environmental point of view in a life cycle perspective.

1.1 Historical background

In a historical perspective, biogas has been produced since the second half of the 19th century. India and China were among the pioneering countries, where biogas produced from manure and kitchen waste for long time has been used as a fuel for gas cookers and lamps. In Sweden, biogas has been produced at municipal waste water treatment plants since the 1960's. The primary incentive was to reduce sludge volumes. However, the oil crises of the 1970's rang alarm bells, leading to research and development of biogas techniques, and construction of new plants in order to reduce environmental problems and dependency on oil. (Swedish Biogas Association, 2004)

Industry was the first to act: sugar refineries and pulp mills started to use anaerobic digestion for waste water purification in the 1970's and 1980's. At this time, several smaller farm-sized plants were also constructed for anaerobic digestion of manure. During the 1980's, several landfill plants started to collect and utilise biogas produced in their treatment areas, an activity that expanded quickly during the 1990's. Several new biogas plants have been constructed since the mid-1990's to digest food industry and slaughterhouse wastes, and kitchen wastes from households and restaurants. (Swedish Biogas Association, 2004)

1.2 Properties of biogas

Biogas consists of 45-85 % methane (CH_4) and 15-45 % carbon dioxide (CO_2), with the exact proportions depending on the production conditions and processing techniques. In addition, hydrogen sulphide (H_2S), ammonia (NH_3) and nitrogen gas (N_2) may be present in small amounts. Biogas is normally saturated with water vapour.

Artificially produced methane, for example from wood products by a process called thermal gasification, is sometimes confusingly called biogas. This is also a renewable source of methane. The amount or volume of biogas is normally expressed in 'normal cubic meters' (Nm^3). This is the volume of gas at 0 °C and atmospheric pressure. The energy value is expressed in joule (J) or watt hours (Wh). Pure methane has an energy value of 9.81 kWh/Nm^3 (9810 Wh/Nm^3). The energy value of biogas varies between 4.5 and 8.5 kWh/Nm^3, depending on the relative amounts of methane, carbon dioxide and other gases present. Thus, if biogas comprises 60 % methane, the energy content is appr. 6.0 kWh/Nm^3. Energy content of biogas compared to other fuels are displayed in Figure 1.

1 Nm^3 biogas (97 % methane) = 9.67 kWh
1 Nm^3 natural gas = 11.0 kWh
1 litre petrol = 9.06 kWh
1 litre diesel = 9.8 kWh
1 litre E85 = 6.6 kWh
1 Nm^3 biogas is equivalent to appr. 1.1 litres of petrol.
1 Nm^3 natural gas is equivalent to appr. 1.2 litres petrol.

Fig. 1. Energy content of different fuels. Source: www.preem.se (petrol, diesel, E85), www.swedegas.se (natural gas)

Both methane and carbon dioxide are odourless. If raw biogas smells, it is usually due to the presence of sulphur compounds. Biogas may ignite at concentrations of about 5-20 % in air, depending on the methane concentration. Methane is lighter than air, whereas carbon dioxide is heavier. This is considered to be advantageous from a safety point of view, since methane easily rises and is quickly diluted by the air. (Swedish Biogas Association, 2004)

1.3 Biogas today and in the future

The global production of biogas is hard to estimate, whereas data on European level is more reliable. Statistics for production and use of biogas is published by EurObserver and Eurostat.

European production of primary energy from biogas reached 7.5 million toe in 2008, i.e. a 4.4 % increase on 2007 (an addition of 318.6 ktoe). Landfill biogas accounted for 38.7 % of the total followed by 13.2 % from waste treatment plants (urban and industrial). The other sources, mainly agricultural biogas units (combining liquid manure with substandard cereals, for instance), and also centralised co-digestion units (liquid manure with other organic matter and/or animal waste) and solid household waste methanisation units, accounted for almost half Europe's biogas production, i.e. 48.2 % in 2008. (Eurobserver, 2009)

Figure 2 illustrates the primary energy production of biogas in Europe in 2007. Unfortunately such map has not been found for 2008 figures. It should be noted that primary energy production estimate of 2008 differs considerably from the estimate for 2007 because of the very significant consolidation in the German statistics. The 2007 data has been consolidated to 3,659.1 ktoe compared to the previous estimate of 2,383.1 ktoe. This major consolidation is justified by taking into consideration from 2008 self-producer heat production, which is essentially the heat produced by farm installations. (Eurobserver, 2009)

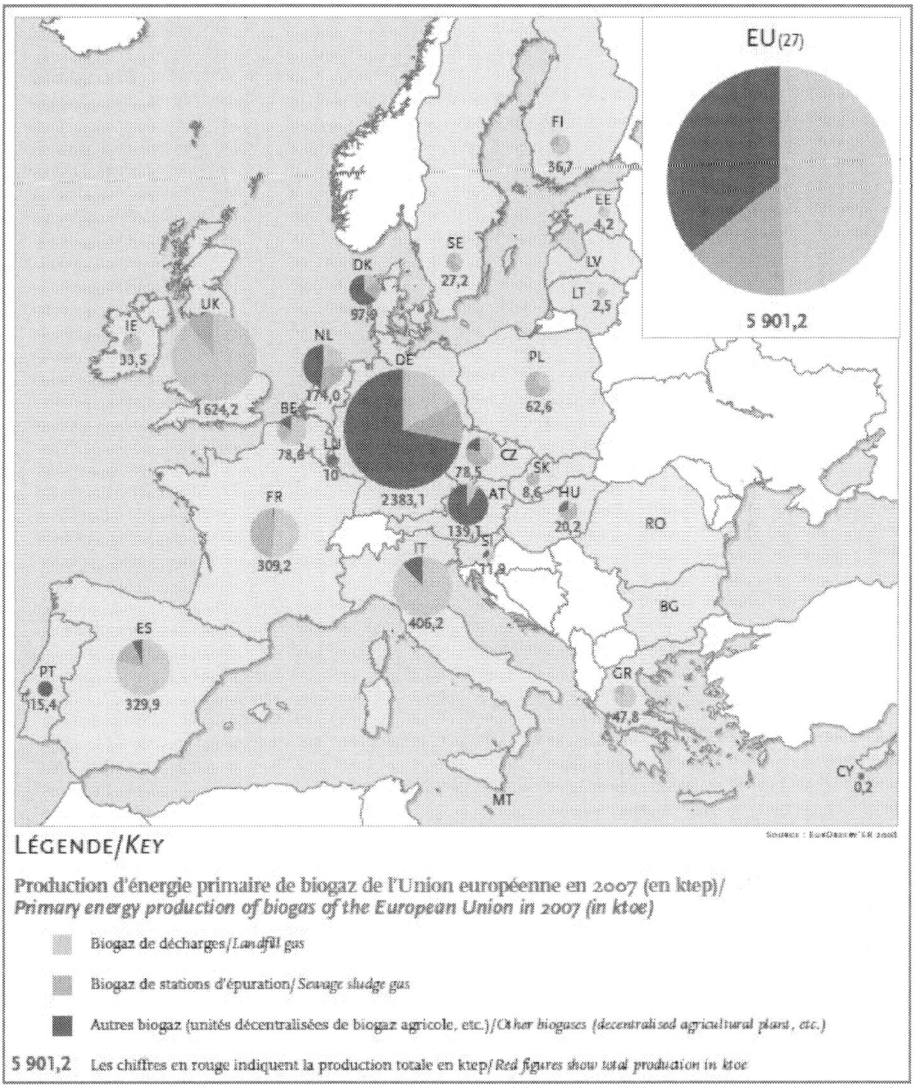

Fig. 2. Estimation of primary energy production of biogas in Europe 2007. Source: Eurobserver, 2008

Electricity production increased in 2008 at a slightly slower rate than that of primary energy production that is up 3.9 % over 2007, or a total of almost 20 TWh. Cogeneration plants generated 18.3 % or nearly 3.7 TWh of this total production. (Eurobserver, 2009)

In order to illustrate the offset for biogas in Europe figures from 2005 have been used as figures from 2008 only covers generated electricity. In 2005 recovered biogas was used for electricity (13 TWh), heat (8 TWh) and vehicle fuel (0.1 TWh). The majority of the heat- and power generation comes from Germany and Great Britain whereas almost all vehicle fuel was generated in Sweden. Figure 3 illustrates the distribution of energy from biogas production in each European country. (AvfallSverige, 2008)

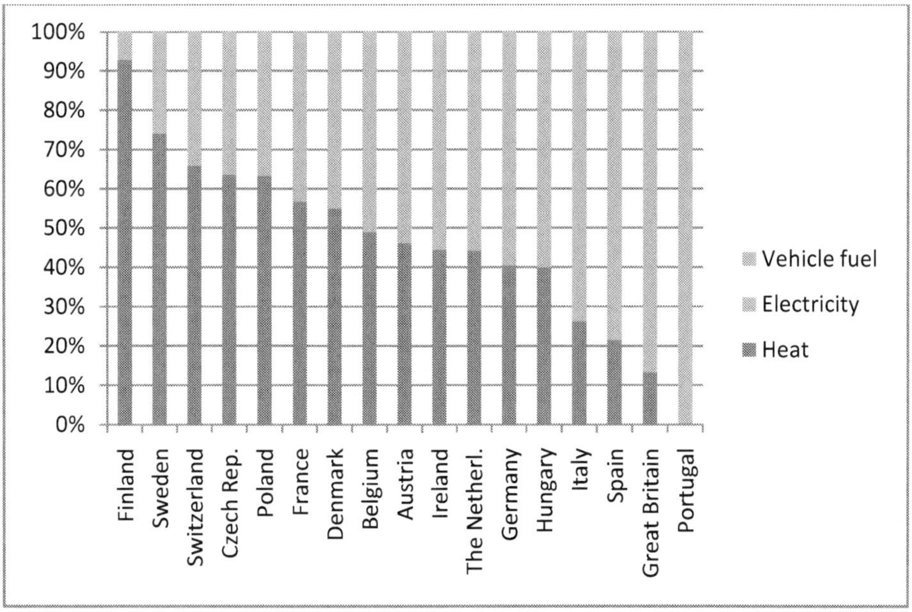

Fig. 3. Distribution for the generation of electricity, heat and vehicle fuel from landfill gas and biogas in each country in 2005. Sources: Switzerland (BFE, 2006), Sweden (Energimyndigheten, 2007), others (Eurobserver, 2007)

What are the trends for 2010? Present growth rates are too low to meet the European Commission's White Paper targets (15 Mtoe in 2010). EurObserv'ER puts production at 8.2 Mtoe in 2010 (mean annual growth rate rising by 4.4% in 2009 and 2010). This production would amount to 5.5% of the European Commission's "Biomass Action Plan" set at 149 Mtoe for 2010. The major price hike in agricultural raw materials should limit the growth of agricultural biogas production, which is the driving force of biogas growth in Europe, to below previous forecast levels.

1.4 General comparison of natural gas, biogas and landfill gas

The composition of biogas depends on a number of factors such as the process design and the nature of the substrate that is digested. A special feature of gas produced at landfills is that it includes nitrogen. The table below lists the typical properties of biogas from landfills, digesters and a comparison with average values for Danish natural gas for 2005. (SGC, 2007)

Property	Unit	Landfill gas	Biogas	Natural gas
Calorific value, lower	MJ/Nm^3	16	23	40
	kWh/Nm^3	4.4	6.5	11
	MJ/kg	12.3	20.2	48
Density	kg/Nm^3	1.3	1.2	0.83
Wobbe index, upper	MJ/Nm^3	18	27	55
Methane number		>130	>135	72
Methane	vol-%	45	65	89
Methane, range	vol-%	35-65	60-70	-
Long-chain hydrocarbons	vol-%	0	0	10
Hydrogen	vol-%	0-3	0	0
Carbon monoxide	vol-%	0	0	0
Carbon dioxide	vol-%	40	35	0.9
Carbon dioxide, range	vol-%	15-50	30-40	-
Nitrogen	vol-%	15	0.2	0.3
Nitrogen, range	vol-%	5-40	-	-
Oxygen	vol-%	1	0	0
Oxygen, range	vol-%	0-5	-	-
Hydrogen sulphide	ppm	<100	<500	3
Hydrogen sulphide, range	ppm	0-100	0-4000	1-8
Ammonia	ppm	5	100	0
Total chlorine as Cl^-	mg/Nm^3	20-200	0-5	0

Table 1. Comparison of properties for landfill gas, biogas and natural gas.
Sources: SGC, 2005; Energinet, 2005

The major difference is of course that natural gas is methane with fossil origin. Emissions of CO_2 from natural gas contributes to global warming, CO_2 from landfill gas and biogas does not. Natural gas is however a less polluting fuel than other fossil fuels, like coal and oil. Especially emissions of greenhouse gases at combustion are lower per unit energy than for coal and oil, but also NO_X emissions are often lower.

1.5 Problem

Natural gas and biogas is essentially the same type of gas, methane. In LCA literature natural gas is compared to other fossil fuels like coal and oil or maybe biomass, e.g. Eriksson et al, 2007. Biogas on the other hand is mostly compared to petrol or diesel, and possibly with system enlargement also with production and use of chemical fertiliser as the biogas process also produces valuable organic fertiliser. Biogas is also compared to other fossil fuels when electricity is generated.

So far, there seem to be few comparisons of natural gas and biogas with respect to environmental performance. A fuel wise comparison (pre combustion) of the two is therefore interesting, regardless of type of energy recovery. Another problem is lack of generic data on biogas as fuel. LCA databases consist of several datasets for natural gas but none or few for biogas.

2. Method and tools

To be able to compare natural gas and biogas, a literature survey has been made for papers on LCA of at least one of the two fuels. Of specific interest are studies showing the contribution from each step of the life cycle from extraction of raw materials to at least gas ready to use, or possibly also combustion with energy recovery as electricity, heat and vehicle fuel. If possible, data on specific emissions have been tracked down, or at least results from impact assessment using a given method.

When performing this meta-study it comes clear that there are many factors or parameters that affect the outcome of the assessment. They are well familiar in LCA as system boundaries, methods for allocation, choice of energy sources etc. The inventory of interesting studies has thus resulted in five papers which have been used to (1) guide the reader of the LCA in what is the environmental impact from each step of the fuel production process and (2) identify crucial factors in LCA of these fuels. The latter is further elaborated in the discussion part.

2.1 Goal and scope definition

The basic idea was to perform the review with a functional unit of 1 MJ of methane gas pre combustion. It is however hard to ignore the fact that the emissions of CO_2 has to be handled in separate ways for the two gases. Therefore utilisation of the methane to end user products as electricity and vehicle fuel has been presented also.

2.2 Inventory analysis and impact assessment

Following studies have been collected:

1. Environmental systems analysis of biogas systems - Part I: Fuel-cycle emissions by Börjesson & Berglund (2006). The study comprises biogas from different substrates. The functional unit is 1 MJ of biogas. Emissions are presented for each step of the process. No impact assessment is made.
2. A life cycle impact of the natural gas in the energy sector in Romania by Dinca et al (2006). The study comprises natural gas with a mix of gas from Russia and Romania. The functional unit is 100 GJ of thermal energy. Emissions are presented for each step of the process. Impact assessment is made using CML 1992.
3. Natural gas and the environmental results of life cycle assessment by Riva et al (2006). The study comprises natural gas from different countries and plants. The functional unit is 1 kWh of electricity. Emissions are not presented for each step of the process. Impact assessment is made for GWP and AP using defined weighting factors with no reference.
4. Life Cycle Assessment of biogas production by monofermentation of energy crops and injection to the natural gas grid by Jury et al (2010). The study comprises both biogas and natural gas. The functional unit is 1 MJ methane injected to the natural gas grid. Emissions weighted to impact categories are not presented for each step of the process (except for GWP). Impact assessment is made using EcoIndicator 1999.
5. Environmental assessment of biogas co- or tri-generation units by life cycle analysis methodology by Chevalier & Meunier (2005). The study comprises both biogas from crop residues and natural gas. The functional unit is 1 MJ of electricity and 1.6 MJ of heat or cold. Emissions are not presented for each step of the process. Impact assessment is made using EcoIndicator 1999.

The most useful study for a stepwise description of the environmental impact from the biogas process is number one in the list above. Studies 2 and 3 describe the whole life cycle for natural gas but with different functional units. It is not possible to find out how allocation between electricity and heat has been made, as the combustion facilities may include co-generation. This problem is further elaborated is the discussion. Study 4 is possible to use for a comparison of the total system using a pre combustion system boundary. Study 5 is possible to use for a comparison of the total system where methane is used for electricity and heat or cold. No study makes a comparison for vehicle fuel which is discussed later on.

2.3 Interpretation and improvement analysis

Interpretation and improvement analysis is made in the Results and conclusions section.

3. Life cycle assessment

Before going into detail of the different biogas production steps a general overview of the biogas system is presented.

Biogas is formed when microorganisms, especially bacteria, degrade organic material in the absence of oxygen. Production of biogas from the remains of dead plants and other organisms is a natural biological process in many ecosystems with a poor oxygen supply, for example in wetlands, rice paddies, lake sediments, and even in the stomachs of ruminating animals. (Swedish Biogas Association, 2004)

The large quantities of organic waste produced by modern society must be treated in some way before being recycled back to nature. Some examples of such organic wastes are sludges from municipal waste water treatment plants, kitchen refuse from households and restaurants, and waste water from the food processing industry. In a biogas process, the natural ability of microorganisms to degrade organic wastes is exploited to produce biogas and a nutrient rich residue which may be used as a fertiliser. The main constituent of biogas, methane, is rich in energy, and has a long history of use by mankind. (Swedish Biogas Association, 2004)

There are several technical solutions for how to recover biogas from organic residues, sewage water and biomass. What they have in common is that a sealed tank, a biogas reactor, is used for the anaerobic degradation of the material. If the gas is to be used as vehicle fuel carbon dioxide, hydrogen-sulphur compounds, ammonia, particles and moisture (steam) must be separated from the gas, making the gas to mainly consist of methane. (IVL, 1999)

Nowadays, production of heat and electricity is one of the major applications. As an environmentally-friendly alternative to diesel and petrol, biogas may also be refined to produce vehicle fuel. (Swedish Biogas Association, 2004)

Landfill gas cannot be used as vehicle fuel due to high concentration of nitrogen. The clean biogas is fuelled to the vehicle in a completely closed system by fast fuelling or slow fuelling. The gas station can be situated close to the production facility or be distributed by pipes or mobile gas tanks. (IVL, 1999)
The production system for biogas is depicted in Figure 4.

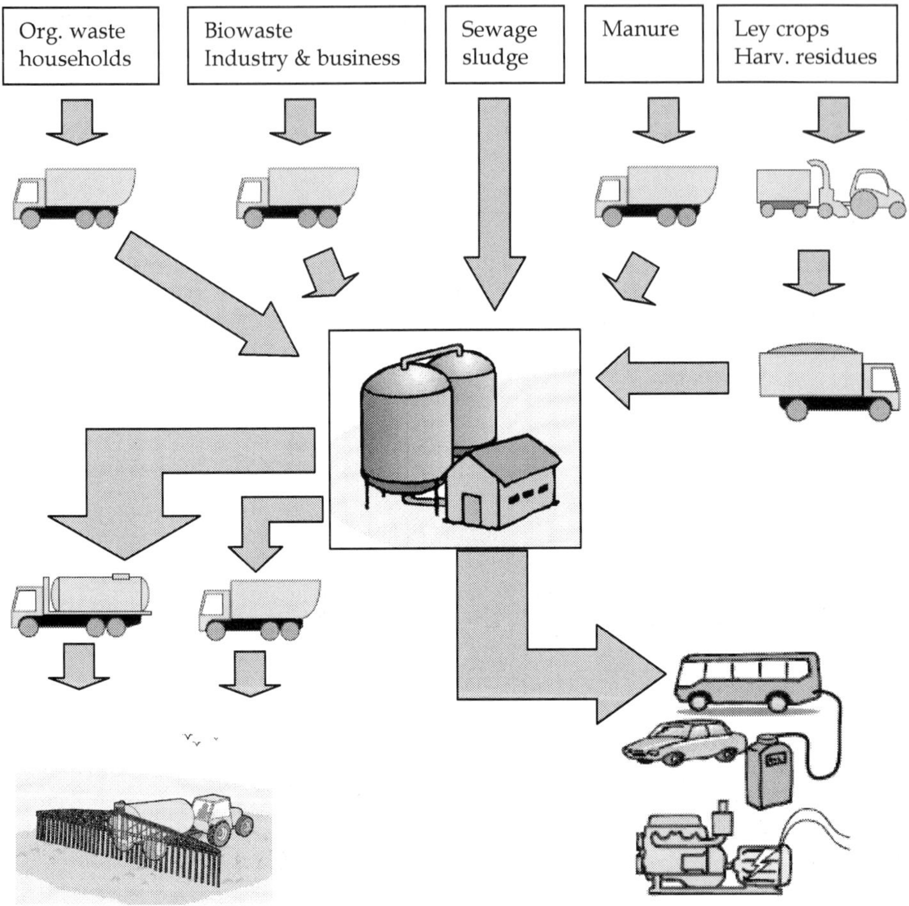

Fig. 4. Biogas system. Source: Eriksson & Hermansson, 2009

As a biogas plant is sealed there are very low losses of methane which does not just affect the energy efficiency but also contributes to global warming. Odour levels are normally lower than for open air composting and similar to reactor composting. The process, if made as wet digestion, uses fresh water for dilution but a large part of the process water is circulated to maintain the bacteria culture in the process. The digestate (the sludge which remains after digestion) is often dewatered leaving a dry digestate which can be used as fertiliser and a wet fraction which is normally sent to a wastewater treatment plant. Some electricity is used for pumping and mixing and heat is needed for hygienisation and heating of the material to the temperature inside the digester. Heat is supplied by a local gas boiler or district heating as to maximise the gas production. (Eriksson & Hermansson, 2009)

Despite energy use and some emissions, the major environmental benefit occurs when biogas substitutes fossil fuels. The digestate reduces the need for chemical fertiliser, but this effect is normally of minor environmental importance. A problem is however that the use of organic fertiliser gives rise to some nutrient leaching compared to mineral fertiliser, in which a much larger share of the nitrogen is plant available which in turn leads to greater precision in fertilising. In a systems perspective, the alternative waste treatment is also of importance. The environmental benefit is larger if the alternative is composting than incineration with energy recovery, in particular if the plant is made as combined heat and power production. (Eriksson & Hermansson, 2009)

3.1 Raw material

The raw material (in thermal applications called the fuel) is called substrate. Biogas can be produced using one or more substrates. The main sources are:

- Municipal organic waste (food waste)
- Biowaste from industry and business activities (e.g. fat, waste from grocery stores, biosludge from pulp and paper industry, dairy by-products, rejected animal food, fishery by-products etc.)
- Raw sewage sludge (produced at wastewater treatment plants)
- Manure
- Harvest residues
- Ley crops

The latter three are more common in small to medium scale plants. Large-scale anaerobic digesters often use a variety of different substrates from one or more sources. What these substrates have in common is that the carbon is present in an easy degradable form (less lignin and cellulose and more carbohydrates, fat and starch) and therefore well suited for anaerobic digestion.

Biogas is also produced in landfill sites due to decomposition of organic material inside the landfill. To facilitate this, the landfill has to be equipped with a gas recovery system. The biogas produced is often more polluted than biogas from an anaerobic digester and therefore mostly used in gas engines or gas boilers for recovery of heat and/or electricity which can be used on site. Landfill gas is not further investigated here.

According to Börjesson & Berglund (2006) (Table 2) the corresponding emissions from this step are as presented in Table 2.

Per tonne raw material	Energy input	Emissions						
Raw material	(MJ)	CO_2 (kg)	CO (g)	NO_x (g)	SO_2 (g)	HC (g)	CH_4 (g)	Particles (g)
Ley crops	440	31	24	270	36	17	9.8	9.9
Straw	230	16	23	150	6.6	11	0.057	2.3
Tops and leaves of sugar beet	100	7.2	7.6	78	4.8	4.7	0.057	1.3
Municipal organic waste	250	17	33	160	5.6	14	0.021	2.2

Table 2. Emissions from and energy input into the cultivation of different crops and collection of municipal organic waste. Source: Börjesson & Berglund, 2006

3.2 Technologies for thermal gasification

Methane gas can be produced from biomass by gasification. The gasification can be thermal or made by anaerobic digestion of easy degradable biomass. A short description on thermal gasification is presented below, but no LCA data have been included in the study as gasification of biomass is rare and still in a developing phase. The information refers to (SGC, 2008).

Gasification is a thermal process that breaks down the chemical bonds in the fuel in order to produce an energy rich gas. The process is an endothermic process which requires external heat. Gasification is divided into two steps; pyrolysis, which is a low temperature process that operates without any oxidation and gasification that needs a gasification agent that contains oxygen such as steam or air. (Bohnet, 2005)

During gasification, it is important to maintain the optimum oxygen input. The maximum efficiency of the gasification is achieved when just enough oxygen is added to allow complete gasification. If more oxygen is added, energy is released as sensible heat in the product stream. If biomass is heated to about 400°C pyrolysis will start to occur. The pyrolysis does not require any oxygen but only the volatile compounds in the biomass will be gasified. Biomass contains ca 60 % volatile compounds compared to coal which contains < 40 % volatile compounds. This makes biomass more reactive than coal. After thermal decomposition the volatile compounds are released as H_2, CO, CO_2, H_2O, CH_4 etc which is also known as pyrolysis gas. The remains after the pyrolysis is char coal. (Bohnet, 2005)

The pyrolysis can not convert all of the biomass into volatile compounds and therefore gasification is required. The gasification requires much higher temperatures than pyrolysis, usually in the range of 800-900°C and with a gasification agent present. The gasification includes partial oxidation and it breaks down most of the feedstock into volatile compounds and the remaining nutrients like alkaline earth metals etc. end up as ash. The produced gas from the gasification contains synthesis gas or syngas which consists of carbon monoxide, CO and hydrogen, H_2. The gas also contains methane, higher hydrocarbons like ethane, tars and inorganic impurities like HCl, NH_3, H_2S and CO_2.

The product gas from the gasifier contains the volatile components from the pyrolysis as well as the syngas. The composition of the gas depends on a number of parameters such as gasification temperature and pressure, feedstock, reactor type and gasification agent. Generally higher temperature favours syngas production while lower temperature yields higher tar and methane rich gases. Increased pressure will increase the methane yield due to the equilibrium of reaction (1). (Bohnet, 2005)

$$CH_4 + H_2O \leftrightarrow CO + 3H_2 \qquad (1)$$

$$CO + H_2O \leftrightarrow CO_2 + H_2 \qquad (2)$$

Because of the endothermic reactions in gasification, heat must be added. This can be achieved either direct, with partial oxidation and/or combustion as in the case with air or pure oxygen as gasification medium or indirect. When air is used as gasification medium in direct gasification, the product gas is nitrogen diluted. This will decrease the lower heating value, LHV, of the gas and increase the cost of the downstream processes as more gas needs to be processed. An alternative is to use pure oxygen as gasification medium. This will eliminate the nitrogen dilution problem but it increases the costs significantly.

3.3 Technologies for biogasification

There are in general two main types of anaerobic digestion, a wet technology where the substrate is diluted with water and a dry or semi-dry technology addressed for dry substrates. First the wet process will be presented, followed by a short description of the less used dry technology. The text refers to Eriksson & Hermansson, 2009.

The substrate enters the biogas plant in a reception hall. The waste is then taken to homogenisation (a mill or screw press) and then by screw transporters to a pulper. In the pulper the waste is mixed with hot water and steam to reach a temperature of 70 ^{o}C with DM 13 % making it a fluid possible to pump. Here hygienisation (pasteurisation) takes place (pathogenic organisms are being killed) during one hour under powerful stirring. Heavy material as stones, gravel and metal is removed from the bottom of the pulper.

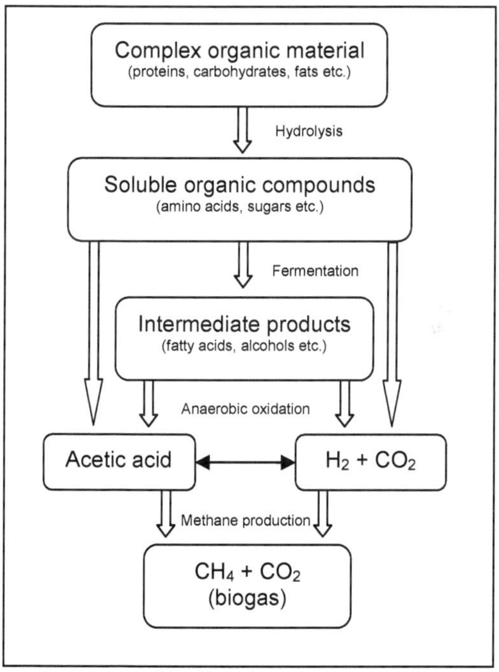

Fig. 5. Degradation process in anaerobic digestion.
Source: Swedish Biogas Association (2004)

After hygienisation the material is pumped to sand- and float filters where light materials as eg. plastic are removed from the surface and heavy material from the bottom. The mix is now called raw sludge and can be compared to dewatered sewage sludge. The raw sludge is pumped to a raw sludge storage equipped with heat recovery and mixer. Reject material is transported to containers for further treatment like landfill and/or incineration.

From the raw sludge storage the sludge is pumped to the digestion chamber where the organic material is degraded due to the microbiological activity. The degradation process is described in Figure 5. There is a mixer in the digestion chamber in order to fulfil complete digestion. The hydraulic retention time and temperature varies between different plants. At temperatures 4-25 °C the process is psychrofilic which is very rare for large scale facilities. At 25-45 °C the process is mesofilic which is the most common process for anaerobic digestion of sewage sludge. Finally at 50-60 °C the process is thermofilic which is the most common process for anaerobic digestion of municipal organic waste. Biogas is released at the top of the digestion chamber and digested material is pumped to a covered digestate storage. Inside the storage some digestion will continue and biogas produced can be recovered.

Another process design is dry digestion. The difference between wet and dry digestion is the DM content in the digestion chamber. In wet digestion the DM is 2-10 w% while dry digestion works at 20-35 w%. Dry digestion should not be mistaken as wet composting as the final products are different. In Germany there are more than 300 plants for dry digestion. The most common method is batch-wise percolation bed in heated digestion reactors where the material is loaded and reloaded by tractors or wheel-loaders through a gas safe port at the short side of the reactor.

According to Börjesson & Berglund (2006) (Tables 3-4) the corresponding emissions from this step comes from transport of raw materials to a centralised biogas plant (Table 3) and plant operation (Table 4).

Per tonne of raw material	Energy input	Emissions						
Raw material	(MJ)	CO_2 (kg)	CO (g)	NO_x (g)	SO_2 (g)	HC (g)	CH_4 (g)	Particles (g)
Ley crops, tops and leaves of sugar beet	11	0.77	0.14	7.1	0.25	0.41	< 0.01	0.12
Straw	29	2.0	0.38	19	0.65	1.1	< 0.01	0.31
Liquid manure	10	0.70	0.13	6.5	0.22	0.37	< 0.01	0.11
Food industry waste	16	1.1	0.21	10	0.36	0.60	< 0.01	0.17
Municipal organic waste	48	3.3	0.62	31	1.1	1.8	< 0.01	0.51

Table 3. Emissions from and energy input into the transport of raw materials to a centralised biogas plant. Source: Börjesson & Berglund, 2006

Note that figures in Table 4 represent a large-scale biogas plant. In Börjesson & Berglund, 2006 data for farm-scale plant is also displayed in Table 4 but have been left out here for space reasons. Large-scale plants have less energy input and less emissions except for electricity input and emissions of CO_2. Electricity and CO_2 are linked to each other as the electricity is assumed to be based on natural gas.

Per tonne of raw material	Energy input		Emissions						
Raw material	Electr. (MJ)	Heat (MJ)	CO_2 (kg)	CO (g)	NO_x (g)	SO_2 (g)	HC (g)	CH_4 (g)	Particles (g)
Ley crops	180	240	12	8.8	36	1.3	2.3	3.2	1.1
Straw	570	870	39	29	120	4.5	8.0	10	3.8
Tops and leaves of sugar beet	160	200	10	7.4	31	1.1	1.9	2.7	1.0
Liquid manure	53	85	3.6	2.7	11	0.4	0.8	1.0	0.4
Food industry waste	53	85	3.6	2.7	11	0.4	0.8	1.0	0.4
Municipal organic waste	230	320	15	11	46	1.7	3	4	1.5

Table 4. Emissions from and energy input into the operation of large-scale biogas plants. Source: Börjesson & Berglund, 2006

3.4 Distribution and use of solid residues

Besides biogas an organic fertiliser is also produced from anaerobic digestion. The digestate can to some extent replace chemical fertiliser and thereby contribute to lower costs and environmental impact. Compared to spreading of manure, methane emissions can be avoided. In a biogas plant the nitrogen is transformed to ammonia which is more easy accessible to the plants than nitrogen bound in organic compounds. Compared to direct use of manure, nitrogen leakage from agriculture can be substantially decreased causing an improved water quality in surrounding watercourses. In addition the digestion reduces odour and as mentioned before kills pathogens.

The digestate contains appr. 95 % water. In order to reduce transport work and problems in finding spreadable land area, the digestate is often dewatered to a water content of less than 70 %. This is performed with a centrifuge. The dry digestate can be stored in a digestate storage and then transported to satellite storages in close connection to the spreading areas. The reject water from the centrifuge contains considerable amounts of nitrogen (ammonia) and can be spread with liquid manure spreaders or diverted to a wastewater treatment plant. If the digestate cannot be used within agriculture it can be mixed with compost, peat and sand to form soil products to be used in gardening. (Eriksson & Hermansson, 2009)

There are also ongoing research which tests drying and pelletisation of the digestate. The pellets can then be used either as fuel or as fertiliser in agriculture or forestry. Pelletisation would mean energy use for drying and pelletisation but also energy savings for transport. Pellets are also easier to handle and store from one season to another. Nutrient pellets could also be more nutrient efficient than dry sludge, as the nitrogen is slower emitted leading to a higher uptake in plants. This is an area for further research and can be of interest in areas with low degree of agriculture.

According to Börjesson & Berglund (2006) (Table 5) the corresponding emissions from this step are as presented in Table 5. It is worth to notice that this study does "not consider potential changes in various emissions from arable soil (such as nitrous oxide, N_2O, and ammonia, NH_3), leakage of nitrate (NO_3^-), or methane emissions, due to changes in the handling of the raw materials."(Börjesson & Berglund, 2006)

Per tonne of digestate	Energy input	Emissions						
Operation	(MJ)	CO_2 (kg)	CO (g)	NO_x (g)	SO_2 (g)	HC (g)	CH_4 (g)	Particles (g)
Transport	16	1.1	0.21	10	0.36	0.60	< 0.01	0.17
Spreading	25	1.7	1.7	15	0.30	0.53	< 0.01	0.23

Table 5. Emissions from and energy input into transport and spreading of digestate.
Source: Börjesson & Berglund, 2006

3.5 Distribution and use of biogas

Biogas contains methane (appr. 60 %), carbon dioxide and minor amounts of hydrosulphurous. The gas can be combusted directly in a gas boiler for heat generation or in a gas engine for electricity generation (occasionally surplus heat is also recovered by water cooling). If the biogas is to be used in vehicles it must be upgraded, i.e. cleaned (removal of carbon dioxide and pollutants) and pressurised. (Eriksson & Hermansson, 2009)

Physical absorption (water wash) is the most commonly used method for upgrading biogas in Sweden. This method makes use of the fact that gases like carbon dioxide, hydrogen sulphide and ammonia are more readily dissolved in water than methane. The solubility of carbon dioxide increases with increasing pressure and decreasing temperature. (Swedish Biogas Association, 2004)

Pressure Swing Adsorption (PSA) is the second commonest method in use. PSA separates out carbon dioxide, oxygen, nitrogen and hydrogen sulphide, trapping molecules according to molecular size using, for example, activated carbon at different pressures. Hence, the method is sometimes called the 'molecular sieve' technique. (Swedish Biogas Association, 2004)

Absorption using Selexol is a method in which carbon dioxide, hydrogen sulphide and ammonia are absorbed by Selexol, a glycol solution. The method is based on the same principle as physical absorption with water. However, Selexol is more effective, since it absorbs three times more carbon dioxide. (Swedish Biogas Association, 2004)

Chemical absorption (chemisorption) uses a chemical to bind carbon dioxide. The advantage of this method is that the chemical only absorbs carbon dioxide and, if present, hydrogen sulphide, whereas virtually no methane is removed. This leads to a very high purity of the upgraded biogas, which contains about 99 % methane. (Swedish Biogas Association, 2004)

One alternative to the conventional technologies is to upgrade biogas with cryogenic technology, which means that the gas is chilled and the differences in condensation temperature for different compounds are used to separate impurities and CO_2 from CH_4. CO_2 condense at -78.5 °C at atmospheric pressure. The technology can be used to upgrade raw biogas by chilling it to the condensation temperature for CO_2 or it can be further chilled to -161 °C (condensation temperature for CH_4 at atmospheric pressure) to produce liquid biogas, LBG. It is more energy intensive to chill the gas to -161 °C but in some situations it results in a more valuable product since LBG is more than 600 times space efficient compared to biogas in its gas phase at atmospheric pressure (around 3 times more space efficient compared to compressed biogas, 200 bar). This makes the biogas available for more customers since the produced LBG can be transported on road in vacuum insulated semi-trailers to remote fuel stations. On a multi-purpose fuel station it is then stored as LBG and fuelled as either LBG or CBG (compressed biogas, 200 bar). LBG can also be produced using one of the conventional upgrading technologies connected with a small-scale liquefaction plant. This small-scale liquefaction plants are either a closed nitrogen cycle or a closed mixed refrigerant cycle. The first has a low efficiency but it is not as complex as the latter since it only use one refrigerant (nitrogen). (Johansson, 2008)

When using cryogenic upgrading technology clean, liquid CO_2, LCO_2, comes as a by-product. This LCO_2 could be used in external processes replacing fossil energy and bring in extra income to the biogas production plant. Two possible applications are cryogenic temperature control and fertilizing of greenhouses. The LCO_2 probably has to be sold directly to the user, and the possibility to get an income from this product is also very site specific. An interesting alternative could be to place a greenhouse close to a biogas production and upgrading plant. In this way the greenhouse could get an organic fertilizer from the digester chamber and heat and CO_2 from the upgrading process. (Johansson, 2008)

The distribution of biogas can also be facilitated by injection to the natural gas grid. The establishment of natural gas grids is therefore very important as it makes the introduction of biogas in the society easier. The economy of biogas production does not allow such heavy investments as gas grids, nor can the delivery safety be high enough. Biogas is a local energy source, whereas natural gas is a transnational energy source.

According to Börjesson & Berglund (2006) (Table 6) the corresponding emissions from this step are as presented in Table 6. The figures for upgrading of biogas have been calculated by subtracting the figures including upgrading from the figures without upgrading.

Per MJ of biogas	Energy input	Emissions						
Raw material	(MJ)	CO_2 (g)	CO (mg)	NO_x (mg)	SO_2 (mg)	HC (mg)	CH_4 (mg)	Particles (mg)
Ley crops	0.11	6.0	3.0	20	0.0	0.4	1.30	0.3
Straw	0.11	6.0	3.0	12	0.2	0.4	1.30	0.3
Tops and leaves of sugar beet	0.11	6.0	2.7	12	0.2	0.4	1.30	0.3
Liquid manure	0.11	6.0	3.2	12	0.2	0.4	1.30	0.3
Food industry waste	0.11	5.6	3.0	12	0.2	0.4	1.33	0.3
Municipal organic waste	0.11	6.0	3.0	12	0.2	0.4	1.30	0.3

Table 6. Emissions from and energy input into upgrading of biogas to vehicle fuel.
Source: Börjesson & Berglund, 2006

4. Results and conclusions

In Table 7 the life cycle inventory from the different steps have been added up to cover the whole biogas system.

Per MJ of biogas	Energy input	Emissions						
Raw material	(MJ)	CO_2 (g)	CO (mg)	NO_x (mg)	SO_2 (mg)	HC (mg)	CH_4 (mg)	Particles (mg)
Ley crops	0.40	21	15	150	16	9.2	5.3	5.0
Straw	0.35	14	12	85	2.9	5.0	1.8	1.7
Tops and leaves of sugar beet	0.27	12	9.3	81	3.7	4.5	1.4	1.6
Liquid manure	0.31	11	7.8	63	1.9	3.3	1.8	1.3
Food industry waste	0.15	5.4	3.5	33	1.0	1.8	0.77	0.67
Municipal organic waste	0.26	12	14	85	2.8	6.1	1.1	1.5

Table 7. Summary of the emissions and energy input in production of biogas in large-scale plant. Source: Börjesson & Berglund, 2006

Even if this is not a complete life cycle inventory it reveals large differences for different substrates. As mentioned above, large-scale or farm-scale plant design also have an influence on the result, as well as choice of system boundaries, electricity generation etc. We will come back to such crucial factors for both biogas and natural gas.

This result now has to be compared to corresponding figures for natural gas. From (IVL, 1999) it was possible to extract figures similar to Table 7. In Table 8 pre combustion as well as different gas applications are presented.

Per MJ of natural gas	Energy input	Emissions						
Step in fuel cycle	(MJ)	CO_2 (g)	CO (mg)	NO_x (mg)	SO_2 (mg)	NMVOC (mg)	CH_4 (mg)	Particles (mg)
Production and distribution	0.047	3.09	3.01	12.7	0.23	1.53	2.8	0.022
Combined cycle	0.001	57	11	11	0.6	0.46	2.0	0.096
Heat station	-	56	10	49	0	1	0.1	-
Power station	-	56	10	58	0	1	0.1	-
Residential service	-	56	10	10	0	1	0.1	-
Light-duty vehicles	-	52	35	28	-	35	-	1.8
Heavy-duty vehicles	-	52	1.7	170	-	4.2	38	1.7

Table 8. Summary of the emissions and energy input in production, distribution and use of natural gas. Source: IVL, 1999

When comparing tables 7 and 8 it is obvious that the pre combustion figures are much lower for natural gas (row 1 "Production and distribution" in Table 8) than for biogas regardless of raw material. For the different gas use applications, emissions of CO_2 are almost the same for all alternatives, whereas the other emissions may vary considerably.

As mentioned earlier it is not easy to find a similar study consistent with Börjesson & Berglund, 2006. The same authors have reported a continued assessment in Börjesson & Berglund, 2007. In the complex assessment three functional units are defined: 1 MJ of heat, 1 MJ of heat and power and 1 MJ of kinetic energy (to reflect vehicle fuel). In fact the assessment also includes functional unit plant nutrients as N and P (this is compensated for by chemical fertiliser) even if this is not put out in words or figures. The system enlargement considers conventional alternatives both for raw material input and energy service output of the biogas system. This means that indirect environmental impact is included. Electricity is assumed to be natural gas in condensing plants, heat comes from fuel oil combustion and petrol and diesel are used for additional transport in light-duty and heavy-duty vehicles. There are also a number of assumptions for alternative waste handling (composting) and cropping systems. The impact assessment includes global warming potential (GWP), acidification potential (AP), eutrophication potential (EP) and photochemical oxidant creation potential (POCP).

This study is, in my eyes, how a proper comparison should be carried out, taking into account primary and secondary upstream and downstream effects of biogas systems. As all energy alternatives are from fossil fuels the biogas system comes out as the least polluting one for all functional units, independent on large-scale or small-scale plants. This is however not a comparison of biogas and natural gas.

From the method section we learned that studies 4 and 5 could be used for pre combustion or total system comparisons. Both studies present results in impact categories using EcoIndicator99. In study 5 the LCI has been weighted with the CML method and EcoIndicator has been used for valuation.

Study 4 (Jury et al, 2010) shows pre combustion results where biogas has a higher impact than natural gas for human health and ecosystem quality. However, for climate change, resources and cumulative energy and exergy demand, the results are opposite. Land occupation and use of fertiliser are used as explanation for this outcome. To produce 1 MJ of biogas about 2.5 MJ energy resources are consumed, whereas natural gas is more efficient with just above 1 MJ. But the non-renewable part is lower for biogas; 0.5 MJ/ MJ in relation to 100 % non-renewable for natural gas.

In study 5 (Chevalier & Meunier, 2005) biogas co- and trigeneration are compared to conventional heat/cold and power production. The conventional generation of heat is from natural gas industrial furnace, cold from vapour compression chiller using electricity and finally electricity from the grid in Germany, Austria and France. Biogas co-generation is by far better than the conventional alternatives and biogas tri-generation is somewhat worse than the French setup, but better than all the rest.

From Riva et al (2006) it is obvious that the environmental performance for natural gas is different depending on country of origin and of course energy conversion technology. In Table 9 the emissions in g/kWh el for natural gas used in Italy are presented. Significant differences between the different alternatives are observed for all emissions.

g/kWh el	ETH Russia	ETH Netherl.	BUWAL	Legislation Steam plant	Legislation Combined Cycle	Gas Russia	Gas Italy
NO_x	1.24	0.88	1.49	0.96	0.61	0.39	0.34
SO_2	0.35	0.009	0.27	0.33	0.22	0.04	0.007
CO_2	742	644	767	635	427	383	356
CH_4	4.07	0.46	1.76	3.87	2.6	1.39	0.17

Table 9. Emissions of natural gas cycle for electricity production. ETH refers to EcoInvent database, BUWAL corresponds to a Western Europe scenario, steam plant and combined cycle use gas from Russia and gas from Russia and Italy are used in a combined cycle. Source: Riva et al, 2006

So, what have we learned from this? From present studies it is possible to make a list of crucial factors or parameters that influence the result.

- From what raw materials are the biogas made of?
- What is the size of the biogas plant, farm-scale or large-scale?
- Which is the alternative use of the raw materials?
- Which is the alternative use of the biogas?
- Are the emissions from use of digestate included and how?
- How are the emissions from the biogas system allocated between digestate and biogas?
- What is the country of origin for natural gas and where is it used?
- What energy conversion technology (heat and/or power) has been applied for the bio/natural gas?
- Is the biogas used in light-duty vehicles substituting petrol or heavy-duty vehicles (busses) substituting diesel oil?
- From what type of driving cycle has the emission factors for vehicles been taken?

These are all crucial factors in the inventory. It is also a fact that the results for CO_2 can point in one direction, but if other impact categories are also included the picture can change. Problems with allocation like heat only (Dinca et al, 2006) or electricity only (Riva et al, 2006) in cogeneration plants has also been identified as an important factor that may influence the results. Some of the papers mentioned above include a variety of sensitivity analyses, such as transport distance from field to biogas plant, as a method to pinpoint these uncertainties. A more thorough comparison of the different studies would probably reveal even more potential key parameters. However it has not been in the scope of this study to perform such a review.

5. Future research

There is definitely a need for a best practice when it comes to LCA of fuels. Life cycle inventories are available in different software as GaBi, Umberto and SimaPro but it takes a great deal of knowledge to grasp what is included and not and what underlying assumptions have been made. There is a risk that biogas, as well as LCA, get negative attention when one supplier of biogas cars states that the emission factor is 22 g CO_2/km at the same time as different websites tells the consumer that the emission factor is 124 g CO_2/km. To common people it is unbelievable that the conclusions can differ that much for the same fuel. A similar discussion is found for bioethanol. How to handle land use issues can be of particular interest as these circumstances has an impact on LCA results and is also an ethical aspect as to whether agricultural fields should be used for food production or for energy purposes. Now famine is much more of a logistic and socio- economic problem, but in a short-term scenario food production may be replaced by energy crops on the margin.

One particular aspect of best practice is to whether gas production and gas use should be separated or not. What is shown above is that it is hard to separate them in a consistent manner. One idea, which is not always used, is to declare emissions or impact from each step, or at least pre combustion and different alternatives for post combustion. It should then be possible to add one or more steps to each other to get the total picture.

In a broader context more research is needed to analyse and optimize both biogasification and thermal gasification. Fiber sludge from pulp and paper industry is a potential substrate that may enter the market if technology and economy allows it.

6. References

AvfallSverige (2008) Energi från avfall ur ett internationellt perspektiv (Energy from waste in an international perspective), report 2008:13, ISSN 1103-4092

BFE (2006) Schweizerische Statistik der erneuerbaren Energien -Ausgabe 2005, Bundesamt für Energie

Bohnet, B. (2005) Ullmann's Encyclopedia of Industrial Chemistry, 7th edition,Wiley-VCH

Börjesson, P. & Berglund, M. (2007) Environmental systems analysis of biogas systems - Part II: The environmental impact of replacing various reference systems, *Biomass and Bioenergy* 31 (2007) 326-344

Börjesson, P. & Berglund, M. (2006) Environmental systems analysis of biogas systems - Part I: Fuel-cycle emissions, *Biomass and Bioenergy* 30 (2006) 469-485

Chevalier, C. & Meunier, F. (2005) Environmental assessment of biogas co- or tri-generation units by life cycle analysis methodology, *Applied Thermal Engineering* 25 (2005) 3025-3041

Dinca, C., Rousseaux, P & Badea, A. (2006) A life cycle impact of the natural gas used in the energy sector in Romania, *Journal of Cleaner Production* 15 (2007) 1451-1462

Energimyndigheten (2007) Produktion och användning av biogas 2005 (Production and use of biogas in 2005), ER2007:05; ISSN 1403-1892

Energinet (2005) Gaskvalitet årsgennemsnit (Gas quality, annual averages) 2005, http://www.energinet.dk

Eriksson, O. & Hermansson, T. (2009) Biogas i Gästrikeregionen - en systemanalys (Biogas in the Gastrike region - a systems analysis)

Eriksson, O., Finnveden, G., Ekvall, T. & Björklund, A. (2007) Life Cycle Assessment of fuels for district heating - A comparison of waste incineration, biomass- and natural gas combustion, *Energy Policy* 35 (2007) 1346-1362

Eurobserver (2009) The state of renewable energies in Europe - 2009 edition, ISSN 2101-9622

Eurobserver (2008) Biogas barometer, SYSTÈMES SOLAIRES le journal des énergies renouvelables N° 186

Eurobserver (2007) Biogas barometer, SYSTÈMES SOLAIRES le journal des énergies renouvelables N° 179

IVL (1999) Miljöfaktabok för bränslen Del 2 Bakgrundsinformation och Teknisk bilaga (Environmental fuel handbook Part 2 Background information and Technical appendix), B 1334 B, Stockholm, Sweden

Johansson, N. (2008) Production of liquid biogas, LBG, with cryogenic and conventional upgrading technology, - Description of systems and evaluations of energy balances, Master Thesis, Department of Technology and Society, Environmental and Energy Systems Studies, Lund university, ISRN LUTFD2/TFEM--08/5032--SE + (1-92)

Jury, C., Benetto, E., Koster, D., Schmitt, B. & Welfring, J. (2010) Life Cycle Assessment of biogas production by monofermentation of energy crops and injection to the natural gas grid, *Biomass and Bioenergy* 34 (2010) 54-66

Riva, A., D'Angelosante, S. & Trebeschi, C. (2006) Natural gas and the environmental results of life cycle assessment, *Energy* 31 (2006) 138-148

SGC (2008) Substitute natural gas from biomass gasification, Swedish Gas Centre, report SGC 187, 1102-7371, ISRN SGC-R-187-SE

SGC (2007) Biogas - basic data on biogas - Sweden, Swedish Gas Centre, broschure

SGC (2005) Energigaser och miljö (Energy gases and the environment), Swedish Gas Centre

Swedish Biogas Association (2004) Biogas – renewable energy from organic waste, broschure

7

Natural gas hydrates

Geir Ersland and Arne Graue
University of Bergen
Norway

1. Introduction

Natural gas hydrates are ice-like materials formed under low temperature and high pressure conditions. Natural gas hydrates consist of water molecules interconnected through hydrogen bonds which create an open structural lattice that has the ability to encage smaller hydrocarbons from natural gas or liquid hydrocarbons as guest molecules. Interest in natural gas hydrates as a potential energy resource has grown significantly in recent years as awareness of the volumes of recoverable gas becomes more focused (Sloan & Koh, 2008). The size of this resource has significant implications for worldwide energy supplies should it become technically and economically viable to produce. Although great efforts are being made, there are several unresolved challenges related to all parts in the process towards full scale hydrate reservoir exploitation. Some important issues are: 1) Localize, characterize, and evaluate resources, 2) technology for safe and economic production 3) safety and seafloor stability issues related to drilling and production. This chapter gives a brief introduction to natural gas hydrate and its physical properties. Some important characteristics of hydrate accumulations in nature are also discussed. Experimental results presented in this chapter emphasis recent work performed by the authors and others where we investigate the possibilities for producing natural gas from gas hydrate by CO_2 replacement. By exposing the hydrate structure to a thermodynamically preferred hydrate former, CO_2, it is shown that a spontaneous conversion from methane hydrate to CO_2 hydrate occured. Several experiments have shown this conversion in which the large cavities of hydrates prefer occupation by CO_2 (Lee et al., 2003; Jadhawar et al. 2005; Ota et al., 2005; Graue et al., 2008). This is also supported by simulations (Phale et al., 2006; Kvamme et al., 2007). Other production schemes proposed in the open literature are also reviewed in this chapter.

2. Structures and Properties

There are three known structures of gas hydrates: Structure I (sI), structure II (sII) and structure H (sH). These are distinguished by the size of the cavities and the ratio between large and small cavities. SI and sII contain both a smaller and a larger type of cavity, but the large type cavity of sII is slightly larger than the sI one. The maximum size of guest molecules in sII is butane. SH forms with three types of cavities, two relatively small ones and one quite large. The symmetry of the cavities leaves an almost spherical accessible volume for the guest molecules. The size and shape of the guest molecule determines which structure is formed due to volumetric packing considerations. Additional characteristics are guest dipole and/or quadropole moments, such as for instance for H_2S and CO_2. The average partial charges related to these moments may either increase the stability of the hydrate (H_2S) or be a decreasing factor in thermodynamic stability (CO_2). SII forms with for instance propane and iso-butane and sH with significantly larger molecules, as for instance cyclo-hexane, neo-hexane. Both methane and carbon dioxide form sI hydrate. SI hydrates forms with guest molecules less than 6 Å in diameter. The cages and the number of each cage per unit cell are shown in Figure 1. SI cages are shown at the top of the figure. The unit cell of sI hydrate contains 46 water molecules and consists of 2 small and six large cages. The unit cell is the smallest symmetric unit of sI. The two smaller cavities are built by 12 pentagonal faces (5^{12}) and the larger of 12 pentagonal faces and two hexagon faces ($5^{12}6^2$). The growth of hydrate adds unit cells to a crystal.

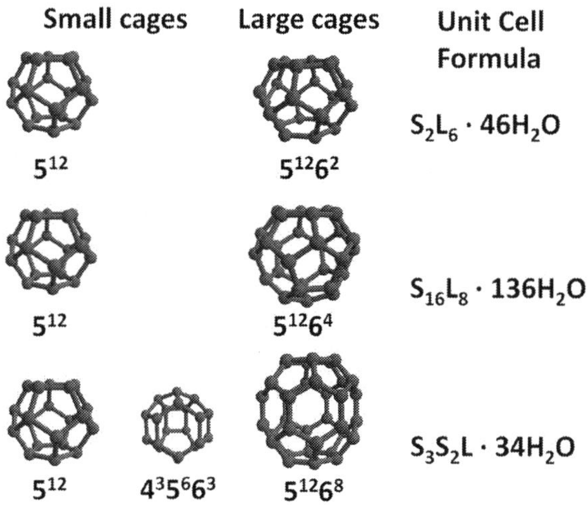

Fig. 1. Hydrate structures, from top: cages of sI, sII, and sH (Husebø, 2008)

3. Hydrates in the Earth

Natural gas occurrences in nature were first recognised in the late 1960s and early 1970s in sub-permafrost in Siberia and the North Slope of Alaska during drilling in established provinces of conventional oil and gas (Makogon et al., 1971). Evidence of natural occurrences of sub-marine gas hydrate accumulation was first found one or two decades later as more deep sea expeditions encountered hydrate bearing sediments. In recent years, a growing number of expeditions have been dedicated to assessing marine gas hydrate accumulations; their nature and their geological settings. Rather comprehensive mapping of hydrate has been devoted to many of the world's continental margins in recent years, as for example in India, Japan and Taiwan. There is no exact information of world wide quantities of gas hydrates. Estimates are based on both indirect (seismic surveys) and direct evidence (drilling) which is very incomplete. Figure 2 shows the world's occurrences of gas hydrate both in permafrost regions and in marine sediments. Estimates of in situ hydrates range from $3053*10^{15}$ m^3 STP (Trofimuk et al., 1973), to $0.2*10^{15}$ m^3 STP (Soloviev et al., 2002). Estimates have generally decreased with time; however, even the most conservative exceed the estimates of conventional gas (Sloan & Koh, 2008). A widely cited estimate is of $20*10^{15}$ m^3, (Kvenvolden, 1988), which exceed the energy in conventional fossil fuels combined.

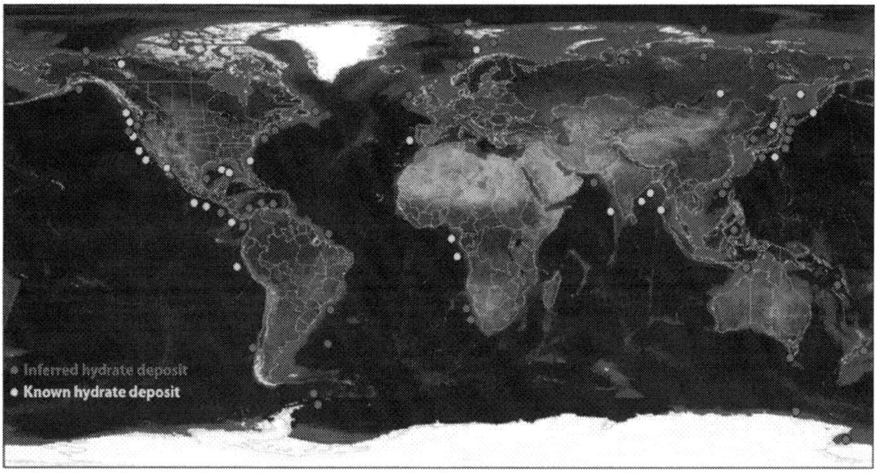

Fig. 2. Map of more than 90 documented hydrate occurrences (Hester and Brewer, 2009). Indirect hydrate markers, as seismic reflectors and pore-water freshening in core samples were used to identify the inferred hydrate deposits. Areas where hydrate samples have been taken are marked as known hydrate deposits. Hydrate deposits are distributed in marine environments and regions of permafrost.

4. Classification of Hydrate Deposits

Boswell and Collett, 2006, proposed a resource pyramid to display the relative size and feasibility for production of the different categories of gas hydrate occurrences in nature (Figure 3). The top resources of the gas hydrates resource pyramid are the ones closest to potential commercialization. According to Boswell and Collett, these are occurrences that exist at high saturations within quality reservoirs rocks under existing Arctic infrastructure. This superior resource type is estimated by US geological survey (USGS) to be in the range of 33 trillion cubic feet of gas-in-place under Alaska's North Slope. Prospects by British Petroleum and the US DOE anticipate that 12 trillion cubic feet of this resource is recoverable. Even more high-quality reservoirs are found nearby, but some distance away from existing infrastructure (level 2 from top of pyramid). The current USGS estimate for total North Slope resources is approximately 590 Tcf gas-in-places. The third least challenging group of resources is in high-quality sandstone reservoirs in marine environments, as those found in the Gulf of Mexico, in the vicinity of existing infrastructure.

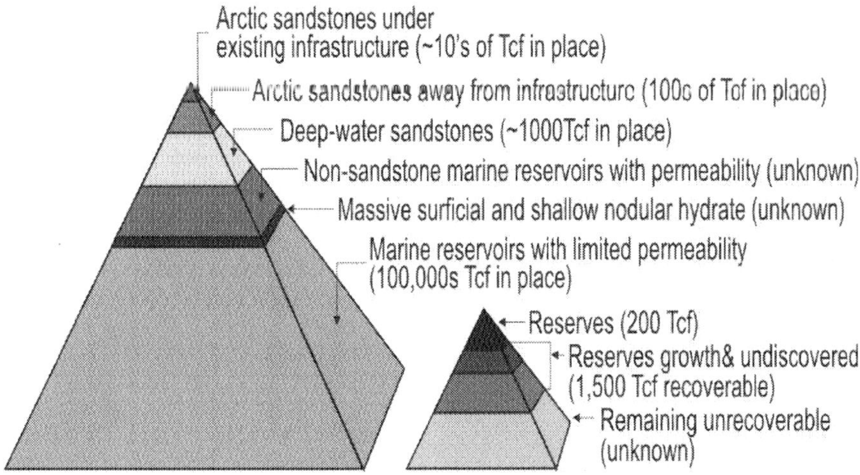

Fig. 3. Hierarchy of production feasibility for gas hydrate resources (left) and conventional gas (right) (Boswell and Collett, 2006)

There is a huge variation in naturally occurring hydrate reservoirs, both in terms of thermodynamic conditions, hosting geological structures and trapping configurations (sealing characteristics and sealing geometry). Hydrates in unconsolidated sand are considered as the main target for production. For the sake of convenience, these types of hydrate occurrences have been further divided into four main classes, as shown in Figure 4 (Moridis and Collett, 2003). Class 1 deposits are characterized with a hydrate layer above a zone with free gas and water. The hydrate layer is composed with either hydrate and water (Class 1W) or gas and hydrate (Class 1G). For both, the hydrate stability zone ends at the bottom of the hydrate interval. Class 2 deposits exist where the hydrate bearing layer, overlies a mobile water zone. Class 3 accumulations are characterized by a single zone of hydrate and the absence of an underlying zone of mobile fluids. The fourth class of hydrate deposits is widespread, low saturation accumulations that are not bounded by confining strata that may appear as nodules over large areas. The latter class is generally not regarded as a target for exploitation.

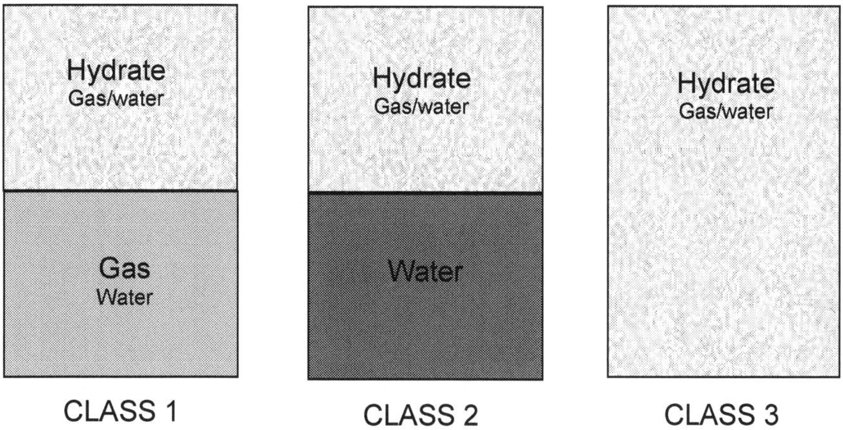

Fig. 4. Schematic over types of hydrate deposits

5. Proposed Production Schemes

The three main methods for hydrate dissociation discussed in the literature are (1) depressurization, where the hydrate pressure is lowered below the hydration pressure P_H at the prevailing temperature; (2) thermal stimulation, where the temperature is raised above the hydration temperature T_H at the prevailing pressure; and (3) through the use of inhibitors such as salts and alcohols, which causes a shift in the P_H-T_H equilibrium due to competition with the hydrate for guest and host molecules. The result of hydrate dissociation is production of water and gas and reduction in the saturation of the solid hydrate phase (Figure 5).

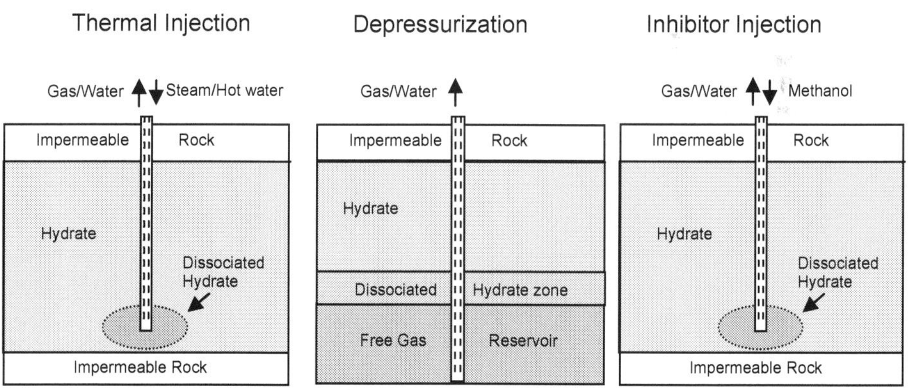

Fig. 5. Gas hydrate production options (after Makogon, 1997)

5.1 Numerical studies

Moridis et al. (2008) report rather comprehensive numerical studies that assess the hydrate production potential for the tree classes of hydrate deposits with the three production options. They found that Class 1 deposits appear to be the most promising target due to the thermodynamic proximity to the hydrate stability zone. That is, the boundary between the free gas zone and the hydrate layer forms the equilibrium line, and hence, only small changes in temperature or pressures will induce dissociation of hydrate. In addition, the free gas zone will secure gas production regardless of the hydrate gas contribution. They found Class 1G to be a more desirable target within Class 1 due to less water production and more evenly distributed pressure fields. Class 2 may attain high rates but are burdened with long lead times with little initial gas production. Class 3 may supply gas earlier, but with lower rates. Moridis et al. (2008), concluded that depressurisation is the favourable production option for all three classes, meaning that the deposit is not a desirable target if depressurisation appears to be ineffective. It is, however, very important to stress that numerical simulations of hydrate exploitation scenarios are still in an early stage, with corresponding challenges at the fundamental level as well as in the parameterisation.

5.2 Field example: the Mackenzie River Delta

The Mackenzie River Delta of Canada was explored mainly for conventional petroleum reserves, but a total of 25 drilled wells have identified possible gas hydrate sites. The gas hydrate research well (JAPEX/JNOC/GSC Mallik 2L-38) drilled in 1998 was designed to investigate the nature of *in situ* hydrates in the Mallik area to explore the presence of sub-permafrost gas hydrate. A major objective was to investigate the gas hydrate zones obtained by well logs in 1972 in a nearby well which was believed to have encountered at least ten significant gas-hydrate stratigrapic units. Drilling and coring gave 37 meters of recovered core in the hydrate interval from depths 878 to 944 meters. Visible gas hydrates were identified in a variety of sediment types, i.e. interbedded sandstone and siltstone. No hydrate was found in the siltstone dominated units, indicating a strong lithological control on gas hydrate occurrence. Well logs suggested the presence of gas hydrates sands from 890-1100 meters depth, with up to 90% gas hydrate saturation. The presence of gas hydrate contributes substantively to the strength of the sediment matrix (Grace et al., 2008). Two production tests were initiated at the Mallik site. The 2007 test was performed without sand controls in order to assess the strength of the sediments. A substantial amount of sand was produced and constrained the test to 24 hours. In March 2008 the test was repeated, this time with sand screen to choke the inflow of sediments. The last Mallik test suggests that a significant gas rate can be achieved by depressurising a sand dominated gas hydrate reservoir (Grace et al., 2008).

6. Environmental Aspects of Gas Hydrates

6.1 Climate change

The natural gas produced from hydrates will generate CO_2 upon combustion, but much less than conventional fuel as oil and coal per energy unit generated. The global awareness of climate change will most likely make it more attractive in relation to oil and coal if fossil fuels, as anticipated, continue to be a major fuel for world economies the next several decades. However, increased global temperatures have the potential of bringing both permafrost hydrates and subsea hydrates out of equilibrium. As a consequence, huge amounts of methane may be released to the atmosphere and accelerate the greenhouse effect due to feedback. In general hydrate is not stable towards typical sandstone and will fill pore volume rather than stick to the mineral walls. This implies that if there are imperfections and leakage paths in the sealing mechanisms the hydrate reservoir will leak. There are numerous small and large leaking hydrate reservoirs which results in methane fluxes into the ocean. Some of these fluxes will be reduced through consumption in biological ecosystems or chemical ecosystems. The net flux of methane reaching the atmosphere per year is still uncertain. Methane is by far a more powerful greenhouse gas than CO_2 (~20 times). Kenneth et al., 2003, hypothesized that major release from methane hydrate caused immense global warming 15 000 years ago. This theory, referred to as "clathrate gun" hypothesis is still regarded as controversial (Sloan & Koh, 2008), but is supported in a very recent paper by Kennedy et al. (2008). The role of gas hydrate in global climate change is not adequately understood. For hydrate methane to work as a greenhouse gas, it must travel from the subsurface hydrate to the atmosphere. Rates of dissociation and reactions/destruction of the methane gas on its way through sediment layers, water and air are uncharted.

6.2 Geomechanical Stability

Gas hydrates will affect the seafloor stability differently for the different types of hydrate occurrences. All of these hydrate configurations may take part of the skeleton framework that supports overlying sediments, which in turn is the fundament for pipelines and installations needed for production. These concerns have already been established for oil and gas exploitation where oil and gas reservoirs that lie below or nearby hydrate bearing sediments. However, geohazards would potentially be far more severe if gas hydrate is to be produced from marine hydrate deposits. During melting, the dissociated hydrate zone may lose strength due to under-consolidated sediments and possible over-pressuring due to the newly released gas (Schmuck and Paull, 1993). If the shear strength is lowered, failure may be triggered by gravitational loading or seismic disturbance that can result in submarine landslides (McIver, 1977). Several possible oceanic landslides related to hydrate dissociation are reported in the literature. Among these are large submarine slides on the Norwegian shelf in the North Sea (Bugge et al., 1988) and massive bedding-plane slides and slumps on the Alaskan Beaufort Sea continental margin (Kayen and Lee, 1993).

7. Production of CH_4 from hydrates by CO_2 exposure

Thermodynamic prediction suggests that replacement of CH_4 by CO_2 is a favourable process. This section reviews some basic thermodynamics and earlier experimental studies of this CH_4-CO_2 reformation process to introduce a scientific fundament for the experimental work presented later in this chapter.

7.1 Thermodynamics of CO_2 and CH_4 Hydrate

CO_2 and CH_4 form both sI hydrates. CH_4 molecules can occupy both large and small cages, while CO_2 molecules will prefer the large $5^{12}6^2$ cage. Under sufficiently high pressures or low temperatures both CO_2 and CH_4 will be stable, but thermodynamic studies suggest that

CH_4 hydrates have a higher equilibrium pressure than that of CO_2 hydrates for a range of temperatures. A summary of these experiments is presented in Sloan & Koh, 2008. Figure 6 shows the equilibrium conditions for CO_2 and CH_4 hydrate in a P-T diagram. This plot is produced using the CSMGem software (Sloan & Koh, 2008), which supplies the most recent thermodynamic predictions.

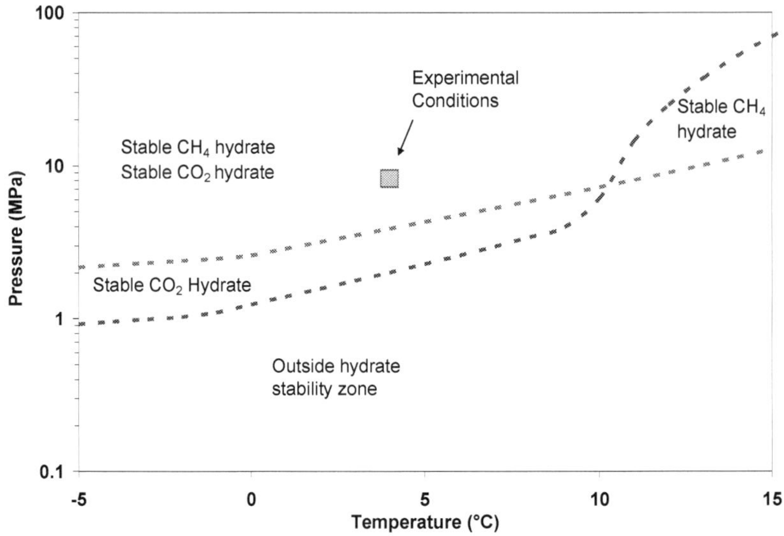

Fig. 6. Stability of CH_4 and CO_2 hydrate (CSMGem software, Sloan and Koh, 2008). Experimental conditions marks the P-T conditions for experiments presented in the next section.

7.2 CO2-CH4 exchange in bulk

Based on the knowledge of increased thermodynamic stability it was hypothesized that CO_2 could replace and recover CH_4 molecules if exposed to CH_4 hydrate (Ohgaki et al., 1994). Several early researchers investigated the CO_2-CH_4 exchange mechanism as a possible way of producing methane from hydrates (Ohgaki et al., 1996; Hirohama et al., 1996). These studies emphasized the thermodynamic driving forces that favour this exchange reaction, though many of the results showed significant kinetic limitations. Many of these early studies dealt with bulk methane hydrate samples placed in contact with liquid or gaseous CO_2, where available surfaces for interaction were limited. Yoon et al., 2004, studied the CO_2-CH_4 exchange process in a high pressure cell using powdered CH_4 hydrate and then exposed it to CO_2. They observed a fairly rapid initial conversion during the first 200 minutes, which then slowed down significantly. Park et al., 2008, found remarkable recovery of methane hydrate by using CO_2 and N_2 mixtures. They found that N_2 would compete with CH_4 for occupancy of the smaller sI cages, while CO_2 would occupy only the larger sI cage - without any challenge of other guests. They also found that sII and sH would convert to sI and yield high recoveries (64-95%) when exposed to CO_2 or CO_2-N_2 mixtures.

An inherent limitation in this experiment is the absence of mineral surfaces and the corresponding impact of liquids that may separate minerals from hydrates. These liquid channels may serve as transport channels as well as increased hydrate/fluid contact areas.

7.3 CO_2-CH_4 Exchange in Porous Media

Lee et al., 2003 studied the formation of CH_4 hydrate, and the subsequent reformation into CO_2 hydrate in porous silica. CH_4 hydrate was formed at 268 K and 215 bar while the conversion reaction was studied at 270 K. The temperatures in the ice stability region could have an impact on the reformation mechanisms since ice may form at intermediate stages of opening and closing of cavities and partial structures during the reformation. Temperatures below zero may also have an impact in the case where water separates minerals from hydrates. Preliminary studies of the CO_2 exchange process in sediments showed slow methane production when the P-T conditions were near the methane hydrate stability and at CO_2 pressure values near saturation levels (Jadhawar et al., 2005). The research presented below revisit the CO_2- CH_4 exchange process in hydrates formed in porous media, this time in larger sandstone core plugs and well within the hydrate stability for both CO_2 and CH_4 hydrate, and outside the regular ice stability zone (Figure 6).

8. MRI of Hydrates in Porous Media

A general schematic of the MRI hydrate forming and monitoring apparatus is shown in Figure 7. The total system consists of the sample, an MRI compatible cell to maintain the sample at high-pressure and low-temperature, high-pressure sources to individually control pore and confining pressures, a sample temperature control system and the MRI to monitor the distribution of water, hydrate and methane. The porous rock sample was sealed with shrink tubing into the centre of the high-pressure MRI cell. This was done so that gases and fluids could flow through the sample while the sample was separated from the confining fluid. One unique yet important feature was employing the confining fluid as the heat transfer medium (Fluorinert FC-40). This allowed accurate and precise control of the sample temperature without the elaborate system that would be required to cool the sample from the outside of the cell. The temperature bath controlled the coolant temperature, which in turn was transferred to the confining fluid by a heat exchanger around the confining-fluid transfer lines. The pressure and temperature were controlled and monitored by computers, which allowed the test to run unattended for extended periods of time. The high magnetic field required that all motors, controllers and pumps had to be several meters from the magnet. MRI images, both 3-D and 2-D, and fast 1D profiles were collected at regular intervals during the hydrate formation process and the CO_2-CH_4 exchange process. The MRI detects gas hydrate as a large drop in intensity between images of liquid water and solid hydrate. Hydrate formation was measured as the loss of MRI intensity as the liquid water converted to solid hydrate. Hydrogen in the solid hydrate has a short relaxation time and is not detected by the MRI by standard spin echo sequences (no signal above the background level). In contrast, the hydrate precursors, water and methane, produce intense MRI images. The images were acquired with a short echo time (< 3ms) and a long recovery time (2-4 sec). CO_2 is insensitive to magnetic resonance at the operating frequency and is therefore, as hydrates, not visible on the images. Two core plug geometries were used in these experiments: The first was a standard cylindrical plug, 3.75 cm diameter and varying lengths between 6 and 10 cm, and the second arrangement had an open fracture down the long axis of the core plug.

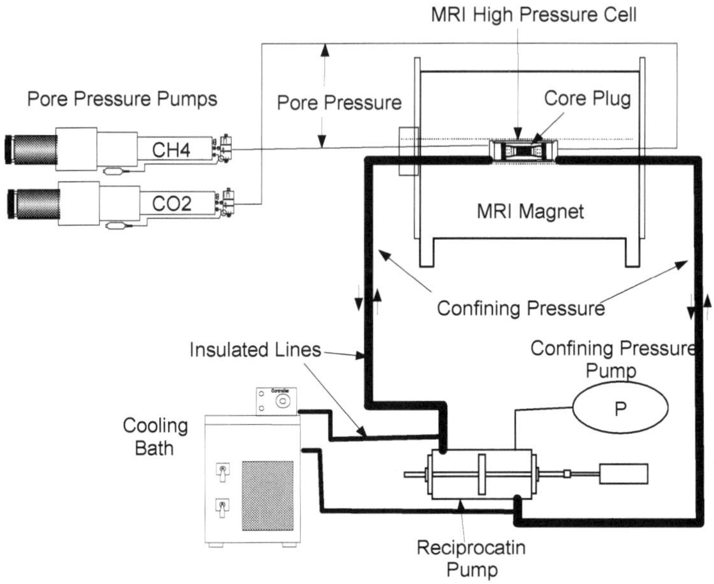

Fig. 7. Design for hydrate experiments

8.1 Core Preparation

The whole core experiments were prepared in one of two ways: 1) the core was dried in a heated vacuum stove and saturated with brine under vacuum. The core was then mounted in the MRI cell and vacuum was pulled from one end to reduce the brine saturation slowly. This procedure secured evenly distributed initial brine saturation. The evacuation valve was closed when the desired saturation was achieved and methane was introduced to the system and pressurized to 1200 psig. 2) The initial water saturation was prepared outside the MRI cell, by spontaneous imbibition. When assembled, several pore volumes of methane were injected through the core to minimize the amount of air in the system. The latter method was chosen in later experiments to keep flow lines dry and to avoid hydrate formation and plugging. Hydrate formed with no distinct difference in induction time or formation rate for both techniques, but the latter method eliminated hydrate formation in the lines. The second arrangement split an original cylinder down the long axis of the plug and inserted a 4 mm thick acetal polyoxymethylene (POM) spacer between the two halves (Figure 8). The spacer had a known volume of free space and small openings in the supporting frame so that fluids could easily enter and leave the spacer. The purpose of the spacer was to simulate a fracture opening in the sample where fluids had enhanced access to the porous media. This fracture increased the surface area for exposing 1) methane to the plug during the hydrate formation stage and 2) liquid carbon dioxide during the methane replacement stage. These experiments were prepared as follows: The high-pressure cell was installed, lines connected and a vacuum applied to the pore space of the core and spacer until approximately 100 millitorr was reached, and then filled with methane gas. After the methane was brought to 1200 psig, with the confining pressure concurrently increased to ca.1700 psig, a pre-determined amount of water was pumped in to the fracture and imbibed into the two core-halves to produce the desired saturation, ranging from 40 to 60% PV. The water was imaged to determine both the quantitative amount and distribution. At 50% PV the water-wet sandstone core imbibed the water, rapidly producing a fairly uniform vertical and horizontal distribution throughout the core.

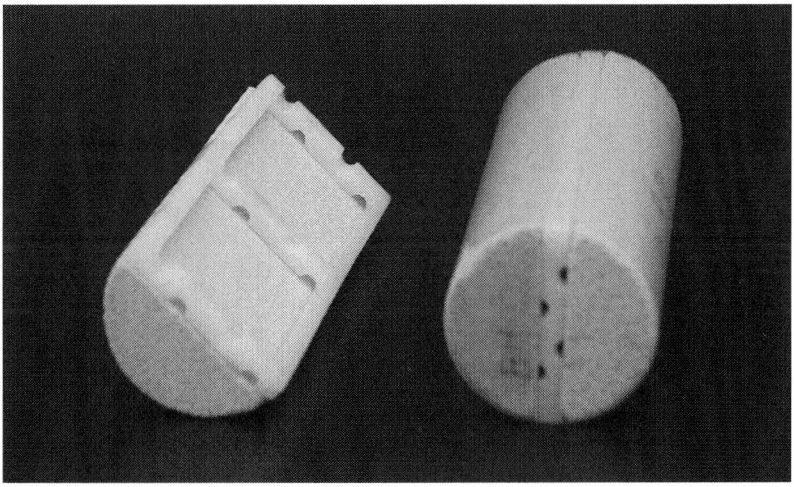

Fig. 8. Core design with spacer

Water salinity varied from 0.1 to 5.0 weight percent NaCl corresponding to values anticipated in permafrost-related hydrate deposits (Sloan and Koh, 2008). The presence of salt, which acts as a hydrate formation inhibitor, ensured that not all of the water was transformed into hydrate.

8.2 Hydrate formation in sandstone

Hydrates were formed in the pore space of a highly permeable sandstone acquired from the Bentheim quarry in Lower Saxony, Germany. The Bentheim sample used in these experiments had a porosity of 23% and a permeability of 1.1 D and was characterized by uniform pore geometry with an average pore diameter of 125 microns. The pore frame consisted of 99.9% quarts. An experiment with a whole sandstone core plug was performed to verify whether hydrate formation in porous media could be formed and detected in the experimental apparatus with the techniques presented in the previous chapter. Formation of methane hydrate within the sandstone pores is shown in the leftmost column in Figure 9. Hydrate growth is identified by the loss of signal between images of the partly water-saturated plug. The core sample was prepared with fairly uniform water saturation (52% average), with pressurized methane (1200 psig) in the remaining pore space. Methane in the core plug did not measurably contribute to the image. The images show the

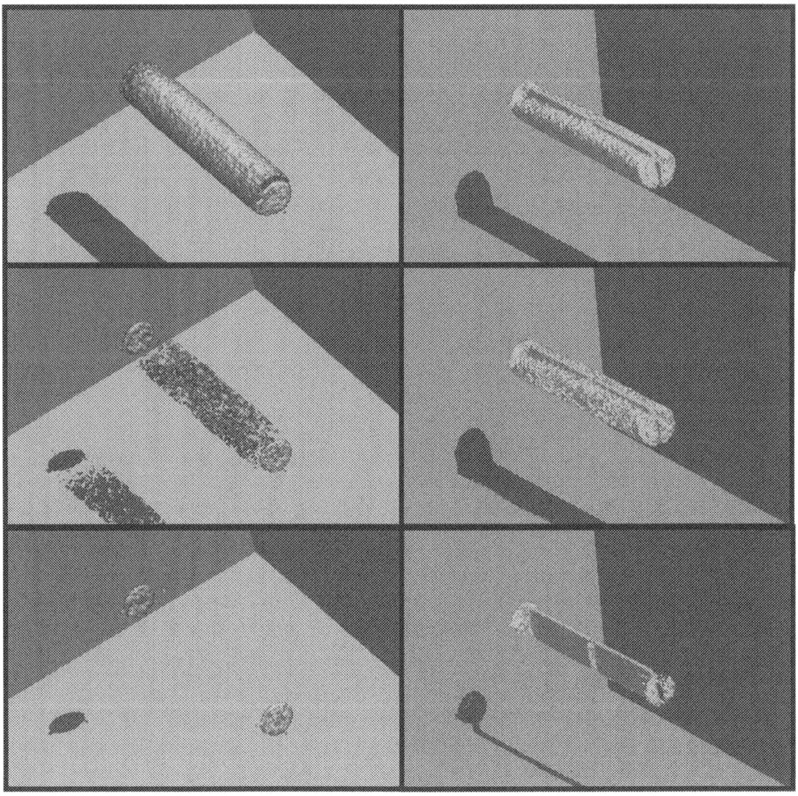

Fig. 9. Hydrate formation in a whole (left) and fractured (right) core plug core

formation of hydrate as a uniform loss of image with time. When cooled, hydrate formation was identified as an abrupt increase of consumed methane and a corresponding drop in the MRI Intensity. The correlation between the two independent measurements of hydrate growth rate was excellent. The core sample was fractured to prepare for the next experiment: measuring methane replacement by carbon dioxide. The right column in Figure 9 shows 3-dimensional MRI images obtained during the formation of methane hydrate in the core halves split by the POM spacer as described in the previous chapter. The first image (uppermost) shows water in the core plug halves and methane in the fracture prior to hydrate formation. The methane in the fracture is visually separated from the water in the plug partly due to the width of the fracture frame and partly due to the more uniform appearance of the methane in the fracture compared to the mottled appearance of water in the porous media. A downward growth pattern in each of the two core halves can be seen from Figure 9. The last image shows that most of the water was converted to hydrates. The open fracture can be seen filled with methane gas

9. Methane Replacement by Carbon Dioxide

To maximize the area of porous media exposed to methane or carbon dioxide a fracture was established along the cylindrical axis of the plug as described in the previous section. This artificial fracture of known volume and orientation provided greater control for introducing gases and/or liquids into the sandstone sample. The fracture frame was used to introduce methane during the initial hydrate formation, expose carbon dioxide to methane hydrate in the porous media and collect the methane expelled from the core plug during the carbon dioxide soak at a confining pressure of ca. 1700 psig and a pore pressure of 1200 psig. When the hydrate formation ceased (see last image in Figure 9) the spacer and connected lines were flushed at constant pressure (1200 psig) with liquid CO_2. Figure 10 shows a series of MRI images collected from the core with spacer after CO_2 was injected to remove methane from the spacer. The system was then closed and CO_2 was allowed to diffuse into the two core halves and methane was allowed to be produced back into the spacer. The first image (A) was acquired after the system was flushed. The region with carbon dioxide reveals no signal because it contains no hydrogen and therefore was not imaged. This suggests that most of the methane was displaced by CO_2. This assumption was confirmed by GC analysis (Gas Chromatography) of the effluent sample. The second image (B) was acquired 112 hours after the flush, at which time the MRI signal reappears in the fracture. C-D show successive images, obtained after 181 and 604 hours respectively, as methane continuously was produced into the spacer. Signal averaging was used in all images. Run time for the images varied from 2 to 9 hours depending on signal/noise ratio and given experimental conditions.

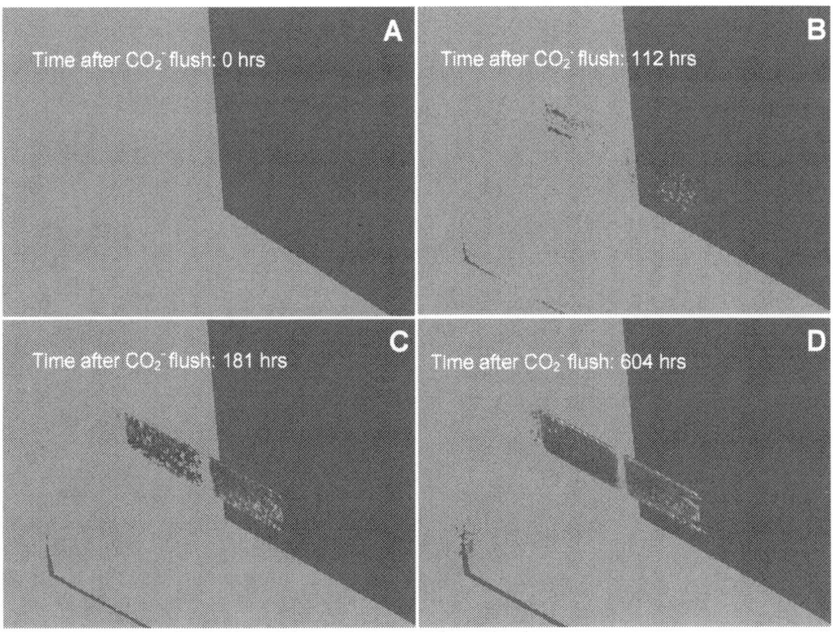

Fig. 10. Methane produced by CO_2 replacement from hydrates

Diffusion processes appeared to be the dominant driving mechanism in supplying CO_2 to the methane hydrate reaction sites and the concomitant increase of methane in the fracture. The exchange process continued over several weeks. When methane production ceased, the spacer was again flushed with CO_2 to accelerate the reaction by supplying fresh and pure liquid CO_2 to the system. The methane production curve found from the average MRI intensity in the fracture is shown for three separate experiments in Figure 11. Two of them are duplicate experiments with initial water saturation of 50 % and 5 wt% NaCl (published in Graue et al., 2008). The agreement between the two is very good. As shown in Figure 11, the methane molar volumes by far exceeded any free methane that might have remained in the pores after hydrate formation (diffusion experiment). Mass balance calculations and the molar production curve from MRI intensities in the fracture suggest that between 50-85 per cent of methane originally in hydrates was recovered by CO_2 replacement. Another observation is the apparent absence of large-scale melting of hydrates during the CO_2-CH_4-exchange. All the experiments run in this system did not detect any significant increase in MRI signal in the hydrate saturated cores that would indicate the presence of free water during CO_2 exchange. This was verified by the evaluation of the MRI signal intensity in the core halves once CO_2 exchange began. MRI intensity remained constant or was even less than the baseline value after the completion of hydrate formation. The exchange process did not cause significant dissociation of the hydrate, at least on the scale of the MRI's spatial resolution of ~0.8 mm^3. These experiments were run at CO_2 partial pressures significantly greater than CO_2 saturation levels, in contrast to earlier studies where the CO_2 levels were only slightly in excess to saturation or were undersaturated. This portion of the work shows that methane can be produced by CO_2 replacement in within sandstone pores.

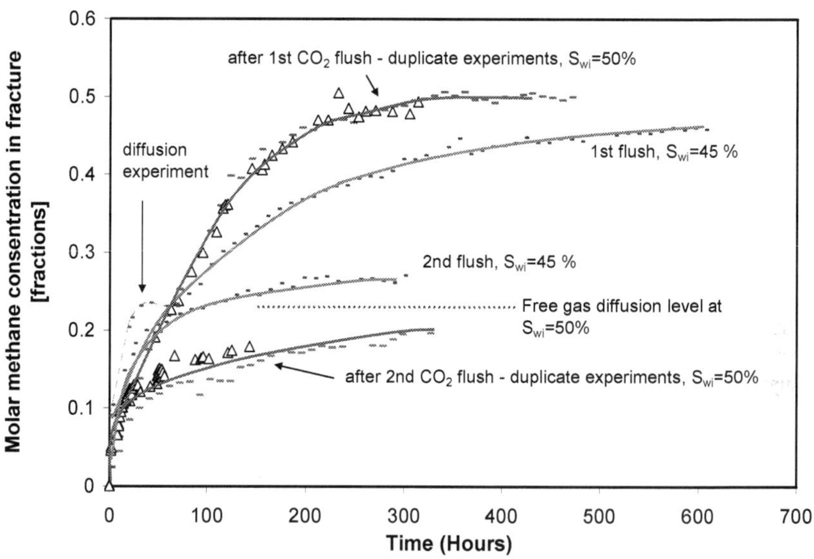

Fig. 11. Methane produced from methane hydrate by CO_2 replacement. Duplicate experiments with S_{wi}=50 % (5wt % NaCl) and one with S_{wi}=50 % (0.1 wt % NaCl).

10. Conclusion

The experimental set-up with the MRI monitoring apparatus was capable of forming large quantities of methane hydrates in sandstone pores and monitor hydrate growth patterns for various initial conditions. Spontaneous conversion of methane hydrate to carbon dioxide hydrate occurred when methane hydrate, in porous media, was exposed to liquid carbon dioxide. The MRI images did not detect any significant increase in signal in the hydrate saturated cores that would indicate the presence of free water during the carbon dioxide replacement.

11. Acknowledgements

The authors are indebted to the Norwegian Research Council and ConocoPhillips for financial support and thank Jim Stevens, James Howard and Bernie Baldwin for their contribution in acquiring the MRI data.

12. References

Sloan ED & Koh, C. (2008). *Clathrate hydrates of natural gases*, 3rd ed. Boca Raton: CRC Press.

Li, B.; Xu, Y. & Choi, J. (1996). Title of conference paper, *Proceedings of xxx xxx*, pp. 14-17, ISBN, conference location, month and year, Publisher, City

Lee H; Seo Y; Seo Y-T; Moudrakovski I. L & Ripmeester J. A. (2003). *Recovering Methane from Solid Methane Hydrate with Carbon Dioxide, Angew. Chem. Int. Ed.*, **42**, 5048 -5051

Jadhawar, P.; Yang, J.; Jadhawar, J.; Tohidi, B. (2005). Preliminary experimental investigation on replacing methane in hydrate structure with carbon dioxide in porous media. *Proceedings of the 5th International Conference on Gas Hydates*, Trondheim, Norway.

Ota, M., Morohashi, K., Abe, Y., Watanabe, M., Smith, J. R. L. & Inomata, H. (2005). *Replacement of CH4 in the hydrate by use of liquid CO2"*. Energy Conversion and Management, **46** (11-12): 1680-1691.

Hester, K. & Brewer, P. G. (2009). *Clathrate Hydrates in Nature*. Annual Reviews of Marine Science, **1** 303-327

Moridis, G.J. & Collett, T. (2003). Strategies for Gas Production From Hydrate Accumulations Under Various Geologic Conditions, LBNL-52568, presented at the *TOUGH Symposium*, Berkeley, CA, May 12-14.

Makogan Y.: *Hydrates of hydrocarbons*, Tulsa, Pennwell Books, 1997.

McIver, R. D. (1977) Hydrates of natural gas - an important agent in geologic processes, In Abstracts with Programs, pages 1089--1090. Geological Society of America.

Boswell, R. & Collett, T.S. (2006) The Gas Hydrate Resource Pyramid, Fire in the ice, NETL Fall Newsletter, 5-7.

Graue A.; Kvamme B.; Baldwin B.A.; Stevens J.; Howard J.; E. Aspenes, Ersland G.; Husebø J. & Zornes D.. (2008). MRI Visualization of Spontaneous Methane Production From Hydrates in Sandstone Core Plugs When Exposed to CO_2. *SPE Journal* (SPE 118851), **13** (2). p. 146-152.

Phale, H. A.; Zhu, T.; White, M. D. & McGrail, B. P. (2006).Simulation study on injection of CO2-Microemulsion for Methane Recovery From Gas-Hydrate Reservoirs. *SPE Gas Technology Symposium*, Calgary, Alberta, Canada

Kvamme B., Graue A., Buanes T., Ersland G.. (2007). Storage of CO_2 in natural gas hydrate reservoirs and the effect of hydrate as an extra sealing in cold aquifers. *International Journal of Greenhouse gas control,* **1** (2) p. 236-246.

Husebø, J. *Monitoring depressurization and CO2-CH4 exchange production scenarios for natural gas hydrates. (2008).* PhD thesis, University of Bergen, Norway.

Makogon, YF., Trebin, F.A., Trofimuk, A.A., Tsarev, V.P., and Cherskiy, N.V. (1971). Detection of a pool of natural gas in a solid (hydrated gas) state," *Doklady Akademii Nauk SSSR,* **196**, pp 203-206 (Translation in *Doklady-Earth Science Section,* **196**, pp 197-200, 1972).

Soloviev, V.A. (2002). Global estimation of gas content in submarine gas hydrate accumulations", *Russian Geology and Geophysics* **43**, pp. 609–624

Trofimuk, A.A.; Cherskiy, N.V. & Tsarev, V.P.(1973) Accumulation of natural gases in zones of hydrate—formation in the hydrosphere. *Doklady Akademii Nauk SSSR,* **212**, pp. 931–934.

Kvenvolden, K.A. (1988). Methane hydrate—a major reservoir of carbon in the shallow geosphere. *Chemical Geology* **71**, pp. 41–51.

Kennedy, M., Mrofka, D., Borch, C. (2008). Snowball Earth termination by destabilization of equatorial permafrost methane clathrate", *Nature,* **453**: 642-645.

Kenneth J.P.; Cannariato G.; Hendy I.L. & Behl R.J. *Methane Hydrates in Quarternary Climate Change: The Clathrate Gun Hypothesis,* Am. Geophys. Union, Washington DC (2003).

Grace, J, Collett, TS, Colwell, F, Englezos, P, Jones, E, Mansell, R, Meekinson, P, Ommer, R, Pooladi-Darvish, M, Riedel, M, Ripmeester, JA, Shipp, C and Willoughby, E. (2008) *Energy From Gas Hydrates: Assessing the Opportunities & Challenges for Canada,* Report in Focus, Council of Canadian Academies.

Bugge, T., Belderson, R. H. and Kenyon, N. H. (1988). *The Storegga slide.* Philos. Trans. R. Soc. London, **325**: 357--388

Kayen, R. E. and Lee, H. J. (1993). Submarine Landslides: Selected Studies in the U.S. Exclusive Economic Zone, *U.S. Geol. Surv. Bull.* **2002**: 97–103

Schmuck, E.A.; and Paull, C.K. (1993) Evidence for gas accumulation associated with diapirism and gas hydrates at the head of the Cape Fear slide. *Geo-Mar. Lett.,* **13**:145-152.

McIver, R. D. (1977) *Hydrates of natural gas - an important agent in geologic processes.* In Abstracts with Programs, pages 1089--1090. Geological Society of America.

Ohgaki, K.; Takano, K.; Moritoki, M. (1994). Exploitation of CH_4 Hydrates under the Nankai Trough in Combination with CO_2 storage", *Kagaku Kogaku Ronbunshu,* **20** 121-123

Ohgaki K; Takano K, Sangawa H; Matsubara T; Nakano S. (1996) Methane exploitation by carbon dioxide from gas hydrates - phase equilibria for CO_2-CH_4 mixed hydrate systems. *J. Chem. Eng. Jpn.* **29** (3): 478-483 (1996)

Hirohama S.; Shimoyama Y.; Tatsuta S.; Nishida N. (1996). Conversion of CH4 hydrate to CO_2 hydrate in liquid CO_2, *J. Chem. Eng. Jpn.* **29**(6): 1014-1020

Yoon J-H.; Kawamura T.; Yamamoto Y.; Komai T. (2004) Transformation of Methane Hydrate to Carbon Dioxide Hydrate: In Situ Raman Spectroscopic Observations *J. Phys. Chem.* A, **108**, 5057-5059.

Park, Y., Cha, M., Cha, J. H., Shin, K., Lee, H., Park, K. P., Huh, D. G., Lee, H. Y., Kim, S. J. & Le, J. (2008) *Swapping Carbon Dioxide for Complex Gas Hydrate Structures.* ICGH, Vancouver, BC, Canada.

8

The effect of H_2S on hydrogen and carbon black production from sour natural gas

[1]M. Javadi, [1]M. Moghiman and [2]Seyyed Iman Pishbin
[1]Ferdowsi University of Mashhad,
[2]Khorasan Gas company
Iran

1. Introduction

Hydrogen is well known as an ideal and clean source of energy which is believed to reduce the emission of carbon dioxide and therefore play a major role in decreasing the global warming problem [Ryu et al, 2007]. Eventual realization of a hydrogen economy requires cheap and readily available hydrogen sources and a technology to convert them into pure hydrogen in an efficient and sustainable manner [Abdel et al, 1998]. In addition to water that is an ideal hydrogen source, CH_4 and H_2S are considered as alternative sources of hydrogen [Jang et al, 2007; T-Raissi, 2003]. On the other hand, there is ample scope for CH_4 and H_2S as the raw source of H_2, because the energy required for CH_4 and H_2S splitting (ΔH_{CH4}=74.9 kJ/mol and ΔH_{H2S}=79.9 kJ/mol) is much less than water splitting (ΔH_{water}= 284.7 kJ/mol) [Jang et al, 2007]. There are several convenient technologies for production of H_2 from CH_4, including steam methane reforming (SMR), partial oxidation, pyrolysis, autothermal pyrolysis, and autothermal SMR [Huang & T-Raissi, 2007a]. Methane decomposition is a moderately endothermic reaction. It requires much less thermal energy (only 37.8 kJ per mol of hydrogen produced) than SMR (69 kJ/mol H_2). Besides, the decrease in the required energy, the CO_2 emission is also decreased in this method. Methane which is the main component of the high quality natural gas can be decomposed to hydrogen and carbon black in pyrolysis reactors [Abanades & Flamant , 2007; Moghima & Bashirnezhad, 2007]. Carbon black is an industrial form of soot produced by subjecting hydrocarbon feedstock to extremely high temperatures in a carefully controlled combustion process. Carbon black is widely used as filler in elastomers, tires, plastics and paints to modify the mechanical, electrical and optical properties of materials in which it is used [Ghosh, 2007; Petrasch et al, 2007].

As the prices of fossil fuel increase, abundant sour natural gas, so called sub-quality natural gas resources become important alternatives to replace increasingly exhausted reserves of high quality natural gases for the production of hydrogen and carbon black [Huang & T-Raissi, 2007b; Abdel et al, 1998]. At oil flow stations it is common practice to flare or vent SQNG, which is produced along with crude oil. This accounts for more than 100 million cubic meters (m^3) world-wide per day, and approximately equals to France's annual gas consumption [Gruenberger et al, 2002]. Clearly this is of considerable concern in terms of global resource utilization and climate change implications. Gas flaring has also been blamed for environmental and human health problems such as acid rain, asthma, skin and breathing diseases [Lambert et al, 2006]. The removal of H_2S from sub-quality natural gas is expensive and not commercially viable for large-scale plants. When H_2S concentration in natural gas is higher than about 1.0%, the high separation cost makes the sour natural gas uneconomical to use [Huang & T-Raissi, 2007b]. As mentioned above, production of hydrogen and carbon black from sour natural gas is one viable option utilizing this untapped energy resource while at the same time reducing carbon oxides and hydrogen sulfide emissions.

There is a massive back ground literature on thermal decomposition of high quality natural gas using different types of reactors. Petrasch & Steinfeld (2007) have studied hydrogen production process using solar reactors with SMR method. Abanades & Flamant (2007) also have investigated the effect of different parameters and system geometry on methane conversion and hydrogen yield using thermal decomposition method in solar reactors. Their results show that the solar reactor producing pure H_2 has high efficiency in CH_4 conversion. Cho et al (2009) have studied on the development of a microwave plasma-catalytic reaction process to produce hydrogen and carbon black from pure natural gas. The direct conversion of methane, using various plasma technologies has widely been studied in order to obtain more valuable chemical products. Gruenberger et al (2002) and Moghiman & Bashirnezhad (2007) have investigated the effect of feedstock parameters on methane decomposition in carbon black furnace.

Although many studies have been carried out on high quality natural gas pyroysis, sour natural gas pyrolysis have received much less attention. Towler & Lynn (1996) introduced thermal decomposition of hydrogen sulfide at high temperature as an alternative of Clause process. The main advantage of the thermal decomposition is reduction of produced tail gas rather than Clauses process. They have investigated the effect of CO_2 presence in feed gas and temperature on decomposition and sulfur compounds production. Also, Huang and T-Raissi et al (2007b, 2007c and 2008) have performed the thermodynamic analyses of hydrogen production from sub-quality natural gas using a Gibbs reactor operation in the AspenPlusTM chemical process simulator. Javadi and Moghiman (2010) have investigated carbon disulfide, hydrogen and solid carbon production from sub-quality natural gas. Their results show that the maximum yield of C(s) is in 1000 °K and then decreases due to increasing of CS2 production.

Based on the importance of sub-quality natural gas pyrolysis, the effects of feedstock parameters, reactor temperature and H_2S/CH_4 molar ratio of feedstock on decomposition process have been studied using the proposed carbon black furnace by Gruenberger et al (2002).

2. Gas furnace carbon black

Hydrogen and carbon black production via thermal decomposition of natural gas have been achieved using a carbon black furnace [Gruenberger et al, 2000 & 2002], plasma [Gaudernack & Lynum 1998], solar radiation [Abanades et al, 2007 & 2008], a molten metal bath and thermal reactors with and without catalyst [Steinberg, 1998; Ishihara et al, 2002; Muradov et al, 1998 & Kim et al, 2004].

Depending on the way that heat is supplied to sour natural gas, carbon black furnaces can be classified as follows:

Type 1: Part of the natural gas or any other fuel burns inside the reactor to provide heat needed to decompose the sour natural gas.

Type 2: Direct heat transfer from inert hot gases introduced into the reactor. This method is known as "Hydrogen Sulfide- Methane Reformation".

The carbon black furnace used in this investigation is a small-scale axial flow reactor identical to that reported previously by Gruenberger et al (2002). The furnace has been designed on the basis of using gaseous fuels as feedstock hydrocarbon, with a maximum output of 10 kg carbon black per hour. The basic geometry of the carbon black furnace is shown in Fig. 1, consisting of a pre-combustor, a mixing zone and a reactor. In the pre-combustor, the axially injected natural gas burns with inlet air introduced through two tangential inlets. Then, the highly swirling hot combustion gases mix with the sub-quality natural gas injected radially into the pre-combustor in the proximity of the mixing zone. A sudden increase in the tube diameter at the exit of the choke promotes vigorous mixing of the SQNG fuel with the hot gases leading to thermal decomposition of CH_4+H_2S and formation of hydrogen, carbon black, sulfur compounds and other precursor species for the formation of carbon black [Lockwood et al, 1995].

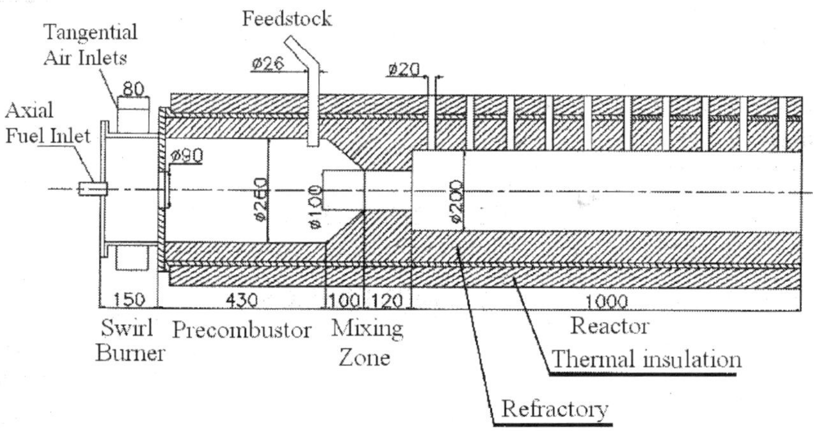

Fig. 1. Carbon black gas furnace[Gruenberger et al, 2002]

3. Chemical reaction modelling

Production of carbon black through thermolysis of SQNG involves a complex series of chemical reactions which control conversion of both CH_4 and H_2S as follows [Huange & T-Raissi, 2007b; Towler & Lynn, 1996]:

$$CH_4 \rightarrow C(S) + 2H_2 \qquad \Delta H^{\circ}_{298} = -74.9\,kJ/mol \qquad \text{Reaction (1)}$$

$$H_2S \rightarrow \frac{1}{2}S_2 + H_2 \qquad \Delta H^{\circ}_{298} = -79.9\,kJ/mol \qquad \text{Reaction (2)}$$

Since reaction 1 is mildly endothermic, it requires temperatures higher than $850^{\circ}K$ to proceed at reasonable rates [Dunker et al, 2006], and, as reaction 2 is highly endothermic, temperatures in excess of $1500^{\circ}K$ is required for achieving reasonable rates [Huang & T-Raissi, 2008].
Under special circumstances including using catalyst H_2S can react with methane producing carbon disulfide (CS_2) and H_2 [Huange & T-Raissi, 2008].

$$2H_2S + CH_4 \Leftrightarrow CS_2 + 4H_2 \qquad \Delta H^{\circ}_{298K} = 232\,kJ/mol \qquad \text{Reaction (3)}$$

A portion of CH_4 and H_2S can oxidize to produce CO, CO_2 and SO_2 [Abdel et al, 1998]:

$$CH_4 + \frac{1}{2}O_2 \rightarrow CO + H_2 + H_2O \text{ and } CO + \frac{1}{2}O_2 \rightarrow CO_2 \qquad \text{Reaction (4)}$$

$$H_2S + \frac{3}{2}O_2 \rightarrow SO_2 + H_2O \qquad \text{Reaction (5)}$$

H_2S can also react with CO_2 producing COS [Sakanishi et al, 2005]:

$$H_2S + CO_2 \Leftrightarrow COS + H_2O \qquad \text{Reaction (6)}$$

4. Turbulence–chemistry interaction

The mixture fraction/PDF method is used to model the turbulent chemical reactions occurring in the diffusion, combustion and thermal decomposition of natural gas in the carbon black furnace. This method, which assumes the chemistry is fast enough for a chemical equilibrium to always exist at molecular level, enables handling of large numbers of reacting species, including intermediate species. Transport equations are solved for the mean mixture fraction $\bar{f}$, its variance $\overline{f'^2}$ and for enthalpy $\bar{h}$. Calculations and PDF integrations are performed using a preprocessing code, assuming chemical equilibrium between 30 different species. The results of the chemical equilibrium calculations are stored in look-up tables which relate the mean thermochemical variables (species mass fractions, temperature and density) to the values of $\bar{f}$, $\overline{f'^2}$ and $\bar{h}$ [Saario & Rebola, 2005].
In non-adiabatic systems, where change in enthalpy, due to heat transfer, affects the mixture state, the instantaneous thermo chemical state of the mixture, resulting from the chemical equilibrium model, is related to a strictly conserved scalar quantity known as the mixture fraction, f, and the instantaneous enthalpy, $H^*, \phi_i = \phi_i(f, H^*)$. The effects of turbulence on the thermo chemical state are accounted for with the help of a probability density function (PDF):

$$\bar{\phi}_i = \int_0^1 \phi_i(f, \overline{H}^*) p(f) df. \tag{1}$$

In this work, the β-probability density function is used to relate the time-averaged values of individual species mass fraction, temperature and fluid density of the mixture to instantaneous mixture fraction fluctuations. The β-PDF in terms of the mean mixture fraction $\bar{f}$ and its variance $\overline{f'^2}$, can be written as:

$$P(f) = \frac{f^{\alpha-1}(1-f)^{\beta-1}}{\int_0^1 f^{\alpha-1}(1-f)^{\beta-1} df}, \quad 0 < f < 1 \tag{2}$$

where:

$$\alpha = \bar{f}\left[\frac{\bar{f}(1-\bar{f})}{\overline{f'^2}} - 1\right], \ \beta = (1-\bar{f})\left[\frac{\bar{f}(1-\bar{f})}{\overline{f'^2}} - 1\right]. \tag{3}$$

Using the unweighted averaging [Jones & Whitelaw, 1982], the values of the two parameters $\bar{f}$ and $\overline{f'^2}$ at each point in the flow domain are computed through the solution of the following conservation equations [Warnatz, 2006]:

$$\frac{\partial}{\partial x}(\rho u_i \bar{f}) = \frac{\partial}{\partial x_i}\left(\frac{\mu_t}{\sigma_t}\frac{\partial \bar{f}}{\partial x_i}\right), \tag{4}$$

$$\frac{\partial}{\partial x}(\rho u_i \overline{f'^2}) = \frac{\partial}{\partial x_i}\left(\frac{\mu_t}{\sigma_t}\frac{\partial \overline{f'^2}}{\partial x_i}\right) + C_g \mu_t \left(\frac{\partial \bar{f}}{\partial x_i}\right)^2 - C_d \rho \frac{\varepsilon}{k}\overline{f'^2}, \tag{5}$$

where the constants σ_t, $C_g(=2/\sigma_t)$ and C_d take the values 0.7, 2.86 and 2.0, respectively. The distribution of the instantaneous enthalpy is calculated from a transport equation as follows:

$$\frac{\partial}{\partial x}(\rho u_i \overline{H^*}) = \frac{\partial}{\partial x_i}\left(\frac{k_t}{c_p}\frac{\partial \overline{H^*}}{\partial x_i}\right) + \tau_{ik}\frac{\partial u_i}{\partial x_k} + S_h \tag{6}$$

where k_t is turbulent thermal conductivity and S_h includes the heat generated by the chemical reaction and radiation. The instantaneous enthalpy is defined as:

$$H^* = \sum_j m_j H_j = \sum_j m_j \left[\int_{T_{ref,j}}^{T} c_{p,j} dT + h_j^{\circ}(T_{ref,j})\right] \tag{7}$$

where m_j is the mass fraction of species j and $h_j^{\circ}(T_{ref,j})$ is the formation enthalpy of species j at the reference temperature $T_{ref,j}$.

5. Numerical solution procedure

Fluent CFD software has been used to model the furnace employing solution-adaptive grid refinement technique to solve the 3D problem. Gambit preprocessor is used for the fully three dimensional geometry creations and unstructured grid generation. The 3D volume grid is represented in Fig 2. The domain is discretized into a grid of 20493 nodes and 82745 tetrahedral cells.

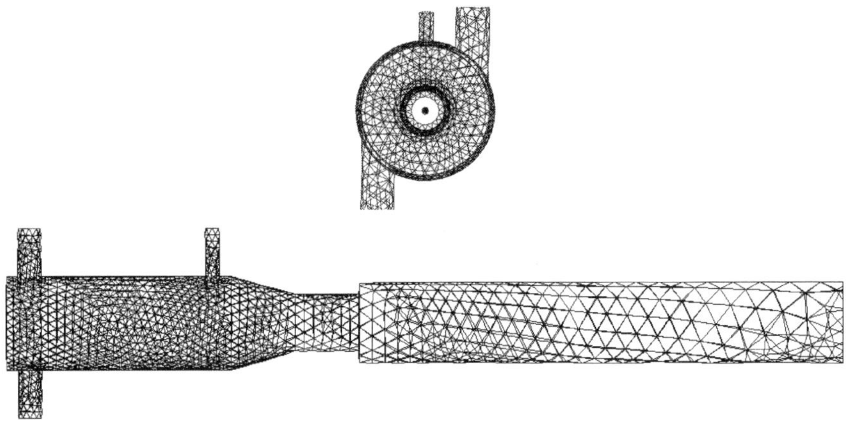

Fig. 2. Three-dimensional tetrahedral grid

The conservation equations for mass, momentum, energy, Reynold's stresses, dissipation rate, mixture fraction and its variance, and concentration of soot are solved by finite-volume analysis, using a second-order upwind scheme for discretisation of the convective terms in the transport equations.

The radiative heat transfer in the absorbing, emitting and scattering medium is calculated by the Discrete Ordinates (DO) radiation model [Murthy & Mathur, 1998]. The RSM (Reynolds stress model) is used for prediction of anisotropic, highly swirling and recirculating flow inside the combustor. Abandoning the isotropic eddy-viscosity hypothesis, the RSM closes the Reynolds-averaged Navier-Stokes equations by solving six differential transport equations for Reynolds stresses, together with an equation for the dissipation rate of turbulence kinetic energy. The conventional wall-function approach is used in the near-wall region. At the inlet boundary, conditions are specified once and did not need updating during the course of the solution procedure. At the outlet boundary, zero gradient conditions are applied. We assumed an isothermal boundary condition at the wall of the furnace.

A grid dependence study was conducted to arrive at the appropriate size of the grid for optimal accuracy and efficiency. The number of grid points was varied from 17231 to 36387 for typical set operating conditions. We observed that the field quantities varied less than 1% after the number of grid points increased beyond 20493. For the radiation model, emissivity coefficient at the flow inlets and outlets were taken to be 1.0 (black body absorption). Wall emissivity was set at 0.6, a typical value for combustion gases.

6. Results and discussion

As mentioned above, the processes of methane pyrolysis differ mainly by the way heat is supplied to the furnace. In this study, sour natural gas decomposition in a carbon black furnace has been investigated for two types of supplying heat. In the first type, the natural gas burns inside the pre-combustor (Fig.1) to provide required heat for decomposing feed sour gas. In this case, the problem is the effect of combustion product (process gases) and excess air which extremely affect on sour natural gas decomposition and furnace product. In the second type the heat transfers from inert hot gases to feed sour gas. In this case only reactions 1 to 3 are involved and there is not the problem of excess air and combustion products. In this study the sour natural gas thermal decomposition inside the axial flow gas furnace designed by Gruenberger et al (2002) has investigated. The results of two types of supplying heat for pyrolysis are as follows:

6.1 Type 1: Pyrolysis by hot combustion gases

The total pre-combustor inlet airflow rate is $19\times10^{-3}\,m^3/s$, at the temperature of 690° K and pressure of 1 bar. The equivalence ratio used for the pre-combustor is 0.92. The accuracy of the quantitative or even the qualitative trends for the combustion and decomposition parameters depend on the accuracy with which the temperature and species concentration fields are determined from the numerical calculation of the present model. To establish the accuracy of our model, we have been calculated and compared the model predictions to the experimental measurements of Gruenberger [Gruenberger et al, 2002] with no H_2S. For comparison purposes, we first conducted computations without H_2S in feed gas.

A comparison of reactor outlet average temperature and carbon black yield (kg carbon black/kg feedstock) predicted by this model and by experimental results is given in Figs. 3 and 4. Results of Fig. 3 depict that the model predicts lower temperatures than the experimental data, especially at high feed flow rates. The discrepancy between the two results might be due to the fundamental assumption made in the combustion model (PDF fast chemistry combustion model), which assumes that chemistry is fast enough for a chemical equilibrium. Results of Fig. 4 show that the predicted and measured carbon black yields are in very good agreement and maximum carbon black yield is reached at the equivalence ratio of 3. The discrepancy between the two results can be attributed to the temperature levels obtained by the two methods (see Fig. 3). The lower temperature levels computed by the model might be due to higher decomposition of CH_4. Fig. 5 presents the calculated distributions for CH_4, H_2S, temperature and mass fraction of soot, carbon black, COS and gaseous sulfur predicted by the model at feed rate of $3\times10^{-3}\,kg/s$. H_2S mass fraction in natural gas is assumed to be 10%. Of particular interest are Figs. 5d-f that show soot formation due to incomplete combustion of inlet methane and production of solid carbon and gaseous sulfur by pyrolysis of methane- hydrogen sulfide jet interaction with hot surroundings. Results from the model calculations seem to indicate that the use of more inlet injection ports for SQNG feed would increase the yield of carbon black and sulfur compounds.

Fig. 6 shows the reactor outlet temperature as a function of inlet mass flow rate for two cases a) with H_2S, b) without H_2S. It can be seen that the results obtained for these two cases are similar. The small discrepancy between the results may be due to CH_4 decomposition reaction that begins at lower temperatures than that of H_2S. Also, Fig. 6 depicts that temperature drops precipitously with increasing flow rate of feed gas due to the endothermic nature of both CH_4 and H_2S decompositions.

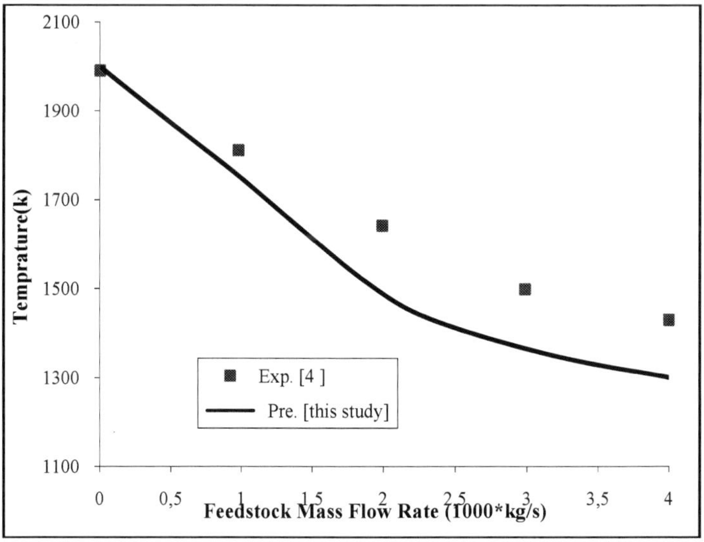

Fig. 3. Comparison of the predicted reactor outlet temperature with the experimental data

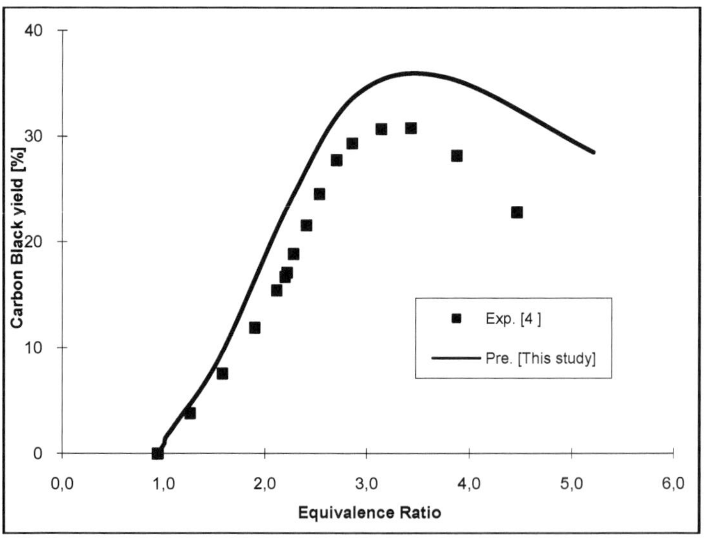

Fig. 4. Comparison of the predicted carbon black yield with the experimental data

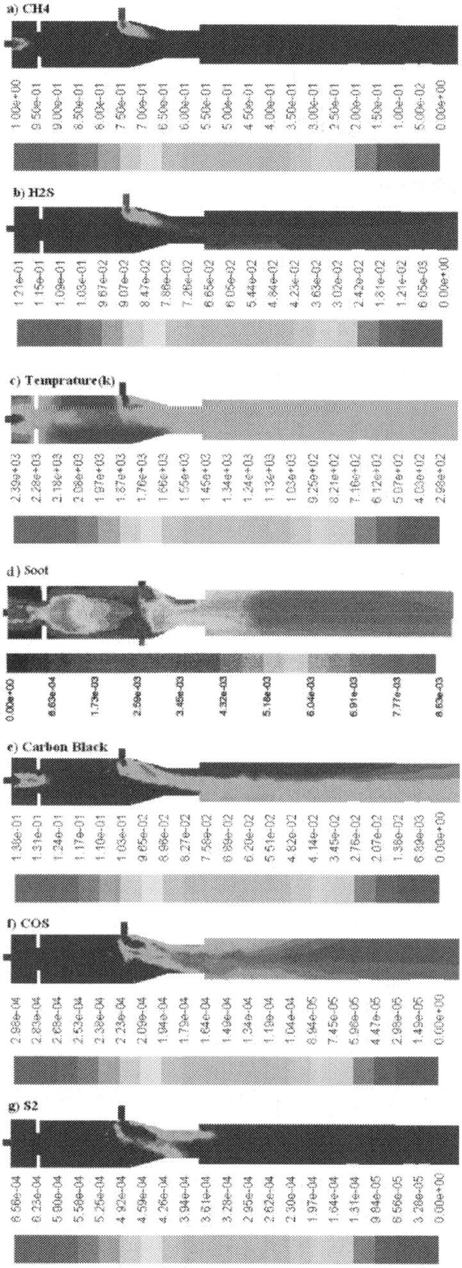

Fig. 5. Contour of species mass fractions and temperature (K)

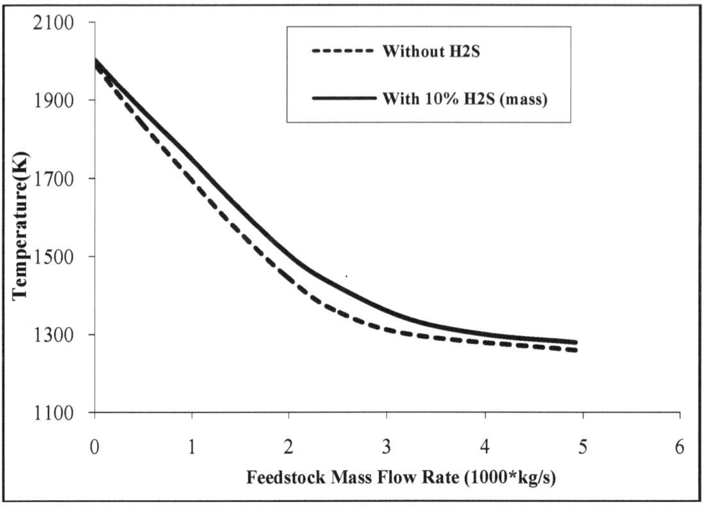

Fig. 6. Effect of feedstock flow rate on calculated outlet temperature

Figs. 7 and 8 show the effect of feed gas flow rate and reactor outlet temperature on CH_4 and H_2S conversions given by [Huang & T-Raissi, 2008]:

$$CH_4 \text{ conversion} = \frac{[CH_4]_0 - [CH_4]}{[CH_4]_0} \times 100, \quad H_2S \text{ conversion} = \frac{[H_2S]_0 - [H_2S]}{[H_2S]_0} \times 100$$

where $[CH_4]_0$ and $[H_2S]_0$ denote the initial (input) concentration of CH_4 and H_2S, respectively. $[CH_4]$ and $[H_2S]$ are equilibrium concentration of CH_4 and H_2S at reactor outlet, respectively. Fig. 7 depicts that the H_2S conversion drops sharply with increased feed gas flow rate, which can be attributed to the endothermic nature of H_2S and CH_4 decomposition reactions. For higher values of feed gas flow rate ($\geq 0.002 \text{kg/s}$) CH_4 conversion decreases with increased feed gas flow rate due to the endothermicity of CH_4 thermolysis. The major factor influencing CH_4 and H_2S conversions appears to be temperature. Fig. 8 shows that CH_4 conversion reaches 100% at temperatures above $1100^\circ K$. Because CH_4 decomposition reaction is mildly endothermic, the temperature must be above $850^\circ K$ for the reaction to proceed at a reasonable rate. This is in accordance with the results of Huang and T-Raissi (2008). At any temperature, H_2S conversion is less than that of CH_4, especially at those below $1300^\circ K$ wherein H_2S conversion is less than 5%. For higher values of reactor temperature ($\geq 1300^\circ K$), H_2S conversion increases sharply with reaction temperature. As the reaction of H_2S decomposition is endothermic, the temperature must be above $1500^\circ K$ for the reaction to proceed at rapid rates.

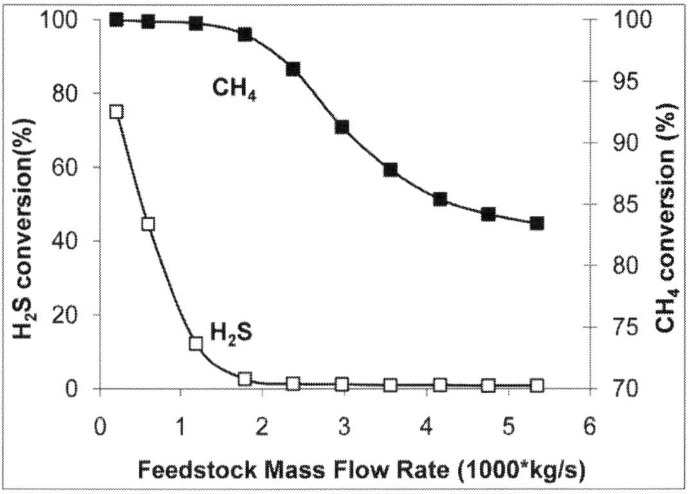

Fig. 7. Effect of feedstock mass flow rate on H_2S and CH_4 conversions

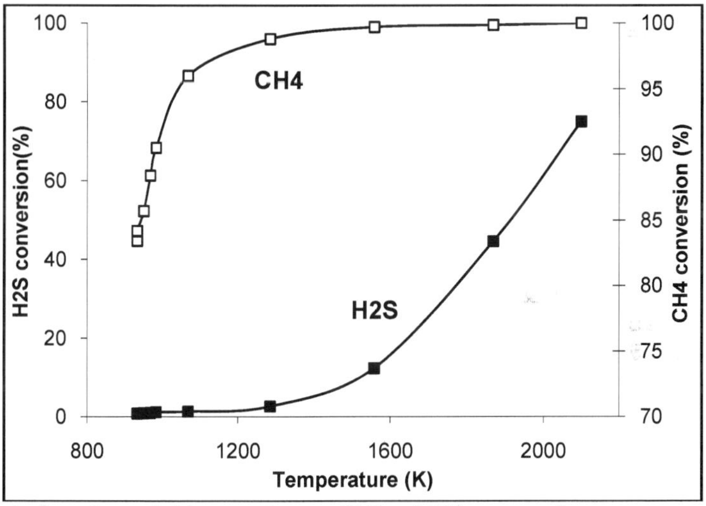

Fig. 8. Effect of reactor outlet temperature on H_2S and CH_4 conversions

Figs. 9 shows the effect of feedstock flow rate on CH_4, carbon black, soot and CO mass fractions at the furnace outlet. It can be seen that for lower values of feed gas flow rates, the very high temperature pre-combustor effluent (see Fig. 6) causes the feedstock methane to convert to CO rather than carbon. For higher values of feedstock flow rate, the formation of carbon black increases, and due to the resulting lower temperatures, the mass fraction of CO and soot decreases. This is so because the soot model strongly depends on the reaction temperature. Fig. 10 shows the effect of feedstock flow rate on H_2 and carbon black yields at the furnace outlet. It can be seen that the yield of H_2 increases with increased feed gas flow rate until it reaches a maximum value, and then drops with further increase in the flow rate. For higher values of feedstock flow rate, the yield of carbon black increases and due to reduction in CH4 conversion (see Fig. 7) the yield of hydrogen decreases.

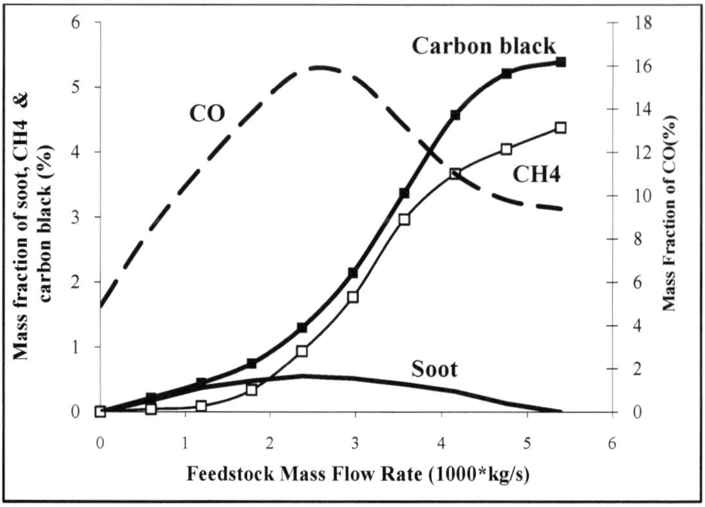

Fig. 9. Effect of feedstock flow rate on CH_4, CO, carbon black, and soot mass fractions

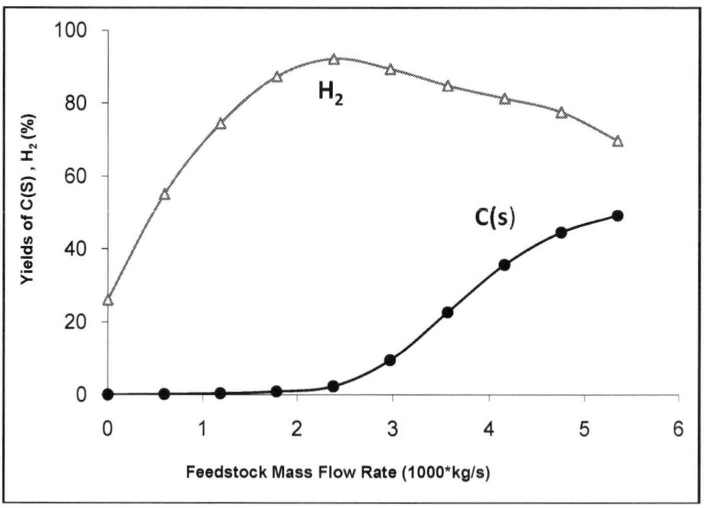

Fig. 10. Effect of feedstock mass flow rate on H_2 and C(S) productions

Fig. 11 depicts the yield of sulfur (due to H_2S decomposition) and SO_2 (due to H_2S combustion) as a function of feedstock flow rate at the outlet of the furnace. S_2 and SO_2 yields are defined as [Huang & T-Raissi, 2008]:

$$S_2(\%) = \frac{2[S_2]}{[H_2S]_0} * 100 \ , \ SO_2(\%) = \frac{[SO_2]}{[H_2S]_0} * 100 .$$

where $[S_2]$ and $[SO_2]$ denote the equilibrium molar concentrations of S_2 and SO_2, respectively. The figure reveals that for low values of feedstock flow rate ($\leq 0.002 kg/s$) that result in high reaction temperatures (see Fig. 3) H_2S converts mostly to S_2 and SO_2. It can be seen that for higher values of feedstock flow rate, yield of S_2 and SO_2 are quite low. This is due to reduced of H_2S conversion (see Fig. 7). Figs. 12 and 13 depict the effects of feedstock mass flow rate and temperature on the yield of COS and CS_2, respectively, as defined by:

$$COS(\%) = \frac{[COS]}{[H_2S]_0} * 100 \ , \ CS_2(\%) = \frac{[CS_2]}{[CH_4]_0} * 100$$

where [COS] and $[CS_2]$ denote the equilibrium molar concentration of COS and CS_2, respectively [Huang & T-Raissi, 2008]. Fig. 12 shows that COS and CS_2 yields increase with increased feedstock flow rate until they reach a peak, and then drop with further increase in feed gas flow rate. Figs. 12 reveals that yield of CS_2 is always low ($\leq 0.0007\%$). This is in accord with results of Huang &T-Raissi (2008) and Towler & Lynn (1996).

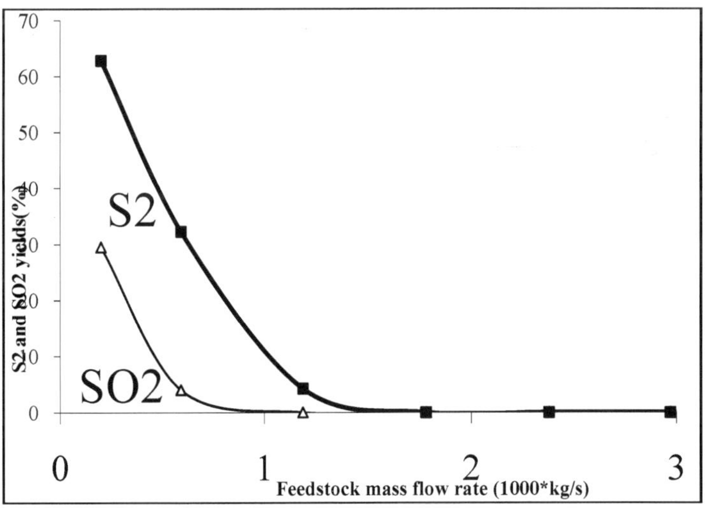

Fig. 11. Effect of feedstock mass flow rate on S_2 and SO_2

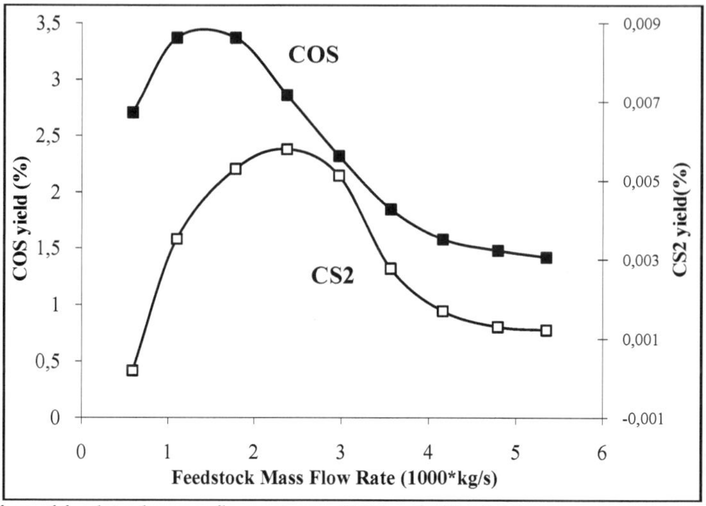

Fig. 12. Effect of feedstock mass flow rate on COS and CS_2 yields

6.2 Type 2: pyrolysis by inert hot gases

In the case of Type 2, the inert hot gas (argon) flow rate introduced into the pre-combustor is 3×10^{-3} kg/s with pressure of 1 bar and the temperature varying from 600° K to 1800° K. The feedstock sour natural gas feed rate is 19×10^{-3} m^3/s with H_2S/CH_4 molar ratio varying from 0 to 4 and the temperature of 300° K. As the reactions of CH_4 and H_2S decomposition are endothermic, in order to avoid the reduction of inert gas temperature some thermal elements have been utilized.

Figs. 13 and 14 illustrate the effects of feedstock H_2S/CH_4 ratio and reaction temperature on CH_4 and H_2S conversions. Results show that the major factor influencing CH_4 and H_2S conversions is temperature. As the temperature increases, CH_4 conversion also increases to reach 100% at temperatures above 1100° K for H_2S/CH_4 feed ratio ranging from 0 to 4. Since the endothermic reaction of CH_4 conversion occurs at lower temperatures compared with H_2S conversion, the reactor temperature is significantly affected by CH_4 molar ratio in feedstock. Based on these results it can be concluded that as the reaction temperature increases up to about 1100° K, methane pyrolysis is the dominant reaction that controls reactor temperature. H_2S decomposition is the limiting step in the H_2S-methane reformation process (Reaction 3) as it proceeds by the sequential reaction involving H_2S pyrolysis to form sulfur diatomic gas (S_2) (Reaction 2), followed by S_2 reaction with CH_4 to produce carbon disulfide (CS_2) and H_2. Fig. 15 depicts that H_2 production increases by increasing molar ratio of CH_4 (decreasing H_2S/CH_4 ratio) in feedstock and temperature. The rate of hydrogen production from pyrolysis of methane is much more than that of hydrogen sulfide. Therefore, any reduction in H_2S/CH_4 feed ratio, for constant feedstock mass flow rate, leads to significant increase in hydrogen production rate.

Figs. 16 and 17 depict the yields of carbon and sulfur as a function of H_2S /CH_4 feed ratio and reaction temperature. Fig. 16 shows that carbon yield increases with increasing temperature until the yield reaches a maximum value (at about 1000°K), and then drops with further increase in the temperature for H_2S/CH_4 ratios more than zero.

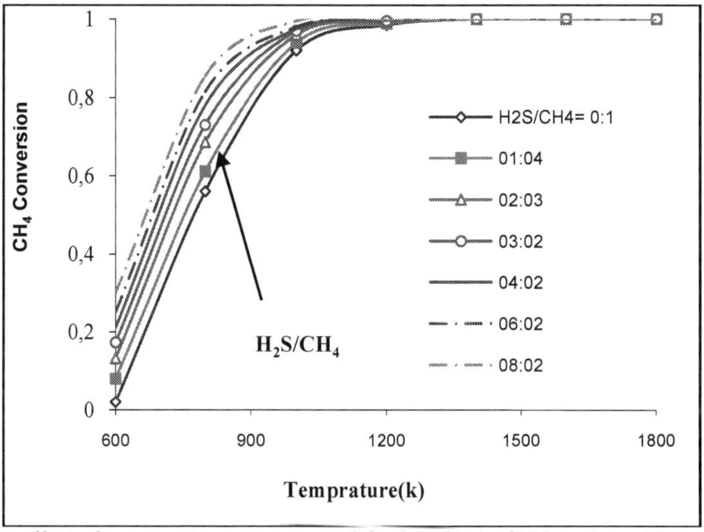

Fig. 13. The effect of reaction temperature and feedstock H_2S/CH_4 ratio on CH_4 conversion for constant feedstock mass flow rate in the case of Type 2

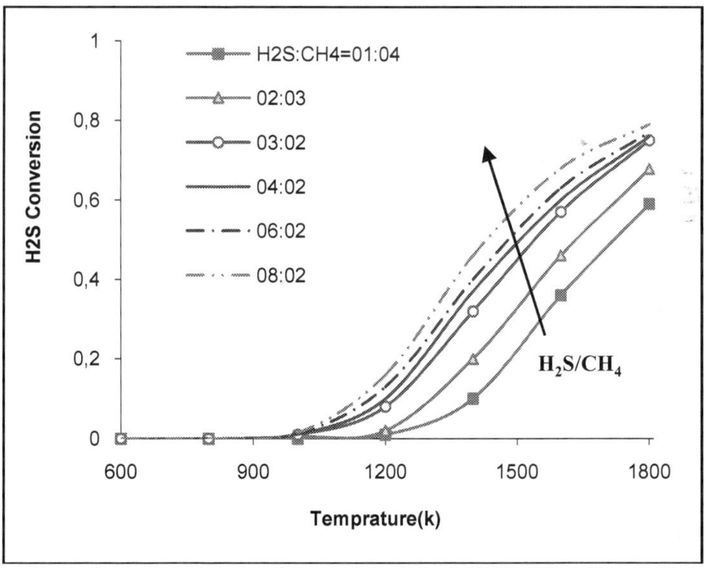

Fig. 14. The effect of reaction temperature and feedstock H_2S/CH_4 ratio on H_2S conversion for constant feedstock mass flow rate in the case of Type 2

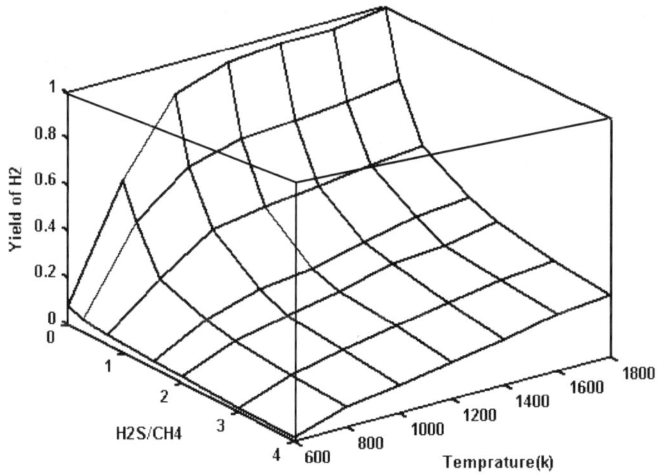

Fig. 15. Yield of H_2 as a function of reaction temperature and H_2S/CH_4 ratio for constant feedstock mass flow rate

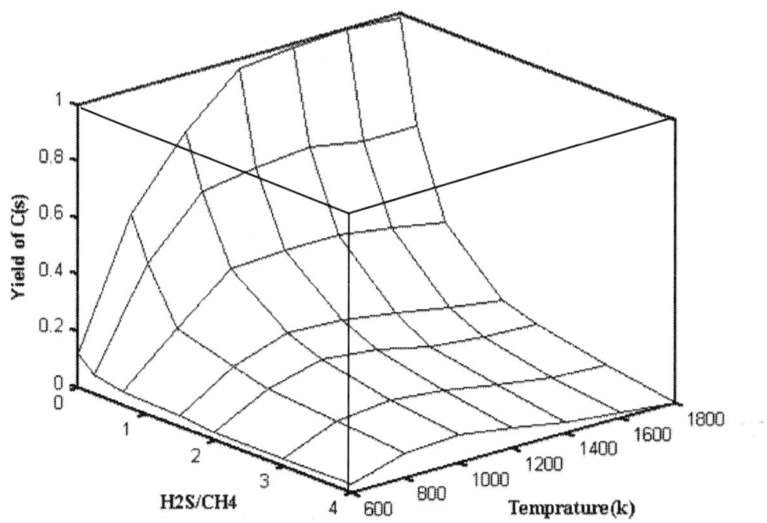

Fig. 16. The effects of temperature and H_2S/CH_4 ratio on C(s) yield for constant feedstock mass flow rate

In temperatures lower than 1000°K, hydrogen and carbon are produced by CH_4 decomposition. But in temperatures more than 1000°K hydrogen is produced from both of methane and hydrogen sulfide, and formed solid carbon from CH_4 decomposition is consumed by CS_2 production (Fig. 18). Fig. 17 depicts that by either increasing H_2S/CH_4 ratio or the reaction temperature, S_2 yield increases. As explained in chemical reactions section, hydrogen sulfide requires high temperatures to be decomposed. Consequently, S_2 yield for temperatures lower than 1000°K is negligible. As shown in Fig .18 , CS_2 yield depend not only on the temperature, but also on the H_2S/CH_4 feed ratio. In temperatures lower than 1000°K, the yield of carbon disulfide (CS_2) approaches zero (Fig. 7). These results are in accordance with results of Huang and T-Raissi (2008).

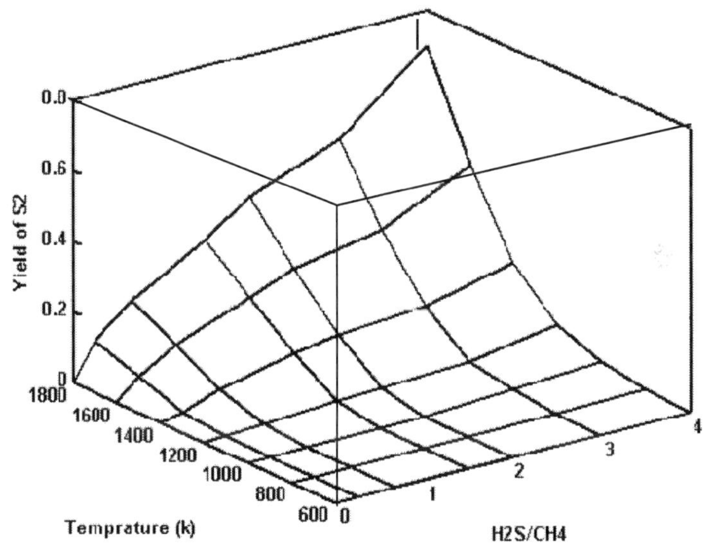

Fig. 17. The effects of reaction temperature and H_2S/CH_4 ratio on S_2 yield for constant feedstock mass flow rate

7. Conclusions

The production of hydrogen and carbon black from sub-quality natural gas containing methane (CH_4) and hydrogen sulfide (H_2S) has been analyzed for two types of supplying heat. Based on the presented results, the following conclusions may be drawn:

- The major factor influencing CH_4 and H_2S conversions is reactor temperature.
- At temperatures above 1100°K, CH_4 conversion is complete.
- At any temperature, H_2S conversion is less than that of CH_4, especially at temperature below 1300°K for which H_2S conversion is less than 5%.
- For lower values of feedstock flow rate, CH_4 is converted to mostly CO and consequently, the yield of carbon black is low. For higher values of feed gas mass flow rates yield of carbon black increases to a maximum value before dropping at much higher feed gas flow rates.

- The yield of hydrogen (in combustion case) increases with increasing feed gas mass flow rate until it peaks and then drops with further increase in the flow rate.
- For temperatures higher than 1200°K, H_2S conversion increases sharply with temperature.
- Since CH_4 conversion occurs at lower temperatures compared with H_2S conversion, the reactor temperature is significantly affected by the amount of CH_4.
- The carbon yield reaches a maximum value about 1000°K and the then drops with further increase in the temperature for sour natural gas (H_2S/CH_4 ratio>0).
- CS2 and COS are minor products in combustion case and can be neglected, while CS2 yield in inert gas case reaches up to 80% as a major product.

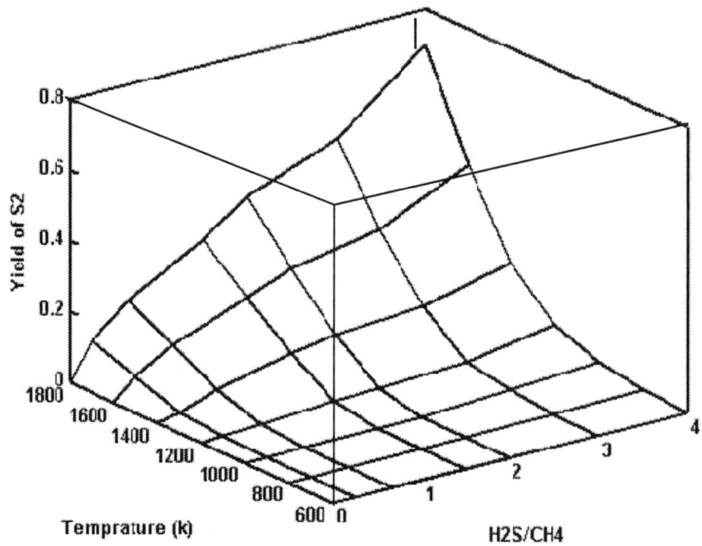

Fig. 18. The effects of reaction temperature and H_2S/CH_4 ratio on CS_2 yield for constant feedstock mass flow rate

8. References

Abanades, S. ; Flamant, G. (2007). Experimental study and modeling of a high-temperature solar chemical reactor for hydrogen production from methane cracking, *International journal of hydrogen energy*, Vol.32, 2007, pp. 1508-1515, ISSN:0360-3199

Abanades, S. & Flamant, G. (2008). Hydrogen production from solar thermal dissociation of methane in a high temperature fluid-wall chemical reactor, *Journal of Chemical engineering and processing* 47, 490–498, ISSN : 0255-2701

Abdel, H. K. ; Shalabi, M. A. ; AL-Harbi D. K. & Hakeem, T. (1998). Non catalytic partial oxidation of sour natural gas , *International journal. of hydrogen energy*, Vol. 23, pp. 457-462, ISSN:0360-3199

Brookes, S. J. and Moss, J. B. (1998). Predictions of soot and thermal radiation properties in confined turbulent jet diffusion flames. Journal of Combustion and Flame, Vol. 116, 1998, pp. 486-503,

Cho, W. ; Lee, S. H. ; Ju, W. S. & Baek, Y. (2004). J. K. Lee, Conversion of natural gas to hydrogen and carbon black by plasma and application of plasma carbon black, *Journal of catalysis today*, Vol. 98, pp. 633–638, ISSN: 0920-5861

Dunker, A. M. ; Kumar, S. & Mulawa, P. A. (2006). Production of hydrogen by thermal decomposition of methane in a fluidized-bed reactor—Effects of catalyst, temperature, and residence time, *International journal of hydrogen energy*, Vol.31, pp. 473-484, ISSN:0360-3199

Gaudernack, B. & Lynum, S. (1998). Hydrogen from natural gas without release of CO_2 to the atmosphere. *International journal of hydrogen energy*, Vol. 23, No. 12, pp. 1087-1093, ISSN:0360-3199

Ghosh, U. (2007). The Role of Black Carbon in Influencing Availability of PAHs in Sediments, *Journal of human and ecological risk assessment*, Vol. 13, pp. 276-285, ISSN: 1080-7039

Gruenberger, T. M.; Moghiman, M.; Bowen, P. J. & Syred, N. (2000). Improving mixing behaviour in new design of carbon black furnace using 3D CFD modeling, *5th European conference on industrial furnaces and boilers*, 972-8034-04-0, Porto, Portugal, April 2000.

Gruenberger, T. M.; Moghiman, M.; Bowen, P. J. & Syred, N. (2002). Dynamic of soot formation by turbulent combustion and thermal decomposition of natural gas, *Journal of Combustion science and technology*, Vol.174, pp.67-86, ISSN:0010-2202

Huang, C. & T-Raissi, A. (2007a). Analyses of one-step liquid hydrogen production from methane and landfill gas, *Journal of power sources*, Vol.173, pp. 950–958, ISSN: 0378-7753

Huang, C. & T-Raissi, A. (2007b). Thermodynamic analyses of hydrogen production from sub-quality natural gas, Part I: Pyrolysis and autothermal pyrolysis, *Journal of power sources*, Vol.163, pp. 645–652, ISSN:0378-7753

Huang, C. & T-Raissi, A. (2007c). Thermodynamic analyses of hydrogen production from sub-quality natural gas, Part II: Steam reforming and autothermal steam reforming, *Journal of power sources*, Vol. 163, pp.637–644, ISSN:0378-7753

Huang, C. & T-Raissi, A. (2008). Liquid hydrogen production via hydrogen sulfide methane reformation, *Journal of power sources*, Vol.175, pp. 464-472, ISSN:0378-7753

Ishihara, T.; Kawahara, A.; Fukunaga, A.; Nishiguchi, H.; Shinkai, H.; Miyaki, M. & Takita, Y. (2002). CH_4 decomposition with a Pd-Ag hydrogen-permeating membrane reactor for hydrogen production at decreased temperature, *Industrial & Engineering Chemistry Research*, Vol. 41, No.14, pp. 3365–3369, ISSN: 0888-5885

Jang, J.S.; Kim, H.G.; Borse, P. H. & Lee, J. S. (2007). Simultaneous hydrogen production and decomposition of H_2S dissolved in alkalinewater over $CdS.TiO_2$ composite photocatalysts under visible light irradiation, *International journal of hydrogen energy*, Vol.32, pp. 4786-4791, ISSN:0360-3199

Javadi, M. Moghiman, M. (2010). Hydrogen and carbon black production from thermal decomposition sour natural gas, *international journal of spray and combustion dynamics, · Vol.2, No.1*, pp. 85–102, ISSN: 1756-8277

Kim, M. H. ; Lee, E. K. ; Jun, J. H. ; Kong, S. J. ; Han, G. Y. ; Lee, B. K. ; Lee, T. J. & Yoon, K. J. (2004). Hydrogen production by catalytic decomposition of methane over activated carbons: kinetic study. *International journal of hydrogen energy,* Vol. 29, No. 2, pp.187–93, ISSN: 0360-3199

Lambert, T. W.; Goodwin, V. M. ; Stefani, D. & Strosher, L.(2006). Hydrogen sulfide (H_2S) and sour gas effects on the eye. A historical perspective, *Journal of science of the total environment,* Vol. 367, pp.1–22, ISSN: 0048-9697

Lockwood, F. C. ; Niekerk, J. E. & Van J. E. (1995). Parametric study of a carbon black oil furnace, *Journal of combustion and flame,* Vol. 103, pp. 76-90, ISSN: 0010-2180

Moghiman, M. & Bashirnezhad, K. (2007). Experimental and numerical studies of carbon black natural gas furnace, *Kuwait journal of science and engineering,* Vol. 34; No. 1B, pp. 167-182, ISSN:1024-8684

Muradov, N. (2001). Catalysis of methane decomposition over elemental carbon. Journal of catalysis communications, Vol. 2, pp.89–94, ISSN: 1566-7367

Murthy, J. Y. & Mathur, S. R. (1998). Radiative heat Transfer in Axisymmetric Geometries using an Unstructured Finite-Volume Method, International journal of numerical heat transfer, Vol.33, No.4, pp.397-416, ISSN: 1040-7790

Petrasch, J. & Steinfeld, A. (2007). Dynamics of a solar thermochemical reactor for steam-reforming of methane, *Journal of chemical engineering science,* Vol. 62, pp. 4214-4228, ISSN: 1307- 6884

Ryu B. H., Lee S. Y., Lee D. H., Han Y., Lee J., Yoon J., Jang, J.S.; Kim, H.G.; Borse, P. H. & Lee, J. S. (2007). Simultaneous hydrogen production and decomposition of H2S dissolved in alkalinewater over CdS.TiO2 composite photocatalysts under visible light irradiation, *International journal of hydrogen energy,* Vol.32, pp. 4786-4791, ISSN: 0360-3199

Sakanishi, K. ; Wu, Z. ; Matsumura, A. & Saito, I.(2005). Simultaneous removal of H_2S and COS using activated carbons and their supported catalysts, *Journal of catalysis today,* Vol.104, pp.94–100, ISSN: 0920-5861

Saario, A. & Rebola, A. (2005). Heavy fuel oil combustion in a cylindrical laboratory furnace: measurements and modeling, *Journal of fuel,* Vol.84, pp.359–369, ISSN: 0016-2361

Steinberg, M. (1998). Production of hydrogen and methanol from natural gas with reduced CO_2 emission. , *International journal of hydrogen energy,* Vol.23, No. 6, pp. 419-425, ISSN: 0360-3199

Towler, G. P. & Lynn, S. (1996). Sulfur recovery with reduced emissions, low capital investment and hydrogen co-production, *Journal of chemical engineering communications,* Vol. 155, 1996, pp. 113-143, ISSN: 0098-6445

T-Raissi, A. (2003). Analysis of Solar Thermochemical water-splitting cycles for hydrogen production, Hydrogen, Fuel Cells, and Infrastructure Technologies, FY 2003 Progress Report Available at : www.fsec.ucf.edu/en/research/hydrogen

Warnatz, J.; Maas, U. & Dibble, R.W., *Combustion,* Springer, ISBN: 3540259929 , Berlin (2006).

9

Soil gas geochemistry: significance and application in geological prospectings

Nunzia Voltattorni[1] and Salvatore Lombardi[2]
[1]*INGV - Rome, Italy*
[2]*Earth Science Dept. - University "La Sapienza" - Rome, Italy*

1. Introduction

Gas geochemistry has been proven to be a reliable and simple technique to apply, at different scales, to many geological scenarios (Annunziatellis et al., 2003; Lewicki et al., 2003; Baubron et al., 2002; De Gregorio et al., 2002; Ciotoli et al., 1998; Ciotoli et al., 1999; Lombardi et al., 1996; Hickman et al., 1995; Duddridge et al., 1991; Durrance and Gregory, 1988; Eremeev et al., 1973). The importance of fluid geochemistry is rooted in the fact that the Earth is an open system and that fluid-releasing crustal phenomena are the major means for the exchange of matter and energy at different depths. As such, fluid-releasing channels like active faults and fractures are actually a "window" on subterranean physical and chemical variations (Ciotoli et al., 2007).

The study of spatial distribution of soil gas anomalies at the surface, can give important and interesting information on the origin and processes involving deep and superficial gas species. This information can be applied and studied in different frameworks, for example:

I) seismic zonation, examining, at the surface, anomalous concentrations of deep gas species that upraise throughout preferential pathways (faults and/or fractures). Soil gas distributions can be directly linked to the evolution of the stress regime and gases migrate preferentially through fractured zones but only along pathways whose permeability has been enhanced by seismic activity and/or through areas of brittle deformation.

II) Environmental protection, such as the monitoring of naturally occurring toxic gases to highlight zones with high health risks for humans. The presence of magmatic chambers can cause an accumulation of gases in the subsurface and local structural features can favour high degassing phenomena. These events are particularly dangerous in populated areas and it is necessary to build risk maps to define the potential health hazard in terms of both short-term and long-term risk.

III) Radionuclide migration, both in the pollution assessment from abandoned uranium mines and in the study of high-level radioactive-waste isolation systems. The main approach is to study the natural migration of radiogenic particles or elements throughout clay formations that are considered an excellent isolation and sealing material due to their ability to immobilize water and other substance over geological timescales. The evaluation of long-term behaviour of clays under normal and extreme conditions is still the main topic in questions relating the role of clays as geological barrier for the permanent isolation of long-lived toxic residues.

Soil gas distribution would be affected by surface features such as pedological, biogenic and meteorological factors: these are supposed to have only a subordinate effect on gas leakage (Hinkle, 1994). However, it is possible to properly interpret soil gas anomalies and recognize influences of surface features studying the association of different gases (having different origin and physical/chemical behaviour), collecting a large number of samples during periods of stable meteorological and soil moisture conditions (e.g., during dry season) and using appropriate statistical treatment of data (experimental variograms to investigate the spatial dependency of gas concentrations).

Soil gas geochemistry involves the study of many gaseous species (radiogenic, trace and diagenetic gases); each of them can give specific information on the conditions that allow their formation, accumulation and/or migration.

Field data can show the usefulness of the soil gas method for detecting, for instance, crustal discontinuities even when faults are buried or cut non-cohesive clastic rocks which makes surface recognition difficult using traditional field methods (Ciotoli et al., 1998; Lombardi et al., 1996; Duddridge et al., 1991; Durrance & Gregory, 1988). These characteristics as well as the rapidity and the low cost of the soil gas survey, make this method a powerful tool for geological investigation which can significantly contribute to hazard assessment and forecasting, especially when continuous monitoring is performed (Klusman, 1993; Reimer, 1990; King et al., 1996; Sugisaki, 1983).

In this chapter, we outline the results from two soil gases: radon, a radiogenic trace gas, and carbon dioxide, which generally acts as carrier for trace gases. We will show data obtained in either prospecting or monitoring case studies.

2. Radon and Carbon Dioxide origin and behaviour

Radon (^{222}Rn) is a rare gas and is probably the gas used the most frequently for mapping and predicting purposes. ^{222}Rn is a naturally occurring radioactive daughter product of the uranium decay chain, with a short half-life (3.8 days). In the geologic environment, it displays a poor intrinsic mobility (Tanner, 1964; Dubois et al., 1995). In diffusive systems, due to its low mobility and its short half-life, radon obviously comes from a short distance below the measuring instrument. Information of a deep origin, however, is expected to be noticed when Rn of a subsurficial origin is extracted by a rising gas/water column. In this latter case, radon being incorporated in the fluid during the last steps of the process, can be used as a tracer, acting as a relative flow meter and velocity meter of the bulk fluid. It gives therefore information about both the steady state conditions and disequilibrium features of a global reservoir, which can be a hydrothermal cell, possibly magma-generated (Pinault & Baubron, 1996). Soil radon activities analyzed in surface conditions depend upon the following main factors: the emanating power of the rock and soil (Morawska & Phillips, 1993), the permeability of the host rock and the flow of the carrying gas (Ball et al., 1991). Generally, radon activities increase with increasing flows (because the gas velocity increases, causing both less time for decay and more extraction). For higher flows, however, dilution of radon by the flux may occur with a subsequent decrease of radon activities measured at the surface.

All these features allow radon to be used as a tool for mapping and determining characteristics of hydrothermal systems (D'Amore et al., 1978; Cox, 1980; Etiope & Lombardi, 1995), for fault detection in volcanic terrains (Crenshaw et al., 1982; Aubert & Baubron, 1988; Baubron et al., 1991), for uranium exploration (Fleischer et al., 1972; Klusman, 1993; Wattananikorn et al., 1995; Charlet et al., 1995) and for groundwater flow characterization (Gascoyne et al., 1993). ^{222}Rn monitoring has long been used for both earthquake (King, 1978; Fleischer & Magro-Campero, 1985; Segovia et al., 1989; Shapiro et al., 1989; Woith et al., 1991) and volcanic prediction purposes (Cox et al., 1980; Del Pezzo et al., 1981; Thomas et al., 1986; Thomas, 1988; Toutain et al., 1992).
Carbon dioxide (CO_2) is the most abundant gas species in hydrothermal to volcanic environments. Kerrick et al. (1995) calculated that non-volcanic CO_2 emissions from high heat flow areas may substantially contribute to the balance of the carbon cycle. Natural discharges of CO_2 have several sources: the mantle, metamorphism of carbonate-bearing rocks, decomposition of organic material and surface biological activity (Irwin & Barnes, 1980). Generally, carbon dioxide in fault zones is a mixture of some of these sources (Sugisaki, 1983). High CO_2 fluxes appear correlated with both high heat flux areas (associated with active and ancient volcanism) and limited areas with deep fracturing (emitting carbon originated from the mantle and from decarbonation processes, with possible mixing of these two sources). Irwin & Barnes (1980) suggested that discharges of CO_2 might indicate areas with high pore pressure at depth, and therefore may serve to identify potential seismic regions. CO_2 is used for fault mapping (Irwin & Barnes, 1980; (Sugisaki et al., 1980; Sugisaki, 1983; Baubron et al., 1990, 1991) as well as for both seismic and volcanic monitoring (Shapiro et al., 1982; Toutain et al., 1992; Rahn et al., 1996).

3. Sampling and analytical procedures

Soil gas surveys can be performed at both regional (e.g., sampling grid: 1 sample/km^2) and local scale (detailed sampling grid including profiles and/or transects) on the basis of the goal of the research. The surveys should be performed during summer or dry periods to avoid climatic factors which may affect soil gas values (Hinkle, 1994).
Shallow soil gas samples are obtained using a 1 m stainless steel probe fitted with a brass valve: this system enables soil gas to be collected and stored in metallic containers (with a vacuum 10^{-2} atm) for laboratory analysis or to be pumped for on-site Rn analysis.
Radon determination is accomplished in the field with an EDA Instrument RDA-200 Radon Detector.
Generally, the studied gases include major (N_2, O_2, CO_2) and trace (^4He, H_2) gases and light hydrocarbons (C_1 to C_4). The determination of helium is performed with a Varian Instrument Mass 4 spectrometer. N_2, O_2, CO_2 and light hydrocarbons concentrations are analyzed using a Fison Instrument GC-8000 Series gas-chromatograph. The used detectors are: Thermal Conductivity Detector (TCD) for N_2, O_2 and CO_2 in order to achieve sensitivity up to percentage and Flame Ionization Detector (FID) for light hydrocarbons with a sensitivity of an order of 0.2 ppm.
A specific technique has been developed to collect submarine samples (Caramanna et al., 2005) in proximity of gas vents. In order to collect free/dry gas samples, a plastic funnel is inverted (30 cm diameter with 12 kg ballast around the lower ring) and placed precisely on the gas vent to be sampled. All of the samplers are stored in a plastic box that is carried

underwater by the divers. The funnel is connected, through a silicon hose, to a Pyrex glass flask with twin valves. This flask is pre-filled with air at a pressure above that of the hydrostatic pressure expected at the sampling depth in order to stop seawater from entering the sampler. Afterwards, collected gas samples are analysed at the laboratory.

4. Results

Results from different geological scenarios are presented highlighting the usefulness of the method in geochemical exploration. In particular, three different topics (seismic zonation, toxic emanation and radionuclide migration) will be treated showing two examples of studies for each of them in order to give a general idea of the soil gas geochemistry applications.

4.1 Seismic zonation

Earthquakes constitute a severe source of human disasters all around the world. Consequently, short-term considerations, through the search for precursory signals, have received great attention in the last several decades. Among the techniques used for the search of precursors, geochemistry has provided some high-quality signals, since the 1960's, mainly as the result of instrumental developments. Focusing interest on geochemical anomalies linked to seismo-tectonic activity is not an unexpected development, owing to the multiple evidence of a genetic link existing between fluid flows and faulting processes (Hickman et al., 1995).

The Colpasquale area is located in the central Italian region of Marche that was devastated by a sequence of shallow earthquakes over a three month-long period (September-December, 1997). The occurrence of this catastrophic event as well as the long duration of the "seismic sequence", presented a unique opportunity to apply a study of gas migration to a zone undergoing active displacement. Soil gas surveys were performed one day, one week, one year and two years after the main shock (Ms 5.6) in this area (Lombardi & Voltattorni, 2010).

Figure 1 shows box plots used to display and compare the distribution characteristics of the Rn and CO_2 soil gases during the different surveys. The graphs indicate that the median values of radon activity (Fig. 1a) are not temporally constant, thus displaying the complex character of this gas whose leakage can vary as a function of many factors including, in particular, the variation in the stress regime. The highest median value (50.8 Bq/l) occurred during the second 1997 survey and increases again in 1999 after a very low 1998 median value. The great variability of mean values during years suggest that gas microseeps can be influenced by many factors, such as fault permeability, fracture width, an increase/decrease of grain surface and porosity, as well as grain comminution by coseismic cracks that produce active surface area and circulation pathways (Holub & Brady, 1981). Median value of CO_2 (Fig. 1b) is quite constant during the first three surveys while doubles during the last campaign.

The use of variogram surface maps has the potential to define phenomena affecting gas distribution along a specific direction (fault-related anisotropy effect). Figure 2 shows the radon variogram surface maps. During the first two surveys the distributions are isotropic, suggesting that there is no preferential direction along which the degassing phenomena occur. A slight anisotropy is evident during the third campaign (Fig. 2c), but only in the last survey is it possible to observe the maximum spatial data continuity (Fig. 2d) along a major NW-SE anisotropy axis. This orientation parallels the direction of the Apennine belt along which the main faults and earthquakes are distributed (Cocco et al., 2000).

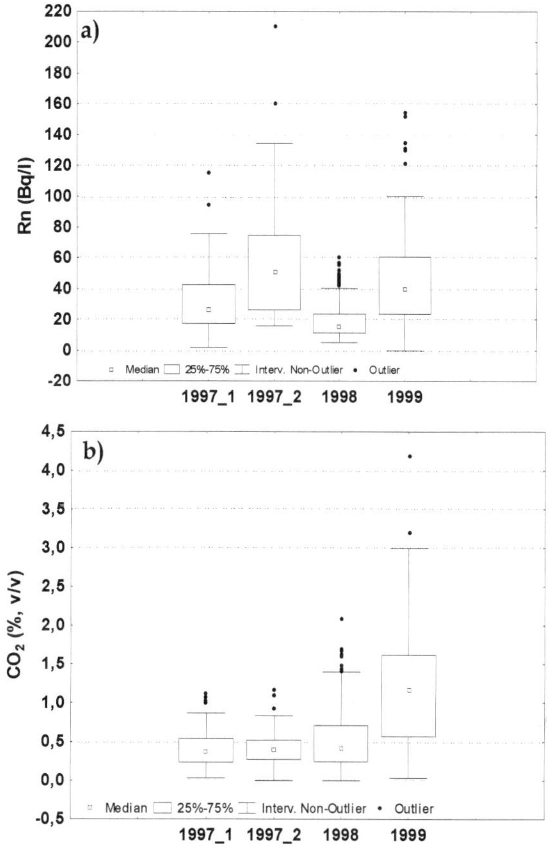

Fig. 1. Box plots of the different soil gas species (Rn and CO_2) during the four surveys at the Colpasquale area (Marche region). The plot a) highlights that Rn concentrations are not temporally constant, but rather the leakages of this gas can vary as a function of many factors, such as the variation of local stress regime.

These results suggest an "evolution" of the radon distribution, starting with an initial radon leak immediately after the first seismic event. This initial phase could have been caused by sudden degassing soon after the earthquake as the result of the opening of numerous fractures, resulting in widespread anomalies and the basic "flooding" of the local soil gas with radon. Once the earthquakes ceased some of the structures began to close and become less permeable to gas flow, allowing the system to slowly return to a state of equilibrium by dissipating the high radon concentrations into the atmosphere. However, where fractures remained open at depth, such as within inferred faults, they provided a steady but reduced flux of radon to the surface.

The CO_2 variogram surface maps (Fig. 3) also show spatial variability over the three years. The anisotropy seems to rotate clockwise from a mostly E-W direction in the first survey to a NE-SW direction in September 1999. The variogram maps imply that the shallow gas distribution may be linked to the variation of the stress regime, as the final NE-SW anisotropy is in accordance with the process of an extensional stress regime of the region (Amato et al, 1998).

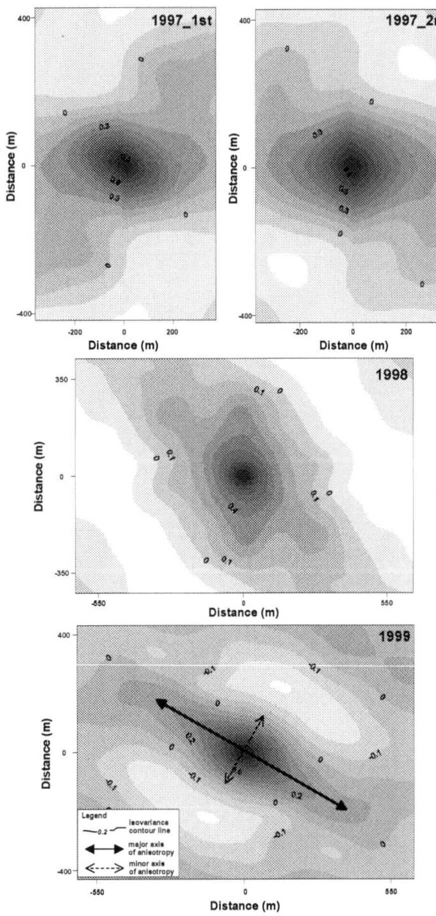

Fig. 2. Variogram surface maps of radon data from the Colpasquale area (Marche region). The first two surveys show an isotropic distribution, meaning that there is no evident preferential anomaly orientation. A slight anisotropy is evident during the third campaign, but only in the last survey it is possible to observe the maximum spatial data continuity with a major anisotropy axis oriented NW-SE.

The Fucino Basin (central Italy) is another studied area characterized by known and inferred structural discontinuities. The Fucino basin (about 250 square kilometers) is an intramontane tectonic depression located in peninsular Italy within the Apennine chain. The basin is affected by a complex fault network (Nijman, 1971; Giraudi, 1989; Blumetti et al., 1993; Galadini & Messina, 1994) due to an intense Quaternary activity whose most evident geomorphic expression are high mountain fronts. The geometry of these faults and the kinematic indicators, mainly normal or oblique slip, confirm that extensional tectonics has been mainly responsible of the evolution of the basin (Blumetti et al., 1988, 1993; Galadini & Messina, 1994)

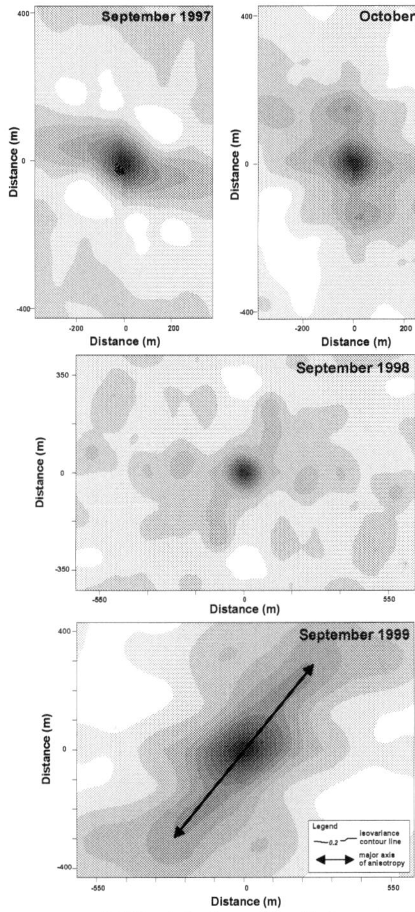

Fig. 3. Variogram surface maps of CO_2 data from the Colpasquale area (Marche region). The maps show a variability of spatial data continuity over the three years. The anisotropy seems to rotate clockwise from a mostly E-W direction at the first survey to a NE-SW direction in September 1999.

Soil gas surveys were performed in two stages: a regional sampling was carried out over the Fucino plain with a density of 4 -6 samples/km^2 (548 samples) and three transects were carried out at a more detailed scale (sample density of 10-20 samples/km^2 for transects 1 and 2, and 80-100 samples/km^2 for transect 3), crossing evident or inferred structural features in order to achieve a better definition of soil gas distribution.

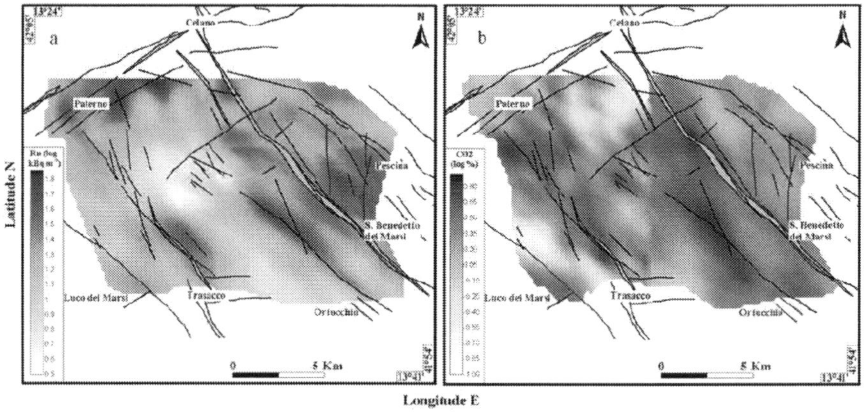

Fig. 4. Contour maps of (a) Rn and (b) CO_2 in the Fucino Plain. Radon and carbon dioxide anomalies well fit with the trend of known structural features both in the eastern and western sectors.

Results obtained from the regional survey show that radon and carbon dioxide anomalies (Figure 4) fit well with the trend of known structural features both in the eastern and western sectors, mimicking the general NW-SE fault orientation of the area.
Higher Rn and CO_2 concentrations characterize the eastern sector of the Fucino Plain where the most active/recently activated faults (OF and SBGMF) occur. The shape of the anomalies confirms the highly anisotropic behaviour and shows the spatial domain of the faults affecting the radon distribution. In the western sector, radon distributions define the southern and the northern segments of the known Trasacco Fault (TF), in agreement with structural observations (Ciotoli et al., 1998). It is quite clear that the anomalies are smeared laterally, confirming the rotation of the anisotropy axis toward the E-W direction. In this sector this phenomenon is probably linked to a more complex fault geometry (wide fracturing) associated with a shallow structural high of carbonate substratum. Furthermore, in the north-western sector of the plain very high radon anomalies clearly define the WSW-ENE ACF, as well as some associated minor Apenninic faults oriented toward the centre of the plain. The CO_2 contour map shows low values with respect to the eastern sector, but an anomalous distribution similar to that of radon. The anomalies are located in correspondence with the TF and with the minor Apenninic faults in the northwester sector of the plain. Minor CO_2 anomalies are located in the centre of the plain where fault-induced liquefaction was recognized during the 1915 Avezzano earthquake, and along a WSW-ENE buried fault.
The results from the three transects yielded anomalies with different features, reflecting the different gas-bearing properties of the eastern seismogenic faults related to the 1915

earthquake (Mb = 7.0) and the hidden structural features occurring in the western side of the plain (Ciotoli et al., 2007). All the achieved results show that gases migrate preferentially through zones of brittle deformation, by advective processes as suggested by the relatively high rate of migration needed to obtain anomalies of short-lived ^{222}Rn in the soil pores.

4.2 Toxic emanation

The sudden and catastrophic, or slow and continuous, release at surface of naturally occurring toxic gases like CO_2 and Rn poses a serious health risk to people living in geologically active regions. In general this problem receives little attention from local governments, although public concern is raised periodically when anomalous toxic gas concentrations suddenly kill humans or livestock.

An area in proximity of Panarea Island (Aeolian Islands, southern Italy) was interested by a huge submarine volcanic-hydrothermal gas burst during November, 2002. The submarine gas emissions chemically modified seawater causing a strong modification of the marine ecosystem and the death of mainly benthonic life forms and serious damage to the sea-grass Posidonia oceanic (Voltattorni et al., 2009).

Gases have been collected from the seafloor at variable depths (depending on gas emission point depth) and the chemical compositions of the submarine gas emissions have displayed a complex combination of temporal and spatial variability from November 2002 to December 2006. The temperatures of leaking fluids are variable at the different gas emission points (Fig. 5): highest temperature measurements refer to Black point ranging between 110 °C and 137 °C, excepting the first measurement (a few days after the gas burst, on November, 13, 2002) being 30 °C. Lower temperatures (mean value: 86.7 °C) have been measured at the Sink point and at Vent 8 (mean value: 52.11 °C). According to Capaccioni et al. (2007), a possible explanation for the temperature variability at the different gas emission points is related to an inferred magmatic system centred on or closer to Black point and whose diameter probably does not exceed a few hundred meters. Vents 1 and 2 have recorded the lowest values (mean values, respectively: 39.75 and 33.8 °C). Being located at the margins of the inferred magmatic system, the latter vents have probably been affected by thermal cooling in the declining stage, because of a rapid inflow of cold seawater from the surroundings.

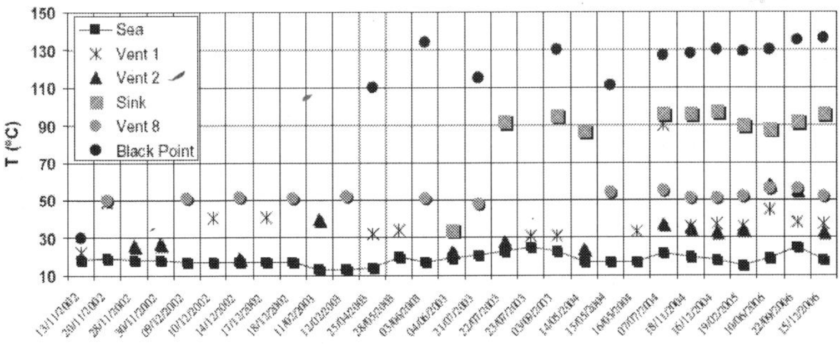

Fig. 5. *T* measurements at Panarea vents. Fluids from vents are very hot (especially from Sink and Black point with temperatures >90 °C) but due to their low flow, they do not affect temperatures of the surrounding seawater.

All of the collected gases are CO_2-dominant (the content varies from a minimum of 83.64 vol. % to a maximum of 98.43 vol. %). Fig. 6 shows a comparison of the CO_2 values from the five monitored vents through a statistical distribution (box plots). The CO_2 leakage varies at the different vents being higher at the Black point and lowest at the Sink point. However, median values are very similar for each vent suggesting a common degassing input linked to local tectonic features. In fact, all the gas emission points are located along N-S, E-W and NE-SW oriented active faults controlling the Aeolian Volcanic District. The main consequence of the presence of high levels of CO_2 in the water chemistry is a generic acidification of the sea with a reduction in pH. This phenomenon affected both the macro and the micro biota. Regarding the macro life-forms in particular, extensive damage to the benthic life-forms was observed; this damage was mainly to the calcareous-shell organisms. Even though the damage to the benthic life-forms seems to be permanent, there is a general healing of the ecosystem with the return of some species of fish. Another organism that was seriously affected by the presence of carbon dioxide is the "Posidonia oceanica" sea-grass. Once the Posidonia was dead, the available substratum was colonized by other species such as more resistant algae. Of the studied micro life-forms, the viral abundance was affected by the presence of the gas vents with a decrease close to the carbon dioxide plumes. From these results it is possible to hypothesize that viruses can be less tolerant than prokaryotes to the carbon dioxide chemistry and this can have consequences on the biota equilibrium in the areas affected by increased levels of CO_2 (Manini et al., 2008).

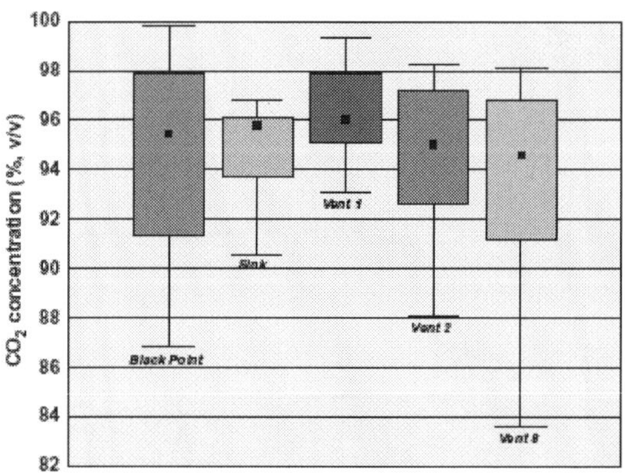

Fig. 6. Box plots of soil gas CO_2 data from the Panarea vents. The median values are very similar for each vent suggesting a common degassing input linked to local tectonics.

Another example of toxic emanation study was performed in the Albani Hills area (a volcano located about 20 km southeast of Rome and extending over an area of about 1500 km) where strong areally diffuse and localised spot degassing processes occur (Annunziatellis et al., 2003). The main structural features which cause the high degassing phenomena are buried highs in the carbonate basement which act as gas traps.

Data were processed in order to build risk maps and highlight areas having a potential health hazard in terms of the short-term risk caused by elevated CO_2 concentrations and the long-term risk caused by high radon concentrations (Beaubien et al., 2003).

Figs. 7 and 8 show the contour maps of radon and carbon dioxide concentrations in soil gas calculated using the kriging method and spherical variograms model estimation. In the surveyed area, the distribution of anomalous radon values (>60 kBq/m^3) shows a maximum anisotropy orientation (N340°–350°), which parallels that of the Apennine mountains. This can be seen both in the western and the eastern sectors along the Appia road (where aligned effervescent water springs occur). Point anomalies occur around the Consorzio Vigna Fiorita (from 75 to 250 kBq/m^3, 1.8–2.4 in log scale), as well as near the village of Cava dei Selci (>100 kBq/m^3) where the major gas release occurs. Background values (i.e. in situ production) occur in the central sector of the area.

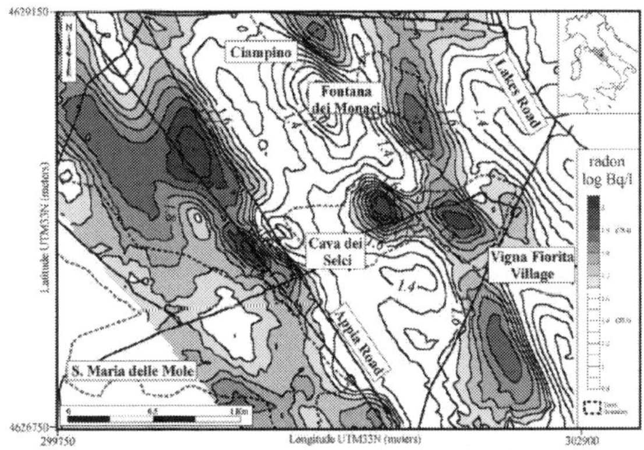

Fig. 7. Map of the radon distribution in soil gas. The radon anomalous values (>60 Bq/l, 1.7 in log scale) shows clear linear trends parallel to the Apennine mountains. The anomalies are located in the western sector where an alignment of sparkling water springs also occur, and in the eastern sector.

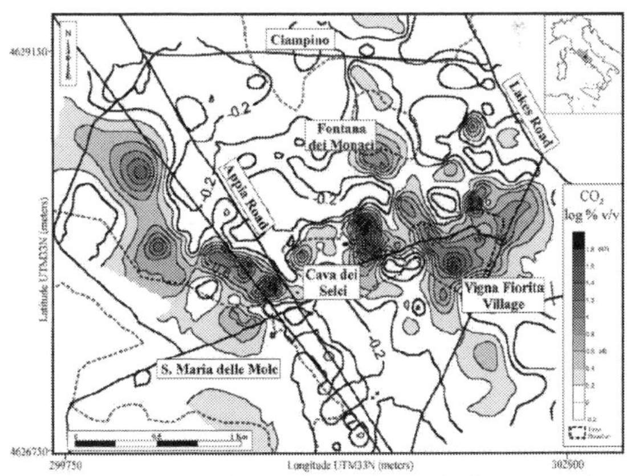

Fig. 8. Map of the carbon dioxide distribution in soil gas. Carbon dioxide concentrations also show a mild anisotropy along a NW–SE major axis, similar to that of radon. Most of the anomalous concentrations (up to 80%, 1.9 in log scale) occur as spots in the eastern sector.

The distribution of radon anomalies in the Ciampino–Marino districts marks the presence of high permeability channels (faults and fractures) along which, due to the action of a carrier gas (such as CO_2), the short-lived Rn is able to migrate quickly and produce soil gas anomalies. Furthermore, the orientation of the anomalies accords with the trend of known structural features, mimicking the general NW–SE trend of the Ciampino high (Di Filippo & Toro, 1995). The anomalies are spatially continuous along the major NW–SE axis, and their width of about 1 km emphasises the spatial domain of the faults which border the Ciampino high structure.

The soil gas CO_2 results (Fig. 8) show a pattern that is similar to that in the radon contour map. Most of the anomalous concentrations (up to 80%, 1.9 in log scale) occur as spots in the eastern sector (Cava dei Selci area and the urbanised area of the Consorzio Vigna Fiorita). The high CO_2 levels in the ground are therefore probably associated with a low enthalpy geothermal system, either metamorphic reactions involving the carbonate substratum or magma degassing, corresponding to faults associated with the Ciampino high.

Generally, the high radon concentration in soils causes high radon concentration indoor: as reported in the literature (Reimer & Gundersen, 1989), indoor radon and soil gas radon show a linear correlation coefficient of 0.77. For this reason, indoor radon measurements (30 samples) were made, using a Genitron Instruments AlphaGuard Radon monitor in random selected private and public dwellings and cellars located in the surveyed area (Cava dei Selci and S. Maria delle Mole villages). Fig. 9 shows a comparison between mean indoor radon values calculated for cellars, ground and first floors and soil gas concentrations. The mean values calculated for the three monitored levels highlight the expected trend, in which cellars show the highest values (in certain sites, measured indoor radon values are extremely high up to 25 kBq/m^3). It is worth noting that the mean soil gas concentration corresponding to the cellar measurements is not the highest. This confirms that enclosed spaces in contact with the ground are more affected by radon and/or toxic gas accumulations.

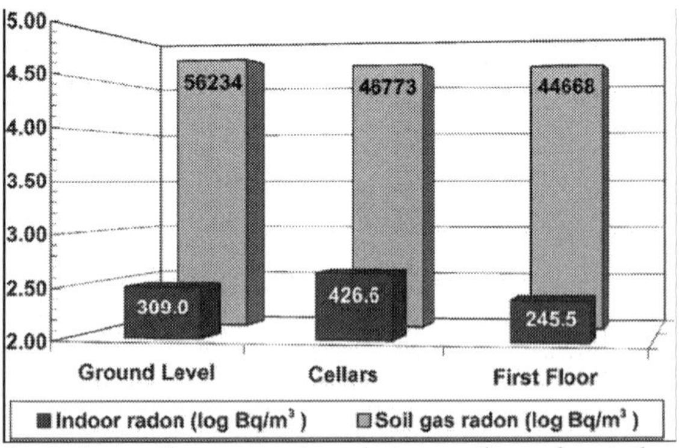

Fig. 9. The bar chart shows the comparison between the radon indoor mean values at different levels (cellars, ground levels and first floor) with the radon concentrations measured in the soil gas samples at the same sites. Numbers in the bars indicate the radon values in Bq/m^3. The figure highlights that cellars show the highest radon values (up to 25,000 Bq/m^3).

4.3 Radionuclide migration

Two different examples of the study of radionuclide migration will be discussed. The first one regards the study of soil gas distributions in clays altered by heating, based on findings at Orciatico site of natural analogue of nuclear waste disposal. The second example is related to the presence of an abandoned uranium mine in proximity of the main natural water resource of Kyrgyzstan (central Asia).

The physical properties of thermally altered clays of the Orciatico area (Tuscany, Central Italy) were studied as argillaceous formations could act as geological barriers to radionuclide migration in high-level radioactive-waste isolation systems. Though available data do not allow exact evaluations of depth, many features of the Orciatico igneous body (widespread glass, highly vesicular peripheral facies etc.) point to a shallow emplacement, comparable with that reasonably forecast for a repository. Not even exact definitions of the temperature of magma at the moment of emplacement are feasible. Only some evaluations can be proposed: from its distinctly femic composition temperatures over 800 °C may be assumed for the alkalitrachytic magma intrusion (Leoni et al., 1984; Hueckel & Pellegrini, 2002). These values are much higher than those expected around a radiowaste container (up to 300°C, according to Dayal & Wilke, 1982); therefore, as to the thermal aspects the Orciatico magmatic body and its metamorphic aureole must be regarded as an extreme condition model of a radiowaste repository and probably it can be mainly used to demonstrate a worst case. The study was performed through detailed soil gas surveys in order to define the gas permeability of the clay unit (Voltattorni et al., 2010). A total of 1086 soil gas samples was collected in the Orciatico area. A first survey was performed collecting 486 samples along a regular grid near the village of Orciatico with a sampling density of about 500 samples/km^2. After that, monthly surveys (from April to September 1998) were performed to monitor possible variations of soil gas concentration due to weather conditions.

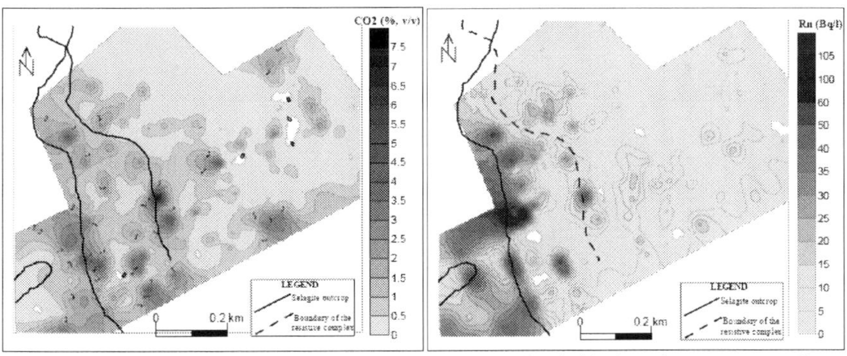

Fig. 10. Carbon dioxide (to the right) and radon (to the left) distributions in soil gases. Anomalous values (CO_2 >2 %,v/v, Rn >25 Bq/l) are in correspondence of the boundary of the resistive complex supposed on geoelectrical results.

The radon, as well as the CO_2 contour line maps, figure 10, show that highest values (^{222}Rn> 25 Bq/l, CO_2>2 %,v/v) occur in the south-western part of the studied area (characterized by the presence of the igneous body outcrop named Selagite) and along a narrow belt, with direction NNW-SSE, where metamorphosed clays (named Termantite) are present.

Furthermore, anomalous values occur in unaltered clays especially in correspondence of the boundary of the resistive complex supposed on previous geoelectrical results (Voltattorni et al., 2010). All over the north-eastern sector, in non metamorphosed clays, radon and carbon dioxide values are very similar to background values reported in literature (Rn: 10-15 Bq/ l, CO_2: 0.5 %,v/v).

As radon and carbon dioxide values seem to decrease gradually from Selagite outcrop towards un-metamorphosed clays, soil gas data set were projected along one longitudinal lines coinciding with a performed geoelectrical profile. Figure 11 shows polynomial regression (3rd degree) of radon and carbon dioxide values plotted against the distance from a reference point. Graphs highlight a slight decreasing trend of radon soil gas values (continuous line) towards the NE, from Selagite outcrop until un-metamorphosed clays.

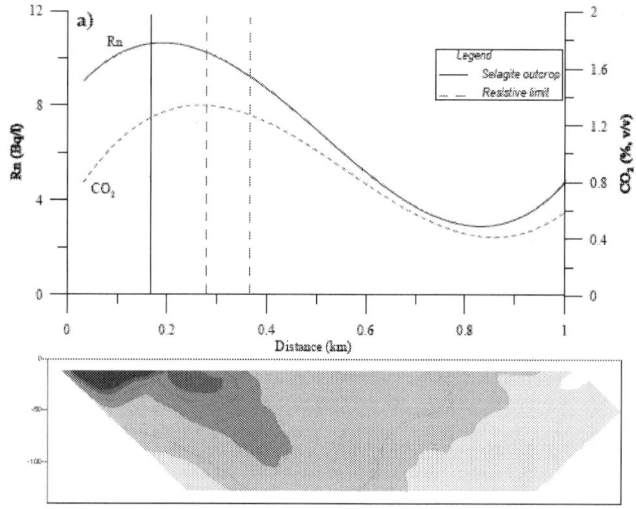

Fig. 11. Comparison between polynomial regression (3° degree) map and geoelectrical profile. Radon graph (continuous line) highlights a general slightly decreasing trend of soil gas values towards the NE, from Selagite outcrop until un-metamorphosed clays. The same behaviour is well evident also for CO_2 polynomial regression (dashed line). Values slightly rise towards un-metamorphosed clays, indicating the presence of structural discontinuities not visible at the surface.

The same behaviour is well evident also for CO_2 polynomial regression (dashed line): the overlapping peaks in the radon-carbon dioxide plots should confirm that the soil gas distribution is linked to clay alteration degree. In fact, highest CO_2 and Rn values were found between Selagite outcrop and the first resistive limit, in a narrow belt characterized by a high alteration degree and, probably, by an intense shallow fracturing (Gregory & Durrance, 1985). On the other hand, after the second resistive limit, where clays did not undergo the effects of the intrusive body, radon and carbon dioxide values are in agreement with the mean values reported in literature excepting in the last 200m of the profile where values slightly increase again.

The results of this study provided specific information about soil gas permeability on the Orciatico clay units characterized by different degrees of thermal alteration. This research represents the first study performed in thermally and mechanically altered clays and results demonstrated that the method gives interesting information also in clays that apparently did not undergo to mineral and geotechnical variations. Radon and carbon dioxide soil gas anomalies are mostly concentrated in zones where the Selagite and thermally altered clays are present. Soil gas distributions are interpreted as being due to intense shallow fracturing of clays along the inferred Selagite boundary: the volcanic intrusion caused thermo-hydro-chemical and thermo-hydro-mechanical stress and contact metamorphism in the clay. Far from Selagite, clays apparently prevent the rising of gases. In fact, small soil gas anomalies were found over the estimated intact Pliocenic clays having permeability due to structural discontinuities not visible at the surface. This study allowed to highlight the role of soil gas technique for the identification of secondary permeability in a clay sequence: clay can strongly modify its characteristics (i.e., reduction of the properties of isolation and sealing material) when affected by even very low thermal alteration although this effect is not visible through traditional investigative methods. The results of this study suggest a review of the role of clays as geological barrier for the permanent isolation of long-lived toxic residues in the radioactive-waste isolation framework.

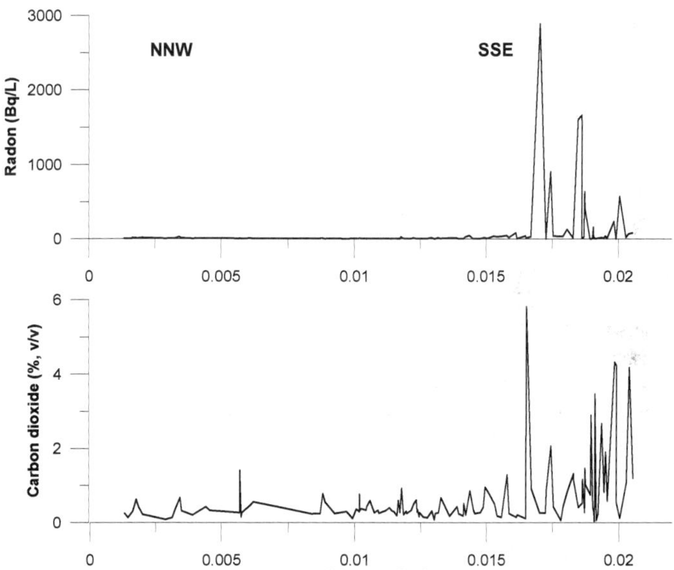

Fig. 12. Radon and carbon dioxide profiles at Djilubulak valley (Kyrgyzstan, central Asia). Graphs highlight a slightly decreasing trend of radon and carbon dioxide soil gas values towards the north, from the waste until the lake.

A different study of radionuclide migration was performed in the Djilubulak ephemeral stream valley on the southern shore of Issyk-Kul (Kyrgyzstan, central Asia), one of the largest and most pristine lakes in the world (Gavshin et al., 2002). The tail storages from the

past mining may pose a pollution hazard to the lake water and sediments. A chain of six protective pools interconnected by drain pipes descend from the abandoned mine and processing plant down the Djilubulak stream valley. To assess the effectiveness of these catch pools and the scale of pollution risk, a soil gas survey was performed from the abandoned mine to the shore of the lake (Giralt et al., 2003; Voltattorni et al., 2004).
In the river bed the soil gas survey was done performing measurements following both profiles perpendicular to the river flow and random distribution. The profiles were carried out approximately every 200 m. In each profile, the measurements were made roughly every 30-40 m. A total of 130 soil gas samples were collected sampling at the lower part of the river valley (close to the lake shore), along the river valley and at the waste.
The highest radon values (>40 Bq/ l) occur in the south-eastern part of the studied area characterised by the presence of the waste. All over the northern sector radon values are very similar to background values reported in literature (10-15 Bq/ l). The CO_2 soil gas distribution shows a greater concentration of anomalous values (> 3%) all over the mine and the waste area. Hypotheses about biogenic and/or thermogenic origin of this gas require isotope analysis. In spite of this, it is reasonable suppose that mine ruins and coal remains influenced soil gas distribution as highest values are present all over the waste and there is a good correspondence between high radon and carbon dioxide values. Fig 12 shows two profiles along which results were projected considering a longitudinal line intersecting the valley. Graphs highlight a slightly decreasing trend of radon and carbon dioxide soil gas values towards the north, from the waste until the lake. The overlapping peaks in the Rn and CO_2 plots imply that the soil gas distribution is linked to the presence of radioactive material in the waste. In fact, highest CO_2 and Rn values were found in the same area. On the other hand, outside the "contaminated" area, where soil did not undergo the effects of the mine activities, radon and carbon dioxide values are in agreement with the mean values reported in literature (Voltattorni et al., 2004).
Soil gas results, therefore, suggest that there has not been a significant down-stream migration of radiogenic particles or elements, either via mass transport during flooding events or via groundwater movement. However, it is worth noting that in case of a catastrophic event such as an intensive flash flood, the deposits of Kadji-Sai could be eroded and distributed in the Djilubulak valley and may reach the shores of Issyk-Kul Lake (Gavshin et al., 2002). These contaminants would then produce high local levels of radioactivity in any area they reach. In the worst case scenario, the exposure rates in the Djilubulak valley and at its confluence with Issyk-Kul Lake may reach values which exceed not only safe exposure rates for general public but even long-term occupational exposure limits. The total amount of radioactive deposits currently at the site would not pose danger to the entire Issyk-Kul Lake and areas further than 10–15 km from the site.

5. Conclusion

The limitation of soil gas investigations lies in weaker crustal gas concentrations in cases of thick sedimentary cover, and in high level of atmospheric dilution in soils (Baubron et al., 2002). However, on the basis of the many achieved results, it can be said that soil gas prospection constitutes a powerful tool to identify complex phenomena occurring within the crust.

The comprehensive approach followed in this study has provided insights on the spatial influence of tectonic discontinuities and geology on gas migration toward the surface. Soil gas measurements, performed at different scales, involved two gaseous species with very different geochemical behaviour. Soil gas surveys yielded different features of the anomalies, reflecting the different gas bearing these properties of the pathways along which gases can migrate.

The association of the two proposed gas species, radon and carbon dioxide, is considered fundamental in the study of gas migration as CO_2 often acts as carrier in transporting the radon trace gas: this mechanism for surface soil gas anomalies is due to advection as suggested by relatively high rate of migration needed to obtain anomalies of short-life ^{222}Rn in the soil pores.

As soil gas distribution can be affected by some phenomena related to the climatic factors, soil moisture and gas behaviour (mobility, solubility and reactivity), a multivariate study including a large number of gaseous species has been considered.

However, independent from gas origin, all the results show that gases migrate preferentially through zones of brittle deformation and enhanced permeability. In order to quantify the spatial influence of fault geometry and geochemical properties on the distribution of soil gases, the geostatistical approach (i.e., variograms) is necessary.

Because of the very high variability of gas concentrations at the surface, soil gas prospection appears necessary in order to select potential optimum sites for surveillance to identify, for example, regional changes of strain fields or variations in toxic emanation. Due to the complex relationship between geology and local phenomena, a network of geochemical stations would be much more useful.

It is hoped that the present study has brought attention to the problems associated with natural gas migration and that there is more awareness of how the soil gas method can be used in these situations, both to plan land-use zoning or to resolve health problems in existing residential areas dealing with the danger of natural toxic gases. In the case of the former, areas defined as high risk can be zoned for agricultural or parkland use and not for residential development, while for the latter modifications can be made on 'high-risk' existing homes or monitoring stations can be installed to improve safety.

Communication of these results to the local government can result in heightened awareness and the initiation of some preventive programmes, such as the development of a continuous monitoring station.

6. References

Amato, A.; Margheriti, L.; Azzara, R.M.; Basili, A.; Chiarabba, C.; Ciaccio, M.G.; Cimini, G.B.; Di Bona, M.; Frepoli, A.; Lucente, F.P.; Nostro, C. & Selvaggi, G. (1998). Passive Seismology and Deep Structure in Central Italy. *Pure and Applied Geophysics*, Special Issue: Geodynamics of the Lithosphere and the Earth's Mantle, 151, 479-493.

Aubert, M. & Baubron, J.C. (1988). Identification of a hidden thermal fissure in a volcanic terrain using a combination of hydrothermal convection indicators and soil atmospheres analysis. *J. Volcanol. Geotherm. Res.*, 35, 217–225.

Annunziatellis, A.; Ciotoli, G.; Lombardi, S. & Nolasco, F. (2003). Short- and long-term gas hazard: the release of toxic gases in the Albani Hills volcanic area (central Italy). *Journal of Geochemical Exploration* 77, 93-108.

Ball, T.K.; Cameron, D.G.; Colman, T.B. & Roberts, P.D. (1991). Behavior of radon in the geological environment: a review. *Q. J. Eng. Geol.*, 24, 169-182.

Baubron, J.C.; Allard, P. & Toutain, J.P. (1990). Diffuse volcanic emissions of carbon dioxide from Vulcano Island, Italy. *Nature*, 344, 51–53.

Baubron, J.C.; Allard, P.; Sabroux, J.C.; Tedesco, D. & Toutain, J.P. (1991). Soil gas emanations as precursory indicators of volcanic eruptions. *J. Geol. Soc. London*, 148, 571–576.

Baubron, J. C.; Rigo, A. & Toutain, J. P. (2002). Soil gas profiles as a tool to characterize active tectonic areas: the Jaut Pass example (Pyrenees, France). *Earth and Planetary Science Lett.*, 196, 69-81.

Beaubien; S.L.; Ciotoli, G. & Lombardi, S. (2002). Carbon dioxide and radon gas hazard in the Alban Hills area (central Italy). *Journal of Volcanology and Geothermal Research*, 123, 63-80

Blumetti, A.M.; Michetti, A.M. & Serva, L. (1988). The ground effects of the Fucino earthquake of Jan. 13th, 1915: an attempt for the understanding of recent geological evolution of some tectonic structure. In: *Historical Seismicity of Central Eastern Mediterranean Region.* C. Margottini and L. Serva Eds., 297-319. Nuove Tecnologie, l'Energie e l'Ambiente, Rome.

Blumetti A, .M.; Dramisa, F. & Michetti, A.M. (1993). Fault-generated mountain fronts in the Central Apennines (Central ltaly): Geomorphological features and seismotectonic implication. *Earth Surf. Processes Landforms*, 18, 203-223.

Capaccioni, B.; Tassi, F.; Vaselli, O. & Tedesco, D. (2007). Submarine gas burst at Panarea Island (southern Italy) on 3 November 2002: A magmatic versus hydrothermal episode. *J. Geophys. Res.*, 112, B05201. doi:10.1029/2006JB0044359.

Charlet, J.M.; Doremus, P. & Quinif, Y. (1995). Radon methods used to discover uranium mineralizations in the lower Devonian of the Ardenne Massif (Belgium). In: *Gas Geochemistry*, C. Dubois Ed., Science Reviews, Northwood, 1–18.

Cox, M.E. (1980). Ground radon survey of an hawaiian geothermal area. *Geophys. Res. Lett.*, 7, 283–286.

Caramanna, G.; Voltattorni, N.; Caramanna, L.; Cinti, D.; Galli, G.; Pizzino, L. & Quattrocchi, F. (2005). Scientific diving techniques applied to the geomorphological and geochemical study of some submarine volcanic gas vents (Aeolian Islands, southern Tyrrhenian sea, Italy). *Proc. 24th Diving for Science Symp. American Academy of Underwater sciences 11-12 March 2005 – Mystic – Connecticut (USA).*

Ciotoli, G.; Guerra, M.; Lombardi, S. & Vittori, E. (1998). Soil gas survey for tracing seismogenic faults: a case-study the Fucino basin (central Italy). *J. Geophys. Res.*, 103B, 23781- 23794.

Ciotoli, G.; Etiope, G.; Guerra, M. & Lombardi, S. (1999). The detection of concealed faults in the Ofanto basin using the correlation between soil gas fracture surveys. *Tectonophysics*, 299 (3–4), 321–332.

Ciotoli, G.; Lombardi, S. & Annunziatellis, A. (2007). Geostatistical analysis of soil gas data in a high seismic intermontane basin: Fucino Plain, central Italy. *J. Geophys. Res.*, 112, B05407, doi:10.1029/2005JB004044.

Cocco, M.; Nostro, C. & Ekström, G. (2000). Static stress changes and fault interaction during the 1997 Umbria-Marche earthquake sequence. *J. of Seism.*, 4, N. 4, 501-516.

Crenshaw, W.B. ; Williams, S.N. & Stoiber, R.E. (1982). Fault location by radon and mercury detection at an active volcano in Nicaragua. *Nature*, 300, 345-346.

Dayal, R. & Wilke, R.J. (1982). Role of clay minerals as backfill in radioactive waste disposal. Proc. Int. Clay Conf. Bologna/Pavia, 1981, pp. 771--787.

D'Amore, F. ; Sabroux, J.C. & Zettwoog, P. (1978). Determination of characteristics of steam reservoirs by radon-222 measurements in geothermal fluids. *Pure Appl. Geophys.*, 117, 253-261.

Del Pezzo, E. ; Gasparini, P. ; Mantovani, M.M. ; Martini, M. ; Capaldi, G. ; Gomes, Y.T. & Pece, R. (1981). A case of correlation between Rn-222 anomalies and seismic activity on a volcano (Vulcano island, southern Thyrrenian Sea). *Geophys. Res. Lett.*, 8, 962-965.

De Gregorio, S.; Diliberto, I.S.; Giammanco, S.; Gurrieri, S. & Valenza, M. (2002). Tectonic control over large-scale diffuse degassing in Eastern Sicily (Italy). *Geofluids*, 2, 273-284.

Di Filippo, M. & Toro, B. (1995). Gravity features. In: *The Volcano of the Alban Hills*, R. Trigila Ed. , 283 pp.

Dubois, C.; Alvarez Calleja, A.; Bassot, S. & Chambaudet, A. (1995). Modelling the 3-dimensional microfissure network in quartz in a thin section of granite. In: *Gas Geochemistry*, C. Dubois Ed., Science Reviews, Northwood, pp. 357-368.

Duddridge, G. A.; Grainger, P. & Durrance, E. M. (1991). Fault detection using soil gas geochemistry, *Q. J. Eng. Geol.*, 24, 427-435.

Durrance, E. M. & Gregory, R .G. (1988). Fracture mapping in clays: Soil gas surveys at Down Ampney, Gloucestershire. DOE Report: DOE/RW/88081, Dep. Of Energy, Washington D.C.

Eremeev, A. N.; Sokolov, V.A. & Solovov, A.P. (1973). Application of helium surveying to structural mapping and ore deposit forecasting. In: *Geochemical Exploration*, 1972, M. J. Jones Ed., pp.183- 192, Inst. of Min. and Metall., London.

Etiope, G. & Lombardi, S. (1995). Soil gases as fault tracers in clay basins: a case history in the Siena Basin (Central Italy). In: *Gas Geochemistry*, C. Dubois Ed., 19-29, Science Reviews, Northwood.

Fleischer, R.L. ; Alter, H.W. ; Furnam, S.C. ; Price, P.B. & Walker, R.M. (1972). Particle track etching. *Science*, 178, 255-263.

Fleischer, R.L. & Magro-Campero, A. (1985). Association of subsurface radon changes in Alaska and the northeastern United States with earthquakes. *Geochim. Cosmochim. Acta*, 49, 1061-1071.

Galadini, F. & Messina, P. (1994). Plio-Quatenary tectonics of the Fucino basin and surrounding areas (Central ltaly), *J. Geol.*,5, 6(2), 73-99.

Gascoyne, M. ; Wuschke, D.M. & Durrance, E.M. (1993). Fracture detection and groundwater flow characterization using He and Rn in soil gases, Manitoba, Canada. *Appl. Geochem.*, 8, 223- 233.

Gavshin, V.M.; Melgunov, M.S.; Sukhorukov, F.V.; Bobrov, V.A.; Kalugin, I.A. & Klerkx, J. (2002). Disequilibrium between uranium and its progeny in the Lake Issyk-Kul system (Kyrgyzstan) under a combined effect of natural and manmade processes. *J.Env. Radioact.*, 83, 1, 61-84.

Giraudi, C. (1989). Lake levels and climate for the last 30,000 years in the Fucino area (Abruzzo, Central Italy): A review. *Palaeogeogr. Palaeoclimatol. Palaeoecol.*, 70, 249-260.

Giralt, S. ; Klerkx, J. ; De Batist, M. ; Beck, C. ; Bobrov, V. ; Gavshin, V. ; Julià, R. ; Kalugin, I. ; Kipfer, R. ; Lignier, V. ; Lombardi, S. ; Matychenkov, V. ; Peeters, F. ; Podsetchine, V. ; Riera, S. ; Romanovsky, V. ; Sukhorukov, F. & Voltattorni, N. (2003). Are environmental changes affecting the natural state of Lake Issyk-Kul ? *Proceedings of NATO Advanced Research Workshop on "Dying and dead seas"*, Liege, 5-10 May 2003.

Gregory, R.G. & Durrance, E.M. (1985). Helium, carbon dioxide and oxygen soil gases: small-scale varations over fractured ground. *J. Geochem. Expl.*, 24, (1), 29-49.

Hickman, S.; Sibson, R. & Bruhn, R. (1995). Introduction to special section: Mechanical involvement of fluids in faulting. *J. Geophys. Res.*, 100, 12,831–12,840.

Hinkle, M. (1994). Environmental conditions affecting concentrations of He, CO_2, O_2 and N_2 in soil gases. *Appl. Geochem.*, 9, 53- 63.

Holub, R. F. & Brady, B. T. (1981). The effect of stress on radon emanation from rock, *J. Geophys. Res.*, 86, 1776–1784.

Hueckel, T. & Pellegrini, R. (2002). Reactive plasticity for clays: application to a natural analog of long-term geomechanical effects of nuclear waste disposal. *Engineering Geology*, 64, 195-215.

Irwin, W.P. & Barnes, I. (1980). Tectonic relations of carbon dioxide discharges and earthquakes. *J. Geophys. Res.*, 85, 3115–3121.

Kerrick, D.M.; McKibben, M.A.; Seward, T.M. & Caldeira, K. (1995). Convective hydrothermal CO_2 emission from high heat flow regions. *Chem. Geol.*, 121, 285–293.

King, C.Y. (1978). Radon emanation on San Andreas fault. *Nature*, 271, 516–519.

King, C.Y.; King, B.S.; Evans, W.C. & Zang, W. (1996). Spatial radon anomalies on active faults in California, *Appl. Geochem.*, 11, 497-510.

Klusman, R.W. (1993). Soil Gas and Related Methods for Natural Resource Exploration. Wiley, Chichester, 483 pp.

Leoni, L.; Polizzano, C.; Sartori F. & Sensi, L. (1984). Chemical and mineralogical transformation induced in Pliocene clays by a small subvolcanic body and consequence for the storage of radioactive wastes. *N. Jb. Mineral. Mh.*, 155-168.

Lewicki, J.L.; Evans, W.C.; Hilley, G.E.; Sorey, M.L.; Rogie, J.D. & Brantley, S.L. (2003). Shallow soil CO_2 flow along the San Andreas and Calaveras Faults, California. *Journal of Geophysical Research*, 108, B4, 14 pp.

Lombardi., S.; Etiope, G.; Guerra, M.; Ciotoli, G.; Grainger, P.; Duddridge, G.A.; Gera, F.; Chiantore, V.; Pensieri, R.; Grindrod, P. & Impey, M. (1996). The refinement of soil gas analysis as a geological investigative technique. Final Report. Work carried out under a cost sharing contract with the European Atomic Energy Community in the framework of its 4th R&D program on Management and Storage of Radioactive Waste (1990-1994), Part A, Task 4: Disposal of Radioactive Waste. EUR 16929 EN.

Lombardi, S. & Voltattorni, N. (2010). Rn, He and CO_2 soil gas geochemistry for the study of active and inactive faults. *Applied Geochemistry*, 25, 1206-1220.

Manini, E., Luna, G., Corinaldesi, C., Zeppilli, D., Bortoluzzi, G., Caramanna, G., Raffa, F. & Danovaro, R. (2008). Prokaryote diversity and virus abundance in shallow hydrothermal vents of the Mediterranean Sea (Panarea Island) and the Pacific Ocean (North Sulawesi-Indonesia). Microbial Ecology, 55, 626-639.

Morawska, L. & Phillips, C.R. (1993). Dependance of the radon emanation coefficient on radium distribution and internal structure of the material. *Geochim. Cosmochim. Acta*, 57, 1783-1797.

Nijman, W. (1971). Tectonics of the Velino-Sirente area, Abruzzi, Central Italy, Proc. K,: Ned. Akad. Wet., Ser. B, 74(2), 156-184.

Pinault, J. L. & Baubron, J. C. (1996). Signal processing of soil gas radon, atmospheric pressure, moisture, and soil temperature data: a new approach for radon concentration modeling, *J. Geophys. Res.*, 101, B2, 3157-3171.

Rahn, T.A.; Fessenden, J.E. & Wahlen, M. (1996). Flux chamber measurements of anomalous CO_2 emission from the flanks of Mammoth Mountain, California. *Geophys. Res. Lett.*, 23, 1861–1864.

Reimer, G.M. & Gundersen, L.C.S. (1989). A direct correlation among indoor Rn, soil gas Rn and geology in the Reading Prong near Boyertown, Pennsylvania. *Health Phys.*, 57, 155-160.

Reimer, G.M. (1990). Reconnaissance techniques for determining soil gas radon concentrations: an example from Prince Geoges County, Maryland. *Geophys. Res. Lett.*, 17, 809- 8012.

Segovia, N. ; De la Cruz Reyna, S. ; Mena, M. ; Ramos, E. ; Monnin, M. & Seidel, J.L. (1989). Radon in soil anomaly observed at Los Azufres Geothermal field, Michoacan: a possible precursor of the 1985 Mexico earthquake (Ms D 8.1). *Natural Hazards*, 1, 319–329.

Shapiro, M.H. ; Melvin, J.D. ; Tombrello, T.A. ; Fong-Liang, J. ; Gui-Ru, L. ; Mendenhall, M.H. & Rice, A. (1982). Correlated radon and CO_2 variations near the San-Andreas fault. *Geophys. Res. Lett.*, 9, 503–506.

Shapiro, M.H. ; Melvin, J.D. ; Copping, N.A. ; Tombrello, T.A. & Whitcombe, J.H. (1989). Automated radon-thoron monitoring for earthquake prediction research. In: *Radon Monitoring in Radioprotection, Environmental Radio-Activity and Earth Sciences*. ICTP, Trieste, pp. 137–153.

Sugisaki, R.; Anno, H.; Aedachi, M. & Ui, H. (1980). Geochemical features of gases and rocks along active faults. *Geochem. J.*, 14, 101–112.

Sugisaki, R. (1983). Origin of hydrogen and carbon dioxide in fault gases and its relation to fault activity. *J. Geol.*, 91, 239-258.

Tanner, A.B. (1964). Radon migration in the ground: A supplementary review. In: *The Natural Radiation Environment,* vol. I, T.F. Gesell and W.M. Lowder Eds., pp. 5-56, Univ. of Tex., Austin.

Thomas, D.M. ; Cox, M.E. & Cuff, K.E. (1986). The association between ground gas radon variations and geologic activity in Hawaii. *J. Geophys. Res.*, 91, 12186–12198.

Thomas, D. (1988). Geochemical precursors to seismic activity. *Pure Appl. Geophys.*, 126, 241–265.

Toutain, J.P. ; Baubron, J.C. ; Le Bronec, J. ; Allard, P. ; Briole, P. ; Marty, B. ; Miele, G. ; Tedesco, D. & Luongo, G. (1992). Continuous monitoring of distal gas emanations at Vulcano, southern Italy. *Bull. Volcanol.*, 54, 147–155.

Voltattorni, N.; Lombardi, S. & Beaubien, S.E. (2004). Evaluation of radioactive elements migration from uranium mines in Kyrghystan (Central Asia). Proceeding oft he *32nd International Geological Congress,* Firenze, Fortezza da Basso, 20-28 Agosto 2004.

Voltattorni, N. ; Sciarra, A. ; Caramanna, G. ; Cinti, D. ; Pizzino, L. & Quattrocchi, F. (2009). Gas geochemistry of natural analogues for the studies of geological CO_2 sequestration. *Applied Geochemistry*, 24, 1339–1346.

Voltattorni, N.; Lombardi, S. & Rizzo, S. (2010). ^{222}Rn and CO_2 soil- gas geochemical characterization of thermally altered clays at Orciatico (Tuscany, Central Italy). *Applied Geochemistry*, 25, 1248-1256.

Wattananikorn, K. ; Techakosit, S. & Jitaree, N. (1995). A combination of soil gas radon measurements in uranium exploration. *Nucl. Geophys.*,9, 643–652.

Woith, H. ; Pekdeger, A. & Zschau, J. (1991). Ground water radon anomalies in space and time: a contribution to the joint Turkish-German earthquake prediction project. In: *Earthquake Prediction: State of the Art*. EUG, Strasbourg.

10

Adsorption of methane in porous materials as the basis for the storage of natural gas

Cecilia Solar, Andrés García Blanco, Andrea Vallone and Karim Sapag
Laboratorio de Sólidos Porosos-Instituto de Física Aplicada-CONICET, Dpto. de Física-Universidad Nacional de San Luis
San Luis, Argentina

1. Introduction

It is well known that the natural gas (NG) is a substance of fossil origin from the decomposition of organic matter. It is found trapped under the terrestrial surface in stratus that avoid the natural release to atmosphere. These underground deposits can be oceanic or terrestrial.

The NG is a homogeneous mixture, having variable proportions of hydrocarbons, being the main constitute the methane (CH_4), which content generally ranges from 55 to 98 % in volume. Also, it contains ethane (C_2H_6), propane (C_3H_8) and heavier constitutes. Although it can be found in gas phase or in solution with oil, under normal atmospheric conditions, remains in gas phase. It may have some impurities or substances that are not hydrocarbons, such as Hydrogen Sulfide, Nitrogen and Carbon Dioxide (Figure 1). According to its origin, natural gas is classified in *associated* and *non-associated,* being the first, the one which remains in contact and/or dissolved with the oil from the deposit. The non-associated gas can be found in deposits lacking oil crude at the initial conditions of pressure and temperature.

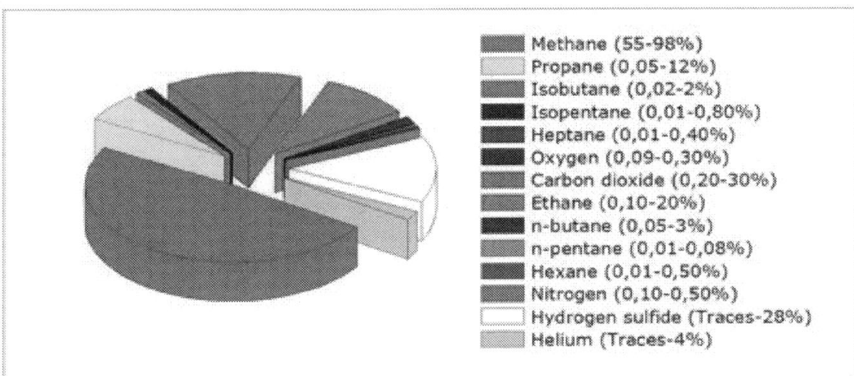

Fig. 1. Principal constitutes of Natural Gas (in percentage).

From the fossil fuels, the cleanest is the natural gas. Its combustion, similarly to other fuels, produces mainly CO_2 and water vapor. The emissions of CO_2 are 25-30% lower than the generated by the fuel-oil and a 40-50% lower than charcoal (Figure 2) per unit of produced energy (Natural Gas and Climate Change Policy, 1998; Comisión Nacional de Energía, 1999).

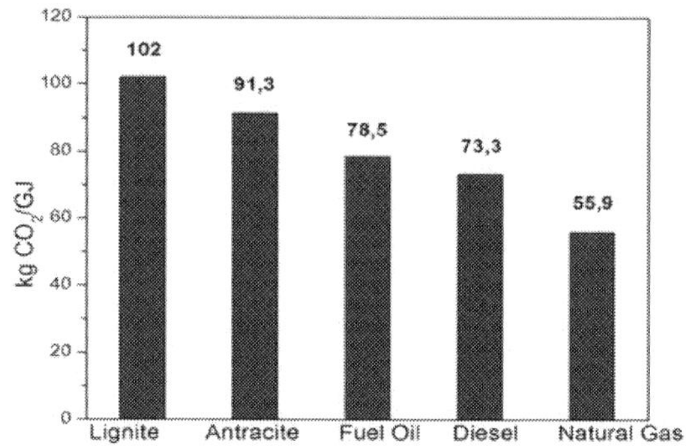

Fig. 2. CO_2 Emissions in the combustion (Kg per GJ).

At worldwide scale, the resources of natural gas are abundant. However, as oil, they are highly concentrated in a reduced number of countries and deposits. Some data reported in the *BP* Statistical Review of World Energy, 2009, revealed interesting information: three countries (Russia, Iran and Qatar), hold the 56% of the world reserves (WR). Almost the 50% of the WR are distributed in 25 deposits around the world and the countries that are members of the OPEC (Organization of the Petroleum Exporting Countries), control the 50% of the WR. The percentage distribution of the WR by the end of 2008 is shown in Figure 3.

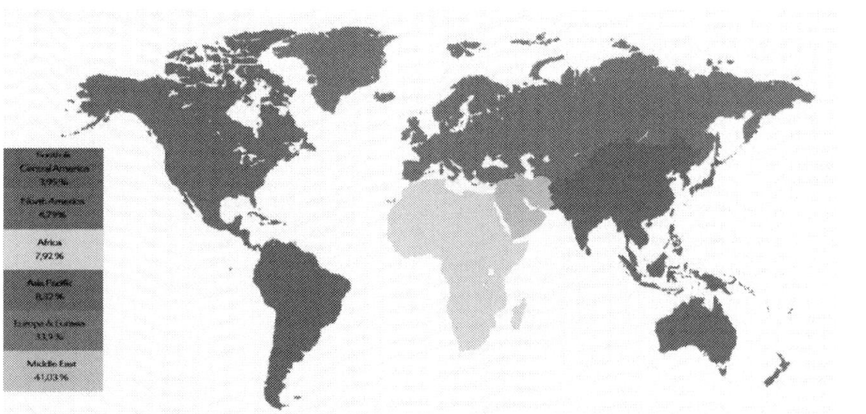

Fig. 3. The percentage distribution of the world reserves of natural gas by the end of 2008 according to the Statistical Review of World Energy, 2009.

As it may be seen from Figure 3, the world reserves of natural gas, although heterogeneously, are distributed throughout the world, constituting an advantage to be able to supply the local requirements. During the last few decades, the volume of discovered gas has been decreasing but it still keeps the necessary volume to ensure their existence for many years. Additionally, the estimations of these reserves are progressing as new techniques of exploitation, exploration and extraction, are discovered. It is estimated that a substantial quantity of natural gas remains undiscovered (World Energy Outlook, 2009).

The NG has vast diversity of applications: in industry, trade, energy generation, residential sector and terrestrial transport, and its use have shown an important growth over the last few years (MacDonald & Quinn, 1998; Inomata et al., 2002; Prauchner & Rodríguez-Reinoso, 2008).

Regarding the particular use as fuel for transport units, such as cars, autobuses and trucks, the natural gas vehicle (NGV) shows diverse environmental benefits. One of them is the reduction of post combustion contaminants, lowering the maintenance costs compared to traditional fuels (Cook et al., 1999; Lozano-Castelló et al., 2002a; Alcañiz-Monge et al., 1997).

The environmental advantages at using the NGV are numerous. However, from the point of view of the combustion products, it can be remarked: i) it does not contain lead or heavy metals traces, avoiding their emission to the atmosphere, ii) lack of suspended solid particles that are present when using gasoline affecting health (increase of respiratory and cardiovascular diseases), iii) absence of sulfur and subsequently no sulfur dioxide (SO_2) emissions, typical contaminant from transport. Compared to liquid fuels, the emissions of the NGV combustion produce up to 76% less CO, 75% less NOx, 88% less hydrocarbons and 30% less CO_2. Furthermore, the physicochemical properties of the natural gas enable the use of catalysts for the combustion of gases, obtaining excellent results and minimizing even more the emissions (Sun et al., 1997).

The advantages of NG have promoted its use in the automotive fleet of many countries, which exceeds six millions of vehicles at present. The advance in the technology for the NGV use has not been standardized throughout the world. This is due to differences regarding the availability of energy resources, contamination levels, fuel pricing policies, applied auditing and, definitely, the set of government actions able to generate expectative among the potential users.

Country	Vehicles
Pakistan	2,000.000
Argentine	1,678.000
Brazil	1,467.000
Italy	433,000
Colombia	251,000
India	225,000
EE.UU	130,200
Germany	54,200
Japan	24,700
France	8,400

Table 1. Estimated Natural Gas Vehicles in different countries.

Table 1 summarizes the number of natural gas vehicles in some representative countries according to the Dirección de Tecnología, Seguridad y Eficiencia Energética, 2006.

In spite of the advantages showed by the NG in comparison to liquid fuels, there is an important disadvantage: its low-energy density (heat of combustion/volume), which constitutes a limitation for some applications. Therefore, under standard conditions of pressure and temperature, the distance traveled by a vehicle per unit of fuel volume, using NG, corresponds to the 0.12% of the trajectory with gasoline. Consequently, the storage of this fuel, whether in quantity or density, plays an important role for its use in diverse kinds of transport.

An alternative is to increase the density, for example, liquefying the NG. The liquefied natural gas (LNG) is stored at the boiling point, 112K (-161°C) in a cryogenic tank at a pressure of 0.1MPa, where the energy density is approximately a 72% of the total gasoline. This means that 1 volume of LNG corresponds to 600 volumes of natural gas under STP (600 v/v) conditions (Cracknell et al., 1993; Menon & Komarneni, 1998). However, this storage method shows multiple inconveniences, mainly because the LNG increases inevitably the temperature within the tank. Thus, the pressure rises and could result in a dangerous situation. Moreover, the filling of the tank must be performed by an expert on cryogenic liquids handling.

A widely used commercial method considered to increase the energy density of the natural gas is to compress and store it as compressed natural gas (CNG). For this case, the NG can be found as a supercritical fluid at room temperature and it becomes compressed at a maximum pressure around 20-25 MPa, reaching a density 230 times higher (230 v/v) than the one obtained for the natural gas under STP conditions (Menon & Komarneni, 1998; Lozano-Castelló et al., 2002b). In this case, the energy density is approximately 25% of the one from gasoline. A disadvantage is the risk of carrying highly compressed gas (20MPa) within the vehicle. Modifications such as thick-walled tanks and complex safety valves would be required.

The use of adsorbent materials, such as activated carbons and zeolites, among others (Rodriguez-Reinoso & Molina-Sabio, 1992; Parkyns & Quinn, 1995; Sircar et al., 1996; Alcañiz-Monge et al., 1997; Lozano-Castello et al., 2002c; Almansa et al., 2004; Marsh & Rodriguez-Reinoso, 2006; Mentasty et al., 1991; Triebe et al., 1996), for the storage of natural gas at low pressures, is known as adsorbed natural gas (ANG). Pressures are relatively low, of the order of 2 to 4 MPa at room temperature, which represents an interesting alternative for the transport and applications at large scale. The technology, in contrast with the other two, is not well developed and is still at scientific level. At this stage, the studies on storage by the ANG method are carried out using the methane, major constituent of the NG. It has been found that the density of the compressed methane at 3.4MPa can be increased in a factor higher than 4 by the use of adsorbents, reaching a relation of methane storage of 180 v/v, which is equivalent to compressed gas at more than 16MPa (Cook et al., 1999; Alcañiz-Monge et al., 1997).

Through this chapter, basic concepts regarding adsorption and adsorbents are reviewed as well as their application for the particular study of methane storage, starting point of the ANG process. In addition, the methodology for the study is described and shows the scientific advance in this field, reporting results from our research group and from other laboratories.

2. Adsorption basics and methodology of study

Adsorption is a phenomenon in which surface plays an important role, unlike absorption where molecules can penetrate the solid structure. The occurrence of this phenomenon in gas-solid interactions is our major focus of interest.

The surfaces of solids, even those homogeneous, have imperfections. These defects are the result of many circumstances, mainly its composition and the interaction that takes place among the molecules that constitute their atmosphere. Figure 4 shows a classical schema of this situation, according to the description made by Somorjai, 1994.

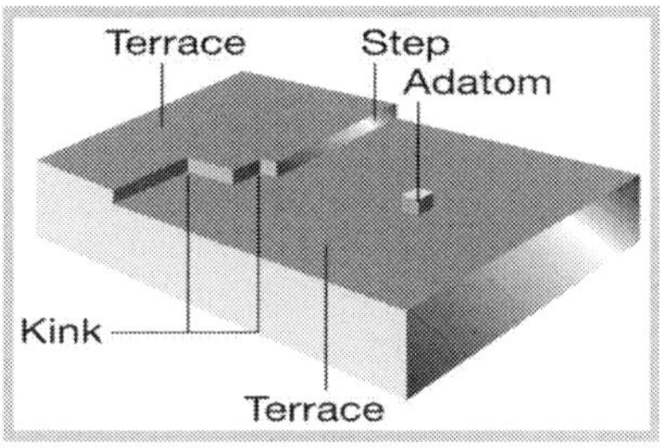

Fig. 4. Scheme of common defects on the apparently homogeneous solid surfaces.

Generally, the properties of the surfaces of the solids differ from their bulk for many reasons. Some of which are enlisted below:

- The perturbation of the superficial electron density is different to the one from the bulk. This is caused by the loss of structural periodicity in the perpendicular direction of the surface.
- The presence of decompensated forces on the surface due to the lack of neighbor atoms (producing potential wells, nearby molecules are attracted).
- Vibrational properties on the surface are different (geometrical and energetic effects, producing curvatures) from the ones on the rest of the solid.
- Some phenomena can occur: *Relaxation* or Superficial *Reconstruction*, which means that the superficial atoms show geometrical and energetic differences to the atoms from the bulk.

These reasons promote the presence of attractive potentials, which are able to attract molecules from the surrounding leaded by thermodynamic parameters, particularly, pressure P and temperature T of the gas-solid system. Moreover, superficial centers can take place showing additional electrostatic effects and creating new attractive or repulsive "sites". Therefore, when one or more molecules from a fluid approach the surface, they could be trapped and nucleation, motion and the formation of layers in the interface, would take place. This process is named Adsorption.

"Adsorption of a gas onto a solid surface" can be defined as the gain of one or more constituents of the gas in the region of the gas-solid interface. Figure 5 shows a schema that represents the process.

The adsorption phenomenon involves an increment of the gas density in the neighborhood of the contact surface and since the process is spontaneous, the change in the free energy of Gibbs is smaller than zero. Given that the entropy change is also below zero (a decrease in the freedom degree of the gas molecules during the process), the enthalpy change is lower than zero. Thus, the process is exothermic (Rouquerol et al., 1999).

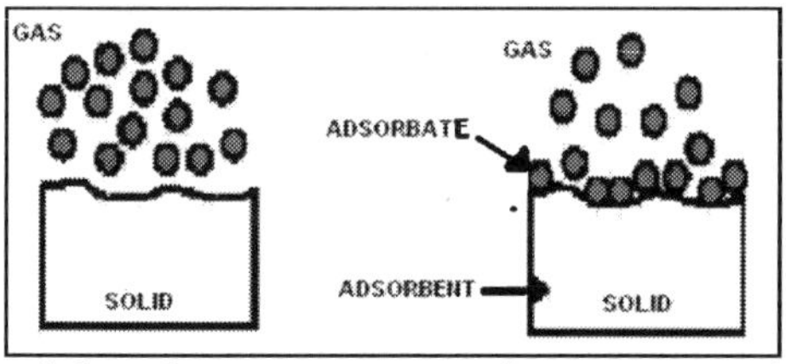

Fig. 5. Representation of the adsorption process of a gas on a solid surface for a given pressure, P and temperature, T.

When the adsorption process is reversible it means physical adsorption or physisorption, our major focus of interest for the study of natural gas storage. In this case, the result of the adsorption heats or enthalpy changes in the process are not elevated values, being for the methane about 16 KJoule/mol (Cook et al., 1999). The interaction forces occurring between the solid surface (adsorbent) and the adsorbed gas (adsorbate) are Van der Waals type, where prior to adsorption, the gas is called adsorbable. Moreover, adsorbate-adsorbate interaction may take place and is neglected in some studies when compared to the adsorbate-adsorbent interaction. It can also be considered that in average, these interactions do not impact the whole process.

The net interaction potential that the molecules surrounding the surface may experience, can be represented as seen in Figure 6, where the energy of interaction of one particle at a distance z of the surface, is the sum of the interaction of each molecule(i) with each atom (j) of the solid, given by equation 1.

$$E(z) = \sum_j \phi_{ij} \tag{1}$$

Figure 6 represents a particle with a kinetic energy E_k approaching to the solid surface.

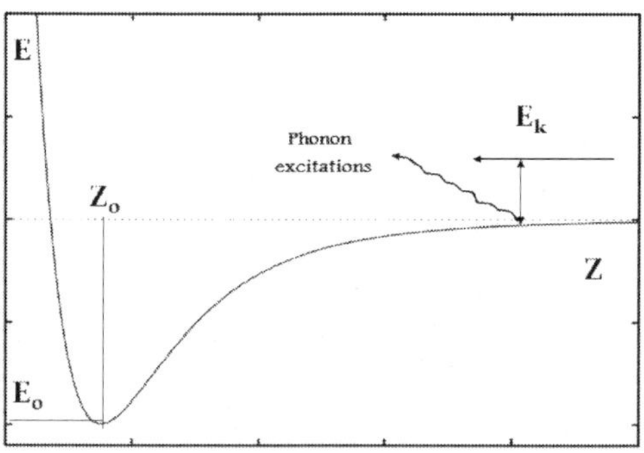

Fig. 6. Representation of the interaction potential that molecules nearby to the surface may sense.

The particle may detect the phonons excitation and subsequently, the potential attraction of the solid, which has a minimum (value) at a distance Z_0, representing the minimal distance of approaching to the solid.
The energy of the adsorbate-adsorbent interaction can be expressed using several terms. Some of them are described in the following equation:

$$E(z) = E_D + E_R + E_P + E_{dip} + E_Q \tag{2}$$

where E_D represents the dispersive potential (attractive); E_R, the repulsive; E_P, the one caused by the polarizability; E_{dip}, the dipolar and E_Q, the quadrupolar interactions (Rouquerol et al., 1999).
Considering only the first two terms, a Lennard-Jones (L-J) potential would take place, which involve the Van der Waals attractive forces and the Pauli repulsive forces.

2.1 Quantification of the Adsorption

Assuming a system set at a given temperature where a gas becomes into contact with a solid surface occupying a volume V at a pressure P_i prior to the adsorption, while a part of the adsorbable gas passes to the adsorbed state, keeping V and T unchanged, it should be noted a pressure decrease, followed by a stabilization of the system to a final equilibrium at pressure P_{eq}. Figure 7 represents the adsorption process at constant V and T.

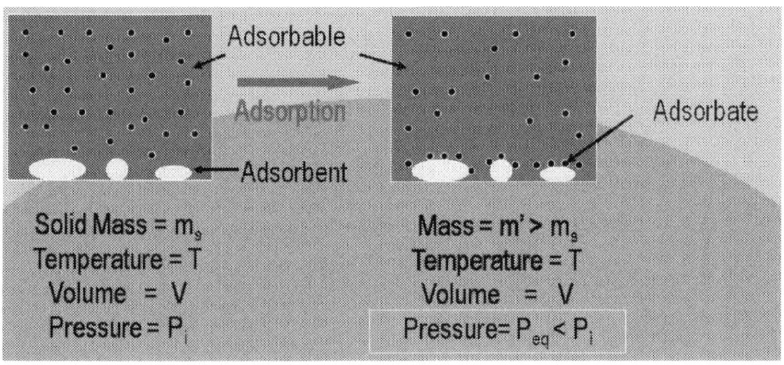

Fig. 7. Scheme of the Adsorption process.

Once the pressure change (P_i-P_{eq}) is determined by an equation of state that represents the gases under study, it is possible to calculate the quantity of moles that are no longer in gas phase but in the adsorbed phase at that pressure. The same can also be expressed in terms of adsorbed volume or grams of adsorbate, which is usually reported in standard conditions of temperature and pressure. Whether P_i is increased, a new P_{eq} is obtained as well as a new adsorbed quantity, maintaining unchanged the temperature and volume of the system. Thereby, the relation between the adsorbed amount and the pressure may be graphically found at constant temperature, reported as *adsorption isotherm*. This method, called volumetric or manometric, is the most widely used to measure the adsorption of gases and was selected for our laboratory to study adsorption processes. By the gravimetric method, the adsorbed quantity is measured from the mass gain during the process.

There is a detail that must be appointed because it would be helpful when interpreting what it is being actually measured. Assuming than n moles of an adsorbable are put into contact with a solid (adsorbent) at a certain volume V and pressure P where the adsorption occurs, once the equilibrium is reached, it is possible to identify three zones with different concentrations $c = dn/dV$, as shown in Figure 8a. Zone I corresponds to the region where the adsorbent is located and none molecule of adsorbable is expected ($c^s=0$). Zone II corresponds to the adsorbed layer, focus of our interest, where the concentration is c^a, which decreases as z increases ($c^a=c(z)$) until $z=t$. The zone III is at c^g concentration, which is the concentration of the adsorbable in absence of the adsorbent and depends only on P and T.
Knowing the area A, where the adsorbed layer is on the surface, as well as the thickness of the adsorbed layer t, the volume of the adsorbed layer can be calculated as $V^a=A.t$, from where the adsorbed quantity in moles, can be deduced.

$$n^a = \int_0^{V^a} c^a dV = A \int_0 c^a dz \qquad (3)$$

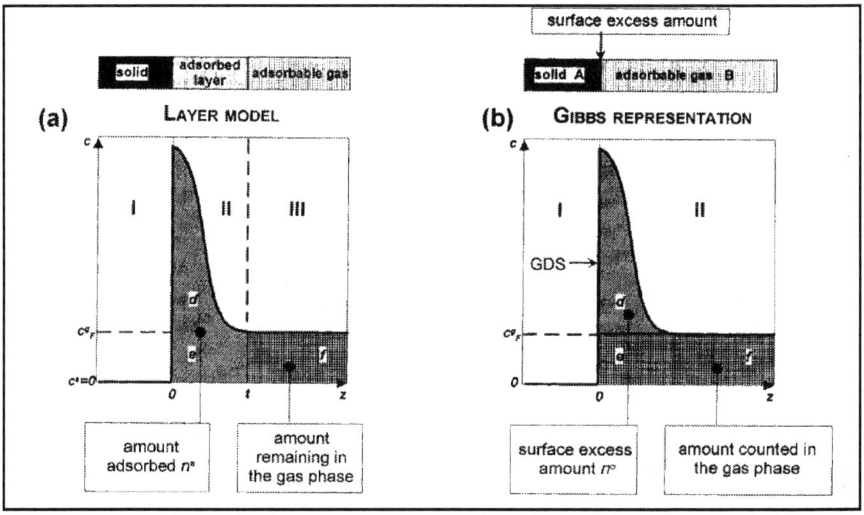

Fig. 8. Variation of the concentration, c, with the distance from the surface, z. a) Adsorbed layer; b) Gibbs representation (from Rouquerol et al., 1999).

The total quantity of moles for the considered volume is:

$$n = n^a + c^g V^g \qquad (4)$$

where V^g is the gas volume that remains at zone III (f region indicated in Fig 8a) after the adsorption process.
Therefore, in order to calculate n^a it must be known the c^a as z function (eq. 3) or V^g and n from eq. 4. However, the concentration profile of the adsorbed zone cannot be determined through an assay, and a measure of the volume V^g is complicated to obtain. This is because when adsorption occurs, the decrease in the system pressure is due to the increase in the

molecules concentration (zone *d* of Figure 8a) at concentrations higher than c^g. On the other hand, the molecules of the zone *e* are at the same concentration than the adsorbable and do not causes a pressure decrease. This would complicate the identification of the molecules that are in the zone *e* and *f*, occupying these latter the volume V^g.
To overcome this inconvenient, the Gibbs representation (Figure 8b) can be used. In this case, the system of reference occupies the same volume than the actual but, at present, it is only divided in two regions: I, the solid and II, the zone where the adsorbable is located. The status of the adsorbable remains unknown (adsorbed or not), while it is separated by a surface that is parallel to the adsorbent, called Gibbs dividing surface (GDS). The actual volume occupies the same volume than the representation, *V*, which is the volume that the molecules (*n*) occupy when put into contact with the solid at an initial pressure P_i. Afterwards, when the equilibrium is reached, a P_{eq} value arises. The entire process follows Figure 7.
Zone II of Figure 8b is the resulting scenario when P_{eq} is reached. Then, the gas molecules can be taken as part of one of two groups: the molecules that maintain the concentration of the gas, c^g, simulating that the adsorption phenomenon does not occur (zones *e* and *f* of Figure 8b), and another group that includes the molecules showing a concentration higher than c^g, that are basically, *"excess"* molecules (zone *e*) called n^σ. These are responsible for the decrease of P_i and the unique measurable molecules in an assay.
In conclusion, the number of *"excess"* molecules is the difference between the total number of molecules and the number of molecules remaining at the same concentration of the gas prior to be adsorbed:

$$n^\sigma = n - c^g V \tag{5}$$

Combining both schematic representations shown in Figure 8, it can be seen that the total volume is the sum of the volume V^g (zone III, Figure 8a) and V^a (zone II Figure 8a). This could be summarized as follows:

$$n^\sigma = n - c^g V^g - c^g V^a \tag{6}$$

From equation 4, it can be obtained the number of molecules from the adsorbed layer (n^a) as a function of the number of total molecules of the studied gas (*n*). Correlating eq. 4 and 6, we obtain:

$$n^a = n^\sigma + c^g V^a \tag{7}$$

At a low pressure assay, c^g corresponds to a small value and $V^a << V^g$. Hence, from equations 6 and 7, we find that:

$$n^a \approx n^\sigma \tag{8}$$

This does not occur at high pressures (pressures higher than the atmospheric), where eq. 7 remains valid.
It can be concluded that the measures that actually can be performed in an assay, are the molecules present in the *"excess"* zone, shown in the Gibbs schema. Therefore, the experimental data that can be graphed correspond to an *excess isotherm*, given by the n^σ molecules. Sometimes the interest is focused on the *absolute isotherm*, particularly for

comparison with theoretical calculations and it is obtained by counting n^a. To conduct assays at subatmospherical pressure, these two isotherms are coincident, but it is not valid for high pressures.
The major interest in this chapter is to use these concepts to achieve adsorption isotherms of methane at low and high pressures. Afterwards, it is possible to obtain information regarding the possibilities of natural gas storage with the adsorbents under study.

2.2 Porous materials

Adsorption is a superficial process and a crucial characteristic for the adsorbents is their high adsorption capacities. Then, the adsorbents require an elevated exposed surface per gram of material, which is called *specific surface area* (S_{esp}) and is expressed in cubical centimeters of adsorbate per gram of adsorbent.
The smaller the elemental constituents of the solid are, the greater the specific surface area is. This characteristic may be shown in fine particles, e.g. powders, as well as solids with small holes, which sizes can range from a few Angstroms to nanometers. These are named *porous solids*.
The IUPAC (Sing et al., 1985), depending on the transversal dimension of the pores in these solids (d), present the following classification:

$$solids \begin{cases} micropores\ d < 20\,\mathring{A} \\ mesopores\ 20\,\mathring{A} < d < 500\,\mathring{A} \\ macropores\ d > 500\,\mathring{A} \end{cases}$$

A solid may exhibit different kinds of pores. Rouquerol et al., 1994 reports diverse possibilities (Figure 9) where the contribution to the specific surface area is variable. The more rough the surface is or the smaller the pores are the greater is the contribution to the S_{esp}.

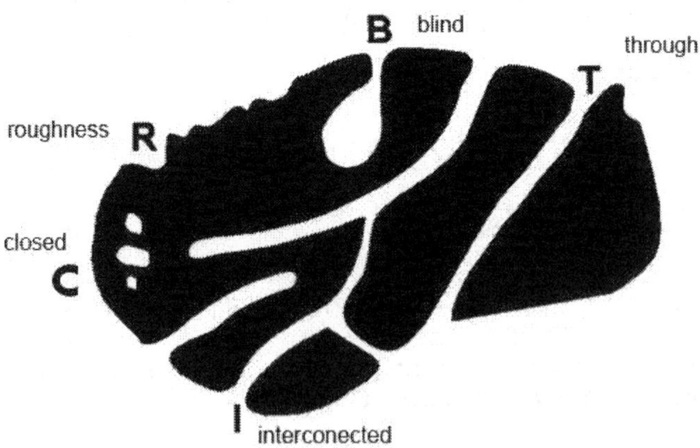

Fig. 9. Types of pores that a solid may exhibit Rouquerol et al., 1994.

Up to present, the adsorption phenomenon has only been studied from the perspective of a plane solid surface and a gas. However, for porous solids the gas molecules are *"surrounded"* by the walls of the pores, being considerably higher the interaction forces. In order to model this interaction, it must be supposed that the potential of the walls has an attractive and a repulsive term, similar to the aforementioned potential style described by Lennard Jones (Figure 6). As the pore becomes smaller, the potentials of the gas-solid interaction of each wall overlap. This, results in further potentiation of the adsorption phenomenon, which turns porous materials into excellent adsorbents. Figure 10 shows a schema of the variation of the potential of the solid-gas interaction for a plane surface and a porous solid while the separation among layers, decreases.

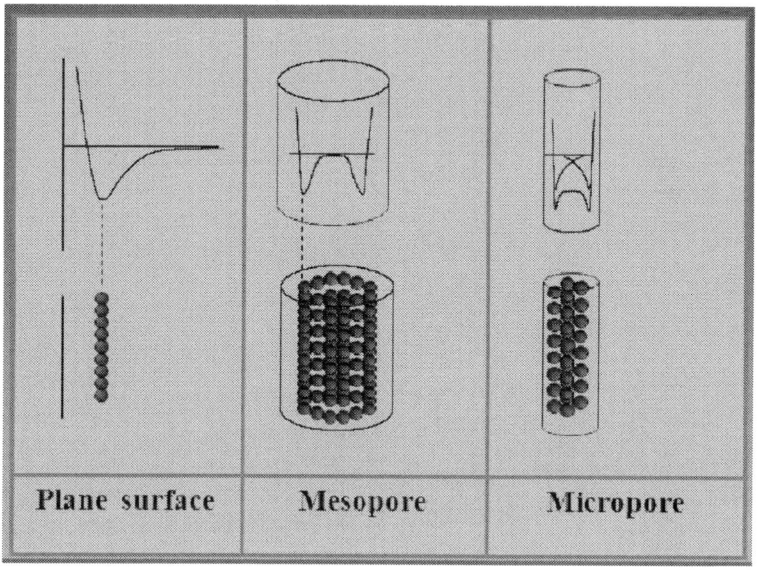

Fig. 10. Potential configuration according to the surface.

Therefore, besides the specific surface, it becomes necessary to study the porosity of the sample to provide comprehensive information related to the adsorption capacity.

2.3 Gas adsorption for the characterization of materials

The textural characteristics of the solids can be studied by gas adsorption, usually with gaseous nitrogen at 77K, at pressures between 10^{-4} Torr to pressures near to the atmospheric. As a result, adsorption isotherms may be obtained and reflect the quantity of adsorbed gas (cm^3/g) as a function of the relative pressure (P/P_0) at constant temperature, where P_0 is the saturation pressure. The appearance of the isotherm is directly related to the characteristics of the solid. An extensive work conducted by Brunauer, Deming, Deming and Teller (Brunauer et al., 1940), reported that a isotherm can be described by one or a combination of the basic shapes illustrated in Figure 11.

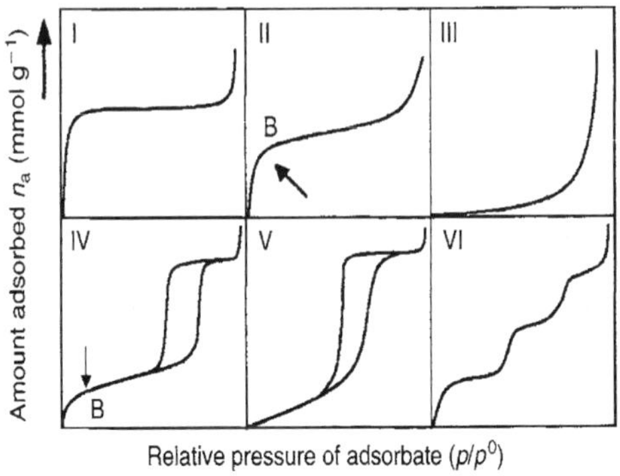

Fig. 11. Types of isotherms representing the most relevant processes taking place in the adsorbate-adsorbent interaction (from Rouquerol et al., 1999).

It should be noted that the shape of the isotherms reveals the diverse processes than can occur as the pressure increases. At low pressures, micropores become filled and a monolayer reaches its capacity at relative pressures of the order of 0.1. From that point, a mono-multilayer filling begins and a capillary condensation is produced at pressures of the order of 0.5 of P/P_0. Afterwards, the mesopores become filled and at pressures near to P_0, the condensation of the N_2 takes place (for N_2 isotherms at 77K). This technique is useful to analyze up to mesopores.
For the storage of methane, the materials of interest must have micropores that show Type I isotherms at the zone of low pressures. However, at high pressures, these materials can have a mono-multilayer filling, typical of mesopores, and hysteresis loops, related to the pore geometry.
From the measured isotherms (Vads vs P/P_0), it is possible to obtain some textural characteristics of the material, such as specific surface area, pores volume, micropores volume, etc. For that purpose are used models that assume the form of the pores and their way of filling, as well as the gases state.
In regards to the models and the manner in which data are obtained for further characterization, notable works have been reported (Gregg & Sing, 1982; Rouquerol et al., 1999).
For the calculation of the specific surface of the solids, the most used method is the one proposed by Brunauer, Emmet and Teller, the BET method (Brunauer et al., 1938). Starting from the thermodynamic equilibrium, at a determined pressure P and temperature T, a series of assumptions are made regarding the events that occur in the gas-solid interface, where molecules reach towards the surfaces. Some of the remarkable assumptions are:
-That the adsorbed molecules are spherical-shaped and gradually accumulate on the surface of the solid. Whether they group side by side, they can form a monolayer.
-That regions exist on the surface of the solid and are covered with 0, 1, 2… m monolayers of molecules.
-That the adsorption energy is E for the first layer and E´ for any other layer and lateral adsorbate-adsorbate interactions do not exist.
-That at equilibrium, the quantity of molecules that enter and exit is the same at a determined region.

From these assumptions, and attending to the kinetic-molecular theory of gases, the following isotherm, known as BET equation, is obtained:

$$\frac{P/P_0}{n_a(1-P/P_0)} = \frac{1}{n_m C} + \frac{C-1}{n_m C}(P/P_0) \tag{9}$$

where P/P_0 is the relative pressure, n_a is the number of adsorbed moles, n_m is the number of moles per gram of solid within a monolayer, and C is a constant related to the energy of adsorbate-adsorbent interaction.

Equation 9 represents the equation of a straight line with slope $\alpha = \frac{C-1}{n_m C}$ and ordinate to the origin $\Omega = \frac{1}{n_m C}$, where n_m and C, may be obtained.

From n_m (mol/g) and having the surface that the gas spheres occupy, A_m *(m²/molecule)*, and the number of molecules that occupy a mol N (*molecules/mol*), it can be stated:

$$S_{BET} = n_m . A_m . N \tag{10}$$

in m^2/g, which represent the specific surface area by the BET method (S_{BET}).
Although this method makes basic suppositions and perhaps unrealistic, it is still a simple method widely used and standardized for the calculation of the "BET specific surface area" (S_{BET}). This method is effective for materials with mono-multilayer formation, particularly when the monolayer is well-formed. This is the case of the type II isotherms from Figure 11, where B points out the zone where the monolayer becomes filled. For microporous materials, caution is necessary because S_{BET} usually overestimates the value of the specific surface area. Rouquerol et al., 1999 detail a series of conditions that must be fulfilled in order to obtain the most accurate calculation.
There are various methods available to calculate the microporosity. One of the most used and accepted is the proposed by Dubinin and collaborators (Dubinin, 1960), which is based on Polanyi's theory, that supposes the existence of adsorption potentials, characteristic of the adsorbents. The adsorbed quantities are a function of this potential and constitute the "characteristic curves". Based on this theory and studying diverse adsorbates on the same surface, Dubinin found that characteristic curves were affined, differing in one constant and with a similar shape to the "tail" of a Gaussian. Hence, he suggested a general shape for these curves and using the Polanyi potential, proposed to calculate special characteristics of the material, particularly the micropores volume. These calculations were performed at the region of low relative pressures, where the process involved in the adsorption is the micropores filling instead of the layer-by-layer adsorption on the pore walls. The following relation, named Dubinin-Raduschevich (Rouquerol et al., 1999), was found:

$$\log(V) = \log(V_o) - D.\log^2\left(\frac{P_o}{P}\right) \tag{11}$$

where V is the adsorbed volume, V_0 is the micropore volume and D is related to the pore size and involves the assay temperature and the affinity of the used adsorbate. The data obtained from the isotherm can be used to determine the micropore volume by the ordinate to the origin, graphing the implicit linear equation (eq. 10).

In consequence two fundamental determinations regarding the characterization of a porous material are S_{BET} and its V_o.

The N_2 at 77K is widely used for the characterization of porous materials. However, for narrowed micropores often called ultramicropores (< 8 Å), this gas has shown diffusion problems. For this reason, CO_2 has been proposed as an alternative characterization gas in this porous region. Since results have been satisfactory (Garrido et al., 1987), studies reporting microporous characterization using CO_2 has become common.

Additionally to experimental data, computational studies are often made in order to obtain information regarding the texture of the materials.

2.4 Computational studies for adsorption

2.4.1 DFT Method

A widely used methodology for the calculation of the pore size distribution is based on the Density Functional Theory, DFT, (Latoskie et al., 1993; Neimark et al., 1997; Neimark et al., 2000; Tarazona, 1985; Murata et al., 2000), which is already incorporated to the software in several equipments. A brief description is given below.

The thermodynamic system chosen to apply the DFT methodology in the adsorption of porous solids is the macrocanonical ensemble. The potential in this ensemble (grand potential) is given by $\Omega(\rho(r))$, which at equilibrium is defined as follows:

$$\Omega(\rho(r)) = A(\rho(r)) + \int dr\rho(r)(V_{ext} - \mu) \tag{12}$$

where $A(\rho(r))$ is the free energy, $\rho(r)$ the density profile, V_{ext} is the wall potential and μ the chemical potential.

The equilibrium density profile is therefore determined by minimizing the grand potential functional with respect to **ρ(r)**. Since ρ(r) is the local density, the adsorbed amount (usually expressed as the surface excess number of adsorbed molecules) must be obtained by the integration over the internal volume of the pore. By repeating this procedure with different values of μ (and hence values of P/P_0) it is possible to construct the adsorption isotherm.

The evaluation of the excess free energy is a more difficult problem. This is because in an inhomogeneous fluid the energy distribution is non-local; it depends on the correlations within the overall density profile. Various attempts have been made to overcome this difficulty by the introduction of weighting or smoothing functions (Gubbins, 1997). This approach has led to the development of the non-local density functional theory (NLDFT), which inter alia has been used for the derivation of the pore size distribution from adsorption isotherm data.

2.4.2 Monte Carlo Method

According to statistical thermodynamics, a system where the chemical potential μ, volume V and temperature T, remain constant while energy and particles are exchanged with the reservoir, is called Grand Canonical ensemble. This kind of ensemble is appropriate for describing an adsorption process of a liquid or a gas on solid surfaces.

Framed in the formal terms characterizing the present ensemble, the Grand Partition Function is set out, from which it is possible to obtain relevant thermodynamic parameters, given in equation 13 (Hill, 1986).

$$\Xi = \sum_N \frac{\lambda^N}{N!\Lambda^{3N}} \int \exp\left(-\beta U_N\right) dr^N \tag{13}$$

where U_N is the total energy of the system, $\lambda = \exp(\beta \mu)$ and $\beta = 1/k_B T$. The probability to find the system at a state i with N molecules in a volume element of the phase spaces; $\delta r^N \delta p^N = \delta r_1 ... \delta r_N \delta p_1 ... \delta p_N$ is $f_i \delta r^N \delta p^N$, where:

$$f_i \propto \frac{(\xi V)^N}{N!} \exp\left(-\beta U_N\right) \tag{14}$$

being $\xi = \frac{\lambda}{\Lambda^3}$ the affinity (for an ideal gas, $\xi = P/k_B T$)

Once the thermodynamic relations of the system are established, the simulated adsorption isotherms can be obtained by studying the situation of the molecules approaching to the surface.

Three elemental processes keeping T and μ unchanged can be considered: Adsorption of a molecule, desorption and displacement (defined as the sum of the desorption and re-adsorption of the same molecule). Using equation 13 and an algorithm, such as the proposed by Metrópolis (Frenkel & Smit, 2002) to calculate the probabilities of transition from an initial state to a final state, it can be obtained:

$$a).\ Displacement: \quad p_{ij} = \min\left\{1, \exp\left[-\beta\left(U_N(i) - U_N(j)\right)\right]\right\} \tag{15}$$

$$b).\ Adsorption: \quad p_{ij} = \min\left\{1, \frac{\xi V}{N+1} \exp\left[-\beta\left(U_{N+1} - U_N\right)\right]\right\} \tag{16}$$

$$c).\ Desorption: \quad p_{ij} = \min\left\{1, \frac{N}{\xi V} \exp\left[-\beta\left(U_{N-1} - U_N\right)\right]\right\} \tag{17}$$

One Monte Carlo step consists on choosing one of the three mentioned processes, assuring equal probabilities. In each case, the displacement, adsorption or desorption for every randomly chosen molecule is performed as described in previous reports (Nicholson & Parsonaje, 1982; Frenkel & Smit, 2002; Sweatman & Quirke, 2006). This process is carried out a sufficient number of times (of the order of 2×10^7 Monte Carlo steps) and the average of N (number of molecules) and U (internal energy) is determined. Subsequently, another μ value (or P value) is set and an adequate number of Monte Carlo steps are performed to reach the N and U average values. Thereby, the quantity of adsorbed molecules is calculated as a function of P (or μ), which is precisely, the adsorption isotherm.

The pressure within the reservoir is related to the chemical potential through:

$$\beta\mu = \beta\mu^{\theta}_{id\ gas} + \ln\left(\beta P_{id\ gas}\right) \tag{18}$$

Whether the work pressure is high enough, the ideal gases equation is no longer valid and a state equation must be used to correlate the chemical potential of the reservoir with the pressure.

$$\beta\mu = \beta\mu^{\theta}_{id\ gas} + \ln\left(\beta P\phi\right) \tag{19}$$

where ϕ is the coefficient of fluid fugacity within the reservoir. In order to calculate ϕ, the state equation is used, i.e. Peng-Robinson (Frenkel & Smit, 2002).

2.4.3 Interaction potentials

To perform a simulation of a gas-solid adsorption process, the interaction potentials gas-gas and gas-solid must be taken into account.

Most methods used for molecular simulation consider gases as interaction sites among the centers of the molecules via Lennard-Jones potential. However, this approximation may be enhanced.

For the calculation of the total energy of the system U, it must be considered the interaction potential among adsorbate molecules (Ugg) and between the adsorbent walls and the adsorbate molecules (Ugs).

The interaction potential among adsorbate molecules is given by the Lennard-Jones potential:

$$U_{gg}(r) = -4\varepsilon_{gg}\left\{\left(\frac{\sigma_{gg}}{r}\right)^{6} - \left(\frac{\sigma_{gg}}{r}\right)^{12}\right\} \tag{20}$$

Where ε_{gg} and σ_{gg} are the energetic parameters of the L-J potential and r is the separation among the molecules. The calculation of this potential is subjected to a cut-off distance beyond which, it is assumed that the Ugg potential is zero ($6\sigma_{gg}$ is usually taken).

In order to calculate the gas-solid potential, Ugs, some aspects must be taken into account such as the chemical composition of the solid surface and more importantly, the pore shape, especially if the pores are small.

For the slit-shaped pores, like two parallel graphite layers as illustrated in Figure 12, the potential proposed by Steele (Steele, 1974) may be applied. The following expression corresponds to this potential to calculate the interaction between an L-J site of an adsorbate molecule i, and the graphite layer of the surface, s:

$$U_{i_\alpha s}(z_{i_\alpha}) = 2\pi\varepsilon_{is}\rho_c\sigma_{is}^2\Delta\left\{\frac{2}{5}\left(\frac{\sigma_{is}}{z_{i_\alpha}}\right)^{10} - \left(\frac{\sigma_{is}}{z_{i_\alpha}}\right)^{4} - \frac{\sigma_{is}^4}{3\Delta\left(z_{i_\alpha} + 0.61\Delta\right)^3}\right\} \tag{21}$$

Where ρ_c is the density of interaction centers within the pore wall and Δ is the separation between the graphite layers of the pore wall. For graphite, ρ_c=114 nm^{-3} and Δ=0.335 nm and z represents the distance between the mass center of the adsorbate molecule and the centers of the carbon atoms from the first layer of the surface.

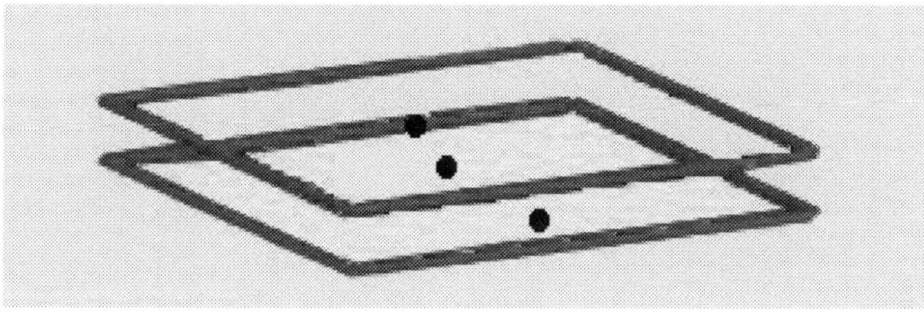

Fig. 12. Geometry of the pore showing parallel layers configuration, widely used to represent activated carbons.

Figure 13 illustrates an example of the interaction potential behavior between a pore wall (graphite) and a CH_4 molecule as a function of the separation distance from the z centers.

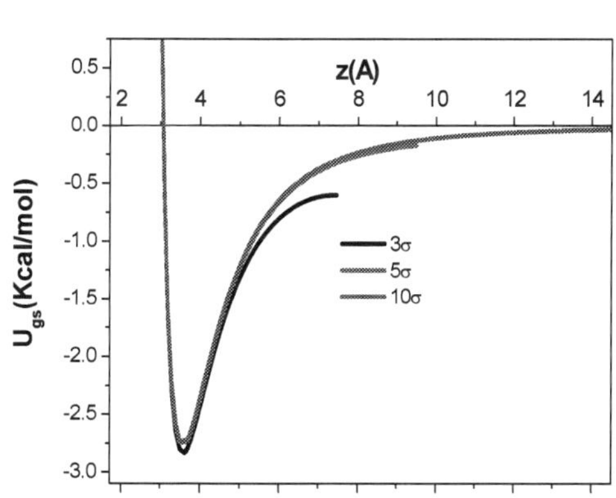

Fig. 13. Gas-solid interaction potential for a graphite layer and a CH_4 molecule for three pore sizes: $3\sigma_{gg}$, $5\sigma_{gg}$ and 10 σ_{gg}.

2.4.4 Characterization – Determination of the pore size distribution

The pore size distribution (PSD) of a porous material is one of the most crucial properties to predict the expected behavior for that material. This is particularly important because the application of the material (gas separation, gas storage, pollutant adsorption, etc.) is based on this characteristic.

The determination of the pore size of non-crystalline materials, such as the activated carbon, is not an easy task considering that materials showing a crystalline arrange, such as zeolites, require a simple DRX analysis to accurately determine the pore size. Gas adsorption probably constitutes the most used tool for PSD calculation and at this point, molecular simulation has played a relevant role during the last few years.

To obtain the PSD, the solution of the generalized equation of adsorption is required:

$$V(P) = A\int f(w)v(w,P)dw \tag{22}$$

where *V(P)* corresponds to the experimental isotherm (volume of excess adsorbate under STP per gram of adsorbent), *f*(w) is the pore size distribution and *v*(w,P) is the average density (in excess) of adsorbate in a pore with size w, obtained by simulation.

Equation 22 is a Fredholm integral equation of the first type and its solution does not constitute a simple problem. Therefore, various methodologies of resolution are known for this equation, among which are remarkable the best fit and matricial methods. In any case, the introduction of regularization parameters is necessary to ensure a "more physical" meaning for the calculated PSD (Sweatman & Quirke, 2006).

The study of the characterization methods for porous materials, in particular by PSD, is an active field, supported by numerous reports. For activated carbons (AC), besides the slit-shape pores, other geometries have been studied, like the squared, rectangular (Davies & Seaton, 1998; Davies & Seaton, 1999, Davies et al., 1999) or triangular (Azevedo et al., 2010) geometries as well as the introduction of heterogeneities on the surface of the graphite layers (Lucena et al., 2010). In general, these approximations are improvements regarding the simulated isotherm fitting and constitute an example of the vast number of factors that have to be considered when simulating the adsorption of this kind of materials.

Another study field is the PSD calculation from the adsorption isotherms obtained for diverse gases, having different sizes and thermodynamic conditions (below or above the Tc) (Quirke & Tennison, 1996; Samios et al., 1997; Ravikovitch et al., 2000; Scaife et al., 2000; Sweatman & Quike, 2001a; Sweatman & Quike, 2001b; Jagiello & Thommes, 2004; Jagiello et al., 2007; Konstantakou et al., 2007; García Blanco et al., 2010). Results have shown discrepancies among the obtained PSDs and have evidenced the convenience of using gases as CO_2 and H_2 for the characterization of materials exhibiting ultramicropores (smaller than 0.7 nm). The N_2 at 77K has shown diffusion limitations at this region but is quite useful for pores with higher sizes, such as the mesopores (50 nm). Therefore, it seems to be clear that a special gas that characterizes with absolute accuracy does not exist. However, several gases can be used in order to obtain an adequate characterization. The characterization of the material under the habitual conditions in which it would be employed can be even more important, for example in the study of the methane storage.

Revealing information has been obtained through simulation techniques regarding the pores required for the storage of methane. For instance, Cracknell et al. 1993 reported a study from a Grand Canonical Monte Carlo (GCMC) simulation, where they compared the methane adsorption on AC with pores showing diverse geometries. It was found that the AC that shows the geometry of parallel plane layers, is also the one that posses the highest adsorption capacity: 166 g/L at 274 K at 34 bar contrasting with results obtained for the zeolites-type geometry, 53.1 g/L, under the same conditions. Additionally, an optimal size of 3 σ_{gg} was reported by Tan & Gubbins, 1990 from data of GCMC and NLDFT simulations. It was concluded that the pore size that maximizes the adsorption of methane falls between 2.9σ_{gg} and 3.9 σ_{gg} (1.1-1.4 nm). Matranga et al., 1992 determined a size of 1.14 nm for a pore showing parallel plane layers configuration intended for a storage system at 34 bar.

2.4.5 Density of the adsorbed phase of methane

As it has been mentioned, the methane overcomes its critical temperature at room temperature and therefore, it should be impossible to condensate under isothermal conditions. This implies that the phase cannot be assumed to remain in liquid phase under these conditions, as occurs for the adsorption of vapors. Consequently, the state of an adsorbed phase for supercritical gases is an unclear subject and for that reason, there are diverse approximates for the calculation of the density or volume of the adsorbed phase (Murata et al., 2001; Zhou et al., 2001; Do & Do, 2003).

Previous studies have reported that the adsorption of supercritical gases tends to accumulate the adsorbate molecules in the neighborhood of the adsorbent surface. Also, it has been observed that a monolayer is usually formed to the distance from the wall that matches the minimum value of the curve of gas-solid potential. This means that only the micropores having sizes of a few molecular diameters are "full" of adsorbate, while higher micropores and mesopores have an adsorbed phase of one or two molecular diameters. This has important consequences on the study of the adsorbed natural gas and agrees with the pore sizes proposed by the bibliography.

Figure 14 illustrates the density profiles of molecules adsorbed into the pore obtained by Monte Carlo simulation. In small pore sizes (3σ_{gg}), there is a vast quantity of molecules distributed into the pore but, as the pore size increases, it can be seen at most, the adsorbed phase composed by a layer of two molecular diameters of thickness. At a pressure of 35 bar, no adsorbed phase was detected in the center of a 7σgg (2.6 nm) pore, instead, the density profile matches the density of the gas at this pressure.

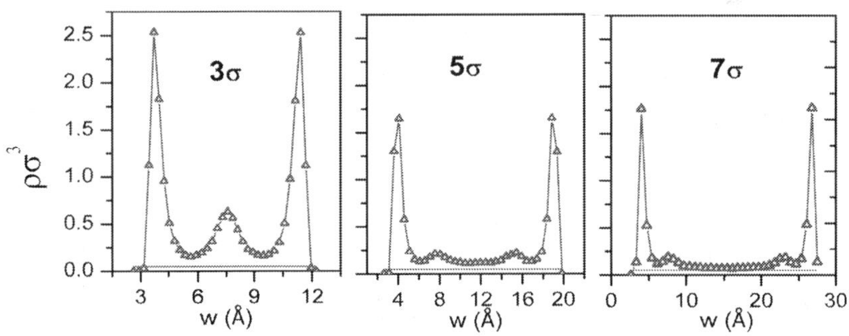

Fig. 14. Density profiles for methane molecules in slit shaped pores with different widths (3,5,7σ_{gg}) at 35 bar obtained from GCMC (Figure supplied by A. de Oliveira, INFAP-CONICET).

3. Adsorbents for the ANG process

Various studies have concluded that the features required by an adsorbent to be adequate for the ANG process are:

a) High adsorption capacity.
b) High adsorption/desorption relations.
c) Micropore sizes of approximate 0.8 nm (bigger than the sizes of two molecules of methane) to facilitate the gas release at room temperature.
d) High packaging density to ensure that the storage capacity and the energetic density are high.
e) Low adsorption heat and high specific heat to minimize the temperature variation in the tank through the adsorption and desorption processes.
f) Suitable properties for the mass transference.
g) Being extremely hydrophobic.
h) Being inexpensive.

To evaluate the quality of the adsorbent for being used in the ANG process for vehicles, there is a parameter called "delivery". It is defined as the gas delivered per unit of stored gas and is expressed by V/V. Specifically delivery is the quantity of gas released from the adsorbent when pressure is reduced to the atmospheric pressure.
During the 90s, the United States Department of Energy of the (USDOE) established an objective of 150 v/v of delivery in vehicles having ANG working at a pressure of 3.5 MPa at 25°C (Cook et al., 1999).
In general, the adsorbent has a porous structure where the molecules that pass through it, can be retained (adsorbed) due to the high affinity that exhibit towards the adsorbent. These adsorbed molecules have a higher density than the one showed at the gas phase.
The suitable adsorbent to be used in the ANG process must be predominantly microporous. Therefore, the storage capacity is optimal when the volume fraction of the deposit corresponding to the micropores, is maximum. Besides, certain contribution of mesoporosity with a size smaller than 5 nm is, to some extent, required to yield the circulation of methane to the micropores interior.

4. Results from our research group

The final section of this chapter presents some of the results obtained in our laboratory regarding ANG studies.

4.1 Powdered activated carbons

In the first place, are shown results using activated carbons (AC) obtained from inexpensive materials as agricultural residues produced by the regional industry. ACs are synthesized by two procedures:
-Chemical activation, using an activating agent such as zinc chloride ($ZnCl_2$).
-Physical activation, using water vapor as activating agent.

Olive and grape lexes were used as precursors. The term lex is used to designate the residue that is left after the oil extraction from the seeds. The other precursor employed was the remains of olive wood resulting from the trees prune.
The synthesis of the chemically activated carbons was carried out as described by Solar et al., 2008 following Tsai et al., 1998. The samples were named by taking into account the parameters of synthesis XZnY, where X is the selected raw material; Zn is the activating agent ($ZnCl_2$) and Y, the impregnation relation. The notation for the used raw material (X) comes from their Spanish name: olive lex (Ac), grape lex (Uv) and olive wood (MO).
Carbonized olive lex and olive wood were used for the physical activation. For the samples nomenclature, the raw material was also considered, for example the carbonized olive lex was named LAC and the carbonized olive wood, MOC. The work conditions were described by Solar et al., 2008.
For the textural analysis of the samples, nitrogen adsorption-desorption isotherms were conducted using an AUTOSORB-1MP (Quantachrome Instruments) and an ASAP 2000 (Micromeritics Instruments Corp). Samples were previously degassed at 250°C and the study was carried out at liquid nitrogen temperature, 77K (-196°C). The specific surface was calculated with the Brunauer, Emmet and Teller, BET, method and the pore size distribution by the Density Functional Theory, DFT.
The adsorption analyses of methane were performed using the HPVA100 volumetric equipment (VTI Corp). The apparatus provides isotherms up to 100 bar of pressure at a wide range of temperatures.
Figure 15 shows the adsorption-desorption nitrogen isotherms at 77K for the prepared carbons.

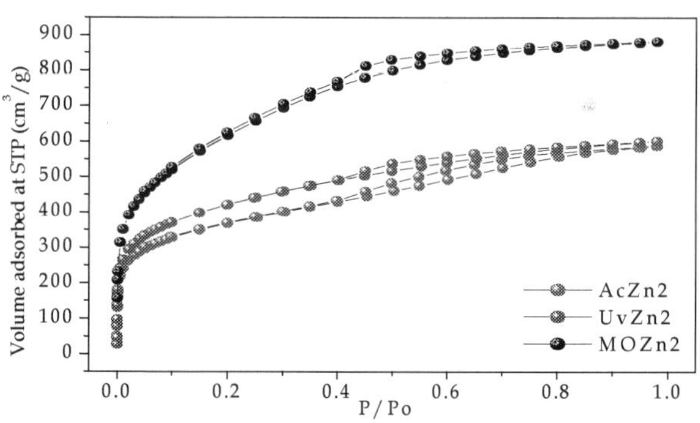

Fig. 15. N_2 adsorption-desorption isotherms obtained from the chemically activated carbons.

The shape of the isotherms is a combination of the Type I at low pressures, characteristic of the microporous solids, and Type IV for higher pressures, according to the Brunauer-Deming-Deming-Teller (BDDT) classification given by Gregg & Sing, 1982. A slight hysteresis loop of the type H2, according to the classification given by Rouquerol et al., 1999 and Martín Martínez, 1990, can be seen, and usually associated to narrow slit-shaped pores.
Table 2 summarizes the data regarding the textural characteristics of the carbons.

Sample	S_{BET} m^2/g	V_T cm^3/g	V_o cm^3/g
AcZn2	1291	0.91	0.54
UvZn2	1470	0.93	0.62
MOZn2	2205	1.36	0.88

Table 2. Textural data from chemically activated carbons.

The specific surface area was calculated by the BET method (S_{BET}) and the micropore volume (V_o) by the Dubinin-Radushkevich method. Finally, the total pore volume (V_T) was estimated from the adsorption of nitrogen at a relative pressure of 0.98 applying the Gurvich rule (Rouquerol et al., 1999). As it can be seen, the sample synthesized from the olive wood is the one showing the better values, that is, higher specific surface area and total volume of pores and micropores.
Figure 16 shows the pore size distribution for the chemically activated samples, where the presence of mesopores can be observed. It was also evidenced by the hysteresis loops of the isotherms. The activated carbon obtained from the olive wood shows a higher quantity of pores, being micro and mesopores. The presence of pores with approximate size of 5Å is more evident for the AcZn2 and UvZn2 samples. The three samples exhibit pores ranging between 10 and 15Å.

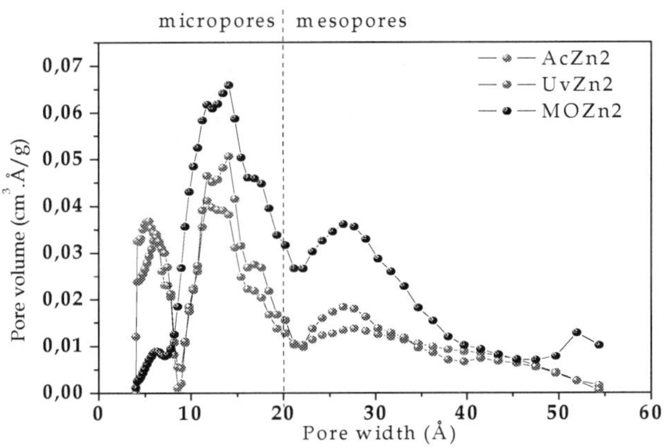

Fig. 16. Pore size distribution (DFT) of the chemically activated carbons.

These materials were assayed in the methane adsorption at high pressure up to 40 bar and at 25°C (Figure 17).

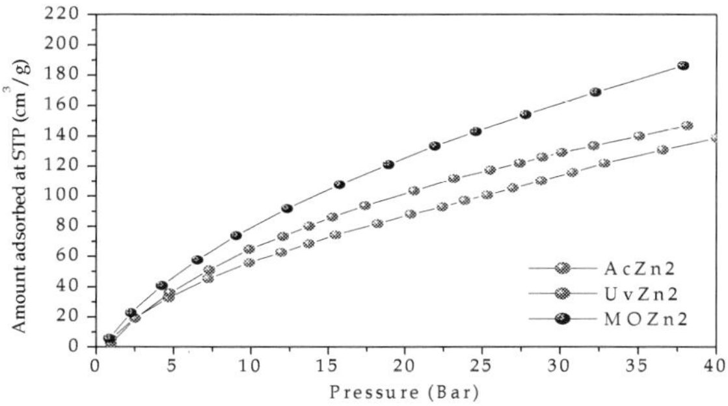

Fig. 17. Methane isotherms from the chemically activated carbons.

Figure 17 shows the isotherms where the samples with higher micropore volumes (Table 2) show higher methane capacity of adsorption in agree with previous reports (Celzard et al., 2005; Lozano-Castelló et al., 2002a; Lozano-Castelló et al., 2002b).
Figure 18 illustrates the isotherms of the physically activated carbon using water vapor. The shape of the isotherms is also a combination of the **Type I** and **IV**, according to the BBDT classification. H2 Hysteresis loops are present.

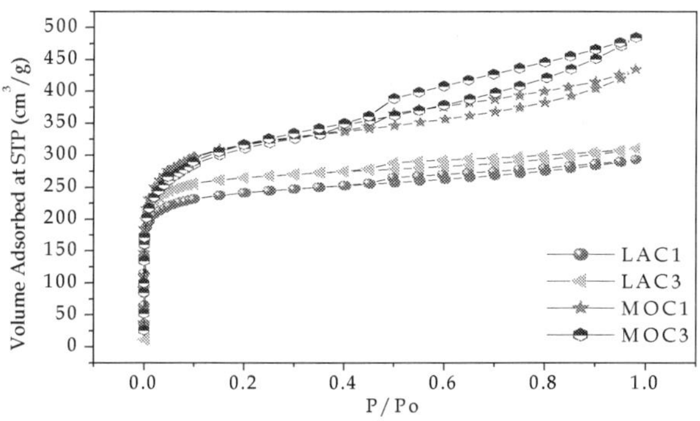

Fig. 18. N_2 adsorption-desorption isotherm from the physically activated carbons.

Table 3 summarizes the textural data from the physically activated carbons using carbonized olive lex (LAC) and carbonized olive wood (MOC) as precursors. By increasing in 30 minutes the activating time for each sample, no significant changes are observed in the specific area, total volume of pores and micropores. Therefore, it seems unnecessary to increase the time by half an hour as the new values are very similar and, at industrial level, the addition of time would increase the costs of the process.

Sample	S_{BET} m^2/g	V_T cm^3/g	V_o cm^3/g	Agent/Material	Temp (ºC)	Time (min)
LAC1	913	0.45	0.37	1g/gh	900	120
LAC3	1015	0.48	0.42	1g/gh	900	150
MOC1	1163	0.67	0.48	1g/gh	900	120
MOC3	1117	0.75	0.46	1g/gh	900	150

Table 3. Textural data from the physically activated carbons.

Comparing both set of samples, certain uniformity can be seen related to the specific surface area values, even though the MOC samples show higher values for the total volume of pores and micropores.

Figure 19 illustrates the pore size distribution (DFT) for the physically activated carbons. A similar behavior may be observed between both of them. For the carbons activated from olive lex, it was found that a longer activation time causes the loss of pores smaller than 10Å. However, the time variation applied to the set of activated carbons obtained from olive wood, does not result in a notable change.

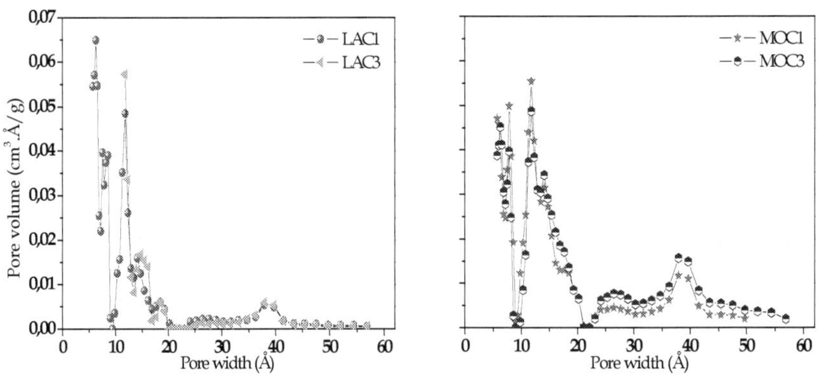

Fig. 19. Pore size distribution (DFT) from the physically activated carbons.

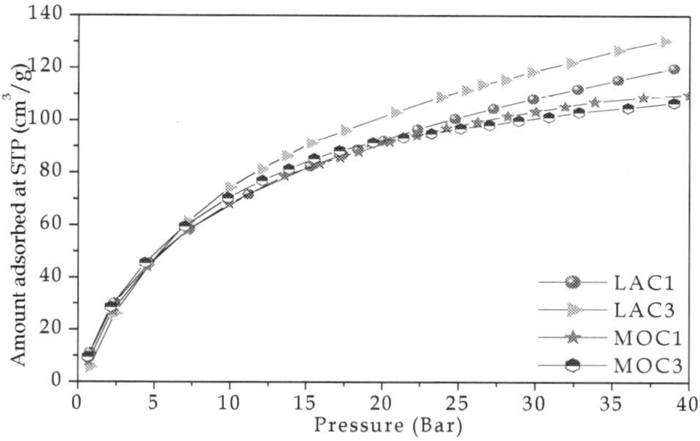

Fig. 20. Methane isotherms from the physically activated carbons.

Regarding the behavior of the physically activated samples in the adsorption of methane, it may be deduced from Figure 20 that they are very similar at low pressures (approximately up to 15 bar). For these samples, a correlation also exists with the nitrogen adsorption that is analogous to the previously observed. However, they show smaller adsorption values than the obtained from the chemically activated samples.

4.2 Monolithic activated carbons

With the aim of increasing the density of the materials and improving their manipulation at technological level, activated carbons in the shape of conglomerates (monoliths) were prepared. Following the methodology described by Almansa et al., 2004, monolithic activated carbons were obtained from coconut shell to study the storage of methane. Figure 21 shows the photographs of the materials.

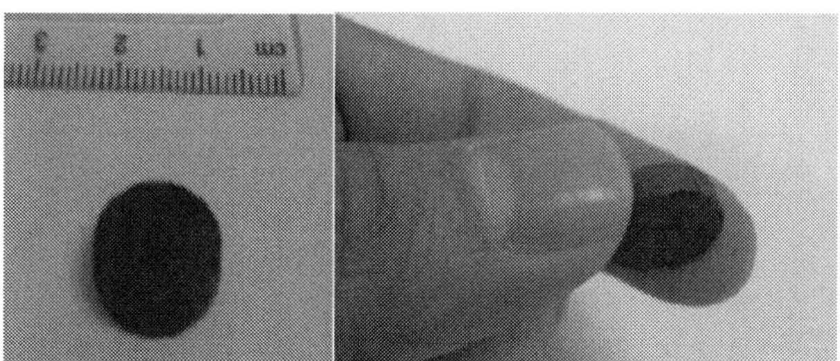

Fig. 21. Photographs of monolithic activated carbons.

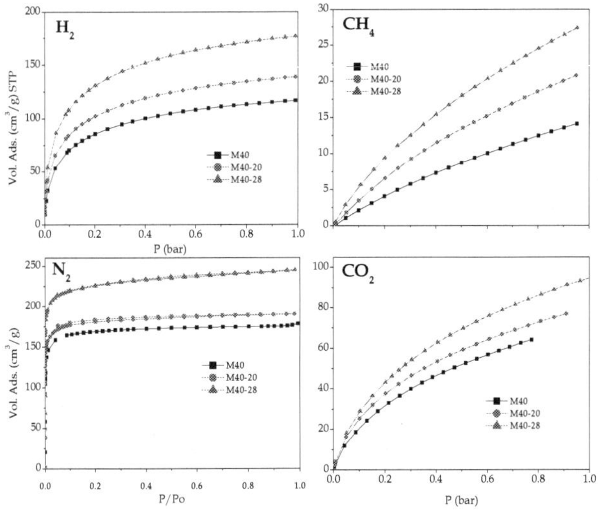

Fig. 22. Adsorption isotherms of N_2 at 77K, H_2 at 77K, CO_2 at 273K and CH_4 at 298K under subatmospheric conditions on the monolithic activated carbons.

The materials were chemically activated with a 40% in weight of $ZnCl_2$ (M40) and subsequently activated using CO_2 to develop higher microporosity up to burn-off percentages of 20 and 28% (M40-20 and M40-28 samples, respectively). For these materials, adsorption isotherms of N_2 at 77K, H_2 at 77K, CO_2 at 273K and CH_4 at 298 K, were measured (Figure 22) at subatmospheric pressures.

From these data, the pore size distribution for each gas was obtained by using a data base of the simulated isotherms through the Monte Carlo method in the Grand Canonical for slit-shaped pores (Figs. 23 to 25). The development of narrow microporosity due to the final activation with CO_2 from the chemically activated monoliths can be observed. Also, the convenience of using various gases for the adequate characterization of the activated carbons becomes evident. The PSDs calculated from the isotherms of CO_2, H_2 and CH_4 can detect narrow porosity that N_2 cannot because, as discussed before, it seems to have diffusion problems. The similitude between the PSDs obtained from CO_2, H_2 and CH_4 at subatmosferic pressures, should be noted (García Blanco et al., 2010).

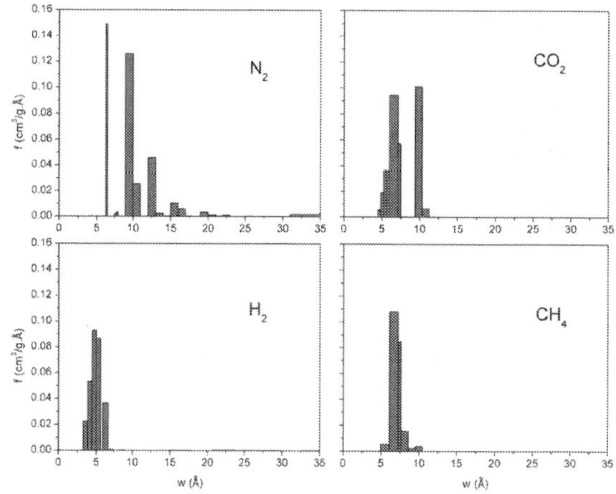

Fig. 23. Pore size distribution from the adsorption isotherms of N_2, H_2, CO_2 and CH_4 for the M40 monolith.

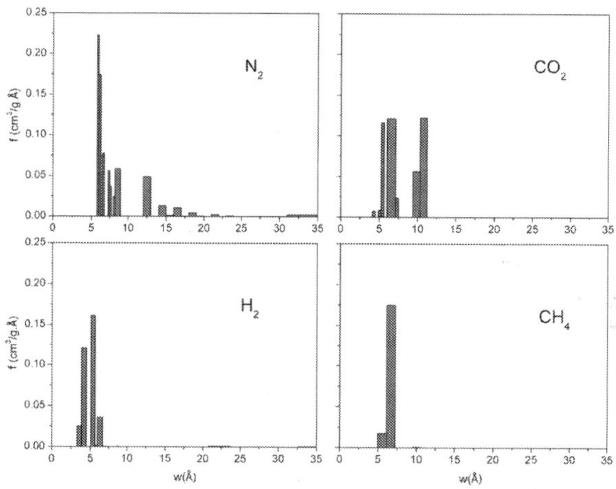

Fig. 24. Pore size distribution from the adsorption isotherms of N_2, H_2, CO_2 and CH_4 for the M40-20 monolith.

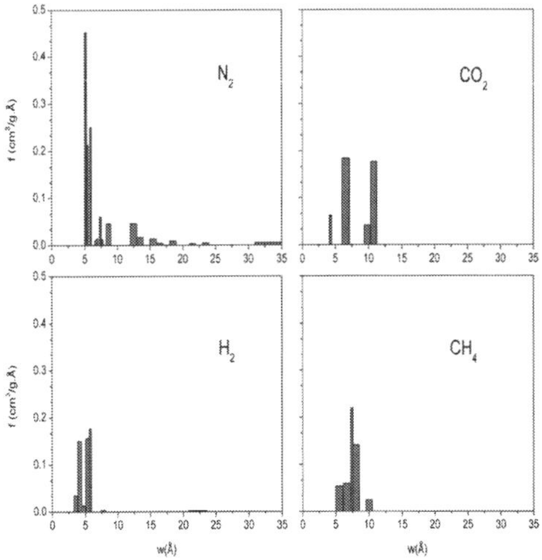

Fig. 25. Pore size distribution from the adsorption isotherms of N_2, H_2, CO_2 and CH_4 for the M40-28 monolith.

Table 4 summarizes the textural data of the samples, comparing diverse methodologies for obtaining the micropore volume. Calculations were made by semi-empirical methods, such as Dubinin-Raduschevich equation and the α-plot method (Gregg & Sing, 1982). The development of the microporosity in the samples and the consistency of the obtained data by the calculated PSDs through Monte Carlo, are remarkable.

	N_2			CO_2		H_2	LP-CH_4
	V_o DR (cm^3/g)	V_o α-plot (cm^3/g)	V_o MC (cm^3/g)	V_o DR (cm^3/g)	V_o MC (cm^3/g)	V_o MC (cm^3/g)	V_o MC (cm^3/g)
M40-0	0.268	0.245	0.259	0.259	0.291	0.175	0.182
M40-20	0.276	0.260	0.291	0.269	0.278	0.204	0.193
M40-28	0.340	0.329	0.378	0.360	0.361	0.269	0.328

Table 4. Textural data of the monolithic activated carbons.

In Figure 26, the adsorption isotherms of CH_4 at 298K and high pressure for the mentioned samples, are shown. The increase in the storage capacity of methane can be seen in accordance to the increase in the microporosity of the samples. This latter was accomplished by the activation with CO_2.

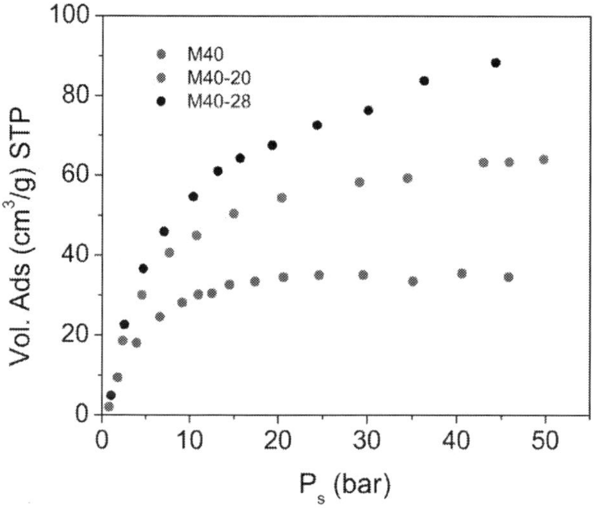

Fig. 26. Methane isotherms from the monolithic activated carbons.

Table 5 present the data obtained for the storage capacity *Q´*, expressed as methane volume stored under STP per stored volume (V/V) calculated at 35 bars with the following equation, given by Celzard et al., 2005:

$$Q' = V/V = Q \cdot M \cdot \mu \cdot \delta ap \tag{23}$$

where *Q* is the molar storage capacity (mol of methane/kg of activated carbon), *M* is the molecular weight of methane (g/mol), μ is the volume occupied by 1 gram of methane under STP conditions (1.5dm^3/g) and δap is the apparent density of activated carbon (g/cm^3).

Sample	Ads. Vol. of CH_4 at 35 bar (cm^3/g)	δap (g/cm^3)	V/V
AcZn2	127	0.38	52
UvZn2	129	0.20	28
MOZn2	178	0.30	57
LAC1	115	0.49	60
LAC3	104	0.50	59
MOC1	126	0.18	24
MOC3	98	0.21	23
M40	35	0.80	30
M40-20	60	0.65	42
M40-28	80	0.60	51

Table 5. Methane storage data.

The physically activated samples, called LACs, show improved values of methane storage (approximately 60 v/v) because of its high apparent density. The MOZn2 sample presents higher methane adsorption than LACs but, because of their lower apparent density, they have similar methane storage capacity. Elevated apparent densities can be seen for the monolithic activated carbons. This enhances the storage capacities compared to a sample showing similar textural properties.

4.3 Adsorption of methane on other porous materials

4.3.1 Zeolites and pillared clays (PILCs)

It was studied the adsorption of methane for zeolites (MS-5A and MS-13X with defined pore sizes of 5 Å and 10 Å respectively) and for aluminium pillared clays (PILC Al).
Figure 27 illustrates the isotherms of N_2 at 77K for these materials. As it can be seen, zeolites are strictly microporous materials, showing N_2 adsorption isotherms of Type I. The pillared clay is a micro-mesoporous material (Sapag & Mendioroz, 2001) and the resulting isotherm corresponds to a combination of the Type I and IIb isotherms (Rouquerol et al., 1999). In Table 6, textural properties of the materials calculated from N_2 isotherms, are shown.

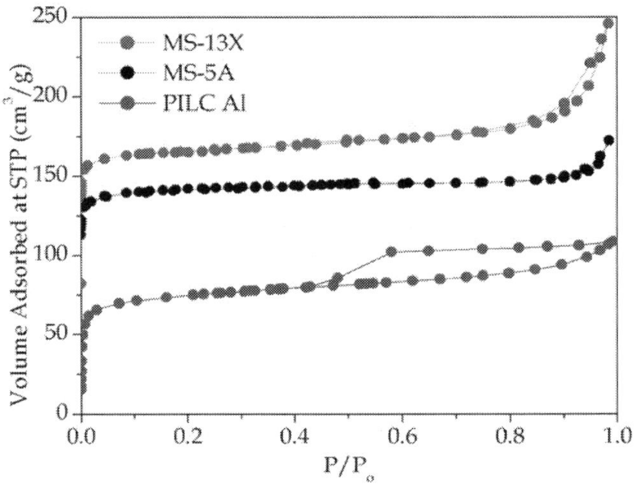

Fig. 27. N_2 adsorption-desorption isotherm for zeolites and PILC.

	S_{BET} (m^2/g)	V_o DR (cm^3/g)	V_T (cm^3/g)
MS-13X	725	0.257	0.320
MS-5A	613	0.221	0.267
PILC Al	283	0.106*	0.170

*Calculated by α-plot

Table 6. Textural data of zeolites and PILC.

In Figure 28 are presented the adsorption isotherms of CH_4 at 298K for zeolites and PILC, at high pressures. For zeolites, the methane adsorption capacity is low due to their pore geometry, among other factors. In addition, the storage capacity of the PILC is particularly low, which is consistent with its lower micropores content in comparison to other materials.

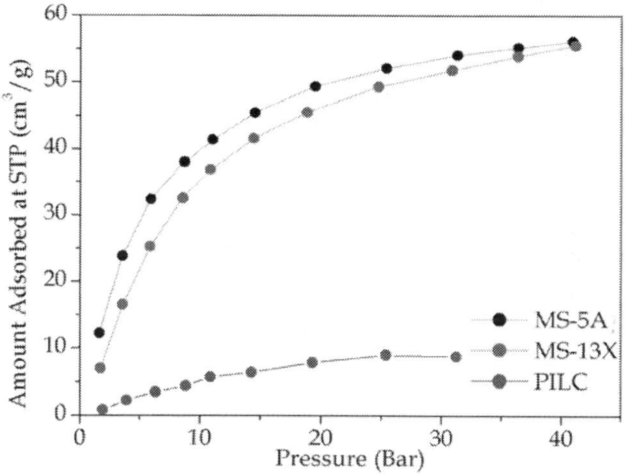

Fig. 28. Methane isotherm for zeolites and PILC.

4.3.2 Carbon nanotubes (NT)

The storage of methane using single-walled carbon nanotubes (SWNT) has been studied. The nanotubes were obtained by chemical vapor deposition (CVD) and commercialized by Carbon Solutions Inc. Since this type of nanotubes usually contain impurities of the catalyst from which they were obtained and from amorphous carbon present with the nanotubes, they are subjected to a purification treatment through the refluxing in concentrated nitric acid (to 65% in weight) at 120°C for 6 hours (NT 6h).

Carbon nanotubes are commonly grouped in bundles of various nanotubes, where the original NT is closed in their end. The treated NT can be opened but they have functional groups at the ends blocking the entrance of the adsorbate molecules (Kuznetsova et al., 2000). Therefore, the adsorption for this kind of materials occur on the IC, G and S sites, indicated in Figure 29, and they have the size of the micropores.

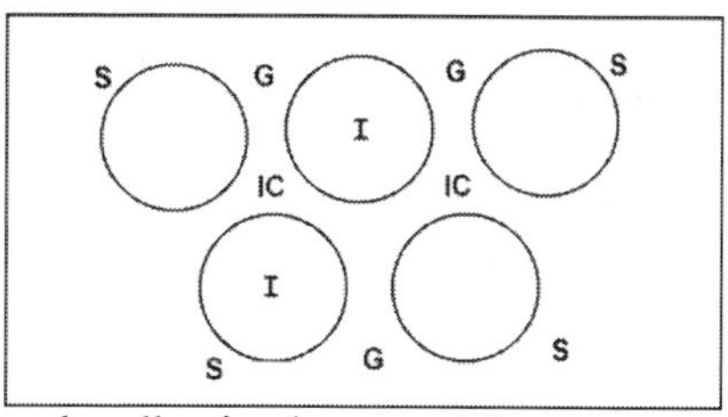

Fig. 29. Adsorption sites in a bundle of carbon nanotubes.

Figure 30 illustrates the N_2 isotherms at 77K of these materials. An important increase in the zone of high relative pressure in the original NT takes place. This is due to the N_2 condensation in the empty sites generated between the nanotubes bundles, corresponding to the meso and macropores. The acid treatment densifies and removes the empty sites (Yang et al., 2005) and the resulting isotherm of the purified nanotubes shows the expected behavior for a microporous material (sites from Figure 29). Table 7 summarizes the textural properties of the materials calculated from the N_2 isotherms.

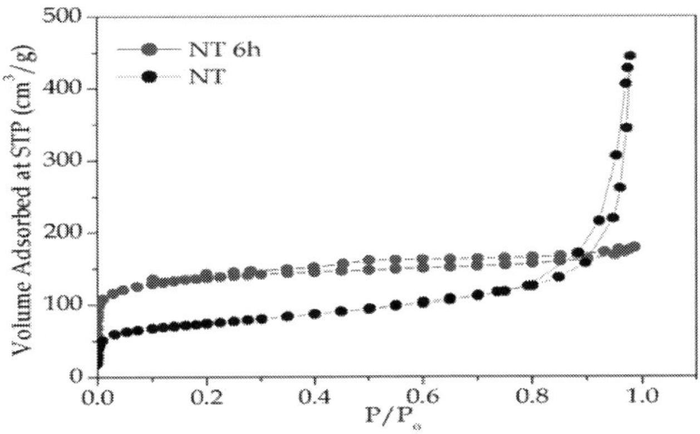

Fig. 30. N_2 adsorption-desorption isotherms of carbon nanotubes.

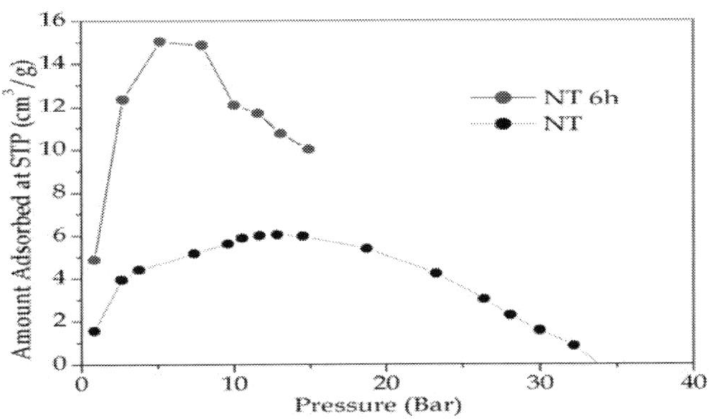

Fig. 31. Methane isotherm of the carbon nanotubes.

	S_{BET} (m^2/g)	V_o DR (cm^3/g)	V_T (cm^3/g)
NT	265	0.11	0.44
NT 6h	510	0.20	0.27

Table 7. Textural data from carbon nanotubes.

Figure 31 corresponds to the CH_4 adsorption at high pressures of the original nanotubes (NT) and the purified nanotubes (NT 6h). For both samples, the CH_4 adsorption is low, indicating that these materials are not suitable for the storage of methane. On the other hand, the decrease in the adsorbed volume along with the pressure increase is due to the saturation of the adsorption sites that are available for methane. Similar observations have been previously reported (Menon, 1968).

4.3.3 Metal Organic Frameworks (MOFs)

The adsorption of methane on MOFs has been studied. MOFs are produced by BASF and commercialized under the denomination of Basolite C300, Basolite A100 and Basolite Z1200. MOFs consist on polymeric framework of metal ions bound one to another by organic ligands. The development during the last few years regarding this type of materials is due to the vast study conducted by the group of Yaghi (Li et al., 1999; Barton et al., 1999). The main characteristics of these materials are the well-arranged pore structure as well as the high pore volume. These features make them attractive for the storage of gases (Lewellyn et al., 2008; Wang et al., 2008; Furukawa & Yaghi, 2009) in spite of their low density.

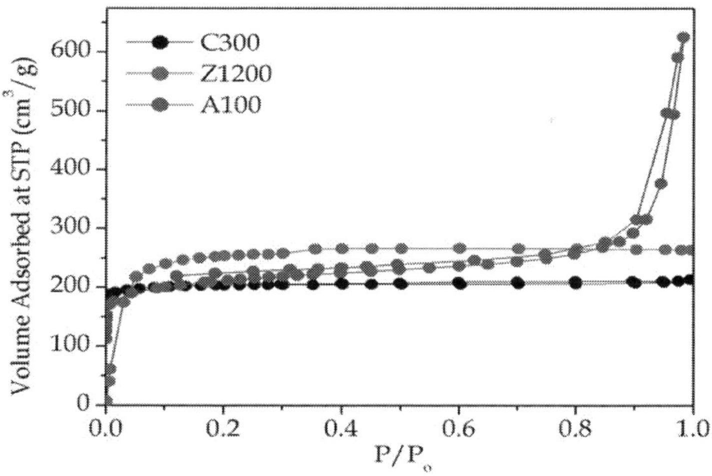

Fig. 32. N_2 adsorption-desorption isotherms of the MOFs.

In Figure 32, an adsorption isotherm of N_2 at 77K for the three studied materials is shown. It is important to note the presence of micropores within the three samples, which is remarked

by the abrupt increase of adsorbed volume at low relative pressures. The isotherms of the C300 and Z1200 samples, present a characteristic plateau of isotherms Type I. In contrast, the growth at high relative pressures of the A100 sample is due to the material flexibility, previously reported by Bourelly et al., 2005.
Table 8 summarizes the data corresponding to the textural characterization of the samples from the N_2 adsorption data, confirming its high microporosity.

	S_{BET} (m^2/g)	V_o DR (cm^3/g)	V_T (cm^3/g)
C300	1059	0.440	0.453
A100	837	0.313	0.969
Z1200	1032	0.421	0.425

Table 8. Textural data of MOFs.

Figure 33 shows the isotherms of CH_4 at 298 K at high pressures. As it can be seen, these samples exhibit a high adsorption capacity for methane, particularly the C300, which almost duplicates the values obtained by the other two samples showing a storage capacity of 70 v/v, evidencing its suitability for the methane storage.
To conclude this chapter, we would like to emphasize the necessity of further research on porous materials, particularly if the purpose of the study is to accomplish the technological application of the ANG process for the storage of methane.

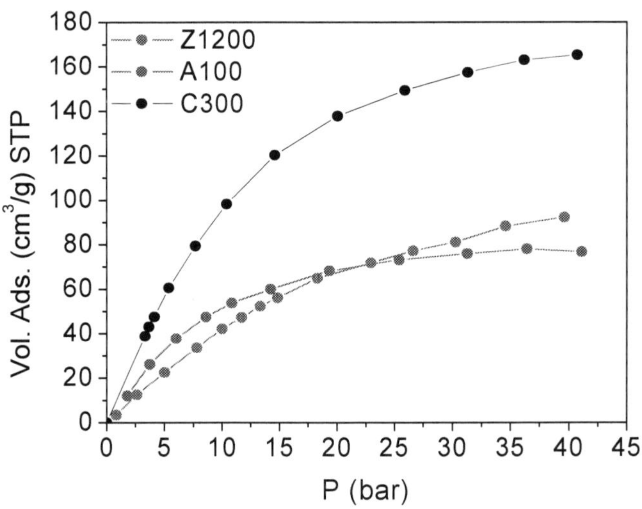

Fig. 33. Methane isotherm for MOFs.

5. Acknowledgements

We want to express our acknowledgement to Universidad Nacional de San Luis, CONICET and FONCyT (Argentine) for the financial support to carry out this work.
Our sincere gratitude to Prof. Aldo Migone, Department of Physics, Southern Illinois University Carbondale, USA, and Prof. Andoni Gil Bravo, Departamento de Química Aplicada, Universidad Pública de Navarra, España for supplying some of the samples reported in this study.

6. References

Alcañiz-Monge, J.; de la Casa-Lillo, M.A.; Cazorla-Amorós, D. & Linares-Solano, A. (1997). Methane storage in activated carbon fibres, *Carbon*, Vol. 35, No. 2, pp. 291-297. ISSN 0008-6223.

Almansa, C.; Molina-Sabio, M. & Rodríguez-Reinoso, F. (2004). Adsorption of methane into ZnCl2-activated carbon derived discs, *Microporous and Mesoporous Materials,* Vol. 76, pp. 185-191. ISSN 1387-1811.

Azevedo, D.C.S.; Rios, R.B.; López R.H.; Torres, A.E.B.; Cavalcante, C.L.; Toso J.P. & Zgrablich G., (2010). Characterization of PSD of activated carbons by using slit and triangular pore geometries. *Applied Surface Science,* Vol. 256, pp. 5191-5197. ISSN 0169-4332.

Barton, T.J.; Buli, L.M.; Klemperer, W.G.; Loy, D.A.; McEnaney, B.; Misono, M.; Monson, P.A.; Pez, G.; Scherer, G.W.; Vartulli, J.C. & Yaghi, O.M. (1999). Tailored porous materials, *Chemistry of Materials,* Vol.11, No.10, pp. 2633-2656. ISSN (electronic): 1089-7690.

Bourelly, S.; Llewellyn, P.L.; Serre, C.; Millange, F.; Loiseau, T & Férey G. (2005). Different adsorption behaviors of methane and carbon dioxide in the isotypic nanoporous metal terephtalates MIL-53 and MIL-47, *Journal the American Chemical Society,* Vol. 127, pp. 13519-13521. ISSN 00027863.

BP Statistical Review of World Energy 2009. (2009). Beyond Petroleum, London. www.bp.com/statisticalreview

Brunauer, S.; Deming, L.S.; Deming, E.W & Teller, E. (1940). On a Theory of the van der Waals Adsorption of Gases, *Journal the American Chemical Society,* Vol. 62, No. 7, pp. 1723-1732. ISSN 00027863.

Brunauer, S.; Emmett, P.H. & Teller, E. (1938). Adsorption of Gases in Multimolecular Layers, *Journal the American Chemical Society,* Vol. 60, No. 2, pp. 309-319. ISSN 00027863.

Celzard, A.; Albiniak, A.; Jasienko-Halat, M.; Mareche, J.F. & Furdin, G. (2005). Methane storage capacities and pore textures of active carbons undergoing mechanical densification, *Carbon,* Vol. 43, pp. 1990-1999. ISSN 0008-6223.

Comisión Nacional de Energía (CNE) (1999). *Información Básica de los Sectores de la Energía.* Edita: CNE, Comisión Nacional de Energía. Publicaciones periódicas anuales. www.cne.es

Cook, T.L.; Komodromos, C.; Quinn, D.F. & Ragan, S. (1999). Adsorbent Storage for Natural Gas Vehicles, In: *Carbon Materials for Advance Technology,* Timothy D. Burchell (Ed.), p. 269-302, Publisher: Pergamon Press Inc, ISBN 0080426832, New York.

Cracknell, R.F.; Gordon, P. & Gubbins, K.E. (1993). Influence of pore geometry on the design of microporous materials for methane storage, *Journal of Physical Chemistry*, Vol. 97, pp. 494-499. ISSN 0022-3654.

Davies, G.M. & Seaton, N.A. (1998). The effect of the choice of pore model on the characterization of the internal structure of microporous carbons using pore size distributions, *Carbon*, Vol. 36, pp. 1473-1490. ISSN 0008-6223.

Davies, G.M. & Seaton, N.A. (1999). Development and validation of pore structure models for adsorption in activated carbons, *Langmuir*, Vol. 15, pp. 6263-6276. ISSN 0743-7463.

Davies, G.M.; Seaton, N.A. & Vassiliadis, V.S. (1999). Calculation of pore size distributions of activated carbons from adsorption isotherms, *Langmuir*, Vol. 15, pp. 8235-8245. ISSN 0743-7463.

Dirección de Tecnología, Seguridad y Eficiencia Energética (2006). *El gas natural vehicular: un combustible con mucho futuro*. http://www.gasnaturalcomercializadora.com

Do, D.D. & Do, H.D. 2003. Adsorption of supercritical fluids in non-porous and porous carbons: analysis of adsorbed phase volume and density. *Carbon*, Vol. 41, No. 9, pp. 1777-1791. ISSN 0008-6223.

Dubinin, D.D. (1960). The Potential Theory of Adsorption of Gases and Vapors for Adsorbents with Energetically Nonuniform Surfaces. *Chemical Reviews*, Vol. 60, No. 2, pp. 235-241. ISSN (electronic) 1520-6890.

Frenkel, D. & Smit, B. (2002). Understanding molecular simulation: From algorithms to applications, Publisher Academic Press, ISBN 0-12-267351, London.

Furukawa, H. & Yaghi, O.M. (2009). Storage of hydrogen, methane, and carbon dioxide in highly porous covalent organic frameworks for clean energy applications, *Journal the American Chemical Society*, Vol. 131, pp. 8875-8883. ISSN 00027863.

García Blanco, A.A.; Alexandre de Oliveira, J.C.; López, R.; Moreno-Piraján, J.C.; Giraldo, L.; Zgrablich, G. & Sapag, K. (2010) A study of the pore size distribution for activated carbon monoliths and their relationship with the storage of methane and hydrogen. *Colloids and Surfaces A: Physicochemical and Engineering Aspects*, Vol. 357, No. 1-3, pp. 74-83. ISSN 0927-7757.

Garrido, J.; Linares-Solano, A.; Martín-Martínez, J. M.; Molina-Sabio, M.; Rodriguez-Reinoso, F. &. Torregrosa, R. (1987). Use of N2 vs. C02 in the Characterization of Activated Carbons, *Langmuir* 1987, Vol. 3, pp. 76-81. ISSN 0743-7463.

Gregg, S.J. & Sing, K.S.W. (1982) *Adsorption, Surface Area and Porosity*. Published Academic Press, ISBN 0123009561, London.

Gubbins, K.E. (1997). Theory and simulation of adsorption in micropores, In: Physical Adsorption: Experiment, theory and applications, Fraissard J. & Conner W.C., (Ed.), pp. 65-103, Kluwer, ISBN 0-7923-4547-9, Dordrecht.

Hill, T.L. (1986). *An Introduction to Statistical Mechanics*, Dover Publications Inc, ISBN 0-486-65242-4, Mineola N.Y.

Inomata, K.; Kanazawa, K.; Urabe, Y.; Hosono, H. & Araki, T. (2002). Natural gas storage in activated carbon pellets without a binder, *Carbon*, Vol. 40, pp. 87-93. ISSN 0008-6223.

Jagiello, J. & Thommes, M. (2004). Comparison of DFT characterization methods based on N2, Ar, CO2 and H2 adsorption applied to carbons with various pore size distributions. *Carbon*, Vol. 42, No. 7, pp. 1227-1232. ISSN 0008-6223.

Jagiello, J.; Ania, C.O.; Parra, J.B.; Jagiello, L.; Pis, J.J. (2007). Using DFT analysis of adsorption data on multiple gases including H2 for the comprehesive characterization of microporous carbons. *Carbon*, Vol. 45, No. 5, pp. 1066-1071 ISSN 0008-6223.

Konstantakou, M.; Steriotis, Th.A.; Papadopoulos, G.K.; Kainourgiakis, M.; Kikkinides, E.S.; & Stubos A.K. (2007). Characterization of nanoporous carbons by combining CO_2 and H2 sorption data with the Monte Carlo simulations. *Applied Surface Science*, Vol. 253, No. 13, pp. 5715-5720. ISSN 0169-4332.

Kuznetsova, A.; Yates, J.T.; Liu, J. & Smalley, R.E. (2000). Physical adsorption of xenon in open single walled carbon nanotubes: Observation of a quasi-one-dimensional confined Xe phase, *Journal of Chemical Physics*, Vol. 112, No. 21, pp. 9590-9598. ISSN (electronic): 1089-7690.

Lastoskie, C.M.; Gubbins, K.E. & Quirke N. (1993). Pore size distribution analysis of microporous carbons : A Density functional approach, *Journal of Physical Chemistry*, Vol. 97, No. 18, pp. 4786-4796. ISSN 0022-3654.

Li, H.; Eddaoudi, M.; O'Keeffe, M. & Yaghi, O.M. (1999) Design and synthesis of an exceptionally stable and highly porous metal-organic framework, *Nature*, Vol. 402, pp. 276-279. ISSN (electronic) 1476-4687.

Llewellyn, P.L.; Bourrelly, S.; Serre, C.; Vimont, A.; Daturi, M.; Hamon, L.; De Weireld. G.; Chang, J.S.; Hong, D.Y.; Hwang, Y.K.; Jhung, S.H. & Férey, G. (2008). High uptakes of CO2 and CH4 in mesoporous metal-organic frameworks MIL-100 and MIL-101, *Langmuir* Vol. 24, pp. 7245-7250. ISSN 0743-7463.

Lozano-Castelló, D.; Alcañiz-Monge, J.; De La Casa-Lillo, M.A.: Cazorla-Amorós, D. & Linares-Solano, A. (2002a). Advances in the study of methane storage in porous carbonaceous materials, *Fuel*, Vol. 81, pp. 1777-1803. ISSN 0016-2361.

Lozano-Castello, D.; Cazorla-Amorós, A.; Linares-Solano, A. & Quinn, D.F. (2002b). Activated carbon monoliths for methane storage: influence of binder, *Carbon* Vol. 40, pp. 2817-2825. ISSN 0008-6223.

Lozano-Castelló, D.; Cazorla-Amorós, D.; Linares-Solano, A. & Quinn, D.F. (2002c). Influence of pore size distribution on methane storage at relatively low pressure: preparation of activated carbon with optimum pore size, *Carbon*, Vol. 40, pp. 989-1002. ISSN 0008-6223.

Lucena, S.M.P.; Paiva, C.A.S.; Silvino, P.F.G.; Azevedo, D.C.S. & Cavalcante Jr., C.L. (2010) The effect of heterogeneity in the randomly etched graphite model for carbon pore size characterization, *Carbon*, Vol. 48, pp. 2554-2565. ISSN 0008-6223.

MacDonald, J.A.F. & Quinn, D.F. (1998). Carbon adsorbents for natural gas storage, *Fuel*, Vol. 77, No. 112, pp. 61-64. ISSN 0016-2361.

Marsh, H. & Rodriguez-Reinoso, F. (2006). *Activated Carbon*. Publisher: Elsevier Science & Technology Books, ISBN 0080444636, Great Britain.

Martín Martínez, J.M. (1990). *Adsorción física de gases y vapores por carbones*, Secretariado de Publicaciones de la Universidad de Alicante (Ed.), Published: Imprenta de la Universidad, ISBN 84-86809-33-9. Universidad de Alicante.

Matranga, K.R.; Myers, A.L. & Glandt, E.D. (1992). Storage of natural gas by adsorption on activated carbon, *Chemical Engineering Science*, Vol. 47, pp. 1569-1579. ISSN 0009-2509.

Menon, P.G. (1968). Adsorption at high pressures, *Chemical Reviews*, Vol. 68, No. 3, pp. 253-373. ISSN (electronic) 1520-6890

Menon, V.C. & Komarnei, S. (1998). Porous adsorbents for vehicular natural gas storage: a review, *Journal of Porous Materials*, Vol. 5, pp. 43-58. ISSN 1380-2224.

Mentasty, L.; Faccio, R.J. & Zgrablich, G. (1991). High Pressure Methane Adsorption in 5A Zeolite and the Nature of Gas-Solid Interactions, *Adsorption Science & Technology*, Vol. 8, pp. 105. ISSN 0263-6174.

Murata, K.; El-Merraoui, M. & Kaneko, K. 2001. A new determination method of absolute adsorption isotherm of supercritical gases under high pressure with a special relevance to density-functional theory study. *Journal of Chemical Physics*, Vol. 114, No. 9, pp. 4196-4205. ISSN (electronic): 1089-7690.

Natural Gas and Climate Change Policy. The European Gas Industry`s View. (1998) EUROGAS, Bélgica. http://www.eurogas.org/publications_environment.aspx

Neimark A.V. & Ravikovitch P.I. (1997). Calibration of pore volume in adsorption experiments and theoretical models, *Langmuir*, Vol. 13, No. 19, pp. 5148-5160. ISSN 0743-7463.

Neimark A.V.; Ravikovitch P.I. & Vishnyakov A. (2000). Adsorption hysteresis in nanopores, *Physical Review E*, Vol 62, No. 2, pp. 1493-1496. ISSN 1550-2376 (online).

Nicholson, D. & Parsonage, N.G. (1982). *Computer simulation and the statistical mechanics of adsorption*, Academic Press, ISBN 0125180608, London.

Parkyns, N.D. & Quinn, D.F. (1995). Natural Gas Adsorbed on Carbon, *Porosity in Carbons*, J. W. Patrick (Ed.), pp. 292-325, ISBN 470-23-454-7, London.

Prauchner, M.J. & Rodríguez-Reinoso, F. (2008). Preparation of granular activated carbons for adsorption of natural gas, *Microporous and Mesoporous Materials*, Vol. 109, pp. 581-584. ISSN: 1387-1811.

Quirke, N. & Tennison, S.R.R. (1996). The interpretation of pore size distributions of microporous carbons. *Carbon*, Vol. 34, No. 10, pp. 1281-1286. ISSN 0008-6223.

Ravikovitch, P. I.; Vishnyakov, A.; Russo, R. & Neimark A.V. (2000). Unified Approach to Pore Size Characterization of Microporous Carbonaceous Materials from N2, Ar, and CO2 Adsorption Isotherms. *Langmuir*, Vol. 16, No. 5, pp. 2311-2320. ISSN 0743-7463.

Rodriguez-Reinoso, F. & Molina-Sabio, M. (1992). Activated carbons from lignocellulosic materials by chemical and/or physical activation: an overview, *Carbon* Vol. 30, No. 7, pp. 1111-1118. ISSN 0008-6223.

Rouquerol, F.; Rouquerol, J. & Sing, K. (1999). *Adsorption by powders and porous solids. Principles, methodology and application,* Published Academic Press, ISBN 0-12-598920-2, London.

Rouquerol, J.; Avnir, D.; Fairbridge, C.W.; Everett, D.H.; Haynes, J.H.; Pernicone, N.; Ramsay, J.D.F.; Sing, K.S.W. & Unger , K.K. (1994). Recommendations For The Characterization Of Porous Solids. *Pure & Applied Chemistry*, Vol. 66, No. 8, pp. 1739-1758. ISSN electronic 1365-3075.

Samios, S.; Stubos, A.K.; Kanellopoulos, N.K.; Cracknell, R.F.; Papadopoulos, G.K. & Nicholson, D. (1997). Determination of micropore size distribution from Gran Canonical Monte Carlo simulations and experimental CO_2 isotherm data. *Langmuir*, Vol. 13, No. 10, pp. 2795-2802. ISSN 0743-7463.

Sapag, K. & Mendioroz, S. (2001). Synthesis and characterization of micro-mesoporous solids: pillared clays, *Colloids and Surfaces A: Physicochemical and Engineering Aspects*, Vol. 187-188, No. 31, pp. 141-149. ISSN 0927-7757.

Scaife, S.; Klusen, P. & Quirke, N. (2000). Characterization of porous materials by gas adsorption: Do different molecular probes give different pore structures?. *Journal of Physical Chemistry B*, Vol. 104, No. 2, pp. 313-318. ISSN (electronic): 1520-5207.

Scaife, S.; Kluson, P. & Quirke, N. (2000), Characterization of porous materials by gas adsorption: Do different molecular probes give different pore structures, *Journal Physical Chemistry B*, Vol. 104, pp. 313-318. ISSN (electronic): 1520-5207.

Sing, K.S.W.; Everett, D.H.; Haul, R.A.W.; Moscou, L.; Pierotti, R.A.; Rouquerol, J. & Siemieniewska, T. (1985). Reporting Physisorption Data For Gas/Solid Systems With Special Reference to the Determination of Surface Area and Porosity, *Pure & Applied Chemistry*, Vol. 57, No. 4, pp. 603-619. ISSN electronic 1365-3075.

Sircar, S.; Golden, T.C. & Rao, M.B. (1996). Activated carbon for gas separation and storage, *Carbon*, Vol. 34, No. 1, p. 1-12. ISSN 0008-6223.

Solar, C.; Sardella, F.; Deiana, C.; Montero Lago, R.; Vallone, A. & Sapag, K. (2008). Natural Gas Storage in Microporous Carbon Obtained from Waste of the Olive Oil Production, *Materials Research*, Vol. 11, No. 4, pp. 409-414. ISSN 1516-1439.

Somorjai, G.A. (1994) *Introduction to Surface Chemistry and Catalysis*. John Wiley & Sons, Inc. (Ed.). ISBN: 978-0-471-03192-5, EEUU.

Steele, W.A. 1974. The interaction of gases with solid surfaces, First Edition, Pergamon, ISBN 0080177247, Oxford.

Sun, J.; Brady, T.A.; Rood, M.J.; Lehmann, C.M.; Rostam-Abadi, M. & Lizzio, A.A. (1997). Adsorbed natural gas storage with activated carbons made from Illinois coals and scrap tires, *Energy & Fuels*, Vol. 11, pp. 316-322. ISSN (electronic): 1520-5029.

Sweatman, M.B. & Quirke, N. (2001a). Characterization of porous materials by gas adsorption at ambient temperatures and high pressure, *Journal of Physical Chemistry B*, Vol. 105, pp. 1403-1411. ISSN (electronic): 1520-5207.

Sweatman, M.B. & Quirke, N. (2001b). Characterization of porous materials by gas adsorption: Comparison of nitrogen at 77K and Carbon Dioxide at 298K for activated carbon. *Langmuir*, Vol. 17, No. 16, pp. 5011-5020. ISSN 0743-7463.

Sweatman, M.B. & Quirke, N. (2006). Modelling gas adsorption in slit-pores using Monte Carlo simulation, In: *Adsorption and transport at the nanoscale*, N. Quirke (Ed.) pp. 15-41. CRC Taylor & Francis, ISBN 041532701-6, Boca Raton FL.

Tan, Z. & Gubbins, K.E. (1990). Adsorption in carbon micropores at supercritical temperatures, *Journal of Physical Chemistry*, Vol. 94, pp. 6061-6069. ISSN 0022-3654.

Tarazona P. (1985). Free-energy density functional for hard spheres, *Physical Review A*, Vol. 31, No. 4, pp. 2672-2679. ISSN 1050-2947.

Triebe, R.W.; Tezel, F.H. & Khulbe, K.C. (1996). Adsorption of methane, ethane and ethylene on molecular sieve zeolites, *Gas Separation & Purification*, Vol. 10, Issue 1, pp. 81-84. ISSN 0950-4214.

Wang, B.; Cote, A.P.; Furukawa, H.; O´Keeffe, M. & Yaghi, O.M. (2008). Colosal cages in zeolitic imidazolate frameworks as selective carbon dioxide reservoirs, *Nature* ,Vol. 453, pp. 207-212. ISSN (electronic) 1476-4687.

World Energy Outlook, 2009. Ed: International Energy Agency (IEA), France ISBN: 978-92-64-06130-9. www.iea.org/about/copyright.asp

Yang, C., Kim, D.Y. & Lee, Y.H. (2005). Formation of densely packed single-walled carbon nanotube assembly. *Chemistry of Materials*, Vol.17, pp. 6422-6429. ISSN 0897-4756.

Zhou, L.; Zhou, Y.; Bai, S.; Lü, C. & Yang, B. (2001). Determination of the Adsorbed Phase Volume and Its Application in Isotherm Modeling for the Adsorption of Supercritical Nitrogen on Activated Carbon. *Journal of Colloid and Interface Science*, Vol. 239, No. 1, pp. 33-38. ISSN 0021-9797.

11

Industrial application of natural gas

Alejandro Sáez
Universidad Técnica Federico Santa María
Chile

1. Introduction to General Aspects of Natural Gas Production and Consumption Worldwide and in Latin America

The world's natural gas reserves are concentrated in a small number of countries representing over 70% of total reserves. Distribution by country is shown below in figure 1.1.

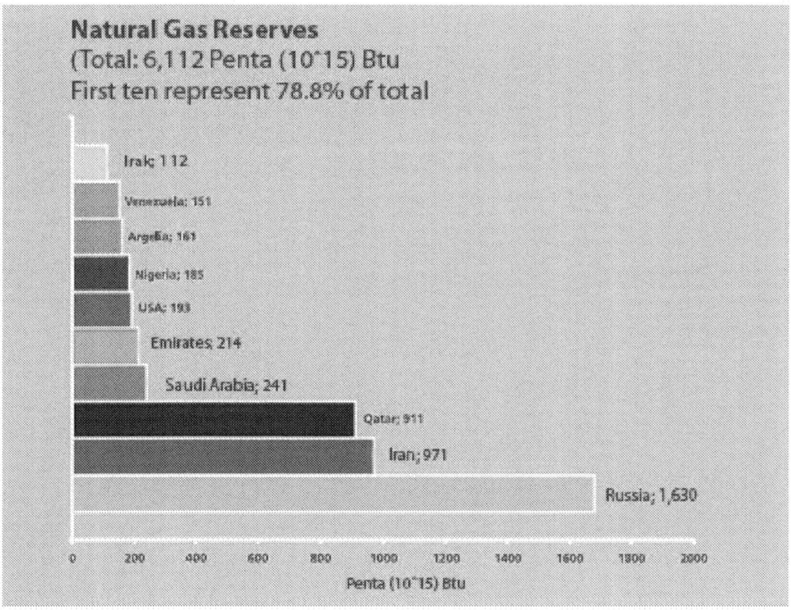

Fig. 1.1 Natural Gas Reserves. Reference (2)

Natural gas reserves are concentrated in the Middle East (40%), primarily Iran (16%) and Qatar (14%). After the Middle East, the world's largest reserves are located in Russia (27%); Africa (8%), above all Nigeria (3%) and Algeria (2.4%), and in the Asia-Pacific Region (8%).

In comparison with the regional distribution of oil reserves, natural gas reserves are more diversely distributed.

Major natural gas producing regions include North America (27%), led by the United States (19%); the Former Soviet Union (26%), where the world's largest production company is located, Russia (20% of total worldwide production in 2008); and the Asia-Pacific Region (13%).

The largest natural gas consumers are the United States and Russia (approximately 22% and 14%, respectively, of total worldwide consumption in 2008). They are followed by Canada, Japan, the United Kingdom, China and Germany, each representing nearly 3% of worldwide consumption.

In terms of the international gas trade, natural gas traded between different countries in 2008 represented over 26% of worldwide production and demonstrates significant potential for growth, particularly as regards LNG (liquefied natural gas). In 2008, 19% of the gas traded internationally was sold through pipelines and 7% as LNG. Japan and Spain are the two largest importers of liquefied natural gas.

As regards South America, the largest reserves are found in Venezuela, Bolivia, Columbia, Argentina and Peru, with a total of 6500 bcm. The countries with the highest consumption are Argentina and Brazil, with a total of 55 bcm/year.

The international natural gas trade in this region is carried out between countries which are interconnected via gas pipelines, such as Argentina with Bolivia, Chile and Uruguay and Bolivia with Brazil and Argentina. For geopolitical reasons and due to differing economic policies, a more universal natural gas transportation system which would allow gas to be traded throughout the continent has not been developed. Multiple ports with liquefied natural gas regasification terminals have been built to meet demand, reaching different regions of the world where it is injected into the local distribution networks of each country. Graph 1.2 shows gas pipelines and liquefied natural gas reception ports in the Southern Cone.

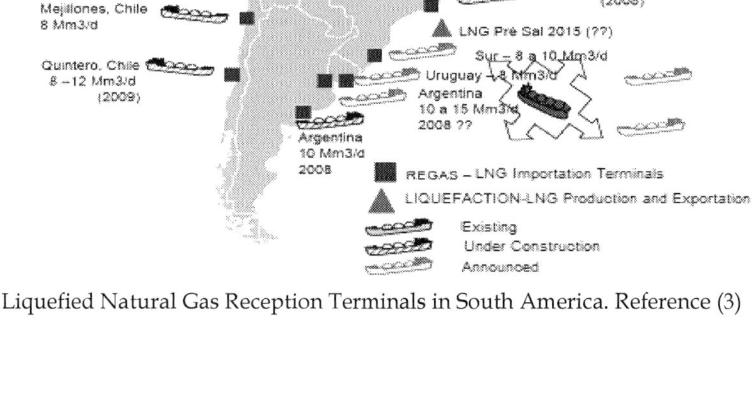

Fig. 1.2. Liquefied Natural Gas Reception Terminals in South America. Reference (3)

2. Properties of Natural Gas and their Impact on Industrial Applications

Natural gas is a fossil fuel found underground, generated by the decomposition of organic matter trapped between rocky strata of the Earth's crust. It is extracted from subterranean deposits of gas, oil and gas or condensate, so it may be obtained alone or together with oil.

2.1 Composition of Natural Gas

Natural gas is a fuel found in deposits in its gas phase. It is colorless and odorless, non-toxic, lighter than air and does not contain olefins (hydrocarbons produced during the process of destructive distillation or reforming). It is constituted primarily by methane (CH4), usually in a percentage of over 85% of volume. The remaining percentage is composed of higher order hydrocarbons such as paraffins or isoparaffins. It also contains water vapor at varying degrees of saturation, or condensed water. It may also contain carbon dioxide, nitrogen, hydrogen sulphide and helium, among others. Table 2.1 shows the typical composition of Argentine natural gas originating from the Northeast Basin.

If the gas contains enough carbon dioxide to cause its calorific value to fall below the values specified in sales contracts it must be subjected to a process to extract this element, in addition to hydrogen sulphide or other sulfur compounds causing it to be highly corrosive and inadmissible for certain industrial applications.

Once extracted, the gas is treated for the purpose of removing undesirable components such as water vapor, carbon dioxide, sulfur compounds, condensable hydrocarbons and solid and liquid particles. This process is known as "gas drying"; therefore, dry gas is gas which has been dehydrated and subjected to a process in which condensable hydrocarbons such as propane, butane pentanes and higher hydrocarbons have been extracted. The amounts which may be recovered from the abovementioned components depend on the original composition of the natural gas and the process used to dry it. Dry gases are also known as lean gases, and wet gases are also known as rich gases.

Component	Identification	Unit	Value
Methane	C_1	mol%	91.15
Ethane	C_2	mol%	5.56
Propane	C_3	mol%	0.16
Carbon Dioxide	CO_2	mol%	2.39
Nitrogen	N_2	mol%	0.72
Other		mol%	0.02
Water	H_2O	mg/Sm^3	14.59
Sulfuric Acid	H_2SO_4	mg/Sm^3	0.65
Total Sulfur	S	mg/Sm^3	1.13

Table 2.1, Typical Elemental Composition of Natural Gas

2.2 Elemental Analysis of Natural Gas

Elemental analysis of a fuel allows the elements composing to be identified, making it possible to determine the stoichiometric ratio as well as the products associated to the combustion process. Table 2.2 shows a comparison of the typical elemental analysis of the main fuels used in the industrial sector.

Fuel	Carbon	Hydrogen	Oxygen	Nitrogen	Sulfur	Ashes
Natural Gas	76.0	23.2	0.0	0.8	0.0	0.0
Liquefied Petroleum Gas	81.0	19.0	0.0	0.0	0.0	0.0
Fuel Oil 6	86.6	10.6	0.4	0.1	2.0	0.1
Fuel Oil 5	86.1	12.0	0.1	0.1	1.7	0.0
Fuel Oil 2	86.3	13.1	0.0	0.0	0.6	0.0
Kerosene	85.9	13.6	0.0	0.0	0.2	0.0
Firewood	45.1	5.6	43.5	0.1	0.1	5.6
Sub-bituminous Coal	49.0	6.1	30.0	0.6	0.3	14.0
Bituminous Coal	72.2	5.6	11.8	1.3	1.7	7.4

Table 2.2. Typical Elemental Analysis of Industrial Fuels (% of mass)

2.3 Properties of Natural Gas

a) Density

Table 2.3 shows the absolute densities of different liquid and gas fuels.

Fuel	Carbon	Hydrogen	Oxygen	Nitrogen	Sulfur	Ashes
Natural Gas	76.0	23.2	0.0	0.8	0.0	0.0
Liquefied Petroleum Gas	81.0	19.0	0.0	0.0	0.0	0.0
Fuel Oil 6	86.6	10.6	0.4	0.1	2.0	0.1
Fuel Oil 5	86.1	12.0	0.1	0.1	1.7	0.0
Fuel Oil 2	86.3	13.1	0.0	0.0	0.6	0.0
Kerosene	85.9	13.6	0.0	0.0	0.2	0.0
Firewood	45.1	5.6	43.5	0.1	0.1	5.6
Sub-bituminous Coal	49.0	6.1	30.0	0.6	0.3	14.0
Bituminous Coal	72.2	5.6	11.8	1.3	1.7	7.4

Table 2.3. Fuel Density

b) Heat Value of Natural Gas

The heat value of a fuel refers to the amount of energy released during complete combustion of one mass unit of the fuel, with the fuel and oxidant at a reference temperature and pressure. The properties of each fuel affect heat value in that the fewer inert elements the gas contains, the greater its heta value will be. As can be observed in Table 2.2, natural gas is a fuel with a high carbon and hydrogen content, making it the fuel with the highest heat value after pure hydrogen. Figure 2.4 shows a comparison of the calorific value of different fuels.

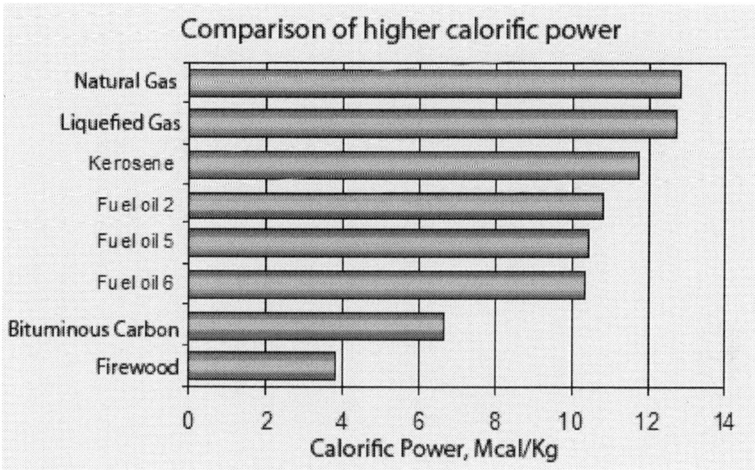

Fig. 2.4 Calorific value or Heat Value of different fuels.

However, the heat value of natural gas normally varies according to its content of inert elements or heavy hydrocarbons, resulting in lower or higher caloric value, respectively.

2.4 Interchangeability of fuel Gases

Two gases are considered interchangeable when, distributed under the same pressure, in the same network, feeding the same burners and without changes to regulation, they produce equivalent combustion results: calorific flow and flame behavior, regardless of the composition of the combustible gases.

The Wobbe index is the most frequently used indicator to establish criteria for the interchangeability of gases. It is defined as a quotient of the gross calorific value based on the square root of the relative density of the gas in relation to air under the same pressure and temperature conditions.

$W = PCS / \sqrt{d}$

Where:

GHV : gross heat value of fuel
d : relative density of gaseous fuel

When gases are interchanged, the value of the Wobbe index must be the same for both gases in order to ensure that the calorific value of the burner remains constant.

2.5 Characteristics of Natural Gas Combustion

Correct burning of natural gas requires the proper proportion of air and gas in order to achieve complete combustion. The stoichiometric amount of air needed for complete combustion is shown in Table 2.4 along with the amount of exhaust gases produced. In addition, characteristic combustion values of other fuels traditionally used in the industry are shown.

Fuel	A_e (a)	G_{ehe} (b)	G_{ese} (c)	G_{er} (d)	H_2O (e)	NHV (f)	W o (g)	Dv (h)	CO_2 (i)	E_a (j)
	Sm 3/kg				Kg/kg	Kcal/kg		[-]	%	
Natural Gas	12.9	14.2	11.6	14.5	2.09	11.500	14.846	0.6	12.2	2
Liquefied Petroleum Gas	12.2	13.3	11.2	13.5	1.71	11.400	8.875	1.65	13.5	2
Fuel oil 6	10.5	11.1	10.0	14.8	0.95	9.550	-	-	16.2	35
Fuel oil 5	10.9	11.5	10.2	14.8	1.08	9.900	-	-	15.7	30
Fuel oil 2	11.1	11.9	10.4	14.1	1.18	10.200	-	-	15.5	20
Kerosene	11.2	12.0	10.5	14.2	1.22	10.300	-	-	15.3	20
Firewood	4.0	4.7	4.0	8.7	0.50	3.500	-	-	20.9	100
Sub-bituminous Coal(m)	5.0	5.5	4.8	9.3	0.55	4.200	-	-	18.9	75
Bituminous Coal	7.6	8.0	7.3	13.6	0.50	6.200	-	-	18.4	75

Am 3: Volume measured in cubic meters under standard or normalized conditions

(a) Stoichiometric Air
(b) Wet Stoichiometric Exhaust Gasses
(c) Dry Stoichiometric Exhaust Gasses
(d) Real Exhaust Gasses considering characteristic operational air
(e) Mass of water generated per kg of burnt fuel
(f) NHV: Net Heat Value
(g) Wobbe Index
(h) Density relative to air
(i) Maximum or stoichiometric CO_2
(j) Excess of characteristic operational air
(k) Liquefied Petroleum gas (LPG) of the type that is commercialized in the Metropolitan Region of Chile
(l) San Pedro de Catamutún Carboniferous type (fuel analysis based on how it is received)

Table 2.4 Combustion Characteristic Values of Different Industrial Fuels

3. Characteristics of Natural Gas Flames and Combustion Products

The flame is the visible and calorific manifestation of the combustion process reaction. In practice, there are different types of flames which vary according to the mixture of fuel and oxidant. Given that the volume of air which participates in combustion is much greater than that of combustible gas, it is ultimately control of air that which defines the shape and dimensions of the flame.

i) Without Prior Mixing (Diffusion).

This refers to a long but low-temperature flame. It is yellow in color, due mainly to the presence of free carbon which has reached only the temperature necessary to become incandescent without oxidizing. The fuel reaches only the first stages of oxidation, with low combustion and burning efficiencies, so it requires a longer reaction time in order to achieve complete combustion. It is appropriate for use in homes or larger combustion chambers.

ii) With Prior Mixing (Premixing).

Premixing improves the homogenization of the fuel and oxidant mixture in order to increase the amount of fuel burned, producing short flames of bluish color, at high temperature and with a highly defined geometry (Figure 3.3). If not enough oxidant is incorporated to ensure complete combustion, a second zone of colorless flame is produced, creating a plume which surrounds the blue flame.

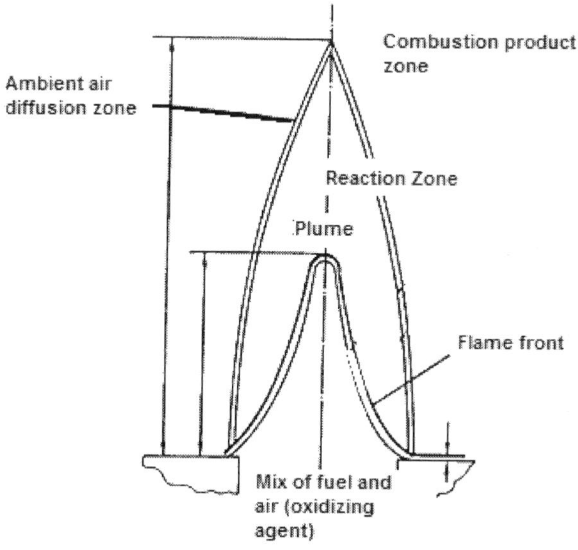

Fig. 3.3. Premixed Laminar Flame

a) Flame Temperature

Theoretical combustion temperature cannot be determined empirically, and corresponds to the temperature that would be reached by combustion products if the heat released during stoichiometric combustion were used exclusively to heat them. This is known as adiabatic flame temperature. However, this situation never occurs in practice, due mainly to:

Heat transfer from combustion products to the surrounding environment.
Reduction of CO2 and H20 (vapor) in CO, H2 and O2, at temperatures above 1,700 °C.

The adiabatic flame temperature of natural gas originating from Argentina is 2,020 °C in the stoichiometric condition. This corresponds to 2,026 °C for ethane and 2,059 °C for propane. This natural gas temperature may increase if operating conditions are altered.

It is important to make the distinction between the adiabatic theoretical temperature of combustion and real flame temperature. The former is a set value (for specific combustion conditions) while the latter corresponds to the actual temperature reached during real combustion, which therefore will vary according to position.

Maximum flame temperature is reached in premixed combustion with air in excess of stoichiometric amounts. Usually, by pre-heating the mixture and/or air, flame temperature increases of 25 °C to 40 °C may be achieved. Figure 3.4 shows the temperature variation experienced by the mixture and combustion products as a function of distance from the burner head.

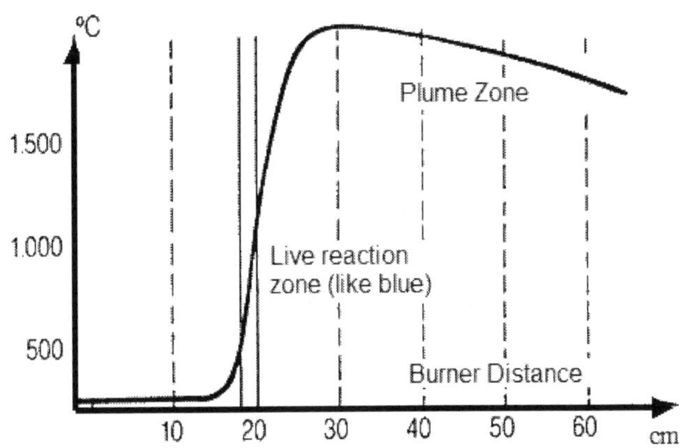

Fig. 3.4. Flame Temperature

b) Flame Speed

The flame front advances as the fuel mixture exits the burner head, producing combustion of the mixture. The reaction between the fuel and air may only occur at a certain speed. This speed depends on the reaction (chemical) and the degree of flame turbulence (physical). Thus, if the flame is turbulent, the mixture will burn more quickly. In addition, flame speed varies according to the fuel-to-oxidant ratio of the premixture known as the primary aeration rate.

Flame speed is an essential factor in proper combustion and better use of energy from fuel. Thus, it is necessary to have a stable flame front, achieved when the transport velocity of the reactants is equal to flame speed. If fuel is supplied at a higher speed than that of the flame, liftoff (movement forward) will occur. On the other hand, if the fuel is supplied at a lower speed than that of the flame, flashback (movement backward) will occur.

Other factors which also affect flame speed are the nature of the fuel, the presence of inert gases (nitrogen, carbon dioxide, etc.) and the temperature of the mixture (combustible gas - combustion air). Figure 3.5 shows flame speed for different combustible gases, as a function of the percentage of primary air in relation to theoretical air (stoichiometric air).

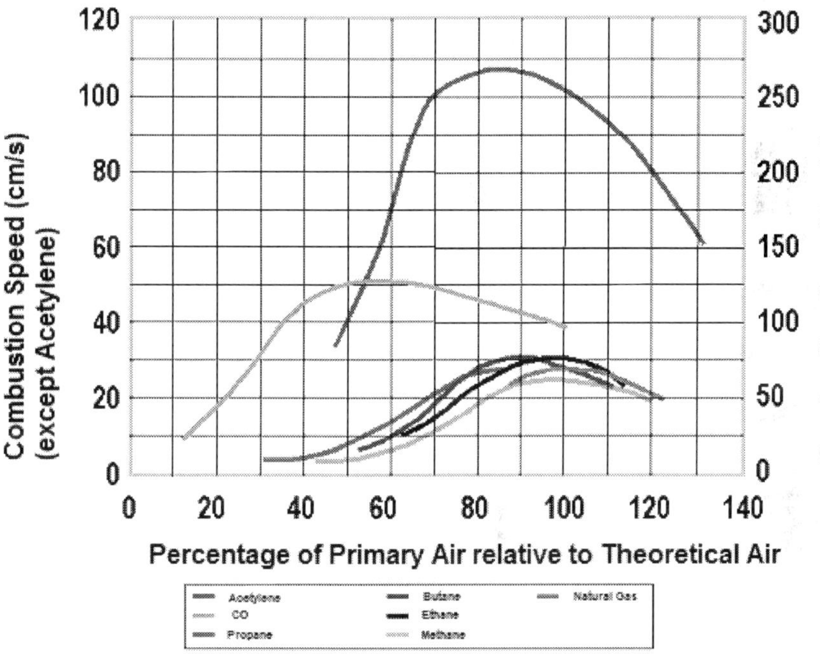

Fig. 3.5. Flame Speed [10]

c) Flame Stability

In order to keep the flame attached to the burner port, equilibrium must be reached between the exit velocity of the fuel-air mixture and flame propagation speed. Frequently, flame instability is the result of pressure variations in fuel supply to the burner or in combustion air:

Increase in fuel pressure. Generates increased pressure on the mixture and increased exit velocity, producing detachment of the flame or liftoff.

Decrease in fuel pressure. Generates a decrease in the exit velocity of the mixture, which produces flashback. This phenomenon is especially relevant to premix burners.

Fluctuations in combustion air supply, producing oscillating flames which cause strong pressure vibrations in the combustion chamber.

d) Inflammability Limit

Inflammability is related to the chemical energy of a mixture of combustible air and the minimum ignition energy; if the former is lower than the latter, it is impossible to produce a flame. This leads to the definition of inflammability limits.

Thus, the lower limit corresponds to a lean mixture situation, in which there is not enough fuel to achieve minimum ignition energy. Conversely, the upper limit (rich mixture) applies when there is not enough air for the combustion reaction to take place. Inflammability limits of some gases are shown in Figure 3.6.

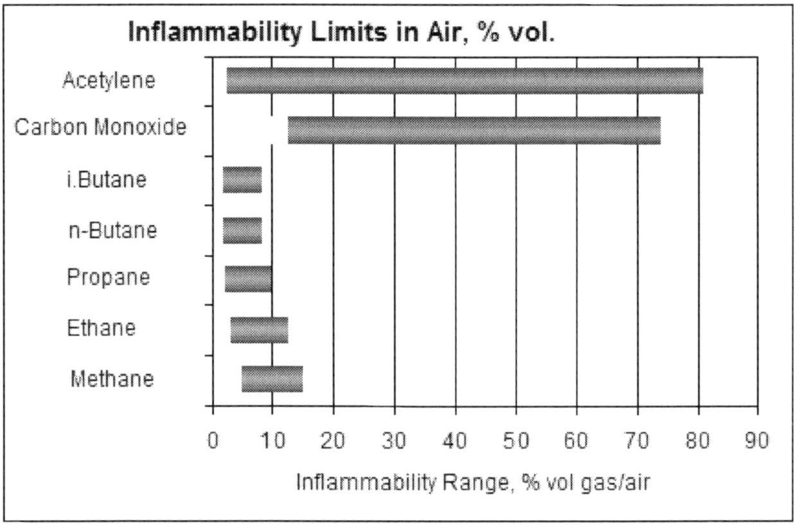

Fig. 3.6. Inflammability Limits of Combustible Gases

e) Flame Color

In complete combustion, the flame should exhibit a nearly transparent plume with a blue or blue-green cone in the middle, depending on the type of gas being burned (Figure 3.7). If combustion is incomplete the flame will be of a yellowish color, also depending on the environment in which the combustion reaction is produced.

Fig. 3.7. Color of a Natural Gas Flame

f) Flame Radiation

When the origin of the radiation is heat, energy is emitted only according to temperature and is known as thermal radiation. Radiation heat transfer occurs when energy is transported from one surface to another as electromagnetic waves, which propagate at the speed of light and do not require a physical medium to be transferred. Radiation heat transfer can be separated into:

Short-wavelength thermal radiation with wavelengths between 0.2 and 3 μm, characteristic of high-temperature radiation sources (T = 6000 °K) such as the sun or artificial lighting, and whose field includes part of ultraviolet radiation (λ<0.4 mm), all of the visible spectrum (0.4< λ <0.7 mm) and near infrared radiation (0.7< λ <3 mm), within which margin it emits 98% of energy.

Long-wavelength thermal radiation, also called irradiation, with wavelengths between 3 and 50mm, characteristic of ambient temperature radiation sources (T = 300°K), such as environmental surfaces, and whose spectrum includes far infrared radiation, within which 97% of energy is emitted.

The radiation that can be produced by a flame depends on how luminous it is. An oil flame can radiate 3 to 4 times as much as a gas flame, due mainly to soot production in the flame which makes it luminescent. Although a gas flame may produce soot under certain conditions of mixing, the amount of radiation attained is still significantly lower than that of an oil flame.

However, given that the emissivity of gases from natural gas combustion is lower (Figure 3.8), the energy transferred in the radiative zone is also lower. Therefore, the gases that reach the convective zone have a lower temperature than the gases produced by the combustion of petroleum-derived fuels.

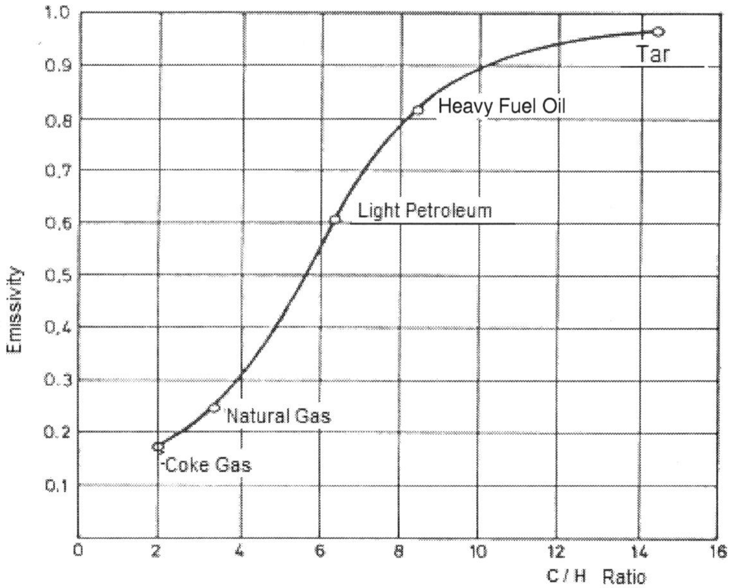

Fig. 3.8 Emissivity of Different Fuel Flames [1]

Thus, if heat transfer in the convective zone remains constant, the gases will exit through the stack at a higher temperature, which results in a loss of thermal efficiency of the equipment (as full use is not made of the energy contained in exhaust gases).

This is very important in terms of converting thermal equipment from oil to natural gas, for which purpose it is necessary to increase the convective heat transfer area. This shift in temperature profile must be taken into consideration for thermal equipment used for mineral fusion and cement furnaces.

4. Type of Burners Used in Industrial Equipment, Boilers and Furnaces

Proper fuel burning is necessarily connected to the technology designed and built in order to make better energetic use of the exothermal process produced.

4.1 Burner Definition and Functions

A burner is a set of mechanisms designed to allow for proper mixing of a fuel and an oxidant in order to produce a combustion chemical reaction with certain flame characteristics. For this reason, the functions of a burner are:

Allow for regulation of air and gas flows.
Ensure mass transfer or mixing of fuel (gas) with oxidant (air) in the correct ratio.

Carry out flame combustion with the dimensions and chemical and physical properties appropriate to the application, maintaining the heat transfer process.
Maintain flame stability.

The most frequently used oxidant is ambient air. However, depending on the application, it is possible to use hot air or oxygen-enriched air. Instead of air, it is also possible to use pure oxygen or combustion products with an elevated oxygen content. In order to make full use of the energy generated during the combustion process, the flame must be stable, without lifting off from the burner or flashing back. To prevent this from happening, there must be an equilibrium between the speed of flame propagation toward the unburned fuel and the exit velocity of this fuel.

Burners which operate using several different gases are called multi-gas, and those which operate using different types of fuel are called "mixed" or "dual", generally using a petroleum derivative (frequently diesel) as the second fuel. If the gas and the other fuel can be burned together, the burner employs what is known as simultaneous operation; if this is not the case, its operating method is known as alternating.

The combustion system used will depend on the type of application. The following are some examples of combustion systems:

Diffusion system, or separate delivery of air and gas.
Premix system, or external mixing of fuel and oxidant.
Atmospheric system, or air drawn into the burner by natural draft.
Forced system, or combustion air supplied by using an additional piece of equipment.

a) Diffusion System or Separate Delivery of Air and Gas

In this combustion system, mixing of fuel and oxidant is carried out in the burner itself, or as they flow toward the combustion chamber. The parameter that indicates the intensity of the mixture is the swirl number (Sx), corresponding to the ratio between the radial moment and the axial moment of the total flow in the burner.

Two types of burners use this combustion system:

When mixing is carried out in the burner head, the gas released into the combustion chamber already has the proper composition. The combustion reaction is quick, and the resulting flame is short and hot. This burner is characterized by a swirl number higher than one.

This system is characterized by generating high turbulence, which causes the reactants (gas and air) to mix quickly. This is induced using vanes or blades, which produce a rotating movement in the air current to encourage turbulence, as shown in Figure 4.1.

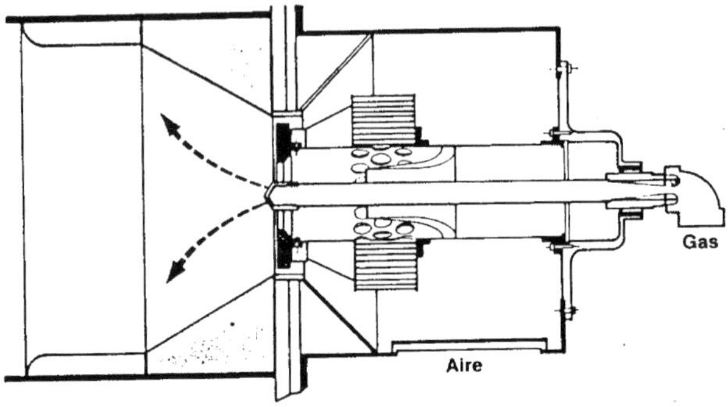

Fig. 4.1, Diffusion Burner with a Set Total Turbulence

In other burners, such as that shown in Figure 4.2, a plate is introduced into the path of air flow in order to stabilize it and encourage mixing with combustible gas.

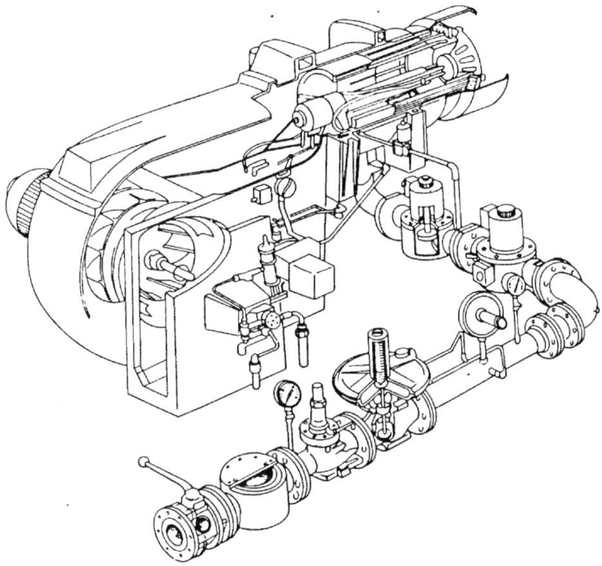

Fig. 4.2. Monoblock Burners with Gas Train and Regulation Unit

In other cases, such as in Figure 4.3, turbulence is not generated using vanes or mixing plates, but through use of converging or divergent flows.

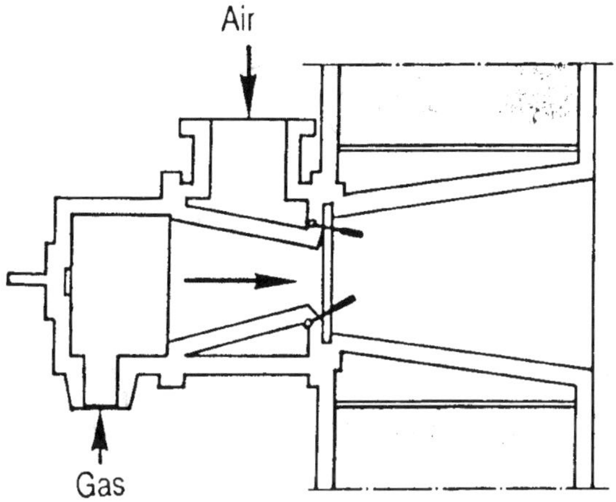

Fig. 4.3. Burner with Converging Air Current

In dysfunctional burners with a swirl number less than one, the flame obtained is long and very luminous due to the longer mixing time needed for fuel and oxidant.

In these low-turbulence burners, air and gas flows are mixed along the combustion chamber (Figure 4.4) and are used for processes in which heat transfer from the flame must be homogenously distributed throughout the space.

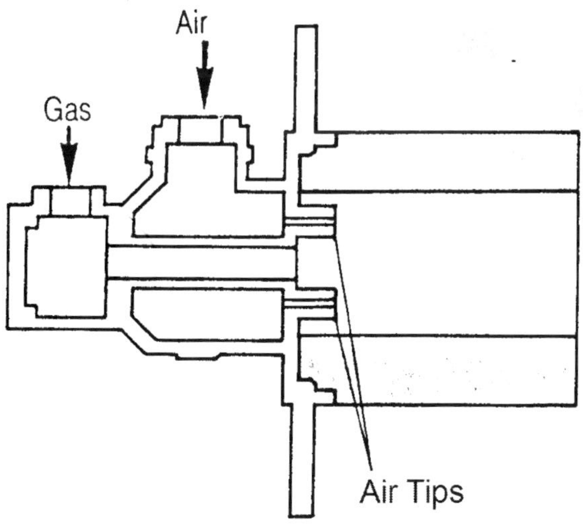

Fig. 4.4. Low-Turbulence Parallel Flow Burner

b) Premix System

In this system, a portion or all of the air required for complete combustion (known as primary air) is mixed with the gas upon entering the burner or immediately before. Thus, better mass transfer (intimate mixture) is achieved between fuel and the oxidant before reaching the burner through an elevated combustion speed and a high volumetric thermal load.

Three types of burners may be identified according to how premixing is carried out:

i) Enclosed Burners with a Mixture of Air and Gas

Air and gas are channeled by pressure through separate tubes with simultaneously controlled progressive regulation valves. The two streams may be unified in a mixing chamber or in the tube itself leading to the burner, as shown in Figure 4.5.

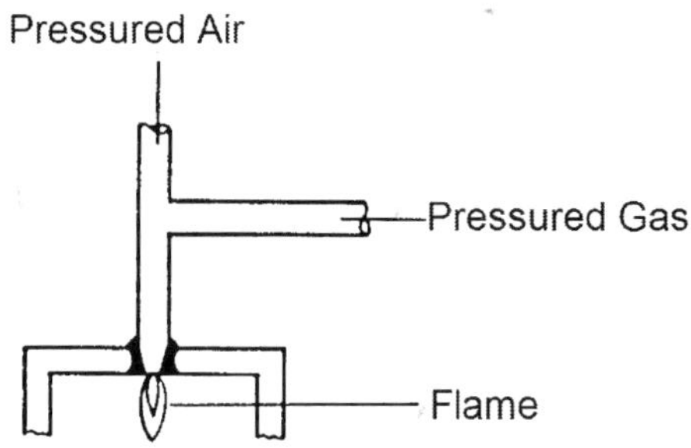

Fig. 4.5. Diagram of Enclosed Burner with a Mixture of Air and Gas

ii) Atmospheric System

An atmospheric burner (Figure 4.6) is comprised by:

A gas and air mixer which uses the kinetic energy of a stream of gas supplied by an injector to suction ambient air and create an inflammable mixture.
A burner head which ensures stable combustion of the air and gas mixture.

These types of mixers may be configured for a unit power of up to 1,000 kW. However, they are designed for unit powers of 30 to 300 kW. Their main advantage is their simplicity and low cost. This type of burner is used when mixing pressures approximate atmospheric pressure and when it is not necessary to obtain the amount of theoretical air in the premixture, as the quantity of air taken in by the gas is not enough to produce complete combustion. The remaining air, known as secondary air, is obtained by diffusion of the ambient air surrounding the flame.

Gas regulation is achieved by varying the pressure in the injector (progressively opening and closing the gas valve). Air is regulated (with gas at a constant pressure) by:

Movement of the injector nozzle in relation to the venturi.
Varying the air entry section by obstructing the orifices where air enters, or by using threaded plates, a moveable ring or a sliding hood.
Constricting the throat of the venturi (not recommended).

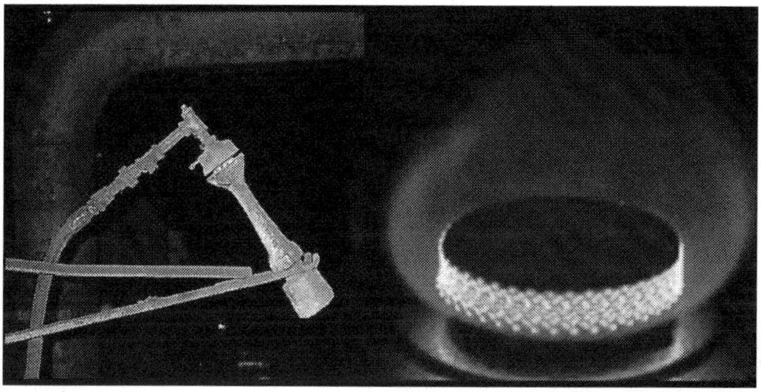

Fig. 4.6 Atmospheric Burners

Atmospheric burners (mixing by atmospheric induction) are virtually the only type used in household applications.

In industrial applications, for feeding enclosed areas such as combustion chambers and furnaces, among others, the air for the mixture is obtained by using special atmospheric induction mixers, with high injection pressure and double induction (Figure 4.7). This improves feed and supplies a combustible mixture which is nearly stoichiometric.

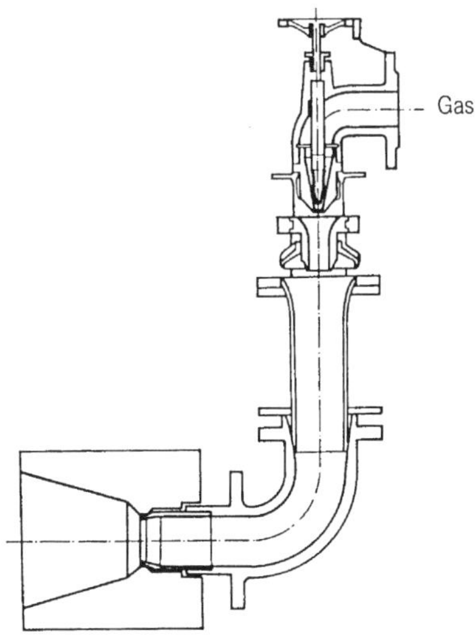

Fig. 4.7. Burner with Pressurized Gas and Double Atmospheric Induction

4.2 Auxiliary Components of Burners

The usage conditions of gas burner equipment make ignition, control and safety devices necessary. These auxiliary components are described as follows.

4.2.1 Ignition Systems

Different systems are used to ignite the combustible mix. The simplest of these consists of a pilot light ignited next to the main burner, which causes combustion when the gas exits the burner. The pilot light, which remains lit while the burner is used, fulfills the functions of initiating combustion of the gas exiting the main burner and preventing liftoff in the main burner once combustion has begun. In order to ignite the pilot light, burners employ several systems such as:

Piezoelectric System (Figure 4.8). This is the most commonly used system in household and low-power burners. It's based on the property had by some materials, such as quartz crystals, of being able to generate an electric potential difference upon being subjected to stress. This potential difference is transferred to two electrodes and produces one or more sparks, which ignite the combustible mixture. Generally, one electrode consists of the mass of the apparatus itself, and the other of a spark plug located near the pilot burner.

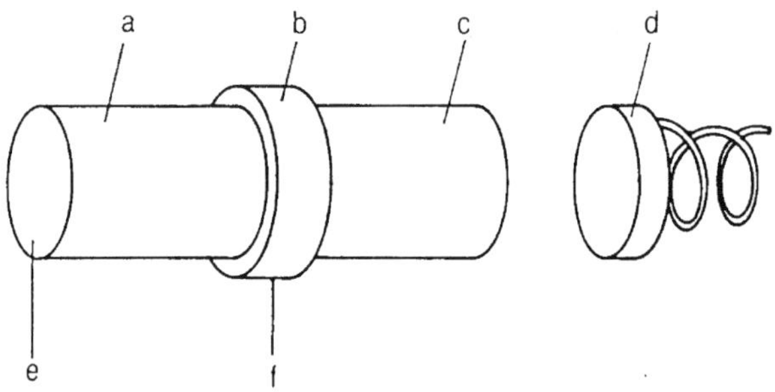

Fig. 4.8. Diagram of a Piezoelectric System

This system consists of a hammer activated manually or by a lever, in which a metal mass (d) strikes two quartz crystals (a and c) with a copper contact between them (b), producing a potential difference between points e and f.
The main advantages of this system are:

It does not require an electrical energy supply.
The energy source does not have to be renewed.
It is a simple system which ignites any type of commercial gas.
It can be automated so that the hammer is activated by a lever.

Electrical Resistance System. This system is based on heat production in an electrical resistor, increasing the temperature enough to ignite the combustible mixture. For this purpose, the resistor must be located near the gas exit port of the pilot light.

Voltage is low (from 2 to 12 V) and obtained from the network via a transformer or batteries. Thus, temperatures of between 700 and 800 °C are reached in the resistor. This procedure offers excellent possibilities for remote ignition.

This ignition device can work directly on the main burner (without a pilot light) as long as there is a flame detection system or device in place.

High Voltage Spark System. This system is essentially used for:

Lighting mixed gas and oil burners (Figure 4.9), whether simultaneous or alternating

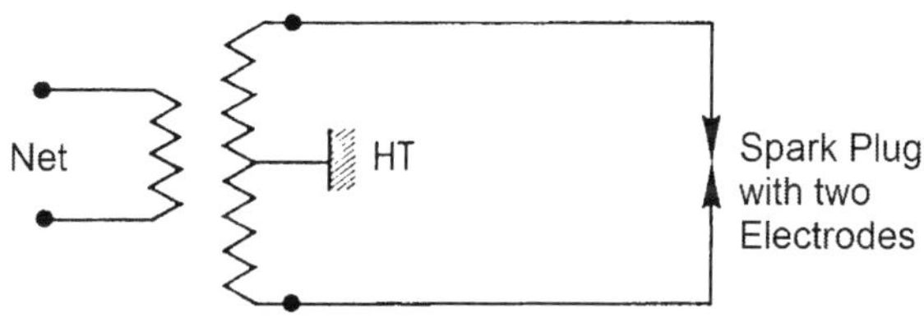

Fig. 4.9. Electrical Diagram for a Spark Plug with two Electrodes.

The no-load voltage is approximately 9,000 V. For a 50,000 Ohm resistor, the voltage should be close to 1,000 V.

Gas ignition only (Figure 4.10)

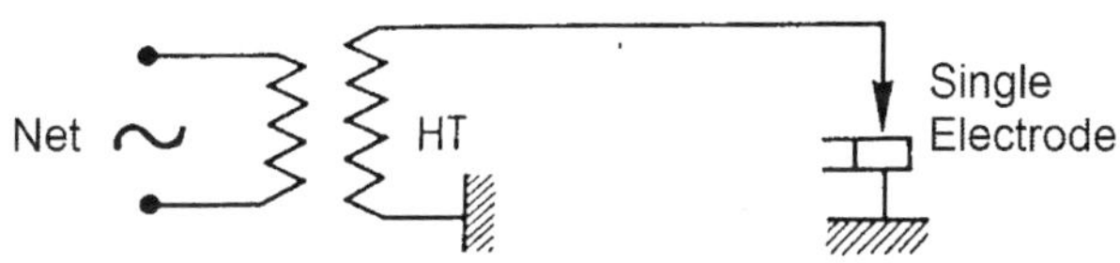

Fig. 4.10. Electrical Diagram for a Spark Plug with a Single Electrode

Automatic Ignition. This type of system, which can operate without a pilot light, consists of the following components:

Ignition device, usually with electrical sparks.
Quick response flame detector, such as those which use flame ionization or a UV photoelectric cell.
Appropriate control components in the event the ignition spark is absent or not produced within a few seconds, blocking the system and shutting off gas flow to the burner.

When the system is connected, the combustion chamber is initially swept with pressurized air to eliminate any potential unburned gas. Subsequently, the gas valve automatically opens and a stream of sparks is produced on the ignition electrode, initiating gas combustion.

If the gas fails to exit the burner, the sparks don't ignite the gas or the spark regeneration system fails, the system will be blocked within a few seconds.

If the burner ignites, this is detected by the ionization electrode, and the gas flow to the burner is automatically kept open and spark production is stopped.

4.2.2 Flame Safety Systems

The flame safety system is made up of a set of components that provide for safe control of burner operation which is appropriate to the application. Its main functions are to:

Provide a safe method for lighting and shutting off the burner (manual or automatic).
Light the burner in the proper order and supervise the flame during the operation.
Protect the system against excessive pressure or temperature.
Control burning capacity.
Maintain the burner ready for operation during the periods when it's turned off.

In order to fulfill these functions, the flame safety system has a programmer, a flame detector, limit and safety controls, operating condition controls, fuel valves and burner power control. The programmer, also called an ignition programmer, coordinates overall operation. In addition to verifying signals from the other components, it must also verify that its own operation is taking place properly. Flame detectors can be classified into three general types:

Thermal. Thermal detectors consist of components which detect the temperature of combustion gases using a thermocouple or a bimetallic strip. Their use is increasingly less common, given their slow response time and unreliability, and they are limited to small burners.
Flame rectification. Flame rectification detectors use the ionization produced by the flame to rectify an alternating current. Despite being simple and reliable, they cannot be used for oil flames, as they can be dirtied by drops of fuel, or in high temperature applications, as they burn too quickly.
Radiation. Radiation detectors are the most versatile, but also the most expensive. Some radiation detectors are sensitive to ultraviolet radiation.

Flame safety systems or flame detectors are devices which automatically interrupt gas supply when ignition has not occurred after a certain period of time, or in the event of accidental extinction of the flame during burner operation. To fulfill this purpose, they detect a distinctive characteristic of the flame such as its temperature, light emitted or its electrical properties. Flame safety devices may consist of:

Complete or total safety: when the safety device cuts off the gas flow to the main burner and the pilot. Simple safety is when gas flow to the pilot is maintained.
Positive safety: when a failure in the safety device produces the same result (gas flow cutoff) as the function of the device itself.

4.2.3 Flashback Protection Systems

In inflammable gas mixtures, the flame propagates from the point of ignition to the rest of the mixture. This phenomenon is known as deflagration, and the velocity of flame propagation or deflagration is about 10cm/sec or even 1m/sec. In some cases, due to low gas or air pressure, fluid at higher pressure flows into the piping with lower pressure fluid. Devices are used to avoid this problem and avoid flashback, fulfilling the following functions (Figure 4.11):

Closing the combustible mixture tube by means of a check valve that closes as a result of the back pressure produced by the downward-flowing gas in combustion.
Stopping the flame front by means of a fine wire mesh for detaining and cooling the flame.
Breakage of a membrane calibrated according to pressure increase, allowing the combustion products which caused the pressure surge to be released into the atmosphere.

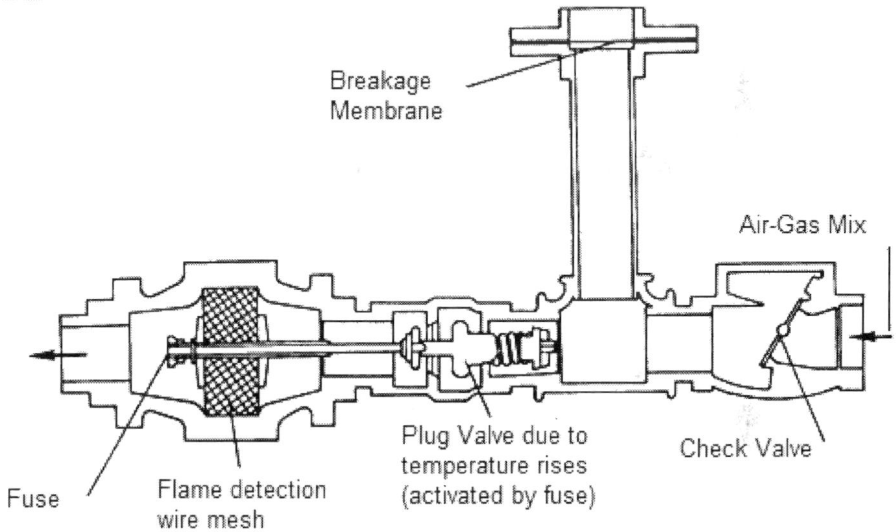

Fig. 4.11. Burner with Flashback Protection and Anti-Explosion Device

Other devices also include a broken membrane detector, which sends a cutoff signal to the main gas flow valve.

4.2.4 Practical Safety Aspects

The following should be considered among the important safety aspects of natural gas use:

Natural gas is lighter than air.
The enclosed spaces where natural gas is used require ventilation.
Flame color is transparent, which complicates visibility in some environments. Thus, burners must be handled with caution.

Gases must be completely swept from the furnace before lighting the burner. Thus, it must be ventilated with air prior to opening the gas flow and ignition. Sweeping time for gas is longer than that required for liquid fuels.
Pressure detection valves and systems must be in place to prevent pressure surges.
Natural gas contains odorizers that permit its detection without instruments and ensure that the operator does not breathe it in. In some cases involving circuit leakage, the operator may lose their sensitivity to the odor; thus, leakage and odor detectors are recommended.
Equipment maintenance must to be performed as indicated in the manual, with the recommended frequency.

5. Process Boiler Applications

Industrial processes requiring vapor are a very important area of natural gas use in practically every country. It's important to consider the characteristics of demand for vapor in processes, as seasonable variability is considerable. Many batch processes require vapor for specific periods of time, after which demand ceases or decreases to a minimal level. This is known as variable vapor demand, and it results in very irregular operation for the boiler, which must transition from zero production, or standby, to maximum production within a short time period. Thus, the efficiency of vapor generation is an average of the efficiencies attained at each demand level, from zero efficiency to the boiler maximum, in the range of 85% to 95% according to design and fuel used. Boilers which use natural gas provide the best efficiencies under these variable demand operating conditions, given that combustion may be efficiently regulated within a wide range of required calorific values. In addition, as natural gas combustion does not result in the generation of particles, heat transfer areas of the boiler remain clean for long periods of operation. This means that the equipment is available for longer periods of time, with less potential for impairing production due to lack of vapor.
The primary characteristics of boilers designed to burn natural gas are that they have a large area for heat transfer by convection and a smaller furnace than boilers using liquid fuels.
Boilers designed to burn combustible liquids have larger furnaces due to the fact that radiation from these fuel flames contains soot particles which emit radiant heat, which is used to generate vapor on the furnace walls. Generally, it is estimated that 50% of the heat required is provided in the furnace by flame radiation, and the other 50% in convection areas (tube bundle). It's very important to consider this situation when boiler fuel is changed. In the case of changing from liquid fuel to natural gas, vapor production will be reduced to maintain efficiency or efficiency will be reduced to maintain vapor production. One way to reduce this impact is to install heat exchangers on the boiler exhaust outlet for preheating boiler feed water.

6. Drying Furnace Applications

Given that its combustion produces very clean gases, natural gas is an optimal fuel for drying processes. This characteristic allows for direct application of hot gases in order to evaporate water from products which need drying, without introducing any type of residue which might alter the composition of the product and impair its quality.

Some examples of where this concept is applied are drying processes involving fish flour, mineral salts, agricultural grains, mineral concentrate, etc. Applying this process, known as direct drying, is more efficient and requires less capital and maintenance costs than indirect drying, which generally uses vapor as a heat transport fluid to heat the walls separating the vapor chamber from the products to be dried.
Experience indicates that fuel savings in the range of 15% to 30% may be obtained over indirect drying when the direct drying process is used.

7. Copper Pyrometallurgy Applications

The use of natural gas in copper pyrometallurgy processes can be divided into its application as a fuel and its application as a product employed in the metal purification process.

Fig. 7.1 Photo of Burners Installed in P&S Furnace

Its use as a fuel takes place in the burners used in the different furnaces involved in the process. Its primary function is to maintain the metal in its liquid state, which requires a temperature of over 1200 °C while copper purification processes are carried out. Just as in Pierce Smith converters, a significant portion of the sulfur and iron contained in the metal is eliminated as air is forced through the liquid metal bath. This process generates heat from the burning of sulfur; however, thermal equilibrium must be maintained with burners, given that as the sulfur is eliminated, its heat contribution diminishes. In these furnaces, it has been proven that the use of natural gas results in a longer service life for refractory mantles, which protect furnace walls by reducing the quantity of soot (carbon) affecting refractory materials and maintaining uniform heat distribution throughout the furnace walls. Figure 7.1 shows the installation of two burners in the mouth of one of these furnaces, in which the color and transparency of the flame and the lack of radiant elements can be noted.
Upon increasing the service life of the furnace, two significant effects are produced:

Maintenance periods are lengthened, resulting in greater productivity.
Maintenance costs for refractory mantles are reduced.

Experience has shown that maintenance periods are extended by up to 50% and even longer in some cases, which signifies an increase in smelting production.
In addition, savings on maintenance of refractory components can amount to several hundred thousand dollars per year.

The next step in copper purification is known as fire refining. In this process, the remaining small amounts of sulfur must be eliminated, so the liquid metal is over-oxidized by injecting air into it until the sulfur is eliminated. However, this leads to oxidation of the metal, which must be treated with a reducing substance which deoxidizes the oxidized metal. During this phase of the process, natural gas is employed as a highly efficient reducing element, as it does not contain sulfur and has a low carbon content (the lowest of all fossil fuels). Experience has shown that black smoke emissions are eliminated during reduction when natural gas is used for this purpose instead of oil.
Finally, natural gas is also used in the burners which maintain the copper in a reductive environment during the process of casting the anodes, the end product of smelting.
Figure 4.6 shows an atmospheric burner used in the casting area of a smelter.

8. Environmental Aspects

All fuels, and particularly those of fossil origin, impact the environment to some degree. This impact is a product not only of their use (combustion processes), but also of extraction processes, as well as construction and operation of the infrastructure needed to make fuel available to consumers.

Most of the installations and infrastructure required for natural gas (gas pipelines and distribution networks, for example) are underground, so they don't substantially affect the esthetic value of the landscape.

Of all the fossil fuels, natural gas today is the cleanest, pollutes the least and has the lowest carbon content. This originates in the molecular composition of its principal component (methane, CH_4), which has four hydrogen atoms for each carbon atom. Natural gas combustion produces 50% to 70% of the carbon oxides produced by other fuels; as a result of its lower carbon dioxide emissions, natural gas contributes to reducing the greenhouse effect.

In addition, its sulfur content is virtually non-existent, so its combustion does not contribute to the generation of acid waste. Under normal conditions of regulation and operation, natural gas combustion does not generate black smoke, ash, slag or other solid waste.

Natural gas combustion generally produces more nitrogen oxides than other fuels. This characteristic has spurred the development of low NOx burners, in which flame temperature is reduced using secondary and tertiary burning techniques, thus preventing formation of these oxides.

9. Economic Aspects of Using Natural Gas as an Industrial Fuel

Natural gas offers some very attractive advantages when it comes to making an economic assessment of its use in industrial processes. These aspects are:

Natural gas does not require additional expense for transportation within the factory; all that's needed is a distribution network and pressure regulation systems for equipment. It should be taken into account that liquid fuels require pumps for transport, implying additional electricity use.
Unlike liquid fuels, natural gas doesn't require storage tanks.
Temperature doesn't have to be controlled, as it does with residual fuels, so additional heating expenses are made unnecessary.
No storage cost is incurred, as natural gas is piped in by the supplier according to demand.
Lower maintenance costs for equipment using natural gas and lower costs for the equipment required when other fuels are used.
Lower costs for meeting local atmospheric emissions standards due to lower costs of treating the gases emitted into the environment.

All of these aspects must be considered when comparative calculations are made of the costs of useful energy supplied by different fuels.
In industry, energy contributions of natural gas are measured based on different physical units of measurement. As this often generates confusion, Table 9.1 provides cost equivalency values based on commonly used physical units.

	US$ / Sm3	US$ / MMBtu	US$ / kJ	US$ / kcal	US$ / kWh
US$ / Sm3	1	28.571	2.719 e-5	1.136 e-4	9.779 e-2
US$ / MMBtu	3.5 e-2	1	9.479 e-7	3.968 e-6	3.421 e-3
US$ / kJ	36,784	1,055,000	1	4.1868	3,600
US$ / kcal	8,800	252,000	0.239	1	859.84
US$ / kWh	10.226	293.07	2.778 e-4	1.163 e-3	1

Table 9.1 Natural Gas Equivalencies in US$, based on Net Calorific Value

Supply contracts may establish certain conditions regarding how gas is delivered. One of the conditions which may create a negative impression is take or pay on an amount of gas. This condition makes paying for gas mandatory under all circumstances, so the manner in which it is proposed must be considered very carefully.

10. Referring

1. J. Stepanke, Industriell Wärmetechnik. Vulkan Verlag, Essen. Deutschland. 1977.
2. International Energy Outlook 2008, World Natural Gas Reserves by Country as of January 1, 2008, page 44
3. www.GasEnergy.com.br

12

Combined operational planning of natural gas and electric power systems: state of the art

Ricardo Rubio-Barros, Diego Ojeda-Esteybar,
Osvaldo Añó and Alberto Vargas
Instituto de Energía Eléctrica, Universidad Nacional de San Juan
Argentina

1. Introduction

The growing installation and utilization of natural gas fired power plants (NGFPPs) over the last two decades has lead to increasing interactions between electricity and natural gas (NG) sectors. From 1990 to 2005, the worldwide share of NGFPPs in the power generation mix has almost doubled, from around 10% to nearly 19%; reaching in 2007, for instance, the 54% in Argentina, the 42% in Italy, the 40% in USA, and the 32% in UK (IEA, 2007; IEA, 2009a). The installation of NGFPPs has been driven by technical, economic and environmental reasons. The high thermal efficiency of combined-cycle gas turbine (CCGT) power plants and combined heat and power (CHP) units, their relatively low investment costs, short construction lead time and the prevailing low natural gas prices until 2004 have made NGFPPs more attractive than traditional coal, oil and nuclear power plants, particularly in liberalized electricity markets. Additionally, burning NG has a smaller environmental footprint and a lower carbon emission rate than any other fossil fuel.

NGFPPs are the link between electric power and NG systems, since they play the role of producers for the former and consumers for the latter. Therefore, the growing use of NGFPPs has had a great impact on the NG market. Power generation accounted for around half of growth in gas use from 1990 to 2004; over the most recent five years, this proportion rose to nearly 80% (IEA, 2007). This fact is especially notable in those countries where large capacities of NGFPPs have been installed.

Two major indexes indicate the level of interrelations between electric power and NG systems. The first one is NG demand for power generation as a share of the total NG consumption, and the second one is the share of electrical energy produced by NGFPPs. Both shares depend not only on the NGFPPs installed capacity, but also on relative fuel prices (NG, coal, oil derivatives products) and the availability of other energy resources (hydroelectricity, wind power). Fig.1 shows year 2007 indexes for several countries and also on a global basis (IEA, 2009a, b). Significant values of both indexes are typically a sign of strong interdependencies. Fig. 1 includes some of the countries with the highest indexes in the world, such as Argentina, Italy, Malaysia, the Netherlands, Russia and Turkey.

The interdependencies between electric power and NG systems can be described from a technical-operational viewpoint. The NGFPPs dispatch determines the total amount of NG consumption and its flows through the pipelines. On the other hand, the NG availability for NGFPPs is constrained by the maximum capacity of gas injection into the pipeline system (from producers, regasification terminals and NG storages), the limited transmission capacity of pipeline network, and priority scheme for the supply of NG in case of shortages, in which residential and commercial customers typically take precedence over large consumers and NGFPPs. Contingencies in NG infrastructure (e.g., interruption or pressure loss in pipelines) may lead to a loss of multiple NGFPPs, and thus jeopardize the security of the power system.

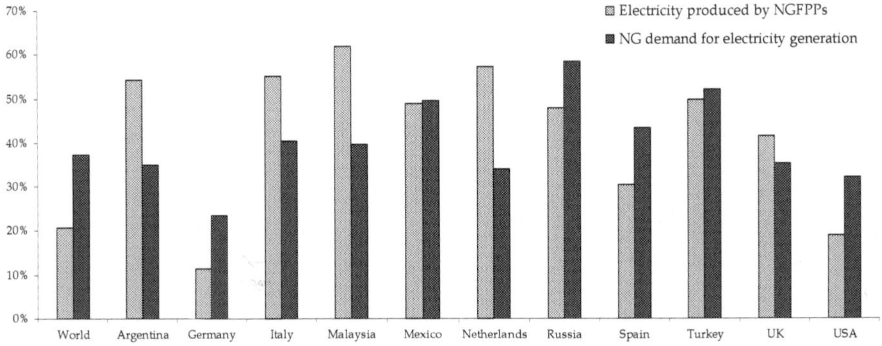

Fig. 1. Indexes of interactions between electric power and NG systems in 2007

These interactions can also be explained from a market perspective. The regulatory frameworks and the type of markets implemented in electric power and NG systems set the extent and the dynamics of the existing interdependencies. Generation companies that own NGFPP participate simultaneously in both markets, therefore, they are best suited for price arbitrage between both commodities. Liberalized and flexible market structures facilitate this practice which is required to reach an electricity and gas *partial economic equilibrium.* According to electricity and NG market prices, and the marginal heat rate of their plants, these companies can decide to use gas and sell electricity in the power market, or resell previously contracted gas on the NG market and purchase electricity to meet their commitments. Therefore, electricity and NG market prices are increasingly interacting between them, which is particularly noticeable when no direct oil indexation is applied to NG pricing (IEA, 2007).

From 2005 to the first half of 2008, the raising NG prices have eroded the competiveness of NGFPPs, decreasing the pace of growth in NG use for electricity generation and reducing the incentives for future investments in these technologies. Nevertheless, as NG prices have converged to lower levels during 2009, the NGFPPs have recovered their investment attractiveness. For the coming decade, NGFPP capacity is estimated to continue to account for the bulk of electricity generation capacity additions. Beyond the factors in favor of NGFPP investments pointed out above, NGFPP could become one of the swing resources utilized to provide flexibility in power systems with large shares of intermittent renewable generation, underpinning the investments in these technologies (IEA, 2009d). On the other hand, from NG market perspective, the electric power sector accounts for 45% of the projected increase in world NG demand by 2030. As a result, the power sector's share of global NG use will rise to 42% in 2030 (IEA, 2009c).

Under the light of all the conditions previously described, there is a strong and rising interdependency between NG and electricity sectors. In this context, it is essential to include NG system models in electric power systems operation and planning. On the other hand, NG system operation and planning require, as input data, the NG demands of NGFPPs, whose accurately values can only be obtained from the electric power systems dispatch. Therefore, several approaches that address the integrated modeling of electric power and NG systems have been presented. These new approaches contrast with the current models in which both systems are considered in a decoupled manner.

Among all the issues that arise from this new perspective, this chapter presents a complete survey of the state of the art in the combined operational planning of NG and electric power systems. The review covers the different time horizons considered in the operational optimization problems, ranging from the long-term (2-3 years) to the single period analysis.

This chapter is organized as follows. Firstly, the general characteristics of electric power and NG systems are described and compared. Secondly, the typical energy systems planning procedure, whose results provide the framework for the operational planning, is introduced. Then, the coordinating parameters used under a decoupled dispatch of electric power and NG systems are explained. Finally, the most relevant economic and market issues associated to this new situation of strong interactions between electric power and NG sectors are described and analyzed.

2. Natural Gas and Electric Power Systems

Energy infrastructure is composed of all the energy systems involved, which provide the energy required by different consumers. The technical energy systems include production, processing, treatment, transportation and storage facilities, which comprise the supply chain from primary energy sources (oil, coal, natural gas, nuclear, solar, wind) to the final energy carriers required by consumers (electricity, natural gas, district heat). In this task, electric power systems play two important roles in energy supply: allows the use of primary energy sources such as nuclear, hydroelectric and wind energy that otherwise were useless, and allows flexibility because most energy sources can be transformed to electricity. Fig. 2 shows a general scheme of the energy system supply focused on the electric power systems and the primary energy resources that converge to them.

In particular, electricity and NG are energy carriers, i.e., a substance or phenomenon that can be used to produce mechanical work or heat or to operate chemical or physical processes (ISO, 1997). While NG is a primary energy because exists in a naturally occurring form and has not undergone any technical transformation, electrical energy is a secondary energy, since it is the result of the conversion of primary energy sources. However, NG and electric power systems have a remarkable common feature which is that extensive networks are used to transport the energy from suppliers to customers. NG can also be carried in the form of Liquefied Natural Gas (LNG), which requires liquefaction trains, LNG ships and regasification terminals to accomplish with transport and re-inject the gas into the network. Pipelines transportation is more cost-effective over short and onshore distances, while LNG is typically used in transcontinental carriage (IEA, 2007).

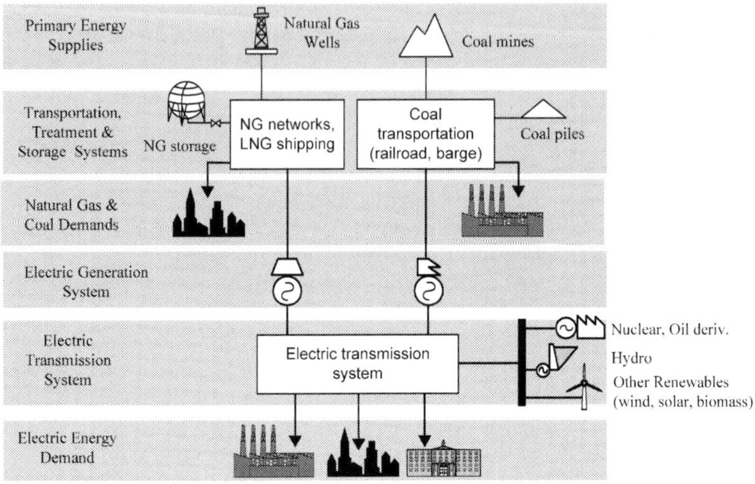

Fig. 2. Electric energy supply system and the converging energy resources

Table 1 shows the similarity in the organizational structures of electric power and NG systems.

Segments	Natural Gas Sector	Electricity Sector
Production (suppliers)	Gas Wells LNG regasification terminals	Electrical power plants (coal, nuclear, gas, hydro)
Transmission	High pressure network	High voltage network
Distribution	Medium/low pressure network	Medium/low voltage network
Consumption	Small consumers (commercial and residential customers) Large consumers (NGFPPs, industries, liquefaction trains)	Small consumers Large consumers

Table 1. Organization of electric power and NG systems

Fig. 3 shows a schematic representation of interconnected NG and electric power systems. The electric power system consists of a 3-bus network connecting NGFPPs, hydroelectric power plants, non-gas thermal power plants to electrical loads. The NG supply system is analogous to the electric power system, with high-, medium- and low-pressure pipelines connecting remote sources - gas wells and LNG regasification terminals - to large/small consumers and NGFPPs.

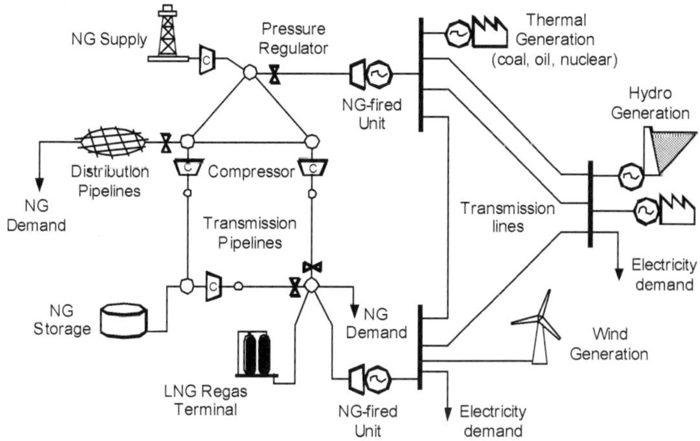

Fig. 3. Natural gas and electricity systems

In electric power systems, large generation units (hydro, nuclear, coal, etc.) are far from consumption centers, so the energy delivery involves several voltage levels (from 500kV to 0.4kV) of transmission and distribution networks to interconnect production areas to consumption centers. The electric power generation system consists of a heterogeneous set of technologies (hydro, CCGT, nuclear, coal/NG fired steam turbine, wind, etc.) with different capacities and operating constraints.

In electrical networks, the steady-state electric power flows are governed by Ohm's and Kirchhoff's laws. These laws can be expressed by means of nodal power balances and line power flows. Mathematically, the power flow through a transmission line depends on the complex voltage difference at its ends and its physical characteristics (serial and shunt resistance-reactance). The maximun capacity of each line is limited to its thermal limit (maximum conductor temperaure) or to stability margins set for the whole electrical network. The energy losses in the electrical networks are due to line resistance (Joule losses) and to a lesser extent to shunt losses (corona effect). The conversion between different voltage levels is performed by means of power transformers. The supervision, control and protection of transmission and distribution networks are performed by complex systems of switchgears and instrumentation equipments.

Like in the electricity production segment, a great diversity of technical characteristics can be found among gas suppliers. Gas wells (commonly located at sites far from load centers) and LNG regasification terminals (harbor locations) have a wide diversity of capacity and operating constraints.

NG transmission and distribution networks provide the same services as their electricity counterparts. Transmission pipelines undertake the responsibility of transporting NG from producers to local distribution companies or directly to large consumers. Distribution networks generally provide the final link in the NG delivery chain, taking it from city gate stations and additional gas supply sources to large and small customers.

Four basic types of facilities are considered in the modelling of NG transmission network: pipelines, compression stations, pressure regulators, and nodes (gathering or interconnection hubs)

A priority scheme for the supply of NG is generally in place for the situation where not enough gas is available to supply all the NG demands. Residential and commercial customers typically take precedence over large consumers and NGFPPs in this allocation.

With some similarities to electric power systems, the steady-state gas flow through a pipeline is a function of the pressure difference between the two ends, gas properties (e.g., compressibility factor, specific gravity) and physical characteristics of the pipe (e.g., diameter, length, friction factor), (Osiadacz, 1987). Therefore, the pressure represents the state variable which its analogous in power systems is voltage.

During transportation of NG in pipelines, the gas flow loses a part of its initial energy due to frictional resistance which results in a loss of pressure. To compensate these pressure losses and maximize the pipeline transport capacity, compressor stations are installed in different network locations. In contrast to electricity networks, where theoretically no significant active power is necessary to maintain a certain voltage, compressors are usually driven by gas turbines. The amount of NG consumed at compressor stations basically depends on the pressure added to the fluid and the volume flow rate through it. However, the operating pressures are constrained by the maximum pressure allowed in pipelines and the minimum pressure required at gate stations. Therefore, the transmission capacity of a gas pipeline is limited.

Valves are protective and control devices whose functions are similar to switchgears in electric power systems. Isolating valves are used to interrupt the flow and shut-off section of a network. Pressure relief valves can prevent equipment damage caused by excessive pressure. Pressure regulators can vary the gas flow through a pipeline and maintain a preset outlet pressure.

Compressor stations and pressures regulators enable a high degree of control of NG flow through the networks. On the other hand, currently it is neither economical nor practical controlling individual power transmission line flows using flexible alternating current transmission system (FACTS).

A comparison between electric power and NG systems is summarized in Table 2.

Characteristic	Power Electric System	Natural Gas System
Energy type	Secondary	Primary
State variables	Voltages	Pressures
Transmission losses (large systems)	Joule effect, up to 3%	Gas consumed in compressors stations, up to 7%
Flow modelling	Steady-state can be assumed for operational simulations	Transient-state is required for operational simulation (time steps shorter than several hours)
Supply hierarchy	Not required in normal operation state	Frequently required in normal operation state. Usually NGFPPs and industries have lower priority
Individual flow controllability	Currently neither economic nor practical at (FACTS)	By means of compressors and compressor stations
Storage facilities	Not yet technically or commercially feasible	Widely used in Europe and USA, not common in Latin America

Table 2. Differences between electric power and natural gas systems

While steady-state operation of power electric systems requires a constant balance between supply and demand, gas storage facilities are typically used to load balancing at any time, on hourly, daily, weekly or seasonally basis, keeping a NG supply as constant as possible. Additionally, large underground storages perform, principally, a supply security (strategic stock) function.

Unlike electricity, which large scale storage is not yet technically or economically feasible, natural gas can be stored for later consumption. There are three major types of NG storage facilities: a) underground storages (depleted gas/oil fields, salt caverns and aquifers), b) LNG tanks and c) pipelines themselves (the amount of gas contained in the pipes is called "line-pack" and can be controlled raising and lowering the pressure). These storage facilities are different in terms of capacities (working volume) and maximum withdrawal rates.

Other important difference between NG and electricity systems is that electricity moves at speed of light, while NG travels through the transmission network at maximum speed always lower than 100 km/h (reference value). These facts imply that the dynamic behaviour of NG systems is much slower than the dynamics of electric power systems. Thus, while steady-state electric power flows are assumed for multi-period simulation with time steps longer than half hour (even up to several minutes), NG flows multi-period simulations with time steps shorter than several hours require pipeline distributed-parameters and transient models (Osiadacz, 1987, 1996). However, many simplified models have been developed to NG flows simulation (Osiadacz, 1987) and transmission system optimization (Osiadacz, 1994; Ehrhardt & Steinbach, 2005).

3. Energy Systems Planning

Nowadays, there is consensus among policy makers that energy sector investment planning, pricing, operation and management should be carried out in an integrated and coordinated manner in order to achieve an economic, reliable and environmentally sustainable energy supply. A hierarchical and sequential procedure is typically used to tackle this huge and complex decision-making problem.

The so-called energy models are the first stage in this hierarchical energy planning procedure. In such models, all (or most) energy carriers are considered in an integrated approach. Several of these energy models have been developed to analyze a range of energy policies and their impacts on the energy system infrastructures and on the environment. Others are focused on the forecast of energy-service demands. An overview and a classification of some of the most relevant energy models, like TIMES (integrated MARKAL-EFOM system) (Loulou et al., 2005), MESSAGE (Messner & Schrattenholzer, 2005), ENPEP-BALANCE (CEESA, 2008) and LEAP (SEI, 2006), are described by Van Beeck (1999). These models are focused on a long term planning horizon (more than 10 years) and can be tailored to cover local, national, regional or world energy systems. The interactions between the energy sectors and the other sectors of the economy (e.g., transport, industry, commerce, agriculture) can be taken into account through model extensions or represented by means of constraints.

Because the dimensions of the problem, energy models are developed neither to represent the characteristics of different transport modes, nor to model the complex physical laws that governs electric power and NG systems. Usually, only nodal energy balances per each energy carrier are considered. Another limitation of these models is related to energy storage facilities, which are typically oversimplified or disregarded.

The results of energy models provide the framework for the following stages, in which each energy carrier system is planned and operated in a decoupled manner. Thus, specific procedures and strategies are implemented according to specific value system, e.g., economic, technical, political and environmental context. Usually, single energy carrier system expansion and operation planning are carried out considering the other energy carriers availabilities and prices as coordinating parameters. Electric power systems are a good example of this approach: they are planned and operated without taking into account the integrated dynamics of the fuel infrastructures and markets, i.e., costs and capacities of fuel production, as well as storage and transportation. The main assumption, in which this decoupled planning and operation approach, is based on the fact that there have not been significant energy exchanges between the energy carriers if they are compared with the total amount of energy supplied by each energy carrier.

More recently, new approaches to energy system planning have been presented. They are focused on a higher technical description of some energy sectors and their transport modes. The model presented by Bakken et al. (2007) includes the topology of several energy systems, and the technical and economic properties of different investment alternatives. Among other energy modes of transport, simplified electricity and NG networks are considered. This approach minimizes total energy system costs (i.e., investments, operating and emissions) for meeting predefined energy demands in a time horizon of 20-30 years. Hecq et al. (2001) and Unsihuay (2007b) propose specific methodologies and tools to address particularly the integrated NG and electric power systems planning. The network models consider not only the electricity and gas nodal balance, but also the loss factors and constrained capacity for each of the pipelines and electric network lines.

4. Coordination of Natural Gas and Electric Power Systems Operations

Nowadays, the operational planning of electric power and NG systems are carried out in a decoupled manner, i.e. different operational optimization problems are performed where each system is self-contained. However, this does not mean that both systems are totally independent. In fact, the existing interactions are modeled by means of fixed coordinating parameters. Typically, three types of parameters can be identified:

a) The NG prices considered in the production cost functions of each NGFPP;
b) The NG availability for the NGFPPs; and
c) NG consumption at each NGFPP

While, the electric power operational planning requires, as input data, the (a) and (b) set of parameters, the NG operational planning needs, as input data as well, the (c) set of parameters.

The decoupled approach consists in two stages. Firstly, the operational planning of electric power system is performed, being the NGFPP's consumption a byproduct of this procedure. Then, the operational planning of the NG systems can be carried out. The results of this last procedure include the NG marginal costs at each NGFPP location and NG actually supplied to each NGFPP.

However, the following situations can occur:

1. The total NG supply is not sufficient to meet the total NG demand, including the NGFPPs' demands. The NG supply to NGFPPs can be curtailed before than other demands, since NGFPPs usually have lower priority of supply.
2. The limited transmission capacity in the NG network can imply that the same situation described in 1) occurs in a specific node.
3. The fixed NG prices, which determine the NGFPPs' production costs, cannot match with the NG marginal costs at nodes where NGFPPs are placed. These marginal costs depend on the NG consumption in the compressor stations (NG network losses) and the binding pipeline's (transmission) capacity constraints..

If any of these situations actually occur, a re-dispatch of the electric power is required updating NG prices and availabilities for each NGFPP according the results obtained from the NG operational planning. Therefore, both operational planning models must be run iteratively. The convergence of procedure is slow and may be hard to reach when NG consumption in NGFPPs is a significant share of the total NG demand.

On the other hand, in a combined operational planning of NG and electric power systems, the described coordinating parameters are endogenous results of the optimization problem. This ensures that the optimal operating schedule for both is achieved simultaneously.

5. Combined Operational Planning of Natural Gas and Electric Power Systems

Several approaches that address the integrated modeling and analysis of energy systems in a more comprehensive and generalized way have been presented. These approaches consider multiple energy carriers; particularly electricity and NG systems interactions and combined operation have been investigated.

An assessment of the impact of NG prices and NG infrastructure contingencies on the operation of electric power systems is presented by Shahidehpour et al. (2005). A security-constrained unit commitment model, in which NG availabilities and prices are external parameters, is used to perform these evaluations. Conversely, Urbina & Li (2008) analyze the effect of pipelines and transmission lines contingencies by means a combined electric power and NG model.

A review of the main approaches and models, which deal with the integrated operational planning of multiple energy carrier systems, is presented in following subsections. This review is based on the survey collected by Rubio et al. (2008). The different approaches are conveniently grouped according to the considered time horizon.

5.1 Long- and Medium-Term

Quelhas et al. (2007) propose a generalized network flow model of an integrated energy system that incorporates the production; storage (where applicable); and transportation of coal, NG, and electricity in a single mathematical framework, for a medium-term operational optimization (several months to 2-3 years).

The integrated energy system is readily recognized as a network defined by a collection of nodes and arcs. Fuel production facilities, electric power plants and storage facilities are also modeled as arcs. A piecewise linear functions are applied to represent all cost and efficiencies. Since the problem is entirely modeled as a network and linear costs, a more efficient generalized network simplex algorithm is applied, than ordinary linear programming. The total costs considered are defined as the sum of the fossil fuel production costs, fuel transportation costs, fuel storage costs, electricity generation costs (operation and maintenance costs), and the electric power transmission costs. The objective of the generalized minimum cost flow problem is to satisfy electric energy demands with the available fossil fuel supplies at the minimum total cost, subject to nodal balances, maximum and minimum flow in each arc and emission (sulfur dioxide) constraints.

Additionally, the hydroelectric systems (hydropower plants and reservoirs) are also taken into account by Gil et al. (2003), but the emission constraints are not considered in this model. Correia & Lyra (1992) present also a generalized network flow model including only hydroelectric, NG and sugar cane bagasse as energy resources.

Bezerra et al. (2006) present a methodology for representing the NG supply, demand and transmission network within a stochastic hydrothermal scheduling model. The NG demand at each node is given by the sum of forecasted non-for-power gas and NGFPPs consumptions. The gas network modeling comprise: a gas balance at each node; maximum and minimum gas production, pipelines flow limits; and loss factors applied to gas flows (to represent the gas consumed by compressor stations). NG storage facilities are not been taking into account in this approach. The stochastic dual dynamic programming (SDDP) algorithm is used to determine the optimal hydrothermal system operation strategy, which minimize the expected value of total operating cost along the time horizon (2-3 years typically). While the total cost includes the fuel and shortage costs relating to electricity supply, the shortage costs associated to non-for-power NG load shedding are not considered. The NG prices are fixed from the outset and they are not results of the optimization process.

5.2 Short-Term

Unsihuay et al. (2007c) present a new formulation in order to include a NG system model in the short-term hydrothermal scheduling and unit commitment. NG wells, pipelines and storage facilities are considered, while nodal balances and pipelines loss factors are taking into account for a simplified gas network modeling. Gas storages are modeled similarly to water reservoirs. A constant conversion factor is used as input-output conversion characteristic for NGFPPSs. A dc power flow modeling without losses is applied to determine electric power flows. The problem is formulated as a multi-stage optimization problem, whose objective function is to minimize the total cost to meet the gas and electricity demand forecast. This total cost is the sum of the non-gas fired generators fuel costs, the startup costs of thermal units and the NG costs calculated at each gas well. The optimization procedure is subject to the following constraints: a) electric power balance at each node, b) hydraulic balance at each water reservoir, c) NG balance at each node and gas storage, d) initial and final water and gas volumes at reservoirs, e) electric power generation limits, f) maximum electric power flow through lines, g) NG withdrawal limits at gas wells, h) pipelines maximum transport capacity, i) bounds on storage and turbined water volumes, j) bounds on storage and outflow gas volumes, k) minimum up and down time of thermal units, and l) minimum spinning reserve requirement.

To solve the integrated electricity-gas optimal short-term planning problem an approach based on dual decomposition, Lagrangian relaxation and dynamic programming is employed.
Li et al. (2008) and Liu, et al. (2009) present the electric power security-constrained unit commitment problem including a NG network model. While in (Li et al., 2008) the NG flows are calculated through a nodal gas balance model, the steady-state physical laws (pressure differences) that govern NG flows are modeled in (Liu, et al., 2009). In both approaches, local NG storages at each NGFPP are considered. Particular and detailed modeling of fuel switching capabilities is described in (Li et al., 2008). Liu, et al. (2009) apply a decomposition method to separate the NG system optimization from the electric power security-constrained unit commitment problem, and treat it as a feasibility check subproblem.
A multi-period combined electricity and NG optimization problem is presented in (Chaudry et al., 2008). The modeling in this approach takes into account not only NG storages facilities, but also the NG contained in the NG network, so-called line pack. The optimization is performed with one month as time horizon with daily time steps. However, the authors include an approximation of the transient NG flows using the finite difference method. A detail model of NG storage injection and withdrawals rates is described.

5.3 Single Period - Snapshot

An et al. (2003) present a combined NG and electricity optimal power flow. The authors deal with the fundamental modeling of NG network, i.e., the steady-state nonlinear flow equations and detailed gas consumption functions in compressor stations. A complete formulation of the NG load flow problem and its similarities with power flows are shown in detail. Ac power flow modeling is applied to determine power flows in the electricity network. The objective function is formulated in terms of social welfare maximization. Thus, the total cost are represented by the generation costs due to non-gas electrical plants and gas supply costs, while the total benefits correspond to the electrical and gas consumers benefits. The benefits that would be allocated to NGFPPs are disregarded since the NGFPPs costs are also not considered.
Unsihuay et al. (2007a) also deal with the integrated NG and electricity optimal power flow. Nonlinear steady-state pipelines flows and compression station are modeled. However, the gas consumption in compressor stations is not considered. The objective function in this approach is to minimize the sum of generation costs due to non-gas electrical plants and costs of gas supply.
Urbina & Li (2007) propose a combined optimization model for electric power and NG systems. The objective is to minimize the electric power production costs subject to the NG transport limitations. A piecewise linear approximation is used to model the NG flows through the pipeline network. Since the steady-state NG flow is a non-convex function, the piecewise approximation is formulated using integer variables. Thus, a mixed integer linear programming is applied to solve the optimization problem.
Mello et al. (2006) and Munoz et al. (2003) present a model to compute the maximum amount of electric power that can be supplied by NGFPPs, subject to NG systems constraints. Nonlinear steady-state NG flows and the effect of compressor stations to enlarge the transmission capacity are included in the NG network modeling. Like in Unsihuay et al. (2007a), the amount of gas consumed in the compressor stations is neglected.

Geidl & Andersson (2007) introduce a comprehensive and generalized optimal power flow of multiple energy carriers. This paper presents an approach for combined optimization of coupled power flows of different energy infrastructures such as electricity, gas, and district heating systems. A steady-state power flow model is presented that includes conversion and transmission of an arbitrary number of energy carriers. The couplings between the different infrastructures are explicitly taken into account based on the new concept of energy hubs. With this model, combined economic dispatch and optimal power flow problems are stated covering energy transmission and conversion. Additionally, the optimality conditions for multiple energy carriers' dispatch are derived, and the approach is compared against the standard method used for electric power systems.

Arnold & Andersson (2008) address the combined electricity and NG optimal power flow (OPF) using the approach proposed by Geidl & Andersson (2007). The OPF problem is solved in a distributed way where each energy hub (combined electric power and NG node), also referred to as control area, is controlled by its respective authority. Applying distribution control techniques, the overall optimization problem is divided into subproblems which are solved iteratively and in a coordinated way. Under this approach different operating targets (e.g., cost minimization, emission caps, security criteria) can be applied at each energy hub.

Hajimiragha et al. (2007) extend the model of Geidl & Andersson (2007) to consider hydrogen as another energy carrier.

Rajabi & Mohtashasmi (2009) present a new model which integrates the NG transport cost in the electric power economic dispatch problem. The NG flows are modeled through the steady-state nonlinear equations and transport cost is defined as the sum of NG consumption in compression stations. The non-for-power NG demand is disregarded.

Ojeda-Esteybar et al. (2009) present a comparison between the decoupled and the combined approach for the optimal dispatch of electric power and NG systems. Rubio-Barros et al. (2009) present a detailed an extensive analysis of the coordinating parameters, which are the reasons for the inefficiencies in the decoupled approach.

6. Economic and Market Issues

Electricity and gas sectors have been liberalized to a certain extent in many countries, introducing competition at varying degrees and at various levels of the value chain. Essentially, these restructures have been attained by unbundling the different segments of the industries. In the electricity sector, the production segment (generation) was separated from the service segments (transmission & distribution). In the same way, the NG sector was split up into a production segment (upstream) and pipeline network services (midstream & downstream). Like in the electricity system, gas transmission and distribution companies provide open pipelines access to other market participants for gas delivery which has permitted producers to sell gas directly to end users and marketers. Different types of markets have been established, allowing the interaction between production sector (suppliers) and consumption sector (demands).

Since a significant share of total NG consumption is used to produce electricity; the market prices of both energy carriers are linked. Therefore, the NGFPPs play a key role in the electricity and gas price dynamic because they are the market participants that allow the arbitrage between the two commodities. Liberalized markets for both commodities promote the arbitrage, and therefore contribute to the price convergence.

The increasing links between gas and electricity also offer both a threat and an opportunity regarding energy supply security. Flexibility facilities, such as energy storage (e.g., gas storage, water reservoirs) and fuel switching (in NGFPPs or steam power plants) are important resources to ensure the gas and electricity supply security and to reduce prices volatility. Additionally, efficient gas and electricity markets tend to reduce gas demand as prices increase, saving gas at times of high demand or low supply.

Different experiences in liberalized electricity markets show that one of the most powerful consumer's mean to avert supplier's market power is the presence of a well-functioning, transparent and liquid wholesale market. Therefore, it is likely that a liquid and competitive wholesale market for NG provides also a powerful tool to counterbalance potential upstream market power in gas. There are numerous policy challenges in establishing well-functioning gas and electricity markets to ensure affordable and reliable energy supply. The short-term price spikes are of paramount importance in order to create resilience to short-term but severe disruptions since these spikes reflect the immediate need for balanced, cost-effective and significant responses. Price caps or other market alterations mitigate these signals and the necessary market response, such as reduced demand, increase supply or storage changes.

6.1 Gas Price Formation

The growth of world oil prices has also produced an increase or readjustment of NG price. This correlation is mainly because both fuels are substitutes of each other; especially in the electricity sector.

The economic theory postulates that in a competitive market, like a mature NG market (e.g., USA, UK), the price maker is defined by short-term prices (spot price on Henry Hub or National Balancing Point) or by standard quotations in a Stock Exchange (NYMEX, ICE). Therefore in markets of this nature, the price reflects the interactions between supply and demand. However, in many markets (e.g. European countries, except UK; Japan; Korea) the linkage between NG and oil prices is still rigidly formalized by contracts which include indexation formulas. In NG monopolies, prices are obtained by subtracting the total costs of transmission and distribution from the final convergent energy market price (electricity price).

In NG markets, like in other commodity markets, exist long-term supply or demand contracts with indexed prices over the time and penalties in any case of lack (called deliver-or-pay or take-or-pay contracts). Usually, a significant share of the NG is traded through this long-term arrangement, thus they establish a price reference for the rest of markets with shorter delivery time.

Until a few years ago, NG markets were considered as regional markets due to the lack of sufficient interchange among them. Currently the NG is becoming an increasingly global commodity due to the rapid growth in the installed LNG infrastructure and the development of LNG markets (IEA, 2007).

The NG prices paid by the consumers are calculated as the sum of the wellhead price (or wholesale market price), the transmission cost and the distribution cost.

It important to point out that the interactions between electric power and NG markets are closely related with NG price foundations and how they respond to the demand variations.

Also, the transport cost allocation methodologies used in electric power and NG systems can have an important impact on the interactions. Morais & Marangon Lima (2003, 2009) analyzes the effects of applying different transmission cost allocations to electricity and NG networks. The authors show that coordination is needed between the applied methods in each network, otherwise wrong economic signal are sent to market players, in particular to NGFPPs.

6.2 NGFPPs Perspective

NGFPPs participate simultaneously in electricity and NG markets. NGFPPs can now purchase NG with great flexibility, through bilateral contracts or through the spot market. On the other hand, the wholesale electricity market is an important part in the decision-making process for NGFPPs. When the market implied marginal heat rate (which is the equivalent heat-rate calculated using the clearing price for electricity divided by the prevailing NG price) is lower than the marginal production heat rate of the NGFPP, the generating company that owns this NGFPP prefers to purchase electricity to meet its commitments instead of generate it itself and resell the previously contracted gas on the spot NG market (Chen & Baldick, 2007).

Another way of looking at the same problem is through the so-called spark spread, which is defined as the difference, at a particular location and time between the fuel cost of generating a MWh of electricity and the price of electricity. As a result, a positive spark spread indicates the power generator should buy electricity rather than produce it.

Other service that can be provided by NGFPPs is called tolling, where a power generator receives fuel from a beneficiary and delivers electric power to the same beneficiary in return for a service fee.

Some aspects of the NGFPPs role in the electric power and NG markets have been addressed in recent studies. Chen & Baldick (2007) propose a short-term NG portfolio optimization for electric utilities that own NGFPPs. This approach considers the financial risk associated with the portfolio and a risk preference function of the electric utility. The portfolio includes base load contracts, intra-day contracts, swing supply and withdrawals from storage facilities as NG supply resource options. Purchasing electricity from the wholesale market, selling NG in the spot market and injections in storage facilities are also the alternatives taken into account to supply a given electricity demand. The approach excludes the option of selling electricity to the wholesale market. Usama & Jirutitijaroen (2009) present a profit-risk maximization model focused on NGFPPs involvement in spot and forward markets. This approach uses the conditional value-at -risk as risk measure within the optimization problem. In both approaches price-taking is assumed, and thus the market fundamental behaviour as result of the price arbitrage are disregarded.

Takriti et al. (2001) discuss the problem of heading between the NG and electricity markets. The problem is addressed from an energy marketer perspective that purchases NG from the open market and sells it to contracted customers, but the marketer also has the option to generate electricity and sell the produced electric power to the wholesale market. A NG storage facility is also another balancing resource considered in the model. Hence, based on multiple forecasts for NG customers' demands, NG prices and electricity prices, a stochastic optimization is performed to find the optimal heading strategy.

NGFPPs might also want to resell their firm (take-or-pay) NG contracts every time the consumption of these amount of gas implies economic losses given certain electricity market conditions. Street et al. (2008) investigates the creation of a secondary market, where

NGFPPs offers flexible NG supply to industrial NG consumers, who would receive the NG originally assigned to the NGFPPs only when the latter are not dispatched. All the periods, in which the NGFPPs are committed, the industrial NG consumers should resort to alternative fuels or NG supply. The success of this secondary market depends on the price of the flexible NG supply contracts. Thus, the authors present a stochastic model to look into a range of prices and the feasibility of this type of market.

6.3 Social Welfare

The objective function of a comprehensive (including the demand response) gas and electric optimal operational planning should be the maximization of social welfare during the considered time horizon. This social welfare is the total gross demand surplus due to gas and electricity consumption minus the total system operating cost (gas supply costs and non-gas electric power plant costs). An et al. (2003) present an assessment of the differences in the social welfare for both, integrated and decoupled gas and electricity optimal power flows. They show that there is a social welfare loss (deadweight) for the decoupled case; except when outset gas prices match the prices obtained in the integrated case.

Ojeda-Esteybar et al. (2009) and Rubio-Barros et al. (2009) present a comprehensive economic impact assessment of decoupled approach for the optimal dispatch of electric power and NG systems. The authors show that a higher economic efficiency is achieved and guarantee only if both energy systems are considered in an integrated manner.

7. Conclusions

Different simplified energy models have been proposed for policy analysis, forecasting, and to support regional or global energy planning. Although economic and physical performances of individual subsystems are well studied and understood, there has been little effort to study the characteristics of integrated systems, especially in the medium- and short-term due to the complexity of the required models.

The interdependencies between electric power and NG systems have shown the need of new approaches and models able to take into account these increasing interactions.

The inclusion NG system model is of paramount importance for the electric power systems planning. New methodologies for the integrated expansion and operation planning of both systems are required. Also, NG infrastructure must be considered in power system reliability assessment.

It is envisioned that energy companies and government agencies must consider an integrated approach for the operation and planning of NG and electricity infrastructures to ensure that the most economical and secure policies are used in the foreseeable future.

8. References

An, S.; Li, Q. & Gedra, T. W. (2003). Natural gas and electricity optimal power flow. *Proceedings of IEEE Power Eng. Soc. Transmission & Distribution Conf.*, Dallas, TX, USA, Sep. 2003.

Bakken, H., Skjelbred, H.I. & Wolfgang, O. (2007). eTransport: investment planning in energy supply systems with multiple energy carriers. *Energy*, Vol. 32, No. 9,pp. 1676-1689.

Arnold, M. & Andersson, G. (2008). Decomposed electricity and natural gas optimal power flow. *Proceedings of 16th Power Systems Computation Conf. (PSCC)*, Glasgow, Scotland, Jul. 2008.

Bezerra, B.; Kelman, R.; Barroso, L. A.; Flash, B.; Latorre, M. L.; Campodónico, N. & Pereira, M. V. F. (2006). Integrated electricity-gas operations planning in hydrothermal systems. *Proceedings of X Symposium of Specialists in Electric Operational and Expansion Planning (SEPOPE)*, Florianópolis, SC, Brazil, May 2006.

Center for Energy, Environmental, and Economic Systems Analysis (CEESA), 2008 Overview of the Energy and Power Evaluation Program (ENPEP-BALANCE). [Online]. Available at: http://www.dis.anl.gov/projects/Enpepwin.html [Accessed 27 November 2009].

Chaudry, M.; Jenkins, N. & Strbac, G. (2008). Multi-time period combined gas and electricity network optimisation. *Elec. Power Syst. Research*, Vol. 78, No. 7, pp. 1265-1279.

Chen, H. & Baldick, R. (2007). Optimizing short-term natural gas supply portfolio for electric utility companies, *IEEE Trans. on Power Syst.*, Vol. 22, No. 1, pp. 232-239.

Correia, P. & Lyra, C. (1992). Optimal scheduling of a multi-branched interconnected energy system. *IEEE Trans. Power Syst.*, Vol. 7, No. 3, pp. 1225-1231.

Ehrhardt, K. & Steinbach, M.C. (2005). Nonlinear Optimization in Gas Networks, in: *Modeling, Simulation and Optimization of Complex Processes*, Editors: Bock, H.G., Kostina, E., Pu, H.X. & Rannacher, R. , pp. 139-148, Springer-Verlag Berlin, Heidelberg, New York.

Geidl, M. & Andersson, G. (2007). Optimal power flow of multiple energy carriers. *IEEE Trans. Power Syst.*, Vol. 22, No. 1, pp. 145-155.

Gil, E. M.; Quelhas, A. M.; McCalley, J. D. & Voorhis, T. V. (2003). Modeling integrated energy transportation networks for analysis of economic efficiency and network interdependencies. *Proceedings of North Amer. Power Symp. (NAPS)*, Rolla, MO, USA, Oct. 2003.

Hajimiragha, A.; Canizares, C.; Fowler, M.; Geidl, M. & Andersson, G. (2007). Optimal energy flow of integrated energy systems with hydrogen economy considerations. *Proceedings of Bulk Power System Dynamics and Control - VII*, Charleston, SC, USA , Aug. 2007.

Hecq, S., Bouffioulx, Y., Doulliez, P. & Saintes, P. (2001). The integrated planning of the natural gas and electricity systems under market conditions. *Proceedings of IEEE Power Eng. Soc. PowerTech*, Porto, Portugal, Sep. 2001.

International Energy Agency (IEA), 2007. *Natural Gas Market Review 2007.* [e-book] France, Paris: OECD/IEA Publications. Available at: http://www.iea.org/publications/free_new_Desc.asp? PUBS_ID=1909 [Accessed 13 March 2010].

International Energy Agency (IEA), 2009a. *Electricity Information 2009.* [e-book] France, Paris: OECD/IEA Publications. Available at: http://www.iea.org/publications/free_new_Desc.asp? PUBS_ID=2036 [Accessed 13 March 2010].

International Energy Agency (IEA), 2009b. *Natural Gas Information 2009.* [e-book] France, Paris: OECD/IEA Publications. Available at: http://www.iea.org/publications/free_new_Desc.asp? PUBS_ID=2044 [Accessed 13 March 2010].

International Energy Agency (IEA), 2009c. *World Energy Outlook 2009.* [e-book] France, Paris: OECD/IEA Publications. Available at: http://www.iea.org/w/bookshop/add.aspx?id=388 [Accessed 13 March 2010].

International Energy Agency (IEA), 2009d. *Natural Gas Market Review 2009*. [e-book] France, Paris: OECD/IEA Publications. Available at: http://www.iea.org/publications/free_new_Desc.asp? PUBS_ID=2102 [Accessed 13 March 2010].

International Organization for Standardization (ISO), 1997. ISO 13600 *Technical energy systems -- Basic concepts*. Geneva: ISO.

Li, T.; Erima, M. & Shahidehpour, M. (2008). Interdependency of natural gas network and power system security. *IEEE Trans. Power Syst.*, Vol. 23, No. 4, pp. 1817-1824, Nov. 2008.

Liu, C.; Shahidehpour, M.; Fu, Y. & Li, Z. (2009). Security-constrained unit commitment with natural gas transmission constraints. *IEEE Trans. Power Syst.*, Vol. 24, No. 3, pp. 1523-1536.

Loulou, R., Uwe, R., Kanudia, A., Lehtila, A. & Goldstein, G. (2005). Documentation for the TIMES Model, Part I. Energy Technology Systems Analysis Programme. [Online]. Available at: http://www.etsap.org [Accessed 27 November 2009].

Mello, O. D. & T. Ohishi, T. (2006). An integrated dispatch model of gas supply and thermoelectric generation with constraints on the gas supply. *Proceedings of X Symp. of Specialists in Electric Operational and Expansion Planning (SEPOPE)*, Florianópolis SC, Brazil, May 2006.

Messner, S. & Schrattenholzer, L. (2005). MESSAGE-MACRO: linking an energy supply model with a macroeconomic model and solving it inter-actively. *Energy*, Vol. 25, No. 6, pp. 145-155.

Morais, M. S. & Marangon Lima, J.W. (2003). Natural gas network pricing and its influence on electricity and gas markets. *Proceedings of IEEE Power Eng. Soc. PowerTech*, Bologna, Italy, Jun. 2003.

Morais, M. S. & Marangon Lima, J.W. (2007). Combiend natural gas and electricity network pricing. *Elec. Power Syst. Research*, Vol. 77, No. 5-6, pp. 712-719.

Munoz, J.; Jimenez-Redondo, N.; Perez-Ruiz, J. & Barquin, J. (2003). Natural gas network modeling for power systems reliability studies. *Proceedings of IEEE Power Eng. Soc. PowerTech*, Bologna, Italy, Jun. 2003.

Ojeda-Esteybar, D.; Rubio-Barros, R.; Añó, O. & Vargas, A. (2009). Despacho óptimo integrado de sistemas de gas natural y electricidad: comparación con un despacho desacoplado y aplicación al sistema argentino. *Proceedings of XIII Encuentro Regional Iberoamericano de (ERIAC)*, Puerto Iguazú, Argentina, May 2009.

Osiadacz, A. J. (1987). *Simulation and Analysis of Gas Networks*. E. & F. N. Spon, ISBN 0-419-12480-2, London.

Osiadacz, A. J. (1994). Dynamic optimization of high pressure gas networks using hierarchical system theory. *Proceedings of 26th Annual Meeting of Pipeline Simulation Interest Group*, San Diego, CA, USA, Oct. 1994.

Osiadacz, A. J. (1996). Different transient models- limitations, advantages and disadvantages. *Proceedings of 28th Annual Meeting of Pipeline Simulation Interest Group*, San Francisco CA, USA, Oct. 1996.

Quelhas, A.; Gil, E.; McCalley, J. D. & Ryan, S. M. (2007). A multiperiod generalized network flow model of U.S. Integrated energy system: part I - model description. *IEEE Trans. Power Syst.*, Vol. 22, No. 2, pp. 829-836.

Rajabi, H. & Mohtashasmi, S. (2009). Economic dispatch problem considering natural gas transportation cost. *Proceedings of World Academy of Science, Engineering and Technology*, Vol. 38, Feb. 2009, pp. 1482-1487, ISSN: 2070-3740.

Rubio, R.; Ojeda-Esteybar, D.; Añó, O. & Vargas, A. (2008). Integrated natural gas and electricity market: a survey of the state of the art in operation planning and market issues. *Proceedings of 2008 IEEE/PES Transmission & Distribution Conf. & Expo: Latin America*, pp. 1-8, Bogotá, Colombia, Aug. 2008.

Rubio-Barros, R.; Ojeda-Esteybar, D.; Añó, O. & Vargas, A. (2009). Identificación de los parámetros para la coordinación de los despachos de los sistemas eléctricos y de gas natural. *Proceedings of VIII Latin-American Congress on Electricity generation and Transmission*, ISBN - 978-85-61065-01-0, Ubatuba, Brasil, Oct. 2009 ISBN - 978-85-61065-01-0.

Shahidehpour, M.; Fu, Y. & Wiedman, T. (2005). Impact of natural gas infrastructure on electric power systems. *Proceedings of IEEE*, Vol. 93, No. 5, pp. 1042–1056.

Street, A.; Barroso, L. A.; Chabar, R.; Mendes, A. T. S. & Pereira, M. V. F. (2008). Pricing flexible natural gas supply contracts under uncertainty in hydrothermal market. *IEEE Trans. Power Syst.*, Vol. 23, No. 3, pp. 1009-1017.

Stockholm Environmental Institute (SEI), 2006 LEAP: User Guide. [Online]. Available at: www.energycommunity.org/documents/Leap2006UserGuideEnglish.pdf [Accessed 27 November 2009].

Takriti, S.; Supatgiat, C. & Wu, L. S.-Y. (2001). Coordination fuel inventory and electric power generation under uncertainty. *IEEE Trans. Power Syst.*, Vol. 16, No. 4, pp. 603-608.

Unsihuay, C.; Marangon-Lima, J. W. & Zambroni de Souza, A. C. (2007a). Modeling the integrated natural gas and electricity optimal power flow. *Proceedings of IEEE Power Eng. Soc. General Meeting*, Tampa, FL, USA, Jun. 2007.

Unsihuay, C.; Marangon-Lima, J. W. & Zambroni de Souza, A. C. (2007b). Integrated power generation and natural gas expansion planning. *Proceedings of IEEE Power Eng. Soc. PowerTech*, Lausanne, Switzerland, Jul. 2007.

Unsihuay, C.; Marangon-Lima, J. W. & Zambroni de Souza, A. C. (2007c). Short-term operation planning of integrated hydrothermal and natural gas systems. *Proceedings of IEEE Power Eng. Soc. PowerTech*, Lausanne, Switzerland, Jul. 2007.

Urbina, M. & Li, Z. (2007). A combined model for analyzing the interdependency of electrical and gas systems. *Proceedings of 39th North American Power Symp.* pp. 468-472, Las Cruces, NM, USA Oct. 2007.

Urbina, M. & Li, Z. (2008). Modeling and analyzing the impact of interdependency between natural gas and electricity infrastructures. *Proceedings of IEEE Power Eng. Soc. General Meeting*, Pittsburgh, PA, USA, Jul. 2007.

Usama, A. & Jirutitijaroen, P. (2009). An optimization model for risk management in natural gas supply and energy portfolio of generation company. *Proceedings of IEEE TENCON*, Singapore, Nov. 2009.

Van Beeck, N. (1999). Classification of energy models, in: Tilburg University and Eindhoven University of Technology: The Netherlands, May 1999.

13

Compressed natural gas direct injection (spark plug fuel injector)

Taib Iskandar Mohamad
Universiti Kebangsaan Malaysia (National University Malaysia)
Malaysia

1. Introduction

The increasing concerns over energy security and the emission of pollutant gases have triggered greater efforts towards developing alternatives to conventional fuels for road vehicles. In the presence of these concerns, automotive engine technology is challenged by the increasing divergence between higher power output, better fuel economy and lower pollutant emission requirements (Stan, 2002).

Several alternatives to gasoline and diesel fuels have been studied on current internal combustion (IC) engines. These include natural gas (NG), which is predominantly methane, liquefied petroleum gas (LPG), hydrogen, as well as ethanol and methanol. They are used either as supplement or replacement to gasoline in spark ignition (SI) engines. For compression ignition (CI) engines, dual fuel operation with diesel fuel providing pilot ignition source has been successful for heavy-duty applications. CI engines have also benefited from the use of various alternative fuels of vegetable origins as diesel replacement. LPG is a promising alternative fuel mainly due to its relatively high energy density, high octane rating and low pollutant emissions. It can be stored as liquid at moderate pressure, which gives it major advantage over most other alternative fuels. Methanol on the other hand has a very high octane rating but low heating value and stoichiometric air fuel ratio (AFR). Thus it leads to higher volumetric fuel consumption when compared to gasoline. Hydrogen fuel for electrically driven fuel cell cars, seen as the future replacement to IC engine technology, is undergoing relatively slower research and development and is expected to be in large scale production at some distance of time. IC engines is therefore will remain the key power source in the 21st century until fuel cell vehicles become widespread (Morita, 2003)

Natural gas use has various advantages over conventional fuels mainly due to its potential for higher thermal efficiency (due to higher octane value that allows the use of higher compression ratios), and lower CO_2 emission (due to lower carbon-to-hydrogen ratio) (Shiga et. al. 2002). From the supply point of view, natural gas has the advantage of energy diversification and the total reserves have been estimated in the same order as petroleum but with only 60% of its production rate (Vuorenkoski, 2004).

According to the statistics by the International Association for Natural Gas Vehicles (IANGV, 2009), there are approximately 11.2 million NGVs in operation worldwide with long establishment record in Europe, North America and South America. Pakistan, Argentina, Iran and Brazil record the highest numbers of NGV with 2.4, 1.8, 1.7 and 1.6 million respectively. The numbers are increasing with mounting interest from other countries like India (725,000 NGV) and Malaysia (42,617 NGV). Most NGV are fuel converted and dual fuel types.

Natural gas is often stored compressed at ambient temperature as compressed natural gas (CNG) in these vehicles but it requires more storage space. NG can also be stored cryogenically at ambient pressure as liquefied natural gas (LNG) in heavy-duty vehicles. For the same energy content, the emission from NG combustion have significantly less harmful combustion products such as CO_2 and NO_x than gasoline and diesel engines (Bradley, 1996). NGV can be categorized into three types, (1) fuel converted, (2) dual fuel operation and (3) dedicatedly developed engine. Most NGV are of type (1) and (2) while type (3) available mainly for heavy duty vehicles. It is well known that when a port injection gasoline engine is converted to NG, with the fuel injected in the intake manifold, power is reduced and upper speed is limited. These are due to reduction of volumetric efficiency and the relatively lower turbulent flame speed of NG-air combustion (Ishii, 1994). The problems can be mitigated by direct injection which increases volumetric efficiency and improves mixing as a result of turbulence induced by high pressure injection. However, to achieve direct fuel injection, a complicated and costly engine modification is required. The cylinder head needs to be redesigned or retrofitted to accommodate the direct fuel injector.

2. Direct injection concepts

Two main characteristics of direct injection are internal mixture formation and closed valve injection. Mixture formation is vital in direct injection because the available time for air-fuel mixing is relatively short compared to indirect port injection or carburetion.

2.1 Internal mixture formations in direct injection spark ignition engines

In spark ignition engines, air and fuel mixing takes place in the cylinder but a premixing process occurs to a certain degrees depending on type of fuel delivery. In a carburetor system, fuel vaporizes and mixes in the air stream prior to entering the combustion chamber. In a port injection system, fuel is injected and the velocity of fuel jet determines atomization and evaporation of fuel in air. In the direct injection method, fuel is directly injected into the combustion chamber as intake valve closes. The turbulence induced by the gas jet and the jet penetration determine the degree of mixing. In general, the mixing process in the direct injection method is restricted to a much shorter time. Furthermore, unlike the carburetion and port injection where mixing starts before air and fuel enter the combustion chamber, the mixing in direct injection mode can only happen in confined cylinder geometry.

The concepts of homogenous and stratified mixture formation are very important when discussing the direct injection in spark ignition engines because they form the basis of a better control of fuel mixture than the one experienced with port fuel injection. In addition, charge stratification can increase thermal efficiency and have the potential of reducing pollutant emissions. However, with direct injection operation, the degree of mixing and mixture uniformity is vital for reliable combustion. A combination of direct injection, high squish, high swirl and optimized piston crown shape can produced fast mixing and a high degree of mixture uniformity, thus turbulent intensity, molecular diffusion and chemical kinetics, which are the main contributors to the establishment and propagation of a turbulent flame (Risi, 1997). Mixture formation in direct injection engines can be classified into homogeneous and stratified charge based on the injection strategies. The concepts of these mixture formations are determined by the engine operation and fuel economy requirements.

2.1.1 Early injection, homogeneous-charge operation

The homogeneous mixture operating mode in the direct injection engine is designed to meet the requirement of medium-to-high engine loads. Depending on the overall air-fuel ratio, the mixture can be homogeneous-stoichiometric or homogeneous lean. Early injection makes it possible to achieve a volumetric efficiency that is higher than port fuel injection, and slightly increased compression ratio operation which contributes to better fuel economy. It also benefits from better emission during cold start and transient operation (Zhao, 1999 & 2002).

2.1.2 Late injection, stratified-charge operation

This operation is mainly to achieve lean burn and unthrottled operations by injecting fuel late during compression stroke. Fuel stratification is achieved by injection strategy such that the air-fuel ratio around the spark gaps yield stable ignition and flame propagation, whereas areas farther from the point of ignition is leaner or devoid of fuel. The advantage of charge stratification includes significant reduction in pumping work associated with throttling, reduced heat loss, reduced chemical dissociation from lower cycle temperatures and increases specific heat ratio for the cycle, which provide incremental gains in thermal efficiency (Zhao, 2002).

2.2 Potential for direct fuel injection in spark ignition engine

Direct injection in spark ignition engines could achieve a number of desirable effects. When direct injection method is applied to gaseous fuel, more achievement in terms of specific power output can be realized due to significant improvement in volumetric efficiency. The advantages of direct injection methods can be summarized as follows (Stan, 2002)

2.2.1 Increased thermal efficiency and lower specific fuel consumption

At part load, avoiding fresh charge throttling results in charge stratification and burned gas in distinct zones. This ideal structure consists of stoichiometric mixture cloud with spark contact, enveloped by fresh air and burned gas that form a barrier against chemical reactions near chamber wall thus avoiding intense heat transfer to the wall during combustion. Thermal efficiency is bettered by increasing compression ratio, as well as turbo charging and supercharging. Knock can be avoided in such cases by different effects: mixture formation just before or during ignition; mixture concentration in central zone of combustion chamber; out of crevice; of mixture cooling by fuel vaporization during injection.

2.2.2 Higher torque due to increase in absolute heating value of mixture

This is achieved by higher scavenging intensity. The more fresh air is captured, the more fuel can be injected. One the other hand, more captured air generally leads to greater scavenging losses. However, fresh air losses without fuel inclusion have no disadvantages in terms of pollutant emissions, leading to only a slight disturbance of thermal efficiency. Such enforced cylinder filling with air forms the basis for efficient downsizing, which involves supercharging of turbo-charging, inter-cooling, and also adapted valve control.

2.2.3 Decreased pollutant emission

Pollutant emission reduction is achieved mainly by the lean burn strategy, which is usually coupled with unthrottled operation. The lean mixture formation however needs a careful calibration of injection timing with respect to ignition timing, as well as proper shape of fuel spray and injection duration.

2.2.4 Improved acceleration behavior

Intensive scavenging and unthrottled operation results in a high charge of fresh air at every load. Thus, inertial effects of the air flow during acceleration can be partially avoided. On the other hand, the possible real-time adaptation of the fuel injection rate to the air flow behavior allows an adaptable correlation of mixture formation and combustion. Such dynamic response on torque demand is well known from advanced car diesel engines with direct injection. The success of a direct injection method depends mainly on the effectiveness of mixture formation and control. This is achieved with precise fuel injection strategies, appropriate cylinder geometry with flow guide and advanced engine management system.
A simple technique to improve engine performance when converting port injection gasoline engines to NG operation using direct fuel injection can be realized using spark plug fuel injector (SPFI). The next section will explore briefly on the development of SPFI and its design specifications.

3. Spark Plug Fuel Injector (SPFI)

The SPFI is a device developed in order to convert any port injection engine to direct injection gaseous fuels (natural gas, hydrogen etc). Normally, converting to direct injection requires modifications to or replacement of the cylinder head to accommodate extra holes for fuel injectors and possibly modifying the piston crowns which incur a high cost. With SPFI, this cost can be reduced because no modification on the original engine structure and spark plug placement are required. It is a technically easier and cheaper conversion system by only replacing the existing spark plug with SPFI. As a result, users will not only benefit from an alternative low cost conversion, but also can enhance the engine performance compared to currently available conversion kits with port fuel injection. In addition, SPFI also can be utilized for dual-fuel systems in the internal combustion (IC) engines such as gasoline-CNG, gasoline-hydrogen and CNG-hydrogen. Fig. 1 shows the Spark Plug Fuel Injector (SPFI) and associated components. The technical drawings and connection to a direct gasoline injector (DGI) encapsulated in a bracket are shown in Fig. 2. Fig. 3 shows the DGI with the bracketing components. The SPFI consists of a spark plug with a 1 mm by 2 mm square cross-section fuel path cut out along the periphery of its threaded section and a steel tube soldered to the end of the cut section. A DGI is connected to it using a specially developed bracket to the end of the fuel path. The distance from the DGI injection nozzle to the SPFI nozzle is 11 cm. The DGI injector is connected to a 230 bar methane bottle through a pressure regulator where methane pressure is reduced to the desired pressure. A specially developed injection control was used to regulate fuel injection by referencing crank angle signals from a camshaft encoder. The length of injection pulse determines fuel mass delivered, therefore air/fuel ratio to the cylinder. The most immediate effect of installation of SPFI on the Ricardo E6 engine is the reduction of motorized engine cylinder peak pressure due to decrease in compression ratio (Mohamad, 2006).

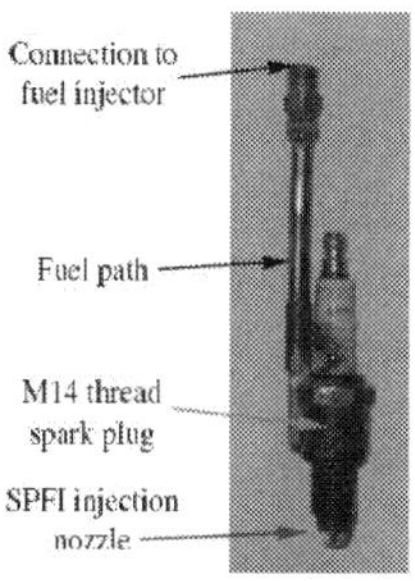

Fig. 1. Spark plug fuel injector

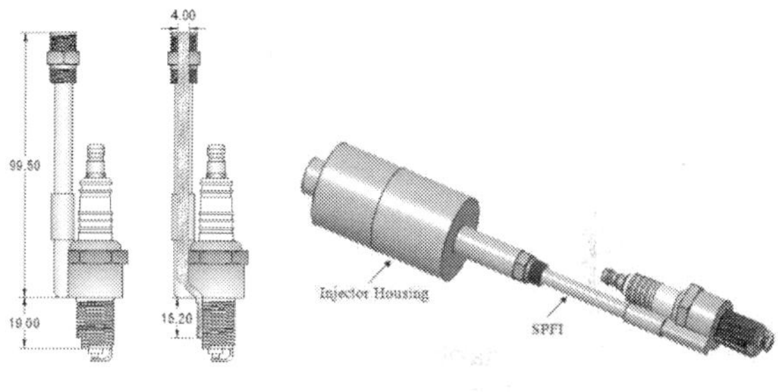

Fig. 2. (Left) Technical drawing of SPFI (units in mm) and (Right) SPFI connected to injector housing

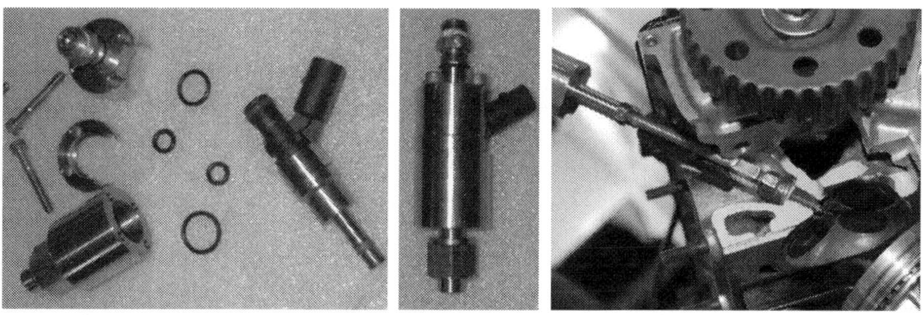

Fig. 3. DGI injector and enclosure component (left), DGI injector in the enclosure (center) and SPFI mounted on an engine (left)

3.1 Injection process

Among the initial SPFI development works is the study on injection process (Mohamad, 2010). It is important to understand the effects of the SPFI fuel path on the injection process which will significantly affect the air-fuel mixing in the combustion chamber. Fig. 4 shows the experimental setup of the injection measurement. The experimental setup can be categorised into four groups; (1.) a laser-optical lens system, (2.) an imaging system, (3.) a fuel supply

system and, (4) a fuel chamber equipped with windows in which the fuel substitute is injected (termed the "bomb"). The fuel injection driver circuit was synchronized with the camera and laser systems. A MOSFET transistor, acting as a gate/bridge between a power supply unit and the SPFI, is excited by the output of a pulse generator which functioned as the main driver to the experiment. At the same time, the pulse generator output was connected to another pulse/delay generator where its signal was sent to the laser source. The laser source simultaneously sent a signal to the camera so that the laser pulse was synchronised with the opening of the shutter. An oscilloscope was connected to both pulse/delay generator to measure the delay between initiation of fuel injection and image capture. For every 0.5-1.0 ms interval (delay with respect to injection signal), 20 images were captured, therefore the sequence of jet development images was independent of the laser frequency.
The plume of fuel emerging from the injector nozzle was imaged. The plume was illuminated with a 100 mm high, 0.5mm thick light sheet from a XeCl excimer laser (308nm) with 100-200mJ/pulse and was imaged with a camera with image intensifier and lens. Laser Induced Fluorescence (LIF) images were taken of fluorescence emitted by the acetone mixed with nitrogen. The LIF is emitted by both droplets and vapour and the signal is proportional to the local mass concentration. Images were taken at several different times after the start of the injector current pulse. At each time, 20 images were taken from successive cycles. The camera settings were f/8, intensifier gate 200ns, intensifier gain x60, camera gain x95. The images were corrected for background intensity and light sheet intensity.

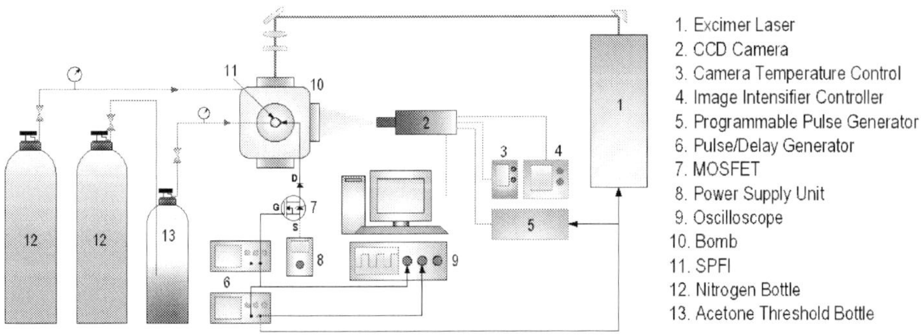

Fig. 4. Schematic of the PLIF imaging of SPFI injection

The nitrogen acetone doping mechanism (Drechsel bottle) is shown in Fig. 5. The nitrogen is bubbled through the acetone and trapped in the upper part of the threshold bottle. The trapped gas increases its pressure and as it reaches the desired injection pressures, the supply valve is closed. An outlet gas pipe is attached to the bottle, positioned above the surface of acetone, thus only acetone-saturated gas is delivered to the SPFI. The amount of seeded acetone is determined based on the saturated vapor pressure of acetone at the room conditions which equals the partial pressure of acetone in the seeded gas. Acetone concentration was kept constant by adjusting the injection frequency to give sufficient time for building up acetone concentration for the next injection. The supply gas and bomb were maintained at room temperature using a water bath. The evaporative cooling in the acetone bottle was kept to a minimum during all pulse durations. The pressure gauge on the gas pipe near SPFI was used to determine the injection pressures.

The experiments were carried out by referencing the motorised cylinder pressure of a single cylinder Ricardo E6 engine (Mohamad 2006). Fig. 6 shows the non-firing cylinder pressure during the compression stroke and the injection timings for stoichiometric air-fuel ratio operations. For 80 bar injection pressure, the optimal injection time is at 215° BTDC of compression stroke (Mohamad 2006). For 50 bar and 60 bar injections, the optimal injections were at 170° BTDC. Fuel injection durations were 6 ms, 10 ms and 12 ms for 80 bar, 60 bar and 50 bar injection pressures respectively in order to supply a stoichiometric quantity of fuel. At 1100 rpm, these injections cover 40°CA, 66°CA and 79°CA respectively, as shown in Fig. 6. Taking into account the fuel delivery delay due to the lengthy fuel path in the SPFI, it was decided to perform injections at three bomb pressures: 1 bar, 3 bar and 10 bar. The 1 and 3 bar pressure represent the range of actual injection cylinder pressure, while the 10 bar pressure was chosen to investigate the effect of gas jet if injection is delayed at later stage of compression stroke.

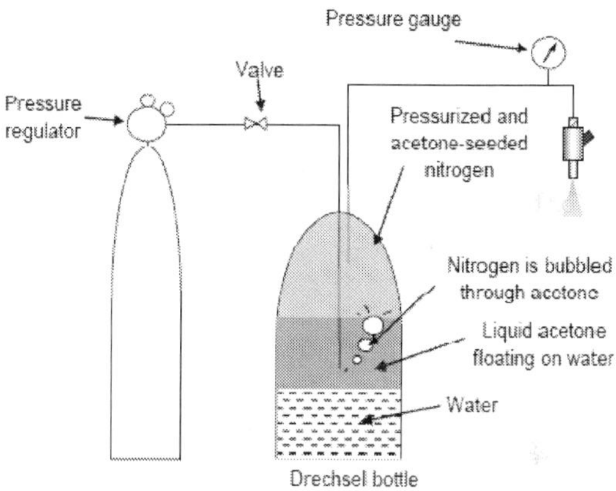

Fig. 5. Nitrogen acetone doping mechanisms (Drechsel bottle) for SPFI spray imaging

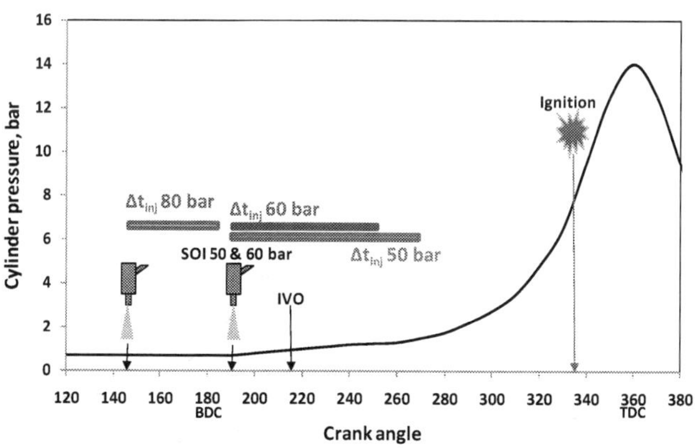

Fig. 6. Cylinder pressure of motorized Ricardo E6 engine (r = 10.5, N = 1100 rpm), starts of injections and injection timings (Δt_{inj}) derived from engine experiments

3.1.1 Image calibration

It is important to quantify the injection penetration and the width of the gas plume from the images. The bomb viewing window allows 110 mm circular diameter visual access to the injection gas flow. Calibration of gas jet dimensions was achieved by imaging a known scale object attached to the fuel injection vertical plane as shown in Fig. 7. Fig. 8 shows a calibration image captured by the CCD camera and compared with a PLIF image with no gas injection. The bright spots at the top of the PLIF image are due to strong elastic scattering of the laser light from the injector and holder breaking through the PLIF filter. The speckled pattern in the bulk of the image is due to thermal and cosmic ray noise in the intensifier and CCD. High intensifier voltage and low level of signal leads to low signal-to-noise ratio. Much effort was applied to avoiding this background noise and improvements were made. In addition, the design of the doping system was limited by a need to operate it at high pressures and limited supply of acetone.

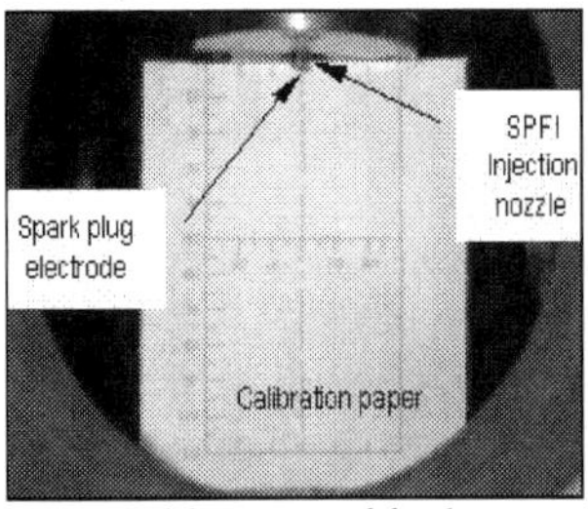

Fig. 7. Calibration of fuel injection measurement

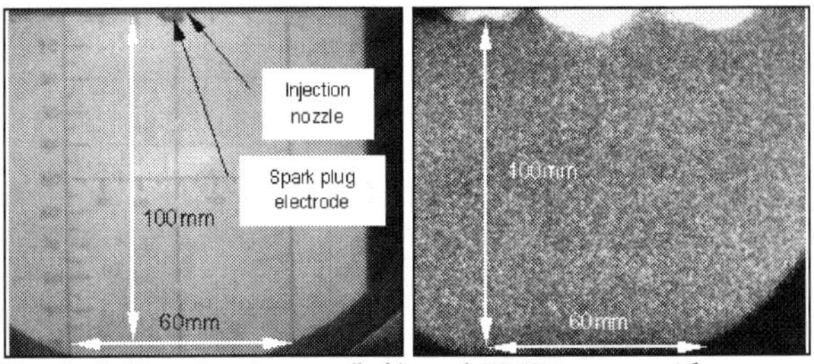

Fig. 8. Calibration image (left) and PLIF image with no gas injection (right)

3.1.2 Injection images

Images of gas jets from SPFI nozzle were obtained successfully with the PLIF method. The images shown in Fig. 10 are the consecutive images of gas jet at various injection pressures. The bright white areas penetrating from the top of each image indicates the injected gas presence. The intensities of brightness can be used to describe qualitatively the gas concentration. It is important to note that because the acetone-doped nitrogen remains static in the injection fuel line due to the slow rate of injection in this experiment (once in every 1 second), it was possible that some of the acetone settled or condensed to the pipe walls, thus reducing the concentration of doping. Nevertheless, the jets are clearly seen in the Fig.s. The shape of the gas jets are in good agreement with the vortex ball model (Turner 1962; Boyan and Furuyama 1998). Fig. 9 shows the vortex ball model which is a theoretical model that assumes the gas transient jet can be characterized as a spherical vortex interacting with a steady-state jet.

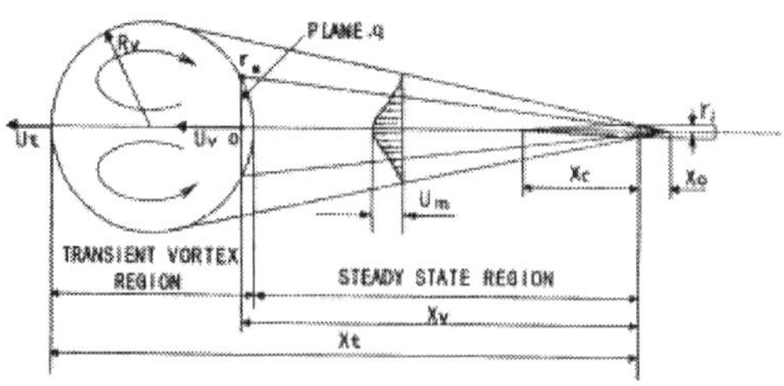

Fig. 9. The vortex ball model (Turner 1962; Boyan and Furuyama 1998)

The injection durations for 50, 60 and 80 bar fuel pressures on the Ricardo E6 engine were 12 ms, 10 ms and 6 ms respectively. These values correspond to the stoichiometric fraction of methane for the engine tests. A fully developed gas jet refers to the gas plume with longest jet tip penetration before detachment from the SPFI nozzle. In general, it can be seen that the fully developed gas jets from the SPFI injection were relatively narrow, with about 23° cone angle and 25 mm of maximum width in a fully developed gas jet. The cone angle is specified by measuring the angle generated by a triangle connecting the center of SPFI nozzle and the widest horizontal span of the gas jet as shown in Fig. 10. The jet tip penetration length and jet width were defined based on the lowest acetone concentration detectable. Fig. 11 shows the visual definition of the tip penetration length and jet width. The penetration length of the fully developed gas jet is between 90 and 100 mm at 8 to 14 milliseconds after the starts of injections. The results show that at all injection pressures, the first appearance of gas jet can be seen at 2.5 ms after SOI which coincides with the first interval of imaging. However the images at this interval were not shown in Fig. 9. It is believed that if the interval is decreased, the first appearance of gas jet can be well defined with respect to different injection pressures. The fully developed jets appear at different times. Higher injection pressure results in faster development of a fully-developed jet which is defined by the furthest distance of jet tip penetration while the plumes were still attached to the injection nozzle. In addition, increasing injection pressure reduces the effective fuel delivery time which is measured from the SOI signal to the time of gas plume detachment from the injection nozzle. Bomb pressure affects the magnitude of tip penetration and effective fuel delivery time. Increasing bomb pressure leads to shorter jet penetration and slower effective delivery time.

The penetrations of gas jet based on the PLIF imaging experiments are shown in Fig. 12. Fig. 13 shows the effect of injection and bomb pressure on tip penetration. Jet penetration is proportional to square root of injection time, except in the early phase. As the bomb pressure increases, jet penetration reduces. However, for the same bomb pressure, the variation in penetration lengths with changing fuel pressure is relatively small due to SPFI injection nozzle exit flow were at sonic conditions for all injection pressures. Therefore gas velocity remains constant except when flow temperature increases. However, the mass flow rate at the sonic conditions not only depends on flow temperature but also depends on the density of the injected gas which is proportional to supply pressure. Therefore mass flow rate determines the length of injection duration and in certain degree affected the length of jet penetration. The penetration lengths are highest at 80 bar and lowest at 60 bar injection. The jet from 60 bar injection is slightly wider but shorter than the one from 50 bar injection.

Measurement of jet tip penetration was performed from 15 images of each injection condition. Error analysis was done with respect to the measured data. The average error was ±0.30mm and the highest error was found to be at ±0.48mm or 3.9% of the measured data. These errors were sourced from a number of factors. Variation of acetone temperature affects the concentration of acetone in the compressed nitrogen. Assuming laser power and camera sensitivity remain constant, the variation of acetone concentration means variation in location of threshold of fluorescent signal. Visual error while measuring the jet parameters from the images was another source. However, the error was proven to be relatively low.

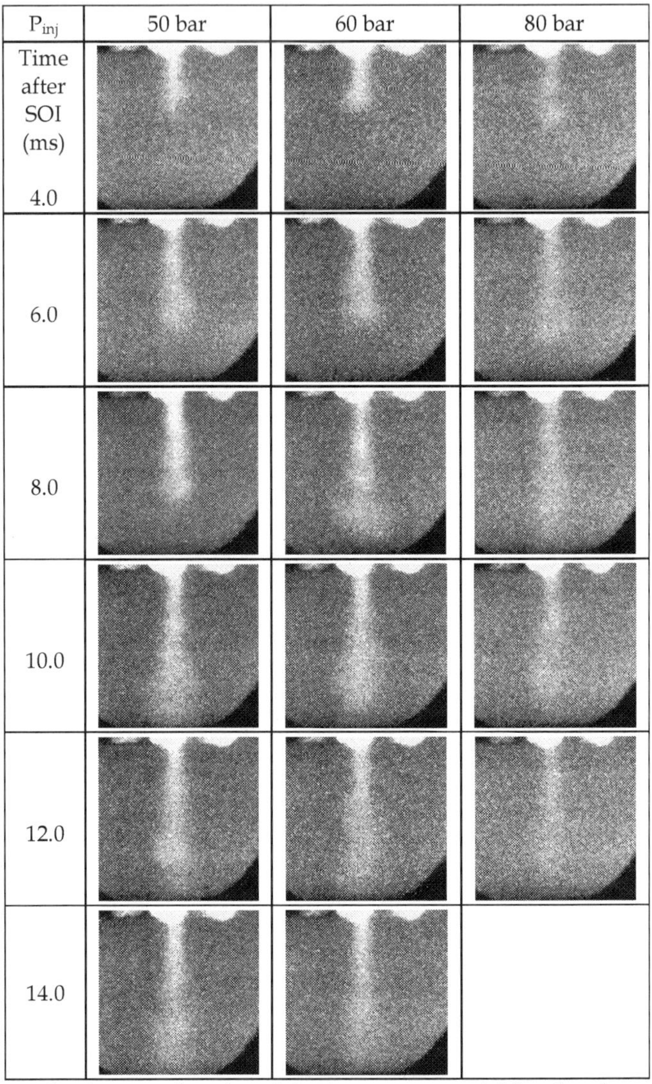

Fig. 10. Consecutive images of various injection pressures at 1 bar bomb pressures. Fully developed gas plume are indicated by the green square backgrounds

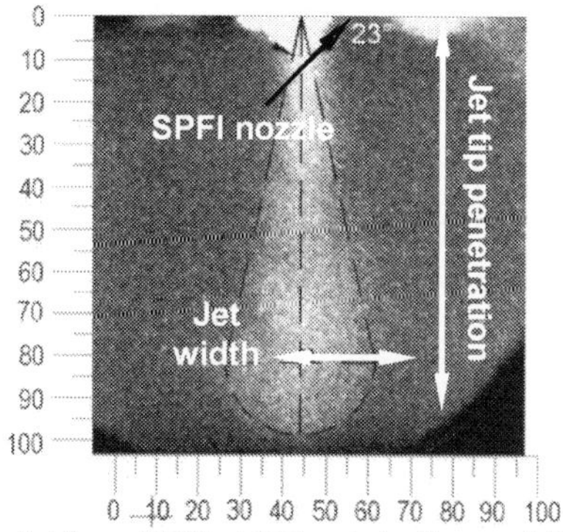

Fig. 11. Fully developed gas jet from SPFI at 60 bar injection and 3 bar bomb pressure

Based on the investigation, SPFI design yields sufficient jet penetration length especially during the later part of compression stroke. However, the width of the jet and the direction of injection away from the point of ignition could be detrimental to the engine performance. SPFI utilizes a fuel injector which is optimised for direct injection and stratified charge operations. However, the optimization of nozzle design and orientation to give best effect on stratified charge direct injection has not been taken advantage of of the SPFI system yet. As a result, the effective fuel injection behaviour is determined mainly by the injection nozzle as well as the fuel path. This work has given a qualitative understanding of the injection and mixing behaviour which is useful for design optimization process of SPFI.

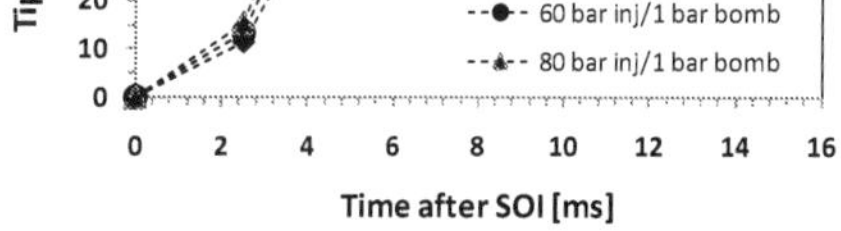

Fig. 12. Effects of injection pressure on the SPFI jet tip penetration

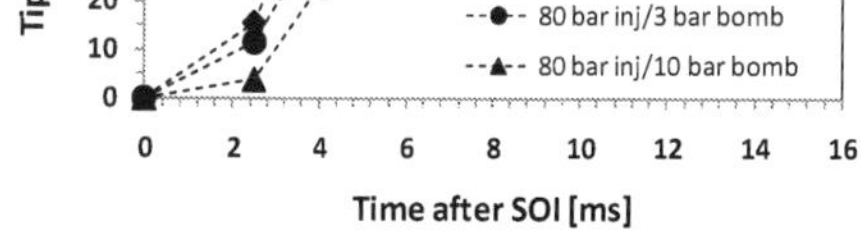

Fig. 13. Effect of bomb pressure on the SPFI jet tip penetration

3.2 Experimental investigation

The SPFI methane direct injection system was designed and tested on a Ricardo E6 engine with gasoline head. The engine is connected and mounted on a common test bed with a direct current electric dynamometer, which functions as motor or brake. Lubricant circulation is driven by an electric motor and water coolant is circulated by separately

driven centrifugal pump. The engine has one intake and one exhaust poppet-type valves. The specifications of the engine are given in Table 1.
Fig. 14 shows the cross sectional area of the engine and the SPFI attachment to the engine. The combustion chamber is disk-shaped with flat cylinder head and flat piston crown. Two 14-mm spark plug holes penetrate from the sides at 60° from vertical axis and pointing to the central axis of the cylinder. The intake manifold arrangement creates both tumble and swirl motion of different degrees. Fig. 15 shows the experimental set up on the Ricardo E6 engine. A shaft encoder was mounted on the camshaft, giving one TTL signal per camshaft rotation which, corresponds to one signal for every two crankshaft rotations. The signal is set as an input to a pulse generator which output signal at changeable pulse length and delay is generated. This secondary signal which determines injector pulse length is then sent to a mosfet that functions as a gate for the high power signal from power supply unit (12 V, 5A) to the GDI injector. Ignition timing varying from 0 to 60° crank angle BTDC can be set using a magnetic strip mechanism attached to the crankshaft and connected to the ignition coil. Engine speed is controlled from the main unit of the electric dynamometer. Crank position is determined from the photodiode signals flashing through 180-rectangular-slotted disk mounted to the crankshaft. The Spark Plug Fuel Injector was mounted through one of the spark plug holes as shown in Fig. 14. Cylinder pressures were measured with an un-cooled type Kistler model 6121 A1 pressure sensor attached to the cylinder head through the other spark plug hole. Pressure signal is amplified through a piezoelectric amplifier. The crank angle and TDC were encoded using the photodiode and slotted disk system. Both crank angle and pressure signals were sent to a data acquisition system at 12000 samples per second rate. The schematic of the experimental control and instrumentation is shown in Fig 4. Methane is supplied from a 230 bar container and a pressure regulator is adjusted to achieved the desired injection pressures. Injection timings were varied to investigate the effects on engine performance. Air/fuel ratio was set to be stoichiometric and ignition timings were set at minimum advance for best torque (MBT). Methane was used as natural gas substitute due to close proximity of properties of these two gases. Methane was injected at 60 bars and 80 bars at various crank angles during the intake or compression stroke at 1100 rpm and mixture lambda value of 1.0. The injection timing are referred to degree crank angle after intake TDC, describe as ATDC in Fig. 16.

Bore (mm)	76.2
Stroke (mm)	111.125
Displacement volume (liter)	0.507
Compression ratio	10.5 : 1
Intake valve open	8° BTDC
Intake valve close	33° ABDC
Exhaust valve open	42° BBDC
Exhaust valve close	8° ATDC
Cooling method	Water cooling
Valve clearance (intake/exhaust)	0.15 mm / 0.20 mm

Table 1. Specification of Ricardo E6 engine with SPFI system

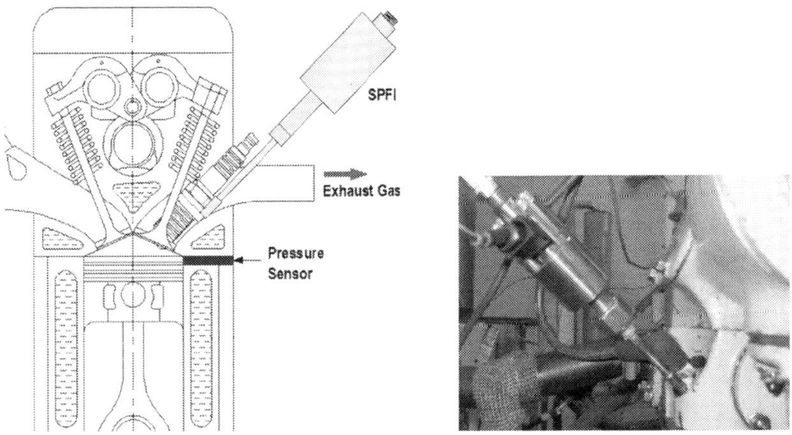

Fig. 14. Cross-sectional view of the combustion chamber (left) and SPFI on the engine cylinder (right).

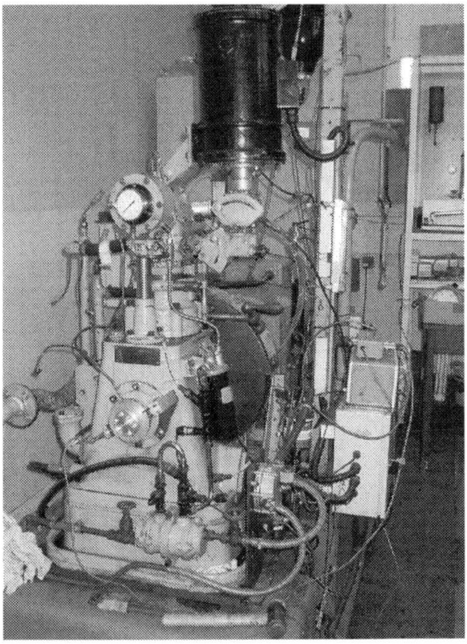

Fig. 15. Experimental set up on a Ricardo E6 engine and an electric dynamometer,

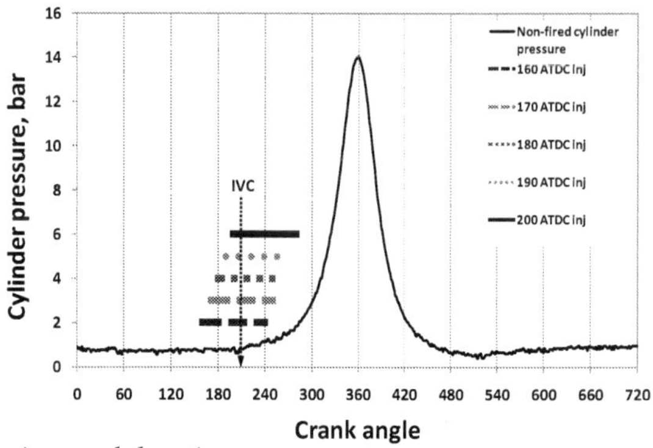

Fig. 16. Injection timings and durations

3.2.1 Engine performance

The outcomes of experimental investigation are shown in Fig. 17 through Fig. 20. In Fig. 17 and Fig. 18, indicated power and indicated mean effective pressure are plotted against injection timings at 1100 rpm engine speed. The graphs show variation in performance at different start of injection timings. When SOI is earlier than 180°BTDC, both power and MEP are the lowest. The best performance was achieved at SOI of 190°BTDC.

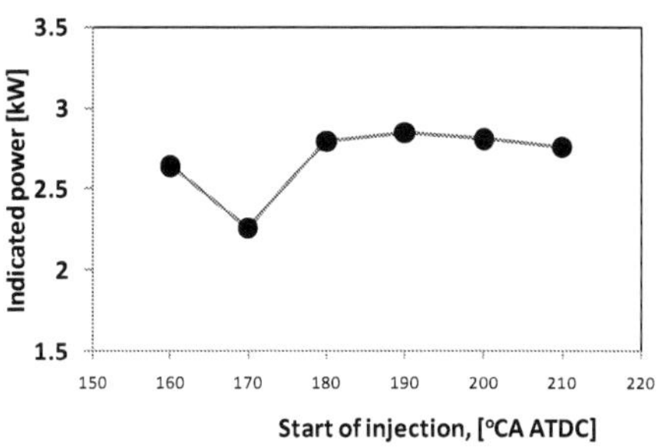

Fig. 17. Indicated power at various injection timings

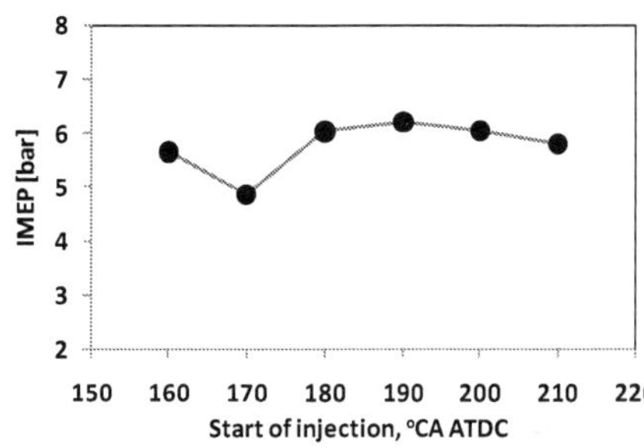

Fig. 18. Indicated Mean Effective Pressure at various injection timings

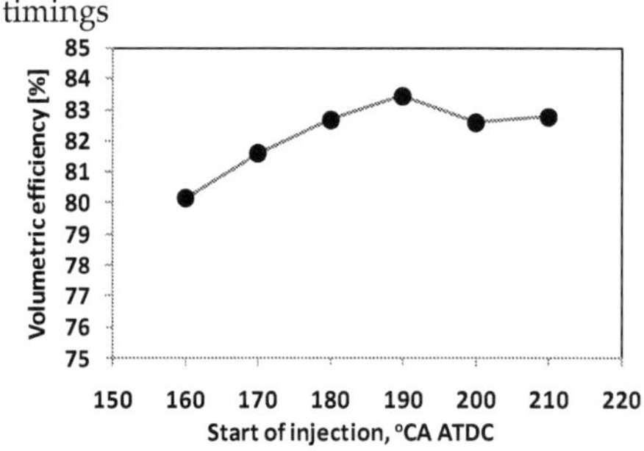

Fig. 19. Volumetric efficiency at various injection timings

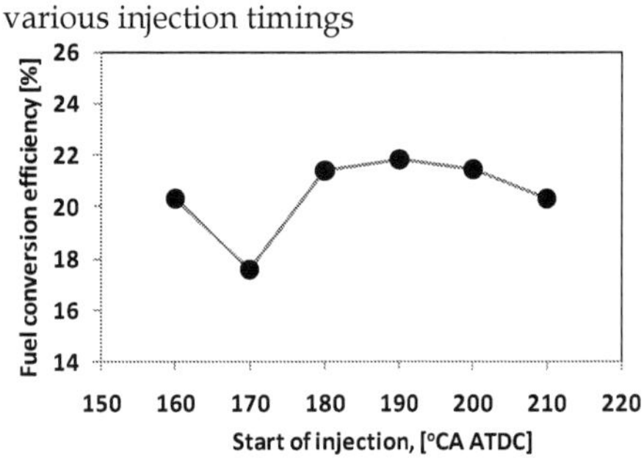

Fig. 20. Fuel conversion efficiency at various injection timings

Fig. 19 shows the volumetric efficiencies are in the excess of 80%, which is significantly high compared to port injection operations. This proved that direct fuel injection using SPFI increases engine ability to inhale more air and as a result, increases the heating value of cylinder charge per engine cycle. Again, the highest volumetric efficiency was achieved at 190°BTDC SOI timing. In terms of fuel conversion efficiency, as shown in Fig. 20, the same pattern as in power and MEP results is shown with the most efficient outcome was achieved at 190°BTDC. SPFI direct injection operation is very sensitive towards injection timing and for any particular engine speed, a proper calibration of SOI timing must be done to get the optimal performance. Ths results are consistent with the findings by Huang et al. (2003) and Zeng et al. (2006).

Table 2 list the comparison of engine performance for the same engine running with different fuel delivery techniques when fuelled with natural gas. In terms of performance (MEP), port injection shows highest results but SPFI betters in terms of volumetric efficiency. Gas mixer (carburetor) operation yielded lowest output in most sectors. The reason for lower performance was due to the fact that SPFI direct injection is not operating at its best engine parameter which should include higher compression ratio and a flow-guided piston crown cavity. Also the air-fuel mixing in the SPFI direct injection operation is subjected to spatial and durational limitation.

Operation	Gas mixer (Simms, 1994)	Port injection	SPFI Direct Injection
MEP, bar	5.52_b	6.63_i	6.20_i
η_v, %	82.5	72.35	83.43
$\eta_{f,\ indicated}$, %	21.3	26.94	21.84
SFC, g/kWh	340_b	267.28_i	329.67_i

Table 2. Engine performance of natural gas operations (subscript i and b are indicated and brake data respectively)

3.2.2 Combustion characteristics

Combustion characteristics in terms of cylinder pressure profile, pressure-volume relations and burning rate were studied. Results, as shown in Fig. 21 through Fig. 23 show the comparison between port injection and SPFI direct injection. In Fig. 21, the pressures versus cylinder volume were plotted. The performance of engine is measured by the area enclosed by the curves, the SPDI operation results in higher peak pressure but smaller area enclosed depicting inferior performance. The importance of peak pressure is most in its location with respect to top dead center (TDC). In Fig. 22, it shows that SPFI operation yields earlier and higher peak pressure as compared to port injection. However, as the combustion proceeds 30°CA ATDC (i.e. 390° ATDC), the cylinder pressure falls below the one of port injection which actually results in slightly lower performance. This was believed due to the weakness in air0-fuel mixing.

The major advantage of SPFI direct injection operation is its burning rate characteristics as shown in Fig. 23. Combustion duration in SPFI direct injection operation is shorter than the one of port injection. However, combustion in SPFI direct injection is slower at the earlier part due to higher charge density but faster at the later part of combustion and subsequently resulted in faster burning rate. Based on the analysis by varying other operational parameters, a number of findings were obtained. Combustion durations were shortened by advancing spark ignition. In addition, by spark advancing, the ratio of the first half to the second half combustion duration is increased resulted in a less uniform burning rate. Combustion durations were not changed with different injection pressures but ignition delay was affected by this variation. However, there is no direct correlation between injection pressure and ignition delay which is most probably due to the effect of charge flow difference. Changing mixture stoichiometry affects the magnitude of ignition delay. Combustion duration, on the other hand increases with leaner mixture. Different load conditions have significant effect on combustion process, Lower loads tend to increase combustion duration but shorten ignition delay.

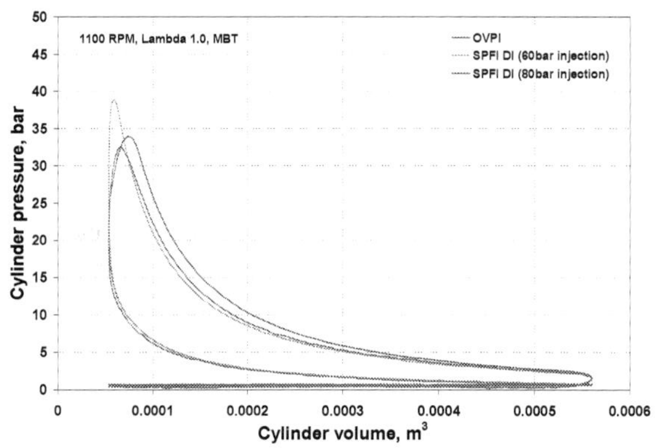

Fig. 21. PV diagrams of open valve port injection (OVPI) and SPFI DI

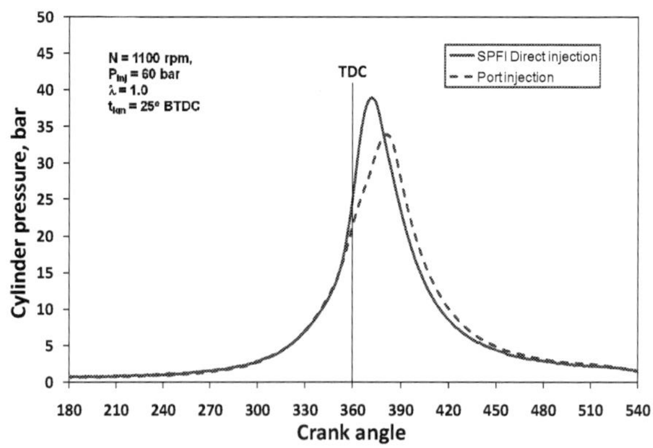

Fig. 22. Cylinder pressure of natural gas port injection and SPFI direct injection

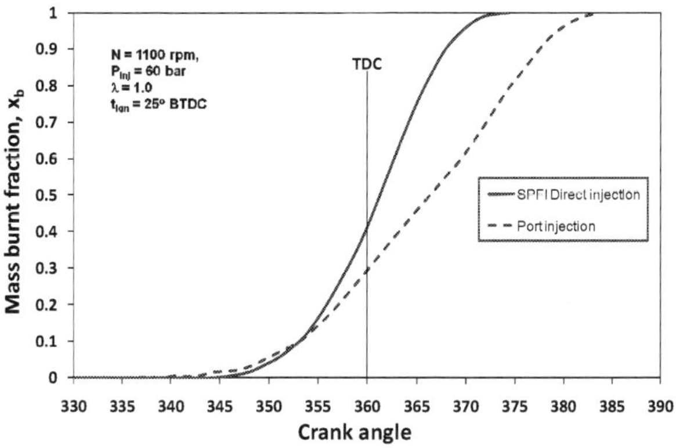

Fig. 23. Normalized mass burnt fraction of natural gas port injection and SPFI direct injection

4. Conclusion

Compressed natural gas direct injection can offer many advantages over port fuel injection but to achieve it requires costly and technically difficult modification to engine structure especially due to the need for extra hole for fuel injector. SPFI offer a simple solution for this. Investigations on injection process and experimental engine performance were carried out on a single cylinder engine. The most immediate effect of installation of SPFI on the Ricardo E6 engine is the reduction of motorized engine cylinder peak pressure due to decrease in compression ratio. Engine experiments were carried out to measure the indicated performance of SPFI direct injection. The engine was run at 1100 rpm and various other operating parameters were varied. The results were compared to the ones from optimally calibrated port injection. At the specified speed, results show that SPFI direct injection performance is slightly lower than those of port injection even though volumetric efficiency is significantly increased. This was mainly because of spatial and temporal limitations in direct injection operation which lead to weak air-fuel mixing as a result of the absence of mixing enhancing geometry within the disk-shaped cylinder. Furthermore, because natural gas is injected away from spark plug electrode, at the time of ignition, the stoichiometry of cylinder charge at the vicinity of spark plug and other areas may not be at optimal conditions. Combustion of methane direct injection has been shown to be faster than of port injection. However, the initial part of combustion is relatively slower in direct injection. This implies the presence of charge stratification. The combustion behavior and indicated performance show that combustion of SPFI direct injection is easily controlled and in good agreement with the data from available literature. Images from PLIF flow visualization from SPFI injection nozzle show narrow gas jets with 9° cone angle and a depth of penetration of 90-100 mm. SPFI methane direct injection is practical, viable and easy way for natural gas conversion but further improvement on its design is required particularly to improve fuel spray to achieve better air-fuel mixing.

5. References

Boyan, X. & Furuyama, M. (1998). *Jet* Characteristics of CNG Injector with MPI system. *JSAE Review*, Vol. 19, 229-234.

Bradley, D.; Gaskell, P.H. & Gu, X.J. (1996). Burning Velocities, Markstein Lengths, and Flame Quenching for Spherical Methane-Air Flames: A Computational Study. *Combustion and Flame*, Vol. 104, 176-198.

Huang, Z.; Shiga, S.; Ueda, T.; Nakamura, H.; Ishima, T.; Obokata, T.; Tsue, M. & Kono, M. (2003). Combustion Characteristics of Natural-gas Direct-injection Combustion under Various Fuel Injection timings. *Proceeding Institute of Mechanical Engineers (Part D): Journal of Automobile Engineering*, Vol. 21, 393 - 401.

Mohamad, T I; Harrison, M.; Jermy, M. & How, H.G. (2010). The Structure of High Pressure Gas Jet from Spark Plug Fuel Injector for Direct Fuel Injection in Spark Ignition Engine. *Journal of Visualization* , Vol. 13, No. 2, 121-131.

Mohamad, T.I.; Harrison, M.; Jermy, M.; Theodoridis, E. & Dolinar, A. (2005). Preliminary Investigation of the Combustion and Performance of Methane Direct Injection in a Single-Cylinder Engine Using Spark Plug Fuel Injector (SPFI) for Low Cost Conversion, in the *Proceedings of the ATCi 2005, Conference on Automotive Technology*, Dec 6-8, Putrajaya, Malaysia.

Morita, K. (2003). Automotive power source in 21st century. *JSAE Review*, Vol. 24, 3-7.

Risi, A.D.; Gajdeczko, B.F. & Bracco, F.V. (1997). A study of H_2, CH_4, C_2H_6 mixing and combustion in a Direct-injection Stratified-Charge engine. *SAE Paper 971710*.

Stan, C. (2002). *Direct Injection Systems: The Next Decade in Engine Technology*, SAE International, Warrendale, PA.

Vourenkoski, A.K. (2004). *Development of a Liquid-phase LPG MPI Conversion System*, PhD Thesis, Cranfield University, Cranfield.

Zeng, K.; Huang, Z.; Liu, B.; Liu, L.; Jiang, D.; Ren, Y. & Wang, J. (2006). Combustion Characteristics of a Direct Injection Natural Gas Engine under various Fuel Injection Timings. *Applied Thermal Engineering*, Vol. 26, 806-813.

Zhao, F. et al. (1999). Automotive spark-ignited direct injection gasoline engines. *Progress in Energy and Combustion Science*, Vol. 25, 437-562.

Zhao, F.; Harrington, D.L. & Lai, M.C. (2002). *Automotive Gasoline Direct-Injection Engines*, SAE International, Warrendale, PA.

14

Hydrogen-enriched compressed natural gas as a fuel for engines

Fanhua Ma, Nashay Naeve, Mingyue Wang, Long Jiang, Renzhe Chen and Shuli Zhao
State Key Laboratory of Automotive Safety and Energy
Tsinghua University
China

1. Introduction

Natural gas is often thought of as the most promising alternative fuels for vehicles. Natural gas is a much more abundant fuel than petroleum and is often described as the cleanest of the fossil fuels, producing significantly less carbon monoxide, carbon dioxide, and non-methane hydrocarbon emissions than gasoline; and when compared to diesel it nearly eliminates the particulate matter. Other advantages of natural gas include a high H/C ratio and a high research octane number (RON) causing the exhaust to be clean and allowing for high anti-knocking properties. Hydrogen is the most abundant element on earth, and is often thought of as the ideal alternative fuel. However, the current infrastructure does not support hydrogen as a wide-spread fuel. In order to expand the role of hydrogen in the near term, one option is to use hydrogen for transportation by mixing it with natural gas and use it for use in ICE engines. This new blended fuel is known as HCNG, or hythane, for which the use will create a basic infrastructure for the use of hydrogen in the future.

2. History

Natural gas-hydrogen mixtures have been used in test engines dating back to as early as 1983 (Nagalim et al, 1983). Nagalingam et al. (1983) conducted experiments with an AVL engine fueled with 100/0, 80/20, 50/50, 0/100. In 1989, HCI (Hydrogen components Inc.) began blending various ratios of hydrogen and natural gas and testing them at Colorado State University (Hythane Company, LLC, 2007). Hythane is a patented mixture of hydrogen and CNG, created by Hydrogen Components, Inc. in Littleton, Colorado. According to US Patient #5,139,002 (Lynch & Marmaro, 1992), Hythane® was invented by Frank Lynch and Roger Marmaro and was granted a US patient in 1992. In this patient, hythane is defined a blend of hydrogen and natural gas provided for burning in an engine without the need for modifications to engine parameters, and is defined as roughly a 15% blend of Hydrogen in CNG fuel. In 1992, the first Hythane® station was built and opened. Since then, there have been engines created specifically for the fuel, and many fleets on the road testing the fuel to further understand the HCNG fuel. There has also been much research relating to the optimization of the in the areas of excess air ratio, hydrogen ratio, and spark timing.

The author of this paper, Dr. Fanhua Ma and his research group at Tsinghua University have been conducting research and development of HCNG (Hydrogen-enriched Compressed Natural Gas) engine and vehicle since 2000. Dr. Ma has published many works, many of which are cited in this chapter and has also succeeded in acquiring several Chinese patients relating to HCNG.

3. Advantages

Enriching natural gas with hydrogen for use in an internal combustion engine is an effective method to improve the burn velocity, with a laminar burning velocity of 2.9 m/s for hydrogen verses a laminar burning velocity of 0.38 m/s for methane. This can improve the cycle-by-cycle variations caused by relatively poor lean-burn capabilities of the natural gas engine. Hydrogen is characterized by a rapid combustion speed, a wider combustion limit and low ignition energy. These characteristics can reduce the exhaust emissions of the fuel, especially the methane and carbon monoxide emissions. The fuel economy and thermal efficiency can also be increased by the addition of hydrogen. The thermal efficiency of hydrogen enriched natural gas is covered in more detail in Ma et al. (2007).

HCNG allows for an initial use of hydrogen while taking advantage of the current CNG infrastructure. This allows for the hydrogen infrastructure to slowly become established until the production and efficiency demands can be met for the hydrogen economy. The research completed for the HCNG engine can be directly applied to a hydrogen engine. The addition of hydrogen to natural gas also greatly reduces the carbon monoxide and carbon dioxide emissions. The HCNG fuel can also help to avoid problems associated with evaporative emissions and cold start enrichment seen in gasoline engines, and the high anti-knock properties of CNG due to the high activation energy helps resists self-ignition.

4. Challenges

There are some challenges when it comes to using the hydrogen-natural gas mixture as a fuel. One of the biggest challenges using HCNG as a fuel for engines is determining the most suitable hydrogen/natural gas ratio. When the hydrogen fraction increases above certain extent, abnormal combustion such as pre-ignition, knock and back-fire, will occur unless the spark timing and air-fuel ratio are adequately adjusted. This is due to the low quench distance and higher burning velocity of hydrogen which causes the combustion chamber walls to become hotter, which causes more heat loss to the cooling water. With the increase of hydrogen addition, the lean operation limit extends and the maximum brake torque (MBT) decreases, which means that there are interactions among hydrogen fraction, ignition timing and excess air ratio. Therefore finding the optimal combination of hydrogen fraction, ignition timing and excess air ratio along with other parameters that can be optimized is certainly a large hurdle.

The emissions levels of fuels are probably the most important factor in determining whether or not the fuel is suitable as an alternative. Although the NOx emissions for CNG are already extremely low compared to traditional fuels, the addition of hydrogen causes increased NOx emissions. The addition of hydrogen has the opposite effect on the hydrocarbon emissions, so it is necessary to compromise at a hydrogen ratio for which the NOx and hydrocarbon emissions are equally low.

Probably most evident challenge for wide-spread use of the new fuel is the current lack of infrastructure. In many countries, however, the infrastructure for natural gas is well developed, which can be further adapted to carry hydrogen for the new fuel. Similar to other gaseous fuels, natural gas and hydrogen are both lighter than air, therefore if there is a leak it will quickly disperse into air with adequate ventilation. Lastly, the currently cost of hydrogen is more expensive than the cost of natural gas resulting in HCNG being more expensive than CNG. Although the cost is likely decrease as the use of hydrogen increases, it will be of great concern to consumers in the near-term.

5. Experimental Apparatus

Unless otherwise stated, the tests in this chapter are all completed using a six-cylinder, single point injection, SI natural gas engine, with the engine specifications shown in table 1.

Displacement	6.234 L
Stroke	120 mm
Bore	105 mm
Compression Ratio	10.0
Rated Power	169 kW / 2800 rpm
Rated Torque	620Nm / 1600 rpm

Table 1. Dongfeng EQD230N engine parameters

The engine is coupled to an eddy-current dynamometer for the measurement and control of speed and load. The exhaust concentration of HC, NO_x, CO, H_2 and the air/fuel ratio are monitored using a HORIBA-MEXA-7100DEGR emission monitoring system and a HORIBA wide-range lambda analyzer, respectively. A high speed YOKOGAWA ScopeCorde is used to record the cylinder pressure from a Kistler 6117B piezoelectric high pressure transducer. Corresponding crankshaft positions were measured by a Kistler 2613B crank angle encoder with a resolution of 1 degree CA.

An online mixing system is used to blend desired amount of hydrogen with natural gas in a pressure stabilizing tank just before entering the engine. The tank is divided into two chambers with a damping line used to improve the mixture uniformity. A schematic of the fuel supply system is shown in figure 1. The flow rate of natural gas and hydrogen are measured using a Micro Motion flow meter that uses the principle of Coriolis force for a direct measure of mass flow and an ALICAT flow control valve is used to adjust the flow rate of the hydrogen according to the flow rate of CNG and obtain the target hydrogen fraction.

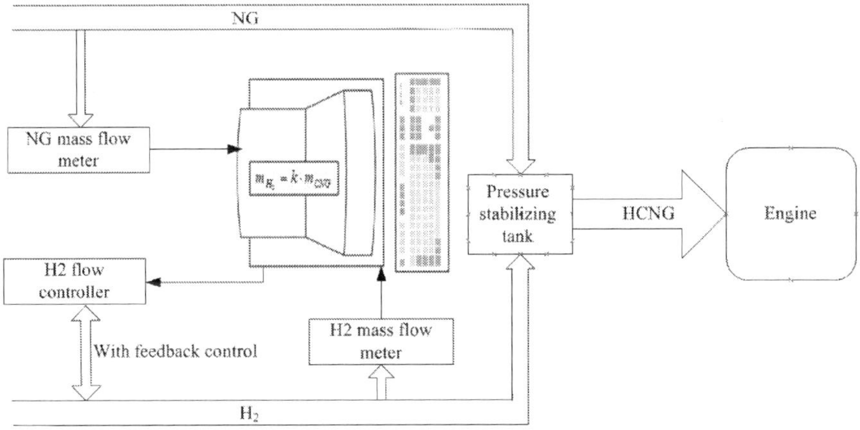

Fig. 1. Control logic of the on-line hydrogen/natural gas mixing system

6. Fundamental Equations

The chemical reactions of the fuel are essential in determining the amount of emissions the fuel will produce. This section will introduce some fundamental equations related to HCNG. Because CNG is made up of primarily methane, the following chemical equations are assumed:

$$CH_4 + 2O_2 \rightarrow (1-x)CO_2 + 2H_2O \tag{1}$$

$$H_2 + \frac{1}{2}O_2 \rightarrow H_2O \tag{2}$$

The fraction of hydrogen is x , and by assuming the molar mass of methane is 16 kg/mol and the molar mass of H_2 is 2 kg/mol, the mass fraction becomes

$$q = \frac{xM_{H_2}}{xM_{H_2} + (1-x)M_{CH_4}} = \frac{2x}{2x + 16(1-x)} = \frac{x}{8-7x} \tag{3}$$

Assuming that the air is 23.2% oxygen, the stoichiometric air-fuel ratio for the hydrogen/methane mixture is

$$l_0 = 34.48\frac{4-3x}{8-7x} \tag{4}$$

In order to calculate the lower heating value, the following equation is used

$$H_u = H_{H_2}q + H_{CH_4}(1 - q) \quad (5)$$

By substituting equation (3) and the relative heating values for hydrogen and methane (assuming the lower heating value of methane is 120 MJ/kg and the lower heating value of hydrogen is 50 MJ/kg), the following equation is formed with the units of MJ/kg.

$$H_u = \frac{120x + 50(8 - 8x)}{8 - 7x} \quad (6)$$

7. Performance Characteristics

Performance plays an important role in the choice of a fuel. HCNG has many advantages when it comes to performance because of the high octane number of hydrogen, the engine performance generally increases with the addition of hydrogen. The transient performance is also very important because in most cases the car will be running in transient conditions, this is covered in more detail by Ma et al. (2009b).

The thermal efficiency of both natural gas and HCNG increases with increasing load, which makes it an ideal fuel for high load applications and heavy-duty vehicles, this relationship can be seen in figure 2. It is also clearly seen in figure 2 that in nearly every case, the HCNG fuel has a higher thermal efficiency than pure natural gas. The results show that the brake effective thermal efficiency increases with an increased percentage of hydrogen at low and medium loads. The increase in thermal efficiency with the hydrogen addition is due to the reduction in the equivalence ratio. The hydrogen addition allows the lean burn limit to be extended because of the fast burn rate of hydrogen. The fast burn rate of hydrogen causes the combustion duration to decreases while the heat release rate and exhaust NOx increase with an increased percentage of hydrogen.

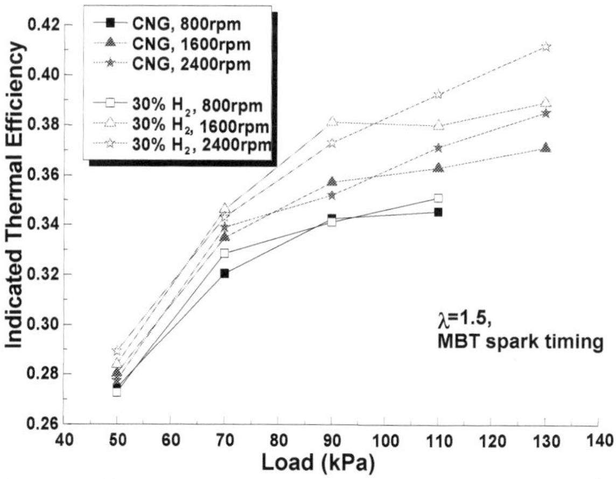

Fig. 2. Indicated thermal efficiency at the maximum brake torque spark timing

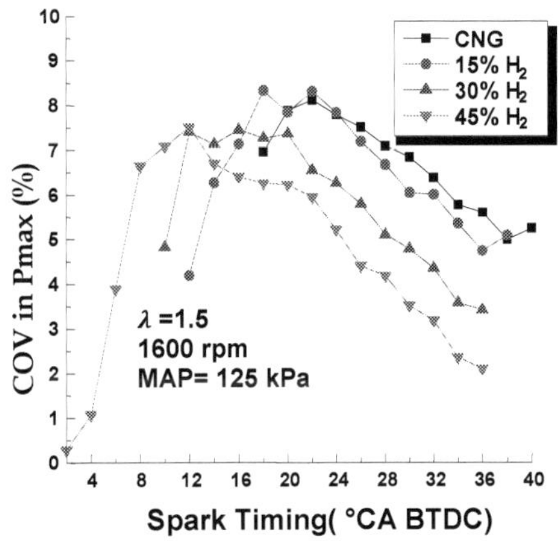

Fig. 3. Variation in the maximum pressure for various spark timings

The cycle-by cycle variations are also reduced with the addition of hydrogen, figures 3 and 4 show graphs of the coefficient of variation in the maximum pressure and indicated mean effective pressure respectively, for different hydrogen ratios. It can be seen that the coefficient of variation is reduced with an increased percentage of hydrogen at lean burn operation. The torque drop caused by retarded spark timing is relatively smaller in the case of HCNG fueling compared to that of CNG fueling, which can be seen in figure 5. This makes it possible to further retard the spark timing in an HCNG engine which results in lower NOx emissions. A higher torque also has other advantages such as resulting in a lower brake specific fuel consumption which is shown in figure 6.

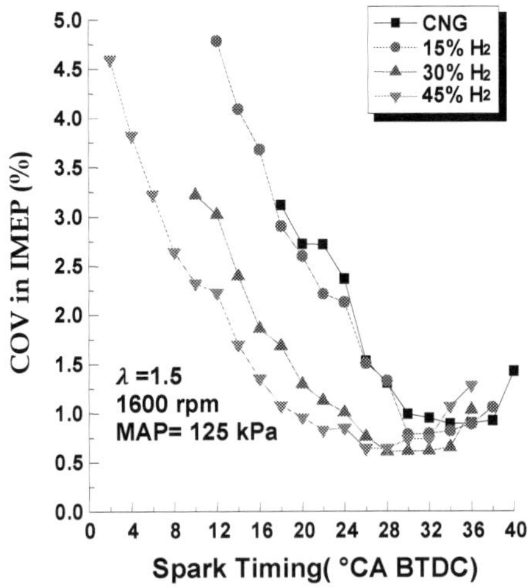

Fig. 4. Variation in the indicated mean effective pressure for various spark timings

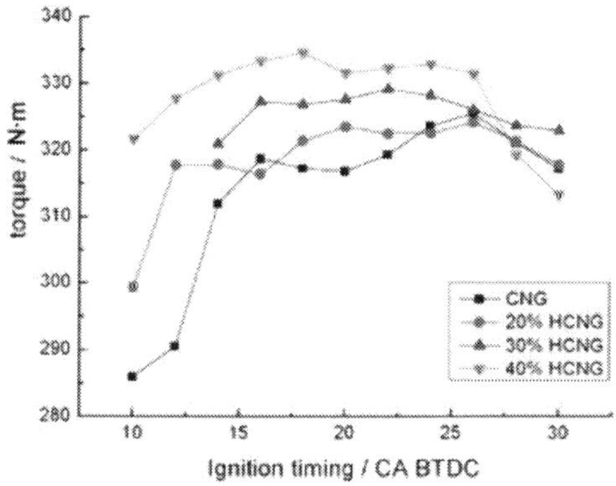

Fig. 5. Torque output for various H/CNG ratios (1200 $rpm, \lambda = 1.3$) (Ma et al., 2009a)

When two fuels with identical lower heating values are used, the fuel with the higher torque output will have lower brake specific fuel consumption (BSFC). According to the rules of lower heating value equivalent transformation, the mass of hydrogen in the HCNG fuel can be converted to a CNG mass with an equal lower heating value; this mass can then be added to the mass of CNG in the HCNG blend, therefore calculating an equivalent CNG mass. Using this equivalent data, the BSFC of HCNG and CNG can be compared and is shown in figure 6. (Ma et al., 2009a) It can be seen that the BSFC of the HCNG fuel is lower than the BSFC of pure CNG in nearly every case.

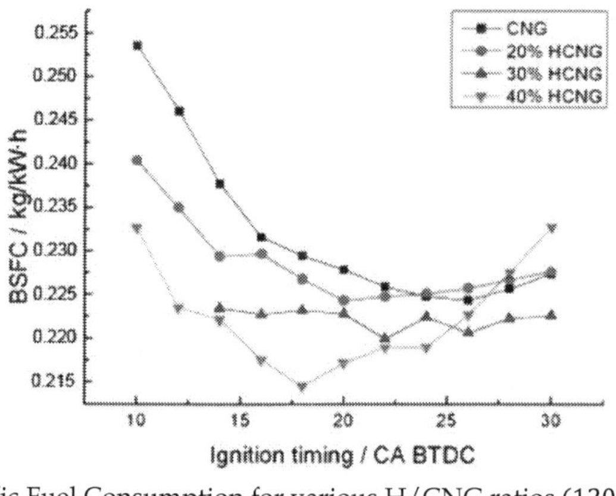

Fig. 6. Brake Specific Fuel Consumption for various H/CNG ratios (1200 $rpm, \lambda = 1.3$) (Ma et al., 2009a)

8. Emission Characteristics

When it comes to alternative fuels, arguably the most important factor in determining the feasibility of the fuel is the exhaust emissions. Because of the strictly controlled emissions regulations, it is not only necessary to find a fuel that has optimum performance, but it is also very important to find a fuel that can meet the respective emissions standards.

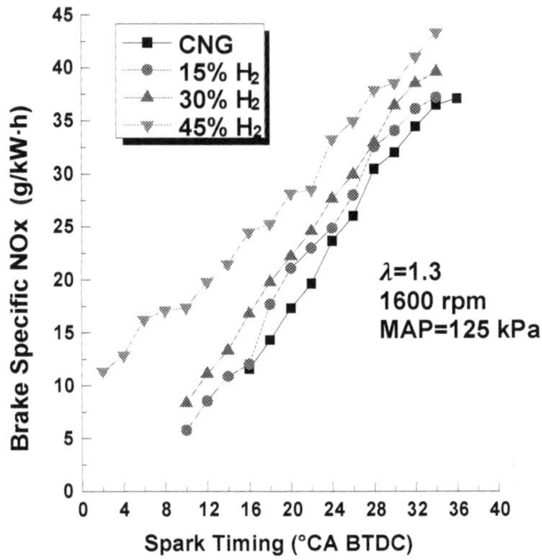

Fig. 7. Brake specific NOx emissions for different hydrogen fractions

Considering emissions, when HCNG fuel is compared with gasoline and diesel it appears to be a very appealing alternative fuel. When compared to gasoline, it produces significantly less nitrous oxide, carbon monoxide, carbon dioxide and non-methane emissions. And when compared with diesel, it nearly eliminates the particulate matter which is often of great concern. Compared to pure natural gas, it has been concluded that the addition of hydrogen increases the NOx emissions while reducing the HC emissions. The combustion stability is also improved by the addition of hydrogen which plays a part in reducing the un-burnt hydrocarbon emissions.

NOx emissions versus ignition timing were plotted in figure 7. As can be seen from the figure, the NOx emissions for the HCNG fuel are greater than the emissions of pure CNG. This is because of the elevated flame temperature due to the hydrogen. However, the NOx emissions of the HCNG are still considered relatively low compared to other fuels, and can be adjusted with further optimization.

Figure 8 indicates the variation of specific brake hydrocarbon emission versus spark timing for HCNG fuel at different ratios of hydrogen. As can be seen from the figure, the hydrocarbon emissions for HCNG fueling are greatly reduced compared to natural gas. The main reason for the decrease in hydrocarbon emissions is that the addition of hydrogen increases the laminar flame speed which decreases the amount of unburned hydrocarbons in the exhaust. Also, methane has a relatively stable chemical structure, therefore making it difficult to reduce emissions by after treatment. For this reason, the engine fueled with

HCNG has a large advantage regarding the hydrocarbon emissions than that of CNG fueling.

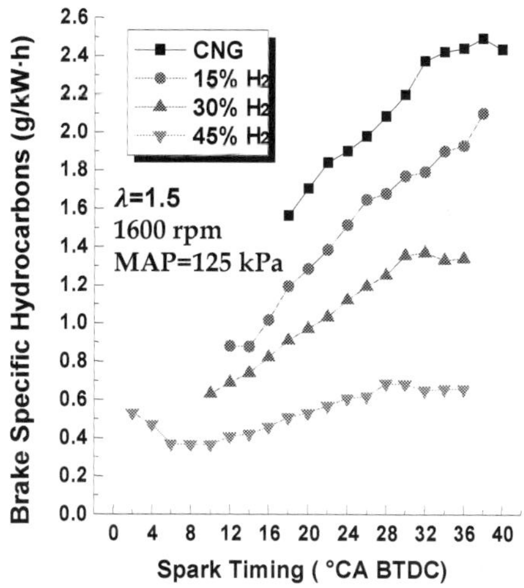

Fig. 8. Brake specific hydrocarbons at different hydrogen fractions

9. Optimization

There are many methods to optimize the engine for performance and emissions based on the properties of the fuel. Although the exhaust emissions from hydrogen-enriched natural gas are already very low, further refinement must be done in order to further reduce emissions and to achieve Enhanced Environmentally Friendly Vehicle (EEV) standards. There are many methods to improve the emission output as well as improving the performance of the engine.

9.1 Lean Burn

Lean burn characteristics are ideal in a fuel, because by running a fuel with a larger excess air ratio can not only reduce the emissions, especially NOx, but can also offers advantages in other areas such as reducing the brake specific fuel consumption. The lean burn limit is increased by the addition of hydrogen because of the faster burn speed as well as the improved laminar burn properties of hydrogen which makes it an ideal fuel to be run on lean-burn conditions. Ma et al. (2008d) specifically investigates the lean burn limit of HCNG.

Probably the largest advantage to running the engine on lean burn, is that it has the ability to greatly reduce the NOx emissions. The reduction in NOx emissions are due to the increased airflow which causes the engine to run at a lower temperature, therefore reducing the NOx emissions. Figure 9 shows how the NOx emissions are reduced at different excess air-ratios. It is very clear from this figure that as the excess air ratio is increased the NOx emissions drop considerably.

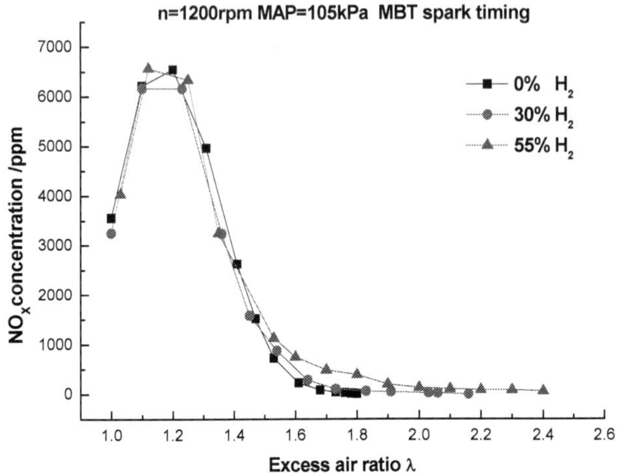

Fig. 9. The effect of excess air ratio vs NOx at MBT spark timing

The effect of the excess air ratio on hydrocarbon emissions can be seen in figure 10. It can be seen from the figure that there is a small reduction at an air-fuel ratio of roughly 1.25, but as the excess air ratio increases even further, the hydrocarbon emissions also increase. The reduction in hydrocarbon emissions at an excess air ratio of around 1.25 is not as evident in the hydrocarbon emissions as it was in the nitrous oxide emissions because as more air is added it can also contribute to unstable combustion which can also contribute to more unburned hydrocarbons. An increased excess air ratio can also increase the cycle-by-cycle variations which causes poor running conditions.

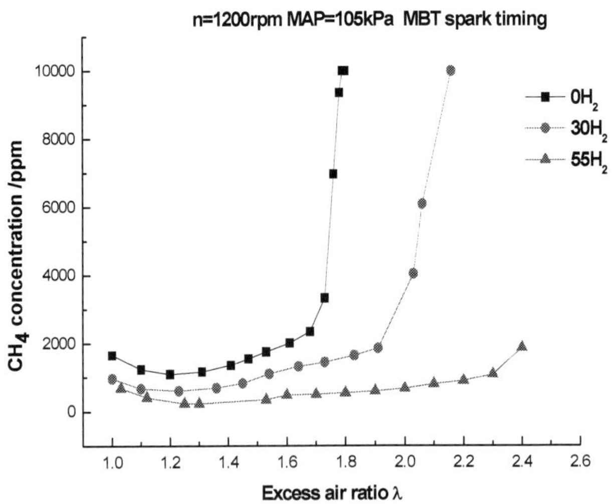

Fig. 10. Hydrocarbon Emissions for different hydrogen fractions and excess air ratios

Carbon monoxide emissions should also be considered when selecting the ideal excess air ratio. As seen in figure 11, by increasing the excess air ratio the carbon monoxide emissions drop dramatically. This occurs because the formation of carbon monoxide is mainly caused by incomplete combustion. However, as the excess air ratio becomes too large the combustion conditions are reduced and the carbon monoxide emissions begin to increase.

Another advantage to lean burn is that as the excess air ratio is increased, the brake specific fuel consumption decreases. This is because as the air-fuel ratio is increased, it usually leaves less unburned fuel. That is true until the excess air ratio reaches a certain limit when the cycle-by-cycle variations begin to increase because of the lack of fuel. Lean operation also reduces the likelihood of knocking, which allows the use of a higher compression ratio. However, there are some difficulties with lean-burn operation including cycle-by-cycle variations. Cycle-by-cycle variations, which increase as the engine is leaned-out, are generally recognized as a limiting factor for the engine's stable operation, fuel efficiency and emissions. Lean operation can decrease the CO and NOx emissions while simultaneously improving engine efficiency. A compromise must be made so that significant reduction in emissions can be made without sacrificing the burn quality of the fuel which may include slow flame propagation, increased cycle by cycle variations and incomplete combustion which may be more clearly explained in Ma et al. (2008a), Ma et al. (2008e) and Ma et al. (2008f).

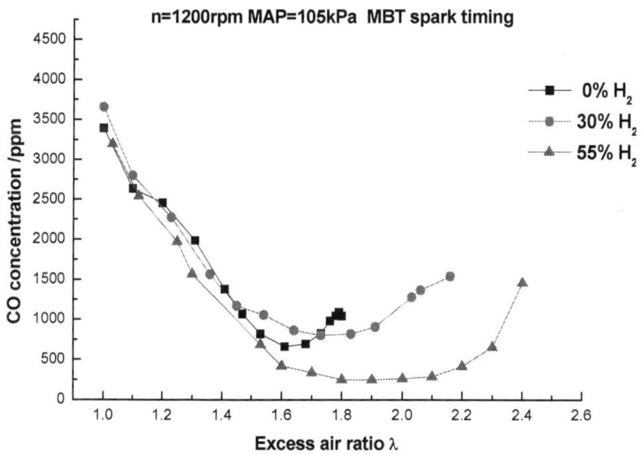

Fig. 11. Carbon Monoxide emissions vs excess air ratio

9.2 Hydrogen Ratio

The addition of hydrogen can greatly improve the performance and emissions of the fuel. There have been many studies completed in efforts to obtain the ideal hydrogen ratio, and the general consensus is that hydrogen/natural gas blends around 20%, results in the best overall combination of emissions and engine performance. According to Wang (2009a), the role of hydrogen in the flame will change from an intermediate species to a reactant when hydrogen fraction in the blends exceeds 20%. (Wang et al., 2009a) Consequently the most suitable hydrogen fraction is significantly related to ignition timing and excess air ratio. According to Akansu et al. (2004) who completed tests on a single cylinder AVL engine at hydrogen/natural gas ratio ranging from 0% to 100%, a 20-30% hydrogen enrichment of natural gas gives the most favorable engine operation. Higher hydrogen contents undermine the knock resistance characteristics of natural gas, lower power output of the engine and increase the fuel cost. Akansu et al. also concludes that, Hydrogen content lower than 20-30% does not make enough use of the performance enhancement potential of hydrogen. (Akansu et al., 2004)

The thermal efficiency of fuel can be improved as seen in figure 2 from a previous section and figure 12 of this section. Also seen in figure 12, the thermal efficiency begins to drop rapidly after reaching a certain excess air ratio, for which this decline in thermal efficiency can be reduced as the hydrogen ratio is increased. This is due to the improvements in burning velocity and improvements in the combustion characteristics which can help extend the lean burn limit and also improve the fuel efficiency. It can be seen in figure 6 that the BSFC of the HCNG fuel can be reduced by increasing the ratio of hydrogen. The minimum BSFC was attained using 40% HCNG, which results in a 5.07% lower BSFC than that of CNG fueling at the same conditions.

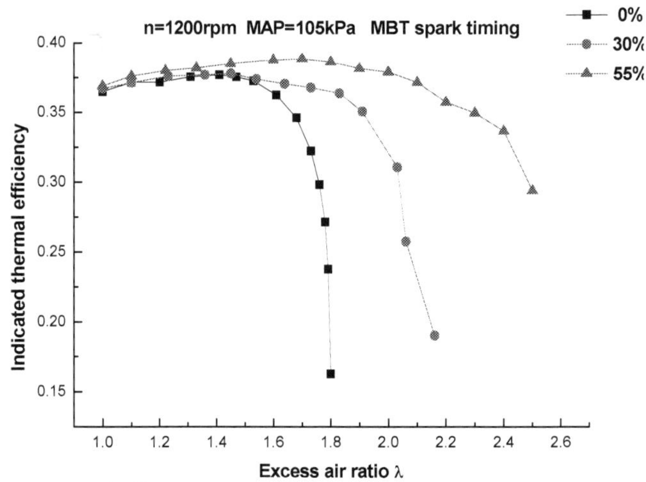

Fig. 12. Indicated thermal efficiency versus excess air ratio

Under idle operation conditions, hydrogen addition is an effective method for improving the power output of the engine and reducing both exhaust emissions and fuel consumption. Furthermore, these results improve as the ratio of hydrogen is increased; however, studies show that under ideal conditions there is not significant improvement when increasing the hydrogen ratio in the HCNG fuel. Under normal operation conditions, the addition of hydrogen is effective at improving the power output of the engine and reducing fuel consumption. The hydrogen-enriched fuel can help improve the burning velocity and improve the incomplete combustion and is seen to increase with the hydrogen ratio. Even though the volumetric calorific value of the HCNG mixture is slightly lower than the calorific value of pure CNG, after the fuel is enriched with hydrogen the combustion efficiency and thermal power conversion efficiency are enhanced resulting in a higher power performance as can be seen in figure 13.

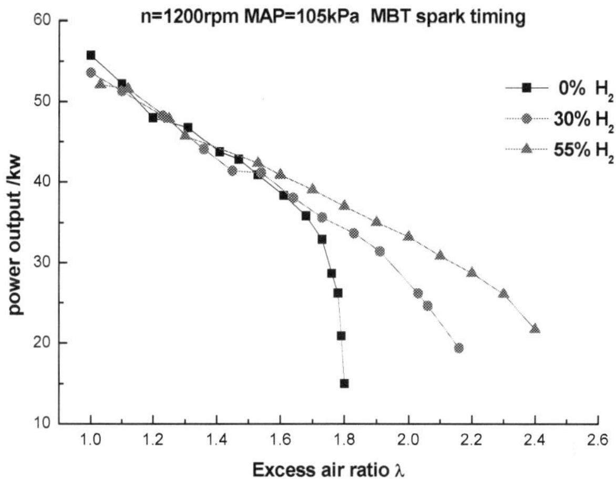

Fig. 13. Engine's power performance versus excess air ratio

Figure 3 and figure 4 from a previous section show that the hydrogen addition can also be an effective method to reduce the as coefficient of variation decreases. Cycle by cycle variations are caused by poor burn quality and have many adverse effects such increasing the emissions and reducing the performance. As the hydrogen fraction is increased, the output torque also increases which can be seen in figure 5. According to Ma, et al. (2009a) this is true at high engine speeds, but for low engine speeds the variation in torque is negligible. Figure 14 shows the coefficient of variation of the indicated mean effective pressure for different hydrogen ratios at different excess air ratios. As can be seen, hydrogen addition can reduce COV_{imep} especially when compared at high excess air ratios due to hydrogen's broader burn limit and its' fast burn speed.

NOx emissions versus ignition timing were plotted in figure 7 of a previous section. As can be seen from the figure, the NOx emissions increase as the hydrogen ratio increase. This is caused by the elevated flame temperature in the cylinder which rises as the hydrogen is added. Carbon monoxide emissions can also be greatly reduced with the addition of hydrogen. Table 2 shows different hydrogen fractions while holding the power constant, it is clearly seen in this table that as the hydrogen fraction is increased the carbon monoxide and unburned hydrocarbon emissions are greatly reduced while the NOx remains at acceptable levels. The reduction in hydrocarbon and carbon monoxide emissions can be attributed to hydrogen's ability to strengthen combustion, especially for lean fuel-air mixtures.

Regarding emissions, the largest advantage to using a higher hydrogen ratio is the reduction in hydrocarbon emissions which can be seen in figure 10 of a previous section. The reduction of hydrocarbon emissions can be explained by the fact that hydrogen can speed up flame propagation and reduce quenching distance, thus decreasing the possibilities of incomplete combustion, and because of the fact that the carbon concentration of the fuel blends is decreased due to hydrogen addition. Hydrogen's ability to strengthen combustion has a large effect on the hydrocarbon emissions, which can be especially evident in lean fuel-air mixtures.

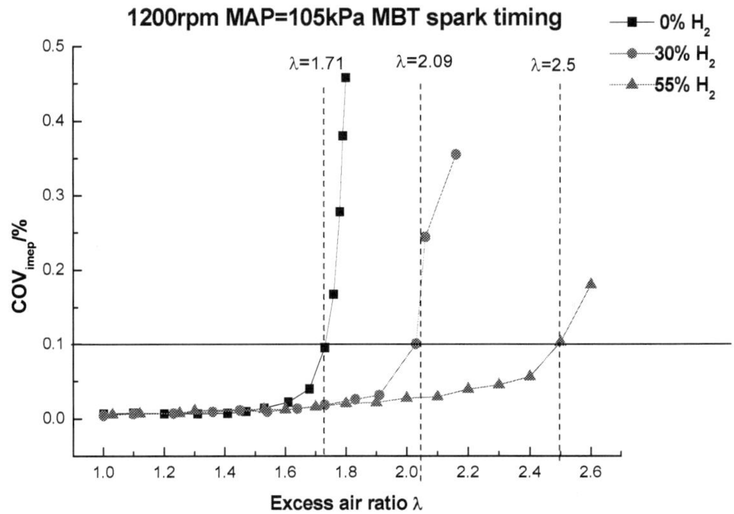

Fig. 14. COV_{imep} versus excess air ratio

Hydrogen fraction (%)	NOx (%)	CH_4 (%)	CO (%)	Economy (%)	Power (%)
0	100	100	100	100	100
10	67.2	84.3	90.4	97	100
20	50.4	71.1	82.7	92	100
30	64.3	65.3	76.5	93	100
40	88.6	60.1	71.3	94	100
50	105	57.3	67.3	94	100

Table 2. The overall performance of different hydrogen fraction at full load 1600r/min

Figures 15 confirms the improvements in flame development speed (characterized as the duration between the spark and 10% mass fraction burned) and propagation speed (characterized as the duration between 10% and 90% mass fraction burned). Fundamentally, the addition of hydrogen provides a large pool of H and OH radicals whose increase makes the combustion reaction much easier and faster, thus leading to shorter burn duration. Engine performance and emissions at different hydrogen ratios are looked at in more detail in Ma et al. (2008h) and Ma et al. (2010).

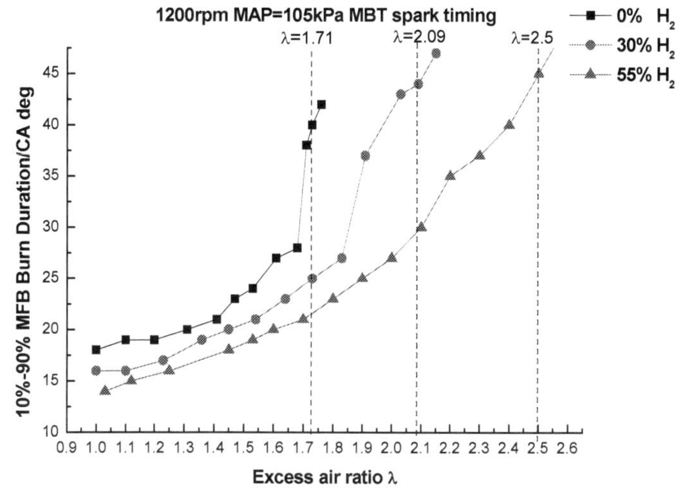

Fig. 15. 10% to 90% MFB burn duration versus excess air ratio

9.3 Spark Timing

Optimizing spark timing can be used as a strategy to avoid knocking and to avoid exceeding the limit of maximum cylinder pressure when operating under lean burn conditions. An effective way to reduce NOx emission is to retard the spark timing which can be seen in figure 7 in a previous section. This is due to the combustion stability and the elevated flame temperature in the cylinder. This figure shows an engine speed of 1600rpm, but at a lower engine speed such as 800 rpm it is found that the NOx emissions also increase slightly when the spark timing becomes closer to TDC beginning around 5 degrees CA BTDC. This is because of the increased engine power loss when the ignition timing is set too close to TDC because the fuel cannot burn completely and the combustion process mainly takes place in the expansion stroke with a relatively low-pressure environment.

The thermal efficiency, shown in figure 16, is also greatly affected by the spark timing. As can be seen in the figure, the thermal efficiency rises as the spark timing is advanced. This is due to the decrease in temperature due to the early ignition timing. The performance and emissions characteristics at different spark timings are more clearly explained in Ma et al. (2008c).

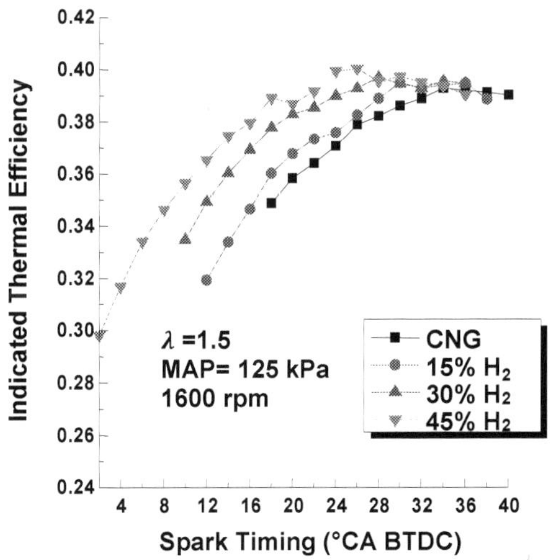

Fig. 16. Indicated thermal efficiency at different spark timings

Maximum brake torque (MBT) spark timing is dependent on flame speed, namely faster flame speed will result in a decrease in the crank angle before TDC at which the spark for maximum torque is applied. With identical fuel energy and equivalent excess air ratios, as the MBT approaches TDC, the torque output increases. This can be explained by the fact that hydrogen addition increases the burning speed of the flame causing the real engine cycle to be similar to the ideal constant volume cycle, thus improving the thermal efficiency of the engine. This relationship can be seen in figure 17. The spark timing also has influence on the coefficient of variation; in general, as hydrogen is added the optimal spark timing should be a few crank angle degrees closer to top dead center. The excess air ratio also has a relatively large effect on the ideal spark timing. In order to degrease the coefficient of variation, the spark timing should be advanced as the excess air ratio is increased.

Figures 7 and 8 from a previous section show the relationship between spark timing and NOx and hydrocarbon emissions, respectively. It is very clear from these figures, that in order to reduce emissions, the spark timing should move closer to top dead center. It can also be observed that as the hydrogen content increases the spark timing should move closer to top dead center. Although there are many other factors that affect the emissions (particularly the air fuel ratio) the spark timing also has a dramatic influence on the emissions.

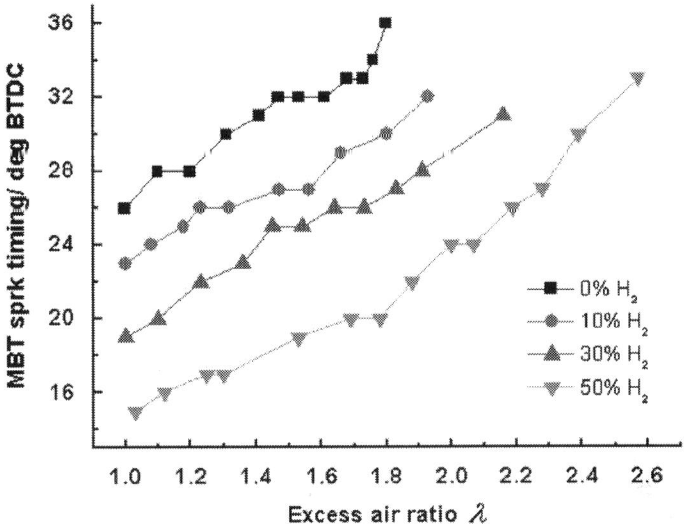

Fig. 17. MBT Spark timing versus Excess air ratio

9.4 Catalytic Converter and European Transient Cycle (ETC) Performance

Although there has not been extensive research completed in this area, a catalytic converter is a common and effective method to reduce the engine emissions, and has also been proven as a suitable method for reducing the emissions in an HCNG engine. ETC performance is a test cycle that has been introduced in the year 2000, in order to receive an emission certification of heavy-duty diesel engines in Europe.

The ETC performance data of the Dongfeng EQD230N engines fuelled with CNG and 20% HCNG without a catalytic converter are shown in figures 18 and 19, and the comparison of CNG and 20%HCNG engine's ETC performance data are listed in table 3. It should be indicated that both the CNG and 20% HCNG engines have been carefully optimized and calibrated. From table 3 it is found that at a 20% hydrogen to natural gas ratio, the engine's NOx emission based on the European transient cycle (ETC) are reduced by nearly 50% compared with CNG engine, which resulted from the addition of 20% HCNG, the engine's larger excess air ratio and increased ignition delay. It can also be found that the engine running on 20% HCNG has about 40% CO reduction, 60% NMHC reduction, 47% CH_4 reduction and 7% BSFC (brake specific fuel consumption) reduction compared with CNG engine, and the peak power maintains unchanged.

A comparison of the ETC performance data for the three different oxidation catalysts running on 20% hydrogen/natural gas are shown in table 4. All three catalysts can obtain enhanced environmentally friendly vehicle (EEV) standards, which are listed in table 5. By implementing a proper oxidation catalyst on a HCNG engine, it allows EEV standards to be more easily achieved. However, by increasing catalytic efficiency, exhaust resistance is increased, engine power is reduced and the brake specific fuel consumption is increased.

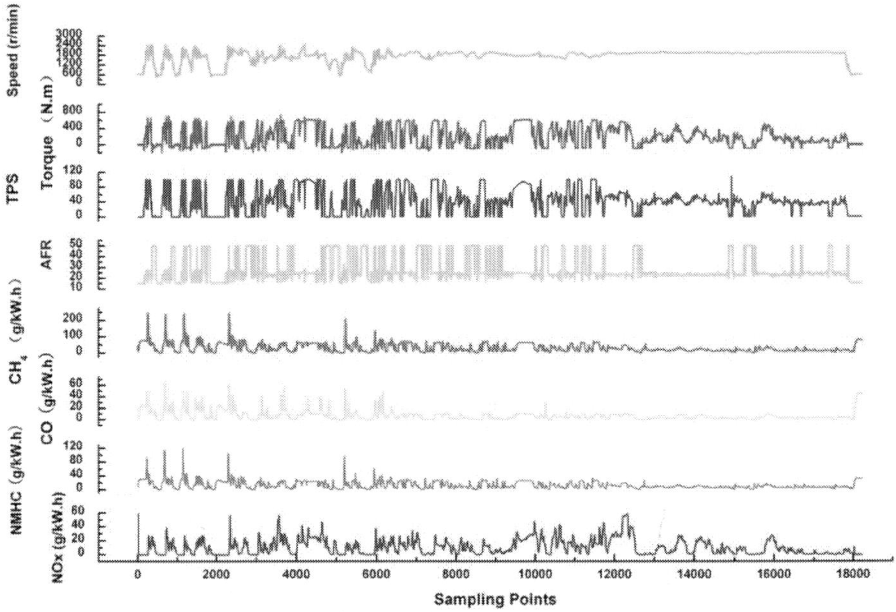

Fig. 18. The ETC performance data for Dongfeng EQD230N CNG engine

A three-way catalyst can be used to reduce the hydrocarbon emissions by oxidizing unburned hydrocarbon, as well as reducing NOx emissions. However, because of the three-way catalyst requires stoichiometric operation, a three-way catalyst may not be the best alternative to reduce emissions in an HCNG or natural gas engine. As the air-fuel ratio approaches the stoichiometric air-fuel ratio, the temperature of the HCNG engine will be elevated resulting in durability problems as well as increasing the emissions and reduces the engine's thermal efficiency.

Fuel type	NOx (g/kW.h)	CO (g/kW.h)	NMHC (g/kW.h)	CH_4 (g/kW.h)	BSFC (g/kW.h)	Peak power (kW)
CNG	4.76	2.45	0.52	4	254	154
20%HCNG	2.31	1.54	0.21	2.1	236	154

Table 3. The comparison of CNG and 20% HCNG engine's ETC performance data

Catalyst type	Nox (g/kW.h)	CO (g/kW.h)	NMHC (g/kW.h)	CH4 (g/kW.h)	BSFC (g/kW.h)	Peak power (kW)
A	1.34	0.25	0.05	0.45	260	150
B	1.42	0.13	0.03	0.42	263	149
C	1.51	0.07	0.02	0.20	272	145

Table 4. The ETC peformance data comparison of three different oxidation catalysts on 20% HCNG engine

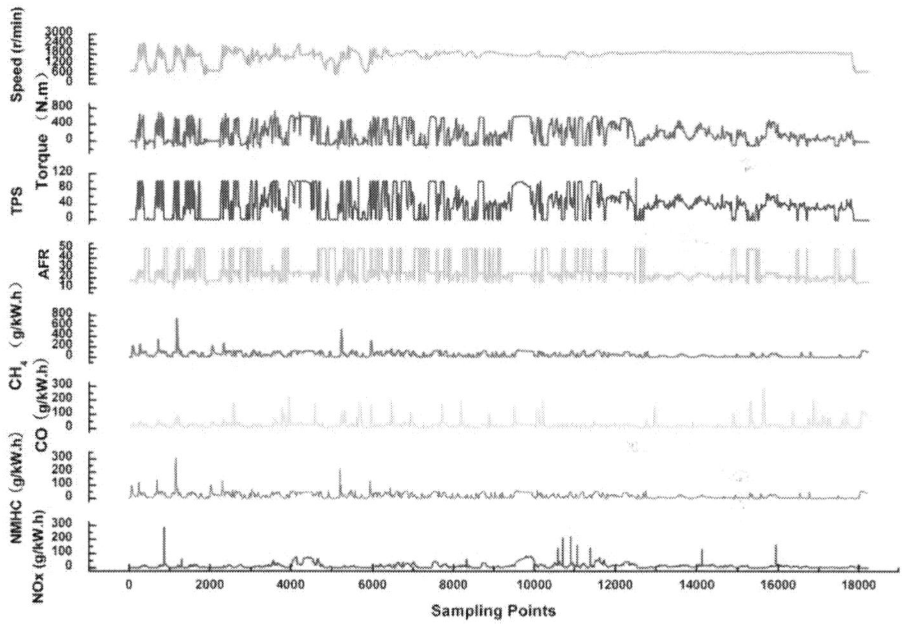

Fig. 19. The ETC performance data for Dongfeng EQD230N engine fuelled with 20% HCNG

Nox (g/kW.h)	CO (g/kW.h)	NMHC (g/kW.h)	CH4 (g/kW.h)
2.0	3.0	0.40	0.65

Table 5. Enhanced environmentally friendly vehicle (EEV) standards

9.5 Exhaust Gas Recycle

Exhaust gas recycle (EGR) is used to reducing emissions in both gasoline and diesels engines. This is another area in which there has not been extensive research in regarding the use of HCNG engines. However, it has been shown in CNG engines to be an effective method in reducing the emissions, especially the NOx emissions.

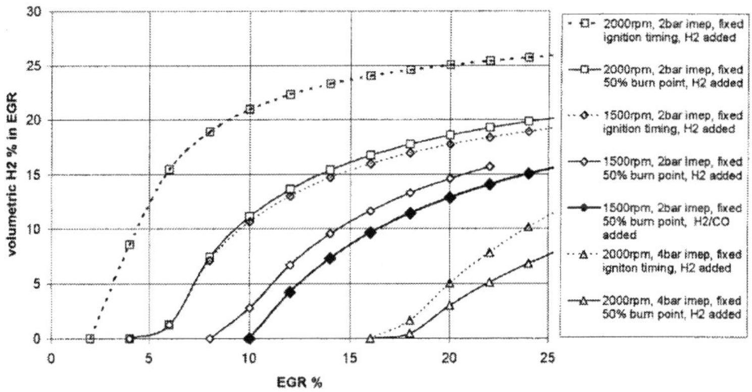

Fig. 20. The percentage hydrogen required in a reformed EGR stream to maintain a COV of iMEP of 5 percent or lower (Allenby et al., 2001)

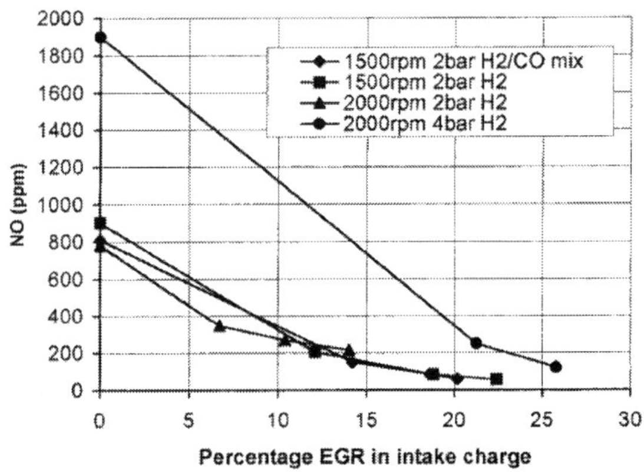

Fig. 21. Dry base engine-out NO emissions versus percentage EGR for four test cases (Allenby et al., 2001)

Research completed by Allenby et al. (2001) on a single cylinder SI engine shows that by adding hydrogen to the EGR of a natural gas engine through a catalyst, the percentage of exhaust gas to be recalculated through this system can be increased; this relationship can be seen in figure 20. In this study, as can be seen in figure 21 the NOx emissions decrease as the percentage of EGR is increased. This study shows that with the addition of hydrogen in the EGR, the engine can tolerate up to 25 percent EGR while maintaining a indicated mean effective pressure, coefficient of variability below 5%, and at this level of EGR, the reduction of NOx emission is greater than 80 percent. (Allenby et al., 2001)

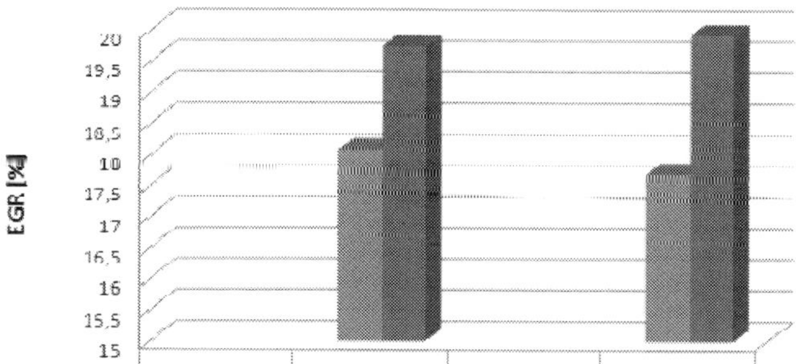

	BMEP [Bar]	2,5		4
Natural gas		18,07		17,69
Hythane		19,74		19,922

Fig. 22. Engine EGR tolerance for different fuels (Kaiadi et al., 2009)

Another study investigating the use of an EGR system was completed by Kaiadi et al. (2009) using a natural gas engine operated a stoichiometric conditions shows that by using an HCNG mixture of 15% as a fuel, the limit for exhaust gas recirculation can be increased as shown in figure 22. This figure shows that by using hydrogen enriched natural gas rather than natural gas alone, the amount of gas that can be recycled is increased by nearly 20%.(Kaiadi et al., 2009)

9.6 Compression Ratio

Because both natural gas and hydrogen are gaseous fuels, they are able to with-stand a higher compression ratio which allows for increased efficiency. Although the studies on the effect of an increased compression ratio on the performance and emission of the fuel are not plentiful, it has been proven to be an effective method to increase performance. According to NRG Tech (2002), by completing tests on an HCNG engine with compression ratio ranged from 9.1 to 15.0 , it was concluded that a desirable compression ratio ranges between 12 and 15. However, care must be taken to avoid engine knock. This can require non-optimal designs for emissions, but will allow knock-free operation. (NRG Tech, 2002)

10. Optimizing the Control System

As can be seen from the previous section, there are many factors that contribute to the performance and emissions of the HCNG fuel. In order to optimize such parameters, there must be the relative software and hardware developed to support this. Simulation can also be useful in determining the ideal parameters. A quasi-dimensional model developed by the author and his research group is presented in Ma et al. (2008b) and Ma et al. (2008g).

There are eight areas of hardware that should be optimized for the HCNG fuel which include: the main chip circuit design, the power management circuit design, the input signal conditioning circuit design, the actuator drive circuit design, the communication circuit design, thermal design, EMI design and access socket design. There are three aspects to the software design, the first is the measurement and modeling of the engine operating parameters. The second is to judge the engine operating conditions and calculation of the state. The third is the implementation of the module results.

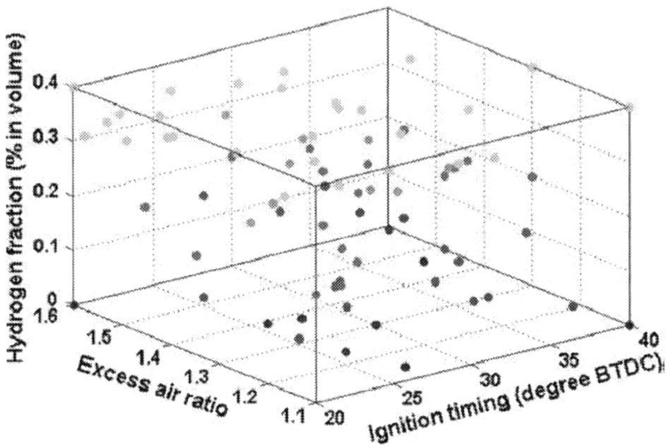

Fig. 23. Control system platform (Ma et al., 2008i)

Finally, the HCNG control unit must with-stand system verification of functionality, reliability and stability. There have already been advances in the electronic control system hardware and software design, additional optimization is necessary to obtain the best combination of parameters. Figure 23 shows an example of data points which were taken in efforts to optimize a control system. The control strategy used by the author is presented in Ma et al. (2008i).

11. Online-Mixing

Another of the many challenges that come with the development and implementation a mixed fuel such as hydrogen-enriched compressed natural gas is the method of mixing the two fuels. In many testing facilities the two fuels are pre-mixed and bottled in high-pressure cylinders, which can be costly, unsafe and constrains the blend ratio; however the use of an online mixing system can not only increase safety and decreases cost, but also allows for a variable blend ratio which can increase efficiency and reduce costs while testing.

The use of premixed, bottled hydrogen/natural gas mixtures restricts the ability to fluctuate the hydrogen ratio, and is especially limiting when doing lab tests. One alternative to having pre-bottled hydrogen-natural gas used for dispensing is by implementing an online mixing system. The relative pressures can be used to control the blend ratio which is described by Dalton's law of additive pressures which states that the pressure of a gas mixture is equal to the sum of the pressure of each gas if it existed alone at the mixture temperature and volume. This can be written to solve for the hydrogen fraction x_1 as follows:

$$x_1 = \frac{P_{H2,1}}{P_1} \tag{8}$$

Where $P_{H2,1}$ is the pressure after the initial charge of hydrogen, and P_1 is the total cylinder pressure at the end of the first charging. After some time, when the cylinders should be recharged, the following equation is used to calculate the hydrogen fraction.

$$x_i = \frac{P_{H2,i} - (1 \quad x_{1-i})P'_{i-1}}{P_i} \qquad (9)$$

In this equation, $P_{H2,i}$ is the hydrogen pressure, x_i is the hydrogen fraction, and P_i is the cylinder pressure after the ith recharge where P_{i-1} represents the pressure before the ith refuel.

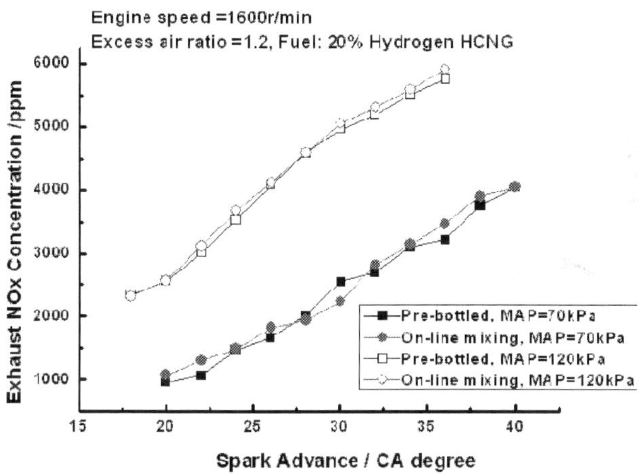

Fig. 24. Online mixing system compared with pre-bottled mixing (Ma et al., 2008j)

Figure 1 from a previous section shows a schematic of an online mixing system that can be used to implement the HCNG fuel. An HCNG dispenser is generally combined with a CNG dispenser for natural gas vehicles as both use the same feed stream from the compressed natural gas grid. In addition the Hydrogen production method differs per station, some stations use on-site generation where other stations use on-site delivery of hydrogen to feed the HCNG dispensers. By using the online mixing system, the power and emissions are nearly identical to those of the pre-bottled HCNG. A comparison of the nitrous oxide emissions can be seen in figure 24. (Ma et al., 2008j)

12. Infrastructure

An established infrastructure is critical for the wide-spread use of any alternative fuel. Currently the infrastructure of natural gas is already well-established and growing in many places, this existing infrastructure could potentially be used as a base for the establishment of infrastructure for the HCNG as a fuel. Furthermore, the HCNG infrastructure can be a good start point to move in the right direction as the world moves closer to a hydrogen economy. According to (Fuel Cells 2000, 2009), There are currently 14 public and R&D

fueling stations around the world including fueling centers in Phoenix, Arizona, Thousand Palms California, Fort Collins Colorado, Las Vegas Nevada, Hempsted New York, University Park Pennsylvania, Montreal Canada, Surrey Canada, Dunkerrque Frace, Toulouse Frace, Faridabad India, Montova Italy, Stavanger Norway and Malmo Sweden. There are also fueling stations planned for Barstow California, Delhi India, Goteborg Sweden, Shanxi Province in China and possibly Grenoble France.

13. Future Research

Future research of the hydrogen enriched compressed natural gas fuel include continuous improvement on performance and emissions, especially to reduce the hydrocarbon emissions (including methane if necessary) which are currently not heavily regulated but will probably be more closely regulated in the future. Additional optimization is also necessary for the HCNG fuel in order to obtain the ideal combination of excess air ratio, hydrogen ratio and spark timing. This should be further followed by the implementation of an adequate control system. Other potential improvements include the reduction of emissions which can be transpire with the addition of a catalytic converter or by implementing an exhaust gas recycle system, lastly there is potential for performance improvements with an increase in the compression ratio.

14. Conclusion

Compared with natural gas, HCNG has many advantages when it comes to performance. Research has shown that the brake effective thermal efficiency increases with an increased percentage of hydrogen. Another effect of the addition of hydrogen is that the brake specific fuel consumption is reduced, the cycle by cycle variations are also reduced, and the thermal efficiency is increased.

Emissions can also be improved with the addition of hydrogen. Compared to pure natural gas, HCNG reduces the HC emissions, which is in part due to the increased combustion stability that comes with the addition of hydrogen. However, due to the increased temperature and combustion duration that accompanies the hydrogen addition, an increase in NOx emissions is observed.

There are many optimization parameters that can be modified to adjust to the HCNG fuel. With the increase of hydrogen addition, the lean operation limit extends which is often used to maximize the thermal efficiency and reduce the nitrous oxide emissions, although due to the increased air running through the engine, at high excess air ratios the combustion becomes more unstable leaving unburned hydrocarbons in the exhaust. Therefore, the excess air ratio should be positioned by finding the best combination of nitrous oxide and the hydrocarbon emissions. Another method to reduce emissions, is to move the spark timing closer to top dead center, however this is greatly dependent on the excess air ratio. The hydrogen ratio can also be increased to extend the lean limit, improve the combustion and reduce the hydrocarbon emissions.

Although the exhaust emissions from hydrogen-enriched natural gas are already very low, further refinement must be done in order to further reduce emissions and to achieve Enhanced Environmentally Friendly Vehicle (EEV) standards. Therefore finding the optimal combination of hydrogen fraction, ignition timing and excess air ratio along with other parameters that can be optimized is certainly a large hurdle. It is not only a challenge to locate the ideal combination of hydrogen fraction, ignition timing, and excess air ratio, but it can also be a large challenge to control these parameters. This requires sufficient control system to be developed for the HCNG engine to maximize the performance simultaneously minimizing the exhaust emissions.

Probably the biggest challenge with the implementation of the fuel comes with developing an infrastructure to support this promising alternative fuel. There HCNG allows for an initial use of hydrogen while taking advantage of the current CNG infrastructure. This allows for the hydrogen infrastructure to slowly become established until the production and efficiency demands can be met for the hydrogen economy. Although there is currently a large amount of research taking place regarding the HCNG fuel, there are certainly many steps to take before wide-spread implementation can occur.

15. References

Akansu, S. O., Dulger, Z., Kahranman, N., & Veziroglu, N. T. (2004). Internal combustion engines fueled by natural gas--hydrogen mixtures. *International Journal of Hydrogen Energy , 29,* 1527-1539.

Allenby, S., Chang, W.-C., Megaritis, A., & Wyszynski, M. L. (2001). Hydrogen enrichment: a way to maintain combustion stability in a natural gas fuelled engine with exhaust gas recirculation, the potential of fuel reforming. *Proceedings Institution of Mechanical Engineers , 215 Part D,* 405-418.

Fuel Cells 2000. (2009). *Worldwide Hydrogen Fueling Stations.* Retrieved 2010 March, from Fuel Cells: www.fuelcells.org/

Hythane Company, LLC. (2007). *Hythane: Blue Sky Clean Air.* Retrieved March 2010, from Hythane: History: http://www.hythane.com/

Kaiadi, M., Tunestal, P., & Johansson, B. (2009). Using Hythane as a Fuel in a 6-0Cylinder Stoichiometric Natural-gas Engine. *SAE International Journal of Fuels and Lubricants, SAE Paper 2009-01-1950 , 2* (1), 932-939.

Lynch, F. E., & Marmaro, R. W. (1992). *Patent No. 5139002.* United States of America.

Ma, F., Ding, S., Wang, Y., Wang, M., Jiang, L., Naeve, N., Zhao, S. (2009a). Performance and Emission Characteristics of a Spark-Ignition (SI) Hydrogen-Enriched Compressed Natural Gas (HCNG) Engine Unver Various Operating Conditions Including Idle Conditions. Energy & Fuels , 23, 2113-3118.

Ma, F., Ding, S., Wang, Y., Wang, Y., Wang, J., Zhao, S. (2008a). Study on combustion behaviors and cycle-by-cycle variations in a turbocharged lean burn natural gas S.I. engine with hydrogen enrichment. *International Journal of Hydrogen Energy,* 33, 7245-7255.

Ma, F., Liu, H., Wang, Y., Wang, J., Ding, S., Zhao, S. (2008b). A Quasi-Dimensional Combustion Model for SI Engines Fuelled by Hydrogen Enriched Compressed Natural Gas. *SAE Paper* 2008-01-1633.

Ma, F., Liu, H., Wang, Y., Li, Y., Wang, J., Zhao, S. (2008c). Combustion and emission characteristics of a port-injection HCNG engine under various ignition timings. *International Journal of Hydrogen Energy*, 33, 816-822.

Ma, F., Wang, M., Jiang, L., Chen, R., Deng, J., Naeve, N., Zhao, S.(2010).Performance and emission characteristics of a turbocharged CNG engine fueled by hydrogen-enriched compressed natural gas with high hydrogen ratio, *International Journal of Hydrogen Energy* , 35, 6438-6447.

Ma, F., Wang, Y. (2008d). Study on the extension of lean operation limit through hydrogen enrichment in a natural gas spark-ignition engine. *International Journal of Hydrogen Energy*, 33, 1416-1424.

Ma, F., Wang, Y., Ding, S., Jiang, L.(2009b).Twenty percent hydrogen-enriched natural gas transient performance research. *International Journal of Hydrogen Energy*, 34, 6523-6531.

Ma, F., Wang, Y., Liu, H., Li, Y., Wang, J., Ding, S. (2008e). Effects of hydrogen addition on cycle-by-cycle variations in a lean burn natural gas spark-ignition engine .*International Journal of Hydrogen Energy*, 33, 823-831.

Ma, F., Wang, Y., Liu, H., Li, Y., Wang, J., Zhao, S.(2007). Experimental study on thermal efficiency and emission characteristics of a lean burn hydrogen enriched natural gas engine. *International Journal of Hydrogen Energy*, 32, 5067-5075.

Ma, F., Wang, Y., Wang, J., Ding, S., Wang, Y., Zhao, S.(2008f). Effects of combustion phasing, combustion duration, and their cyclic variations on Spark-Ignition (SI) engine efficiency. *Energy&Fuels, 22, 3022-3028.*

Ma, F., Wang, Y., Wang, J., Zhao, S. , Yin, Y., Cheng, W., Zhou, M.(2008g). Development and validation of a quasi-dimensional combustion model for SI engine fuelled by HCNG with variable hydrogen fractions. *SAE Paper* 2008-01-1580.

Ma, F., Wang, J., Wang, Y., Wang, Y., Li, Y., Liu, H., Ding, S.(2008h). Influence of different volume percent hydrogen/natural gas mixtures on idle performance of a CNG engine.*Energy&Fuels*, 22, 1880-1887.

Ma, F., Wang, J., Wang, Y., Wang, Y., Zhong, Z., Ding, S., Zhao, S. (2008i). An investigation of optimum control of a spark ignition engine fueled by NG and hydrogen mixtures, *International Journal of Hydrogen Energy*, 33, 7592-7606.

Ma, F., Wang, Y., Wang, J., Zhao, S., Yin, Y., Cheng, W., Zhou, M. (2008j). Development and Validation of an On-line Hydrogen-Natural Gas Mixing System for Internal Combustion Engine Testing. *SAE Paper 2008-02-1508.*

Nagalim, B., Duebel, F., & Schmillen, K. (1983). Performance study using natural gas, hydrogen-supplemented natural gas and hydrogen in AVL research engine. *International Journal of Hydrogen Energy , 8* (9), 715-720.

NRG Tech. (2002). Hydrogen/Natural Gas Blends for Heavy and Light-Duty Applications. *2002 U.S. DOE Hydrogen Program Review*. Reno, NV.

15

Looking for clean energy considering LNG assessment to provide energy security in Brazil and GTL from Bolivia natural gas reserves

Miguel Edgar Morales Udaeta, Jonathas Luiz de Oliveira Bernal, Geraldo Francisco Burani and José Aquiles Baesso Grimoni
IEE/USP (Instituto de Eletrotécnica e Energia da Universidade de São Paulo)
Av. Prof. Luciano Gualberto, 1289
CEP 05508-010
São Paulo – SP
Brazil

This work aims to identify market opportunities for LNG in Brazil as a complement to natural gas supply, characterizing it both on its production and transportation side and analyzing costs and prospects for expanding the supply and use infrastructure so as to ensure that the growing demand of this energy input is met and allow greater flexibility for the natural gas industry and the power sector, with an idea about risk sources and protection mechanisms used to ensure the reliability of power generation, gas market flexibility and the need for LNG supply planning considering the spot and/or long-term market.

Another aspect approached is the economics and the viability of Natural Gas Industrialization in Bolivia, by producing secondary fuels such as GTL-diesel from natural gas (cleaner than the oil byproduct), seeking a clean development from the environmentally correct use of power by this GLT process. Bolivia has resources that could supply these secondary energy resources as from GTL. It is possible to process 30MCMpd of gas obtaining profits from the gas and also from the liquid hydrocarbons that are found in it. The Bolivian GTL would present the following advantages: It would export diesel and/or gasoline and would not have to import it anymore; the GTL-FT exports could reach 35Mbpy, acquiring competitive prices; It would increase productive jobs not only due to GTL itself, but also due to secondary economy linked to the GTL market; The use of GTL-FT diesel would provide a "cleaner" environment especially in the urban areas; Finally, from the macroeconomic perspective, the investment in the plant construction and supporting works would generate a great number of jobs.

Energy resources play a fundamental strategic role in the development of a country and its economic activity. The worldwide tendency of increasing demand for energy, especially in developing countries - such as Brazil and or Bolivia - made energy resources achievement and supply a critical issue for the continuity of socioeconomic growth. To assure the availability of these resources so as to guarantee the future of the economy, of the

environment and of the society as a whole is a constant challenge. In this sense, all the factors involved that determine or make security and availability exist over time must be highly relevant, as well as the planning of energy supply and use.
Natural gas (NG) is a fossil fuel, formed by hydro-carbons that can be found in nature in an isolated way or associated to oil. Formed mainly of methane, its advantage is to have low amounts of contaminants, such as nitrogen, carbon dioxide, sulfur composites and particulates, which make its combustion be considered clean. Moreover, it has high calorific power, allowing its direct use, without the need of refining.
In that sense, a technical and economic analysis of secondary fuels from hydrocarbons is conducted in order to identify the possibility of manufacturing the Bolivian Natural Gas using GTL to produce byproducts. The energy economics analysis of LNG is also made viewing energy security in Brazil. Technically and economically speaking, it is important to identify the industrialization, technology investment amounts and production costs of a GTL and or LNG project. Commercially speaking, the aim is to identify the current and future situations of GTL and LNG in the market and what that represents. From the energy economic point of view, the intention is to determine the financial issues (amounts, interest fees, benefit periods) linked to investments in GTL and/or LNG. Legally and politically speaking, the present legal situation will be compared to the most appropriate regulatory issues related to natural gas conversions, both chemical (GTL) an physical (LNG).

Keywords:
LNG, natural gas, energy, energy planning, power, GTL, Diesel, secondary Fuels, Reserves, Clean Development, Energy Resources, Generation, Gas-Chemical, Gas Byproducts

1. LNG and its Production Chain

A LNG project basically has four stages, also called production chain: natural gas exploration and production (E&P); next come the liquefying process, transportation to the import terminals and, finally, regasification.

1.1 Natural Gas Production

Liquefied natural gas (LNG) is essentially natural gas (NG), cooled at a certain temperature below its vaporization point. Thus, the LNG productive chain starts in the exploration and production of natural gas.
At this initial exploration phase, there is a close relation between the NG and petroleum industries. This occurs because usually, in the same basin, there may be gas together with petroleum, either dissolved or as a gas layer formed in the upper part of the deposit. In this case, it is said that natural gas is "associated" to petroleum. In turn, the so-called "non-associated" gas is the one found in fields where there is very little or no petroleum, allowing only the exploration of gas. This way, the geological research efforts to locate these fields, as well as the drilling, development and exploration technologies may be shared between the two industries.
The exploration process is divided into geological and geophysical research and drilling. In the research phase, an analysis is made on the rocky structures and on the underground of the region where petroleum and/or gas is being sought, which allows selecting the drilling

sites. Drilling is part of proving the existence of compounds (oil and/or gas) and its economic viability for later exploration.
After the discovery of a basin, and the analysis of the economic viability of the field, comes the production process. With similar characteristics and technologies, petroleum and NG prospections are jointly conducted, so as to provide the exploration of the two compounds. During NG production, the primary purification process of the gas also occurs, when liquids (water and others), particulate matter and contaminants (sulfur) are separated, so as to make NG adequate to be conveyed to the processing unit.

1.2 Liquefying

The natural gas liquefying plant is the main stage in the LNG production chain. In it, the temperature of natural gas is reduced to -162° C, which is below the vaporization point of methane. Hence, the methane gas turns liquid and its volume is reduced to 1/600 of the original volume.
The liquefying plant is usually built in coastal areas, in bays, so that it facilitates the production outflow by vessels, thus making it also desirable for the plant to be close to the NG producing fields, as the transportation price via gas pipelines is considerable and, depending on the distance to be covered, it may increase the global costs of the project.
The premises composing the liquefying plant are: a gas processing unit (UPGN) in case the gas has not been previously processed with the separation of components of greater commercial value and the standardization of the product global composition. The gas is then dehydrated and broken down, so that hydrocarbons are separated: processed or dry gas (essentially methane), ethane, GLP (propane and butane) and C5+ components (especially natural gasoline). This way, the natural gas processed is led to the liquefying stage in a set of heat exchangers and LNG storage tanks.
The liquefaction of NG is conducted at several stages of gas cooling until the cooled liquid is obtained in a process similar to that of a conventional refrigerator. A cooling gas extracts heat from the NG by means of heat exchangers in parallel sets, forming liquefying trains until this gas is cooled at a temperature of -162°C.
Propane is the main cooling gas, leading the NG temperature to -30°C; the gas will go through other cooling trains in which nitrogen, associated to other hydrocarbons, act as secondary coolers, making NG go below the vaporization temperature.
The technology that uses propane as initial cooling gas is the most commonly used and gained the market along the evolution and diffusion of LNG in the world market, incorporating several technological improvements, mainly concerning cooling compression turbines, which account for a large share of the plants operational cost and their efficiency, allied to increase in power and environmental improvement in the use of cooling gases, besides the development of much more efficient thermal insulating materials, which revest the storage tanks, were essential for the growth in the insertion of LNG as a viable option to natural gas.
The storage of liquefied NG is made in tanks with compression and re-liquefying systems to recover the gases that leak from stocking and resume the gas state; the logistics of liquefying, shipping and transportation forecasts is necessary for minimizing the stored volume, maximizing the LNG production and therefore mitigating losses from re-liquefying and storage.

1.3 Shipping

In order to convey the LNG between the liquefying and regasification plants, specially built vessels for storing gas in its liquid form are used, which count on large reservoirs capable of keeping the gas temperature during transportation. However, losses occur in this process varying from 1% to 3% of the initial volume, according to the distance to be covered, besides the consumption of the gas employed as fuel for the LNG Carrier Ship.
Figure 1 below shows the two types of vessels that convey LNG: the ones that store gas in spherical tanks and those counting on tanks in longitudinal positions; the costs between the two types is similar both in construction and in operation.

Fig. 1. LNG Carrier Ships

In function of its great meaningfulness for the world LNG industry, Japan concentrates a large share of the shipyards that build these types of vessels, and today it has European and Korean shipyards as competitors in this sector. The major producing companies are Daewoo Shipbuilding, Hyundai Heavy Industries, Mitsui Engineering & Shipbuilding, Samsung Heavy Industries, Kawasaki Shipbuilding and Mitsubishi Heavy Industries.
Besides LNG Carrier Ships, LNG can also be conveyed by smaller tanks, by means of trucks or trains generally used to supply peak, temporary or isolated demands when the development cost of a gas pipeline makes the gas supply too expensive.

1.4 Regasification

Regasification plants constitute the importation side in the LNG chain. They are usually located close to the natural gas consumer centers and harbor LNG Carrier Ships in especially built terminals. The plants are formed by LNG storage tanks and heat exchangers where LNG is again transformed into gas for distribution.

1.5 State of the Art of LNG in the World

The greatest world natural gas consumers count on a mature market and fully established infrastructure with maintenance characteristics in the development of both its infrastructure and demand making it necessary to transport natural gas from other producing countries up to the consumer countries. Hence, the LNG technology emerges an alternative to cover great distances.
These facts make an increment in LNG production by competitors be expected, owing to an increase in natural gas prices and to the reduction in LNG costs managed with improvements in the liquefying, storage and transportation technology, making business

with natural gas fields far away from the consumer centers economically interesting, which makes the construction of gas pipelines too expensive.

LNG consumers may use it as a logistic alternative to natural gas in countries that do not count on reserves or physical links with producing regions via gas pipelines, as is the case of Japan, the greatest LNG consumer in the world. Thus, it can be used as a guarantee of power supply in demand peaks, known as *'peak-shaving'* as is the case of the Unites States.

1.6 Major LNG Producers in the World

The EIA (*Energy Information Administration*) divided the LNG exportation industry into three geographical sectors: Pacific basin, Atlantic basin and Middle East basin[1].

1.6.1 Pacific Basin

The producing countries of this Basin are Indonesia, Malaysia, Australia, Brunei, Unites States and Russia, and Indonesia is the world leader in LNG production and export. Table 1 details each country export and its sales markets.

This Basin accounts for approximately 49% of the LNG world production.

Producer	Exports 2002 (TCF)	Exports 2003 (TCF)	Exports 2007 (TCF)	Major Consumers
Indonesia	1.1	1.4	1.4	Japan, Taiwan and South Korea
Malaysia	0.741	0.741	1.1	Japan, Taiwan and South Korea
Australia	0.367	0.572	0.747	Japan
Brunei	0.351	0.351	0.351	Japan, South Korea
Unites States	0.068	0.068	0.068	Japan
Russia			0.234	Unites States
TOTAL	**2.6**	**3.1**	**3.9**	

Table 1. Pacific Basin: LNG Production and Sales Markets

1.6.2 Atlantic basin

The main producing countries in this basin are Algeria, Nigeria, Trinidad and Tobago, Libya, Egypt and Norway, Algeria being the most important country in the LNG production. In Table 2, details of LNG exports in this basin can be observed.

The exporters in this basin produce about 29% of the LNG world production.

1 The division of the LNG industry was made by EIA by means of *The Global Liquefied Natural Gas Market*, site: http://www.eia.doe.gov/oiaf/analisispaper/global/

Producer	Exports 2002 (TCF)	Exports 2003 (TCF)	Exports 2007 (TCF)	Major Consumers
Algeria	0.935	1.1	1.1	France, Belgium, Spain, Turkey and the Unites States
Nigeria	0.394	0.463	0.863	Turkey, Italy, France, Portugal, Spain and Unites States
Trinidad & Tobago	0.189	0.482	0.735	Unites States, Puerto Rico, Spain, and Dominican Republic.
Libia	0.021	0.021	0.021	-
Egypt			0.594	Italy and the Unites States
Norway			0.200	Spain, France, and the Unites States.
TOTAL	**1.5**	**2.1**	**3.5**	

Table 2. Atlantic basin: LNG Production and Sales Markets

1.6.3 Middle East

The producing countries in this basin are Qatar, Oman and the United Arab Emirates, and Qatar is the most important, as shown in Table 3.
The Middle East exporters produce about 23% of the LNG production.
Today, with the largest gas reserves ever found, Iran has great potential to export gas to markets in Europe, Asia, and India, both by gas pipelines and LNG.

Producer	Exports 2002 (TCF)	Exports 2003 (TCF)	Exports 2007 (TCF)	Major Consumers
Qatar	0.626	0.726	1.184	Japan, South Korea, Unites States and Europe
Oman	0.356	0.356	0.517	South Korea, Japan
Arab Emirates	0.178	0.278	0.278	Japan
TOTAL	**1.2**	**1.4**	**2.0**	

Table 3. Middle East: Production and LNG Sales Markets

1.7 Major LNG Consumers in the world

The LNG world market can be divided into two large zones; the Pacific and the Atlantic Basins.
Japan, South Korea and Taiwan are the main consuming countries in the Pacific Basin, which means about 68% of global imports; LNG is used to supply about 90% of the natural gas needs in these countries, making this type of fuel of vital importance for energy supply and security.

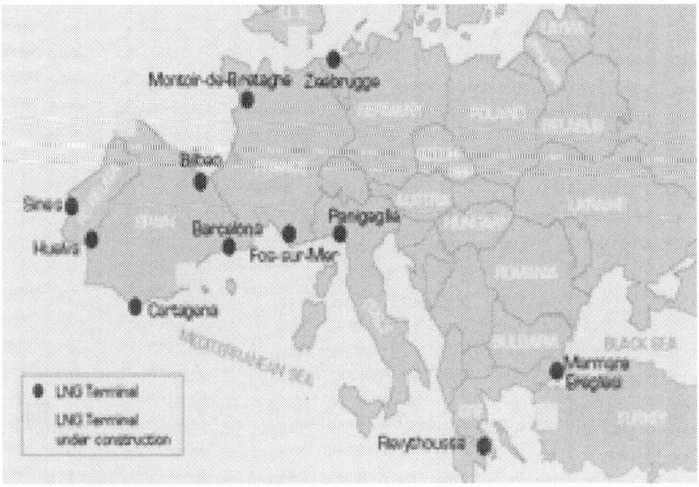

Fig. 2. Europe: Regasification Terminals for LNG Imports
Source: Energy Information Administration (EIA), The Global Liquefied Natural Gas Market

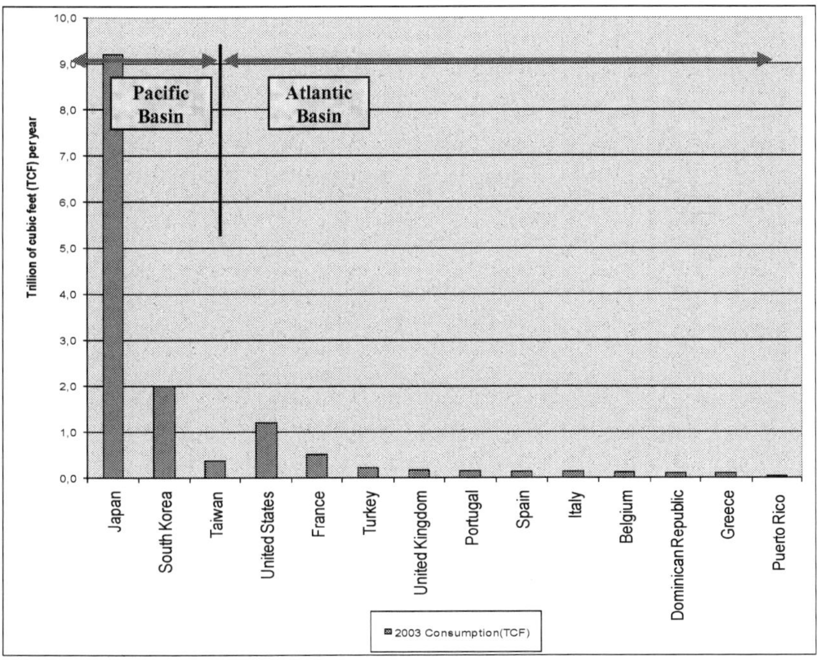

Fig. 3. World: LNG Consumers per Basin

Japan is the largest world consumer, importing around 48% of the world production, counts on 23 LNG regasification terminals, which represent 12% of the power and 95% of the natural gas used by the country. South Korea is the second largest world importers with 3 regasification terminals.
In the Atlantic basin, seven European countries share 28% of the world imports, including the European Union. In this set, the number rises to 32% of the global imports with 11 regasification terminals (Figure 2).

2. Analysis of the Production Costs in the LNG Chain

2.1 Liquefying

Along the last two decades, the specific cost (cost per unit of LNG produced) of the natural gas liquefying plants has significantly been reduced. This cost reduction was possible mainly owing to the technological improvement of the process and to the scale gains, obtained after the manufacturing of trains with greater capacity, but also in function of a greater competition among the companies acting in the liquefying plants design and construction.

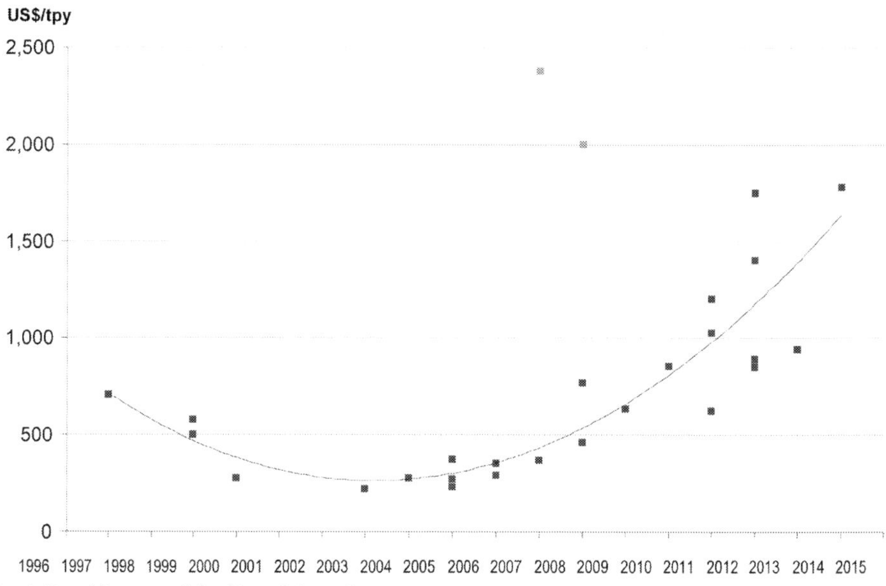

Fig. 4. Specific cost of the liquefying plants

While the first LNG producing plants, built in Algeria in the 1970s, counted on six liquefying trains to reach a capacity of 8 million tons of LNG per year (mtpa), the new projects for expanding the Qatargas and RasGas plants - the major LNG exporters in Qatar and two of the major in the world have about the same production level from a single train.

Since 1964, when the first liquefying plant started operating in Algeria with a single 1.1 mtpa train, the average size of the trains has increased along the years, reaching fourfold greater levels now.

Plant	Location	Started Operation	Capacity (millions of tons per year)	Estimated cost (billions of dollars)	Specific cost
Gladstone	Australia	2014	3.5	6.2	1780
Sunrise LNG	Australia	2013	5.0	4.7	940
Gassi Touil	Algeria	2012	4.0	7.0	1750
Brass LNG	Nigeria	2012	10.0	8.5	850
Soyo	Angola	2012	5.0	7.0	1400
Olokola	Nigeria	2012	11.0	9.8	891
South Pars	Iran	2011	10.0	12.0	1200
Gorgon	Australia	2011	16.0	19.2	1200
Qatargas 4	Qatar	2011	7.8	8.0	1026
Skikda	Algeria	2011	4.5	2.8	622
Pluto	Australia	2010	4.3	9.6	2233
P. Malchorita	Peru	2010	4.5	3.8	854
Qatargas 3	Qatar	2009	7.9	5.0	634
Sakhalin-2	Russia	2008	9.6	19.2	2000
Qatargas 2	Qatar	2008	15.6	12.0	769
NW Shelf T5	Australia	2008	4.4	2.0	460
Snohvit	Norway	2007	4.0	9.5	2375
EGLNG	Equatorial Guinea	2007	3.8	1.4	368
Rasgas 2T345	Qatar	2006	14.1	5.0	355
Darwin	Australia	2006	3.7	1.1	292
Atlantic LNG T4	Trinidad e Tobago	2005	5.2	1.2	231
Egyptian LNG1	Egypt	2005	3.6	1.4	375
Segas	Egypt	2005	4.8	1.3	271
Rasgas 2T3	Qatar	2004	4.7	1.3	277
MLNG Tiga	Malaysia	2004	6.8	1.5	221
Oman LNG	Oman	2003	7.3	2.0	274
NLNG 1-2	Nigeria	2000	6.6	3.8	576
Rasgas	Qatar	1999	6.6	3.3	500
Qatargas 1	Qatar	1997	9.9	7.0	707
MLNG Dua	Malaysia	1995	7.8	1.6	205

Table 4. Existing and planned gas liquefying projects

Despite the reduction in the LNG production costs observed in the last decades, recent contracts for building new liquefying plants have shown an inversion in this trend. Figure 4 shows the evolution in the capital cost per ton per year of some existing liquefying projects and under planning for the next years. The projects are listed in Table 4 further on.
Some factors have to be analyzed in order to understand the rising cost for building LNG plants observed since 2003. Firstly, in the last years, the global demand for energy has grown at record levels, partly due to the strong economic growth in China and in India, but also due to the development of other emerging markets - in which Brazil is included - and this growing demand has generated a world boom in the building sector for energy infrastructure. Moreover, few companies have expertise for LNG designs and, with so many enterprises under planning to start operating in five years, the costs for hiring these companies has significantly risen.
Another aspect is that of the labor costs and the raw material used in LNG designs, such as steel, cement and nickel, the prices of which have substantially and systematically risen in the foreign market along the years.

Some projects are particularly more expensive, most of the time for being located in regions in which there are difficulties to conduct works due to climate conditions, as is the case of the Snohvit projects in Norway and Sakhalin-2 in Russia. Both are located in regions of extreme cold, which implies greater design costs.

2.2 Shipping

After the natural gas liquefying process, large LNG Carrier Ships are filled and convey it to regasification plants, making the shipping by LNG Carrier Ships a crucial element for their flexibility in serving diversified markets all over the world.
The LNG commercial transportation started in 1964, taking LNG from Algeria to the United Kingdom, and since then, the LNG industry has substantially developed, presenting great reliability in terms of security, process technology and operational procedures.
The LNG transportation by LNG Carrier Ships represents from 10% to 30% of the total cost considering the chain from natural gas prospection up to the regasification in the import terminals. The value of the freight represents nearly 70% of the total LNG transportation cost, the rest being related to fuel price, insurance, among others, and the costs for building LNG Carrier Ships exert great influence on the total value for LNG transportation.
Despite the important representativity on the total costs of the LNG chain, shipping becomes more competitive in relation to gas pipelines since the distance run increases. Figure 5 presents a cost comparison between the transportation of natural gas via gas pipeline and the transportation by LNG Carrier Ships, in relation to the distance; it can be verified that the transportation of natural gas via LNG is more advantageous than by sea gas pipeline for distances longer than 700 miles and more advantageous than by land gas pipeline as from 2,200 miles.

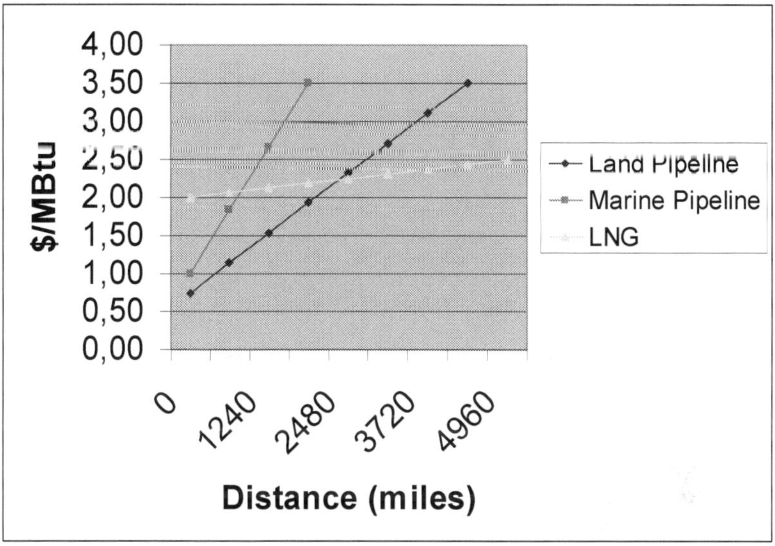

Fig. 5. Comparison between gas pipeline and LNG, in function of distance

In the 1990s, when the LNG industry started to grow more significantly, the number of shipyards with capacity to build LNG Carrier Ships grew and the competition between shipyards led to technological improvements and price reduction, aided by the devaluation of the Japanese and South-Korean currencies, major builders of these LNG Carrier Ships, as related to the American dollar.

However, in more recent years, with so many LNG projects under development, there has been a significant increase in demand for LNG Carrier Ships, leading to an increase in their price.

In the last decade, LNG production grew more than 50% all over the world. During this period, China and India started to import LNG, the United Kingdom resumed LNG imports after 40 years and other LNG markets, such as Spain, South Korea and Taiwan presented an expressive growth.

Due to this growing demand, the fleet of LNG conveying vessels grew from 130 in the early 2002 to about 250 in the late 2007, and by 2011, the number of LNG Carrier Ships may reach 380.

Figure 6 presents the number of LNG Carrier Ships delivered per year since 1993. Figure 7 shows the price evolution per capacity of these LNG Carrier Ships in the delivery year. The graphs show how the demand for LNG Carrier Ships is increasing and the combination of this increase and also the growing cost of raw material and labor led to a strong rise in LNG Carrier Ship prices.

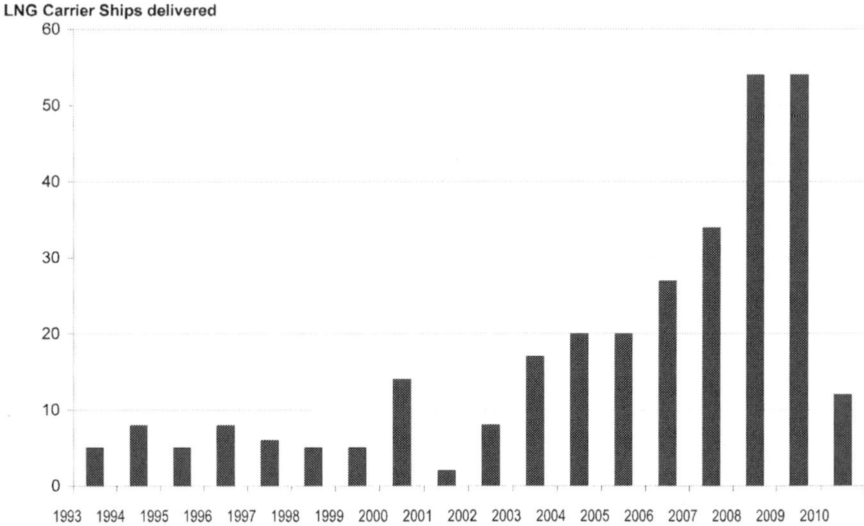

Fig. 6. Number of LNG Carrier Ships delivered

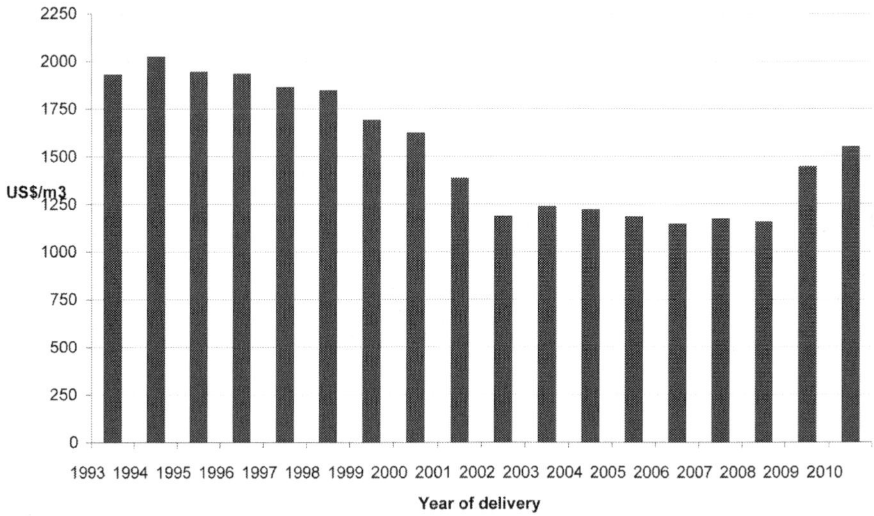

Fig. 7. average price per LNG Carrier Ship capacity

The standard size of the tanks has also changed along the years, in function of the growing demand for greater efficiency and cost reduction in LNG transportation, with scale gains. Whereas in the 1970s and 1980s the average capacity of the LNG Carrier Ships was 125,000 m^3, in the 1990s this average increased to almost 135,000 m^3 and is still growing.

The average capacity of the LNG Carrier Ships delivered in the last years was around 150,000 m^3, and there are at least 40 vessels with a capacity over 200,000 m^3. Figure 8 shows how the average capacity of the LNG Carrier Ships increased since the 1970s.

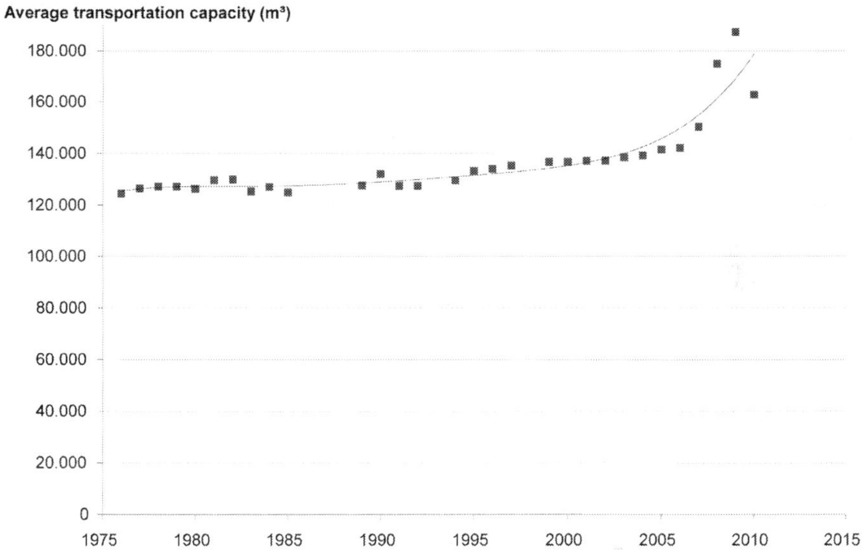

Fig. 8. Evolution in the average capacity of LNG Carrier Ships

Although there is a clear trend for increasing the LNG Carrier Ships capacity, some issues have to be considered to assess whether the LNG Carrier Ships size will keep growing. There is a limiting factor, since only some terminals are able to harbor vessels with capacity higher than 200,000 m^3. Hence, this increase in the number of large vessels may affect the design of new liquefying plants and regasification terminals.

Conversely, these new very large vessels are dedicated to specific projects, exerting their cost advantage with scale gains when sailing in regular routes, carrying regular LNG volumes. In the last years, the markets that have mostly grown are the ones of short and medium-terms, which use standard LNG Carrier Ships from 145,000 m^3 to 160,000 m^3.

2.3 Regasification

Regasification is the final process in the LNG chain, when it is unloaded from the LNG Carrier Ships, reheated and again transformed into gas. By the end of the 1990s, 75% of the LNG regasification capacity was found in Asia, mainly in Japan, where the LNG industry developed early owing to the limited access to gas pipelines and the lack of other local natural resources. However, in the last years, LNG imports have grown especially in Europe and in North America.

Since 2000, most of the LNG new import terminals projects were built in Europe, where the dropping North Sea reserves, the high production costs and liberalization of the power and natural gas markets generated new opportunities for LNG. There are now more than 100 new regasification terminals or expansion projects to start operating in the next years in the world, as a response to the increase in the demand for LNG, and at least 70 of these are in Europe or North America. According to Table 5, only 16% of the new regasification projects will be built in Asia.

	Europe	Asia	America (Atlantic)	America (Pacific)
2000	15%	75%	11%	0%
2000 to 2007	45%	37%	19%	0%
2007 -	30%	16%	47%	8%

Table 5. Regasification capacity in different periods

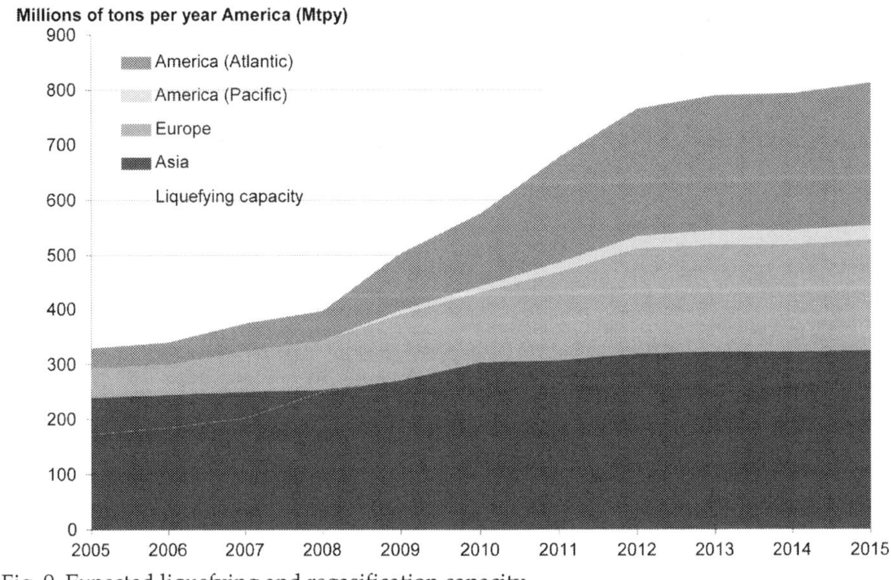

Fig. 9. Expected liquefying and regasification capacity

Figure 9 shows that the new terminals will substantially increase the global regasification capacity that will keep growing until it reaches 800 mtpa in 2015. The graph also shows how the regasification capacity operates with low use rate; it can absorb practically twice the total LNG produced and is growing faster than the liquefying capacity. This disparity reflects a smaller cost as compared to the other segments in its production chain.

The costs for building regasification terminals are very specific for each project, depending on several factors, such as location, cost of the land, type and number of storage tanks, harbor infrastructure, among others. Thus, the costs for building these terminals present great variation and do not follow any trend along the years, as shown in Figure 10.

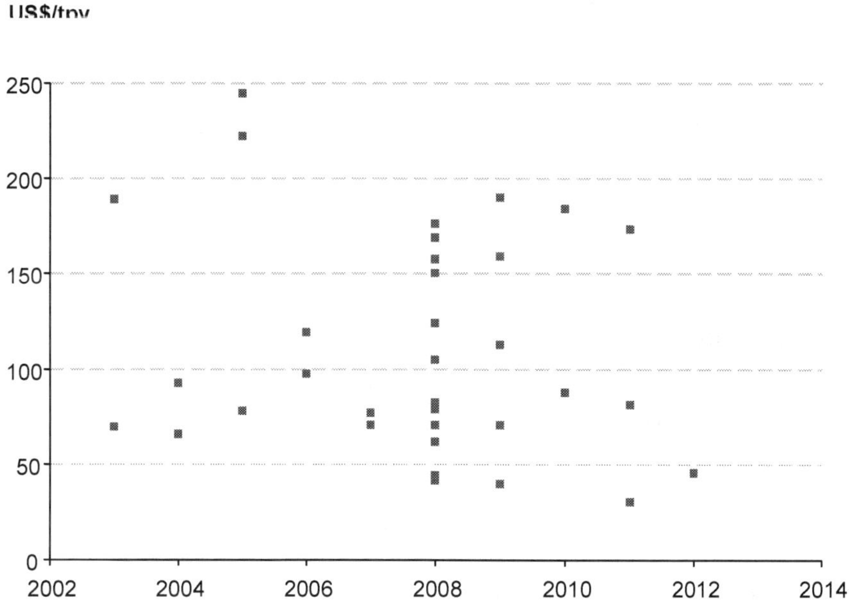

Fig. 10. Specific cost for building regasification terminals

2.4 Investments at each stage of the LNG production chain

Projects in the LNG area require heavy investments and the capital invested at each stage of the production chain may vary significantly. The information found in the media is usually sparse; hence, concrete data and specific details of contracts are difficult to obtain. It is possible, however, to estimate an average of the costs per unit at each stage of the chain in a period.

Figure 11 presents how the costs are divided at each stage, from natural gas prospection to LNG regasification for the projects that started operating recently and for those that are to start operating by 2011. From 2002 to 2007, the gas liquefying process showed to be more expensive, accounting for 44% of the total cost. This fact represents an expressive increase in the chain total cost for the next years, seeing the growing costs of the liquefying plants. The graph also shows that in the period analyzed, the average costs of gas exploration did not suffer great alterations and regasification has a smaller impact on the LNG chain total cost.

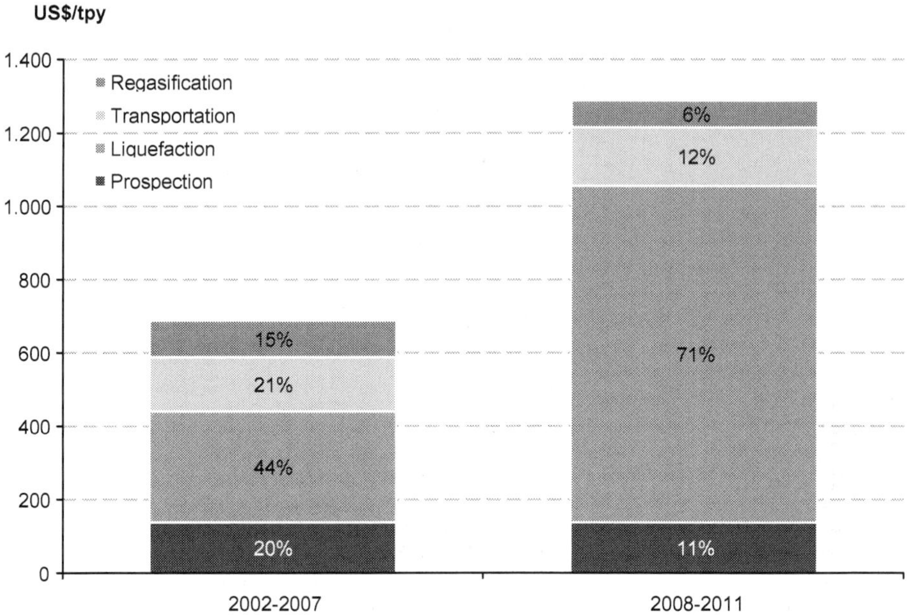

Fig. 11. Average cost per unit of the production chain in different periods

3. Perspectives for LNG in Brazil

3.1 Natural Gas Context

Today the gas volume available is of about 65.5 MMm^3 a day; 30 MMm^3 come from Bolivia, 21.5 MMm^3 from the Southeast and 14 MMm^3 the Northeast. The gas produced in Brazil comes mainly from off-shore platforms (61%), from depths varying between zero and 300m (33%), 1500m (62%) and only a small parcel from depths greater than 1500m (5%). Brazil presents three main and independent production and transportation networks, depicted in Figure 12: one in the North region, one in Northeast region and one in the South cone (S-SE-CO). The South cone network connects Southeast Center-West and South States, and the Northeast network connects the Northeast States from Ceará to Bahia. These two networks will be connected by Gasene, foreseen for the first half of 2010.

The Petrobras main expansion programs to supply natural gas are Plangas, mainly destined to the industrial market in the Southeast and TC - Term of Agreement - a program based mainly on LNG and guided towards thermoelectric consumption.

Fig. 12. Natural Gas Networks in Brazil

The expansion of supply in the short-term occurs mainly with the natural gas from the Espírito Santo Basin and in medium and long-term, the gas used comes from the Santos Basin.

3.2 Recent History

Even though natural gas thermal generation has been stimulated by the government since 2000, by the PPT (Priority Program for Thermoelectric Plants), aiming at the consumption of large volumes for paying the take-or-pay (contracting modality in which, in case the consumption is smaller than the volume contracted, the contracted value is paid; in case it is higher, the measured volume is paid) of the Bolivian gas, the investments foreseen did not materialize. This occurred owing to indefinition in the trading rules and to uncertainties concerning the natural gas price or even the very conception of the Plan.

As a result, Petrobras stimulated the consumption of natural gas in the industrial and transportation sectors. This policy worked very well and, coupled with the rise in petroleum byproduct prices and environmental constraints, resulted in a growth from 10 to 20% per year in gas consumption in Brazil. This growth even affected the operations of natural gas thermoelectric plants which, due to the lack of fuel, started failing to deliver the whole power expected when called to operate.

In April 2005, there was a reduction of 2300 MW on average in the importation coming from Argentina and the Uruguaiana Thermoelectric Plant for unavailability of fuel to meet the demand for power generation. In the late 2006, the National System Operator - ONS conducted availability tests of natural gas thermoelectric plants, which resulted in a cut of 2700 MW on average in the power supply in the South and Southeast regions. Besides, in 2007 Bolivia interrupted the gas supply for the Cuiabá Thermoelectric Plant, resulting in an additional cut of 200 MW on average.

These cuts in availability of power generation by the thermoelectric plants for lack of natural gas caused a structural unbalance in the power supply and demand, leading to a greater dependence on the electric system in relation to the hydrologic regime. As a consequence, Petrobras was forced to sign a Term of Agreement (TC) with ANEEL. The TC was signed to cover the present deficit and is a fixed commitment, subject to penalties. Besides, it contemplates LNG implementation.

3.2.1 Term of Agreement

Insufficient investments to follow the demand for natural gas in the industry and the growing yield of thermoelectric plants composed the present scenario of gas rationing risk. As early as 2005, a gas supply deficit from 20 to 30 MMm^3 per day was verified; the crisis was not anticipated only due to a smaller consumption of gas by the thermoelectric plants (7.1 MMm^3 a day).

In order to have knowledge on the real consumption and generation capacity of the natural gas thermoelectric plants, ANEEL asked the National System Operator (ONS) to conduct availability tests in the 2004 to 2006 thermal generation. In 2004, all the natural gas thermoelectric plants in the Northeast participating in the Thermoelectric Priority Program (PPT) were summoned. The yield registered by ONS observed availability lower than that verified by ANEEL in about 757 MW on average. This deficit led to the signature of the Reserve Recomposition Agreement approved by the ANEEL instruction 1.090/2004.

In December 2006, ANEEL asked ONS to test the natural gas thermoelectric plants. This time the thermoelectric plants of the South-Southeast were called, and effective availability also inferior to that verified by ANEEL was observed, resulting in a cut of reserves of about 2,700 MW on average.

As from these tests results, the Term of Agreement (TC) was signed between Petrobras and ANEEL. In it, Petrobras committed to make natural gas available for thermoelectric plants in the South, Southeast and Northeast, according to a previously established schedule, to be concluded by 2011. Table 6 presents the schedule in which the infrastructure events associated to the evolution in availability of simultaneous generation in the NG thermoelectric plants considered in the TC are listed.

Nº	Subsystem	Events	Mark
1	SE/CW	Increase in the ES production and gas pipeline	1st half 2008
2	SE/CW	LNG in SE (Rio de Janeiro)	1st half 2009
3	SE/CW	GASBEL	2nd half 2009
4	NE	Backup Hiring	2nd half 2007
5	NE	LNG in NE (Pecém)	April 2008
6	NE	Interconnection works (Southern NE and Northern NE)	2nd half 2008
7	NE	GASENE	1st half 2009
8	S	Additional compression in the Paulinia-Araucaria gas pipeline	1st half 2008
9	S-SE/CW	NG from South to Southeast	1st half 2008

Table 6. Schedule of events in the TC

Nevertheless, in 2007 there was gas unavailability for the thermoelectric plants which resulted in an ANEEL penalty of R$ 84 million to Petrobras and in the temporary reduction in the gas supply for the utilities in October 2007. The utilities most affected by the reduction were CEG in the State of Rio de Janeiro and COMGAS in the State of São Paulo; the supply was re-established by a temporary restraining order.
Owing to the risk of natural gas scarcity, in 2006, Petrobras announced the Natural Gas Production Anticipation Plan (Plangas). The plan includes expansion projects in all the natural gas supply stages, from production distribution by gas pipelines.

3.2.2 Plangas

Before approaching the expansion plan schedule, it is worth explaining that the natural gas production has three main stages:

- Exploration and Production (E&P): stage at which the removal of the natural gas from the reservoirs is considered;
- Processing: gas treatment in the so-called UPNG (Natural Gas Processing Units) where liquids and impurities are removed so as to deliver the gas within the composition standards provided by the Brazilian Petroleum Agency (ANP) law;
- Transportation: stage at which the natural gas transportation by pipelines is considered.

Plangas was divided into two stages. At the first stage, natural gas availability in the Southeast had an increment of 24.2 MMm^3 per day. At the second stage, more than 15 MMm^3 per day were made available, totaling an increment of 39.2 MMm^3 per day for the Plangas.
The discovery of new fields (São Mateus, Juruá-Araracanga and Jaraqui) in the North region, will meet the increase in demand and the fall in the Urucu field production. This gas has been available for Manaus from 2009, with the completion of the works in the Coari-Manaus gas pipeline.
In 2008, the arrival of LNG in Pecém in the Northeast accounted for a large part of the addition in gas supply. The gas pipelines network in this region is also in expansion and was recently connected to the Southeast network through the Cacimbas-Catu gas pipeline, a

GASENE branch. The recent increase in gas availability for the Northeast was caused by the beginning of operations in the Manati field, in Bahia.

Still viewing the expansion in the Brazilian natural gas production, Petrobras conducts investments in the Santos Basin. The short-term production expectations in the Santos Basin are 30 MMm^3 per day of natural gas, with excellent perspectives for continuous growth, mainly after the finding of the of the Tupi and Júpiter mega-fields with an estimated reserve of 5 to 8 billion barrels of petroleum equivalent and, more recently, the announcement of the finding of field BM-S-9, known as Carioca, with an estimated reserve of 33 billion barrels of petroleum equivalent.

Besides the distance of these new fields from the Brazilian coast, another great difficulty of these recent findings is the thickness of the water blade and well depth - the sum of the parts results in total depths of over 5,000 meters. This is because the E&P cost considerably increases with the depth of the fields due to the need of using more resistant materials and more adequate to the pre-salt environment, as presented in Figure 13. Hence, the exploration of these wells is only viable and attractive with the increase in the petroleum barrel price.

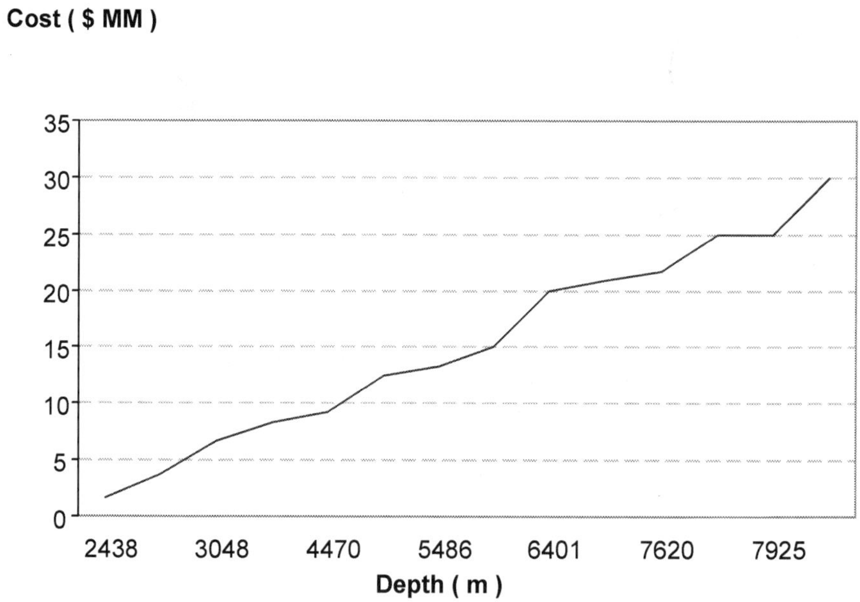

Fig. 13. E&P Cost x Depth (Source: British Petroleum)

4. Market Opportunity for LNG in Brazil

4.1 The need of flexibility for the Brazilian natural gas market and its relation with LNG

Albeit incipient, the Brazilian natural gas industry needs great flexibility. In the 1990s, the conduction of liberalizing reforms changed the economic context in Brazil, causing the industrial organization and the contracts traditionally used in the incipient stage of the natural gas industry not be the best instruments to reduce the investments risks in relation to the infrastructure of this industry. The present context derives from different factors that transformed the basic conditions of the Brazilian natural gas industry, such as: liberalization of the prices of fuels competing with natural gas; exhaustion of the developmentist model of traditional financing by the public sector and foreign credits to state companies; partial privatization of power companies; formation of large international groups capable of competing in the world market, as from the privatization process and the introduction of competition in the power industry and natural gas sector in the developed countries; regional energy integration, also in the natural gas industry; technological evolution, growing technological and business convergence in the power sector and in the natural gas industry.

In the present economic context, the prices of fuels competing with natural gas are given in a liberalized market environment. Therefore, these prices present greater volatility, varying according to the international market, climate conditions and the demand in the Brazilian power sector. As a consequence, the value of natural gas has undergone more changes more often, and a greater flexibility is necessary in the natural gas industry for the gas price to vary, aiming to keep its competitivity with the competing fuels.

An important factor that also contributed to the need of flexibility in the Brazilian natural gas industry is the one related to its power sector. Power generation in Brazil is basically conducted by the hydropower plants, generating about 80% of the Brazilian electric power. The hydropower plants have an installed capacity for generating 77.4 GW, which corresponds to 70.2% of the total power generated. In turn, the thermoelectric plants have an installed capacity of 24.7 GW, 11.8 GW of which from gas thermoelectric plants. This respectively represents 22.4% and 10.7% of the whole supply of domestic capacity for power generation in the country.

Besides the installed capacity, the Brazilian hydropower plants also have large reservoirs, the water storage capacity of which is among the greatest in the world. This great storage capacity allows stocking water, increasing hydropower plants generation capacity and power is generated at very low costs for practically the whole of its market in abundant rain periods. Thanks to the reservoirs system, to the country geographical size and to the interconnection of the Brazilian power system, even if one region in the country is undergoing a period of low rain, another region with abundant rains and full reservoirs may see to the power demand of the of the "dry" region, thus creating a compensation mechanism among the hydropower plants in Brazil and minimizing the risk caused by the lack of rains. As a consequence of this characteristic of the Brazilian power sector, the economic value of natural gas destined to power generation in abundant rain periods is drastically reduced, and may fall to zero.

Despite the important role of the hydropower plants at the base of the Brazilian power generation, the thermoelectric plants have a complementary role, yet fundamental, of guaranteeing a greater security to the national generation system, diversifying the energy source. The yield of these thermoelectric plants depends on variations in the rain regime and on demand peaks. This way, traditional instruments used in the natural gas industry, such as long-term contracts with take-or-pay clauses, would not be adequate to the natural gas thermoelectric plants in Brazil, which need greater flexibility.

The Brazilian natural gas industry does not have flexibility, despite the great need. On the demand side, only as from 2007 did Petrobras provide the possibility of signing interruptible natural gas contracts, yet there is not a secondary market for this input. Predominantly the gas supply contracts used are long-term and have take-or-pay clauses. On the supply side, the existing flexibility is very small, owing to the specificities of the Brazilian natural gas industry, such as: the lack of natural gas storage capacity out of the transportation network; the fact that 75% of the domestic production of this gas is associated, making a variation in gas production aiming at greater flexibility also affect petroleum production; since practically the whole domestic natural gas production derives from off-shore reservoirs there is, therefore, a high development opportunity cost of these gas fields; the on-shore Brazilian natural gas production lies basically in the isolated system of Amazon, and cannot meet the needs of flexibility in the Northeast and Centro-Southeast-South regions and, finally, since Brazil already uses practically the whole total transportation capacity of the Gasbol, there is little surplus capacity to conduct an increase in supply to meet the need of flexibility of the Brazilian natural gas industry.
In this context, opportunities for LNG are identified in in Brazil, in the sense of diversifying the power supply sources while allowing greater supply flexibility for the natural gas industry and for the electric power sector

4.2 The LNG importation project by Petrobras

Since the late 1990s, LNG has been the object of studies for Petrobras, as an alternative to complement the natural gas supply in Brazil. In 2004, Petrobras started studies to import this input flexibly, in order to adapt the supply to the volatile demand of the gas thermoelectric plants. This natural gas import alternative gained momentum after the nationalization of natural gas in Bolivia in May 2006, when a greater uncertainty scenario concerning the future supply of this gas was generated. Therefore, due to the expected growth in domestic demand for natural gas and the risk of the country not being able to meet it with greater flexibility, the Ministry of Mines and Energy (MME), together with the Petrobras projects, in its Resolution nº 4 of November 21, 2006, established the option of using LNG as a way of meeting such needs, as presented below:
"Article 1 - Declaring a priority and an emergence the implementation of Liquefied Natural Gas – LNG Projects, consisting in the importation of natural gas in cryogenic form, storage and regasification, as well as the necessary infrastructure, aiming to:

> I - Ensure the availability of natural gas for the domestic market viewing to prioritize the supply to thermoelectric plants;
> II - facilitate the adjustment in the natural gas supply to the characteristics of the domestic market through flexible supply;
> III - mitigate risks of failure in the natural gas supply due to hazards;
> IV - diversify the imported natural gas supply sources; and
> V - reduce the implementation deadline of the Natural Gas Supply Projects.

Article 2 - Aiming at the full conduction of the activities provided in Article 1, the implementation of mechanisms for abiding by this Resolution is assured, as well as the articulation of the institutional means to overcome possible problems in the implementation of LNG projects."

As can be seen in Figure 14 below, Petrobras expects the Brazilian natural gas demand to grow nearly 90% between 2007 and 2012 (a greater value than that considered by EPE in the Decennial Plan 2007/2016). To meet this growth in demand, the state company intends to increase its national production to nearly 73 million m³/day, use the maximum capacity of Gasbol and import 31.1 million m³/day of LNG. The main reasons that led Petrobras to opt for the use of LNG as an instrument to complement the Brazilian natural gas supply are its smaller implementation time and smaller fixed cost as related to other options, such as the development of new natural gas fields and the construction of new gas pipelines for importing this gas; the diversification in the natural gas supply; and the possibility of purchasing LNG through short or long-term, fixed or flexible contracts.

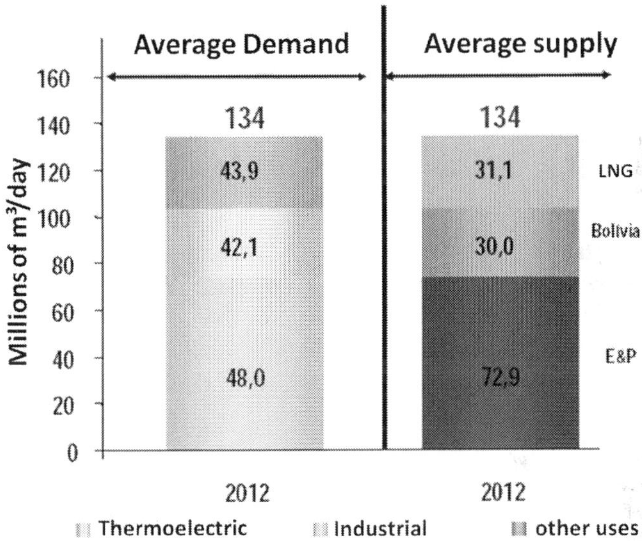

Fig. 14. Expectation of Natural Gas Supply and Demand in Brazil in 2012.

In the late 2007, the Petrobras LNG importation project foresaw investments in infrastructure of about US$ 152 million, for building two flexible LNG regasification terminals, located in the Guanabara Bay (US$ 112 million), in Rio de Janeiro and in Pecém, in Ceará (US$ 40 million). Besides these two terminals, Petrobras also studied four more projects for LNG flexible terminals, located in Suape (PE), São Francisco (SC), Aratu (BA) and São Luis (MA).

The Pecém terminal was inaugurated in August 20, 2008 at a total cost of R$ 380 million, which includes the pier adaptation and the construction of a 22.5 km gas pipeline. The terminal, operated by Transpetro - a Petrobras logistics company - has the capacity to regasify 7 million cubic meters a day, the equivalent to about half of the present consumption of natural gas guided towards the Brazilian thermal market and a 129,000 m³ storage capacity. The Guanabara Bay terminal, in turn, has the capacity to regasify 14 million cubic meters a day and store 138,000 m³. The regasification at the two terminals is

conducted in LNG Carrier Ships, which are used for storage, too. Petrobras contracted the Golar Spirit (Pecém) and Golar Winter (Guanabara Bay) vessels of the Norwegian company Golar LNG at a total cost of US$ 900 million for 10 years, already including operational expenses.
In order to obtain the LNG supply, Petrobras signed a Master Agreement (intent agreement) for importing this commodity with the companies Nigerian LNG, from Nigeria, and Sonatrach, from Algeria. This agreement foresees purchases in the LNG market spot without fixed volume and price based on the natural gas quotation at Henry Hub[2] at the moment of purchase. Petrobras also signed a confidentiality agreement with Oman LNG for negotiating a potential LNG supply, besides negotiating with other vendors. According to Petrobras, the travelling time for LNG to reach Brazil, after the purchase is conducted in the market spot, would be of at most 18 days, depending on the origin. Table 7Table presents the estimated travelling time for LNG to arrive in Brazil.

Destination (simple trip - 19 knots)	Nigeria (Bonny)	Algeria (Skikda)	Algeria (Arzew)	Trinidad & Tobago (Point Fortin)	Qatar (Ras Laffan)
Baia de Guanabara (RJ)	7d 10h	10d 12h	9d 18h	6d 20h	17d 21h
Pecém (CE)	6d 4h	7d 23h	7d 5h	3d 15h	17d 12h

Table 7. travelling time for LNG to reach Brazil

Petrobras means to import LNG so that there is a flexible natural gas supply source, directed to meet mainly the thermoelectric plants demand. It intends to purchase LNG in the market spot and pass it on to the thermoelectric plants according to their needs. The hiring modality of this natural gas with the thermoelectric plants will be of "preferential supply". In this new modality, the consumer (in this case, thermoelectric plants) has the prerogative of interrupting supply, which is interruptible only by the client, being the supplier obliged to provide the supply of gas available when demanded. The gas price in this contract will be composed of two parcels: one concerning the remuneration of investments in infrastructure of the gas transportation (capacity) and the other concerning energy, which will depend on the value of natural gas at Henry Hub. Moreover, the contract will provide the antecedence and the nomination conditions of the gas.
The yield of the thermoelectric plants is determined by the National Power System Operator (ONS), which seeks to optimize the Brazilian power generation, so as to minimize the system operation cost, taking into consideration, among other variables, the level of water storage in the reservoirs of the interlinked system, the occurrence of rains, the fuel costs and the demand for power. Hence, the thermoelectric plants only operate when there is not enough water for the hydropower plants or when it is convenient to reduce hydropower production to save the water in the reservoirs. It is worth noting that the period in which rains are less abundant in Brazil, causing a lower water level in the reservoirs, goes from May to October, which corresponds with the period in which the cold countries of the North hemisphere are experiencing their hottest seasons. Thus, during the dry period in Brazil, the world demand for LNG tends to be reduced, also resulting in a lower price of this

[2] Point in the transportation network of the American State of Louisiana, where there is an interconnection of 9 interstate and 4 inner state gas ducts. The prices negotiated at this point are a reference for the spot and future market prices.

commodity at Henry Hub. Therefore, Petrobras will probably conduct most of its LNG purchases in the lower prices period, reducing the cost of generating power with LNG.
It also is worth stressing that, according to Administrative Rule n° 253 of September 2007, of the Ministry of Mines and Energy, the ONS will give instruction notice to the thermoelectric plants that use re-gasified natural gas, two months prior to its effective instruction. This deadline respects the LNG supply logistics, allowing Petrobras to import LNG in market spot, with enough time to meet the demand of the thermoelectric plants.

4.3 Risks associated to market spot

As seen in the previous sections, the natural gas industry in Brazil has little flexibility and Petrobras, for some years, has been studying LNG projects aiming to meet the growing national demand for gas, and also allowing a greater flexibility to see to the fluctuations of this demand, especially concerning thermoelectric generation. As stated before, the "preferential" contract modality will allow Petrobras to offer the thermoelectric plants the flexibility obtained in the LNG market spot.
Nonetheless, although the LNG market spot offers a flexibilized supply of natural gas to Petrobras, it also presents greater price risks, once the spot contracts of the Atlantic basin, in which Brazil lies, are based on the Henry Hub quotation, which is highly volatile, as can be seen in Figure 15 below.

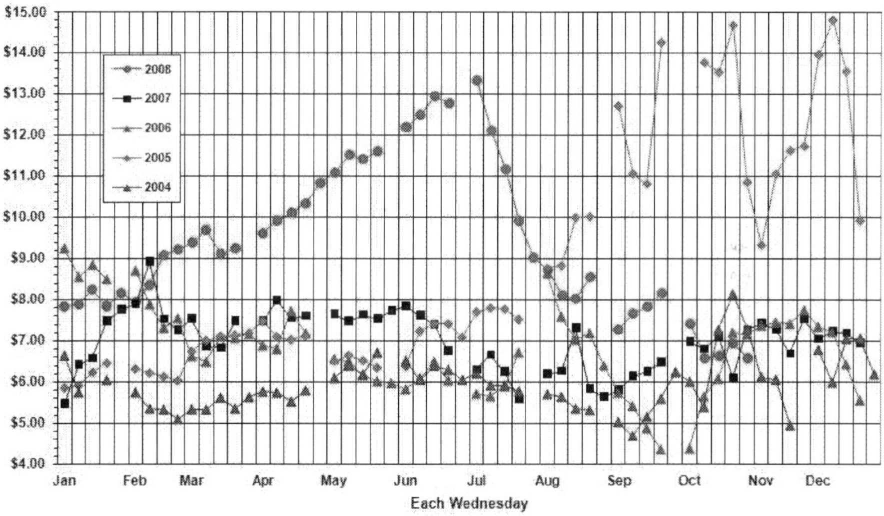

Fig. 15. Spot price of natural gas in Henry Hub in US$/MMBtu

Such a fact generates uncertainty in relation to the future price of LNG paid by Petrobras, making the natural gas Brazilian industry be influenced by events in the American market. Besides, this uncertainty may influence investments in the national gas industry which uses LNG as a basic input, due to the difficulty in foreseeing future prices of this input and, consequently, its use economic viability.

The purchase of LNG only in the market spot also has risks concerning the volume available, generating uncertainties as to its future supply. Albeit growing, the LNG market spot is still incipient, accounting for only 13% of the total. Thus, if there are any contingency in LNG supply, its sellers prioritize meeting the obligations provided in their fixed and long-term contracts, leaving the market spot aside. A way of mitigating this risk would be storing LNG, so as to use it in a high-price period and purchasing when prices were lower. However, Brazil does not count on storage infrastructure out of the transportation network, already reduced.

Today, as a way of reducing the uncertainty generated by the price and volume risks of purchasing LNG in the market spot, the Brazilian natural gas industry counts on the possibility of conducting a combination between the purchase of this commodity by means of spot contracts and of strict long-term contracts. Hence, the guarantee of supplying LNG with a price already determined in a strict long-term contract would reduce the uncertainty generated by the risks of purchasing LNG in the market spot. In turn, the LNG purchases in the market spot would reduce the uncertainty deriving from a strict long-term contract.

5. About GTL Production with Natural Gas from Bolivia

Bolivia intends to industrialize natural gas in different ways; one of them is the conversion to liquid process (especially diesel), also known as Gas To Liquids (GTL) process, which is based on obtaining syngas by the Fischer Tropsch method (F-T). The conversion efficiency is of the order of 60% but it is foreseen to reach up to 70%.

Today presenting a small energy industry based on natural gas and practically no project of massive use of this resource, the implementation of this project and other large-scale ones is a huge challenge for Bolivia allowing, by the implementation of a GTL-FT project, generating added value for the natural gas reserves and allowing access to scale economies.

5.1 Technical Specificities of the Bolivian Natural Gas

The major characteristics of natural gas in Bolivia are non-associated gas and very rich in methane, making the exploitation and use of this resource very attractive. Table 8 details the Bolivian natural gas composition.

Main components	Chemical formula	Percentile in volume (*) [%]
Methane	CH_4	89.10
Ethane	C_2H_6	5.83
Propane	C_3H_8	1.88
Butanes	C_4H_{10}	0.74
Pentates	C_5H_{12}	0.23
Hexanes	C_6H_{14}	0.11

Table 8. Chemical composition of the Bolivian natural gas

5.2 Natural Gas Petrochemistry

The hydrocarbons that come with methane in Natural Gas, such as ethane, propane and butane (n-butane and iso-butane), could be applied in byproduct production, by means of a traditional petrochemical plant, because this industry uses, among others, the same hydrocarbons above; however, those are obtained in the extraction of crude oil (the condensed propane and butane are generically named LPG, "Liquefied Petroleum Gas"), which is distributed in steel containers for residential consumption. Figure 3 presents a summarized diagram with the processes and some of the products associated to the traditional petrochemical industry of crude oil refinement.

Particularly, the area within the dotted line in Figure 16 is associated to petrochemistry (gas-chemistry), based specially on the transformation of ethane, propane and butane deriving from crude oil refinement, in a process called "steam cracking". This process allows obtaining oils, such as ethylene and propylene, from which it is possible to get, for example, polypropylene and polyethylene, highly used and known plastic materials.

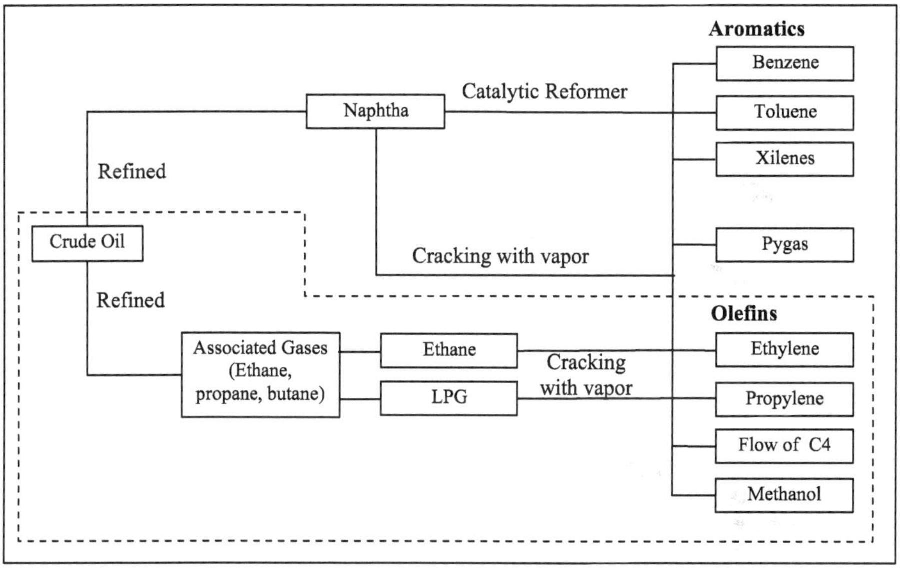

Fig. 16. Traditional Petrochemical Industry

In a similar way, ethane, propane and butane, companions of methane found in the Bolivian NG, can be applied in traditional petrochemical processes, here named gas-chemistry. However, since the companions mentioned are found in low quantities in Bolivian most important cas reserves, it can be concluded that only the massive exportation of methane will allow obtaining sufficient amounts of the "liquids of natural gas" to generate a gas-chemical industry in the country.

In summary, the creation of a gas-chemical industry in Bolivia depends on the LNG project, since this is a methane massive exportation project and, with that, it will be possible to count on great amounts of liquids from NG and then develop a Bolivian gas-chemical industry.

5.3 The CH_4 Industrialization

Methane industrialization, as well as the petrochemical (gas-chemical) industry and energy strategy should be considered fundamental for the Bolivian industrialization. As the Bolivian NG, in the most important gas fields, is mostly constituted of methane, it is important to talk about the methane industrialization, and, on this basis, the other components that come with it (GTL, GTO, GTM, etc.).

The first stage in the industrialization of methane is to obtain the syngas. The synthesis gas is a mixture of carbon monoxide and hydrogen, obtained from chemical reactions of methane with substances easily found in nature, such as carbon dioxide, oxygen and water. As its name shows, the synthesis gas is the basis to synthesize many compounds that are both economically and industrially important. Depending on the desired compounds, the synthesis gas is prepared with different proportions of carbon monoxide and hydrogen, as shown in Table 9.

Reacting Compounds	Chemical Reactions (under adequate conditions of pressure and temperature)	Proportion (mol to mol) of carbon monoxide and hydrogen in syngas
Methane with carbon dioxide	$CH_4 + CO_2 \rightarrow 2CO + 2H_2$	1:1
Methane with air oxygen	$2CH_4 + O_2 \rightarrow 2CO + 4H_2$	1:2
Methane with water	$CH_4 + H_2O \rightarrow CO + 3H_2$	1:3
As an example, to obtain a synthesis gas in which the carbon monoxide and hydrogen are in a proportion from 1 to 2, respectively, a partial combustion of the methane with the oxygen of the air is made, reaction additionally generates considerable amounts of thermal energy.		

Table 9. Methane reactions in order to form synthesis gas

5.4 Synthesis Gas as Vector for secondary Fuels

From the reaction of the syngas (synthesis gas) components, using different catalysts, many products can be made (see figure 17); among the most important products, depending on the proportion of carbon monoxide/hydrogen in the syngas, it is possible to have:

- LPG, petrol, diesel, jet fuel and ultra-pure paraffin, all of those with the Fischer-Tropsch process. The Natural Gas transformation into the products above, all of them liquid, is denominated GTL (Gas to Liquids) process.
- Hydrogen, denominated the Fuel of the Future.
- Ammonia, basis of the fertilizing industry, which is the product of the reaction of the nitrogen in the air with the hydrogen from methane.
- Dimethyl ether, a diesel and LPG substitute, which can also be used in the electricity industry.
- Methanol, from which it is possible to synthesize olefins, such as ethylene and propylene, and, from these, the products in Figure 17

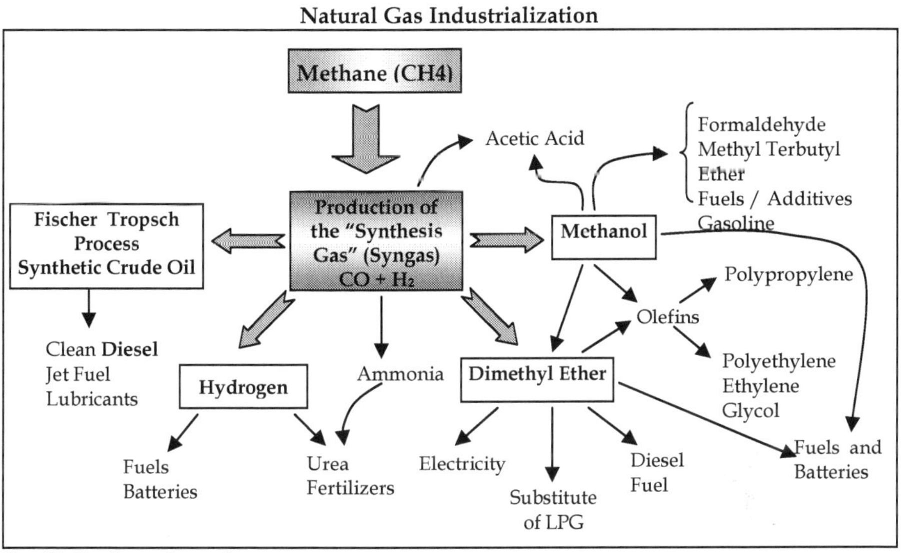

Fig. 17. Products from the syngas

Fundamentally, it is necessary to consider these general aspects:

- The technology;
- The present and future markets;
- The possibility of getting to these markets;
- The amount of investments;
- The advantages;
- And the specifically Bolivian aspects, such as:
- Benefits to the country and areas of production;
- Mediterranean Climate;
- Legal security.

Considering all the general and specific aspects mentioned above, it is necessary to carefully decide about the best industrialization route or routes to be taken.

5.5 GTL Production Factors

A project of Gas to Liquids (GTL) production by the Fischer-Tropsch process - GTL-FT - consists in obtaining syngas from the partial combustion of methane with oxygen in oxygen-poor stoichiometric proportion. The syngas obtained can thus be transformed into liquid fuel of massive use, such as gasoline, diesel and jet fuel from different catalyzers and syngas reaction times.

The basic GTL-FT process starts with the methane separated from its liquid companions (dry). Compounds such as ethane, propane, butanes and pentanes, can be industrialized independently of the GTL-FT project, originating polymers, and synthetic oils.

The F-T process is a multiple- step process, with great power consumption, which separates the natural gas molecules, predominantly methane, joins them again to produce larger

molecules. The first step requires the incoming of oxygen (O_2) separated from air. Oxygen is blown into a reactor to extract the methane hydrogen (CH_4) atoms. The products are synthetic hydrogen (H_2) gases and carbon monoxide (CO), called syngas (Figure 18).
The second step uses a catalyzer to recombine hydrogen and carbon monoxide, leading to liquid hydrocarbons. In the last stage, liquid hydrocarbons are converted and broken down into products that can be immediately used or be mixed to other products. The most widely known product is the extremely pure diesel, also known as ***gas oil***. The diesel obtained from the F-T process, as opposed to that deriving from petroleum distillation, has practically null sulfur oxide and nitrogen oxide content, virtually does not present aromatics, its combustion produces little or no particulate matter emission, and it has a high rate of cetane. Kerosene, ethanol and dimetil ether (DME) can also be produced. Another product of the reaction is naphta, which has high paraffin content. Waxes deriving from the GTL process may be pure enough to be used in the cosmetic industry and that of canned food.

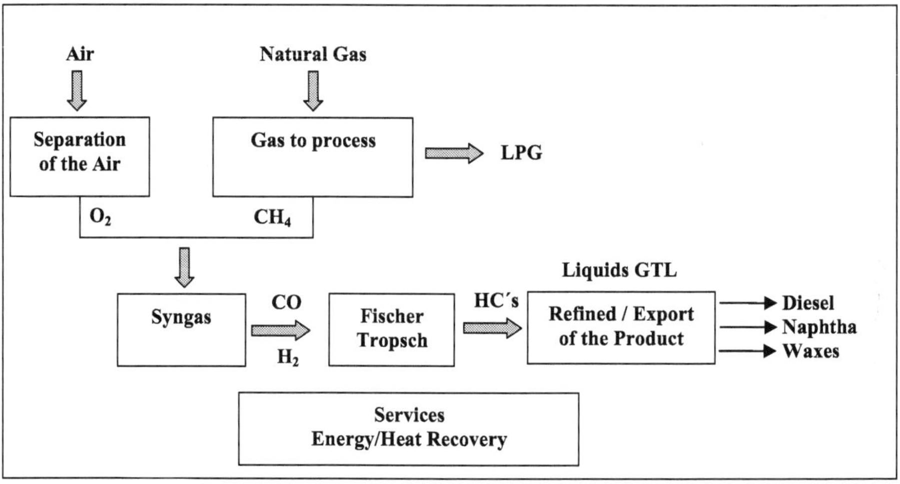

Fig. 18. Conversion of natural gas into liquid fuels
Source: Authors' elaboration, 2006.

For all this, it may be concluded that the GTL-FT technology has to be developed on industrial scale, for the range of products that may be obtained and for the amount of existing gas. In this sense, Bolivia may produce investment costs and commercial production scenarios that allow having a general view of the project and analyze its benefits.

6. Market for products of a GTL-FT Industry

If Bolivia were to process 30 million m^3 of Natural Gas daily (the same amount agreed with Brazil and the same expected to be exported to the USA), approximately 100 thousand bpd of mostly liquid hydrocarbons, gasoline and diesel would be produced. This amount, as Table 10 shows, is minimal if compared to the worldwide demand.

Products	Estimated Demand in 2005* (million bpd)	Estimated Demand in 2010* (million bpd)
Naphtha	5.2	5.7
Gasoline	20.9	22.3
Kerosene	6.6	7.7
Diesel	22.2	25.1
Fuel Oil	9.2	9.1
Others	8.8	9.6

Table 10. Projected Hydrocarbons Demand

In Bolivia, there is an internal market for 14 thousand bpd, from which 5 thousand bpd are imported. Considering the price of US$35.00 per diesel barrel, the money spent on this activity is US$64 million per year. There are two very tempting markets available to Bolivia for selling diesel: Chile and Brazil.

6.1 Pacific Market – CHILE

Chile would be a great buying potential for Bolivia's diesel-GTL since it consumes 250 thousand bpd of oil and their byproducts and 95% of what it consumes come from imports. Due to the high environmental degree of contamination it presents, Chile would become a guaranteed purchaser of GTL diesel. Figure 19 shows the distribution of the oil import in Chile according to origin.

In order to export eco-diesel (diesel-GTL) to Chile, also a South-American Country, and countries on the other side of the Pacific Ocean, the port of Arica (normally used for Bolivian exportations) would be used for exporting to other countries, such as China.

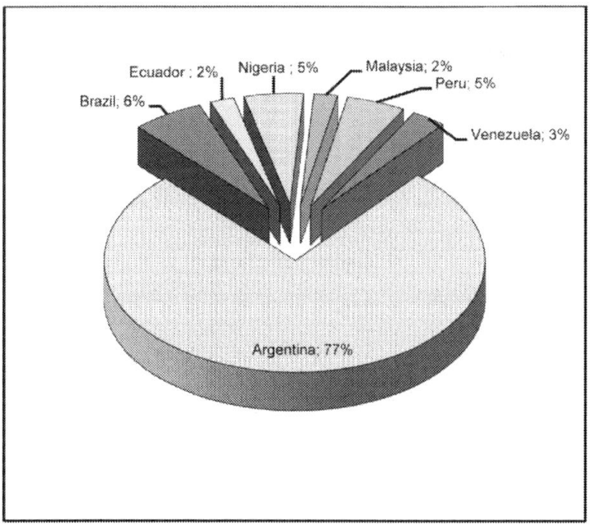

Fig. 19. Oil importation for Chile in 2001

6.2 Atlantic Market – BRAZIL

Brazil is another potential purchaser of Bolivian eco-diesel because its refinement capacity will be insufficient, according to the National Oil Agency (ANP).

The diesel consumed in the States of Rondônia, Mato Grosso do Sul, Mato Grosso and Goiás is mainly obtained from the neighboring State of São Paulo, though it is possible to obtain fuel from Bolivia (diesel-GTL). As seen in Figure 20, the eco-diesel produced in Bolivia would be more economic for the mentioned states, better still if the project is jointly implemented by Brazil and Bolivia.

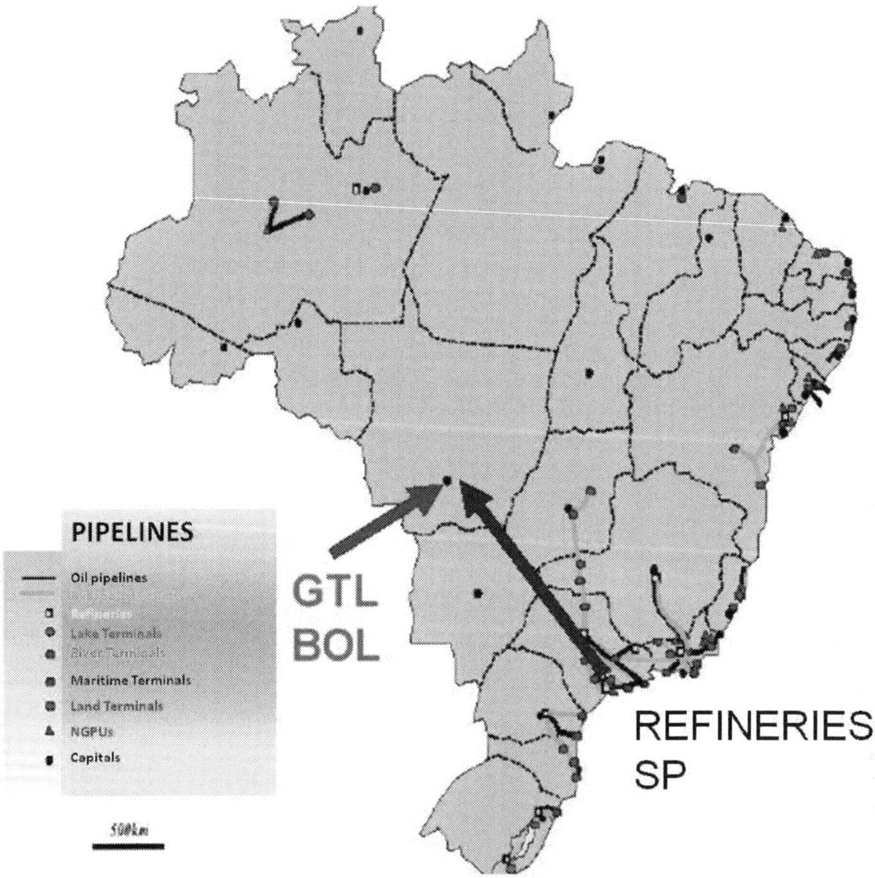

Fig. 20. Eco-diesel from Bolivia to Brazil (Atlantic side)

Brazil needs to invest great amounts in refineries (Figure 21) to avoid importing diesel for diesel engines. The Brazilian state oil enterprise, Petrobras, has gas reserves without market in Bolivia, which makes it possible to manufacture eco-diesel from the gas reserves and export it to Brazil. It is worth mentioning that both countries would gain from that.

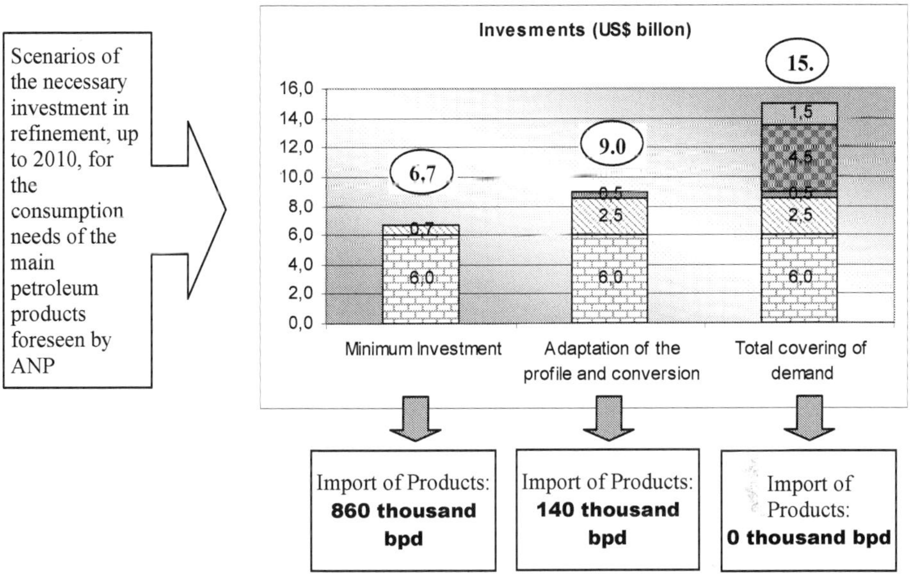

Fig. 21. Oil byproducts 2010 ANP scenarios

6.3 Market for liquid hydrocarbons

The price and the quality of GTL-FT products will determine, as for any other product, their capacity to gain access and to compete in the world hydrocarbon market in favorable conditions. Their nearest competitors are the products of the oil distillation process. In a GTL-FT diesel projected production of 100 thousand bpd, only 0.3% would be covered by the international market (Table 10).

In Figure 22, for example, there is a comparison between the qualities of GTL diesel and the conventional one, also referred to as 'dirty' diesel. Compared to the conventional diesel, GTL-FT diesel contains much reduced amounts of aromatic hydrocarbons and sulfur, and for that reason it emits reduced amounts of detrimental compounds into the atmosphere. For the same reason, GTL-FT diesel fully meets the strictest specifications of developed countries legislation. Figure 9 corresponds to a standard proposal of a rule in the United States on the maximum amounts or limits of aromatic compounds (10%) and sulfur (15 ppm) contained in the diesel.

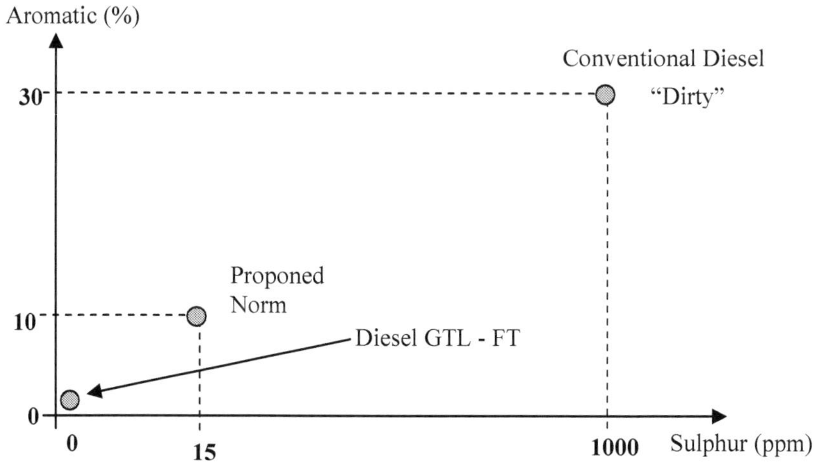

Fig. 22. GTL-FT diesel is a clean fuel

Regarding the price, it is important to stress that the possibilities for this to be competitive are going to depend on the cost structure of the company. In Figure 23, an attempted cost structure is shown, elaborated by the Foster Wheeler company for the production of clean diesel on commercial scale. It can be observed that the GTL-FT diesel production cost is of approximately US$ 19.80 per barrel, far below even the present international price of the diesel, of around US$ 35.00 per barrel, the reason why the profits would be very attractive, considering that the ecological clean diesel has an additional benefit in quality.

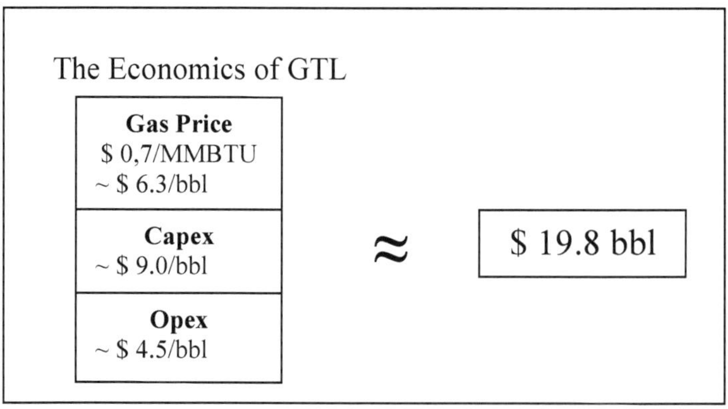

Fig. 23. GTL-FT diesel production cost

If the production cost of a barrel of GTL-FT 'clean' diesel with the present sale price is compared to the diesel originating from oil distillation (around US$ 35.00 by barrel), it is possible to conclude that GTL-FT diesel is a very competitive product in price.
As seen in figure 10, the production cost of the clean diesel is of US$ 18.00 per barrel. As an illustration, this cost was calculated considering that:

- 9 thousand cubic feet of gas are required to produce a barrel of diesel.
- The cost of one thousand cubic feet of dry natural gas is US$ 0.70.
- The capital expenses are about US$ 9.00 per barrel.
- The operation expenses are about of US$ 4.50 per barrel.

Table 11 displays a production cost sensibility analysis for a barrel of GTL-FT diesel, which uses the price of the raw material (of the Natural Gas) to be used in the GTL plant clean diesel production as a control variable. The price was calculated so that the plant could buy the gas, and is expressed in dollars per thousand cubic feet and their equivalent in dollars per barrel of diesel. In addition, the costs of capital and the costs of operation shown in figure 11 have remained unaltered, and are expressed in dollars per barrel.

Gas Price (US$/1000cf)	Gas Price (US$/bbl)	Capital Costs (US$/bbl)	Operation Costs (US$/bbl)	Approximated GTL-diesel Costs (US$/bbl)
0.7	6.3	9.0	4.5	19.8
1.0	9.0	9.0	4.5	22.5
1.4	12.6	9.0	4.5	26.1
1.8	16.2	9.0	4.5	29.7
2.0	18.0	9.0	4.5	31.5
2.2	19.8	9.0	4.5	33.3
2.5	22.5	9.0	4.5	36.0
2.8	25.2	9.0	4.5	38.7
3.0	27.0	9.0	4.5	40.5

Table 11. Gas price and the diesel GTL costs

6.4 Assessment of GTL project

In order to have an idea of how financially attractive GTL-FT projects are, a part of the manuscript "Banks Endorse Qatar GTL Project", is transcribed. This text was published in the Petroleum Economist of March, 2003:
Bankers consider that GTL schemes provide a triple advantage on the traditional projects of oil and gas, which is true in the oil product market as well as for pipelines or LNG.

- The products are commercialized globally in a mature market, so that the sponsor does not have to depend on a specific buyer for long-term agreements.
- The gain margins are much greater than those for the traditional oil refinement that seems to be kept under pressure in the near future.
- The GTL Gas chain is much smaller than in the traditional Gas schemes.

Furthermore, N. White, Director of Energy Economy (of the Arthur D. Little company), says: "the advantage of GTL is that there is not an obligation to construct a new logistic system. It is possible to use the existing distribution system to bring products to the markets".

6.5 Existing Natural Gas to implement a GTL-FT project

The Bolivian Natural Gas fields are appropriate to implement one or more GTL project from the qualitative and quantitative point of view:

From the quantitative point of view, to produce around 100 thousand bpd of GTL byproducts, for 25 years, is necessary to process 30 million m^3 of gas per day, which demands 10 TCF of the economic gas reserves. Bolivia had, among proven and probable reserves, about 50 TCF in the beginning of 2005, fully meeting this requirement.

From the qualitative point of view, the Bolivian reserves are of non-associated gas; this means that it does not have many accompanying liquid hydrocarbons, which allows minimum investments to separate the methane from other hydrocarbons. Bolivia has the greatest non-associated Gas reserves in South America: greater than, for example, the ones in the 226 TCF of total gas reserves in Venezuela (the greatest gas reserves of South America), but only 14 TCF are of non-associated gas.

Another important aspect is that the sulfur contents of Bolivian hydrocarbons are generally low, which avoids investments in desulphurization plants and, furthermore, avoids the poisoning of the catalysts, fundamental aspect in the process.

Figure 24 illustrates the flow of the potential projects and exportation and industrialization products of the Bolivian natural gas.

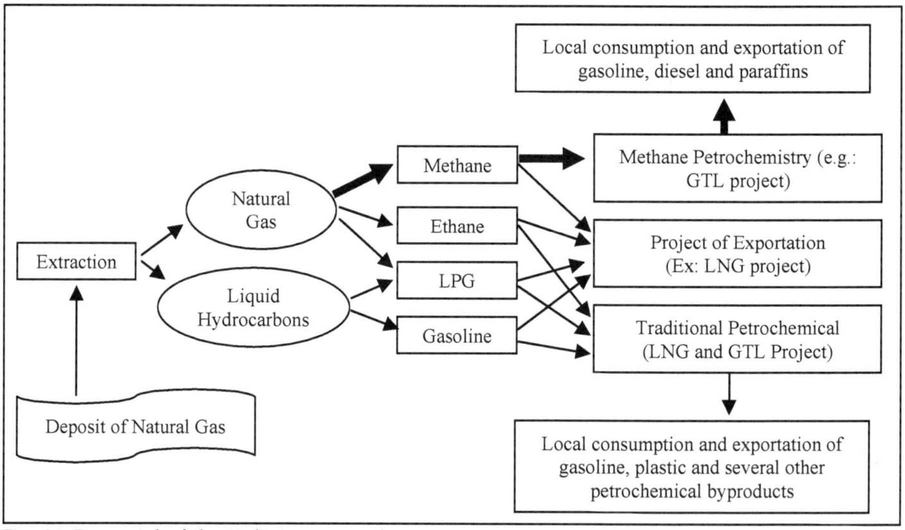

Fig. 24. Potential of the Bolivian natural gas

7. References

[1] International Forum: "Industrialización Del Gás Boliviano: Sueño o Realidad?" La Paz, December, 2003.

[2] SH - Superintendência de Hidrocarburos de Bolívia; www.superhid.gov.bo. 2005.

[3] SIRESE - Superintendencia General, "Sistema de Regulación Sectorial"; ww.sirese.gov.bo. 2005.

[4] Código Petroleiro Davenport de 1995-2000.
[5] DORIA, M.S., 2003. "Gas Bolivia"; La Paz - Bolivia.
[6] KINN, L.C., 2004. "Política Energética Integral"; Santa Cruz - Bolivia.
[7] MDE - Ministerio de Desarrollo Económico, Comisión Política de Estado sobre el Gas Natural, 2002. "Política de Estado sobre la Utilización del Gas Natural"; Bolivia.
[8] Udaeta, M.E.M.; Lafuente, R.J.O., june 2003. "Perspectiva del rubro energético en Bolivia y gas natural". Cochabamba - Bolivia. Acta Nova - UCB Magazine of Science and tecnology. Vol. 2 N° 2.
[9] Udaeta, M.E.M; Reis, L.B; Lafuente, R.J.O; Zurita, R.O.R; Burani, G.F. "Análisis de Ia Industria Energética en Bolivia en el Marco del Mercado Competitivo". Rio de Janeiro -Brasil, 2001. Magazine "Revista Brasileira de Energia" -Vol. 8 No 1, SBPE.
[10] Boarati, J.H.; Galvão, L.C.R.; Udaeta, M.E.M. "Full Cost Account for Electricity From Gas or Hydroelectric". In: Gas And Electricity Networks: Complementarity Or Competition? Brasília - DF. Symposium Papers. França: CIGRÉ, 2002.
[11] Oliva, R.C.R., "Exeqüibilidade da industrialização do gás natural na Bolívia e a sustentabilidade de abastecimento a mercados além das suas fronteiras", São Paulo - Brasil, 2006.
[12] California Energy Commission, April, 2005.
[13] Geology College, November, 2003.
[14] Alaskan Natural Gas To Liquid, August, 2000.
[15] Foster Wheeler, Gas to Market Technology, 2002.
[16] Gas To Liquids brochure - www.gassolutions.conoco.com
[17] Udaeta, M.E.M. "Gestão de Energia", Apostila. Target Engenharia e Consultoria. São Paulo -Brasil, 2001. 183 p.
[18] Udaeta, M.E.M.; Galvão, L.C.R.; Lafuente, R.J.O. Capitulo Bolivia. In: Paula, Ericson de. (Org.). Energía para el Desarrollo de América del Sur. Mackenzie. São Paulo, 2002, p. 69-110.
[19] Dubrovsky, H.; Udaeta, M.E.M.; Paula, E.; Gonzáles, M.I.; Giraldo, A.E.B.; Manzoni, G.B.L.; Oxilia, V.; Becerra, A.; Abreu, A.T.; Cordeiro, J.L.. Energía para el Desarrollo de América del Sur. São Paulo. Mackenzie, 2002. v. 1. 410 p.
[20] Vega, F.F. de.; "El Gas Natural en América Latina y el Caribe". Buenos Aires - Argentina, 2004. PennWell.
[21] Galvão, L.C.R.; Reis, L.B.; Udaeta, M.E.M. "A Opção pelo Desenvolvimento Sustentável com a Proposta do Planejamento Integrado de Recursos Energéticos". In: III Encontro De Economistas De Língua Portucuesa, Macau. 1998, Anais.
[22] Boarati, J.H.; Galvão, L.C.R.; Udaeta, M.E.M. "Full Cost Account for Electricity From Gas or Hydroelectric". In: Gas And Electricity Networks: Complementarity Or Competition? Brasília - DF. Symposium Papers. França: CIGRÉ, 2002.
[23] Udaeta, M.E.M; Reis, L.B; Lafuente, R.J.O; Zurita, R.O.R; Burani, G.F. "Análisis de Ia Industria Energética en Bolivia en el Marco del Mercado Competitivo". Rio de Janeiro -Brasil. Periódico "Revista Brasileira de Energia" -Vol. 8 No 1- 2001, SBPE. ISSN OlO4-303X.
[24] Udaeta, M.E.M.; Lafuente, R.J.O. "Perspectiva del rubro energético en Bolivia y gas natural". Cochabamba - Bolivia. Acta Nova Revista Semestral de ciencias y tecnología de la UCB. Vol. 2 N° 2, Junho de 2003, p. . ISSN - 1683 - 0768.

[25] Carvalho, C.E.; Gimenes, A.L.V.; Reis, L.B.; Grimoni, I.A.B.; Udaeta, M.E.M "Diagnóstico Energético e Gestão da Energia em Indústrias de Pequeno e Médio Porte". In. XVI SNPTEE, Campinas - SP, 2001. CD-ROM, Anais.

[26] Galvão, L.Cr.; Carvalho, C.E.; Gimenes, A.L.V.; Udaeta, M.E.M. "Power Production With Natural Gas Under The Concept Of The Local IRP Application" In: 25TH International Conference Of The IAEE, Aberdeen - Escócia: IAEE -International Association for Energy Economics, 2002. CD-ROM, Proceedings.

[27] Galvão, L.C.R.; Udaeta, M.E.M. "Aspectos Relevantes do Gás Natural Visando o Planejamento Energético". In: III CBPE, São Paulo -SP. Junho de 1998. SBPE, Anais.

[28] Udaeta, M.E.M.; Zurita, R.O.R.; Lafuente, R.J.O.; Galvão, L.C.R. "La Industria Energética en Bolivia y su Vocación Integradora Através del Gas Natural" In. IV Encontro Brasileiro Dos Profissionais Do Gás, São Paulo, Gasbrasil, 2003, Cd-Rom, Anais.

[29] Institute of Americas, El Gasoducto Sudamericano "Mesa Redonda Ejecutiva Sobre La Integración Regional Energética". Presentada por: Ministerio de Industria, Energía y Minería del Uruguay y el Instituto de las Américas. Montevideo, Uruguay, August of 2005.

[30] Carrera G. A. Zamalloa "Avaliação de Alternativas Tecnológicas (GNL e GTL) para a Viabilização de Jazidas de Gás Natural Remotas em Países em Desenvolvimento" - Estudo de Caso: Jazida de Camisea no Peru" 2004.

16

Practical results of forecasting for the natural gas market

Primož Potočnik and Edvard Govekar
University of Ljubljana
Slovenia

1. Introduction

The need for natural gas consumption forecasting is rooted in the requirements to balance the supply and consumption of natural gas. For daily operation of natural gas suppliers and distributors, short-term forecasting with the forecasting horizon of several days is required. Forecasting resolution is required on daily and also on hourly basis. Since the natural gas market is very dynamic, many factors influence the consumption and consequently, the natural gas demand forecasting becomes very challenging. The restrictive economic policies that drastically penalize the forecasting errors only increase the forecasting challenge.

Various approaches to energy consumption forecasting have been investigated in the literature. Forecasting methods include time series and regression methods (Ediger et al., 2006; Ediger & Akar, 2007), nonlinear regression (Vondráček et al., 2008), expert systems (Smith et al., 1996; Chandrashekara et al., 1999; Petridis et al., 2001; Tzafestas & Tzafestas, 2001), stochastic models (Hubele & Cheng, 1990, Vajk & Hetthéssy, 2005), artificial neural networks (Mihalakakou et al., 2002; Beccali et al., 2004; Gonzalez & Zamarreno, 2005; Karatasou et al., 2006; Hamzaçebi, 2007), wavelets (Benaouda, 2006) and support vector machines (Pai & Hong, 2005a; Pai & Hong, 2005b).

Based on our experience, the forecasting solution can be considerably improved by incorporating the proper influential variables into the solution, and by properly understanding the underlying principles of energy consumption. Consequently, the forecasting approach was developed based on the understanding the underlying natural gas consumption cycles (Potočnik et al., 2007a) and the forecasting system for the Slovenia energy market was developed (Potočnik et al., 2005; Potočnik et al., 2007b; Potočnik et al., 2008). The proposed forecasting approach was embedded into stand-alone forecasting applications for various companies and natural gas distributors in Slovenia. This chapter presents an overview of practical results for a larger gas distributing company, obtained during the last three years of online operation. The forecasting requirements for the Slovenia natural gas market are explained in section 2, section 3 presents data for the case study, development and validation of the model are presented in section 4, section 5 is devoted to the presentation and discussion of practical forecasting results, obtained through the several years of operation, and the last section 6 draws some conclusions.

2. Forecasting requirements

The forecasting requirements that motivate the forecasting efforts of energy distribution companies vary from country to country and are also subject to modifications according to the legislation. Slovenia's economic incentive model is briefly described in this section. Currently, companies in Slovenia are motivated to accurately forecast their daily gas consumption for the next "gas consumption" day. Slovenia's gas consumption day is defined as the period from 8.00 AM until 8.00 AM the following day. The forecast should be delivered approx. at 9.00 AM for the gas consumption day starting tomorrow, therefore the forecasting horizon is H = 2 days. Currently, the incentive model only refers to forecasting on daily resolution, therefore hourly forecasting is only required by the company for online optimization of their resources and operation strategies. Both, daily and hourly forecasting results are considered in this chapter. Daily forecasting requirements are shown in Fig. 1, and expanded hourly forecasting requirements are plotted in Fig. 2.

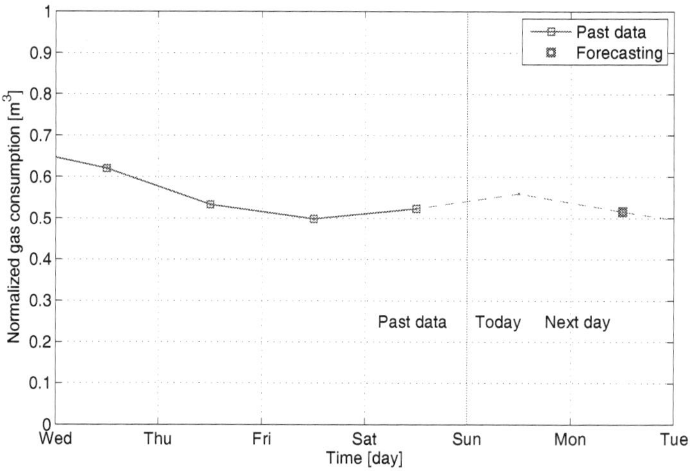

Fig. 1. Forecasting requirements for Slovenia's gas market on daily resolution.

The economic incentive model in Slovenia currently regulates the transfer of savings based on daily forecasting results. Daily natural gas consumption forecast y_F with the forecasting horizon H = 2 days is compared with actual (measured) gas consumption y_M, and the forecasting error E is expressed through a percentage of the maximum transport capacity (MTC) of the distribution system:

$$E = 100(y_F - y_M)/MTC \quad [\%] \tag{1}$$

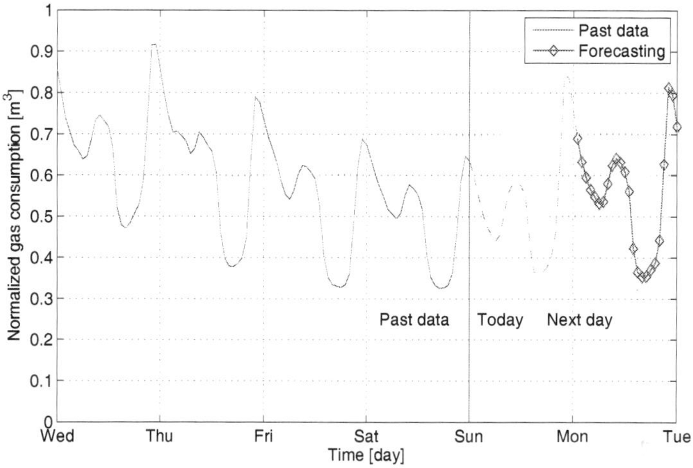

Fig. 2. Forecasting requirements for Slovenia's gas market on hourly resolution.

Two error levels are particularly important for the economic evaluation of the forecasting model:

E3: defining the relative forecasting error E = 3% MTC
E8: defining the relative forecasting error E = 8% MTC

The economic incentive model is schematically shown in Fig. 3 and is defined by the following rules:

$E < E3$:	distributor receives a positive daily profit,
$E > E3$ and $E < E8$:	daily profit is 0,
$E > E8$:	daily profit is negative.

For negative errors (forecasted consumption less than actual) of more than *E8*, the negative daily profit increases more rapidly than for positive errors of more than *E8*. Absolute values of daily profits depend linearly on the size of the distribution system (defined by MTC value). In addition to the economic evaluation relative to the *E3* and *E8* levels, distributors that decide to forecast their gas consumption also obtain an extra stimulation directly proportional to the daily gas consumption. This is an important measure that further encourages distributors to take the risk of natural gas consumption forecasting.

For the case study considered in this chapter, the customer's forecasting requirements were defined as follows:

1. forecasting is required only during the winter season, since consumption during the summer season is quite simple,
2. forecasting is required both in daily, and in hourly resolution.

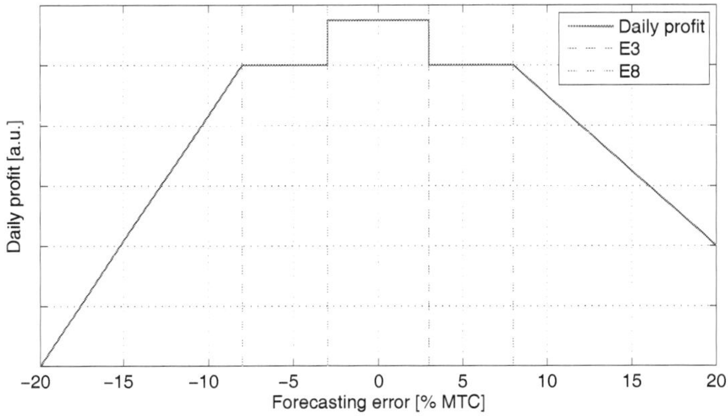

Fig. 3. Economic incentive model for natural gas consumption forecasting

According to the confidentiality policy of the company, its identity and the absolute values of the natural gas consumption data should not be revealed. Consequently, relative gas consumption data, normalized to the maximum transfer capacity (MTC) of the distribution system, are considered throughout this chapter.

3. Data

The initial forecasting model was developed based on natural gas consumption data, as shown in Fig. 4. Only data for incomplete two winter seasons were available. The consumption data are shown in relative units, expressed as a percentage of MTC. Fig. 4. shows daily gas consumption data, and Fig. 5. shows the 2007 portion of the data in hourly resolution. The hourly data exhibit wild fluctuations, therefore the zoom of the hourly natural gas consumption data is displayed in Fig. 6 for the period of three weeks of March 2007.

Beside the past consumption data, various meteorological data and past weather forecasts were also available. Based on the expert knowledge, it was known that the outside temperature heavily influences the natural gas consumption, therefore the temperature data were collected as the most important influential inputs. But when forecasting for the future, weather data are not available and the forecasting model must rely on the weather forecast. For the geographical region of Slovenia, the ALADIN weather model is applied daily by the The Environmental Agency of the Republic of Slovenia (ARSO). The model provides daily forecasts of the most important meteorological parameters, including the temperature, for the forecasting horizons up to 72 hours which suffices for the short-term natural gas consumption forecasting requirements.

Fig. 7 shows comparison of measured temperature and ALADIN weather forecast. Forecasting horizons are defined according to the requirements for the natural gas forecasting. Discrepancy (E_T) between measured and forecasted temperatures on daily basis for the winter season 2008-2009 are as follows: mean(E_T) = -0.93 ^{o}C, std(E_T) = 1.68 ^{o}C. Fig. 7

also reveals several data errors both on measured temperature as well on the temperature forecast. Such obvious outliers can be easily detected and removed but more subtle errors are much more difficult to detect by automated signal processing methods.

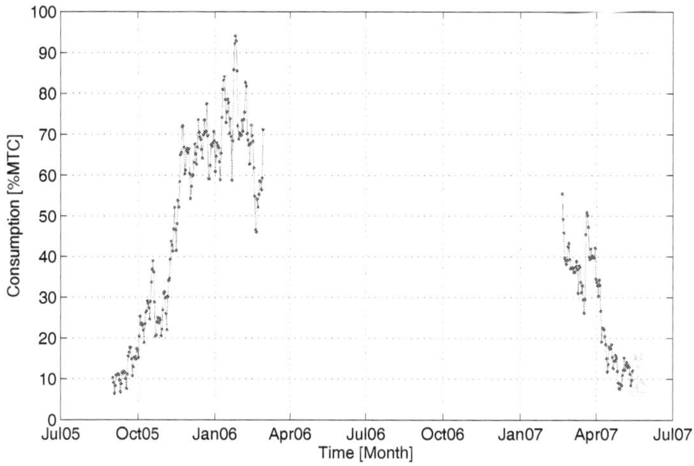

Fig. 4. Daily past natural gas consumption data, available for the development of the model

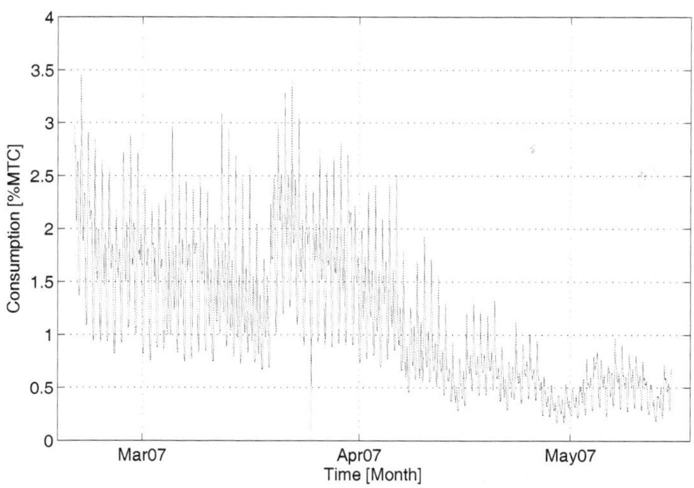

Fig. 5. Hourly past natural gas consumption data for the 2007 period

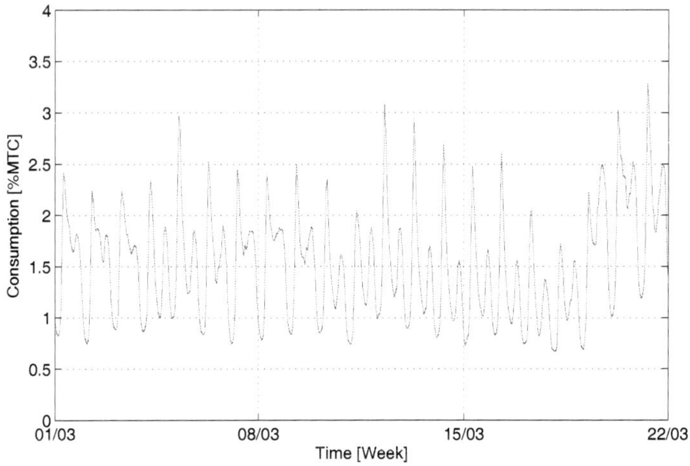

Fig. 6. Hourly past natural gas consumption data for March 2007

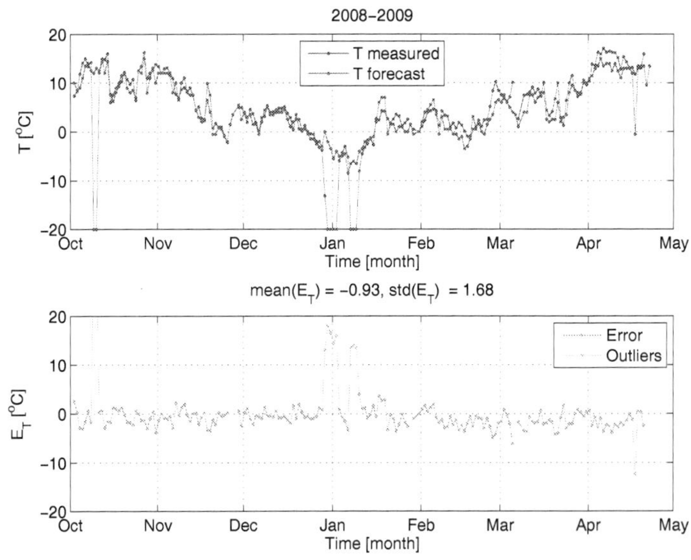

Fig. 7. Comparison of measured temperature and ALADIN weather forecast

4. Model formation and validation

The natural gas consumption data exhibit complex dynamics both on daily, and even more on hourly scale. Therefore it is advantageous to decompose the forecasting task into subdomains (daily and hourly resolution) and construct the forecasting models separately for daily and for hourly forecasting. This results in decreased complexity of the forecasting submodels compared to the single forecasting model (Potočnik et al., 2008). The proposed approach comprises a daily forecaster and a series of hourly forecasters, each for the particular hour of the day. The output of the daily forecaster is a single forecast for the next gas consumption day. The outputs of hourly forecasters are relative hourly consumptions (hourly profiles) that can be combined with a daily forecast to obtain absolute values of natural gas consumption forecast in hourly resolution. The distributed forecasting approach is schematically shown in Fig. 8.

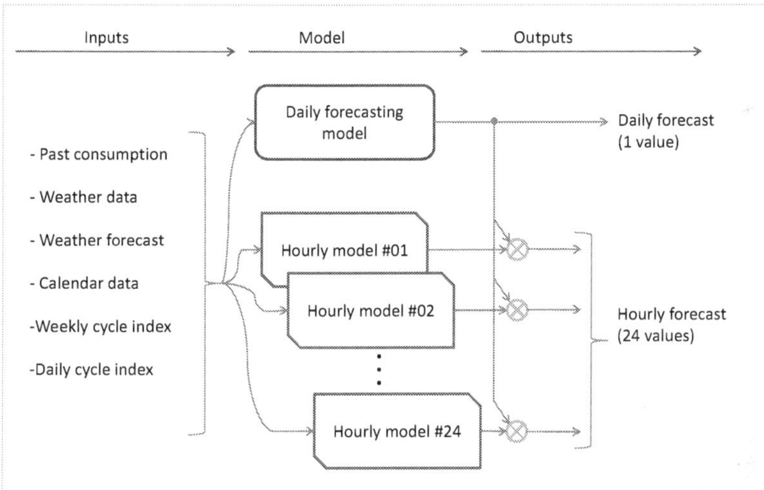

Fig. 8. Complete forecasting solution for combined daily and hourly forecasting

The absolute natural gas consumption values are only forecast by the daily model. The task for the hourly models is simplified by substituting the absolute requirement by the relative one, therefore hourly models are designed to provide normalized daily profiles which are then multiplicatively combined with the daily forecast to obtain final absolute gas consumption forecasts in hourly resolution. The development of daily and hourly submodels are described in the following subsections.

4.1 Development of a daily forecasting model

Natural gas consumption during the winter season is highly correlated with the outside temperature, since the major natural gas consumption in cities is governed by the heating patterns of a residential and business part of the population. The correlation with the temperature can be captured by an ordinary linear regression.

Some portions of the consumption depend nonlinearly on the activity pattern of the population and consequently, nonlinear modelling structures should be applied for proper modelling of the consumption. An interesting solution to avoid nonlinear modelling is to extract features that capture the nonlinear relationships and therefore enable further simplification of the modelling procedure. As proposed in (Potočnik et al., 2007a), natural gas consumption cycles can be efficiently calculated from the available data and offer powerful solution to feature extraction of nonlinear patterns of calendar-based consumption activity.

For the application in a daily forecasting model, a weekly gas consumption index (*WGCI*) was calculated in order to obtain a typical consumption pattern through the days of the week. The weekly cycle is mainly governed by the proportions of residential, business and industrial consumption. It can be calculated through the normalisation of daily consumption data $y(t)$ by the mean of the current week (starting at the current day minus three days, and ending at the current day plus three days):

$$WGCI(d) = \frac{1}{N_W} \sum_{n=1}^{N_W} \frac{y(d+7n)}{\frac{1}{7}\sum_{k=-3}^{3} y(d+7n+k)}, \quad d = 1,2,\ldots,7 \tag{2}$$

Weekly gas consumption index is calculated from daily data for each day of the week (d=1, Monday; d=2, Tuesday; ...; d=7, Sunday) by averaging through the number of available weeks N_W. When the weekly-normalised daily consumption data are collected for each day of the week, a weekly cycle with confidence intervals can be obtained, as shown in Fig. 9.

Characteristically lower consumption can be observed on Friday, Saturday and Sunday. The variations of *WGCI* within each day of the week are small enough compared to the variation of *WGCI* across the week, therefore a complete weekly cycle represents an important nonlinear feature that can be helpful in building the daily forecasting model. WGCI can be included into the forecasting model as an input either directly:

$$WGCI(t+H), \tag{3}$$

or indirectly by weighting the last measured daily gas consumption $y_M(t)$ by the ratio of weekly gas consumption cycle for the forecast day $WGCI(t+H)$ and the current day $WGCI(t)$:

$$\frac{WGCI(t+H)\, y_M(t)}{WGCI(t)}. \tag{4}$$

H denotes the forecasting horizon (H=2) and t the current day (the last day with available measured data). Both possible regressors are defined by Eq. (3-4) with respect to the day $t+H$ for which the forecast should be given.

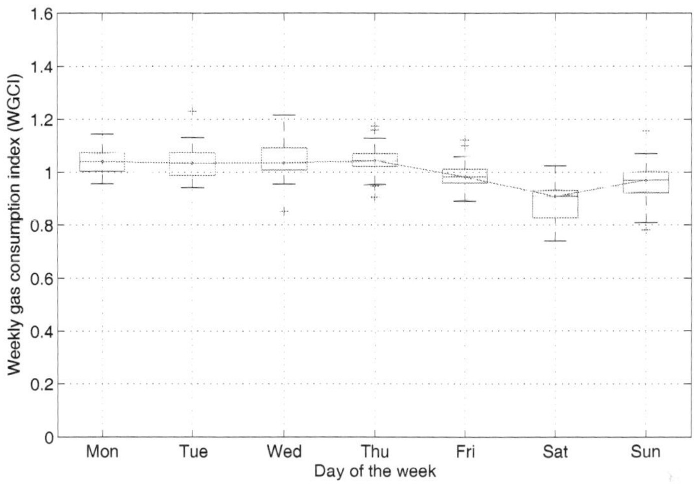

Fig. 9. Weekly gas consumption index (*WGCI*)

Population activity that depends on calendar data, such as holidays, can also be expressed through *WGCI* indexes where the preliminary study should be performed to obtain the appropriate clusters of similar days. For example: holidays can be encoded as Sundays, the day after holiday as Monday, etc.

The other possible regressors for the forecasting model include:

- past natural gas consumption $y_M(t)$, $y_M(t-1)$, ...,
- past weather data, such as temperatures $T_M(t)$, $T_M(t-1)$, ...,
- weather forecasts, such as temperature forecasts $T_F(t+H)$, $T_F(t+H-1)$, ...

The regressors can be specified on various resolution scales, e.g. hourly or daily resolution, therefore the list of possible regressors can grow very large.

Consequently, an optimization approach should be applied to select the most informative subset of regressors that yield the best model structure. We constructed the forecasting model as a liner combination of possible regressors, and the best subset of regressors was determined through the stepwise regression procedure. The final version of the daily forecasting model that was a result of the feature selection optimization based on the available data was defined in the following form:

$$\begin{aligned} y_F(t+H) = \quad & a_0 + a_1 WGCI(t+H) + \\ & + a_2 \frac{WGCI(t+H)\, y_M(t)}{WGCI(t)} + \\ & + a_3 T_F(t+12h) + a_4 T_F(t+30h) + \\ & + a_5 T_F(t+36h) + a_6 T_F(t+42h) + \\ & + a_7 T_F(t+54h), \end{aligned} \tag{5}$$

Subscripts F denote the forecasted value and M the measured value. Weather forecasts such as T_F are obtained from the corresponding Environmental Agency. Both weekly gas consumption features, as defined by Eq. (3-4), were included in the forecasting model. Past weather data were surprisingly not included in the model but forecasted temperatures T_F were included on the hourly resolution, as specified by the forecasting horizons $T_F(t+12h)$, ..., $T_F(t+54h)$.

Daily forecasting model, optimized by the stepwise regression procedure as described above, was tested on available past data. The results on past data are shown in Fig. 10. Note that past weather forecasts and not past weather data were used in order to estimate the industrial applicability of the model. In order to compare various forecasting results, we express the forecasting error of the model with the mean absolute error (*MAE*):

$$MAE = \frac{100}{MTC} \frac{1}{N} \sum_{t=1}^{N} \left| y_F(t) - y_M(t) \right| \quad [\%] \tag{6}$$

For the training results, shown in Fig. 10, the forecasting error amounts to <u>$MAE = 2.4$ %</u>. This result is very good from the practical perspective and helps the company to generate extra savings due to forecasting accuracy. Taking into consideration various unknown uncertainties and unknown future natural gas consumption network dynamics, the estimate about the future online performance was proposed as $MAE \approx 3$ %.

For better alignment with online forecasting conditions, the complete model formation procedure was accomplished on past weather forecasts and not on past measured data. However, by using the past measured weather data, the model's forecasting capacity can be estimated. If past weather data and not past weather forecasts were applied to the natural gas consumption forecasting model (Eq. 5), the forecasting error of $MAE = 1.5$ % is obtained. This shows high forecasting capacity of the proposed model but is of little help for online application due to the fact that weather data for the future are not available.

The final step in preparing the forecasting model for online application was fine-tuning of model parameters according to the policy that favours higher forecasting accuracy toward the newest data. This is specially important when several years of available data exist. Usually the oldest data are less relevant but still useful for the construction of the forecasting model.

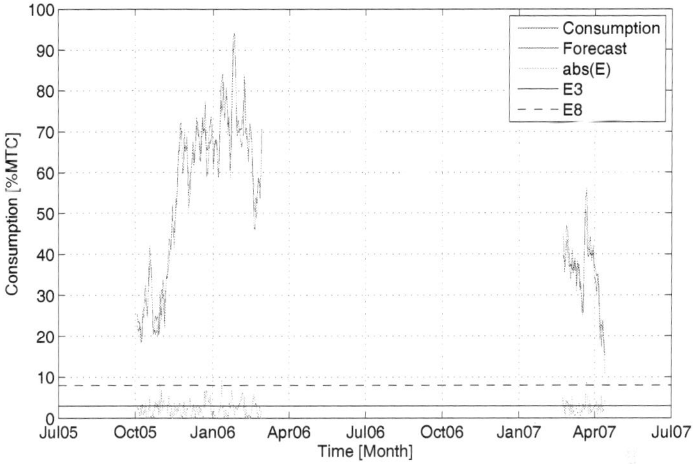

Fig. 10. Daily forecasting model applied on past training data

We propose the following fine-tuning strategy that can be realised through the following two steps:

1. Amplification of the error function, as obtained by the basic forecasting model, according to the desired amplification function (linear, quadratic, sigmoid, ...).
2. Numerical optimization of the basic forecasting model in order to minimize the amplified error function.

By using any of the amplified error functions, as shown in Fig. 11, the original error function is transformed into an amplified error function. The second step is then to numerically optimize model parameters to minimize the transformed error function.

We applied the linear error function that linearly increases the importance of newer data and obtained the fine-tuned forecasting error $\underline{MAE = 2.6\ \%}$. This error is higher compared to the basic model (2.4 %) but the optimized model is expected to be more relevant for online application on new data. Fig. 12 shows the comparison of initial and optimized model parameters for the daily model (Eq. 5) according to the fine-tuning strategy.

By performing the described steps, the daily model is prepared for online application. An online application of the daily model should be supported by periodical model updates based on newest data, followed by fine-tuning as described above.

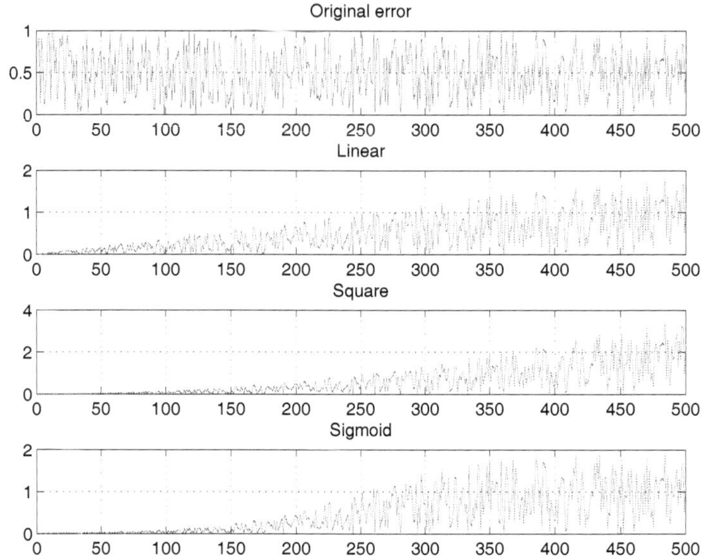

Fig. 11. Examples of amplified error functions for model fine-tuning

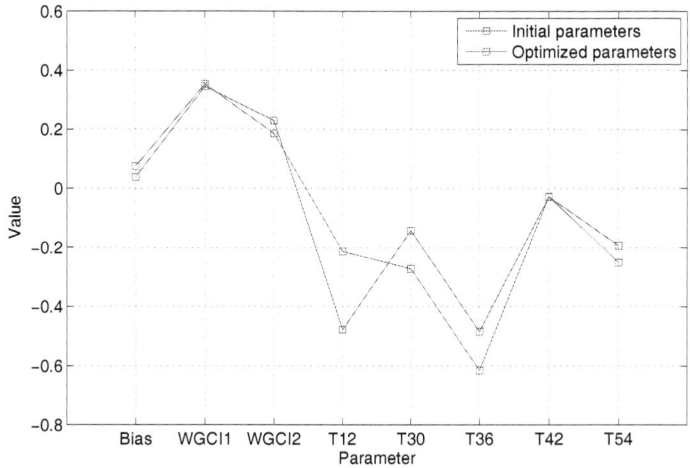

Fig. 12. Initial model parameters and optimized model parameters

4.2 Model formation for hourly forecasting

In this section, the development of an hourly forecasting model is described. The hourly model is designed to forecast the relative hourly profile and this information is then multiplicatively combined with daily forecast as shown in Fig. 8 to obtain final forecasting solution for combined daily and hourly forecasting.

In accordance with the daily model strategy, where a weekly gas consumption index (*WGCI*) was calculated in order to obtain a typical consumption pattern through the days of the week, we derive a similar feature for the hourly model. A daily gas consumption index (*DGCI*) describes a typical consumption pattern through the hours of the day and can be calculated by normalisation of hourly consumption data by the mean of hourly consumption for the current day:

$$DGCI(h) = \frac{1}{N}\sum_{n=1}^{N} \frac{y(h+24n)}{\frac{1}{24}\sum_{k=-12}^{11} y(h+24n+k)}, \quad h = 1,2,\ldots,24 \tag{7}$$

Daily gas consumption index is calculated from hourly data for each hour of the day (h=1, 2, ..., 24) by averaging through the number of available data days *N*. When the daily-normalised hourly consumption data are collected for each hour of the day, a daily cycle with confidence intervals can be obtained, as shown in Fig. 13. Morning and evening consumption peaks, as well as the low-consumption night period, can easily be observed.

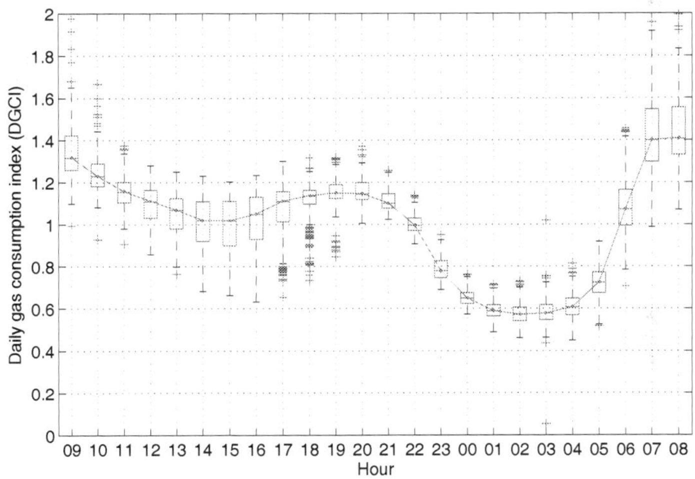

Fig. 13. Daily gas consumption index (*DGCI*)

Daily gas consumption index can be further specialized for particular days of the week by applying the basic equation (Eq. 7) to particular days or groups of the days only, as shown in Fig. 14. If daily cycles are similar enough across various days of the week, it is not necessary to specialize the *DGCI* for particular days of the week and a basic equation (Eq. 7) can be utilized. Daily gas consumption index represents an important nonlinear feature that facilitates the construction of a simplified hourly model which is linear in the parameters.

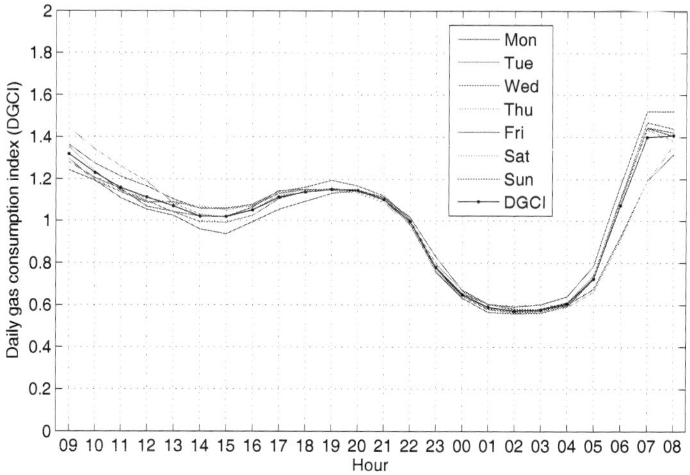

Fig. 14. Daily gas consumption index for particular days of the week

Hourly model can be constructed either as a single model or as a set of 24 submodels for each hour separately. In our case study, the later approach is applied in order to slightly increase the forecasting accuracy of hourly forecasting. The hourly model is composed of 24 submodels, each forecasting normalized consumption for a specific hour of the day. Inputs and parameters for each hourly submodel are optimized through a stepwise regression from a pool of a larger set of available regressors. This results in optimized model structures for each hourly submodel. The pool of possible regressors includes the variables available to the daily model and additional daily gas consumption index.

The stepwise regression procedure automatically selects the most appropriate set of relevant regressors for each hourly submodel, as shown in the following few examples denoting the hourly forecasts for the hours hrs = {09, 15, 21}:

Hourly model, hrs = 09:

$$\begin{aligned} y_F(t+hrs) = \; & b_0 + b_1 DGCI(hrs) + b_2 y_M(t-1, hrs) + \\ & + b_3 T_F(t+30h) + b_4 T_F(t+36h) + \\ & + b_5 T_F(t+42h) + b_6 T_F(t+54h) \end{aligned} \quad (8)$$

Hourly model, hrs =15:

$$\begin{aligned} y_F(t+hrs) = \; & b_0 + b_1 DGCI(hrs) + \\ & + b_2 T_F(t+12h) + b_3 T_F(t+30h) + \\ & + b_4 T_F(t+36h) + b_5 T_F(t+42h) + \\ & + b_6 T_F(t+48h) + b_7 T_F(t+54h) \end{aligned} \quad (9)$$

Hourly model, *hrs* =21:

$$
\begin{aligned}
y_F(t+hrs) = \; & b_0 + b_1 y_M(t-1, hrs) + \\
& + b_2 T_F(t+12h) + b_3 T_F(t+18h) + \\
& + b_4 T_F(t+36h) + b_5 T_F(t+48h) + \\
& + b_6 T_F(t+54h)
\end{aligned}
\tag{10}
$$

Automatic selections of regressors for each hourly submodel shows that the most informative regressors include:

- *DGCI(hrs)*: daily gas consumption index for the particular (forecasted) hour,
- $y_M(t-1,hrs)$: past normalized hourly consumption for the forecasted hour,
- $T_F(t+Xh)$: temperature forecast for various forecasting horizons.

A complete hourly forecast is composed from values of all 24 hourly submodels and thus an output of an hourly model is a normalized hourly profile, specified for each hour of the day.

4.3 Complete forecasting model

The complete natural gas consumption forecast is finally obtained by a daily forecast, multiplicatively combined with an hourly normalized profile. This results in a final forecast in an hourly resolution, as shown in Fig 8. An example of a complete forecast for the training data is shown in Figures 15-16 for two weeks of January 2006 and March 2007.

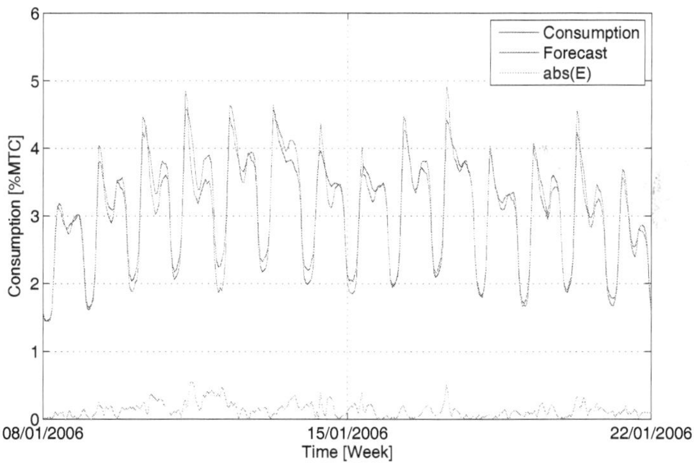

Fig. 15. Complete forecast (daily + hourly), January 2006, training data

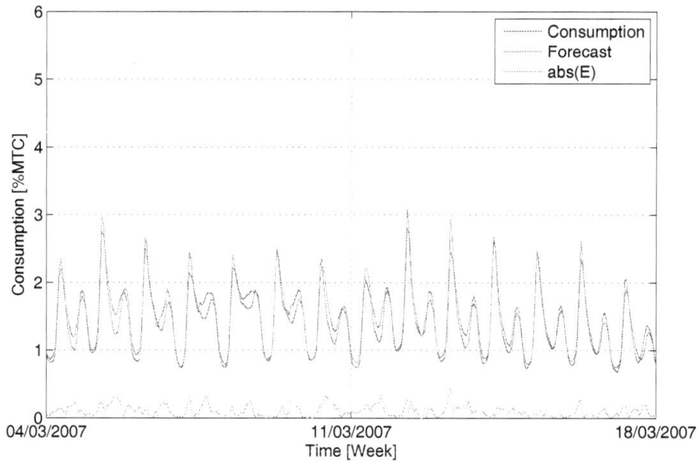

Fig. 16. Complete forecast (daily + hourly), March 2007, training data

5. Practical forecasting results

The proposed forecasting solution was installed in September 2007 and operates online since then. The system is designed as a software package composed from the data acquisition module, the forecasting module and the output module. Input data are parsed from various data sources, such as weather forecasts and autoregressive time series data. The forecasting module is responsible for data preprocessing and for daily generation of natural gas consumption forecasts. The output module provides for integration of forecasting results with the company's SCADA (supervisory control and data acquisition) system. During the last three seasons of operation (2007/2008, 2008/2009, 2009/2010), the forecasts failed few times due to system errors or missing values in input data. The forecasting results, obtained through online operation, are shown and discussed in the following sections.

5.1 Daily forecasting results

Forecasting results based on online operation of the forecasting system are shown in Fig. 17-19 for last three seasons of operation. Results are obtained in daily resolution by forecasting with the horizon H=2 days. Daily forecasting results are summarized in Table 1.

Season	Forecasting days	MAE [%MTC]	E<E3 [%]	E<E8 [%]
2007/2008	191	3.60	51	94
2008/2009	199	2.82	64	96
2009/2010	188	3.05	54	97
Totals	578	3.15	56	96

Table 1. Daily forecasting results

The average mean absolute error for all seasons amounts to <u>MAE = 3.15 %</u> for a complete set of 578 forecasting days. On average, 56% of forecasting errors were below E3 margin (error < 3% MTC), and 96% of forecasting errors were below E8 margin (error < 8% MTC). The result is in accordance with the initially estimated $MAE \approx 3$ %. The obtained forecasting result is beneficial to the company and is therefore considered as very positive. There were several days without forecasts which is mainly due to errors in input data or missing inputs.

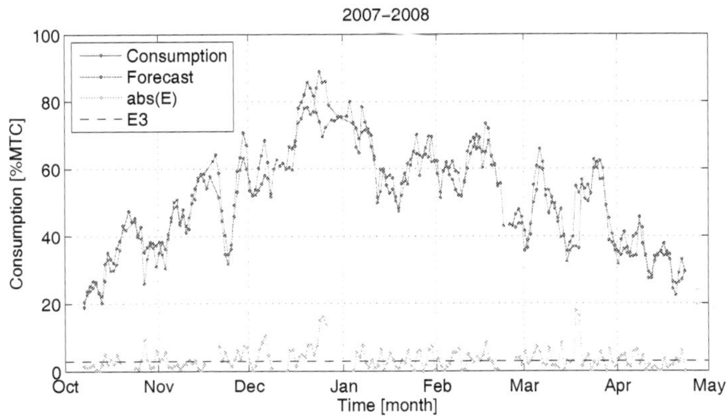

Fig. 17. Daily forecasting results for winter season 2007/2008

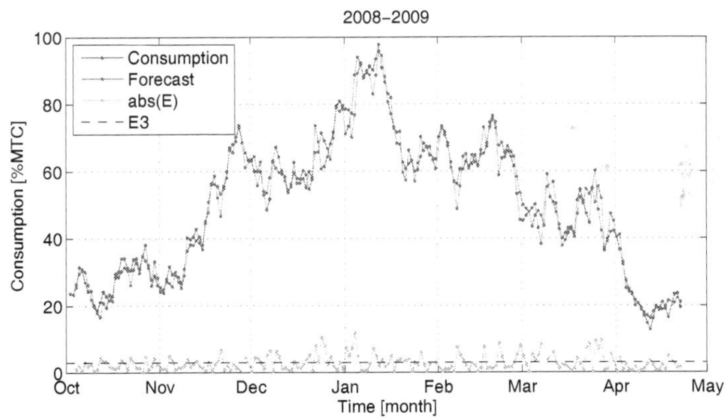

Fig. 18. Daily forecasting results for winter season 2008/2009

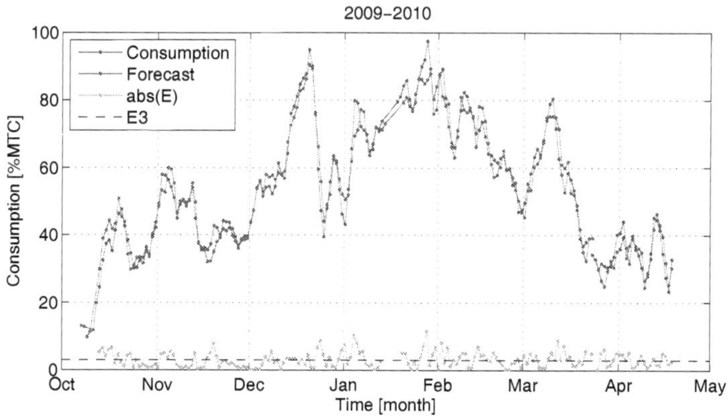

Fig. 19. Daily forecasting results for winter season 2009/2010

5.2 Hourly forecasting results

Whereas daily forecasting results are reflected directly in the cash flow due to economic incentive model, the hourly forecasting results are only utilized by the company's internal resource optimization strategies. Fig. 20-22 show several weeks of hourly forecasting for each forecasting season. Hourly forecasting results are collected in Table 2.

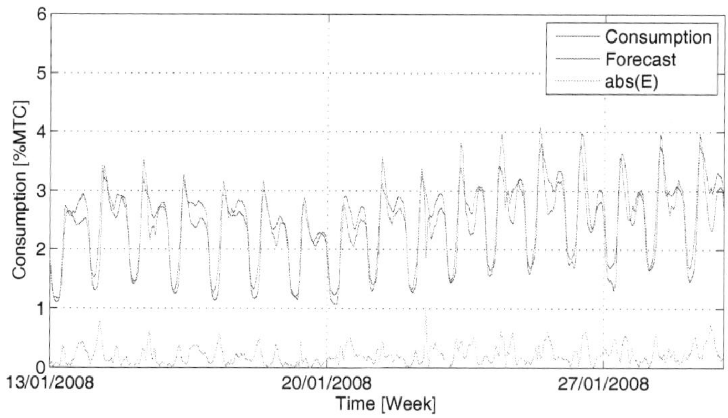

Fig. 20. Hourly forecasting results for winter season 2007/2008 (detail)

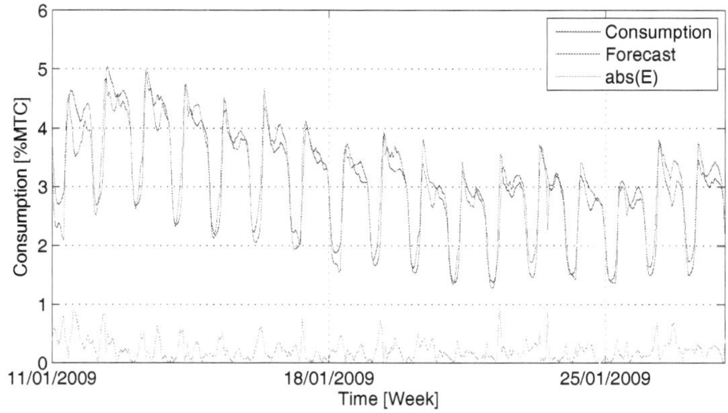

Fig. 21. Hourly forecasting results for winter season 2008/2009 (detail)

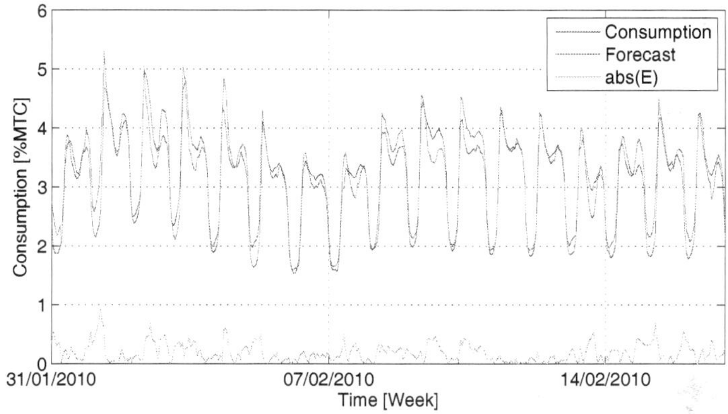

Fig. 22. Hourly forecasting results for winter season 2009/2010 (detail)

Season	Forecasting days	MAE (hourly) [%MTC]	MAE (daily sum) [%MTC]
2007/2008	191	0.24	5.85
2008/2009	199	0.20	4.69
2009/2010	188	0.17	4.05
Totals	578	0.20	4.87

Table 2. Hourly forecasting results

The third column in Table 2 shows hourly forecasting accuracy averaged per hour and expressed in percentage of MTC. The last column scales the hourly error to the daily resolution for comparison with the daily forecasting model. The average daily error amounts to 3.15% and the average hourly forecasting amounts to 4.87%. Consequently, this means that the hourly forecasting model increases the daily model error by 1.72% due to expansion of the daily forecast into hourly forecast, and this result can be interpreted as expected and appreciated. Such results are very useful to the company for planning in advance proper operating actions and optimization strategies of heating and natural gas distributing resources.

6. Conclusions

Natural gas consumption forecasting is required to balance the supply and consumption of natural gas. Companies and natural gas distributors are motivated to forecast their consumption by the economic incentive model that dictates the cash flow rules corresponding to the forecasting accuracy. The rules are quite challenging but enable the company to gain positive cash flow by forecasting accurately their short-term natural gas consumption.

In this chapter, some practical forecasting results for the Slovenia natural gas market are presented. In 2007, an online forecasting system was developed and installed for a larger gas distributing company in Slovenia (according to the confidentiality policy of the company, its identity should not be revealed). The chapter presents the development of the forecasting models for both daily and hourly forecasting and summarizes the practical forecasting results, obtained during the last three years of online operation (winter seasons 2007/2008, 2008/2009, and 2009/2010).

Average daily forecasting result, expressed over three successive forecasting winter seasons as a mean absolute error, amounts to $\underline{MAE = 3.15\ \%}$. The result is considered as very successful and confirms the applicability of such an approach. The results obtained also enable the company to gain benefits according to the economic incentive model. Based on the presented case study, some practical conclusions can be drawn:

- An initial small data set of two incomplete winter seasons (Fig. 4) suffices for a construction of an adequate forecasting model.
- Extraction of specific natural gas consumption features, such as *WGCI* (Eq. 2) and *DGCI* (Eq. 7) is highly recommended in order to construct simple (linear in the parameters), robust and effective forecasting models.
- Stepwise regression procedure proved to be successful in extraction of informative inputs for both daily and hourly forecasting models. The simple forecasting models, as presented by Eq. 5 for the daily forecasting and Eq. 8-10 for hourly forecasting, proved to be quite robust and successful for the considered case study.
- During the online operation of the forecasting model, its performance can be successfully improved by regular model adaptation on new data. Applying the proposed weighting error function (Fig. 11) is essential for proper adaptation of the model through longer operation periods.
- Finally, it should be mentioned that natural gas forecasting results rely heavily on available weather forecasts. Therefore, the accuracy of the forecasts of influential data should be taken into account for each case study and the eventual natural gas forecasting efficiency estimated in order to gain the full perspective of possible natural gas forecasting outcomes.

7. References

Beccali, M.; Cellura, M.; Lo Brano, V. & Marvuglia, A. (2004). Forecasting daily urban electric load profiles using artificial neural networks. *Energy Conversion and Management*, 45, 2004, 2879-2900.

Benaouda, D.; Murtagh, F.; Starck J.-L. & Renaud, O. (2006). Wavelet-based nonlinear multiscale decomposition model for electricity load forecasting, *Neurocomputing, 70,* 1-3, 2006, 139-154.

Chandrashekara, A.S.; Ananthapadmanabha, T. & Kulkarni, A.D. (1999). A neuro-expert system for planning and load forecasting of distribution systems. *Electrical Power and Energy Systems*, 21, 1999, 309-314.

Ediger, V.Ş.; Akar, S. & Uğurlu, B. (2006). Forecasting production of fossil fuel sources in Turkey using a comparative regression and ARIMA model, *Energy Policy, 34,* 18, 2006, 3836-3846.

Ediger, V.Ş. & Akar, S. (2007). ARIMA forecasting of primary energy demand by fuel in Turkey, *Energy Policy, 35,* 3, 2007, 1701-1708.

Gonzalez, P.A. & Zamarreno, J.M. (2005). Prediction of hourly energy consumption in buildings based on a feedback artificial neural network. *Energy and Buildings,* 37, 2005, 595-601.

Hamzaçebi, C. (2007). Forecasting of Turkey's net electricity energy consumption on sectoral bases, *Energy Policy, 35,* 3, 2007, 2009-2016.

Hubele, N.F. & Cheng, C.S. (1990). Identification of seasonal short-term forecasting models using statistical decision functions. *IEEE Transactions on Power Systems*, 5, 1990, 40-45.

Mihalakakou, G.; Santamouris, M. & Tsangrassoulis, A. (2002). On the energy consumption in residential buildings, *Energy and Buildings, 34,* 7, 2002, 727-736.

Karatasou, S.; Santamouris, M. & Geros, V. (2006). Modeling and predicting building's energy use with artificial neural networks: Methods and results. *Energy and Buildings, 38,* 8, 2006, 949-958.

Pai, P.F. & Hong W.-C. (2005a). Support vector machines with simulated annealing algorithms in electricity load forecasting, *Energy Conversion and Management, 46,* 17, 2005, 2669-2688.

Pai, P.F. & Hong W.-C. (2005b). Forecasting regional electricity load based on recurrent support vector machines with genetic algorithms, *Electric Power Systems Research, 74,* 3, 2005, 417-425.

Petridis, V.; Kehagias, A.; Petrou, L.; Bakirtzis, A.; Kiartzis, S.; Panagiotou, H. & Maslaris, N. (2001). A Bayesian multiple models combination method for time series prediction. *Journal of Intelligent and Robotic Systems*, 31, 2001, 69-89.

Potočnik, P.; Thaler, M., Poredoš, A. & Grabec, I. (2005). Risk estimation of excess energy demand in a distribution system based on energy consumption prediction, In: *Physics of risk,* Second annual meeting COST Action P10, Toledo, Spain, April 23-26, 2005.

Potočnik, P.; Govekar, E. & Grabec, I. (2007a). Short-term natural gas consumption forecasting, *Proceedings of the 16th IASTED International Conference on Applied Simulation and Modelling,* Palma de Mallorca, Spain, August 29-31, 2007, 353-357.

Potočnik, P.; Thaler, M.; Govekar, E.; Grabec, I. & Poredoš, A. (2007b). Forecasting risks of natural gas consumption in Slovenia. *Energy Policy*, 35, 2007, 4271-4282.

Potočnik, P.; Govekar, E. & Grabec I. (2008). Building forecasting applications for natural gas market, In: *Natural gas research progress*, Nathan David (Ed.), Theo Michel (Ed.), New York, Nova Science Publishers, 2008, 505-530.

Smith, P.; Husein, S. & Leonard, D.T. (1996). Forecasting short term regional gas demand using an expert system, *Expert Systems with Applications, 10,* 2, 1996, 265-273.

Tzafestas, S. & Tzafestas, E. (2001). Computational intelligence techniques for short-term electric load forecasting. *Journal of Intelligent and Robotic Systems*, 31, 2001, 7–68.

Vajk, I. & Hetthéssy, J. (2005). Load forecasting using nonlinear modelling, *Control Engineering Practice, 13, 7*, 2005, 895-902.

Vondráček, J.; Pelikán, E.; Konár, O.; Čermáková, J.; Eben, K.; Malý, M. & Brabec, M. (2008). A statistical model for the estimation of natural gas consumption. *Applied Energy,* 85, 5, May 2008, 362-370.

17

Statistical model of segment-specific relationship between natural gas consumption and temperature in daily and hourly resolution

Marek Brabec, Marek Malý, Emil Pelikán and Ondřej Konár
Department of Nonlinear Modeling, Institute of Computer Science,
Academy of Sciences of the Czech Republic
Czech Republic

1. Introduction

In this chapter, we will describe a statistical model which was developed from first principles and from empirical behavior of the real data to characterize the relationship between the consumption of natural gas and temperature in several segments of a typical gas utility company's customer pool. Specifically, we will deal with household and small+medium (HOU+SMC) size commercial customers. For several reasons, consumption modeling is both challenging and important here. The essential fact is that these segments are quite numerous in terms of customer numbers. It leads to three practically significant consequences.

- First, their aggregated consumption constitutes an important part of the total gas consumption for a particular day.
- Secondly, their consumption depends strongly on the ambient temperature. Hence, the temperature lends itself as a nice and cheap-to-obtain, exogeneous predictor. The temperature response is nonlinear and quite complex, however. Traditional, simplistic approaches to its extraction are not adequate for many practical purposes.
- Further, the number of customers is high, so that their individual follow-up in fine time resolution (say daily) is not feasible from financial and other points of view. Routinely, their individual data are available only at a very coarse (time-aggregated) level, typically in the form of approximately annual consumption totals obtained from more or less regular meter readings. When daily consumption is of interest, the available observations need to be disaggregated somehow, however.

Disaggregation is necessary for various practical purposes - for instance for the routine distribution network balancing, for billing computations related to the natural gas price changes (leading to the need for pre- and post-change consumption part estimates), etc. As required by the market regulator, the resulting estimates need to be as precise as possible, and hence they need to use available information effectively and correctly. Therefore, they should be based on a good, formalized model of the gas consumption. Since the main driver of the natural consumption is temperature, any useful model should reflect the consumption response to temperature as closely as possible. It ought to follow basic qualitative features of the relationship (consumption is a decreasing function of temperature having both lower and upper asymptotes), but it needs to incorporate also much finer details of the relationship observed in empirical data.

Our model tries to achieve just this and a bit more, as we will describe in the following paragraphs. It is based on our analyses of rather large amounts of real consumption data of unique quality (namely of fine time resolution) that was obtained during several projects our team was involved in during the last several years. These include the Gamma project, Standardized load profiles (SLP) projects in both the Czech Republic and Slovakia, as well as the Elvira project (Elvira, 2010). Consumption-to-temperature relationships were analyzed there in order to be able to model/describe them in a practically usable way.

Our resulting model is built in a stratified way, where the strata had been defined previously via formal clustering of the consumption dynamics profiles (Brabec at al., 2009). The stratification concerns the values of model parameters only, however. The form of the model is kept the same in all strata, both in order to retain simplicity advantageous for practical implementation and for saving the possibility of a relatively easy (dynamic) model calibration (Brabec et al., 2009a). Model parameters are estimated from data in a formalized way (based on statistical theory). The data consist of a sample of consumption trajectories obtained through individualized measurements (obtained in rare and costly measurement campaigns for nationwide studies mentioned above).

Construction of the model keeps the same philosophy as our previous models that have been in practical use in Czech and Slovak gas utility companies (Brabec et al., 2009), (Vondráček et al., 2008). It is modular, stressing physical interpretation of its components. This is useful both for practical purposes (e.g. the ability to estimate certain latent quantities that are not accessible to direct measurement but might be of practical interest) and for model criticism and improvement (good serviceability of the model).

The model we present here is substantially different from the standardized load profile (SLP) model we published previously (Brabec et al., 2009) and from other gas consumption models (Vondráček et al., 2008) in that it has no standard-consumption (or consumption under standard conditions) part. It is advantageous that the model is more responsible to the temperature changes, especially in years whose temperature dynamics is far from being "standard" and in transition (spring and fall) periods even during close-to-normal years. Absence of the smooth standard-consumption part also simplifies the interpretation of various model parts. It calls for expansion of the temperature response function. Here, we start from the approach (Brabec at al., 2008), but we expand it substantially in three important ways:

- Shape of the temperature response is estimated in a flexible, nonparametric way (so that we let the empirical data to speak for themselves, without presupposing any a priori parametric shape).
- Dynamic character of the temperature response and mainly its lag structure is captured in much more detail.

- The model now allows for temperature*(type of the day) interaction. In plain words, this means that is allows for different temperature responses for different day of week.

Numerous papers have discussed various aspects of modeling, estimation and prediction of natural gas consumption for various groups of customers such as residential, commercial, and industrial. Similar tasks are solved in the context of electricity load. Load profiles are typically constructed using a detailed measurements of a sample of customers from each group. Other, methods include dynamic modeling (historical load data are related to an external factor such as temperature) or proxy days (a day in history is selected which closely matches the day being estimated). The optimal profiling method should be chosen based on cost, accuracy and predictability (Bailey, 2000). Close association between gas demand and outdoor temperature has been recognized long time ago, so the first approaches to modeling were typically based on regression models with temperature as the most important regressor. Among such models, nonlinear regression approaches to gas consumption modeling prevail (Potocnik, 2007). The concept of heating degree days is sometimes used to suppress the temperature dependency during the days when no heating is needed (Gil & Deferrari, 2004).

In addition to the temperature, weather variables like sunshine length or wind speed are studied as potential predictors. Among other important explanatory variables mentioned in the literature one can find calendar effects, seasonal effects, dwelling characteristic, site altitude, client type (residential or commercial customer), or character of natural gas end-use. Economical, social and behavioral aspects influence the energy consumption, as well. Data on many relevant potential predictors are not available. Regression and econometric models may include ARMA terms to capture the effects of latent and time-varying variables. Another large group of models is based on the classical time series approach, especially on Box-Jenkins methodology (Lyness, 1984), or on complex time series modifications.

In the following, we will first describe the model construction in a formalized and general way, having in mind its practical implementation, however. Then, we will illustrate its performance on real data.

2. Model description and estimation of its parameters

2.1 Segmentation

As mentioned in the Introduction already, we will deal here only with customers from the household and small+medium size commercial segments (HOU+SMC). The segmentation is considered as a prerequisite to the statistical modeling which will be stratified on the segments. In the gas industry (at least in the Czech Republic and Slovakia), the tariffs are not related to the character of the consumption dynamics, unlike in the (from this point of view, more fortunate) electricity distribution (Liedermann, 2006). Therefore, the segmentation has to be based on empirical data. In order to be practical, it has to be based on time-invariant characteristics of customers which are easily obtainable from routine gas utility company databases. These include character of customer (HOU or SMC), character of the consumption (space heating, cooking, hot water or their combinations; technological usage). Here, we used hierarchical agglomerative clustering (Johnson & Wichern, 1988) of weekly standardized consumption means averaged across customers having the same values of selected time-invariant characteristics. Then, upon expert review of the resulting clusters, we used them as segments, similarly as in (Vondráček et al., 2008). This way, we have $K = 8$ segments (4 HOU + 4 SMC in the Czech Republic and 2 HOU + 6 SMC in Slovakia).

2.2 Statistical model of consumption in daily resolution

Here we will formulate a fully specified statistical model describing natural gas consumption Y_{ikt} of a particular (say the i-th, $i = 1,\ldots,n_k$) customer of the k-th segment ($k = 1,\ldots,K$) on during the day $t = 1,2,\ldots$ (using julian date starting at a convenient point in the past). In fact, in order to deal with occasional zero consumptions (that would produce mathematically troublesome results in the development later), we define Y_{ikt} as the consumption plus a small constant (we used 0.005 m^3 when consumption was measured in m^3/100). Another, more complicated possibility is to model zero consumption process more explicitly is described in (Brabec et al., 2008).

We stress that the model is built from down to top (from individual customers) and it is intended to work for large regions, or even on a national level. It has been implemented in the Czech Republic and Slovakia separately. They are of the same form but they have different parameters, reflecting differences in consumption, gas distribution, measurement etc. Then we have:

$$Y_{ikt} = p_{ik}.f_{kt} + \varepsilon_{ikt} = \\ p_{ik}.\exp\left(\sum_{j=1}^{5}\alpha_{jk}.I_{t\in D_j} + \chi_k.I_{t\in Christmas} + \phi_k.I_{t\in Easter} + \zeta_{kt}\right) + \varepsilon_{ikt} \qquad (1)$$

where $I_{condition}$ is an indicator function. It assumes value of 1 when the *condition* in its argument is true and 0 otherwise. The model (1) has several unknown parameters (that will have to be estimated from training data somehow).

We will now explain their meaning. α_{jk} is the effect of the j-th type of the day ($j = 1,\ldots,5$). Note that different segments have different day type effects (because of the subscripting by k). The notation is similar to the so called textbook parametrization often used in the ANOVA and general linear models' context (Graybill, 1976; Searle, 1971). We haste to add that, for numerical stability, the model is actually fitted in the so called sum-to-zero (or contr.sum) parametrization

$$\mu' = \sum_{j=1}^{5}\alpha_{jk},\ \alpha'_{jk} = \alpha_{jk} - \sum_{j=1}^{5}\alpha_{jk},\ j = 1,\ldots,5 \qquad (2)$$

(Rawlings, 1988). In other words, we reparametrize the model (1) to the sum-to-zero for numerical computations and then we reparametrize the results back to the textbook parametrization for convenience. Table 1 shows how different types of the day $D_1,\ldots,D_5$ are defined by specifying for which particular triplet ($t-1,t,t+1$) a particular day type holds. Non- working days are the weekends and (generic) bank holidays of any kind. On the other hand, χ_k and ϕ_k are effects of special Christmas and Easter holidays. Note that these effects act on the top of the generic holiday effect, so that the total holiday effect e.g. for 25th of December is (on the log scale) the sum of generic holiday (given by the day type 4, from Table 1) and Christmas effects. *Christmas* period is (in the Central European implementations of the model) defined to consist of days of December, 23, 24, 25, 26, while *Easter* period is defined to consist form the Wednesday, Thursday, Friday, Saturday of the week before the Easter Monday. ζ_{kt} is the temperature correction which is the most important part of the model with quite rich internal structure that we will explain in detail in the next section. p_{ik} is a multiple of the so called expected annual consumption (scaled as a daily consumption average) for the i-th customer. It is estimated from past consumption record (typically 3 calendar years) of the particular customer. For instance, if we have m roughly annual consumption readings $Y_{ik,\Im_{i1}},\ldots,Y_{ik,\Im_{im}}$ in the intervals $\Im_{i1} = [t_{i1},t_{i2}],\ldots,\Im_{im} = [t_{i,2m-1},t_{i,2m}]$, we compute

$$\hat{p}_{ik} = \frac{Y_{ik,\Im_{i1}} + \ldots + Y_{ik,\Im_{im}}}{t_{im} - t_{i1} + 1} \tag{3}$$

and then condition on that estimate (i.e., we take the $\hat{p}_{ik}$ for the unknown p_{ik}) in all the development that follows. That way, we buy considerable computational simplicity, compared to the correct estimation based on nonlinear mixed effects model style estimation (Davidian & Giltinan, 1995; Pinheiro & Bates, 2000) at the expense of neglecting some (relatively minor) part of the variability in the consumption estimates. It is important, however that the integration period for the $\hat{p}_{ik}$ estimation is long enough.

Note that (1) immediately implies a particular separation

$$\mu_{ikt} \equiv p_{ik}.f_{kt} \tag{4}$$

of substantial practical importance. In fact, (4) achieves multiplicative separation of the individual-specific but time-invariant and common across individuals but time-varying terms. Obviously, the separation is additive on the log scale.

ε_{ikt} is an additive random error term (independent across i,k,t) which describes variability of individual customers around a central tendency of the consumption dynamics. In accord with the heteroscedasticity of the consumptions observed in practice, we assume that $\varepsilon_{ikt} \sim N\left(0,\sigma_k^2.\mu_{ikt}\right)$, i.e. that the error is distributed as a normal (or Gaussian) random variable with zero expected value and variance $\sigma_k^2.\mu_{ikt}$ (which means that variance to mean ratio is allowed to differ across segments). This means that also the observable consumption Y_{ikt} has a normal distribution, $Y_{ikt} \sim N\left(\mu_{ikt},\sigma_k^2.\mu_{ikt}\right)$, with expected value μ_{ikt} (i.e. the true consumption mean for a situation given by calendar effects and temperature is given by μ_{ikt}), variance $\sigma_k^2.\mu_{ikt}$, and coefficient of variation $\frac{\sigma_k}{\sqrt{\mu_{ikt}}}$. This is a bit milder variance-to-mean relationship than that used in (Brabec et al., 2009). The distribution is heteroscedastic (both over individuals and over time). Specifically, variability increases for times when the mean consumption is higher and also for individuals with higher average consumption (within the same segment). These changes are such that the coefficient of variation decreases within a segment, but its proportionality factor is allowed to change among segments to reflect different consumption volatility of e.g. households and small industrial establishments.

Taken together, it is clear that the model (1) has multiplicative correction terms for different calendar phenomena which modulate individual long term daily average consumption and a correction for temperature.

Type of the day code, j	Previous day $(t-1)$	Current day (t)	Next day $(t+1)$
1	working	working	working
2	working	working	non- working
2	non- working	working	non- working
3	non- working	working	working
4	working	non- working	non- working
4	non- working	non- working	non- working
5	non- working	non- working	working
5	working	non- working	working

Table 1. Type of the day codes

2.3 Temperature response function

Temperature response function ζ_{kt} is in the core of model (1). Here, we will describe how it is structured to capture details of the consumption to temperature relationship:

$$\zeta_{kt} = \left(\sum_{j=1}^{5} \psi_{jk}.I_{t \in D_j} \right) . \left(1 + \exp\left(\beta_k . \frac{\sum_{j=0}^{9} T_{t-j}}{10} \right) \right) . \left(\rho_k(T_t) + \omega_k . \sum_{j=1}^{7} \delta_k^{j-1} . \rho_k(T_{t-j}) \right), \quad (5)$$

where T_t is a daily temperature average for day t. We use a nation-wide average based on official met office measurements, but other (more local) temperature versions can be used. Even though a more detailed temperature info can be obtained in principle (e.g. reading at several times for a particular day, daily minima, maxima, etc.), we go with the average as with a cheap and easy to obtain summary.

$\rho_k(.)$ is a segment-specific temperature transformation function. It is assumed to be smooth and monotone decreasing (as it should to conform with principles mentioned in the Introduction). Since it is not known a priori, it has to be estimated from the data. Here we use a nonparametric formulation. In particular, we rely on loess smoother as a part of the GAM (generalized additive model) specified by (1) and (5), (Hastie & Tibshirani, 1990, Hastie et al., 2001).

It is easy to see that the right-most term in the parenthesis represents a nonlinear, but time invariant filter in temperature. In the transformed temperature, $\widetilde{T}_{kt} \equiv \rho_k(T_t)$, it is even a linear time invariant filter. In fact, it is quite similar to the so called Koyck model used in econometrics (Johnston, 1984). It can be perceived as a slight generalization of that model allowing for non-exponential (in fact even for non-monotone) lag weight on nonlinear temperature transforms $\widetilde{T}_{kt}$. $\omega_k > 0$ and $\delta_k^j > 0, j = 1,\ldots,7$ are the parameters which characterize shape of the lag weight distribution. The behavior is somewhat more complex than geometrical decay dictated by the Koyck scheme. While the weights decay geometrically from ω_k at lag 1 (with the rate given by δ_k), they allow for arbitrary (positive) lag-zero-to- lag-one weight ratio (given by ω_k). In particular, they allow for local maximum of the lag distribution at lag one, which is frequently observed in empirical data. The parametrization uses weight of 1 for zero lag within the right-most parenthesis in order to assure identifiability (since the general scaling is provided by the two previous parentheses).

The term in the middle parenthesis essentially modulates the temperature effect seasonally. The moving average in temperature modifies the effect of left and right parentheses terms slowly, according to the "currently prevailing temperature situation", that is differently in year's seasons. In a sense, this term captures (part of) the interaction between the season and temperature effect - we use the word "interaction" in the typical linear statistical models' terminology sense of the word here (Rawlings, 1988). The impact is controlled by the parameter β_k. Note that the weighing in the 10-day temperature average could be non-uniform, at least in principle. Estimation of the weights is extremely difficult here so that we stick to the uniform weighting.

The left-most parenthesis contains an interaction term. It mediates the interaction of nonlinearly transformed temperature and type of the day. In other words, the temperature effect is different on different types of the day. This is a point that was missing in the SLP model formulation (Brabec et al., 2009) and it was considered one of its weaknesses - because the empirical data suggest that the response to the same temperature can be quite different if it occurs on a working day than in it occurs on Saturday, etc. The (saturated) interaction is described by the parameters $\psi_{jk}, j = 1,\ldots 5$. For numerical stability, they are estimated using a similar reparametrization as that mentioned in connection with α_{jk} after model (1) formulation in the section 2.2.

Consumption estimate $\hat{Y}_{ikt}$ (we will denote estimates by hat over the symbol of the quantity to be estimated) for day t, individual i of segment k is obtained as

$$\hat{Y}_{ikt} = \hat{\mu}_{ikt} = \hat{p}_{ik} . \hat{f}_{kt} . \quad (6)$$

Therefore, it is given just by evaluating the model (1), (5) with unknown parameters being replaced by their estimates.
This finishes the description of our gas consumption model (GCM) in daily resolution, which we will call GCMd, for shortness.

2.4 Hourly resolution

The GCMd model (1), (5) operates on daily basis. Obviously, there is no problem to use it for longer periods (e.g. months) by integrating/summing the outputs. But when one needs to operate on finer time scale (hourly), another model level is necessary. Here we follow a relatively simple route that easily achieves an important property of "gas conservation". In particular, we add an hourly sub-model on the top of the daily sub-model in such a way that the daily sum predicted by the GCMd will be redistributed into hours. That will mean that the hourly consumptions of a particular day will really sum to the daily total. To this end, we will formulate the following working model:

$$\log\left(\frac{q_{kth}}{1-q_{kth}}\right) = \lambda_{kth} = I_{t\in work} \cdot \sum_{j=1}^{24} I_{j=h} \cdot \varphi_{jk}^{w} + I_{t\in nonwork} \cdot \sum_{j=1}^{24} I_{j=h} \cdot \varphi_{jk}^{n} + \varepsilon_{kth} \quad (7)$$

where we use $\log(.)$ for the natural logarithm (base e). Indicator functions are used as before, now they help to select parameters (φ) of a particular hour for a working (w) and nonworking (n) day. This is an (empirical) logit model (Agresti, 1990) for proportion of gas consumed at hour h of the day t (averaged across data available from all customers of the given segment k):

$$q_{kth} = \frac{\sum_{i\in k} Y_{ikth}}{\sum_{i\in k} \sum_{h'} Y_{ikth'}} \quad (8)$$

with Y_{ikth} being consumption of a particular customer i within the segment k during hour h of day t. The logit transformation assures here that the modeled proportions will stay within the legal (0,1) range. They do not sum to one automatically, however. Although a multinomial logit model (Agresti, 1990) can be posed to do this, we prefer here (much) simpler formulation (7) and following renormalization. Model (7) is a working (or approximative) model in the sense that it assumes iid (identically distributed) additive error ε_{kth} with zero mean and finite second moment (and independent across k,t,h). This is not complete, but it gives a useful and easy to use approximation.

Given the φ_{hk}^{w} and φ_{hk}^{n}, it is easy to compute estimated proportion consumed during hour h and normalize it properly. It is given by

$$\widetilde{q}_{kth} = \frac{\dfrac{1}{1+\exp(-\lambda_{kth})}}{\sum_{h'\in t} \dfrac{1}{1+\exp(-\lambda_{kth'})}} \tag{9}$$

Amount of gas consumed at hour h of day t is then obtained upon using (1) and (9). When we replace the unknown parameters (appearing implicitly in quantities like μ_{ikt} and $\widetilde{q}_{kth}$) by their estimates (denoted by hats), as in (6), we get the GCM model in hourly resolution, or GCMh:

$$\hat{Y}_{ikth} = \hat{\mu}_{ikt} . \hat{\widetilde{q}}_{kth} \tag{10}$$

In the modeling just described, the daily and hourly steps are separated (leading to substantial computational simplifications during the estimation of parameters). Temperature modulation is used only at the daily level at present (due to practical difficulty to obtain detailed temperature readings quickly enough for routine gas utility calculations).

3. Discussion of practical issues related to the GCM model

3.1. Model estimation

Notice that real use of the model described in previous sections is simple both in daily and hourly resolution, once its parameters (and the nonparametric functions $\rho_k(.)$) are given. For instance, its SW implementation is easy enough and relies upon evaluation of a few fairly simple nonlinear functions (mostly of exponential character). Indeed, the implementation of a model similar to that described here in both the Czech Republic and Slovakia is based on passing the estimated parameter values and tables defining the $\rho_k(.)$ functions (those need to be stored in a fine temperature resolution, e.g. by 0.1 °C) to the gas distribution company or market operator where the evaluation can be done easily and quickly even for a large number of customers.

The separation property (4) is extremely useful in this context. This is because that the time-varying and nonlinear consumption dynamics part f_{kt} needs to be evaluated only once (per segment). Individual long-term-consumption-related p_{ik}'s enter the formula only linearly and hence they can be stored, summed and otherwise operated on, separately from the f_{kt} part.

It is only the estimation of the parameters and of the temperature transformations that is difficult. But that work can be done by a team of specialists (statisticians) once upon a longer period. We re-estimate the parameters once a year in our running projects.

For parameter estimation, we use a sample of customers whose consumption is followed with continuous gas meters. There are about 1000 such customers in the Czech Republic and about 500 in Slovakia. They come from various segments and were selected quasi-randomly from the total customer pool. Their consumptions are measured as a part of large SLP projects running for more than five years. Time-invariant information (important for classification into segments) as well as historical annual consumption readings are obtained from routine gas utility company databases. It is important to acknowledge that even though the data are obtained within a specialized project, they are not error-free. Substantial effort has to be exercised before the data can be used for statistical modeling (model specification and/or parameter estimation). In fact, one to two persons from our team work continuously on the data checking, cleaning and corrections. After an error is located, gas company is contacted and consulted about proper correction. Those data that cannot be corrected unambiguously are replaced by "missing" codes. In the subsequent analyses, we simply assume the MCAR (missing at random) mechanism (Little & Rubin, 1987).

As we mentioned already, the model is specified and hence also fitted in a stratified way – that is separately for each segment. Parameter estimation can be done either on original data (individual measurements) or on averages computed across customers of a given segment. The first approach is more appropriate but it can be troublesome if the data are numerous and/or contain occasional gross errors. In such a case the second might be more robust and quicker.

For the functions ρ_k, we assume that they are smooth and can be approximated with loess (Cleveland, 1979). Due to the presence of both fixed parameters and the nonparametric ρ_k's, the model GCMd is a semiparametric model (Carroll & Wand, 2003). Apart from the temperature correction part, the structure of the model is additive and linear in parameters, after log transformation, therefore it can be fitted as a GAM model (Hastie & Tibshirani, 1990), after a small adjustment. Naturally, we use normal, heteroscedastic GAM with variance being proportional to the mean, logarithmic link and offset into which we put $\log(p_{ikt})$ here. The estimation proceeds in several stages, in the generalized estimating equation style (Small & Wang, 2003). We start the estimation with estimation of the function ρ_k. To that end, we start with a simpler version of the model GCMd which formally corresponds to a restriction with parameters $\psi_{jk} \equiv 1, \beta_k = -\infty, \omega_k = 0$ being held. The $\hat{\rho}_k$ obtained from there is fixed and used in the next step where all parameters are re-estimated (including $\psi_{jk}, \beta_k, \omega_k$). The β, ω, δ parameters that appear nonlinearly in the temperature correction (5) are estimated via profiling, i.e. just by adding an external loop to the GAM fitting function and optimizing the profile quasilikelihood (McCullagh & Nelder, 1989) $Q_P(\beta, \omega, \delta) = \max_{others} Q(\beta, \omega, \delta, others)$ across β, ω, δ, where "others" denotes all other parameters of the model. This is analogous to what had been suggested in (Brabec et al., 2009).

Hourly sub-model needed for GCMh is estimated by a straightforward regression. Alternatively, one might use weighting and/or GAM (generalized linear model) approach.

For practical computations, we use the R system (R Development Core Team, 2010), with both standard packages (gam, in particular) and our own functions and procedures.

3.2 Practical applications of the model and typical tasks which it is used for

The model GCM (be it GCMd or GCMh) is typically used for two main tasks in practice, namely redistribution and prediction. First, it is employed in a retrospective regime when known (roughly annual) total consumption readings need to be decomposed into parts corresponding to smaller time units in such a way that they add to the total. In other words, we need to estimate proportions corresponding to the time intervals of interest, having the total fixed. When the total consumption $Y_{ik,[t_{1i},t_{2i}]}$ over the time interval $[t_{1i},t_{2i}]$ is known for an i-th individual of the k-th segment and it needs to be redistributed into days $t \in [t_{1i},t_{2i}]$, we use the following estimate:

$$\hat{Y}_{ikt}^{R} = \frac{Y_{ik,[t_{1i},t_{2i}]}\hat{Y}_{ikt}}{\sum_{t'=t_{1i}}^{t_{2i}} \hat{Y}_{ikt'}} = \frac{Y_{ik,[t_{1i},t_{2i}]}\hat{f}_{kt}}{\sum_{t'=t_{1i}}^{t_{2i}} \hat{f}_{kt'}} \tag{11}$$

where $\hat{Y}_{ikt}$ has been defined in (6). Disaggregation into hours would be analogous, only the GCMh model would be used instead of the GCMd. Such a disaggregation is very much of interest in accounting when the price of the natural gas changed during the interval $[t_{1i},t_{2i}]$ and hence amounts of gas consumed for lower and higher rates need to be estimated. It is also used when doing a routine network mass balancing, comparing closed network inputs and amounts of gas measured by individual customers' meters (for instance to assess losses). The disaggregated estimates might need to be aggregated again (to a different aggregation than original readings), in this context. The estimate of the desired consumption aggregation both over time and customers is obtained simply by appropriate integration (summation) of the disaggregated estimates (11):

$$\hat{Y}_{I,[T_1,T_2]}^{R} = \sum_{i,k \in I} \sum_{t=T_1}^{t=T_2} \hat{Y}_{ikt}^{R} \tag{12}$$

where I is a given index set. It might e.g. require to sum consumptions of all customers of two selected segments, etc.

Secondly, one might want to have prospective estimates of consumption over the interval which lies, at least partially, in future. Redistribution of the known total is not possible here, and the estimates have to be done without the (helpful) restriction on the total. They will have to be based on $\hat{Y}_{ikt}$ alone. It is clear that such estimates will have to be less precise and hence less reliable, in general. This is even more true in the situation when the average annual consumption changes systematically, e.g. due to the external economic conditions (like crisis) which the GCM model does not take into account. At any rate, the disagreggated estimates can then be used to estimate a new aggregation in a way totally parallel to (12), i.e. as follows:

$$\hat{Y}_{I,[T_1,T_2]} = \sum_{i,k \in I} \sum_{t=T_1}^{t=T_2} \hat{Y}_{ikt} \tag{13}$$

It is important to bear on mind that the estimates (both $\hat{Y}_{ikt}^{R}$ and $\hat{Y}_{ikt}$, as well as their new aggregations) are estimates of means of the consumption distribution. Therefore, they are not to be used directly e.g. for maximal load of a network or similar computations (mean is not a good estimate of maximum). Estimates of the maxima and of general quantiles (Koenker, 2005) of the consumption distribution are possible, but they are much more complicated to get than the means.

3.3 Model calibration

In some cases, it might be useful to calibrate a model against additional data. This step might or might not be necessary (and the additional data might not be even available). One can think that if the original model is good (i.e. well calibrated against the data on which it was fitted), it seems that there should be no space for a further calibration. It might not be necessarily the case at least for two reasons.

First, the sample of customers on which the model was developed, its parameters fitted, and its fit tested might not be entirely representative for the total pool of customers within a given segment or segments. The lack of representativity obviously depends on the quality of the sampling of the customer pool for getting the sample of customers followed in high resolution to obtain data for the subsequent statistical modeling (model "training" or just the estimation of its parameters). We certainly want to stress that a lot of care should be taken in this step and the sampling protocol should definitely conform to principles of the statistical survey sampling (Cochran, 1977). The sample should be definitely drawn at random. It is not enough to haphazardly take a few customers that are easy to follow, e.g. those that are located close to the center managing the study measurements. Such a sample can easily be substantially biased, indeed! Taking the effort (and money) that is later spent in collecting, cleaning and modeling the data, it should really pay off to spend a time to get this first phase right. This even more so when we consider the fact that, when an inappropriate sampling error is made, it practically cannot be corrected later, leading to improper, or at least, inefficient results. The sample should be drawn formally (either using computerized random number generator or by balloting) from the list of all relevant customers (as from the sampling frame), possibly with unequal probabilities of being drawn and/or following stratified or other, more complicated, designs. It is clear, that to get a representative sample is much more difficult than usual, since in fact, we sample not for scalar quantities but for curves which are certainly much more complicated objects with much larger space for not being drawn representatively in all of their (relevant) aspects. It might easily happen that while the sample is appropriate for the most important aspects of the consumption trajectory, it might not be entirely representative e.g. for summer consumption minima. For instance, the sample might over-represent those that do consume gas throughout the year, i.e. those that do not turn off their gas appliances even when the temperature is high. The volume predicted error might be small in this case, but when being interested in relative model error, one could be pressed to improve the model by recalibration (because the small numerators stress the quality of the summer behavior substantially).

Secondly, when the model is to be used e.g. for network balancing, it can easily happen that the values which the model is compared against are obtained by a procedure that is not entirely compatible with the measurement procedure used for individual customer readings and/or for the fine time resolution reading in the sample. For instance, we might want to compare the model results to amount of gas consumed in a closed network (or in the whole gas distribution company). While the model value can be obtained by appropriate integration over time and customers easily, for instance as in (13), obtaining the value which this should be compared to is much more problematic than it seems at first. The problem lies in the fact that, typically there is no direct observation (or measurement) of the total network consumption. Even if we neglect network losses (including technical losses, leaks, illegal consumption) or account for them in a normative way (for instance, in the Czech Republic, there are gas industry standards that describe how to set a (constant) loss percentage) and hence introduce the first approximation, there are many problems in practical settings. The network entry is measured with a device that has only a finite precision (measurement errors are by no means negligible). The precision can even depend on the amount of gas measured in a complicated way. The errors might be even systematic occasionally, e.g. for small gas flows which the meter might not follow correctly (so that summer can easily be much more problematic than winter). Further, there might be large customers within the network, whose consumption need to be subtracted from the network input in order to get HOU+SMC total that is modeled by a model like GCM. These large customers might be followed with their own meters with fine time precision (as it is the case e.g. in the Czech Republic and Slovakia), but all these devices have their errors, both random and systematic. From the previous discussion, it should be clear now that the "observed" SMC+HOU totals

$$Z_{..t} = (input)_t - (sum\,of\,nonHOUSMC\,customers)_t - (normative\,losses)_t \tag{14}$$

have not the same properties as the direct measurements used for model training. It is just an artificial, indirect construct (nothing else is really feasible in practice, however) which might even have systematic errors. Then the calibration of the model can be very much in place (because even a good model that gives correct and precise results for individual consumptions might not do well for network totals).

In the context of the GCM model, we might think about a simple linear calibration of $Z_{..t}$ against $\sum_{i,k} \hat{Y}_{ikt}$ (where it is understood that the summation is against the indexes corresponding to the HOU+SMC customers from the network), i.e. about the calibration model described by the equation (15) and about fitting it by the OLS, ordinary least squares (Rawlings, 1988) i.e. by the simple linear regression:

$$Z_{..t} = \kappa_1 + \kappa_2 . \sum_{i,k} \hat{Y}_{ikt} + error_t . \tag{15}$$

Conceptually, it is a starting point, but it is not good as the final solution to the calibration. Indeed, the model (15) is simple enough, but it has several serious flaws. First, it does not acknowledge the variability in the $\sum_{i,k} \hat{Y}_{ikt}$. Since it is obtained by integration of estimates obtained from random data, it is a random quantity (containing estimation error of $\hat{Y}_{ikt}$'s). In particular, it is not a fixed explanatory variable, as assumed in standard regression problems that lead to the OLS as to the correct solution. The situation here is known as the measurement error problem (Carroll et al., 1995) in Statistics and it is notorious for the possibility of generating spurious regression coefficients (here calibration coefficients) estimates. Secondly, the (globally) linear calibration form assumed by (15) can be a bit too rigid to be useful in real situations. Locally, the calibration might be still linear, but its coefficients can change smoothly over time (e.g. due to various random disturbances to the network).

Therefore, we formulate a more appropriate and complete statistical model from which the calibration will come out as one of its products. It is a model of state-space type (Durbin & Koopman, 2001) that takes all the available information into account simultaneously, unlike the approach based on (15):

$$
\begin{aligned}
&Y_{ikt} = \mu_{ikt} + \varepsilon_{ikt} \qquad\qquad k = 1,\cdots,K \\
&Z_{..t} = \exp(\gamma_t).\sum_{k=1}^{K} \tau_k.\sum_{i=1}^{n_k} Y_{ikt} + \eta_t \\
&\gamma_t = \gamma_{t-1} + \xi_t \\
&\varepsilon_{ikt} \sim N\left(0, \sigma_k^2.\mu_{ikt}\right), \eta_t \sim N\left(0, \sigma_\eta^2\right), \xi_t \sim N\left(0, \sigma_\xi^2\right)
\end{aligned}
\qquad (16)
$$

Here, we take the GCMd parameters as fixed. Their unknown values are replaced by the estimates from the GCMd model (1), (5) fitted previously (hence also μ_{ikt} appearing explicitly in the first K equations, as well as in the error specification and implicitly in the $(K+1)$-th equation are fixed quantities). Therefore, we have only the variances σ_k^2, σ_η^2, σ_ξ^2 as unknown parameters, plus we need to estimate the unknown γ_t's. In the model (16), the first $K+1$ equations are the measurements equations. In a sense they encompass simultaneously what models (1), (5) and (15) try to do separately. There is one state equation which describes possible (slow) movements of the linear calibration coefficient $\exp(\gamma_t)$ in the random walk (RW) style (Kloeden & Platen, 1992). The RW dynamics is imposed on the log scale in order to preserve the plausible range for the calibration coefficients (for even a moderately good model, they certainly should be positive!). The random error terms are specified on the last line. We assume that ε, η and ξ are mutually independent and that each of them is independent across its indexes (t and i,k). For identifiability, we have to have a restriction on τ_k's (that is on the segment-specific changes of the calibration). In general, we prefer the multiplicative restriction $\prod_{k=1}^{K} \tau_k = 1$, but in practical applications of (16), we took even more restrictive model with $\tau_k \equiv 1$.

Although the model (16) can be fitted in the frequentist style via the extended Kalman filter (Harvey, 1989), in practical computations we prefer to use a Bayesian approach to the estimation of all the unknown quantities because of the nonlinearities in the observation operator. Taking suitable (relatively flat) priors, the estimates can be obtained from MCMC simulations as posterior means. We had a good experience with Winbugs (2007) software.

Advantage of the model (16) is that, apart from calibration, it provides a diagnostic tool that might be used to check the fitted model. For instance, comparing the results of the GCMd model (1), (5) alone to the results of the calibration, i.e. of (1), (5), (15), we were able to detect that the GCMd model fit was OK for the training data but that it overestimated network sums over the summer, leading to further investigation of the measurement process at very low gas flows.

4. Illustration on real data

In this paragraph, we will illustrate performance of the GCMd on real data coming from various projects we have been working with. Since these data are proprietary, we normalize the consumptions deliberately in such a way, that they are on 0-1 scale (zero corresponds to the minimal observed consumption and one corresponds to the maximal observed consumption). This way, we work with the data that are unit-less (while the original consumptions were measured in $m^3/100$).

Figure 1 illustrates that the gas consumption modeling is not entirely trivial. It shows individual normalized consumption trajectories for a sample of customers from HOU4 (or household heaters') segment that have been continuously measured in the SLP project. Since considerable overlay occurs at times, the same data are depicted on both original (left) and logarithmic (right) consumption scale. Clearly, there is a strong seasonality in the data (higher consumption in colder parts of the year), but at the same time, there is a lot of inter-individual heterogeneity as well. This variability prevails even within a single (and rather well defined) customer segment, as shown here. Some individuals show trajectories that are markedly different from the others. Most of the variability is concentrated to the scale, which justifies the separation (4). Due to the normalization, we cannot appreciate the fact that the consumptions vary over several orders of magnitude between seasons, which brings further challenges to a modeler. Note that model (1) deals with these (and other) complications through the particular assumptions about error behavior and about multiplicative effects of various model parts.

Figure 2 plots logarithm of the normalized consumption against the mean temperature of the same day for the data sampled from the same customer segment as before, HOU3. Here, the normalization (by subtracting minimum and scaling through division by maximum) is applied to the ratios $\frac{Y_{ikt}}{p_{ik}}$ as to the quantities more comparable across individuals. Clearly, the asymptotes are visible here, but there is still substantial heterogeneity both among different individual customers and within a customer, across time (temperature response is different at different types of the day, etc., as described by the model (1)). This second, within individual variability is exactly where the model (5) comes into play. All of this (and more) needs to be taken into account while estimating the model.

After motivating the model, it is interesting to look at the model's components and compare them across customer segments. They can be plotted and compared easily once the model is estimated (as described in the section 3.1). Figure 3 compares shapes of the nonlinear temperature transformation function $\rho_k(.)$ across different segments, k . It is clearly visible that the shape of the temperature response is substantially different across different segments - not only between private (HOU) and commercial (SMC) groups, but also among different segments within the same group. The segments are numbered in such a way that increasing code means more tendency to using the natural gas predominantly for heating. We can observe that, in the same direction, the temperature response becomes less flat. When examining the curves in a more detail, we can notice that they are asymmetric (in the sense that their derivative is not symmetric around its extreme). For these and related reasons, it is important to estimate them nonparametrically, with no pre-assumed shapes of the response curve. The model (5) with nonparametric ρ_k formulation brings a refinement e.g. over previous parametric formulation of (Brabec at al., 2009), where one minus the logistic cumulative distribution function (CDF) was used for temperature response as well as over other parametric models (including asymmetric ones, like 1-smallest extreme value CDF) that we have tried. Figure 4 shows $\exp(\alpha_{1k}),\ldots,\exp(\alpha_{5k})$'s of model (1), which correspond to the (marginal) multiplicative change induced by operating on day of type 1 though 5. Indeed, we can see that HOU1 consisting of those customers that use the natural gas mostly for cooking have more dramatically shaped day type profile (corresponding to more cooking over the weekends and using the food at the beginning of the next week, see the Table 1). Figure 5 shows a frequency histogram for normalized p_{ik}'s from SMC2 segment (subtracting minimum p_{ik} and dividing by maximum p_{ik} in that segment).

One could continue in the analysis and explore various other effects or their combinations. For instance, there might be considerable interest in evaluating μ_{ikt} for various temperature trajectories (e.g. to see what happens when the temperature falls down to the coldest day on Saturday versus shat happens when that is on Wednesday). This and other computations can be done easily once the model parameters are available (estimated from the sample data). Similarly, one can be interested in hourly part of the model. Figure 6 illustrates this viewpoint. It shows proportions of the daily total consumed at a particular hour for the HOU1 segment. They are easily calculated from (9), when parameters of model (7) have been estimated. For this particular segment of those customers that use the gas mostly for cooking, we can see much more concentrated gas usage on weekends and on holidays (related to more intensive cooking related to lunch preparation).

How does the model fit the data? Figure 7 illustrates the fit of the model to the HOU4 (heaters') data. This is fit on the same data that have been used to estimate the parameters. Since the model is relatively small (less than 20 parameters for modeling hundreds of observations), signs of overfit (or of adhering to the training data too closely, much more closely than to new, independent data) should not be too severe. Nevertheless, one might be interested in how does the model perform on new data and on larger scale as well. The problem is that the new, independent data (unused in the fit) are simply not available in the fine time resolution (since the measurement is costly and all the available information should be used for model training). Nevertheless, aggregated data are available. For instance, total (HOU+SMC) consumptions for closed distribution networks, for individual gas companies and for the whole country are available from routine balancing. To be able to compare the model fit with such data, we need to integrate (or re-aggregate) the model estimates properly, e.g. along the lines of formula (13). When we do this for the balancing data from the Czech Republic, we get the Figure 8. The fit is rather nice, especially when considering that there are other than model errors involved in the comparison (as discussed in the section 3.3) - note that the model output has not been calibrated here in any way.

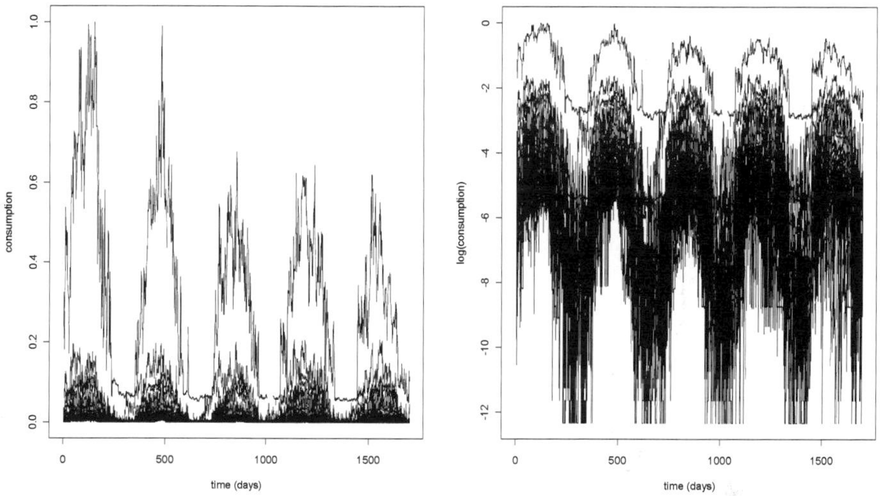

Fig. 1. Overlay of individual consumption trajectories (left - normalized untransformed, right - logarithmically transformed normalized consumptions). Day 1 corresponds to starting point of the SLP projects (October 1, 2004).

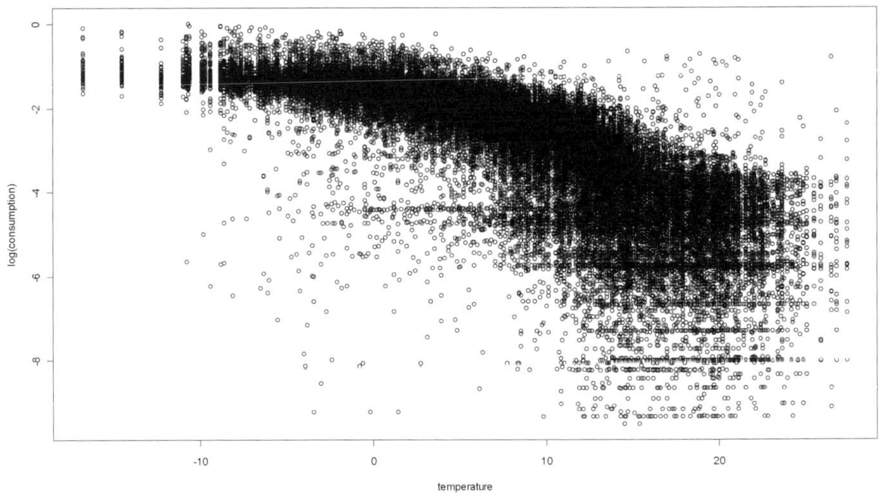

Fig. 2. Logarithmically transformed normalized consumption against current day average temperature.

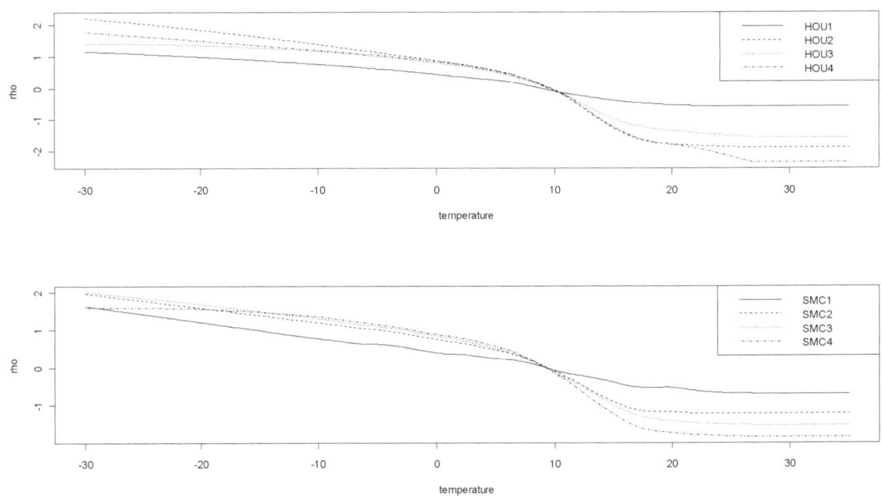

Fig. 3. Temperature response function $\rho_k(.)$ of (5), compared across different HOU and SMC segments.

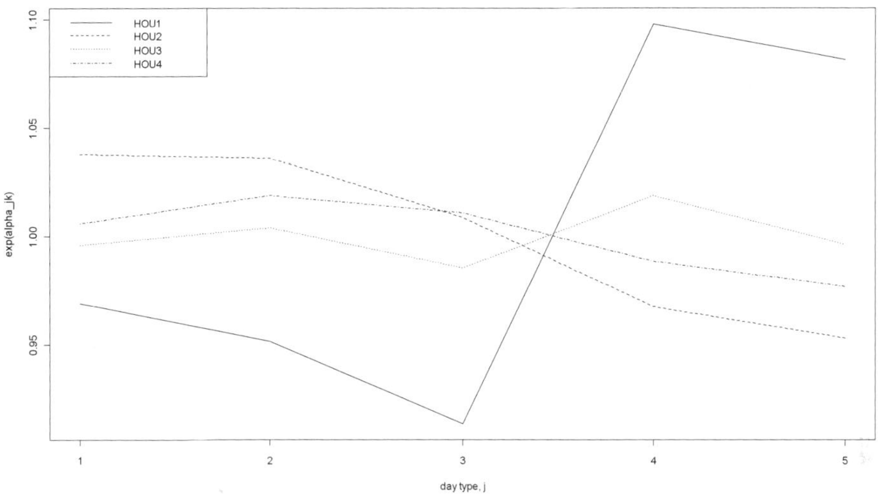

Fig. 4. Marginal factors of day type, $\exp(\alpha_{jk})$ from model (1).

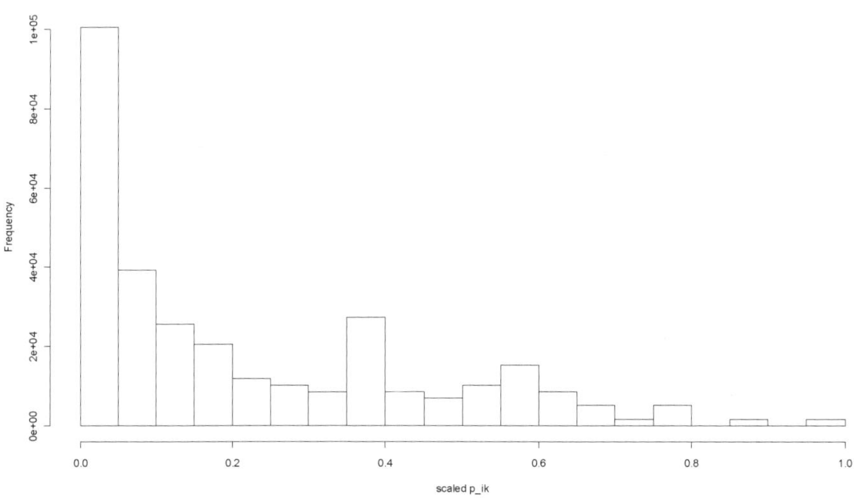

Fig. 5. Histogram of normalized p_{ik}'s for SMC2 segment.

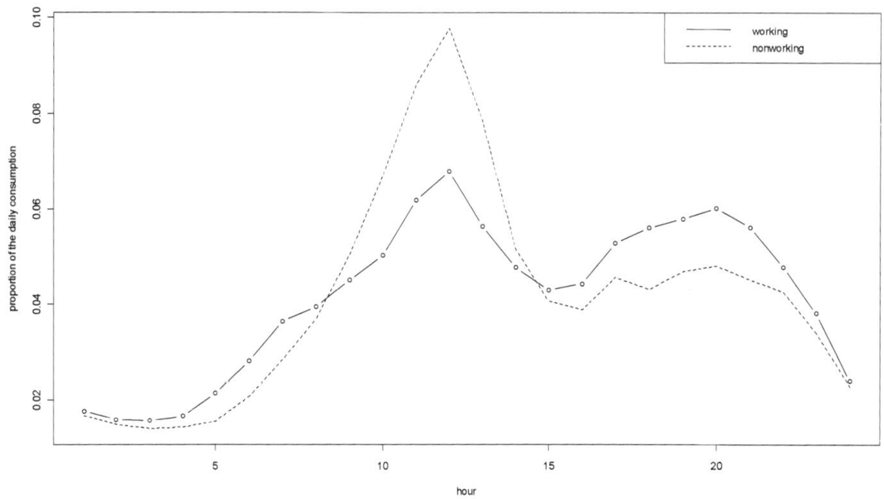

Fig. 6. Proportions of daily consumption totals consumed in a particular hour of the day, i.e. $\widetilde{q}_{kth}$'s from (9), compared between working and nonworking day for HOU1 segment (i.e. for „cookers").

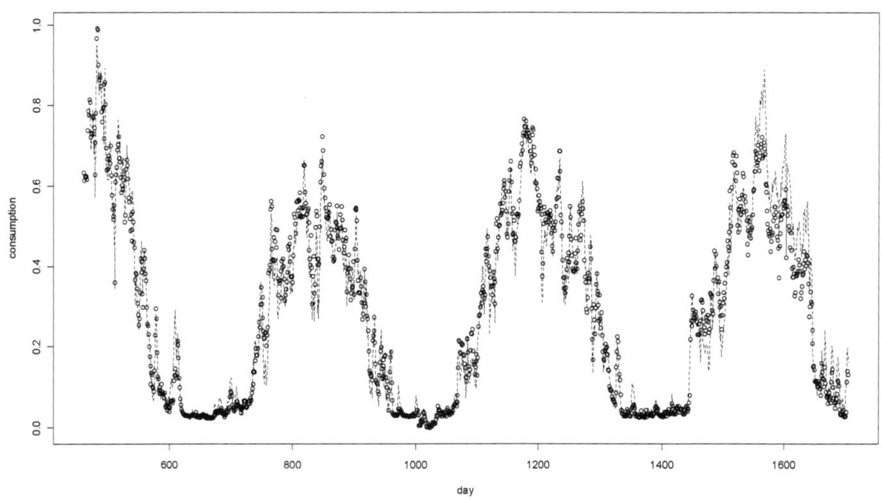

Fig. 7. Fit of the model (1) to the HOU4 data (normalized consumptions as dots and normalized model output as a dotted line).

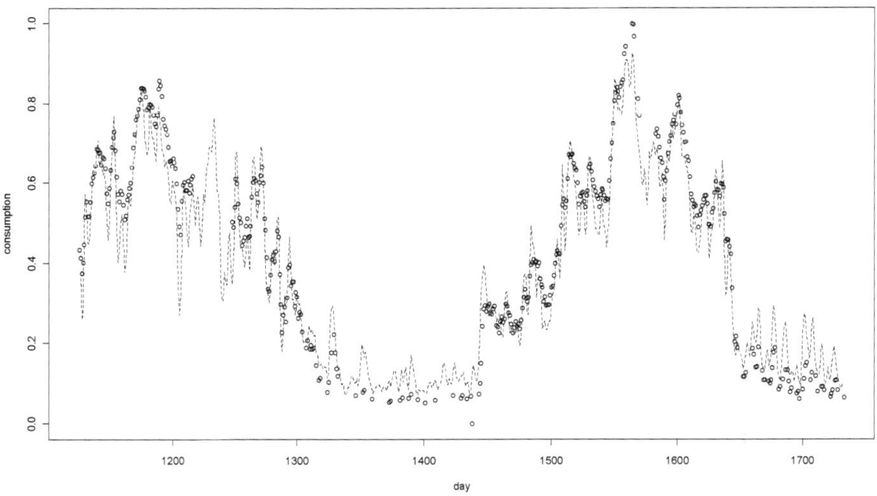

Fig. 8. Fit of the model (1) after disaggregation and re-aggregation of normalized model output, according to (13) on the CR total HOU+SMC consumption data over a period of more than a year.

5. Future work and discussion of some open problems

The model GCM, as described in previous sections uses a single temperature average for all customers. That might be perfectly appropriate if it is employed within a gas company operating over a relatively small and homogeneous region. Even for larger and less homogeneous regions, it can provide good approximations - as we know from its nationwide implementations in both the Czech Republic and Slovakia (both of them being relatively small countries, admittedly). For large and climatologically heterogeneous countries, it might be useful to "regionalize" the GCM in the sense that the global temperature T_t entering the formula (5) would be replaced by the local temperature relevant for the i,k -th customer, i.e. by T_{ikt} . Obviously, it would not be practical to require temperature measurements for each individual customer. Therefore, T_{ikt} would on the i,k index only through the relation of being included in some more local region for which the temperature daily average would be available separately (e.g. county). Technically, this is very simple indeed. Nevertheless, such an improvement requires appropriate (regionally) stratified sample.

The calibration model (16) can be expanded to cover not only proportional but also additive biases. Note that, compared to the full linear calibration of (15), model (16) assumes that the additive bias is zero. The assumption is in line with what we experienced in practice, but for other situations, the model (16) can be expanded by one more state equation to have time-varying intercept as well.

Another useful way of expanding (16) is to drop the $\tau_k \equiv 1$ restriction (while keeping the multiplicative identifiability-related restriction). That might be useful in case when different segments would show very different proportional biases. In our experience, the segment-specific multipliers are very difficult to estimate, however.

The GCM model is very efficient computationally and easy to comprehend conceptually because it implies the relation (4), i.e. the multiplicative separation of the individual-specific but time-invariant and common but time-varying parts. Lack of interaction between the two parts (i.e. between the individual and dynamical parts) is important in practice because it eases implementation substantially. Sums of the p_{ik}'s and sums of the f_{kt}'s can be formed separately when doing the integrations like (12) and (13). The log-additive GCM model certainly captures substantial part of the consumption behavior. If more detailed modeling is attempted, p_{ik}'s might be allowed to follow a time trend (e.g. in connection with changes in economy or with building insulation trends, etc.). Willingness to expand the model along these lines might be hampered by the fact that the impact of this should not be overwhelming however, when the GCM model parameters are re-estimated periodically in relatively short periods (e.g. annually), as suggested. Furthermore, trend common to everybody (within a segment) might not be strong enough to matter at all. More useful would be to assume a trend, but to allow the trend to change the trend from individual to individual. In other words, to allow the individual*dynamics interaction (where the * and the word interaction are used in the statistical sense, as explained before) instead of the additivity of the two terms currently assumed. Obviously, full (or saturated) interaction in the analysis of variance (ANOVA) model style (Graybill, 1976) style is out of question here (since it would not be even estimable). Nevertheless, it is possible to attempt for a more parsimonious model where only part of the interaction (with less degrees of freedom than the saturated interaction) would be specified. Particularly promising route is to allow for time-varying p_{ikt}, i.e. for individual p_{ikt}'s to follow time series models implying slow, but individual-specific dynamics. This is an interesting topic, we have been working on recently (Brabec et al., 2007; Brabec et al., 2008a).

6. Conclusion

In this chapter, we have introduced a gas consumption model GCM for household and small medium customers in daily and hourly resolution and showed how it can be used for various practical tasks, including estimations of consumption aggregates integrated over time and/or customers as well as network related balancing. A model similar to the implementation described here has been running in nationwide system in the Czech Republic and Slovakia for several years already.

The model has a moderately rich structure but it has been built with very strong accent on easy and efficient practical implementation in a gas company or energy market operator environment. It is built in a modular way, enhancing serviceability and making local adjustments to somewhat different conditions rather easy. For more complicated

adjustments, we might be a help a new user with the statistical modeling part if it would result in an interesting project.
The GCM model is built from the first principles, in close contact with empirical behavior of the observed consumptions. It is specified in formal terms as a full blown statistical model (not only mean behavior but also variability assumptions and distributional behavior are given by the model). Our practical experience in natural gas modeling has been strongly supporting the idea that rigorous statistical formulation always pays off here and that it is to be preferred to a haphazard ad hoc or even black box type approaches. There is a lot of structure and many systematic features that a good gas consumption model should follow closely in order to be useful.

7. Acknowledgement

The work was partly supported by the grant 1ET400300513 of the Grant Agency of the Academy of Sciences of the Czech Republic as well as by the Institutional Research Plan AV0Z10300504 'Computer Science for the Information Society: Models, Algorithms, Applications'. We would like to acknowledge important support from the M100300904 project of the Academy of Sciences of the Czech Republic. We also would like to thank to the people from the RWE GasNet, formerly West Bohemian Gas Distribution Company (J. Bečvář, J. Čermáková and others) and to V. Jilemnický from RWE Plynoprojekt for their help and willingness to discuss gas distribution background problems and issues.

8. References

Agresti, A. (1990). *Categorical data analysis*. John Wiley. New York.

Bailey, J. (2000). Load profiling for retail choice: Examining a complex and crucial component of settlement. *Electricity Journal*. 13, 69-74

Brabec, M.; Konár, O.; Malý, M.; Pelikán, E.; Vondráček, J. (2009). A statistical model for natural gas standardized load profiles. *JRSS C - Applied Statistics*. 58, 1, 123-139

Brabec, M.; Malý, M.; Pelikán, E.; Konár, O. (2009a). Statistical calibration of the natural gas consumption model. *WSEAS transactions on systems*. 8, 7, 902-912

Brabec, M.; Konár, O.; Pelikán, E.; Malý, M. (2008). A nonlinear mixed effects model for prediction of natural gas consumption by individual customers. *International Journal of Forecasting*. 24, 659-678

Brabec, M.; Konár, O.; Pelikán. E.; Malý, M. (2008a). Hierarchical model for estimation of yearly sums from irregular longitudinal data. *Book of abstracts, ISF symposium on forecasting*, Nice, France, page 139

Brabec, M.; Konár, O.; Malý,M.; Pelikán, E.; Vondráček, J. (2007). State space model for aggregated longitudinal data. *Abstract Book, 27th International Symposium on Forecasting*, New York 24.-27.6.2007, page 46, ISF.

Carroll, R. J. D.; Ruppert, L. A.; Stefanski. (1995). *Measurement error in nonlinear models*. Chapman & Hall/CRC. London.

Carroll, R. J. & Wand, M. P. (2003). *Semiparametric regression*. Cambridge University Press. Cambridge.

Cleveland, W. S. (1979). Robust Locally Weighted Regression and Smoothing Scatterplots. *Journal of the American Statistical Association*. 74, 829-836

Cochran, W. G. (1977). *Sampling techniques*. John Wiley. New York.
Davidian, M. & Giltinan, D. M. (1995). *Nonlinear models for repeated measurement data*. Chapman and Hall. London.
Durbin, J. & Koopman, S. J. (2001). *Time series analysis by state space methods*. Oxford University Press. Oxford.
Elvira (2010). Elvira project webpage, http://www.cs.cas.cz/nlm/elviraindex-en.htm
Gil S.; Deferrari J. (2004). Generalized model of prediction of natural gas consumption. *Transactions of the ASME*. 126, 90-97
Graybill, F. A. (1976). *Theory and application of the linear model*. Wadsworth & Brooks-Cole. Pacific Grove.
Harvey, A. C. (1989). *Forecasting, structural time series models and the Kalman filter*. Cambridge University Press. Cambridge.
Hastie, T. & Tibshirani, R. (1990). *Generalized additive models*. Chapman and Hall. New York.
Hastie, T.; Tibshirani, R.; Friedman, J. (2001). *The elements of statistical learning*. Springer. New York.
Johnson, R. A. & Wichern, D. W. (1988). *Applied multivariate statistical analysis*. Englewood Cliffs: Prentice Hall. New Jersey.
Johnston, J. (1984). *Econometric Methods*. McGraw Hill. New York.
Kloeden, P. E. & Platen, E. (1992). *Numerical solution of stochastic differential equations*. Springer. New York.
Liedermann, P. (2006). Standardized load profiles for electricity supply—surrogate method for billing customers without continuous measurement (in Czech). *Energetika*, 56, 402–405
Lyness F.K. (1984). Gas demand forecasting. *Statistician*. 33, 9-21
Koenker, R. (2005). *Quantile regression*. Cambridge University Press. Cambridge.
Little, R. J. A. & Rubin, D. B. (1987). *Statistical analysis with missing data*. John Wiley. New York.
McCullagh, P. & Nelder, J. A. (1989). *Generalized linear models*. Chapman & Hall. London.
Pinheiro, J. C. & Bates, D. M. (2000). *Mixed-effects models in S and S-plus*. Springer. New York.
Potočnik,P.; Thaler, M.; Govekar, E.; Grabec, I.; Poredoš, A. (2007). Forecasting risks of natural gas consumption in Slovenia. *Energy Policy*. 35, 4271-4282
Rawlings, J. O. (1988). *Applied regression analysis: A research tool*. Wadsworth & Brooks Cole. Pacific Grove.
R Development Core Team (2010). *R: A Language and Environment for Statistical Computing*, R Foundation for Statistical Computing, Vienna, Austria. http://www.r-project.org/. Accessed on 23 March 2010.
Searle, S. R. (1971). *Linear models*. John Wiley. New York.
Small, C. G. & Wang, J. (2003). *Numerical methods for nonlinear estimating equations*. Clarendon Press. Oxford.
Vondráček, J.; Pelikán, E.; Konár, O.; Čermáková, J.; Eben, K.; Malý, M.; Brabec, M. (2008). A statistical model for the estimation of natural gas consumption. *Applied Energy*. 85, 5, 362-370
Winbugs (2007). *Winbugs with Doodle Bugs*. Version 1.4.3. (6th August 2007) http://www.mrc-bsu.cam.ac.uk/bugs/winbugs/contents.shtml. Medical Research Council. United Kingdom.

18

Molecular dynamics simulations of volumetric thermophysical properties of natural gases

Santiago Aparicio[1] and Mert Atilhan[2]
[1]*Department of Chemistry. University of Burgos*
Spain
[2]*Department of Chemical Engineering. Qatar University*
Qatar

1. Introduction

The accurate knowledge of thermophysical properties of natural gas mixtures is of great importance for practical purposes for the gas industry from exploration stages to final customer use (Jaescke et al., 2002; Wagner & Kleinrahm, 2004; Gallagher, 2006). Two main properties are required by the oil and natural gas industry: *i*) phase equilibria and *ii*) pressure-density-temperature ($P\rho T$) data. The large impact of $P\rho T$, volumetric, data on production, processing and transportation of natural gas is well-known (Hall & Holste, 1990; Husain, 1993; Wagner & Kleinrahm, 2004; Bluvshtein, 2007). Although the required accuracy for the considered properties varies depending on the purpose for which they are used (performance analysis or gas sales, Mokhatab et al. 2006), a high degree of accuracy is frequently required for most of the applications. Thermophysical properties of natural gas systems must be accurately known for national and international custody transfer considering that flowmeters measurements applied to custody transfer are used to buy and sell natural gas between pipeline companies. It should be remarked that custody transfer operations relies on accurate density data, and thus, the economic impact of accurate density measurements is very large both for the calculation of energy content of natural gas and for the flow rate obtained usually from orifice meters (which are used in about 90 % of the metering stations). Inaccuracies in density measurements may lead to very large economical losses or profits for gas producers or buyers, and thus, to conflicts between companies or even countries. Therefore, density values with accuracies below the 0.1 % level are commonly required.
Experimental measurements using highly accurate densimeters, mainly magnetic suspension type, may lead to measurements with a level of uncertainty below the required 0.1 % limit (Wagner & Kleinrahm, 2004; Patil et al., 2007), and thus, this would be the best and most reliable option to obtain density data. Nevertheless, natural gas composition changes remarkably from one to other reservoir (because of age and deep), moreover, the advances in exploration and drilling technologies allow to explore and produce from non-conventional reservoirs (such as ultra-deep ones) which characteristics are very different to the traditional ones. Therefore, considering that to carry out experimental density

measurements is very costly, both in time and economical resources, and having in mind that it is not possible to measure all possible compositions of natural gas in the wide pressure–temperature ranges required for production, transportation and processing purposes, thus, accurate predictive models are required by the gas industry. The current industrial standard model for custody transfer purposes is AGA8-DC92 (Starling & Savidge, 1992); this is a complex multiparametric equation of state which has been used for years as an international standard. Nevertheless, in spite of the common use of AGA8-DC92 in the gas industry it presents several problems. Any predictive model must be validated with reliable experimental data obtained on a limited number of samples that have well defined compositions. None of the samples used for the AGA8-DC92 validation are in the extended region, and thus, the application of this model for mixtures with large concentrations of CO_2, N_2, or long alkanes (as the natural gases from non-conventional reservoirs; Babusiaux, 2004) may be problematic. Moreover, the accuracy of AGA8-DC92 model decreases remarkably on going from region 1 (265 to 335 K, 0 to 12 MPa, deviations < 0.1 %) to regions 2 (12 to 17 MPa, 211 to 394 K, deviation < 0.3 %) and 3 (17 to 70 MPa, 144 to 477 K, deviation < 0.5 %) of model validity, and thus the model performance decreases remarkably when pressure and temperature increases. Recent studies have reported larger deviations than the claimed ones for AGA8-DC92 (Patil et al., 2007). Therefore, two main conclusions may be extracted: *i*) performance of AGA8-DC92 has to be systematically analyzed using a collection of carefully selected mixtures and *ii*) probably a new model has to be proposed as international standard for custody transfer and transmission purposes.

Properties of natural gases, including $P\rho T$ behavior, are a reflection of intermolecular forces rising by the simultaneous presence of very different types of molecules in these complex multicomponent fluids, and their evolution with pressure and temperature. Therefore macroscopic properties of these fluids, such as density, are a consequence of the microscopic structure and behavior of the involved molecules in the considered phases. These relationships between microscopic and macroscopic behavior may be developed through the principles of statistical mechanics if an accurate knowledge of the forces acting between the involved molecules were available. Hence, if we want to develop reliable and accurate models to predict natural gas density, we should get a deeper insight into the microscopic structure for complex gas mixtures and their relationships with $P\rho T$ behavior. For this purpose, computational chemistry methods, mainly classical molecular dynamics and Monte Carlo approaches, are very useful tools allowing *i*) to infer microscopic structural and energetic features and *ii*) to predict macroscopic relevant properties, such as density, as a function of pressure and temperature.

We report in this work a computational study in which the ability of classical molecular dynamics simulation methods to predict $P\rho T$ behavior of complex natural gas mixtures is analyzed. Computational predictions are compared with available highly accurate experimental density data. The possibility of using this approach to predict density values with an acceptable degree of accuracy at moderate computational costs is analyzed, the weaknesses and strengths of the method together with possible future directions are considered.

2. Literature Review

A detailed analysis of the literature shows that the available studies on the use of molecular modelling with predictive purposes for thermophysical properties of natural gas like mixtures are scarce. We will not report here the available literature on the use of Gibbss ensemble Monte Carlo methods for the prediction of phase equilibria, both for pure compounds or for the involved binary mixtures, because this is not the object of this work. Moreover, most of the studies use Monte Carlo approach instead of the classical molecular dynamics methods proposed in this work. In this section, we will analyze the relevant studies available in the open literature in which computational methods, both classical molecular dynamics and Monte Carlo approaches, are used to predict thermophysical properties of natural gas mixtures. Results of literature analysis are reported in a chronological ordering.

Saager and Fischer, 1989, reported a study on NVT molecular dynamics simulations of PVT and thermal properties of pure methane up to 1000 MPa.

Duan et al., 1992, reported a wide study in which PVT properties of pure methane are predicted in the 273 - 2000 K and 100 - 20000 bar ranges using NVE molecular dynamics simulation methods together with a united atom approach. Results show deviations within the 1.5 % range.

Yoshida & Uematsu, 1996, published a study reporting the results for the prediction of PVT properties of natural gases by Monte Carlo molecular simulation. They studied light mixtures composed by methane, ethane, propane, CO_2 and N_2 in the conditions of natural gas transportation in pipelines. They reported deviations lower than 1.5 % in pressure, although density prediction analysis is not carried out in a straightforward manner.

Duan et al., 1996, used molecular dynamics to simulate the PVT properties of the ternary mixture methane + CO_2 + N_2, leading to results with low deviations compared with the general equation of state proposed by the authors.

Neubauer et al., 1999, reported a study in which NPT Monte Carlo method was applied for the simulation of volumetric properties of natural gas mixtures, both in the single phase and two - phase conditions. A united atom approach was used leading to density deviations up to 5 % in the high pressure region, decreasing with decreasing pressure and increasing temperature. These too large deviations, that obviously hinder the application of the proposed method with purely predictive purposes, are a consequence of the poor representation of the composition of the gas during the simulation.

Errington et al., 1999, reported a NPT Monte Carlo study on the properties of the n-alkanes homologous series, developing a united atom intermolecular potential and showing its validity for alkanes up to C78. Saturated densities are predicted within a 2 % limit and liquid densities for the longer n-alkanes to 1 %.

Dysthe et al., 1999, used equilibrium molecular dynamics together with the Green-Kubo formalism to predict transport coefficients of multicomponent natural gas like mixtures including alkanes up to C4 , N_2 and He, both in the gas and liquid phases. Simulations were performed in the NVT ensemble with a united atom approach for alkanes leading to viscosity deviations of 7 % and 11 % for the gaseous and liquid states, respectively.

Escobedo and Chen, 2001, developed a NPT Monte Carlo study for the prediction of Joule - Thomson inversion curves for several fluids, including pure methane and a gas condensate mixture (with alkanes up to C7), leading to reliable predictions.

Lagache et al., 2001, reported a study in which NPT Monte Carlo simulations were performed to compute second order derivatives of the Gibbs energy for simple alkanes up to butane and for the methane - ethane binary mixture. The authors used a united atom potential. Results show that predicted data are in fair agreement with experimental values, even for complex properties such as Joule - Thomson coefficient for which deviations below the 10 % are obtained.

Ungerer, 2003, reported a wide review in which the use of Monte Carlo and molecular dynamics methods is analyzed for several relevant fields in the petroleum and gas industry.

Lagache et al., 2004, reported a study in which NPT Monte Carlo was used for the prediction of density and other relevant properties (thermal expansivity, isothermal compressibility, isobaric heat capacity and Joule - Thomson coefficient) of methane, ethane and two mixtures including heavy components up to 35 carbon atoms. The authors use a united atom approach to model the involved molecules. Density, and compressibility factor, show deviations up to 3 %, whereas for the remaining studied properties deviations are lower than 10 %. The authors show the importance of the mixtures characterization and representation to obtain accurate results.

Ungerer et al., 2004, reported a NPT Monte Carlo study on the prediction of relevant properties, including density, for H_2S - rich gases, showing that although reported results provide valuable information on the understanding of the complex mixed fluids, this computational approach does not lead to the high accuracy required for process design purposes for the studied acid gases.

Ungerer et al. (Ungerer et al., 2006; Ungerer et al., 2006) reported a wide and useful study on the application of Monte Carlo methods for oil and gas production and processing purposes. They showed some of the results previously reported by Lagache et al., 2004, and claimed again to the remarkable importance of an adequate characterization of studied mixtures to obtain accurate results for single phase and equilibria properties.

Bessieres et al., 2006, reported a study in which NPT Monte Carlo simulations were used to predict the Joule-Thomson inversion curve of pure methane using and united atom approach. Results show very accurate predictions with deviations below the 1 % limit.

Vrabec et al., 2007, carried out a study on the performance of molecular simulation methods for the prediction of Joule-Thomson inversion curves for light natural gas mixtures. Reported results show deviations usually within the 5 % range, larger for high temperatures, but being competitive with most state-of-the-art EOS in predicting Joule - Thomson inversion curves.

In a review work, Ungerer et al., 2007, analyzed the weaknesses and strengths of using molecular simulation for the predication of thermophysical properties of complex fluids, including natural gas like mixtures. The authors claim that one of the main limitations of the computational approach is the availability of potentials and force field parameters tested in wide pressure-temperature ranges.

The main conclusions obtained from the analysis of the available open literature are:

i) Monte Carlo approach is used in an exclusive basis when thermophysical properties of natural gas mixtures are under study.

ii) United atom potentials are the most common option.

iii) Studies in wide pressure-temperature ranges and for multicomponent mixtures are very scarce, and thus the performance of the proposed approaches is not clear.

Thus, in this work we report results using the molecular dynamics approach, all-atoms potential, and analysis in wide *P-T* ranges for selected pure and mixed fluids. This methodology was considered because of the absence of similar studies in the open literature to analyze its validity for natural gas industry production, transportation and processing purposes.

3. Computational Methods

Classical molecular dynamics simulations were carried out using the TINKER molecular modeling package (Ponder, 2004). All simulations were performed in the NPT ensemble; the Nosé–Hoover method (Hoover, 1985) was used to control the temperature and pressure of the simulation system. The motion equations were solved using the Verlet Leapfrog integration algorithm (Allen & Tildesley, 1989). Long-range electrostatic interactions were treated with the smooth particle mesh Ewald method (Essmann, 1995).

The simulated systems consist of cubic boxes, with the number of molecules and compositions, for mixed fluids, reported in Table 1, to which periodic boundary conditions were applied in the three directions to simulate an infinite system. The composition of the mixed fluid selected to test the performance of the computational approach for multicomponent natural gas–like mixtures resembles the one reported by Patil et al., 2007, for which very accurate and reliable density data obtained through magnetic suspension densitometers are reported. The number of molecules used for the simulation of pure compounds was selected to obtain systems with 4000 - 5000 total atoms leading to reasonable computing times. For the mixture we have selected a total number of molecules (1000) that allow the representation of all the involved species, even those that appear at very low mole fraction but which effect on the mixed fluid behavior is important.

The simulations were performed using a cutoff radius of $L/2$ Å for the non bonded interactions, L being the initial box side. Initial boxes generated using the PACKMOL program (Martínez & Martínez, 2005) were minimized according to the MINIMIZE program in TINKER package to a 0.01 kcal mol^{-1} Å^{-1} *rms* gradient. Long simulation times are needed for computing the properties of these fluids and procedures have to be designed carefully to avoid the presence of local minima. Therefore several heating and quenching steps in the NVT ensemble up to 600 K were performed after which a 100 ps NVT equilibration molecular dynamics simulation was run at the studied temperature; finally, from the output NVT simulation configuration, a run of 1 ns (time step 1 fs) in the NPT ensemble at the studied pressure and temperature was run, from which the first 0.5 ns were used to ensure equilibration (checked through constant energy) and the remaining 0.5 ns for data collection. n-Alkanes were described according to the so called Optimized Potential for Liquid Simulations (*all atom* version) OPLS–AA (Jorgensen et al., 1996), eqs. 1-4. Parameters for CO_2 and N_2 were obtained from the literature (Shi & Maginn, 2008; Lagache et al., 2005), with bonds, angles and non-bonded interactions treated using eqs. 1,2 and 4. A Lennard- Jones 6-12 potential, eq. 4, was used to describe the interactions between sites which are separated more than three bonds, if they are in the same molecule (intramolecular interactions), and for interactions between sites belonging to different molecules (intermolecular interactions). Non-bonded interactions between 1-4 sites are scaled with a 0.5 factor. Lorentz-Berthelot mixing rules are applied for Lennard–Jones terms between different sites, eqs. 5-6. The used forcefield parameters are reported in Table 2.

$$E_{bond} = \sum_{bonds} k_r \left(r - r_{eq}\right)^2 \qquad (1)$$

$$E_{angle} = \sum_{angles} k_\theta \left(\theta - \theta_{eq}\right)^2 \qquad (2)$$

$$E_{bond} = \frac{V_1}{2}(1+\cos\phi) - \frac{V_2}{2}(1-\cos 2\phi) + \frac{V_3}{2}(1+\cos 3\phi) \qquad (3)$$

$$E_{nonbonded} = \sum_i \sum_j \left[\frac{q_i q_j e^2}{r_{ij}} + 4\varepsilon_{ij}\left(\frac{\sigma_{ij}^{12}}{r_{ij}^{12}} - \frac{\sigma_{ij}^{6}}{r_{ij}^{6}} \right) \right] \qquad (4)$$

$$\sigma_{ij} = \frac{\left(\sigma_i + \sigma_j\right)}{2} \qquad (5)$$

$$\varepsilon_{ij} = \sqrt{\varepsilon_i \varepsilon_j} \qquad (6)$$

Thermophysical properties were obtained from the statistical analysis of the results in the last 0.5 ns together with fluctuation formulae for the NPT ensemble, eqs. 7-9. We have selected five relevant properties to analyze the performance of the proposed computational approach (Lagache et al., 2004): density (ρ, because of its importance for natural gas engineering purposes), thermal expansion coefficient (α_P), isothermal compressibility (β_T), isobaric heat capacity (C_P) and Joule - Thomson coefficient (μ, calculated according to eq. 10).

$$\alpha_P = \frac{1}{k_B T^2 V} \left\langle \delta V \cdot \delta\left(K + \mathrm{P} + PV\right) \right\rangle \qquad (7)$$

$$\beta_T = \frac{1}{k_B T V} \left\langle \delta V^2 \right\rangle \qquad (8)$$

$$C_P = \frac{1}{k_B T^2} \left\langle \delta\left(K + \mathrm{P} + PV\right)^2 \right\rangle \qquad (9)$$

$$\mu = \frac{v}{C_P}\left(T\alpha_P - 1\right) \qquad (10)$$

In eqs. 7-9, <> stands for ensemble averages, *K* for kinetic energy, P for potential energy, *P* for pressure, *T* for temperature, *V* for volume and k_B for the Boltzmann constant. The notation δX (where X stands for any of the quantities reported in eqs. 7-9) denotes X-<X>. In eq.10, v stands for the molar volume.

Pure fluids simulations		Mixed fluid simulations		
Component	N	Experimental mole fraction (Patil et al., 2007)	N	Mole fraction
Methane	1000	0.90991	910	0.91000
Ethane	500	0.02949	29	0.02900
Propane	500	0.01513	15	0.01500
i-Butane	300	0.00755	8	0.00800
n-Butane	300	0.00755	8	0.00800
i-Pentane	300	0.00299	3	0.00300
n-Pentane	300	0.00304	3	0.00300
CO_2	1000	0.02031	20	0.02000
N_2	1000	0.00403	4	0.00400

Table 1. Molecular representation of studied systems used for molecular dynamic simulations. N stands for the number of molecules used in the simulation. For mixed fluid simulations the experimental compositions reported by Patil et al., 2007, are reported together with the composition used in our simulations for comparative purposes

alkanes			
	r_{eq} / Å	k_r / kcal mol^{-1}	
C-C	1.529	268.0	
H-C	1.090	340.0	
	θ_{eq} / deg	k_θ / kcal mol^{-1}	
H-C-H	107.8	33.00	
H-C-C	110.7	37.50	
C-C-C	112.7	58.35	
	V_1 / kcal mol^{-1}	V_2 / kcal mol^{-1}	V_3 / kcal mol^{-1}
H-C-C-H	0.000	0.000	0.318
H-C-C-C	0.000	0.000	0.366
C-C-C-C	1.740	-0.157	0.279
	q / e$^-$	σ / Å	ε / kcal mol^{-1}
C (in CH_4)	-0.240	3.500	0.066
C (in RCH_3)	-0.180	3.500	0.066
C (in R_2CH_2)	-0.120	3.500	0.066
C (in R_3CH_3)	-0.060	3.500	0.066
H	0.060	2.500	0.030
CO_2			
	r_{eq} / Å	k_r / kcal mol^{-1}	
C-O	1.160	1030.0	
	θ_{eq} / deg	k_θ / kcal mol^{-1}	
O-C-O	180.0	56.00	
	q / e$^-$	σ / Å	ε / kcal mol^{-1}
C	0.700	3.143	0.053
O	-0.350	3.424	0.157
N_2			
	r_{eq} / Å		
N-N	1.098		
	q / e$^-$	σ / Å	ε / kcal mol^{-1}
N	0.000	3.798	0.142

Table 2. Forcefield parameters used along this work obtained from the literature (Jorgensen et al., 1996; Shi & Maginn, 2008; Lagache et al., 2004)

4. Molecular Dynamics Simulations Results

4.1 Methane

As a first proof of the ability of the computational approach proposed in this work to reproduce accurately the properties of pure methane (the main component of natural gas mixtures) we have calculated the phase equilibria diagram of this fluid. Although the objective of this work is not devoted to the analysis of phase equilibria, these results may be considered as a previous test of the model. Phase equilibria was calculated using Monte Carlo method from the coupled-decoupled (Martin & Siepmann, 1999), dual cut-off (Vlugt et al., 1998), configurational bias (Siepmann & Frenkel, 1992) simulations, using the MF-CPN strategy (Martin & Frischknecht, 2006), which were performed using MCCCS Towhee code in the NVT-Gibbs ensemble (Errington & Panagiotopoulos, 1998). Simulated phase equilibria in comparison with literature experimental results are reported in Fig. 1. Critical properties were calculated using the Ising scaling law (Smit and Williams, 1990) with a scaling exponent of 0.32.

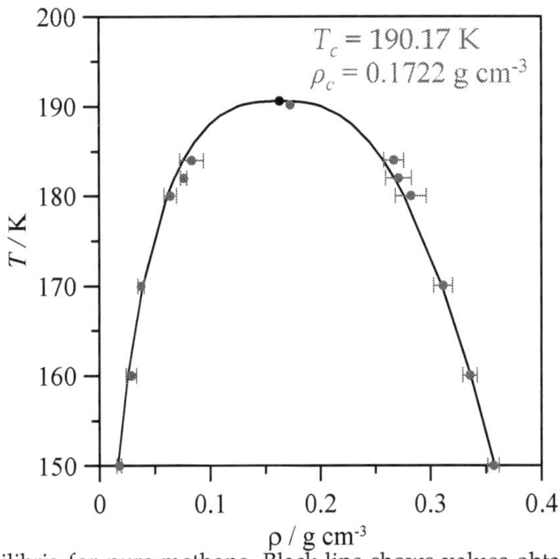

Fig. 1. Phase equilibria for pure methane. Black line shows values obtained from the NIST webbook reference data (Linstrom & Mallard, 2009), red points show values calculated in this work using Monte Carlo simulations (together with error bars), black point shows reference critical point from NIST webbook (Linstrom & Mallard, 2009), the calculated point at the highest temperature shows the simulated critical point obtained from Monte Carlo results and Ising scaling law (with numerical values reported in red within the Figure)

Results reported in Fig. 1 show a fair agreement between experimental reference data (NIST webbook, Linstrom & Mallard, 2009) and simulated values. Critical temperature is just a 0.23 % lower and critical density a 5.53 % larger than the experimental values. Therefore the proposed forcefield parameterization reproduces accurately the phase equilibria of pure methane.

To obtain reliable density values from molecular dynamics simulations, the evolution of calculated densities with simulation time was analyzed (together with energy values) to assure that stable configurations are obtained and to check if longer simulation times are required. Therefore in Fig. 2 we report some representative results of these analyses for pure methane as a function of pressure and temperature, we report selected results for the lowest, medium and larger densities studied.

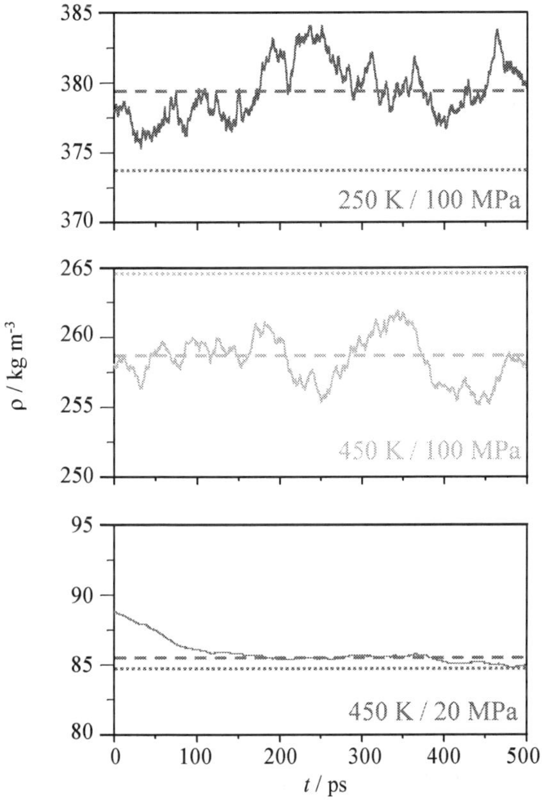

Fig. 2. Representation of calculated densities for pure methane, ρ, obtained from NPT molecular dynamics simulations as a function of simulation time. The results are obtained for the last 500 ps of the simulations. Symbols: continuous lines show density values, dashed lines average values obtained from the simulation and dotted line experimental values obtained from NIST webbook reference data (Linstrom & Mallard, 2009)

Results reported in Fig. 2 show that density values converge rapidly, and thus no longer simulation times are required (total simulation time is 1000 ps). In some cases, as it is reported in Fig. 2 for 450 K / 20 MPa simulations (and usually appears for the lower temperature simulations), density convergence is slower, nevertheless density also converge for this temperature, and thus statistical analysis was performed with the data once convergence is reached. Once density convergence is reached, fluctuations are in the ± 1.5 % range of the average density value.

Calculated densities of pure methane are reported in Table 3 in comparison with available reference data. Simulations are carried out in wide pressure-temperature ranges (250 to 450 K and 20 to 100 MPa), relevant for natural gas engineering purposes, to analyze the global performance of the approach. These pressure-temperature ranges were used for all the simulations along this work. Deviations between experimental and simulated densities are in the 1 - 2 % range for the studied pressure - temperature range, and thus, slightly larger than the accuracy required for many natural gas engineering purposes.

T / K	P / MPa	ρ_{ref} / kg m^{-3}	ρ_{sim} / kg m^{-3}	standard deviation of ρ_{sim} / kg m^{-3}	$100\frac{(\rho_{ref}-\rho_{sim})}{\rho_{ref}}$
250	20	223.23	219.30	0.84	1.76
250	60	332.38	336.69	1.98	-1.30
250	100	373.51	379.78	2.01	-1.68
350	20	119.41	117.94	0.64	1.23
350	60	255.39	251.25	1.93	1.62
350	100	312.26	308.73	1.19	1.13
450	20	84.72	85.36	0.30	-0.76
450	60	202.23	198.83	1.19	1.68
450	100	264.73	257.96	1.10	2.56

Table 3. Comparison between reference, ρ_{ref} (Linstrom & Mallard, 2009), and NPT molecular dynamics simulated, ρ_{sim}, density values, for methane as a function of pressure and temperature

T / K	P / MPa	$10^3\,\alpha_P$ / K^{-1}		$10^3\,\beta_T$ / MPa^{-1}		C_P / J mol^{-1} K^{-1}		µ / K MPa^{-1}	
		ref	*sim*	*ref*	*sim*	*ref*	*sim*	*ref*	*sim*
250	20	7.8345	7.05	29.943	27.72	66.246	64.42	1.0400	0.87
250	60	2.6297	2.93	4.1285	4.66	50.537	48.85	-0.32718	-0.26
250	100	1.8203	1.99	2.1330	2.40	48.036	47.05	-0.48724	-0.45
350	20	4.4143	4.85	46.945	46.09	50.698	49.73	1.4443	1.91
350	60	2.5420	2.55	7.6472	8.49	50.924	49.92	-0.13605	-0.14
350	100	1.7380	1.88	3.4090	3.58	49.162	48.84	-0.40934	-0.36
450	20	2.7308	2.52	46.794	40.89	49.556	48.21	0.87455	0.52
450	60	2.1109	1.91	10.145	10.96	52.622	50.95	-0.075505	-0.22
450	100	1.5561	1.51	4.5467	4.03	52.218	50.02	-0.34786	-0.40

Table 4. Comparison between reference, *ref* (Linstrom & Mallard, 2009, and Lemmon et al., 2002), and NPT molecular dynamics simulated properties, *sim*, for methane as a function of pressure and temperature. Properties: thermal expansion coefficient (α_P), isothermal compressibility (β_T), isobaric heat capacity (C_P) and Joule-Thomson coefficient (µ)

The remaining properties obtained from molecular dynamics simulations are reported in Table 4. Average absolute deviation, between reference and simulated data, are 7.70 %, 9.19 %, 2.53 % and 37.62 % for thermal expansion coefficient, isothermal compressibility, isobaric heat capacity and Joule - Thomson coefficient, respectively. Therefore, deviations obtained for these properties obtained from fluctuation analysis of simulations are too large, especially for Joule - Thomson coefficient. We report in Fig. 3 a comparison between experimental and simulated values of isobaric heat capacity and Joule - Thomson coefficient, from the reported results it may be inferred that trends of the properties with pressure and temperature are properly reproduced by simulations, with the larger deviations obtained for Joule - Thomson coefficient at the lower pressures. Deviations obtained in this work are in agreement with those reported by Lagache et al. (2004) using Monte Carlo approach with a different force field parameterization. Therefore, the main problems rising from molecular dynamics simulations are for derived properties, especially for Joule - Thomson coefficient, as we may expect.

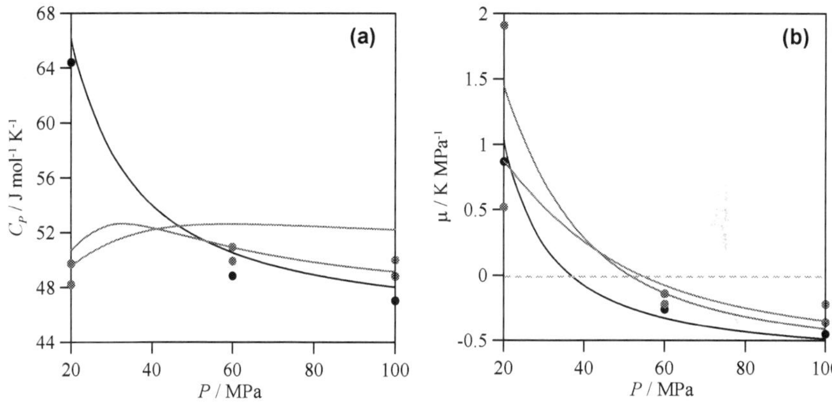

Fig. 3. Isobaric heat capacity (C_P) and Joule-Thomson coefficient (μ) of pure methane as a function of pressure and temperature. Symbols, NPT molecular dynamics simulated properties, and lines, reference data (Linstrom & Mallard, 2009). Color code: black (250 K), blue (350 K) and red (450 K)

4.2 Ethane

Density predictions for ethane are reported in Table 5, deviations between experimental and simulated data are larger for ethane than for methane (Tables 3 and 5), and for methane these deviations are larger than the ones required for natural gas engineering purposes. Thus inaccuracies of simulations increase with increasing chain length of n-alkane, in spite of using an all-atoms force field approach, this could lead to increasing inaccuracies for multicomponent mixtures rich in large n-alkanes. Derived properties are reported in Table 6, large deviations are obtained for all the studied properties, especially for Joule-Thomson coefficient, as we obtained in the previous section for methane. Average absolute deviation, between reference and simulated data, are 8.86 %, 10.62 %, 3.06 % and 17.798 % for thermal expansion coefficient, isothermal compressibility, isobaric heat capacity and Joule-Thomson coefficient, respectively.

T / K	P / MPa	ρ_{ref} / kg m^{-3}	ρ_{sim} / kg m^{-3}	standard deviation of ρ_{sim} / kg m^{-3}	$100\frac{(\rho_{ref}-\rho_{sim})}{\rho_{ref}}$
250	20	486.74	471.42	1.22	3.15
250	60	529.91	510.23	1.81	3.71
250	100	557.16	537.81	1.95	3.47
350	20	343.55	329.85	0.85	3.99
350	60	445.96	436.74	1.74	2.07
350	100	489.08	478.33	1.39	2.20
450	20	202.36	195.12	0.94	3.58
450	60	370.02	360.81	1.22	2.49
450	100	429.48	420.75	1.40	2.03

Table 5. Comparison between reference, ρ_{ref} (Linstrom & Mallard, 2009), and NPT molecular dynamics simulated, ρ_{sim}, density values, for ethane as a function of pressure and temperature

T / K	P / MPa	$10^3\,\alpha_P$ / K^{-1}		$10^3\,\beta_T$ / MPa^{-1}		C_P / J mol^{-1} K^{-1}		μ / K MPa^{-1}	
		ref	*sim*	*ref*	*sim*	*ref*	*sim*	*ref*	*sim*
250	20	2.5063	2.84	3.1016	2.66	77.586	75.04	-0.29735	-0.25
250	60	1.6200	1.39	1.5345	1.63	71.927	70.09	-0.46941	-0.55
250	100	1.2873	1.32	1.0391	1.02	70.138	68.85	-0.52185	-0.54
350	20	4.8378	4.35	16.779	14.61	98.734	94.76	0.61455	0.50
350	60	1.8187	1.67	3.1585	3.56	81.503	78.87	-0.30069	-0.36
350	100	1.3102	1.35	1.7358	1.47	78.897	76.32	-0.42192	-0.43
450	20	4.6465	5.24	44.489	38.01	99.313	96.23	1.6323	2.17
450	60	1.8816	2.06	5.5582	6.23	91.100	88.78	-0.13673	-0.07
450	100	1.2809	1.36	2.5866	2.44	88.511	85.24	-0.33508	-0.33

Table 6. Comparison between reference, *ref* (Linstrom & Mallard, 2009, and Lemmon et al., 2002), and NPT molecular dynamics simulated properties, *sim*, for ethane as a function of pressure and temperature. Properties: thermal expansion coefficient (α_P), isothermal compressibility (β_T), isobaric heat capacity (C_P) and Joule-Thomson coefficient (μ)

4.3 CO_2 and N_2

Results for carbon dioxide and nitrogen are reported in Tables 7 - 10. Although these gases are minor component of common natural gas mixtures, their effect on the thermodynamic properties of natural gas mixtures is relevant, and thus, the accurate reproduction of their properties through simulations is required to obtain reliable results. Density is simulated for both gases with low deviations, average absolute deviation, between reference and simulated data, are 1.70 % and 1.71 % for carbon dioxide and nitrogen respectively. Nevertheless, these deviations are larger than 1 %, and thus, as for the results reported for n-alkanes in previous sections, simulations lead to deviations too large for engineering purposes. Derived properties reported in Tables 8 and 10, show too large deviations, leading to average absolute deviations, between reference and simulated data, of 9.52 and 9.11 % (for thermal expansion coefficient, for carbon dioxide and nitrogen, respectively), 13.09 and 13.17 % (for isothermal compressibility), 3.54 and 3.32 % (for isobaric heat capacity) and 70.57 and 40.3 % (for Joule-Thomson coefficient).

T / K	P / MPa	ρ_{ref} / kg m^{-3}	ρ_{sim} / kg m^{-3}	standard deviation of ρ_{sim} / kg m^{-3}	$100\frac{(\rho_{ref}-\rho_{sim})}{\rho_{ref}}$
250	20	1105.5	1082.45	1.40	2.09
250	60	1183.6	1161.16	1.10	1.90
250	100	1235.6	1215.04	1.55	1.66
350	20	614.18	603.76	1.04	1.70
350	60	923.92	910.12	1.47	1.49
350	100	1027.7	1013.22	1.33	1.41
450	20	285.14	280.48	1.25	1.63
450	60	689.59	677.97	1.01	1.69
450	100	847.00	832.01	1.38	1.77

Table 7. Comparison between reference, ρ_{ref} (Linstrom & Mallard, 2009), and NPT molecular dynamics simulated, ρ_{sim}, density values, for carbon dioxide as a function of pressure and temperature

T / K	P / MPa	$10^3\, a_P$ / K^{-1}		$10^3\, \beta_T$ / MPa^{-1}		C_P / J mol^{-1} K^{-1}		μ / K MPa^{-1}	
		ref	*sim*	*ref*	*sim*	*ref*	*sim*	*ref*	*sim*
250	20	3.1348	2.83	2.3550	2.04	83.318	80.11	-0.10335	-0.15
250	60	2.1531	1.96	1.2889	1.13	76.168	73.35	-0.22540	-0.26
250	100	1.7459	1.60	0.9062	0.79	73.521	70.81	-0.27302	-0.31
350	20	10.574	9.49	37.489	32.62	115.34	111.89	1.6780	1.51
350	60	2.7954	2.54	3.7943	3.31	74.349	72.10	-0.013846	-0.07
350	100	1.9221	1.75	1.9368	1.70	69.370	67.18	-0.20202	-0.25
450	20	4.5519	4.04	54.620	47.14	64.820	62.38	2.4962	2.06
450	60	2.8988	2.60	8.1235	7.02	69.816	67.12	0.27833	0.16
450	100	1.9076	1.75	3.3676	2.92	66.107	63.57	-0.11128	-0.18

Table 8. Comparison between reference, *ref* (Linstrom & Mallard, 2009, and Lemmon et al., 2002), and NPT molecular dynamics simulated properties, *sim*, for carbon dioxide as a function of pressure and temperature. Properties: thermal expansion coefficient (a_P), isothermal compressibility (β_T), isobaric heat capacity (C_P) and Joule-Thomson coefficient (μ)

T / K	P / MPa	ρ_{ref} / kg m^{-3}	ρ_{sim} / kg m^{-3}	standard deviation of ρ_{sim} / kg m^{-3}	$100\frac{(\rho_{ref}-\rho_{sim})}{\rho_{ref}}$
250	20	268.10	262.99	1.62	1.91
250	60	523.04	513.14	1.51	1.89
250	100	631.37	621.81	1.44	1.51
350	20	178.03	174.46	1.00	2.01
350	60	400.90	394.65	1.74	1.56
350	100	520.38	513.04	1.25	1.41
450	20	136.50	134.03	1.04	1.81
450	60	325.58	320.51	1.91	1.56
450	100	442.22	434.70	1.58	1.70

Table 9. Comparison between reference, ρ_{ref} (Linstrom & Mallard, 2009), and NPT molecular dynamics simulated, ρ_{sim}, density values, for nitrogen as a function of pressure and temperature

T / K	P / MPa	$10^3\,\alpha_P$ / K^{-1}		$10^3\,\beta_T$ / MPa^{-1}		C_P / J mol^{-1} K^{-1}		μ / K MPa^{-1}	
		ref	*sim*	*ref*	*sim*	*ref*	*sim*	*ref*	*sim*
250	20	5.4355	5.04	41.524	36.13	40.966	39.42	0.91540	0.70
250	60	2.9791	2.73	7.0050	6.11	40.680	39.20	-0.33604	-0.44
250	100	2.0939	1.93	3.2581	2.81	39.685	38.29	-0.53276	-0.61
350	20	3.1609	2.86	43.877	38.05	34.185	33.13	0.48940	0.00
350	60	2.3437	2.13	9.6497	8.39	36.591	35.54	-0.34318	-0.51
350	100	1.7733	1.60	4.4900	3.92	36.671	35.73	-0.55688	-0.67
450	20	2.2621	2.02	44.586	38.50	32.184	31.14	0.11451	-0.61
450	60	1.8502	1.66	11.083	9.61	34.323	33.16	-0.41970	-0.67
450	100	1.4925	1.35	5.3610	4.66	34.852	33.53	-0.59686	-0.75

Table 10. Comparison between reference, *ref* (Linstrom & Mallard, 2009, and Lemmon et al., 2002), and NPT molecular dynamics simulated properties, *sim*, for nitrogen as a function of pressure and temperature. Properties: thermal expansion coefficient (α_P), isothermal compressibility (β_T), isobaric heat capacity (C_P) and Joule–Thomson coefficient (μ)

4.4 Lean Natural Gas Mixtures

Results for the multicomponent natural gas mixture studied in this work are reported in Table 11. Deviations are in the 2–4 % range with an average absolute deviation from experimental results of 2.55 %, which is too large in comparison with reference methods used by the natural gas industry for density predictions. There are not available experimental data for derived properties for the studied mixture. Therefore, although thermal expansion coefficient, isothermal compressibility, isobaric heat capacity and Joule–Thomson coefficient were calculated using molecular dynamics simulations for the studied mixture, these results are not reported here because it is not possible to infer their accuracy. Nevertheless, we may expect that considering that density deviations reported in Table 11 for the studied mixture are larger than those reported in previous sections for methane, ethane, carbon dioxide and nitrogen, deviations of derived properties for the mixture should be also larger than for the studied pure fluids.

T / K	P / MPa	ρ_{ref} / kg m^{-3}	ρ_{sim} / kg m^{-3}	standard deviation of ρ_{sim} / kg m^{-3}	$100\frac{(\rho_{ref}-\rho_{sim})}{\rho_{ref}}$
270	3.483	31.673	32.39	5.87	-2.26
270	17.261	206.727	201.32	1.22	2.62
270	34.543	298.051	288.81	0.89	3.10
290	3.453	28.415	29.05	6.02	-2.23
290	17.271	175.201	171.27	1.07	2.24
290	34.506	274.016	265.64	0.70	3.06
305	3.484	26.894	27.38	6.24	-1.81
305	20.707	185.026	181.45	0.38	1.93
305	34.472	257.481	251.40	0.97	2.36
340	3.450	23.264	24.27	5.98	-4.32
340	20.699	151.983	147.85	0.94	2.72
340	34.486	224.033	219.71	0.96	1.93

Table 11. Comparison between reference, ρ_{ref} (Patil et al., 2007), and NPT molecular dynamics simulated, ρ_{sim}, density values, for the natural gas like mixture (composition reported in Table 1) as a function of pressure and temperature

5. Conclusions

Results reported in this work show preliminary conclusions obtained in the first stages of a wide computational study that we are carrying out on the performance of molecular simulation approaches for the prediction of thermophysical properties of multicomponent natural gas like mixtures. The main conclusions that may be inferred from these initial results may be resumed in:

i) Density for the main components of natural gases is predicted with deviations in the 1 - 2 % range for the 250 - 450 K / 20 - 100 MPa studied ranges. These deviations are low considering the purely predictive character of the considered approach but are too large for natural gas engineering purposes.

ii) Derived properties (thermal expansion coefficient, isothermal compressibility, isobaric heat capacity and Joule-Thomson coefficient), for the main components of natural gases are predicted with deviations up to 10 % (even larger for Joule-Thomson coefficient for some fluids). Therefore, although these are the deviations commonly obtained using molecular dynamics approach for many complex fluids, they are too large to use this approach for industrial purposes. Fortunately, deviations for isobaric heat capacity, which is a property with remarkable importance for natural gas engineering purposes, is predicted with deviations usually lower than 5 %.

iii) Predictions for the studied multicomponent natural gas like mixture lead to analogous results than those mentioned in conclusions *i* and *ii*: low deviations for density which lead to probably larger deviations for the remaining properties, but for all of them too large for industrial purposes. Moreover, it should be remarked that the studied mixture is composed of n-alkanes only up to C5, and thus, heavier mixtures containing larger alkanes should lead to even larger deviations.

Therefore, considering the results reported in this work, molecular dynamics approach is not able to lead to predictions that may be used for natural gas production, processing or transportation purposes in the present situation. Nevertheless, we think that results reported in this work, and by other authors in the literature (Lagache et al., 2001; Lagache et al. 2004), are very promising, and the use of molecular simulation approach to predict complex natural gas like mixtures properties should not be discarded. Therefore studies are being carried out by the authors with the next main objectives:

i) Improvement of forcefield parameterizations.

ii) Comparison of the performance of several force fields.

iii) Improvements of all atoms force fields approach in comparison with united atoms approach, does the increase of computational time lead to a remarkable improvement in the accuracy of predictions?

iv) Decrease the errors in density predictions below the 1 % limit, in wide pressure - temperature ranges, and thus, leading to results competitive with the methods commonly used by the gas industry nowadays.

v) Analyze the predictions for heavy mixtures containing long alkanes.

vi) Comparison of the performance of Monte Carlo and molecular dynamics approaches.

vii) Analyze from a molecular viewpoint the structural factors that govern these complex mixtures behavior using these computational tools.

6. References

Allen, M. P. & Tildesley, D. J. (1989). *Computer Simulation of Liquids,*Clarendon Press, ISBN 0198556454, Oxford.

Babusiaux, D. (2004). *Oil and Gas Exploration and Production: Reserves, Costs, Contracts,* Editions Technip, ISBN 2710808404, Paris.

Bessieres, D.; Randzio, S. L.; Piñeiro, M. M.; Lafitte, Th. & Daridon, J. L. (2006). A Combined Pressure-controlled Scanning Calorimetry and Monte Carlo Determination of the Joule−Thomson Inversion Curve. Application to Methane. *J. Phys. Chem. B,* 110, 11, February 2006, 5659-5664, ISSN 1089-5647.

Bluvshtein, I. (2007). Uncertainties of gas measurement. *Pipeline & Gas Journal,* 234, 5, May 2007, 28-33, ISSN 0032-0188.

Bluvshtein, I. (2007). Uncertainties of measuring systems. *Pipeline & Gas Journal,* 234, 7, July 2007, 16-21, ISSN 0032-0188.

Duan, Z.; Moller, N. & Weare, J. H. (1992). Molecular dynamics simulation of PVT properties of geological fluids and a general equation of state of nonpolar and weaklyu polar gases up to 2000 K and 20000 bar. *Geochim. Cosmochim. Acta,* 56, 10, October 1992, 3839-3845, ISSN 0016- 7037.

Duan, Z.; Moller, N. & Weare, J. H. (1996). A general equation of state for supercritical fluid mixtures and molecular dynamics simulation of mixture PVTx properties. *Geochim. Cosmochim. Acta,* 60, 7, April 1996, 1209-1216, ISSN 0016- 7037.

Dysthe, D. K., Fuch, A. H.; Rousseau, B. & Durandeau, M. (1999). Fluid transport properties by equilibrium molecular dynamics. II. Multicomponent systems. *J. Chem. Phys.,* 110, 8, February 1999, 4060-4067, ISSN 0021-9606.

Errington, J.R. & Panagiotopoulos, A. Z. (1998). A Fixed Point Charge Model for Water Optimized to the Vapor−Liquid Coexistence Properties. *J. Phys. Chem. B,* 102, 38, September 1998, 7470-7475, ISSN 1089-5647.

Errington, J. & Panagiotopoulos, A. Z. (1999). A New Intermolecular Potential Model for the *n*-Alkane Homologous Series. *J. Phys. Chem. B,* 103, 30, July 1999, 6314-6322, ISSN 1089-5647.

Escobedo, F. A. & Chen, Z. (2001). Simulation of isoenthalps curves and Joule - Thomson inversion of pure fluids and mixtures. Mol. Sim., 26, 6, June 2001, 395-416, ISSN 0892-7022.

Essmann, U. L.; Perera, M. L.; Berkowitz, T.; Darden, H.; Lee,H. & Pedersen, L. G. (1995) *J. Chem. Phys., 103,* 19, November 2005, 8577-8593, ISSN 0021-9606.

Gallagher, J. E. (2006). *Natural Gas Measurement Handbook,* Gulf Publishing Company, ISBN 1933762005, Houston.

Hall, K. R. & Holste, J. C. (1990). Determination of natural gas custody transfer properties. *Flow. Meas. Instrum.,* 1, 3, April 1990, 127-132, ISSN 0955-5986.

Hoover, W. G. (1985). Canonical dynamics: Equilibrium phase-space distributions. *Phys. Rev. A, 31,* 3, March 1985, 1695-1697, ISSN 1050-2947.

Husain, Z. D. (1993). Theoretical uncertainty of orifice flow measurement, *Proceedings of 68th International School of Hydrocarbon Measurement,* pp. 70-75, May 1993, publ, Oklahoma City.

Jaescke, M.; Schley, P. & Janssen-van Rosmalen, R. (2002). Thermodynamic research improves energy measurement in natural gas. *Int. J. Thermophys.,* 23, 4, July 2002, 1013-1031, ISSN 1572-9567.

Jorgensen, W. L.; Maxwell, D. S. & Tirado–Rives, J. (1996). Development and testing of the OPLS All-Atom force field on conformational energetics and properties of organic liquids. *J. Am. Chem. Soc., 118,* 45, November 1996, 11225-11236, ISSN 0002-7863.

Lagache, M.; Ungerer, P., Boutin, A. & Fuchs, A. H. (2001). Prediction of thermodynamic derivative properties of fuids by Monte Carlo simulation. *Phys. Chem. Chem. Phys.*, 3, 8, February 2001, 4333-4339, ISSN 1463-9076.

Lagache, M. H.; Ungerer, P.; Boutin, A. (2004). Prediction of thermodynamic derivative properties of natural condensate gases at high pressure by Monte Carlo simulation. *Fluid Phase Equilibr.*, 220, 2, June 2004, 211-223, ISSN 0378-3812.

Lemmon, E. W.; McLinden, M. O.; Huber, M. L. NIST Standard Reference Database 23, Version 7.0, National Institute of Standards and Techcnology, Physical and Chamical Properties Division, Gaithersburg, MD, 2002.

Linstrom, P. J. & Mallard, W.G. Eds. (2009). *NIST Chemistry WebBook, NIST Standard Reference Database Number 69,* National Institute of Standards and Technology, Gaithersburg. Available at http://webbook.nist.gov

Martínez, J. M. & Martínez, L. (2003). Packing optimization for automated generation of complex system's initial configurations for molecular dynamics and docking. *J. Comput. Chem., 24,* 7, May 2003, 819-825, ISSN 0192-8651.

Martin, M.G. & Frischknecht, A. L. (2006). Using arbitrary trial distributions to improve intramolecular sampling in configurational-bias Monte Carlo. *Mol. Phys.*, 104, 15, July 2006, 2439-2456, ISSN 0026-8976.

Martin, M.G. & Siepmann, J.I. (1999). Novel Configurational-Bias Monte Carlo Method for Branched Molecules. Transferable Potentials for Phase Equilibria. 2. United-Atom Description of Branched Alkanes. *J. Phys. Chem. B,* 103, 21, May 1999, 4508-4517, ISSN 1089-5647.

Mokhatab, S.; Poe, W. A. & Speight, J. G. *Handbook of Natural Gas Transmission and Processing*, Gulf Professional Publishing, ISBN 0750677767, Burlington.

Neubauer, B.; Tavitian, B.; Boutin, A.; Ungerer, P. (1999). Molecular simulations on volumetric properties of natural gas. *Fluid Phase Equilibr.*, 161, 1, July 1999, 45-62, ISSN 0378-3812.

Patil, P.; Ejaz, S.; Atilhan, M.; Cristancho, D.; Holste, J. C. & Hall, K. R. (2007). Accurate density measurements for a 91 % methane natural gas-like mixture. *J. Chem. Thermodyn.*, 39, 8, August 2007, 1157-1163, ISSN 0021-9614.

Ponder, J. W. (2004). *TINKER: Software tool for molecular design.* 4.2 ed, Washington University School of Medicine.

Saager, B. & Fischer, J. (1990). Predictive power of effective intermolecular pair potentials: MD simulation results for methane up to 1000 MPa. *Fluid Phase Equilibr.*, 57, 1-2, July 1990, 35-46, ISSN 0378-3812.

Shi, W. & Maginn, E. (2008). Atomistic Simulation of the Absorption of Carbon Dioxide and Water in the Ionic Liquid 1-n-Hexyl-3-methylimidazolium Bis(trifluoromethylsulfonyl)imide ([hmim][Tf2N]. *J. Phys. Chem. B,* 112, 7, January 2008, 2045-2055, ISSN ISSN 1089-5647.

Siepmann, J.I. & Frenkel, D. (1992). Configurational bias Monte Carlo: a new sampling scheme for flexible chains. *Mol. Phys.*, 75, 1, January 1992, 59-70, ISSN 0026-8976.

Smit, B. & Williams, C. P. (1990). Vapour-liquid equilibria for quadrupolar Lennard-Jones fluids. *J. Phys. Condens. Matter*, 2, 18, May 1990, 4281-4288, 0953-8984.

Starling, K.E. & Savidge, J.L. (1992) *Compressibility Factors of Natural Gas and Other Related Hydrocarbon Gases*, AGA transmission Measurement Committee Report 8, American Gas Association, 1992.

Ungerer, P. (2003). From Organic geochemistry to statistical thermodynamics: the development of simulation methods for the petroleum industry. *Oil & Gas Science and Technology – Rev. IFP*, 58, 2, May 2003, 271-297, ISSN 1294-4475.

Ungerer, P.; Wender, A.; Demoulin, G.; Bourasseau, E. & Mougin, P. (2004). Application of Gibbs Ensemble and NPT Monte Carlo Simulation to the Development of Improved Processes for H_2S-rich Gases. *Mol. Sim.*, 30, 10, August 2004, 631-648, ISSN 0892-7022.

Ungerer, P.; Lachet, V. & Tavitian, B. (2006). Properties of natural gases at high pressure. In: *Applications of molecular simulation in the oil and gas industry. Monte Carlo methods.*, 162-175, Editions Technip, ISBN 2710808587, Paris.

Ungerer, P.; Lachet, V. & Tavitian, B. (2006). Applications of molecular simulation in oil and gas production and processing. *Oil & Gas Science and Technology – Rev. IFP.*, 61, 3, May 2006, 387-403, ISSN 1294-4475.

Ungerer, P.; Nieto-Draghi, C.; Rousseau, B.; Ahunbay, G. & Lachet, V. (2007). Molecular simulation of the thermophysical properties of fluids: From understanding toward quantitative predictions. J. Mol. Liq., 134, 1-3, May 2007, 71-89, ISSN 0167-7322.

Vlugt, T. J. H.; Martin, M.G.; Smit, B.; Siepmann, J.I. & Krishna, R. (1998). Improving the efficiency of the configurational-bias Monte Carlo algorithm. *Mol. Phys.*, 94, 4, July 1998, 727-733, ISSN 0026-8976.

Vrabec, J.; Kumar, A. & Hasse, H. (2007). Joule–Thomson inversion curves of mixtures by molecular simulation in comparison to advanced equations of state: Natural gas as an example. *Fluid Phase Equilibr.*, 258, 1, September 2007, 34-40, ISSN 0378-3812.

Wagner, W. & Kleinrahm, R. (2004). Densimeters for very accurate density measurements of fluids over large ranges of temperature, pressure, and density. *Metrologia*, 41, 2, March 2004, S24-S29, ISSN 0026-1394.

Yoshida, T.; Uematsu, M. (1996). Prediction of PVT properties of natural gases by molecular simulation. *Transactions of the Japan Society of Mechanical Engineers, Series B*, 62, 593, , 278-283, ISSN 03875016.

19

Static behaviour of natural gas and its flow in pipes

Ohirhian, P. U.
University of Benin, Petroleum Engineering Department, Benin City, Nigeria.
Email: peter@ohirhian.com, okuopet@yahoo.com

Abstract

A general differential equation that governs static and flow behavior of a compressible fluid in horizontal, uphill and downhill inclined pipes is developed. The equation is developed by the combination of Euler equation for the steady flow of any fluid, the Darcy–Weisbach formula for lost head during fluid flow in pipes, the equation of continuity and the Colebrook friction factor equation. The classical fourth order Runge-Kutta numerical algorithm is used to solve to the new differential equation. The numerical algorithm is first programmed and applied to a problem of uphill gas flow in a vertical well. The program calculates the flowing bottom hole pressure as 2544.8 psia while the Cullender and Smith method obtains 2544 psia for the 5700 ft (above perforations) deep well

Next, the Runge-Kutta solution is transformed to a formula that is suitable for hand calculation of the static or flowing bottom hole pressure of a gas well. The new formula gives close result to that from the computer program, in the case of a flowing gas well. In the static case, the new formula predicts a bottom hole pressure of 2640 psia for the 5790 ft (including perforations) deep well. Ikoku average temperature and deviation factor method obtains 2639 psia while the Cullender and Smith method obtaines 2641 psia for the same well.. The Runge-Kutta algorithm is also used to provide a formula for the direct calculation of the pressure drop during downhill gas flow in a pipe. Comparison of results from the formula with values from a fluid mechanics text book confirmed its accuracy. The direct computation formulas of this work are faster and less tedious than the current methods. They also permit large temperature gradients just as the Cullender and Smith method.

Finally, the direct pressure transverse formulas developed in this work are combined wit the Reynolds number and the Colebrook friction factor equation to provide formulas for the direct calculation of the gas volumetric rate

Introduction

The main tasks that face Engineers and Scientists that deal with fluid behavior in pipes can be divided into two broad categories - the computation of flow rate and prediction of pressure at some section of the pipe. Whether in computation of flow rate, or in pressure transverse, the method employed is to solve the energy equation (Bernoulli equation for liquid and Euler equation for compressible fluid), simultaneously with the equation of lost head during fluid flow, the Colebrook (1938) friction factor equation for fluid flow in pipes and the equation of continuity (conservation of mass / weight). For the case of a gas the equation of state for gases is also included to account for the variation of gas volume with pressure and temperature.

In the first part of this work, the Euler equation for the steady flow of any fluid in a pipe/ conduit is combined with the Darcy - Weisbach equation for the lost head during fluid flow in pipes and the Colebrook friction factor equation. The combination yields a general differential equation applicable to any compressible fluid; in a static column, or flowing through a pipe. The pipe may be horizontal, inclined uphill or down hill.

The accuracy of the differential equation was ascertained by applying it to a problem of uphill gas flow in a vertical well. The problem came from the book of Ikoku (1984), "Natural Gas Production Engineering". The classical fourth order Runge-Kutta method was first of all programmed in FORTRAN to solve the differential equation. By use of the average temperature and gas deviation factor method, Ikoku obtained the flowing bottom hole pressure (P_{wf}) as 2543 psia for the 5700 ft well. The Cullender and Smith (1956) method that allows wide variation of temperature gave a P_{wf} of 2544 psia. The computer program obtaines the flowing bottom hole pressure (P_{wf}) as 2544.8 psia. Ouyang and Aziz (1996) developed another average temperature and deviation method for the calculation of flow rate and pressure transverse in gas wells. The average temperature and gas deviation formulas cannot be used directly to obtain pressure transverse in gas wells. The Cullender and Smith method involves numerical integration and is long and tedious to use.

The next thing in this work was to use the Runge-Kutta method to generate formulas suitable for the direct calculation of the pressure transverse in a static gas column, and in uphill and downhill dipping pipes. The accuracy of the formula is tested by application to two problems from the book of Ikoku. The first problem was prediction of static bottom hole pressure (P_{ws}). The new formula gives a P_{ws} of 2640 psia for the 5790ft deep gas well. Ikoku average pressure and gas deviation factor method gives the

P_{ws} as 2639 psia, while the Cullender and Smith method gives the P_{ws} as 2641 psia. The second problem involves the calculation of flowing bottom hole pressure (P_{wf}). The new formula gives the P_{wf} as 2545 psia while the average temperature and gas deviation factor of Ikoku gives the P_{wf} as 2543 psia. The Cullender and Smith method obtains a P_{wf} of 2544 psia. The downhill formula was first tested by its application to a slight modification of a problem from the book of Giles et al.(2009). There was a close agreement between exit pressure calculated by the formula and that from the text book. The formula is also used to calculate bottom hole pressure in a gas injection well.

The direct pressure transverse formulas developed in this work are also combined wit the Reynolds number and the Colebrook friction factor equation to provide formulas for the direct calculation of the gas volumetric rate in uphill and down hill dipping pipes.

A differntial equation for static behaviour of a compressible fluid and its flow in pipes

The Euler equation is generally accepted for the flow of a compressible fluid in a pipe. The equation from Giles et al. (2009) is:

$$\frac{dp}{\gamma}+\frac{vdv}{g} \pm d\ell \sin\theta + dh_l = 0 \tag{1}$$

In equation (1), the plus sign (+) before $d\ell \sin\theta$ corresponds to the upward direction of the positive z coordinate and the minus sign (-) to the downward direction of the positive z coordinate.

The generally accepted equation for the loss of head in a pipe transporting a fluid is that of Darcy-Weisbach. The equation is:

$$H_L = \frac{fLv^2}{2gd} \tag{2}$$

The equation of continuity for compressible flow in a pipe is:

$$W = A\gamma v \tag{3}$$

Taking the first derivation of equation (3) and solving simultaneously with equation (1) and (2) we have after some simplifications,

$$\frac{dp}{d\ell} = -\frac{\left[\dfrac{fW^2}{2\gamma A^2 dg} \mp \gamma \sin\theta\right].}{\left[1-\dfrac{W^2}{\gamma^2 A^2 g}\dfrac{d\gamma}{dp}\right]} \tag{4}$$

All equations used to derive equation (4) are generally accepted equations No limiting assumptions were made during the combination of these equations. Thus, equation (4) is a general differential equation that governs static behavior compressible fluid flow in a pipe. The compressible fluid can be a liquid of constant compressibility, gas or combination of gas and liquid (multiphase flow).

By noting that the compressibility of a fluid (C_f) is:

$$C_f = \frac{1}{\gamma}\frac{d\gamma}{dp} \tag{5}$$

Equation (4) can be written as:

$$\frac{dp}{d\ell}=-\frac{\left[\frac{fW^2}{2\gamma A^2 dg}\mp\gamma\sin\theta\right]}{\left[1-\frac{W^2 C_f}{\gamma A^2 g}\right]} \tag{6}$$

Equation (6) can be simplified further for a gas.
Multiply through equation (6) by γ, then

$$\frac{\gamma dp}{d\ell}=-\frac{\left[\frac{fW^2}{2gA^2 dg}\mp\gamma^2\sin\theta\right]}{\left[1-\frac{W^2 C_f}{\gamma A^2 g}\right]} \tag{7}$$

The equation of state for a non-ideal gas can be written as

$$\gamma=\frac{p\,M}{zRT} \tag{8}$$

Substitution of equation (8) into equation (7) and using the fact that

$$\frac{pdp}{d\ell}=\frac{1}{2}\frac{dp^2}{d\ell}, \text{ gives}$$

$$\frac{dp^2}{d\ell}=-\frac{\left[\frac{fW^2\ zRT}{A^2 dMg}\mp\frac{2p^2 M\sin\theta}{zRT}\right]}{\left[1-\frac{W^2\ zRTC_f}{MA^2 g\,p}\right]} \tag{9}$$

The cross-sectional area (A) of a pipe is

$$A^2=\left(\frac{\pi\,d^2}{4}\right)^2=\frac{\pi^2\,d^4}{16} \tag{10}$$

Then equation (9) becomes:

$$\frac{dP^2}{d\ell} = -\left[\frac{1.621139\frac{fW^2}{d^5}\frac{zRT}{M\,g} \mp \frac{2\,M\sin\theta\,P^2}{zRT}}{1-\frac{1.621139W^2\;zRTC_f}{M\,g\,d^4\,P}}\right]. \tag{11}$$

The denominator of equation (11) accounts for the effect of the change in kinetic energy during fluid flow in pipes. The kinetic effect is small and can be neglected as pointed out by previous researchers such as Ikoku (1984) and Uoyang and Aziz(1996). Where the kinetic effect is to be evaluated, the compressibility of the gas (C_f) can be calculated as follows:
For an ideal gas such as air,

$$C_f = \frac{1}{p}.\ \text{For a non ideal gas, } C_f = \frac{1}{p} - \frac{1}{z}\frac{\partial z}{\partial p}.$$

Matter et al. (1975) and Ohirhian (2008) have proposed equations for the calculation of the compressibility of hydrocarbon gases. For a sweet natural gas (natural gas that contains CO_2 as major contaminant), Ohirhian (2008) has expressed the compressibility of the real gas (C_f) as:

$$C_f = \frac{K}{p}$$

For Nigerian (sweet) natural gas K = 1.0328 when p is in psia
The denominator of equation (11) can then be written as

$$1-\frac{KW^2zRT}{M\,g\,d^4P^2}\text{, where K = constant.}$$

Then equation (11) can be written as

$$\frac{dy}{d\ell} = \frac{(A \pm By)}{(1-\frac{G}{y})} \tag{12}$$

where

$$y = p^2,\ A = \frac{1.621139fW^2zRT}{gd^5M},\ B = \frac{2M\sin\theta}{zRT},\ G = \frac{KW^2zRT}{gMd^4}.$$

The plus (+) sign in numerator of equation (12) is used for compressible uphill flow and the negative sign (-) is used for the compressible downhill flow. In both cases the z coordinate is taken positive upward. In equation (12) the pressure drop is $\sqrt{y_1 - y_2}$, with $y_1 > y_2$ and incremental length is $l_2 - l_1$. Flow occurs from point (1) to point (2). Uphill flow of gas occurs in gas transmission lines and flow from the foot of a gas well to the surface. The pressure at

the surface is usually known. Downhill flow of gas occurs in gas injection wells and gas transmission lines.
We shall illustrate the solution to the compressible flow equation by taking a problem involving an uphill flow of gas in a vertical gas well.

Computation of the variables in the gas differential equation
We need to discuss the computation of the variables that occur in the differential equation for gas before finding a suitable solution to it.. The gas deviation factor (z) can be obtained from the chart of Standing and Katz (1942). The Standing and Katz chart has been curve fitted by many researchers. The version that was used in this section of the work that of Gopal(1977). The dimensionless friction factor in the compressible flow equation is a function of relative roughness ($\in$ / d) and the Reynolds number (R_N). The Reynolds number is defined as:

$$R_N = \frac{\rho v d}{\mu} = \frac{W d}{\Lambda g \mu} \tag{13}$$

The Reynolds number can also be written in terms of the gas volumetric flow rate. Then

$$W = \gamma_b Q_b$$

Since the specific weight at base condition is:

$$\gamma_b = \frac{p_b M}{z_b T_b R} = \frac{28.97 G_g p_b}{z_b T_b R} \tag{14}$$

The Reynolds number can be written as:

$$R_N = \frac{36.88575 G_g P_b Q_b}{R g d \mu_g z_b T_b} \tag{15}$$

By use of a base pressure (p_b) = 14.7psia, base temperature (T_b) = 520°R and R = 1545

$$R_N = \frac{20071 Q_b G_g}{\mu_g d} \tag{16}$$

Where d is expressed in inches, Q_b = MMSCF / Day and μ_g is in centipoises.
Ohirhian and Abu (2008) have presented a formula for the calculation of the viscosity of natural gas. The natural gas can contain impurities of CO_2 and H_2S. The formula is:

$$\mu_g = \frac{0.0109388 - 0.0088234xx - 0.00757210xx^2}{1.0 - 1.3633077xx - 0.0461989xx^2} \tag{17}$$

Where

$$xx = \frac{0.0059723p}{z\left(16.393443 - T/p\right)}$$

In equation (17) μ_g is expressed in centipoises(c p) , p in (psia) and Tin (°R)

The generally accepted equation for the calculation of the dimensionless friction factor (f) is that of Colebrook (1938). The equation is:

$$\frac{1}{\sqrt{f}} = -2\log\left(\frac{\in}{3.7d} + \frac{2.51}{R_N\sqrt{f}}\right) \tag{18}$$

The equation is non-linear and requires iterative solution. Several researchers have proposed equations for the direct calculation of f. The equation used in this work is that proposed by Ohirhian (2005). The equation is

$$f = \left[-2\log\left(a - 2b\ \log\ (a + bx_1)\right)\right]^{-2} \tag{19}$$

Where

$$a = \frac{\in}{3.7d},\ b = \frac{2.51}{R_N}.$$

$$x_1 = -1.14\log\left(\frac{\in}{d} + 0.30558\right) + 0.57\log R_N\left(0.01772\log R_N + 1.0693\right)$$

After evaluating the variables in the gas differential equation, a suitable numerical scheme can be used to it.

Solution to the gas differential equation for direct calculation of pressure transverse in static and uphill gas flow in pipes.

The classical fourth order Range Kutta method that allows large increment in the independent variable when used to solve a differential equation is used in this work. The solution by use of the Runge-Kutta method allows direct calculation of pressure transverse.. The Runge-Kutta approximate solution to the differential equation

$$\frac{dy}{dx} = f(x,y) \qquad \text{at } x = x_n \tag{20}$$

given that $y = y_o$ when $x = x_o$ is

$$y = y_o + \frac{1}{6}(k_1 + 2(k_2 + k_3) + k_4) \tag{21}$$

where

$$k_1 = Hf(x_o, y_o)$$

$$k_2 = Hf(x + \frac{1}{2}H, y_o + \frac{1}{2}k_1)$$

$$k_3 = Hf(x_o + \frac{1}{2}H,\ y_o + \frac{1}{2}k_1)$$

$$k_4 = Hf(x_o + H,\ y + k_3)$$

$$H = \frac{x_n - x_o}{n}$$

n = number of applications

The Runge-Kutta algorithm can obtain an accurate solution with a large value of H. The Runge-Kutta Algorithm can solve equation (6) or (12). The test problem used in this work is from the book of Ikoku (1984), "Natural Gas Production Engineering". Ikoku has solved this problem with some of the available methods in the literature.

Example 1

Calculate the sand face pressure (p_{wf}) of a flowing gas well from the following surface measurements.

Flow rate (Q) = 5.153 MMSCF / Day
Tubing internal diameter (d) = 1.9956in
Gas gravity (G_g) = 0.6
Depth = 5790ft (bottom of casing)

Temperature at foot of tubing (T_{wf}) = 160 °F

Surface temperature (T_{sf}) = 83 °F

Tubing head pressure (p_{tf}) = 2122 psia
Absolute roughness of tubing ($\in$) = 0.0006 in
Length of tubing (l) = 5700ft (well is vertical)

Solution

When length (ℓ) is zero, p = 2122 psia
That is (x_o, y_o) = (0, 2122)
By use of 1 step Runge-Kutta.

$$H = \frac{5700 - 0}{1} = 5700\text{ft.}$$

The Runge-Kutta algorithm is programmed in Fortran 77 and used to solve this problem. The program is also used to study the size of depth(length) increment needed to obtain an accurate solution by use of the Runge-Kutta method. The first output shows result for one-step Runge-Kutta (Depth increment = 5700ft). The program obtaines 2544.823 psia as the flowing bottom hole pressure (P_{wf}).

```
TUBING HEAD PRESSURE =        2122.0000000 PSIA
SURFACE TEMPERATURE =         543.0000000 DEGREE RANKINE
TEMPERATURE AT TOTAL DEPTH =     620.0000000 DEGREE RANKINE
GAS GRAVITY =    6.000000E-001
GAS FLOW RATE =            5.1530000 MMSCFD
DEPTH AT SURFACE =            .0000000 FT
TOTAL DEPTH =       5700.0000000 FT
INTERNAL TUBING DIAMETER =          1.9956000 INCHES
ROUGHNESS OF TUBING =  6.000000E-004 INCHES
INCREMENTAL DEPTH =        5700.0000000 FT

    PRESSURE PSIA          DEPTH FT

       2122.000                .000
       2544.823            5700.000
```

To check the accuracy of the Runge-Kutta algorithm for the depth increment of 5700 ft another run is made with a smaller length increment of 1000 ft. The output gives a p_{wf} of 2544.823 psia. as it is with a depth increment of 5700 ft. This confirmes that the Runge-Kutta solution can be accurate for a length increment of 5700 ft.

```
TUBING HEAD PRESSURE =        2122.0000000 PSIA
SURFACE TEMPERATURE =         543.0000000 DEGREE RANKINE
TEMPERATURE AT TOTAL DEPTH =      620.0000000 DEGREE RANKINE
GAS GRAVITY =    6.000000E-001
GAS FLOW RATE =            5.1530000 MMSCFD
DEPTH AT SURFACE =            .0000000 FT
TOTAL DEPTH =       5700.0000000 FT
INTERNAL TUBING DIAMETER =          1.9956000 INCHES
ROUGHNESS OF TUBING =  6.000000E-004 INCHES
INCREMENTAL DEPTH =        1000.0000000 FT

    PRESSURE PSIA          DEPTH FT
     2122.000                  .000
     2206.614              1140.000
     2291.203              2280.000
     2375.767              3420.000
     2460.306              4560.000
     2544.823              5700.000
```

In order to determine the maximum length of pipe (depth) for which the computed P_{wf} can be considered as accurate, the depth of the test well is arbitrarily increased to 10,000ft and the program run with one step (length increment = 10,000ft). The program produces the P_{wf} as 2861.060 psia..

```
TUBING HEAD PRESSURE =          2122.0000000 PSIA
SURFACE TEMPERATURE =           543.0000000 DEGREE RANKINE
TEMPERATURE AT TOTAL DEPTH =       687.0000000 DEGREE RANKINE
GAS GRAVITY =     6.000000E-001
GAS FLOW RATE =              5.1530000 MMSCFD
DEPTH AT SURFACE =                .0000000 FT
TOTAL DEPTH =       10000.0000000 FT
INTERNAL TUBING DIAMETER =             1.9956000 INCHES
ROUGHNESS OF TUBING =   6.000000E-004 INCHES
INCREMENTAL DEPTH =       10000.0000000 FT

    PRESSURE PSIA          DEPTH FT

     2122.000                  .000
     2861.060             10000.000
```

Next the total depth of 10000ft is subdivided into ten steps (length increment = 1,000ft). The program gives the P_{wf} as 2861.057 psia for the length increment of 1000ft.

```
TUBING HEAD PRESSURE =          2122.0000000 PSIA
SURFACE TEMPERATURE =           543.0000000 DEGREE RANKINE
TEMPERATURE AT TOTAL DEPTH =        687.0000000 DEGREE RANKINE
GAS GRAVITY =     6.000000E-001
GAS FLOW RATE =              5.1530000 MMSCFD
DEPTH AT SURFACE =                .0000000 FT
TOTAL DEPTH =       10000.0000000 FT
INTERNAL TUBING DIAMETER =             1.9956000 INCHES
ROUGHNESS OF TUBING =   6.000000E-004 INCHES
INCREMENTAL DEPTH =        1000.0000000 FT

    PRESSURE PSIA          DEPTH FT
     2122.000                  .000
     2197.863              1000.000
     2273.246              2000.000
     2348.165              3000.000
     2422.638              4000.000
     2496.680              5000.000
     2570.311              6000.000
     2643.547              7000.000
     2716.406              8000.000
     2788.903              9000.000
     2861.057             10000.000
```

The computed values of P_{wf} for the depth increment of 10,000ft and 1000ft differ only in the third decimal place. This suggests that the depth increment for the Range - Kutta solution to the differential equation generated in this work could be a large as 10,000ft. By neglecting the denominator of equation (6) that accounts for the kinetic effect, the result can be compared with Ikoku's average temperature and gas deviation method that uses an average value of the gas deviation factor (z) and negligible kinetic effects. In the program z is allowed to vary with pressure and temperature. The temperature in the program also varies with depth (length of tubing) as

$$T = GTG \times \text{current length} + T_{sf}\text{, where, } GTG = \frac{(T_{wf} - T_{sf})}{\text{Total Depth}}$$

The program obtains the P_{wf} as 2544.737 psia when the kinetic effect is ignored. The output is as follows:

```
TUBING HEAD PRESSURE =          2122.0000000 PSIA
SURFACE TEMPERATURE =           543.0000000 DEGREE RANKINE
TEMPERATURE AT TOTAL DEPTH =         620.0000000 DEGREE RANKINE
GAS GRAVITY =    6.000000E-001
GAS FLOW RATE =              5.1530000 MMSCFD
DEPTH AT SURFACE =              .0000000 FT
TOTAL DEPTH =       5700.0000000 FT
INTERNAL TUBING DIAMETER =            1.9956000 INCHES
ROUGHNESS OF TUBING =   6.000000E-004 INCHES
INCREMENTAL DEPTH =          5700.0000000 FT

    PRESSURE PSIA          DEPTH FT

    2122.000                   .000
    2544.737               5700.000
```

Comparing the P_{wf} of 2544.737 psia with the P_{wf} of 2544.823 psia when the kinetic effect is considered, the kinetic contribution to the pressure drop is 2544.823 psia - 2544.737psia = 0.086 psia.The kinetic effect during calculation of pressure transverse in uphill dipping pipes is small and can be neglected as pointed out by previous researchers such as Ikoku (1984) and Uoyang and Aziz(1996)

Ikoku obtained 2543 psia by use of the the average temperature and gas deviation method. The average temperature and gas deviation method goes through trial and error calculations in order to obtain an accurate solution. Ikoku also used the Cullendar and Smith method to solve the problem under consideration. The Cullendar and Smith method does not consider the kinetic effect but allows a wide variation of the temperature. The Cullendar and Smith method involves the use of Simpson rule to carry out an integration of a cumbersome function. The solution to the given problem by the Cullendar and Smith method is p_{wf} = 2544 psia.

If we neglect the denominator of equation (12), then the differential equation for pressure transverse in a flowing gas well becomes

$$\frac{dy}{dl} = A + By \tag{22}$$

where

$$A = \frac{1.621139 fW^2 zRT}{g d^5 M} \tag{23}$$

$$B = \frac{2M \sin\theta}{zRT} = \frac{2 \times 28.79 G_g \sin\theta}{RTz} \tag{24}$$

The equation is valid in any consistent set of units. If we assume that the pressure and temperature in the tubing are held constant from the mid section of the pipe to the foot of the tubing, the Runge-Kutta method can be used to obtain the pressure transverse in the tubing as follows.

$$\frac{dy}{d\ell} = \frac{\dfrac{46.9643686\, GgQ_b^2 fzRT}{gd^5} + \dfrac{59.940Gg\ \mathrm{Sin}\ \theta\, y}{zRT}}{1 - \dfrac{46.9643686KzGgQ_b^2}{gRd^4}\left(\dfrac{p_b}{T_b}\right)^2\left(\dfrac{T}{y}\right)} \tag{25}$$

The weight flow rate (W) in equation (12) is related to Q_b (the volumetric rate measurement at a base pressure (P_b) and a base temperature (T_b)) in equation (25) by:

$$W = \gamma_b Q_b \tag{26}$$

Equation (25) is a general differential equation that governs pressure transverse in a gas pipe that conveys gas uphill. When the angle of inclination (θ) is zero, $\sin\theta$ is zero and the differential equation reduces to that of a static gas column. The differential equation (25) is valid in any consistent set of units. The constant K = 1.0328 for Nigerian Natural Gas when the unit of pressure is psia.

The classical 4th order Runge Kutta alogarithm can be used to provide a formula that serves as a general solution to the differential equation (25). To achieve this, the temperature and gas deviation factors are held constant at some average value, starting from the mid section of the pipe to the inlet end of the pipe. The solution to equation (25) by the Runge Katta algorithm can be written as:

$$p_1 = \sqrt{p_2^2 + \overline{y}}. \tag{27}$$

Where

$$\bar{y} = \frac{aa}{6}\left(1 + x + 0.5x^2 + 0.36x^3\right) + \frac{p_2^2}{6}\left(4.96x + 1.48x^2 + 0.72x^3\right) + \frac{u}{6}\left(4.96 + 1.96x + 0.72x^2\right)$$

$$aa = \left(\frac{46.9643686\, G_g Q_b^2 f_2 z_2 R T_2}{g d^5} + \frac{57.94 G_g \sin\theta\, P_2^2}{z_2 T_z R}\right) L$$

$$u = \frac{46.9643686\, G_g Q_b^2 f_2 z_{av} T_{av} L}{g d^5}$$

$$x = \frac{57.940\, G_g \sin\theta\, L}{z_{av} T_{av} R}$$

When $Q_b = 0$, equation (27) reduces to the formula for pressure transverse in a static gas column.

In equation (27), the component k_4 in the Runge Kutta method given by $k_4 =$ $H f(x_o + H, y + k_3)$ was given some weighting to compensate for the fact that the temperature and gas deviation factor vary between the mid section and the inlet end of the pipe.

Equation (27) can be converted to oil field units. In oil field units in which L is in feet, R = 1545, temperature is in °R, g = 32.2 ft/sec², diameter (d) is in inches, pressure (p) is in pound per square inch (psia), flow rate (Q_b) is in MMSCF / Day, P_b = 14.7 psia and T_b = 520° R.,the variables aa, u and x that occur in equation (25) can be written as:

$$u = \frac{25.130920 G_g Q_b^2 f_2 z_{av} T_{av} L}{d^5}$$

$$x = 0.03749 \times \frac{G_g L \sin\theta}{z_{av} T_{av}}$$

The following steps are taken in order to use equation (27) to solve a problem.

1. Evaluate the gas deviation factor at a given pressure and temperature. When equation (27) is used to calculate pressure transverse in a gas well, the given pressure and temperature are the surface temperature and gas exit pressure (tubing head pressure).
2. Evaluate the viscosity of the gas at surface condition. This step is only necessary when calculating pressure transverse in a flowing gas well. It is omitted when static pressure transverse is calculated.
3. Evaluate the Reynolds number and dimensionless friction factor by use of surface properties. This step is also omitted when considering a static gas column.
4. Evaluate the coefficient aa in the formula. This coefficient depends only on surface properties.

5. Evaluate the average pressure (p_{av}) and average temperature (T_{av}).
6. Evaluate the average gas deviation factor.(z_{av})
7. Evaluate the coefficients x and u in the formula. Note that u = 0 when $Q_b = 0$.
8. Evaluate $\bar{y}$ in the formula.
9. Evaluate the pressure p_1. In a flowing gas well, p_1 is the flowing bottom hole pressure. In a static column, it is the static bottom hole pressure.

Equation (27) is tested by using it to solve two problems from the book of Ikoku(1984), "Natural Gas Production Engineering". The first problem involves calculation of the static bottom hole in a gas well. The second involves the calculation of the flowing bottom hole pressure of a gas well.

Example 2

Calculate the static bottom hole pressure of a gas well having a depth of 5790 ft. The gas gravity is 0.6 and the pressure at the well head is 2300 psia. The surface temperature is 83°F and the average flowing temperature is 117°F.

Solution

Following the steps that were listed for the solution to a problem by use of equation (27) we have:

1. Evaluation of z – factor.

   The standing equation for P_c and T_c are:

$$P_c\,(\text{psia}) = 677.0 + 15.0\,G_g - 37.5\,G_g^2$$

$$T_c\,(^\circ R) = 168.0 + 325.0\,G_g - 12.5\,G_g^2$$

Substitution of $G_g = 0.6$ gives, $P_c = 672.5$ psia and $T_c = 358.5$°R. Then $P_r = 2300/672.5 = 3.42$

and $T_r = 543/358.5 = 1.52$

The Standing and Katz chart gives $z_2 = 0.78$.

Steps 2 and 3 omitted in the static case.

4. $$aa = \left(\frac{25.13092\,G_g Q_b^2 f z_2 T_2}{d^5} + \frac{0.037417\,G_g p_2^2 \sin\theta}{z_2 T_2}\right) L$$

Here, $G_g = 0.6$, $Q_b = 0.0$, $z_2 = 0.78$, d = 1.9956 inches, $p_2 = 2300$ psia,

$T_2 = 543$ °R and L = 5700 ft. Well is vertical, $\theta = 90°$, $\sin\theta = 1$. Substitution of the given values gives:

$aa = 0.0374917 \times 0.6 \times 2300^2 \times 5790 / (0.78 \times 543) = 1626696$

5. $$p_{av} = \sqrt{2300^2 + 0.5 \times 1626696} = 2470.5 \text{ psia}$$

Reduced $p_{av} = 2470.5 / 672.5 = 3.68$

$T_{av} = 117\,°F = 577\,°R$

Reduced $T_{av} = 577/358.5 = 1.61$

From the standing and Katz chart, $z_{av} = 0.816$

7. In the static case u = 0, so we only evaluate x

$$x = \frac{0.0374917 \times 0.6 \times 5790 \operatorname{Sin} 90^{o}}{0.816 \times 577} = 0.2766$$

8. $$\bar{y} = \frac{aa}{6}\left(1 + x + 0.5x^2 + 0.36x^3\right) + \frac{p_2^2}{6}\left(4.96x + 1.48x^2 + 0.72x^3\right)$$

Substitution of a = 1626696, x = 0.2766 and $P_2 = 2300$ gives

$$\bar{y} = 358543 + 1322856 = 1681399$$

9. $$p_1 = \sqrt{p_2^2 + \bar{y}} = \left(2300^2 + 1681399\right)^{0.5} = 2640.34 \text{ psia} \approx 2640 \text{psia}$$

Ikoku used 3 methods to work this problem. His answers of the static bottom hole pressure are:

Average temperature and deviation factor = 2639 psia
Sukkar and Cornell method = 2634 psia
Cullender and Smith method = 2641 psia

The direct calculation formula of this work is faster.

Example 3

Use equation (27) to solve the problem of example 1 that was previously solved by computer programming.

Solution

1. Obtain the gas deviation factor at the surface. From example 2, the pseudocritical properties for a 0.6 gravity gas are, $P_c = 672.5$ psia. and $T_c = 358.5$, then

$P_r = 2122 / 672.5 = 3.16$

$T_r = 543 / 358.5 = 1.52$

From the Standing and Katz chart, $\mathbf{Z}_2 = 0.78$

2. Obtain, the viscosity of the gas at surface condition. By use of Ohirhian and Abu equation, $$xx = \frac{0.0059723\,p}{z\left(16.393443 - \frac{T}{p}\right)} = \frac{0.0059723 \times 2122}{0.78\left(16.393443 - \frac{543}{2122}\right)} = 0.9985$$

Then $$\mu_g = \frac{0.0109388 - 0.008823(0.9985) - 0.0075720(0.9985)^2}{1.0 - 1.3633077(0.9985) - 0.0461989(0.9985)^2} = 0.0133 \text{ cp}$$

3. Evaluation of the Reynolds number and dimensionless friction factor

$$R_N \frac{20071 Q_b G_g}{\mu g d} = \frac{20071 \times 5.153 \times 0.6}{0.0133 \times 1.9956} = 2.34 \times 10^6$$

The dimensionless friction factor by Ohirhian formula is

$$f = \left[-2\log\left(a - 2b\log\left(a + bx_1\right)\right)\right]^{-2}$$

Where

$$a = \in /3.7d,\ b = 2.51 / R_N$$

$$x_1 = -1.14\log\left(\frac{\in}{d} + 0.30558\right) + 0.57\ \log R_N \left(0.01772\ \log R_N + 1.0693\right)$$

Substitute of $\in = 0.0006,\ d = 1.9956,\ R_N = 2.34\times10^6$ gives $f = 0.01527$

4. Evaluate the coefficient aa in the formula. This coefficient depends only on surface properties.

$$aa = \left(\frac{25.13092\,G_g Q_b^2 f\, z_2 T_2}{d^5} + \frac{0.037417 G_g p_2^2 \sin\theta}{z_2 T_2}\right)L$$

Here, $G_g = 0.6$, $Q_b = 5.153$ MMSCF/Day, $f = 0.01527$, $z_2 = 0.78$, $d = 1.9956$ inches, $p_2 = 2122$ psia, $T_2 = 543\,^\circ R$, $z = 5700$ ft

Substitution of the given values gives;

aa = (81.817446 + 239.14594) × 5700 = 1829491

5. Evaluate P_{av}

$$p_{av} = \sqrt{p_2^2 + 0.5aa} = \sqrt{2122^2 + 0.5\times 1829491} = 2327.6\ psia$$

6. Evaluation of average gas deviation factor.

Reduced average pressure = $p_{av} / p_c = 2327.6 / 672.5 = 3.46$

$$T_{av} = T_2 + \alpha L / 2$$

Where α is the geothermal gradient.

$$\alpha = \left(T_1 - T_2\right)/L = \left(620 - 543\right)/5700 = 0.01351$$

T_{av} at the mid section of the pipe is 2850 ft. Then, $T_{av} = 543 + 0.01351 \times 2850 = 581.5\,^\circ R$

Reduced $T_{av} = 581.5 / 358.5 = 1.62$

Standing and Katz chart gives $z_{av} = 0.822$

7. Evaluation of the coefficients x and u

$$x = \frac{0.0374917 G_g L}{z_{av} T_{av}} = \frac{0.0374919\times 0.6\times 5700}{0.822\times 581.5} = 0.26824$$

$$u = \frac{25.13092 G_g Q_b^2 f\ z_{av} T_{av}\ L}{d^5}$$

$$= \frac{25.13092\times 0.6\times 5.153^2\times 0.01527\times 0.822\times 581.5\times 5700}{1.9956^5} = 526662$$

8. Evaluate $\bar{y}$

$$\bar{y} = \frac{aa}{6}\left(1 + x + 0.5x^2 + 0.36x^3\right) + \frac{p_2^2}{6}\left(4.96x + 1.48x^2 + 0.72x^3\right) + \frac{u}{6}\left(4.96 + 1.96 + 0.72x^2\right)$$

Where u = 526662, x = 0.26824, p_2 = 2122 psia and aa = 1829491. Then,

$$\bar{y} = 399794 + 1088840 + 485752 = 1974386\, psia^2$$

9. Evaluate p_1 (the flowing bottom hole pressure)

$$p_1 = \sqrt{p_2 + \bar{y}}$$
$$= \sqrt{2122^2 + 1974386} = 2545.05 \text{ psia}$$
$$\approx 2545 \text{ psia}$$

The computer program obtains, the flowing bottom hole pressure as 2544.823 psia. For comparison with other methods of solution, the flowing bottom hole pressure by:

Average Temperature and Deviation Factor, P_1 = 2543 psia

Cullender and Smith, P_1 = 2544

The direct calculating formula of this work is faster. The Cullendar and Smith method is even more cumbersome than that of Ikoku.t involves the use of special tables and charts (Ikoku, 1984) page 338 - 344.

The differential equation for static gas behaviour and its downhill flow in pipes

The problem of calculating pressure transverse during downhill gas flow in pipes is encountered in the transportation of gas to the market and in gas injection operations. In the literature, models for pressure prediction during downhill gas flow are rare and in many instances the same equations for uphill flow are used for downhill flow.

In this section, we present the use of the Runge-Kutta solution to the downhill gas flow differential equation.

During downhill gas flow in pipes, the negative sign in the numerator of differential equation (12) is used..The differential equation also breaks down to a simple differential equation for pressure transverse in static columns when the flow rate is zero. The equation to be solved is:

$$\frac{dy}{d\ell} = \frac{(A - By)}{(1 - \frac{G}{y})} \qquad (28)$$

Where $y = p^2$,

$$A = \frac{1.621139\, f\, W^2 zRT}{gd^5 M}, \quad B = \frac{2M \sin\theta}{zRT}, \quad G = \frac{KW^2 zRT}{gMd^4}$$

Also, the molecular weight (M) of a gas, can be expressed as M = 28.97Gg.

Then, the differential equation (28) can be written as:

$$\frac{dP^2}{d\ell} = \frac{\left[\frac{0.0559592\, fzRTW^2}{gd^5G_g} - \frac{59.940\, G_g \sin\theta\, P^2}{zRT}\right]}{\left[1 - \frac{0.0559592\, zRTW^2K}{gd^5G_g\, P^2}\right]} \tag{29}$$

The differential equation (29) is valid in any consistent set of units. The relationship between weight flow rate (W) and the volumetric flow rate measured at a base condition of pressure and temperature (Q_b) is;

$$W = \gamma_b Q_b \tag{30}$$

The specific weight at base condition is:

$$\gamma_b = \frac{p_b M}{z_b T_b R} = \frac{28.97 G_g p_b}{z_b T_b R} \tag{31}$$

Substitution of equations (30) and (31) into differential equation (29) gives:

$$\frac{dp^2}{dl} = \frac{\left[\frac{46.9583259 f zTG_g P_b^2 Q_b^2}{gd^5 z_b^2 T_b^2 R} - \frac{59.940 G_g \sin\theta\, P^2}{zRT}\right]}{\left[1 - \frac{46.95832593 Gg Q_b^2 K}{gRd^4}\left(\frac{P_b}{T_b}\right)^2\left(\frac{T}{P^2}\right)\right]} \tag{32}$$

The differential equation (32) is also valid in any consistent set of units.

Solution to the differential equation for downhill flow

In order to find a solution to the differential equation for downhill flow (as presented in equation (29) and (32) we need equations or charts that can provide values of the variables z and f. The widely accepted chart for the values of the gas deviation factor (z) is that of Standing and Katz (1942). The chart has been curve fitted by some researchers. The version used in this section is that of Ohirhian (1993). The Ohirhian set of equations are able to read the chart within $\pm 0.777\%$ error. The Standing and Katz charts require reduced pressure (Pr) and reduced temperature (T_r). The Pr is defined as Pr = P/Pc and the T_r is defined as $T_r = T / T_c$; where Pc and T_c are pseudo critical pressure and pseudo critical temperature, respectively.

Standing (1977) has presented equations for Pc and T_c as functions of gas gravity (G_g). The equations are:

$$Pc = 677 + 15.0\, G_g - 37.5 G_g^2 \tag{33}$$

$$T_c = 168 + 325\, G_g - 12.5\, G_g^2 \tag{34}$$

The differential equation for the downhill gas flow can also be solved by the classical fourth order Runge-Kutta method.
The downhill flow differential equation was tested by reversing the direction of flow in the problem solved in example 3.

Example 4

Calculate the sand face pressure (p_{wf}) of an injection gas well from the following surface measurements.

Flow rate (Q) = 5.153 MMSCF / Day
Tubing internal diameter (d) = 1.9956 in
Gas gravity (G_g) = 0.6
Depth = 5790ft (bottom of casing)
Temperature at foot of tubing (T_{wf}) = 160°F
Surface temperature (T_{sf}) = 83°F
Tubing head pressure ($P_{sf'}$) = 2545 psia
Absolute roughness of tubing ($\in$) = 0.0006in
Length of tubing (L) = 5700ft (well is vertical)

Solution

Here, $(x_o, y_o) = (0, 2545)$
By use of I step Runge-Kutta.

$$H = \frac{(5700 - 0)}{1} = 5700$$

The Runge-Kutta algorithm is programmed in Fortran 77 to solve this problem. The output is as follows.

```
TUBING HEAD PRESSURE =    2545.0000000 PSIA
SURFACE TEMPERATURE =   543.0000000 DEGREE RANKINE
TEMPERATURE AT TOTAL DEPTH = 620.0000000 DEGREE RANKINE
 GAS GRAVITY = 6.000000E-001
GAS FLOW RATE = 5.1530000 MMSCFD
DEPTH AT SURFACE = .0000000 FT
TOTAL DEPTH = 5700.0000000 FT
INTERNAL TUBING DIAMETER = 1.9956000 INCHES
ROUGHNESS OF TUBING = 6.000000E-004 INCHES
INCREMENTAL DEPTH = 5700.0000000 FT

    PRESSURE PSIA          DEPTH FT
        2545.000               .000
       2327.930            5700.000
```

The other outputs from the program (not shown here) indicates that the contribution of kinetic effect to pressure transverse during down hill flow is also negligible. The program also shows that an incremental length as large as 5700 ft can yield accurate result in pressure transverse calculations.

Neglecting the kinetic effect, the Runge-Kutta algorithm can be used to provide a solution to the differential equation (32) as follows

$$p_1 = \sqrt{p_2^2 - |\bar{y}|} \tag{35}$$

Here

$$\bar{y} = \frac{aa}{6}\left(1 - x + 0.5x^2 + 0.3x^3\right) + \frac{p_1^2}{6}\left(-5.2x + 2.2x^2 - 0.6x^3\right) + \frac{u}{6}\left(5.2 - 2.2x + 0.6x^2\right)$$

$$aa = \left(\frac{46.958326\, f_1\, z_1\, T_1\, G_g\, P_b^2 Q_b^2}{g d^5 z_b^2 T_b^2 R} - \frac{57.940\, G_g \sin\theta\, p_1^2}{z_1 T_1 R}\right) L$$

$$u = \frac{46.958326 f_1 z_{av} T_{av} G g P_b^2 Q_b^2 L}{g d^5 z_b^2 T_b^2 R}, \quad x = \frac{57.940\, G g \sin\theta\, L}{z_{av} T_{av} R}$$

f_1 = Moody friction factor evaluated at inlet end pipe
T_1 = Temperature at inlet end of pipe
T_{av} = Temperature at mid section of pipe = $0.5(T_1 + T_2)$
p_1 = Pressure at inlet end of pipe
z_1 = Gas deviation factor evaluated with p_1 and T_1
z_{av} = Gas deviation factor calculated with temperature at mid section (T_{av}) and pressure at the mid section of pipe (p_{av}) given by $p_{av} = \sqrt{p_1^2 - 0.5|aa|}$
p_2 = Pressure at exit end of pipe, psia
p_1 = Pressure at inlet end of pipe
Note that $p_1 > p_2$ and flows occurs from point (1) to point (2)
Equation (35) is valid in any consistent set of units.
Equation (35) can be converted to oil field units. In oil field units in which L is in feet, R = 1545, temperature(T) is in ^{o}R, g = 32.2 ft/sec^2, diameter (d) is in inches, pressure (p) is in pound per square inche (psia), flow rate (Q_b) is in MMSCF / Day and P_b = 14.7 psia, T_b = 520° R. The variables aa, u and x that occur in equation (35) can be written as:

$$aa = \left(\frac{25.1472069\, G_g f_1 z_1 T_1 Q_b^2}{d^5} - \frac{0.0375016\ G_g \sin\theta\ p_1^2}{z_1 T_1}\right) L$$

$$u = \frac{25.1472069\ G_g f_1 z_{av} T_{av} Q_b^2 L}{d^5}, \quad x = \frac{0.0375016\, G_g \operatorname{Sin}\theta\ L}{z_{av} T_{av}}$$

Example 5

Use equation (35) to solve the problem of example 4

Solution

Step 1: obtain the gas deviation factor at the inlet end

$T_1 = 83^oF = 543^o R$

$P_1 = 2545$ psia

Gg = 0.6

By use of equation (33) and (34)

Pc (psia) = 677 + 15 x 0.6 - 37.5 x 0.6^2 = 672.5 psia

Tc (^{o}R) = 168 + 325x0.6 - 12.5 x 0.6^2 = 358.5 ^{o}R

Then, P_{1r} = 2545/672.5 = 3.784, T_{1r} = 543/358.5 = 1.515

The required Ohirhian equation is

$$z = \left(z_1 + \left(1.39022 + Pr\left(0.06202 - 0.02113 \times Pr\right)\right) \times \log Tr\right) Fc$$

Where

$$z_1 = 0.60163 + Pr\left(-0.06533 + 0.0133 Pr\right)$$

$$Fc = 20.208372 + Tr(-44.0548 + Tr(37.55915 + Tr(-14.105177 + 1.9688 Tr)))$$

Substitution of values of Pr = 3.784 and Tr = 1.515 gives z = 0.780588

Step 2

Evaluate the viscosity of the gas at inlet condition. By use of Ohirhian and Abu formula (equation 17)

$$xx = \frac{0.0059723 \times 2545}{0.780588\left(16.393443 - 543/2545\right)} = 1.203446$$

$$\mu_g = \frac{0.0109388 - 0.008823(1.203446) - 0.0075720(1.203446)^2}{1.0 - 1.3633077(1.203446) - 0.0461989(1.203446)^2} = 0.015045 \text{ cp}$$

Step 3

Evaluation of Reynolds number (R_N) and dimensionless friction factor (f). From eqn. (26)

$$R_N = \frac{20071 \times 5.153 \times 0.6}{0.015045 \times 1.9956} = 2066877$$

The dimensionless friction factor can be explicitly evaluated by use of Ohirhian formula (equation 19)

$$\in/d = 0.0006/1.9956 = 3.066146E-4$$

$$a = 3.066146E-4/3.7 = 8.125985E-5$$

$$b = 2.51/2066877 = 1.21393E-6$$

$$x_1 = -1.14 \log(3.066146E-4 + 0.30558) + 0.57 \times \log 2066877 (0.01772 \log 2066877 + 1.0693)$$
$$= 4.838498$$

Substitution of values of a, b and x_1 into $f = \left[-2\log\left(a - 2b\log\left(a + bh\right)\right)\right]^{-2}$ gives

f = 0.01765

Step 4

Evaluate the coefficient aa in the formula. This coefficient depends only on surface (inlet) properties. Note that the pipe is vertical $\theta = 90^{o}$ and sin $90^{o} = 1$

$$aa = \left(25.147207 \times 0.6 \times 0.017650 \times 0.780588 \times 543 \times 5.153^2 \times 5700\right) / 1.9956^5$$
$$- \left(0.037502 \times 0.6 \times 1 \times 2545 \times 5700\right) / \left(0.780588 \times 543\right)$$
$$= 539803 - 1959902 = -1420099$$

Step 5

Evaluate the average pressure (p a v) at the mid section of the pipe given by

$$p_{av} = \sqrt{p_1^2 - 0.5|aa|} = \sqrt{2545^2 - 0.5 \times 1420099} = 2401.5$$

Step 6:

Evaluate the average gas deviation factor (za v). Reduced average pressure(p a v r) = 2401.5/672.5 = 3.571. $T_{av} = T_1 + \alpha L/2$ where α = geothermal gradient given by:

$$\alpha = (T_2 - T_1)/L = (620 - 543)/5700 = 0.013509$$

T a v at mid section of pipe (2850 ft) then, is: $T_{av} = 543 + 0.013509 \times 2850 = 581.5^{o}R$

Reduced $T_{av} = 581.5/358.5 = 1.622$

Substitution into the Ohirhian equation used in step 1, gives z = 0.821102

Step 7:

Evaluate the coefficients x and u

$$x = \frac{0.0375016 \times 0.6 \times 1 \times 5700}{0.821102 \times 581.5} = 0.268614$$

$$u = \left(25.147207 \times 0.6 \times 0.017650 \times 0.821102 \times 581.5 \times 5.153^2 \times 5700\right) / 1.9956^5 = 608079$$

Step 8: Evaluate $\bar{y}$

$$\bar{y} = \frac{u}{6}\left(5.2 - 2.2x + 0.6x^2\right) + \frac{p_1^2}{6}\left(-5.2x + 2.2x^2 - 0.6x^3\right) + \frac{aa}{6}\left(1 - x + 0.5x^2 - 0.3x^3\right)$$

Substitution of u = 608079, p_1 = 2545 psia, aa = -1420099 and x = 0.268614 gives

$$\bar{y} = 471499 - 1349039 - 180269 = -1057809$$

Step 9: Evaluate p_2, the pressure at the exit end of the pipe

$$P_2 = \sqrt{\left(2545^2 - 1057809\right)} = 2327.92\,\text{psia} \approx 2328\,\text{psia}$$

Pressure drop across 5700 ft of tubing is 2545 psia - 2328 psia = 217 psia

This pressure drop may be compared with the pressure drop across the 5700 ft of tubing when gas flows uphill against the force of gravity. From example 3, tubing pressure at the surface = 2122 psia when the bottom hole pressure (inlet pressure) = 2545 psia. Then pressure drop = 2545 psia - 2122 psia = 423 psia. The pressure drop during down hill flow is less than that during up hill flow.

The general solution (valid in any system of units) to the differential equation for downhill flow was tested with slight modification of a problem from the book of Giles et al (2009). In the original problem the pipe was horizontal. In the modification used in this work, the pipe was made to incline at 10 degrees from the horizontal in the downhill direction. Other data remained as they were in the book of Giles et al.. The data are as follows:

Example 6

Given the following data,

Length of pipe (L) = 1800ft

$z_2 = z_1 = z_{av} = 1$ (air is flowing fluid)

$p_1 = 49.5$ psia = 49.5×144 psf = 7128 psf

W = 0.75 lb/sec

$Q_b = 9.81937$ ft³/sec

$P_b = 14.7$ psia = 2116.8 psf

$T_b = 60\ ^oF = 520\ ^oR$

$T_1 = T_{av} = 90\ ^oF = 550\ ^oR$

$G_g = 1.0$ (air)

R = 1544

$\mu = 390 \times 10^{-9}$ lb sec/ft²

d = 4 inch = 0.333333 ft

Absolute Roughness ($\in$) = 0.0003 ft

calculate the exit pressure (p_2) at 1800 ft of pipe.

Solution

Step 1: Obtain the gas elevation factor at inlet end, $z_1 = 1.0$, air in flowing fluid

Step 2: Obtain the viscosity of the gas at inlet condition. Viscosity of gas is 390×10^{-9} lb sec / ft ^{2} (given)

Step 3: Evaluate the Reynolds number and friction factor.

$$R_N = \frac{36.88575\, G_g\, P_b Q_b}{g\, R\, d\, \mu\, z_b T_b}$$

Here, $G_g = 1.0(\text{air})$, $P_b = 2116.8$ psf, $Q_b = 9.81937\ \text{ft}^3/\text{sec}, g = 32.2\ \text{ft}/\text{sec}^2$, $d = 0.3333$ft,

$R = 1544, \mu = 390 \times 10^{-9}$, $z_b = 1.0$ (air), $T_b = 520^\circ R$

Then

$$R_N = \frac{36.88575 \times 1 \times 2116.8 \times 9.81937 \times 10^9}{32.2 \times 1544 \times 0.3333 \times 390 \times 1 \times 520} = 2281249$$

$$\frac{\in}{d} = \frac{0.0003}{0.33333} = 0.0009$$

From Moody chart, $f_1 = 0.0205$

Step 4: Evaluate the coefficient aa in the formula

$$aa = \left(\frac{46.958326 f_1 z_1 T_1 G g P_b^2 Q_b^2}{g d^5 z_b^2 T_b^2 R} - \frac{57.940 G g \text{Sin}\theta P_1^2}{z_1 T_1 R} \right) L$$

$$\frac{46.958326\times 0.0205\times 1\times 550\times 1\times 2116.8^2\times 9.81937^2}{32.2\times 0.33333^5\times 520^2\times 1544}=4134.70$$

$$\frac{57.94\times 1\times 0.173648\times 7128^2}{1\times 550\times 1544}=\frac{511191540.9}{849200}=601.97$$

Then aa = (4134.70 - 601.97) × 1800 = 6358914

Step 5

Evaluate the average pressure (p_{av}) at the mid section of the pipe

$$p_{av}=\sqrt{p_1^2-0.5\times 6358914}=6901.4\text{ psf}$$

Step 6: Evaluate average gas derivation factor (Z_{av}) for air, z_{av} = 1.0

Step 7: Evaluate the coefficients x and u

$$x=\frac{57.94G_g\text{Sin}\theta L}{z_{av}T_{av}R}=\frac{57.94\times 1\times 0.173648\times 1800}{1\times 550\times 1544}=0.021326$$

$$u=\frac{46.958326f_1z_{av}T_{av}GgP_b^2Q_b^2L}{gd^5z_b^2T_b^2R}$$

$$=\frac{46.958326\times 0.0205\times 1\times 550\times 1\times 2116.8^2\times 9.81937^2\times 1800}{32.2\times 0.33333^5\times 1\times 520^2\times 1544}=7442642.4$$

Step 8: Evaluate $\bar{y}$

$$\bar{y}=\frac{u}{6}\left(5.2-2.2x+0.6x^2\right)+\frac{p_1^2}{6}\left(-5.2x+2.2x^2-0.6x^3\right)+\frac{aa}{6}\left(1-x+0.5x^2-0.3x^3\right)$$

Where x = 0.021326, $(5.2 - 2.2x + 0.6x^2)$ = 5.153356, $(-5.2x + 2.2x^2-0.6x^3)$ = - 0.10990 $(1 - x + 0.5x^2-0.3x^3)$ = 0.9789.

$$\text{Then,}\quad \bar{y}=\frac{7442642.4}{6}\times 5.153356+\frac{7128^2}{8}\left(-0.1099\right)+\frac{6358914}{6}\left(0.9789\right)$$

$$= 6392431 - 930644 + 1037457 = 6499244$$

Step 9: Evaluate p_2, the pressure at the exit end of the pipe

$$p_2=\sqrt{7128^2-6499244}=6656.5\text{psf}=\frac{6656.5\text{ psia}}{144}=46.2\text{ psia}$$

Pressure drop = 49.5 psia - 46.2 psia = 3.3 psia

When the pipe is horizontal, p_2 (from Fluid Mechanics and Hydraulics) is 45.7 psia. Then, pressure drop = 49.5 psia - 45.7 psia = 3.8 psia

Direct calculation of the gas volumetric rate

The rate of gas flow through a pipe can be calculated if the pipe properties, the gas properties, the inlet and outlet pressures are known. The gas volumetric rate is obtained by solving an equation of pressure transverse simultaneously with the Reynolds number and the Colebrook friction factor equation.

Direct calculation of the gas rate in uphill pipes

Ohirhian(2002) combined the Weymouth equation with the Reynolds number and the Colebrook friction factor equation to arrive at an equation for the direct calculation of the gas volumetric rate during uphill gas flow. In this section, the formula type solution to the differential equation for horizontal and uphill gas flow is combined with the Reynolds number and the Colebrook equation to arrive at another equation for calculating the gas volumetric rate during uphill gas flow.

Combination of the pressure transverse formula for uphill gas flow (equation (27) in oil field units, with the Reynolds number, equation (16) and equation (18) which is the Colebrook friction factor equation, leads to:

$$\frac{1}{\sqrt{f}} = -2\log\left(\frac{\in}{3.7\,d} + \frac{2.51}{\alpha}\right) \quad \text{..... (36)}$$

$$R_N = -2\alpha\log\left(\frac{\in}{3.7\,d} + \frac{2.51}{\alpha}\right) \quad (37)$$

$$Q_b = \frac{-\mu_2\, d\, R_N}{20071\, G_g} = \frac{-2\mu_2\, d\,\alpha}{20071\, G_g}\log\left(\frac{\in}{3.7\,d} + \frac{2.51}{\alpha}\right) \quad (38)$$

$$\alpha = \frac{20071\, G_g\left[p_1^2 - p_2^2\left(1 + \frac{x_b}{6}\right) - \frac{1}{6}\left(S\,Lx_a\right)\right]^{0.5}}{\mu_2\, d\sqrt{B}\left[z_2 T_2\, x_a + z_{av}\, T_{av}\, x_c\right]^{0.5}}$$

Where,

$$x = \frac{0.0375016\, G_g \sin\theta\, L}{z_{av}\, T_{av}}$$

$$B = \frac{4.191201\, G_g\, L}{d^5}$$

$$S = \frac{0.03075016\, G_g \sin\theta\, p_2^2}{z_2\, T_2}$$

z_{av} is computed with $T_{av} = 0.5(T_1 + T_2)$ and $p_{av}^2 = 0.5(p_1^2 + p_2^2)$.

μ_2 is viscosity of gas at p_2 and T_2. The subscript 2 refers to surface condition.

$$x_a = 1 + x + 0.5x^2 + 0.36x^3$$

$$x_b = 4.96x + 1.48x^2 + 0.72x^3$$

$$x_c = 4.96 + 1.96x + 0.72x^2$$

The above equations are in oil field units in which d is expressed in inches, L in feet, μ in centipoises, T in degrees rankine and pressure in pounds per square inches. In this system of units, $p_b = 14.7$ psia, $T_b = 520\ ^\circ R$, $z_b = 1.0$. The subscript 2 refers to surface condition in the gas well.

Example 7

A gas well has the following data:
$L = 5700$ft, $G_g = 0.6$, $\theta = 90^\circ$, $Z_2 = 0.78$, $Z_{av} = 0.821$, $f = 0.0176$, $T_2 = 543\ ^\circ R$, $T_{av} = 581.5\ ^\circ R$, $d = 1.9956$ in, absolute roughness of pipe= 0.0006 in., $p_1 = 2545$ psia, $p_2 = 2122$ psia, μ_2, viscosity of gas at p_2 and $T_2 = 0.0133$ cp. Calculate the flow rate of the well (Q_b) in MM SCF / Day,

Solution
Substituting given values,

$$x = 0.0375016 \times \frac{0.6 \times 1 \times 5700}{0.822 \times 581.5} = 0.268250$$

$$B = 4.191201 \times \frac{0.6 \times 5700}{1.9956^5} = 452.6012689$$

$$S = 0.0375016 \times \frac{0.6 \times 1 \times 2122^2}{0.78 \times 543} = 238.680281$$

$$x_a = 1 + x + 0.5x^2 + 0.36x^3 = 1+0.268250 + 0.5x0.268250^2 +0.36x0.268250^3$$
$$= 1.311178$$

$$x_b = 4.96x + 1.48x^2 + 0.72x^3 = 4.96x0.268250+1.48x0.268250^2+0.72x0.268250^3$$
$$= 1.45096$$

$$x_c = 4.96 + 1.96x + 0.72x^2 = 4.96+ 1.96x0.268250+0.72x0.268250^2=5.53758$$

$$\alpha = \frac{20071 \times 0.6 \left[2545^2 - 2122^2\left(1 + \frac{1.45096}{6}\right) - \frac{1}{6}(238.680281 \times 5700 \times 1.311178)\right]^{0.5}}{0.0133 \times 1.9956 \sqrt{452.6012689}\left[1.45096 \times 0.78 \times 543 + 0.822 \times 581.5 \times 5.53758\right]^{0.5}}$$

$$= 288996.2$$

From example 3, actual Reynolds number is 2.34E06 and f = 0.01527. Then,

$$\alpha = R_N\sqrt{f} = 234000 \times \sqrt{0.01527} = 289158.1$$

$$Q_b = \frac{-2\mu_2 \, d\,\alpha}{20071\, G_g}\log\left(\frac{\in}{3.7\,d} + \frac{2.51}{\alpha}\right) = \frac{-2\times 0.0133\times 1.9956\times 288996.2}{20071\times 0.6}\times \log\left(\frac{0.0006}{3.7\times 1.9956} + \frac{2.51}{288996.2}\right)$$

$$= 5.154 \text{ MMSCF / Day}$$

Direct calculation of the gas volumetric rate in downhill flow

Combination of equation (35) in oil field units, with the Reynolds number, equation (16) and equation (17) which is the Colebrook friction factor equation, leads to:

$$\frac{1}{\sqrt{f}} = -2\log\left(\frac{\in}{3.7\,d} + \frac{2.51}{\alpha_1}\right)\ldots \qquad \ldots.(39)$$

$$R_N = -2\alpha_1 \log\left(\frac{\in}{3.7\,d} + \frac{2.51}{\alpha_1}\right)\ldots \qquad \ldots.(40)$$

$$Q_b = \frac{-\mu_{av}\, d\, R_N}{20071\, G_g} = \frac{-2\mu_{av}\, d\,\alpha_1}{20071\, G_g}\log\left(\frac{\in}{3.7\,d} + \frac{2.51}{\alpha_1}\right) \qquad (40)$$

$$\alpha_1 = \frac{20071\, G_g\left[J + \frac{1}{6}(S\,Lx_d)\right]^{0.5}}{\mu_1\, d\sqrt{B}\left[z_1 T_1 x_d + z_{av} T_{av} x_f\right]^{0.5}} \qquad 41)$$

$$J = p_2^2 - p_1^2\left(1 + \frac{x_e}{6}\right), \text{ if } B \geq S \qquad (42a)$$

$$J = p_1^2\left(1 - \frac{x_e}{6}\right) - p_2^2, \text{ if } B < S \qquad (42b)$$

$$x_d = 1 - x + 0.5x^2 + 0.3x^3$$

$$x_e = -5.2x + 2.2x^2 - 0.6x^3$$

$$x_f = 5.2 - 2.2x + 0.6x^2$$

$$B = \frac{4.191201\, G_g\, L}{d^5}$$

$$S = \frac{0.03075016\, G\, g \sin\theta\, p1^2}{z_1 T_1}$$

$$x = \frac{0.0375016\, G_g \sin\theta\, L}{z_{av}\, T\, av}$$

z_{av} and μ_{av} are evaluated with $T_{av} = 0.5(T_1 + T_2)$ and p_{av} $= (2p_1p_2) / (p_1 + p_2)$

The subscript 1 refers to surface condition and 2 to exit condition in the gas injection well. The above equations are in oil field units in which d is expressed in inches, L in feet, μ in centipoises, T in degrees rankine and pressure in pounds per square inches. In this system of units, p_b = 14.7 psia, T_b = 520 °R, z_b = 1.0. The subscript 2 refers to surface condition in the gas well.

Example 8

A gas injection well has the following data:

L = 5700ft, G_g = 0.6, θ = 90 °, sin90 °= 1, z_1= 0.78059, Z_{av} = 0.821, T_1= 543 °R, T_{av}= 543 °R, d = 1.9956 in, p_1= 2545 psia, p_2= 2327.92 psia, absolute roughness of tubing(∈) = 0.0006 in. Calculate the gas injection rate in MMSCF / Day. Take the viscosity of the at surface condition(μ_1) as 0.015045 cp and the average viscosity of the gas(μ_{av}) as 0.0142 cp.

Solution

$$x = 0.0375016 \times \frac{0.6 \times 1 \times 5700}{0.821 \times 581.5} = 0.26865$$

$$B = 4.191201 \times \frac{0.6 \times 5700}{1.9956^5} = 452.89458$$

$$S = 0.0375016 \times \frac{0.6 \times 1 \times 2545^2}{0.78059 \times 543} = 343.83795$$

$$x_d = 1 - x + 0.5x^2 + 0.3x^3 = 1 - 0.26865 + 0.5 \times 0.26865^2 + 0.36 \times 0.26865^3$$
$$= 0.76162$$

$$x_e = -5.2x + 2.2x^2 - 0.6x^3$$

$$= -5.2 \times 0.26865 + 2.2 \times 0.26865^2 - 0.6 \times 0.26865^3$$
$$= -1.24982$$

$$x_f = 5.2 - 2.2x + 0.6x^2 =$$

$$5.2 - 2.2 \times 0.26865 + 0.60. \times 26865^2 = 4.65228$$

$$\text{Since } B > S,\ J = p_2^2 - p_1^2\left(1 + \frac{x_e}{6}\right) \text{ is used}$$

$$J = 2327.92^2 - 2545^2\left(1 - \frac{1.24982}{6}\right) = 291372.42$$

$$\frac{1}{6}(S\,Lx_d) = \frac{1}{6}(343.83795 \times 5700 \times 0.76162) = 248780.17$$

$$\left[z_1 T_1 x_d + z_{av} T_{av} x_f\right]^{0.5} =$$

$$(0.78 \times 543 \times 0.76162 + 0.821 \times 581.5 \times 4.65228)^{0.5}$$
$$= 50.43440$$

$$\alpha = \frac{20071\,G_g\left[J + \frac{1}{6}(S\,Lx_d)\right]^{0.5}}{\mu_1\, d\sqrt{B}\left[z_1 T_1 x_d + z_{av} T_{av} x_f\right]^{0.5}}$$

$$= \frac{20071 \times 0.6\,[291372.42 + 248780.17]^{0.5}}{0.015045 \times 1.9956 \times \sqrt{452.89458} \times 50.43440} = 274641.7$$

From example 5, actual Reynolds number is 2066877 and f = 0.01765. Then,

$$\alpha = R_N\sqrt{f} = 2066877 \times \sqrt{0.01765} = 274591.4$$

$$Q_b = \frac{-2\mu_{av}\, d\,\alpha}{20071\,G_g}\log\left(\frac{\epsilon}{3.7\,d} + \frac{2.51}{\alpha}\right) = \frac{-2\times0.0142\times1.9956\times274641.7}{20071\times0.6}\times\log\left(\frac{0.0006}{3.7\times1.9956} + \frac{2.51}{274641.7}\right)$$

= 5.227 MMSCF / Day

By use of an average viscosity of 0.0140 cp, the calculated Q_b = 5.153 MMSCF / Day. Taking 0.0140 cp as the accurate value of the average gas viscosity, the absolute error in estimated average viscosity is (0.0002 / 0.014 = 1.43 %) . The equation for the gas volumetric rate is very sensitive to values of the average gas viscosity. Accurate values of the average gas viscosity should be used in the direct calculation of the gas volumetric rate.

Conclusions

1. A general differential equation that governs static behavior of any fluid and its flow in horizontal, uphill and downhill pipes has been developed.
2. classical fourth order Runge-Kutta numerical method is programmed in Fortran 77, to test the equation and results are accurate. The program shows that a length ncrement as large as 10,000 ft can be used in the Runge-Kutta method of solution to differential equation during uphill gas flow and up to 5700ft for downhill gas flow
3. The Runge-Kutta method was used to generate a formulas suitable to the direct calculation of pressure transverse in static gas pipes and pipes that transport gas uphill or downhill. The formulas yield very close results to other tedius methods available in the literature.

4. The direct pressure transverse formulas developed are suitable for wells and pipelines with large temperature gradients.
5. Contribution of kinetic effect to pressure transverse in pipes that transport gas is small and ca n be neglected..
6. The pressure transverse formulas developed in this work are combined with the Reynolds number and Colebrook friction factor equation to provide accurate formulas for the direct calculation of the gas volumetric rate. The direct calculating formulas are applicable to gas flow in uphill and downhill pipes.

Nomenclature

p = Pressure
γ = Specific weight of flowing fluid
v = Average fluid velocity
g = Acceleration due to gravity in a consistent set of umits.
$d\ell$ = Change in length of pipe
θ = Angel of pipe inclination with the horizontal, degrees
dh_l = Incremental pressure head loss
f = Dimensionless friction factor
L = Length of pipe
d = Internal diameter of pipe
W = Weight flow rate of fluid
C_f = Compressibility of a fluid
C_g = Compressibility of a gas
K = Constant for expressing the compressibility of a gas
M = Molecular weight of gas
T= Temperature
R_N = Reynolds number
ρ = Mass density of a fluid
μ =Absolute viscosity of a fluid
z = Gas deviation factor
R = Universal gas constant in a consistent set of umits.
Q_b = Gas volumetric flow rate referred to P_b and T_b,
γ_b = specific weight of the gas at p_b and T_b
p_b = Base pressure, absolute unit
T_b = Base temperature, absolute unit
z_b = Gas deviation at p_b and T_b usually taken as 1
G_g = Specific gravity of gas (air = 1) at standard condition
μ_g = Absolute viscosity of a gas
$\in$ = Absolute roughness of tubing
GTG = Geothermal gradient
f_2 = Moody friction factor evaluated at outlet end of pipe.

z_2 = Gas deviation factor calculated with exit pressure and temperature of gas

Z_1 = Gas deviation factor calculated with exit pressure and temperature of gas

p_2 = Pressure at exit end of pipe.

p1 = Pressure at inlet end of pipe $p_1 > p_2$
T_2 = Temperature at exit end of pipe
T_1 = Temperature at inlet end of pipe

$T_{av} = 0.5(T_1 + T_2)$

z_{av} = Gas deviation factor evaluated with T_{av} and average pressure (p_{av}) given by $p_{av} = \sqrt{p_2^2 + 0.5aa}$ in uphill flow, and $p_{av} = \sqrt{p_1^2 - 0.5|aa|}$ in downhill flow

SI Metric Conversion Factors

$(°F - 32) / 18 = °C$

ft × 3.048000 *E-01 = m

m × 2.540 * E + 00 = cm

lbf × 4.448222 E + 00 = N

lbm × 4.535924 E - 01 = kg

psi × 6.894757 E + 03 = Pa

lb sec / ft^2 × 4.788026 E + 01 = Pa.s

cp × 1.0 E - 3 = Pas

foot^3 / second (ft^3 / sec) x 2.831685 E - 02 = metre^3 / sec (m^3 / s)
foot^3 / second x 8.64 *E - 02 = MMSCF / Day
MMSCF / Day x 1.157407 E + 01 = foot^3 / second (ft^3 / sec)
MMSCF / Day x 3.2774132E + 01 = metre^3 / sec (m^3 / s)
MMSCF / Day = 4.166667E + 04 ft^3/hr
ft^3/hr = 7.865792E - 06m^3/s
* Conversion factor is exact.

References

1. Colebrook, C.F.J. (1938), Inst. Civil Engineers, 11, p 133
2. Cullender, M.H. and R.V. Smith (1956). "Practical solution of gas flow equations for wells and pipelines with large temperature gradients". Transactions, AIME 207, pp 281-89.
3. Giles, R.V., Cheng, L. and Evert, J (2009).. Schaum's Outline Series of Fluid Mechanics and Hydraulics, McGraw Hill Book Company, New York.
4. Gopal, V.N. (1977), "Gas Z-factor Equations Developed for Computer", Oil and Gas Journal, pp 58 - 60.
5. Ikoku, C.U. (1984), Natural Gas Production Engineering, John Wiley & Sons, New York, pp. 317 - 346

6. Matter, L.G.S. Brar, and K. Aziz (1975), Compressibility of Natural Gases", Journal of Canadian Petroleum Technology", pp. 77-80.
7. Ohirhian, P.U.(1993), "A set of Equations for Calculating the Gas Compressibility Factor" Paper SPE 27411, Richardson, Texas, U.S.A.
8. Ohirhian, P.U.: "Direct Calculation of the Gas Volumeric Rates", PetEng. Calculators, Chemical Engineering Dept. Stanford University, California, U.S.A, 2002
9. Ohirhian, P.U. and I.N. Abu (2008), "A new Correlation for the Viscosity of Natural Gas" Paper SPE 106391 USMS, Richardson, Texas, U.S.A.
10. Ohirhian, P.U. (2005) "Explicit Presentation of Colebrook's friction factor equation". Journal of the Nigerian Association of Mathematical physics, Vol. 9, pp 325 - 330.
11. Ohirhian, P.U. (2008). "Equations for the z-factor and compressibility of Nigerian Natural gas", Advances in Materials and Systems Technologies, Trans Tech Publications Ltd, Laubisrtistr. 24, Stafa - Zurich, Switzerland..
12. Ouyang, I and K. Aziz (1996) "Steady state Gas flow in Pipes", Journal of Petroleum Science and Engineering, No. 14, pp. 137 - 158.
13. Standing, M.B., and D.L. Katz (1942) "Density of Natural Gases", Trans AIME 146, pp. 140–9.
14. Standing, M.B. (1970), Volumetric and Phase Behavior of Oil Field Hydrocarbon Systems, La Habra, California.

Steady State Compressible Fluid Flow in Porous Media

Peter Ohirhian
University of Benin, Petroleum Engineering Department
Benin City, Nigeria

Introduction

Darcy showed by experimentation in 1856 that the volumetric flow rate through a porous sand pack was proportional to the flow rate through the pack. That is:

$$\frac{dp}{d\ell_p} = K^{/} Q = K^{/} v \qquad \text{(i)}$$

(Nutting, 1930) suggested that the proportionality constant in the Darcy law (K/) should be replaced by another constant that depended only on the fluid property. That constant he called permeability. Thus Darcy law became:

$$\frac{dp}{d\ell_p} = \frac{k\,v}{\mu} \qquad \text{(ii)}$$

Later researches, for example (Vibert, 1939) and (LeRosen, 1942) observed that the Darcy law was restricted to laminar (viscous) flow.
(Muskat, 1949) among other later researchers suggested that the pressure in the Darcy law should be replaced with a potential (Φ). The potential suggested by Muskat is:

$$\Phi = p \pm \rho g z$$

Then Darcy law became:

$$-\frac{dp}{d\ell_p} = \frac{k\,v}{\mu} \pm \rho g \qquad \text{(iii)}$$

(Forchheimer, 1901) tried to extend the Darcy law to non laminar flow by introducing a second term. His equation is:

$$-\frac{dp}{d\ell_p} = \frac{kv}{\mu} \pm \rho g - \beta \rho v^2 \qquad (iv)$$

(Brinkman, 1947) tried to extend the Darcy equation to non viscous flow by adding a term borrowed from the Navier Stokes equation. Brinkman equation takes the form:

$$-\frac{dp}{d\ell_p} = \frac{kv}{\mu} \pm \rho g + \frac{\mu' d^2 v}{d\ell_p^2} \qquad (v)$$

In 2003, Belhaj et al. re- examined the equations for non viscous flow in porous media. The authors observed that; neither the Forchheimer equation nor the Brinkman equation used alone can accurately predict the pressure gradients encountered in non viscous flow, through porous media. According to the authors, relying on the Brinkman equation alone can lead to underestimation of pressure gradients, whereas using Forchheimer equation can lead to overestimation of pressure gradients. Belhaj et al combined all the terms in the Darcy , Forchheimer and Brinkman equations together with a new term they borrowed from the Navier Stokes equation to form a new model. Their equation can be written as:

$$\frac{dp}{d\ell_p} = \frac{\mu}{\phi} \frac{d^2 v}{d\ell_p^2} - \frac{\mu v}{k} \quad \beta\rho\, v^2 + \rho g - \frac{\rho v dv}{d\ell_p} \qquad (vi)$$

In this work, a cylindrical homogeneous porous medium is considered similar to a pipe. The effective cross sectional area of the porous medium is taken as the cross sectional area of a pipe multiplied by the porosity of the medium. With this approach the laws of fluid mechanics can easily be applied to a porous medium. Two differential equations for gas flow in porous media were developed. The first equation was developed by combining Euler equation for the steady flow of any fluid with the Darcy equation; shown by (Ohirhian, 2008) to be an incomplete expression for the lost head during laminar (viscous) flow in porous media and the equation of continuity for a real gas. The Darcy law as presented in the API code 27 was shown to be a special case of this differential equation. The second equation was derived by combining the Euler equation with the a modification of the Darcy-Weisbach equation that is known to be valid for the lost head during laminar and non laminar flow in pipes and the equation of continuity for a real gas.

Solutions were provided to the differential equations of this work by the Runge- Kutta algorithm. The accuracy of the first differential equation (derived by the combination of the Darcy law, the equation of continuity for a real gas and the Euler equation) was tested by data from the book of (Amyx et al., 1960). The book computed the permeability of a certain porous core as 72.5 millidracy while the solution to the first equation computed it as 72.56 millidarcy. The only modification made to the Darcy- Weisbach formula (for the lost head in a pipe) so that it could be applied to a porous medium was the replacement of the diameter of the pipe with the product of the pipe diameter and the porosity of the medium. Thus the solution to the second differential equation could be used for both pipe and porous medium. The solution to the second differential equation was tested by using it to calculate the dimensionless friction factor for a pipe (f) with data taken from the book of (Giles et al., 2009). The book had f = 0.0205, while the solution to the second differential equation obtained it as 0.02046. Further, the dimensionless friction factor for a certain core (f_p) calculated by the solution to the second differential equation plotted very well in a graph of f_p versus the Reynolds number for porous media that was previously generated by (Ohirhian, 2008) through experimentation.

Development of Equations

The steps used in the development of the general differential equation for the steady flow of gas pipes can be used to develop a general differential equation for the flow of gas in porous media. The only difference between the cylindrical homogenous porous medium lies in the lost head term.

The equations to be combined are;

(a) Euler equation for the steady flow of any fluid.
(b) The equation for lost head
(c) Equation of continuity for a gas.

The Euler equation is:

$$\frac{dp}{\gamma}+\frac{vdv}{g} \pm d\ell_p \sin\theta + dh_l = 0 \tag{1}$$

In equation (1), the positive sign (+) before $d\ell_p \sin\theta$ corresponds to the upward direction of the positive z coordinate and the negative sign (-) to the downward direction of the positive z coordinate. In other words, the plus sign before $d\ell_p \sin\theta$ is used for uphill flow and the negative sign is used for downhill flow.

The Darcy-Weisbach equation as modified by (Ohirhian, 2008) (that is applicable to laminar and non laminar flow) for the lost head in isotropic porous medium is:

$$dh_L = \frac{c^/ v\mu d\ell_p}{k\gamma} \tag{2}$$

The (Ohirhian, 2008) equation (that is limited to laminar flow) for the lost head in an isotropic porous medium is;

$$dh_L = \frac{32c\, v\, \mu\, d\ell_p}{\gamma\, d_p^{\,2}} \tag{3}$$

The Darcy-Weisbach equation as modified by (Ohirhian, 2008) (applicable to laminar and non-laminar flow) for the lost head in isotropic porous medium is;

$$dh_L = \frac{f_p v^2 d\ell_p}{2 g d_p} \tag{4}$$

The Reynolds number as modified by (Ohirhian, 2008) for an isotropic porous medium is:

$$R_{Np} = \frac{\gamma v d_p}{g \mu} = \frac{4 \gamma Q}{\pi g \mu d_p}$$

$$= \frac{4W}{\pi g \mu d_p} \tag{5}$$

In some cases, the volumetric rate (Q) is measured at a base pressure and a base temperature. Let us denote the volumetric rate measured at a base pressure (P_b) and a base temperature (T_b) then,

$$W = \gamma_b Q_b$$

The Reynolds number can be written in terms of γ_b and Q_b as

$$R_{Np} = \frac{4 \gamma_b Q_b}{\pi g \mu d_p} \tag{6}$$

If the fluid is a gas, the specific weight at P_b and T_b is

$$\gamma_b = \frac{p_b M}{z_b T_b R} \tag{7}$$

Also, $M = 28.97\, G_g$, *then:*

$$\gamma_b = \frac{28.97 G_g p_b}{z_b T_b R} \tag{8}$$

Substitution of γ_b in equation (4.8) into equation (4.6) leads to:

$$R_{NP} = \frac{36.88575 G_g P_b Q_b}{R g d_p \mu_g z_b T_b} \tag{9}$$

Example 1

In a routine permeability measurement of a cylindrical core sample, the following data were obtained:

Flow rate of air = 2 cm^2 / sec

Pressure upstream of core = 1.45 atm absolute

Pressure downstream of core = 1.00 atm absolute

Flowing temperature = 70 °F

Viscosity of air at flowing temperature = 0.02 cp

Cross sectional area of core = 2 cm^2

Length of core = 2 cm

Porosity of core = 0.2

Find the Reynolds number of the core

Solution

Let us use the pounds seconds feet (p s f) consistent set units. Then substitution of values into

$$\gamma_b = \frac{p_b M}{z_b T_b R}$$

gives:

$$\gamma_b = \frac{14.7 \times 144 \times 28.97}{1 \times 530 \times 1545} = 0.0748\ \ell b / ft^3$$

$$Q_b = 2\ cm^3 / sec = 2 \times 3.531467\ E-5\ ft^3 / sec$$

$$= 7.062934\ e-5\ ft^3 / sec$$

$$W = \gamma_b Q_b = 0.0748\ \ell b / ft^3 \times 7.062934\ e-5\ ft^3 / sec$$

$$= 5.289431\ E-6\ \ell b / sec$$

$$\mu = 0.02\ cp = 0.02 \times 2.088543\ \ell b / sec / ft^2 \qquad A_p = \frac{\pi d_p^2}{4}, \text{ then, } d_p = 1.128379\sqrt{A_p}$$

$$= 4.177086\ E-7\ \ell b / sec / ft^2$$

$$= 1.128379\sqrt{2 \times 0.2} = 0.713650\ cm$$

$$= 0.\ 023414\ ft$$

$$\text{Then } R_{NP} = \frac{4\ W}{\pi\, g\, \mu\, d_p} = \frac{4 \times 5.289431\text{E}-6}{\pi \times 32.2 \times 4.177086\text{E}-7 \times 0.02341} = 21.385242$$

Alternatively

$$R_{NP} = \frac{36.88575 G_g P_b Q_b}{R g d_p \mu_g z_b T_b} = \frac{36.885750 \times 1 \times 14.7 \times 144 \times 7.052934\text{E}-5}{32.2 \times 4.177086\text{E}-7 \times 1 \times 530 \times 0.023414}$$

$$= 21.385221$$

The equation of continuity for gas flow in a pipe is:

$$W = \gamma_1\ A_1\ v_1 = \gamma_2\ A_2\ v_2 = \text{Constant} \tag{10}$$

Then, $W = \gamma\ A\, v$.

In a cylindrical homogeneous porous medium the equation of the weight flow rate can be written as:

$$W = \gamma\, A_p\, v. \tag{11}$$

Equation (11) can be differentiated and solved simultaneously with the lost head formulas (equation 2, 3 and 4), and the energy equation (equation 1) to arrive at the general differential equation for fluid flow in a homogeneous porous media.

Regarding the cross sectional area of the porous medium (A_p) as a constant, equation (11) can be differentiated and solve simultaneously with equations (2) and (1) to obtain.

$$\frac{dp}{d\ell_p} = \frac{\left(\dfrac{c^{/}\, v\, \mu}{k} \mp \gamma\, \sin\theta\right)}{\left(1 - \dfrac{W^2}{\gamma^2\, A_p^2\, g}\dfrac{d\gamma}{dp}\right)} \tag{12}$$

Equation (12) is a differential equation that is valid for the laminar flow of any fluid in a homogeneous porous medium. The fluid can be a liquid of constant compressibility or a gas. The negative sign that proceeds the numerator of equation (12) shows that pressure decreases with increasing length of porous media.

The compressibility of a fluid (C_f) is defined as:

$$C_f = \frac{1}{\gamma}\frac{d\gamma}{dp} \tag{13}$$

Combination of equations (12) and (13) leads to:

$$\frac{dp}{d\ell_p} = \frac{\left(\dfrac{c^{/} v\,\mu}{k} \mp \gamma \sin\theta\right)}{\left(1 - \dfrac{W^2}{\gamma A_p^2 g}\right)} \tag{14}$$

Differentiation of equation (11) and simultaneous solution with equations (2), (1) and (13) after some simplifications, produces:

$$\frac{dp}{d\ell_p} = \frac{\left(\dfrac{32\,c\,v\,\mu}{d_p^2} \mp \gamma \sin\theta\right)}{\left(1 - \dfrac{W^2 C_f}{\gamma A_p^2 g}\right)} \tag{15}$$

Differentiation of equation (6) and simultaneous solution with equations (4), (1) and (1 3) after some simplifications produces:

$$\frac{dp}{d\ell_p} = \frac{\left(\dfrac{f_p W^2}{2\gamma A_p^2 d_p} \mp \gamma \sin\theta\right)}{\left(1 - \dfrac{W^2 C_f}{\gamma A_p^2 g}\right)} \tag{16}$$

Equation (16) can be simplified further for gas flow through homogeneous porous media. The cross sectional area of a cylindrical cross medium is:

$$A_p = \frac{\pi d_p^2}{4} \tag{17}$$

The equation of state for a non ideal gas is:

$$\gamma = \frac{p\,M}{z\,T\,R} \tag{18}$$

Where

p = Absolute pressure

T = Absolute temperature

Multiply equation (11) with γ and substitute A_p in equation (17) and use the fact that:

$$\frac{p\,dp}{d\ell_p} = \frac{1}{2}\frac{dp^2}{d\ell_p}$$

Then

$$\frac{dP^2}{d\ell_p} = \left[\frac{1.621139\dfrac{f_p W^2\,zRT}{d^5\,M\,g} \mp \dfrac{2\,M\sin\theta\,P^2}{zRT}}{1-\dfrac{1.621139W^2\,zRTC_f}{M\,g\,d^4\,P}}\right] \tag{19}$$

The compressibility of ideal gas (C_g) is defined as

$$C_g = \frac{1}{p} - \frac{1}{z}\frac{z}{p}\,. \tag{20}$$

For an ideal gas such as air,

$$C_g = \frac{1}{p}\,. \tag{21}$$

(Matter et al, 1975) and (Ohirhian, 2008) have proposed equations for the calculation of the compressibility of hydrocarbon gases. For a sweet natural gas (natural gas that contains CO_2 as major contaminant), (Ohirhian, 2008) has expressed the compressibility of the real gas (C_g) as:

$$C_f = \frac{\mathbf{K}}{p} \tag{22}$$

For Nigerian (sweet) natural gas K = 1.0328 when p is in psia. Then equation (19) can then be written compactly as:

$$\frac{d p^2}{d\ell_p} = \frac{(AA p \pm B_p\ p^2)}{(1 - \frac{C_p}{p^2})} \tag{23}$$

Where

$$AA_p = \frac{1.621139 f_p W^2 zRT}{g d_p^5 M}, B_p = \frac{2M \sin\theta}{zRT},$$

$$C_p = \frac{KW^2 zRT}{gMd_p^4}$$

The denominator of the differential equation (23) is the contribution of kinetic effect to the pressure drop across a given length of a cylindrical isotropic porous medium. In a pipe the kinetic contribution to the pressure drop is very small and can be neglected. What of a homogeneous porous medium?

Kinetic Effect in Pipe and Porous Media

An evaluation of the kinetic effect can be made if values are substituted into the variables that occurs in the denominator of the differential equation (23)

Example 2

Calculate the kinetic energy correction factor, given that 0.75 pounds per second of air flow isothermally through a 4 inch pipe at a pressure of 49.5 psia and temperature of 90 0 F.

Solution

The kinetic effect correction factor is $1 - \frac{C}{p^2}$

Where C for a pipe is given by, $C = \frac{KW^2 zRT}{gMd^4}$

Here

$$W = 0.75 \quad lb/sec,\ d = 4 inch = 4/12\ ft = 0.333333\ ft,$$

$$p = 45.5\ psia = 49.5 \times 44\ psf = 7128\ psf,\ T = 90\ ^{o}F = (90 + 460)\ ^{\circ}R = 550\ ^{\circ}R$$

$K = 1$ for an ideal gas, $z = 1.0$ (air is the fluid), $R = 1545, g = 32.2\ ft/sec^2$, $M = 28.97$. Then,

$$C = \frac{1 \times 0.75^2 \times 1 \times 1545 \times 550}{32.2 \times 28.97 \times 0.333333^4} = 41504.58628$$

The kinetic effect correction factor is

$$1 - \frac{C}{p^2} = 1 - \frac{41504.58628}{7128^2} = 0.999183$$

Example 3

If the pipe in example 1 were to be a cylindrical homogeneous porous medium of 25 % porosity, what would be the kinetic energy correction factor?

Solution

Here, $d_p = d\sqrt{\phi} = 0.333333\sqrt{0.25} = 0.1666667$ ft

$$C_p = \frac{1 \times 0.75^2 \times 1 \times 1545 \times 550}{32.2 \times 28.97 \times 0.166667^4}$$

$$= 344046.0212$$

Then,

$$1 - \frac{C_p}{p^2} = 1 \frac{3441046.0212}{7128^2} = 0.993221$$

The kinetic effect is also small, though not as small as that of a pipe. The higher the pressure, the more negligible the kinetic energy correction factor. For example, at 100 psia, the kinetic energy correction factor in example 2 is:

$$1 - \frac{3441046.0212}{(100 \times 144)^2} = 0.998341$$

Simplification of the Differential Equations for Porous Media

When the kinetic effect is ignored, the differential equations for porous media can be simplified. Equation (14) derived with the Darcy form of the lost head becomes:

$$\frac{dp}{d\ell_p} = \left(\frac{c^{/} v \mu}{k} \mp \gamma \sin\theta \right) \tag{24}$$

Equation (15) derived with the (Ohirhian, 2008) form of the lost head becomes:

$$\frac{dp}{d\ell_p} = \left(\frac{32 c v \mu}{d_p^2} \mp \gamma \sin\theta \right) \tag{25}$$

Equation (16) derived with the (Ohirhian, 2008) modification of the Darcy- Weisbach lost head becomes:

$$\frac{dp}{d\ell_p} = \left(\frac{f_p W^2}{2\gamma A_p^2 d_p} \mp \gamma \sin\theta \right) \tag{26}$$

In terms of velocity (v) equation (26) can be written as:

$$\frac{dp}{d\ell_p} = \left(\frac{f_p v^2}{2\gamma d_p} \mp \gamma \sin\theta \right) \tag{27}$$

In certain derivations (for example, reservoir simulation models) it is required to make v or W subject of equations (24) to (27)

Making velocity (v) or weight (W) subject of the simplified differential equations
When v is made subject of equation (24), we obtain:

$$v = \frac{-k}{c/\mu} \left(\frac{dp}{d\ell_p} \mp \gamma \sin\theta \right) \tag{28}$$

When v is made subject of equation (25), we obtain:

$$v = \frac{-d_p^2}{32c\mu} \left(\frac{dp}{d\ell_p} \mp \gamma \sin\theta \right) \tag{29}$$

When v^2 is made subject of equation (27), we obtain:

$$v^2 = \frac{-2g d_p}{f_p \gamma} \left(\frac{dp}{d\ell_p} \mp \gamma \sin\theta \right) \tag{30}$$

When W^2 is made subject of equation (26), we obtain:

$$W^2 = \frac{-2g d_p A_p^2}{f_p \gamma} \left(\frac{dp}{d\ell_p} \mp \gamma \sin\theta \right) \tag{31}$$

Let S be the direction of flow which is always positive, then equation (28) can be written as:

$$v_s = \frac{-k}{\mu} \left(\frac{dp}{ds} - \frac{\gamma}{1.01325} \frac{dz}{ds} \times 10^6 \right) \tag{32}$$

Where:

v_s = Volumetric flux across a unit area of porous medium in unit time along flow path, S cm / sec

$\gamma = \rho g$ = Specific weight of fluid, gm weight / cc

ρ = Mass Density of fluid, gm mass / cc

g = Acceleration due to gravity, 980.605 cm / sec^2

$\frac{dp}{ds}$ = Pressure gradient along S at the point to which v_s refers, atm / cm

μ = Viscosity of the fluid, centipoises

z = Vertical coordinate, considered positive downwards, cm

k = Permeability of the medium, darcys.

1.01325×10^6 = dynes / sq cm atm

According to (Amyx et al., 1960), this is "the generalized form of Darcy law as presented in APT code 27 ".

Horizontal and Uphill Gas Flow in Porous Media

In uphill flow, the + sign in the numerator of equation (23) is used. Neglecting the kinetic effect, which is small, equation (23) becomes

$$\frac{dp^2}{d\ell_p} = AA_p + B_p p^2 \tag{33}$$

$$AA_p = \frac{1.621139 f_p zTRW^2}{g d_p^5 M},$$

$$B_p = \frac{2M\sin\theta}{zTR}$$

An equation similar to equation (33) can also be derived if the Darcian lost head is used. The horizontal / uphill gas flow equation in porous media becomes.

$$\frac{dp^2}{d\ell_p} = AA_p^{/} + B_p p^2 \tag{34}$$

Where

$$AA_p^{/} = \frac{2c'\mu zTRW}{A_p Mk} = \frac{8c'\mu \, zTRW}{\pi d_2^{\,2} Mk}$$

$$= \frac{2.546479 \, c'\mu zTRW}{d_p^{\,2} Mk}$$

Solution to the Horizontal/Uphill Flow Equation

Differential equations (33) and (34) are of the first order and can be solved by the classical Runge - Kutta algorithm. The Runge - Kutta algorithm used in this work came from book of (Aires, 1962) called "Theory and problems of Differential equations". The Runge - Kutta solution to the differential equation

$$\frac{dy}{dx} = f(x, y) \text{ at } x = x_n \text{ given that}$$

$$y = y_0 \text{ at } x = x_0 \text{ is}$$

$$y = y_0 + \frac{1}{6}\left(k_1 + 2(k_2 + k_3) + k_4\right) \tag{35}$$

where

$$k_1 = Hf(x_o, y_o)$$

$$k_2 = Hf(x_0 + \frac{1}{2}H, y_o + \frac{1}{2}k_1)$$

$$k_3 = Hf(x_o + \frac{1}{2}H, y_o + \frac{1}{2}k_1)$$

$$k_4 = Hf(x_o + H, y_0 + k_3)$$

$$H = \frac{x_n \; x_o}{n}$$

$$n = subintervals\ (steps)$$

Application of the Runge - Kutta algorithm to equation (33) leads to:

$$p_1^2 = p_2^2 + \bar{y}_a \tag{36}$$

Where

$$\bar{y}_a = aa_p^a\left(1 + x_a + 0.5x_a^2 + 0.36x_a^3\right) + \frac{p_2^2}{6}\left(4.96x_a + 1.48x_a^2 + 0.72x_a^3\right) + \frac{u_p^a}{6}\left(4.96 + 1.96x_a + 0.72x_a^2\right)$$

$$aa_p^a = (AA_{p2} + S_2)L$$

$$AA_{p2} = \frac{1.621139\, f_p z_2 T_2 RW^2}{g d_p^5 M},$$

$$S_2 = \frac{2M \sin\theta p_2^2}{z_2 T_2 R}$$

$$u_p^a = \frac{1.621139 f_p z_{av} T_{av} RW^2}{g d_p^5 M},$$

$$x_a = \frac{2M \sin\theta L}{z_{av}^a T_{av} R}$$

Where:

p_1 = Pressure at inlet end of porous medium

p_2 = Pressure at exit end of porous medium

f_p = Friction factor of porous medium.

θ = Angle of inclination of porous medium with horizontal in degrees.

z_2 = Gas deviation factor at exit end of porous medium.

T_2 = Temperature at exit end of porous medium

T_1 = Temperature at inlet end of porous medium

z_{av} = Average gas deviation factor evaluated with T_{av} and p_{av}

T_{av} = Arithmetic average temperature of the porous medium given by $0.5(T_1 + T_2)$ and

$$p_{av}{}^{a} = \sqrt{p_2^{\,2} + aa_p^{\,a}}$$

In equation (36), the component k_4 in the Runge - Kutta algorithm was given some weighting to compensate for the variation of temperature (T) and gas deviation factor (z) between the mid section and the inlet end of the porous medium. In isothermal flow where there is little variation of the gas deviation factor between the mid section and the inlet end of the porous medium, the coefficients of x_a change slightly, then,

$$\bar{y}_a = aa_p^{\,a}\left(1 + x_a + 0.5x_a^{\,2} + 0.25x_a^{\,3}\right) + \frac{p_2^{\,2}}{6}\left(5x_a + 2x_a^{\,2} + 0.5x_a^{\,3}\right) + \frac{u_p^{\,2}}{6}\left(5 + 2x_a + 0.5x_a^{\,2}\right)$$

Application of the Runge-Kutta algorithm to equation (34) produces.

$$p_1^{\,2} = p_2^{\,2} + \bar{y}_b \tag{37}$$

$$\bar{y}_b = aa_p^{\,b}\left(1 + x_b + 0.5x_b^{\,2} + 0.36x_b^{\,3}\right) + \frac{p_2^{\,2}}{6}\left(4.96x_b + 1.48x_b^{\,2} + 0.72x_b^{\,2}\right) + \frac{u_p^{\,2}}{6}\left(4.96 + 1.96x_b + 0.72x_b^{\,2}\right)$$

Where $aa_p^b = \left(AA_{p2}^{/} + S_2 \right) L$

$$AA_{p2}^{/} = \frac{2c'\mu z_2 T_2 RW}{A_p M k} = \frac{8c'\mu\, z_2 T_2 RW}{\pi d_2^2\, M k}$$

$$= \frac{2.546479\; c'\mu z_2 T_2 RW}{d_p^2\; M k}$$

$$S_2 = \frac{2M \sin\theta p_2^2}{z_2 T_2 R},\; u_p^b = \frac{2c'\mu z_{av}^b\, T_{av} RW}{A_p Mk}$$

$$= \frac{2.546479c'\mu z_{av}^b\, T_{av} RW}{d_p^2\; Mk}$$

$$x_b = \frac{2M \sin\theta L}{z_{av}^b\, T_{av} R}$$

Where

z_{av}^b = Average gas deviations factors evaluated with T_{av} and p_{av}^b

T_{av} = Arithmetic average Temperature of the porous medium = $0.5(T_1+T_2)$,

$p_{av}^b = \sqrt{p_2^2 + 0.5 aa_p^b}$

All other variables remain as defined in equation (36). In isothermal flow where there is not much variation in the gas deviation factor (z) between the mid section and inlet and of the porous medium there is no need to make compensation in the k_4 parameter in the Runge Kuta algorithm, then equation (37) becomes:

$$p_1^2 = p_2^2 + \bar{y}_{bT} \tag{38}$$

Where:

$$\bar{y}_{bT} = aa_p^{\,b}\left(1 + x_b + 0.5x_b^{\,2} + 0.25x_b^{\,2} + 0.5x_b^{\,3}\right) + \frac{p_2}{6}\left(5x_b + 2x_b^2 + 0.5x_b^3\right) + \frac{u_p^{\,b}}{6}\left(5 + 2x_b + 0.5x_b^2\right)$$

Equation (36) can be arranged as:

$$W^2 f_p BB_p^{\,a}\left[z_2T_2\left(1 + x_c + 0.5x_c^2 + 0.36x_c^3\right) + PU\right] = \left[\begin{array}{l} p_1^2 - \frac{p_1^2}{6}\left(4.96x_c + 1.48x_c^2 + 0.72x_c^2\right) - p_2^2 \\ - S_2\left(1 + x_c + 0.5x_c^2 + 0.36x_c^3\right) \end{array}\right] \quad (39)$$

Where

$$PU = z_{av}T_{av}\left(4.96x_c + 1.48x_c^2 + 0.72x_c^2\right)$$

$$BB_p^{\,a} = \frac{1.621139RL}{6g\,d_p^5\,M}, \quad S_2 = \frac{2M\sin\theta p_2^2}{z_2T_2R}, \quad x_c = \frac{2M\sin\theta L}{z_{av}^{\,c}T_{av}R}$$

$z_{av}^{\,c}$ = Average gas deviations factors evaluated with T_{av} and $p_{av}^{\,c}$ and

$$p_{av}^{\,c} = \sqrt{\frac{p_1^2 + p_2^2}{2}}$$

All other variables remain as defined in previous equations.
In isothermal flow where there is no significant change in the gas deviation factor (z), equation (39) becomes:

$$W^2 f_p BB_p^{\,a} z_2T_2\left[\begin{array}{l} \left((1 + x_c + 0.5x_c^{\,2} + 0.25x_c^{\,3}\right) \\ +(5 + 2x_c + 0.5x_c^{\,2})\right) = \end{array}\right]$$

$$p_1^2 - \frac{p_1^2}{6}\left(5x_c + 2x_c^2 + 0.5x_c^3\right) - p_2^2 - \frac{S_2L}{6}\left(1 + x_c + 0.5x_c^2 + 0.25x_c^3\right) \tag{40}$$

When the porous medium is horizontal, $S_2 = 0$ and $x_c = 0$ then from equation (40),

$$f_p = \frac{p_1^2 - p_2^2}{W^2 BB_p^a \left(z_2 T_2 + 4.96 z_{av} T_{av}\right)} \tag{41}$$

In an isothermal flow where there is no variation in z,

$$f_p = \frac{p_1^2 - p_2^2}{6W^2 BB_p^a z_2 T_2} \tag{42}$$

Example 4

The following data came from the book of (Giles et al., 2009) called "theory and problem of fluid mechanics and hydraulics"

W = 0.75 1b/sec of air, R = 1544, L = 1800ft, d = 4inch = 0.333333ft,

$g = 32.2 ft/sec^2$, $z_2 = Z_{av}^a = 1$ (air is fluid), $T_2 = T_{av} = 90^0 F = 550^0 R$

(Isothermal flow), p_1 = 49.5psia = 7128psf, P_2 = 45.73 psia = 6585.12 psf.

Pipe is horizontal.

(a) Calculate friction factor of the pipe (f)
(b) If the pipe were to be filled with a homogenous porous material having a porosity of 20% what would be the friction factor (fp)?

Solution

(a) Let BB^a the equivalent BB_p^a by use of a pipe then.

$$BB^a = \frac{`1.621139RL}{6gd^5M}$$

$$= \frac{1.621139 \times 1544 \times 1800}{6 \times 32.2 \times 0.333333^5 \times 28.97}$$

$$= 195610.8241$$

$$f = \frac{p_1^2 - p_2^2}{6W^2 BB^a z_2 T_2} = \frac{7128^2 - 6585.12^2}{6 \times 0.75^2 \times 195610.8241 \times 1 \times 550}$$

$$= 0.20463$$

The calculated f agrees with f = 0.0205 obtained by Giles et al., who used another equation.

(b) $d_p = 0.333333 \times \sqrt{0.2\text{ft}} = 0.149071\text{ft}$

$$BB_p^{\ a} = \frac{`1.621139RL}{6gd_p^{\ 5}M}$$

$$= \frac{1.621139 \times 1544 \times 1800}{6 \times 32.2 \times 0.1490751^{5} \times 28.97}$$

$$= 10934995.62$$

$$f_p = \frac{p_1^{\ 2} - p_2^{\ 2}}{6W^2 BB_p^{\ a} z_2 T_2}$$

$$= \frac{7128^{\ 2} - 6585.12^{\ 2}}{6 \times 0.75^{\ 2} \times 10934995.6 \quad 2 \times 1 \times 550}$$

$$= 3.667626 \quad E - 4$$

The equation for pressure transverse in a porous medium by use of Darcian lost head (equation (37) can be arranged as:

$$\frac{W^2 BB_p^{\ b}}{k}\left[z_2 T_2 \left(1 + x_c + 0.5x_c^{\ 2} + 0.36x^3{}_c\right) + z_{av}^{\ \ b} T_{av}\left(4.96 + 1.96x_c + 0.72x_c^{\ 2}\right)\right]$$
$$= p_1^{\ 2} - \frac{p_1^{\ 2}}{6}\left(4.96x_c + 1.48x_b^{\ 2} + 0.36x_c^{\ 3}\right) - p_2^{\ 2} - \frac{S_2 L}{6}\left(1 + x_c + 0.5x_c^{\ 2} + 0.3x_c^{\ 2}\right) \tag{43}$$

Where

$$BB_p^{\ b} = \frac{2c'\mu RL}{6A_p M} = \frac{2.576479c'\mu RL}{6\ d_p^{\ 2} M},$$

$$S_2 = \frac{2M \sin\theta\, P_2^{\;2}}{z_2 T_2 R}$$

$$x_c = \frac{2M \sin\theta\, L}{z_{av}^{\;c}\; T_{av} R}$$

$z_{av}^{\;c}$ = Average gas deviation factor calculated with $p_{av}^{\;c}$ and T_{av}

$$p_{av}^{\;c} = \sqrt{\frac{p_1^{\;2} + p_2^{\;2}}{2}}$$

When the porous medium is horizontal, $S_2 = 0$, , and $x_c = 0$, then ,

$$k = \frac{W^2 BB_p^{\;b}\left[z_2 T_2 + 4.96\, z_{av}^{\;c}\, T_{av}\right]}{\left(p_1^{\;2} - p_2^{\;2}\right)} \tag{44}$$

When the flow is isothermal and there is no significant variation in the gas deviation factor (z) equation (44) becomes.

$$\frac{W^2 BB_p^{\;b}}{k}\left[\begin{array}{l} z_2 T_2\left(1 + x_c + 0.5x_c^{\;2} + 0.25x_c^{3}\right) \\ + z_2 T_2\left(5 + 2x_c + 0.5x_c^{\;2}\right) \end{array}\right]$$

$$= \left[p_1^{\;2} - \frac{p_1^{\;2}}{6}\left(5x_c + 2x_c^{\;2} + 0.5x_c^{\;3}\right) - p_2^{\;2} - \frac{S_2 L}{6}\left(1 + x_c + 0.5x_c^{\;2} + 0.25x_c^{\;2}\right)\right] \tag{45}$$

When the porous medium is horizontal, equation (45) becomes

$$k = \frac{6WBB_p^{\;b} z_2 T_2}{\left(p_1^{\;2} - p_2^{\;2}\right)} \tag{46}$$

Example 5

The following problem came from the book of (Amyx et al., 1960). During a routine permeability test, the following data were obtained.

Flow rate (Q) = 1,000cc of air in 500sec.
Pressure down stream of core (p_2) = 1 atm. absolute
Flowing temperature (T) = 70 ^{0}F
Viscosity or air at test temperature (μ) = 0.02c_p
Cross-sectional area of core (A_p) = 2cm^2
Pressure upstream of core (p_1) = 1.45 atm absolute
Length of core (L_p) = 2cm
Compute the permeability of the core in millidarcy

Solution

In oil field units in which pressure is in atmospheres and temperature is expressed in degree Kelvin, R = 82.1

$$\text{Here, } T = 70^0F = (70 + 460)^0R = \frac{530^0 R}{1.8}$$

$$= 294.4^0 K$$

$$Q = 1000cc / 500\,\text{sec} = 2cc / \text{sec}$$

$$z_1 = z_2 = z_{av}^{\;c} = 1\ (air\ is\ fluid)$$

The volumetric flow rate can be converted to weight flow rate by:

$$W = \gamma Q \text{ where } \gamma = \frac{pM}{zTR}$$

Substituting given values

$$W = \frac{1 \times 28.97 \times 2}{1 \times 82.1 \times 294.4}$$

$$= 0.002397163\ gm / \text{sec}$$

Taking the core to be horizontal

$$k = \frac{6wBB_P^{\;b} z_2 T_2}{\left(p_1^{\;2} - p_2^{\;2}\right)} \text{ where}$$

$$BB_p^{\;b} = \frac{2\mu RL}{6A_p M}, (c' = 1 \text{ in a consistent set of units})$$

$$= \frac{2 \times 0.02 \times 82.1 \times 2}{6 \times 2 \times 28.97} = 1.889311E^{-}2$$

Then

$$k = \frac{6 \times 2.397163E \quad 3 \times 1.889311E \quad 2 \times 1 \times 294.4}{1.45^2 - 1^2}$$

$$= 0.07256\text{darcy} = 72.56 \text{ millidarcy}$$

Amyx , et al obtained the permeability of this core as 72.5md with a less rigorous equation.

Horizontal and Downhill Gas Flow in Porous Media

In downhill flow, the negative (-) sign in the numerator of equation (23) in used. Neglecting the kinetic effect, equation (23) becomes:

$$\frac{dp^2}{d\ell_p} = AA_p - B_p p^2 \tag{47}$$

Where

$$AA_p = \frac{1.621139 f_p zTR}{g d_p^5 M},$$

$$B_p = \frac{2M \sin\theta}{zTR}$$

By use of the Darcian lost head, the differential equation for downhill gas flow in porous media becomes.

$$\frac{dp^2}{d\ell_p} = AA'_p - B_p p^2 \tag{48}$$

Where

$$AA'_p = \frac{2c'\mu zTRW}{A_p Mk} = \frac{2.546479\ c'\mu zTRW}{d_p^2 Mk}, \quad B_p = \frac{2M \sin\theta}{zTR}$$

Solution to the differential equation for horizontal and downhill flow

The Runge-Kutta numerical algorithm that was used to provide a solution to the differential equation for horizontal and uphill flow can also be used to solve the differential equation for horizontal and downhill flow. Application of the Runge - Kutta algorithm to equation (47) produces.

$$p_2 = \sqrt{p_1^2 - \left| \overline{y}_c \right|} \tag{49}$$

Where

$$\bar{y}_c = aa_p^{\ c}(1 - x_d + 0.5x_d^{\ 2} - 0.3x_d^{\ 3})$$

$$+\frac{p_1^{\ 2}}{6}\left(-5.2x_d + 2.2x_d^{\ 2} - 0.6x_d^{\ 3}\right)$$

$$+\frac{u_p^{\ c}}{6}\left(5.2 - 2.2x_d - 0.6x_d^{\ 3}\right)$$

$$aa_p^{\ c} = (AAp_1 - S_1)L$$

$$AA_{p1} = \frac{1.621139 f_p z_1 T_1 RW^2}{gd_p^{\ 5} M},$$

$$S_1 = \frac{2M\sin\theta p_1^{\ 2}}{z_1 T_1 R}$$

$$u_p^{\ c} = \frac{1.621139 f_p z_{av}^{\ d} T_{av} RW^2}{gd_p^{\ 5} M},$$

$$x_d = \frac{2M\sin\theta L}{z_{av}^{\ d} T_{av} R}$$

$z_{av}^{\ d}$ = Gas deviation factor (z) calculated

with $T_{av} = 0.5(T_1 + T_2)$

and $p_{av}^{\ d} = \sqrt{p_1^{\ 2} - \left| aa_p^{\ c} \right|}$

Other variables remain as defined in previous equations.
In equation (49), the parameter k_4 in the Runge-Kutta algorithm is given some weighting to compensate for the variation of the temperature (T) and the gas deviation factor between the mid section and the exit end of the porous medium. In isothermal flow in which there is no significant variation of the gas deviation factor (z) between the midsection and the exit end of the porous medium, equation (49) becomes.

$$p_2 = \sqrt{p_1^{\ 2} - \left| \bar{y}_{cT} \right|} \tag{50}$$

Where

$$\bar{y}_{cT} = aa_p{}^c (1 - \dot{x}_d + 0.5x_d{}^2 - 0.35x_d{}^3) + \frac{p_1{}^2}{6}\left(-5.0x_d + 2.0x_d{}^2 - 0.7x_d{}^3\right) + \frac{u_p{}^c}{6}\left(5.0 - 2.0x_d - 0.7x_d{}^2\right)$$

Other variables in equation (50) remain as defined in equation (49).
Application of the Runge -Kutta algorithm to the down hill differential equation by use of Darcian lost head (equation (48)) gives

$$p_2{}^2 = p_1{}^2 - |\bar{y}_d| \qquad (51)$$

Where

$$\bar{y}_d = aa_p{}^d (1 - x_e + 0.5x_e{}^2 - 0.3x_e{}^3) + \frac{p_1{}^2}{6}\left(-5.2x_e + 2.2x_e{}^2 - 0.6x_d{}^3\right) + \frac{u_p{}^d}{6}\left(5.2 - 2.2x_e - 0.6x_e{}^2\right)$$

$$aa_p{}^d = (AAp'_1 - S_1)L$$

$$AAp'_1 = \frac{2c'\mu z_1 T_1 RW}{A_p Mk} = \frac{2.54679\, c'\mu z_{av}{}^e\, T_{av} RW}{d_p{}^2\, Mk},$$

$$S_2 = \frac{2M\sin\theta\, P_1^2}{z_1 T_1 R}$$

$$x_e = \frac{2M\sin\theta\, L}{z_{av}{}^{e}\, T_{av} R},$$

$$u_p{}^{d} = \frac{2c'\mu z_{av} T_{av} RW}{A_p Mk} = \frac{2.546479 c'\mu z_{av}{}^{e} T_{av} RW}{d_p{}^{2} Mk}$$

$z_{av}{}^{e}$ = Gas deviation factor (z) calculated

with T_{av} and $p_{av}{}^{e}$ $\quad T_{av} = 0.5(T_1 + T_2)$

$$p_{av}{}^{e} = \sqrt{p_1{}^{2} - \left| aa_p{}^{d} \right|}$$

Equation (49) can be written as:

$$\frac{f_p W^2 \left[J_p + \frac{S_1 L}{6}(1 - x_f + 0.5^2 x_f - 0.3 x_f{}^3) \right]}{BB_p^a \left[Z_1 T_1 (1 - x_f + 0.5 x_f{}^2 - 0.3 x_f{}^3) + XX \right]} \tag{53}$$

Where $XX = z_{av}{}^{f} T_{av}(5.2 - 2.2x_f + 0.6x_f{}^2)$

$$J_p = P_1^2 - \frac{p_1^2}{6}\left(-5.2x_f + 2.2x_f{}^2 - 0.6x_f{}^3\right) - p_2{}^2 \text{ if } BB_p{}^a \geq S_1$$

$$J_p = P_2^2 - p_1^2 - \frac{p_1^2}{6}\left(5.2x_f + 2.2x_f{}^2 - 0.6x_f{}^3\right) - p_2{}^2 \text{ if } BB_p{}^a < S_1$$

$$BB_p{}^a = \frac{1.621139RL}{6gd_p{}^5 M} = \frac{0.270110RL}{gd_p{}^5 M}$$

$$S_1 = \frac{2M\sin\theta\, p_1^2}{6gd_p{}^5 M}, \quad x_f = \frac{2M\sin\theta L}{z_{av}{}^{f} T_{av} R}$$

$z_{av}{}^{f}$ = Gas deviation factor at the midsection of the porous medium calculated with

$$T_{av} \text{ and } p_{av}{}^{f}, \text{ where } T_{av} = 0.5(T_1 + T_2)$$

$$\text{and } p_{av}{}^{f} = \frac{2p_1p_2}{p_1+p_2}$$

During isothermal flow in which there is no significant variation of the gas deviation factors (z) between the mid section and the exit end of the porous medium, equation (52) can be written as:

$$f_p w^2 = \frac{\left[J_p + \frac{S_1 L}{6}\left(1 - x_f + 0.5x_f{}^2 - 0.35x_f{}^3\right)\right]}{BB^a{}_p\left[z_1T_1\left(1 - x_f + 0.5x_f{}^2 - 0.35x_f{}^3\right)\right] + z_1T_1\left(5.0 - 2.0x_f + 0.7x_f{}^2\right)} \qquad (54)$$

The variables in equation (53) remain as defined in equation (52)

Example 6

Suppose the porous medium of example 3b was vertical what would be the dimensionless friction factor by use of the same pressure as they were in example 3b?

Solution

Here, $P_1 = 7128$psf, $P_2 = 6585.12$psf, $T_1 = T_{av} = 550^0$R, W=0.75 ℓb / sec, R= 1544, L_p= 1800ft, g= 32.2ft / sec ^{2}, d_p =0.066667ft, $z_1 = z_{av}{}^f = 1$, since $\theta = 90^0$, $\sin 90^0 = 1$

$$x_f = \frac{2M \sin \theta L}{z_{av}{}^f\, T_{av} R} = \frac{2 \times 28.97 \times 1 \times 1800}{1 \times 550 \times 1544}$$

$$= 0.122812$$

The flow is isothermal; z is constant at 1.0 so equation (52) is used.

$$1 - x_f + 0.5x_f{}^2 - 0.35x_f{}^3 = 0.884081$$

$$5.0 - 2.0x_f + 0.7x_f{}^2 = 4.764934$$

$$-5.0x_f + 2.0x_f{}^2 + 0.7x_f{}^3 = 0.585191$$

$$BB_p{}^a = \frac{0.270110RL}{gd_p{}^5 M}$$

$$= \frac{0.270110 \times 1544 \times 1800}{32.2 \times 0.066667^5 \times 28.97} = 61128325.$$

$$S_1 = \frac{2M\sin\theta p^2{}_1}{z_1 T_1 R} = \frac{2 \times 28.97 \times 1 \times 7128^2}{1 \times 550 \times 1544}$$

$$= 3466.601$$

$$BB_p{}^a > S_1, then$$

$$J_p = P_1^2 - \frac{p_1^2}{6}\left(-5.0x_f + 2.0x_f{}^2 - 0.7x_f{}^3\right) - p_2^2$$

$$= 7128^2 - \frac{7128^2}{6}\left(-0.585191\right)$$

$$-6585.12^2 = 12400014$$

$$\frac{S_1 L}{6}\left(1 - x_f + 0.5x_f{}^2 - 0.35x_f{}^3\right)$$

$$= \left(3466.601 \times 1800 \times 0.884081\right) / 6 = 919426.8859$$

Then

$$f_p = \frac{12400014 + 919426.8859}{0.75^2 \times 611283825\left[\left(1 \times 550 \times 0.884081\right) + \left(1 \times 550 \times 4.764934\right)\right]}$$

$$= 1.977439E6$$

There is a drastic reduction in the f_p as compared to f_p= 6.560860 E-6 when the porous medium was horizontal. The effect of inclination becomes more severe as the porous medium gets longer.

Equation (40) can be written as:

$$k = \frac{BB_p{}^b W\left[z_1 T_1\left(1 - x_f + 0.5x_f{}^2 - 0.3x_f{}^3\right) + Z_{av}{}^f T_{av}\left(5.2 - 2.2x_f + 0.6x_f{}^2\right)\right]}{\left[J_p + \frac{S_1 L}{6}\left(1 - x_f + 0.5x_f{}^2 - 0.3x_f{}^3\right)\right]}.$$

Where

$$J_p = P_1^2 - \frac{p_1^2}{6}\left(-5.2x_f + 2.2x_f{}^2 - 0.6x_f{}^3\right) - p_2^2, \text{if } BB_p{}^b \geq S_1$$

$$J_p = P_2^2 - \frac{p_1^2}{6}\left(-5.2x_f + 2.2x_f^2 - 0.6x_f^3\right) - p_2^2 \quad \text{if } BB_p^b < S_1$$

$$BB_p^b = \frac{2c^{/}\mu RL}{6A_pM} = \frac{2.546479c^{/}\mu RL}{6d_p^2M}$$

All other variables remain as defined in equation (52). During isothermal flow in which there is no significant variation of the gas deviation factor (z) between the mid section and the exit end of the porous medium, equation (54) can be written as:

$$k = \frac{BB_p^b W z_1 T_1 [\,(1 - x_f + 0.5x_f^2 - 0.3x_f^3) + (5.0 - 2.0x_f + 0.7x_f^2)}{J_p + \frac{S_1L}{6}\left(1 - x_f + 0.5x_f^2 - 0.35x_f^3\right)} \quad (55)$$

Beside the coefficients of x_f all other variables in equation (55) remain as defined in equation (54).

Example 7
Compute the permeability of the core of example 4 assuming that the case was vertical.

Solution
From example 4, W= γ Q
Substituting the given values, W=0.00239716gm/sec
Sin θ = sin 90^0 =1.0, M =28.97, L_p=2cm

$$z_{av}^{f} = z_1 = 1,\ T_1 = T_{av} = 294.4\ ^oK,\ A_p = 2cm^2$$

$$p_1 = 1.45atm,\ p_2 = 1.0atm,\ \mu = 0.02cp.$$

$$x_f = \frac{2M \sin \theta L^{'}}{z_{av}^{f}\, T_{av} R} = \frac{2 \times 28.97 \times 2}{1 \times 294.4 \times 82.1}$$

$$= 0.004794$$

The flow is isothermal so equation (55) is used.

$$1 - x_f + 0.05x_f^2 - 0.35x_f^3 = 0.995217$$

$$5.0 - 2x_f + 0.7x_f^2 - 0.70x_f^3 = 4.990428$$

$$-5.0x_f + 2.0x_f^2 - 0.70x_f^3 = -0.023941$$

$$BB^b_p = \frac{2c^1\mu RL}{6A_pM} = \frac{0.02 \times 82.1 \times 2}{3 \times 2 \times 28.97} = 0.018893$$

$$S_1 = \frac{2M \sin\theta\, p_1^2}{z_1T_1R} = \frac{2 \times 28.97 \times 1 \times 1.45^2}{1 \times 294.4 \times 82.1}$$

$$= 0.005040$$

$BB_p^b > S_1$, therefore,

$$J_p = p_1^2 - \frac{p_1^2}{6}\left(-5.0x_f + 2.0x_f^2 - 0.7x_f^3\right) - p_2^2 = 1.45^2 - \frac{1.45^2}{6}\left(-0.023941\right) - 1^2 = 1.110883$$

Substitution of given values into equation (54) gives

$$k = \frac{0.018893 \times 0.002397 \times 1 \times 2944\left[0.99521 + 4.990428\right]}{1.11088 + \dfrac{0.005040 \times 2 \times .0995217}{6}}$$

$$= \frac{0.018893 \times 0.00239716 \times 1762.173888}{1.110883 + 0.001672}$$

=0.071734 darcy = 71.734 millidarcy.

Comparing 71.734 md with 72.562md obtained when the core was considered horizontal, it is seen that inclination has reduced, the calculated permeability (k) by (72.562-71.734)/72.564 = 1.141093 percent
The longer the core, the more, the effect of inclination.

Example 8
Use the data of example 4 to calculate the dimensionless friction factor (f_p). Because of simplicity assume that the core is horizontal.

Solution

$$p_1 = 1.45\text{atm} = 1.45 \times 14.7 \times 144\text{psf}$$

$$= 3069.36\text{psf}$$

$$p_2 = 1\text{atm} = 14.7 \times 144\text{psf} = 2116.80\text{psf}$$

$$z_2 = 1, T_2 = 530^0 R, \theta = .0.2$$

$$L_p = 2\text{cm} = 2/2.54\text{in} = 2/(2.54 \times 12)\ \text{ft}$$

$$M = 28.97, g = 32.2\text{ft}/\sec^2$$

$$A_p = 2 \times 0.2\text{cm}^2 = 0.4\text{cm}^2, R = 1545$$

$$d_p = 1.128379\sqrt{A_p} - 0.713650\text{cm}$$

$$= 0.023414\text{ft}$$

$$\gamma_b = \frac{p_b M}{z_b T_b R} = \frac{1 \times 14.7 \times 28.97}{1 \times 530 \times 1545}$$

$$= 0.074890\ \ell b/\text{ft}^3$$

$$Q_b = 2\text{cm}^3/\sec = 2 \times 3.531467E^-5\text{ft}^3/\sec$$

$$W = \gamma_b Q_b = 5.289431E^-6\ \ell b/\sec$$

$$\mu = 0.02 \times 2.088543E^-5\ \ell b \sec/\text{ft}^2$$

$$= 4.177086E^-7\ \ell b \sec/\text{ft}^2$$

$$BB_p^a = \frac{1.621139RL}{6gd_p^5 M} = \frac{1.621139 \times 1545 \times 0.0656168}{6 \times 32.2 \times 0.023414^5 \times 28.97}$$

$$= 4172824$$

$$f_p = \frac{p_1^2 - p_2^2}{6W^2 BB_p^a z_2 T_2}$$

$$= \frac{3069.36^2 - 2116.80^2}{6 \times \left(5.289431E-6\right)^2 \times 4172824 \times 1 \times 530}$$

$$= 0.133065E8$$

The coordinate (R_{NP}, f_p) = (21.385242 , 0.0133065E8) locates very well in a previous graph of f_p versus R_{NP} that was generated by (Ohirhian, 2008). The points plotted in the graph were obtained by flowing water through synthetic tight consolidated cores. The plot is reproduced here as follows.

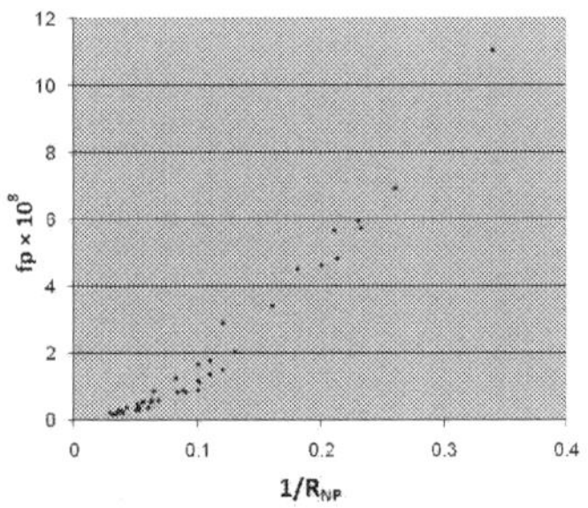

Plot of f_p versus R_{Np} for Porous Media

Assignment

Use the data of example 4 to calculate the dimensionless friction factor (f_p) considering the core to be vertical

Conclusions

(1) The Darcy law as presented in API code 27 has been derived from the laws of fluid mechanics
(2) New general differential equations applicable to horizontal, uphill and downhill flow of gas through porous media have been developed.
(3) The Runge-Kutta algorithm has been used to provide accurate solutions to the differential equations developed in this work.
(4) The solution to the differential equation shows that inclination has the effect of reducing laboratory measured values of gas permeability and dimensionless friction factor- the longer a core the more the reduction of measured permeability / dimensionless friction factor.

Nomenclature

$dp =$ Incremental pressure drop

$d\ell_p$ = Incremental length of porous medium

Q = Volumetric flow rate

V = Average velocity flowing fluid

$K^{/}$ = Proportionality constant that is dependent on both fluid and rock properties

k = Permeability of porous medium

$\mu =$ Absolute viscosity of flowing fluid

$\rho =$ Mass density of flowing fluid

g = Acceleration due to gravity

z = Elevation of the porous medium above a datum. The + sign is used where the point of interest is above the datum the - sign is used where the chosen point is and below the datum

$\mu^{/} = \text{Effective viscosity of flowing fluid} = \frac{\mu}{\phi}$

p = Pressure

γ = Specific weight of flowing fluid

v = Average fluid velocity

g = Acceleration due to gravity in a consistent set of units.

$d\ell_p$ = Incremental length of porous medium

θ = Angel of porous medium inclination with the horizontal, degrees

dh_L=Incremental lost head

c = Dimensionless constant which is dependent on the pone size disribution of porous medium

c_1 = Constant used for conversion of units. It is equal to 1 in a consistent set of units

d_p = Diameter of porous medium = $d\sqrt{\phi}$

d= Diameter of cylindrical pipe

ϕ = Porosity of medium

f_p = Dimensionless friction factor of porous medium that is dependent on the Reynolds number of porous medium.

R_{Np}= Reynolds number of isotropic porous medium.

A_p= Cross-sectional area of porous medium

W = Weight flow rate of fluid

γ_b= Specific weight of fluid at P_b and T_b

Q_b = Volumetric rate of fluid, measured at P_b and T_b

P_b = Base pressure, absolute unit
T_b = Base Temperature, absolute unit
z_b = Gas deviation factor at p_b and T_b usually taken as 1
G_g = Specific gravity of gas (air = 1) at standard condition
M = Molecular weight of gas
R = Universal gas constant
A_1 = Pipe cross sectional area at point 1
v_1 = Average fluid velocity at point. 1
γ_1 = Specific weight of fluid at point 1
A_2 = Pipe cross-sectional area at point 2
v_2 =Average fluid velocity at point 2
γ_2 = Specific weight of fluid at point 2
T = Absolute temperature
K = Constant for calculating the compressibility of a real gas
p_1 = Pressure at inlet end of porous medium
p_2 = pressure at exit end of porous medium
θ = Angle of inclination of porous medium with horizontal in degrees.
z_2 = Gas deviation factor at exit end of porous medium.
T_2 = Temperature at exit end of porous medium
T_1 =Temperature at inlet end of porous medium
z_{av} = Average gas deviation factor evaluated with T_{av} and p_{av}
T_{av} = Arithmetic average temperature of the porous medium given by $0.5(T_1 + T_2)$ and p_{av} = Average pressure

References

Amyx, J.W., Bass, D.M. & Whitting, R.L. (1960). Petroleum Reservoir Engineering - Physical Properties, Mc Graw Hill Book Company, pp73-78, New York.

Aires, F, (1962).Differential Equations, McGraw Hill Book Company, New York.

Belhj, H.A., Agha, K.R., Butt, S.D.& Islam, M.R. (2003). Journal of tech. papers, 27th Nigeria Annual Int. Conf. of Soc. of Pet. Engineers.

Brinkman, H.C. (1947) A calculation of the viscous force exerted by a flowing fluid on a dense swarm of particles. Appl. Soc. Res. A1, pp. 27-34.

Darcy, H. (1856). Les fontainer publigues de la ville de Dijoin, Dalmont,

Forchheimer, P.Z . (1901). Wasserbewegung durch Boden, ZVDI, 45, PP. 1781

Giles, R.V., Cheng, L. & Evert, J. (2009). Schaum's Outline Series ofFluid Mechanics and Hydraulics, McGraw Hill Book Company, New York.

LeRosen, A. L. (1942). Method for the Standardization of Chromatographic Analysis, J. Amer. Chem. Soc., 64, 1905-1907

Matter, L.G., Brar, S.& Aziz, K. (1975). "Compressibility of Natural Gases, Journal of Canadian Petroleum Technology", pp. 77-80.

Muskat, M. (1949). "Physical Principles of Oil Production", p. 142, McGraw-Hill Book Company, Inc., New York.

Nutting , P.G. (1930). Physical Analysis of Oil Sands, Bulletin of American Association of Petroleum Geologists, 14, 1337.

Ohirhian, P.U. (2008). A New Dimensionless Friction Factor for Porous Media, Journal of Porous Media, Volume 11, Number 5.

Ohirhian, P.U. (2009). Equations for the z-factor and compressibility of Nigerian Natural gas, Advanced Materials Research, Vols. 62-64, pp. 484-492, Trans Tech Publications, Switzerland

Vibert. T. (1939). Les gisements de bauxite de I'Índochine, Gênie Civil 115, 84.

21

Natural gas properties and flow computation

Ivan Marić and Ivan Ivek
Ruđer Bošković Institute
Croatia

1. Introduction

Precise measurement of fluid flow rate is essential in commercial and in process control applications. The flow rate can be measured using different principles and devices (Baker, 2000, Miller, 1996): Orifice, Turbine, Venturi, Nozzle, Target, V-cone, Pitot, Multiport averaging, Elbow, Wedge, Laminar flow, Gilfo, Positive displacement, Thermal mass, Ultrasonic-time of flight, Variable area, Vortex, Coriolis. The measurement accuracy varies from ±5% of rate (Pitot) down to ±0.2% of rate (Coriolis). The Coriolis mass flowmeters are generally used to measure the mass flow of liquids but have been also used for the measurement of flow of high density gases. The turbine meters are widely used for the measurement of the volumetric flow rate of clean gases (±0.5% of rate) and liquids (±1% of rate). The flow rate measurements based on orifice meters are less accurate (1-2% URV) but the orifice plates are the most widely used devices in natural gas flow rate measurements due to their simplicity and robustness. We will here illustrate the thermodynamic effects that may cause significant error in measurements of natural gas flow rate based on orifice meters. We will also demonstrate how they could be efficiently compensated.

In measurements based on orifice plates the temperature of the fluid measured upstream of the orifice plate is used for the calculation of the flow rate but the fluid temperature is preferably measured downstream of the orifice plate (ISO-5167-1, 2003). When a gas is forced to flow through an orifice its temperature is changed due to the Joule-Thomson (JT) effect. The effect can be generally neglected for low flow rates i.e. for low differential pressures measured across the orifice meter (ISO-5167-1, 2003). At higher differential pressures and at lower temperatures the flow rate error increases and generally needs to be compensated (Marić, 2007). The precise compensation of flow rate error implies double calculation of natural gas properties and the flow rate, which extends the calculation time significantly and may become impractical for implementation in low-computing-power embedded systems. To avoid the computational burden the original high complexity models of natural gas properties can be replaced by the corresponding low-complexity surrogate models (Marić & Ivek, IEEE, Marić & Ivek, 2010) with no significant deterioration of flow rate accuracy.

Comprehensive presentation of modern methods of estimating the physical properties of gases and liquids can be found in (Poling at al., 2000). Formulations explicit in the Helmholtz energy have been widely used to represent the properties of natural gas because of the ease of calculating all other thermodynamic properties by mathematical

differentiation (Lemmon & Starling, 2003, Span & Wagner, 1996, Span & Wagner, 2003). The Helmholtz energy is a fundamental thermodynamic property from which all other thermodynamic properties can be calculated as derivatives with respect to molar density or temperature. The detailed procedure for the calculation of thermodynamic properties based on formulations explicit in Helmholtz energy (Lemmon & Starling, 2003) and on AGA-8 detail characterization equation (Starling & Savidge, 1992) is given in (ISO-207651-1, 2005). Here we will elaborate an alternative procedure for the calculation of properties of a natural gas that was originally published in the Journal Flow Measurement and Instrumentation (Marić, 2005 & 2007). The procedure is derived using fundamental thermodynamic equations (Olander, 2007), DIPPR AIChE (DIPPR® Project 801, 2005) generic ideal heat capacity equations, and AGA-8 (Starling & Savidge, 1992) extended virial-type equations of state. The procedure specifies the calculation of specific heat capacities at a constant pressure c_p and at a constant volume c_v, the JT coefficient μ_{JT}, and the isentropic exponent κ of a natural gas. The effect of a JT expansion on the accuracy of natural gas flow rate measurements will be pointed out.
The possibilities of using the computational intelligence methods - Artificial Neural Networks - ANNs (Ferrari & Stengel, 2005, Wilamowski et al., 2008) and machine learning tools - Group Method of Data Handling - GMDH (Ivakhnenko, 1971, Nikolaev & Iba, 2003) for meta-modeling the effects of natural gas properties in flow rate measurements (Marić & Ivek, 2010) will be illustrated. The practical examples of ANN and GMDH surrogate models for the compensation of natural gas flow rate measurement error caused by the thermodynamic effects, with the corresponding accuracies and execution times will be given. The models are particularly suitable for implementation in low computing power embedded systems.

2. A procedure for the calculation of thermodynamic properties of natural gas

This section summarizes the procedure (Maric, 2007) for the calculation of specific heat capacity at constant pressure c_p and at constant volume c_v, JT coefficient μ_JT and isentropic exponent κ of a natural gas based on thermodynamic equations, AGA-8 extended virial type characterization equation (Starling & Savidge, 1992, ISO-12213-2, 2006) and DIPPR generic ideal heat capacity equations (DIPPR® Project 801, 2005). First, the relation of the molar heat capacity at constant volume to equation of state will be derived. Then the relation will be used to calculate a molar heat capacity at constant pressure, which will be then used for the calculation of the JT coefficient and the isentropic exponent. The total differential for entropy (Olander, 2007), related to temperature and molar volume, is:

$$ds = \left(\frac{\partial s}{\partial T}\right)_{v_m} dT + \left(\frac{\partial s}{\partial v_m}\right)_T dv_m, \tag{1}$$

where s denotes entropy, T denotes temperature and v_m is a molar volume of a gas. By dividing the fundamental differential for internal energy $du = T \cdot ds - p \cdot dv_m$ by dT while holding v_m constant the coefficient of dT in Eq. (1) becomes $c_{m,v}/T$ since the molar heat at constant volume is defined by $c_{m,v} = (\partial u/\partial T)_{v_m}$. The Maxwell relation

$(\partial s/\partial v_m)_T = (\partial p/\partial T)_{v_m}$, is used to substitute the coefficient of dv_m. Finally, the Eq. (1) becomes:

$$ds = \frac{c_{m,v}}{T} dT + \left(\frac{\partial p}{\partial T}\right)_{v_m} dv_m . \tag{2}$$

Similarly, starting from a total differential for entropy related to temperature and pressure (Olander, 2007) $ds = (\partial s/\partial T)_p dT + (\partial s/\partial p)_T dp$ and by dividing the fundamental differential for enthalpy $dh = T \cdot ds + v_m \cdot dp$ by dT while holding p constant, the coefficient of dT in total differential becomes $c_{m,p}/T$ since the molar heat capacity at constant pressure is defined by: $c_{m,p} = (\partial h/\partial T)_p$. The Maxwell relation $(\partial s/\partial p)_T = (\partial v_m/\partial T)_p$ is used to substitute the coefficient of dp and the following relation is obtained:

$$ds = \frac{c_{m,p}}{T} dT + \left(\frac{\partial v_m}{\partial T}\right)_p dp , \tag{3}$$

Subtracting Eq. (2) from Eq. (3), then dividing the resulting equation by dv_m while holding p constant and finally inverting the partial derivative $(\partial T/\partial v_m)_p$ the following equation is obtained:

$$c_{m,p} - c_{m,v} = T\left(\frac{\partial v_m}{\partial T}\right)_p \left(\frac{\partial p}{\partial T}\right)_{v_m} . \tag{4}$$

A total differential of thermodynamic property, Eqs. (2) and (3), must be the exact differential i.e. the order of forming the mixed second derivative is irrelevant. The partial derivative of the first coefficient with respect to the second variable equals to the partial derivative of the second coefficient with respect to the first variable. By applying this property to Eq. (2) and by assuming T to be the first variable with the corresponding coefficient $c_{m,v}/T$ and v_m the second variable with the corresponding coefficient $(\partial p/\partial T)_{v_m}$ we obtain:

$$\left(\frac{\partial c_{m,v}}{\partial v_m}\right)_T = T\left(\frac{\partial^2 p}{\partial T^2}\right)_{v_m} , \tag{5}$$

The Eq. (5) can be rewritten in the following integral form:

$$c_{m,v} = c_{m,vI} + T \int_{v_{mI}\to\infty(T=const)}^{v_m} \left(\frac{\partial^2 p}{\partial T^2}\right)_{v_m} dv_m \,, \tag{6}$$

where $c_{m,vI}$, v_{mI} and v_m denote the ideal molar heat capacity at constant volume and the corresponding molar volume of ideal and real gas at temperature *T*. Real gases behave more like ideal gases as pressure approaches zero or $v_{mI} \to \infty$. After substituting $v_m = 1/\rho_m$, $p = RTZ\rho_m$ and $c_{m,vI} = c_{m,pI} - R$ the Eq. (6) transforms to:

$$c_{m,v} = c_{m,pI} - R - RT \int_{\rho_{mI}\to 0(T=const)}^{\rho_m} \frac{1}{\rho_m}\left(2\left(\frac{\partial Z}{\partial T}\right)_{\rho_m} + T\left(\frac{\partial^2 Z}{\partial T^2}\right)_{\rho_m}\right) d\rho_m \,, \tag{7}$$

where $c_{m,pI}$ denotes the temperature dependent molar heat capacity of ideal gas at constant pressure, *R* is the universal gas constant, *Z* is the compression factor and ρ_{mI} and ρ_m are the corresponding molar densities of ideal and real gas at temperature *T*. After substituting the first and the second derivative of the AGA-8 compressibility equation (Starling & Savidge, 1992, ISO-12213-2, 2006)

$$Z = 1 + B\rho_m - \rho_r \sum_{n=13}^{18} C_n^* + \sum_{n=13}^{58} C_n^* \left(b_n - c_n k_n \rho_r^{k_n}\right) \rho_r^{b_n} e^{-c_n \rho_r^{k_n}} \,, \tag{8}$$

into the Eq. (7) and after integration we obtain

$$c_{m,v} = c_{m,pI} - R + RT\rho_r\left(2C_0 + TC_1 - C_2\right), \tag{9}$$

with

$$C_0 = \sum_{n=13}^{18} C_n^{*\prime} - \frac{B'}{K^3} \,, \tag{10}$$

$$C_1 = \sum_{n=13}^{18} C_n^{*\prime\prime} - \frac{B''}{K^3} \,, \tag{11}$$

$$C_2 = \sum_{n=13}^{58} \left(2C_n^{*\prime} + TC_n^{*\prime\prime}\right) \rho_r^{b_n - 1} e^{-c_n \rho_r^{k_n}} \,, \tag{12}$$

where ρ_r is the reduced density ($\rho_r = K^3 \rho_m$), B is the second virial coefficient, $\{C_n^*\}$ are the temperature dependent coefficients, K is the mixture size parameter while $\{b_n\}$, $\{c_n\}$ and $\{k_n\}$ are the equation of state parameters. The mixture size parameter K is calculated using the following equation (ISO-12213-2, 2006):

$$K^5 = \left(\sum_{i=1}^{N} y_i K_i^{5/2} \right)^2 + 2 \sum_{i=1}^{N-1} \sum_{j=i+1}^{N} y_i y_j (K_{ij}^5 - 1)(K_i K_j)^{5/2}, \tag{13}$$

where y_i denotes the molar fraction of the component i, while $\{K_i\}$ and $\{K_{ij}\}$ are the corresponding size parameters and the binary interaction parameters given in [ISO-12213-2, 2006]. According to (ISO-12213-2, 2006) the second virial coefficient is calculated using the following equation:

$$B = \sum_{n=1}^{18} a_n T^{-u_n} \sum_{i=1}^{N} \sum_{j=1}^{N} y_i y_j B_{nij}^* E_{ij}^{u_n} (K_i K_j)^{3/2}, \tag{14}$$

and the coefficients $\{B_{nij}^*\}$, $\{E_{ij}\}$ and $\{G_{ij}\}$ are defined by

$$\begin{aligned} B_{nij}^* = & (G_{ij} + 1 - g_n)^{g_n} (Q_i Q_j + 1 - q_n)^{q_n} (F_i^{1/2} F_j^{1/2} + 1 - f_n)^{f_n} \cdot \\ & \cdot (S_i S_j + 1 - s_n)^{s_n} (W_i W_j + 1 - w_n)^{w_n} \end{aligned}, \tag{15}$$

$$E_{ij} = E_{ij}^* (E_i E_j)^{1/2}, \tag{16}$$

and

$$G_{ij} = G_{ij}^* (G_i + G_j)/2, \tag{17}$$

where T is temperature, N is the total number of gas mixture components, y_i is the molar fraction of the component i, $\{a_n\}$, $\{f_n\}$, $\{g_n\}$, $\{q_n\}$, $\{s_n\}$, $\{u_n\}$, and $\{w_n\}$ are the equation of state parameters, $\{E_i\}$, $\{F_i\}$, $\{G_i\}$, $\{K_i\}$, $\{Q_i\}$, $\{S_i\}$ and $\{W_i\}$ are the corresponding characterization parameters while $\{E_{ij}^*\}$ and $\{G_{ij}^*\}$ are the corresponding binary interaction parameters. The main symbols and units are given in Table 1.
The temperature dependent coefficients $\{C_n^*; n = 1,...,58\}$ and the mixture parameters U, G, Q and F are calculated using the equations (ISO-12213-2, 2006):

$$C_n^* = a_n (G + 1 - g_n)^{g_n} (Q^2 + 1 - q_n)^{q_n} (F + 1 - f_n)^{f_n} U^{u_n} T^{-u_n}, \tag{18}$$

Symbols and units		
Symbol	**Description**	**Unit**
B	Second virial coefficient	m^3*$kmol^{-1}$
B^*_{nij}	Mixture interaction coefficient	-
C	Coefficient of discharge	-
$c_{m,p}$	Molar heat capacity at constant pressure	J/(mol·K)
$c_{m,v}$	Molar heat capacity at constant volume	J/(mol·K)
C^*_n	Temperature and composition dependent coefficients	-
c_n	AGA-8 equation of state parameter	-
c_p	Specific heat capacity at constant pressure	J/(kg·K)
$c_{m,pI}$	Ideal molar heat capacity of the natural gas mixture	J/(mol·K)
$c^j_{m,pi}$	Ideal molar heat capacity of the gas component j	J/(mol·K)
D	Upstream internal pipe diameter	m
d	Diameter of orifice	m
h	Specific enthalpy	J/kg
K	Size parameter	-
M	Molar mass of the gas mixture	kg·$kmol^{-1}$
p	Absolute pressure	Pa
q	Mass flow rate	kg/s
R	Molar gas constant 8314.51	J/(kmol·K)
s	Specific entropy	J/(kg·K)
T	Absolute temperature	K
v_m	Molar specific volume	m^3/kmol
v_{mI}	Molar specific volume of ideal gas	m^3/kmol
y_i	Molar fraction of i-th component in gas mixture	-
Z	Compression factor	-
β	Diameter ratio d/D	-
Δp	Differential pressure	Pa
$\Delta\omega$	Pressure loss	Pa
κ	Isentropic exponent	-
μ_{JT}	Joule-Thomson coefficient	K/Pa
ρ_m	Molar density	kmol/m^3
ρ_{mI}	Molar density of ideal gas	kmol/m^3
ρ_r	Reduced density	-

Table 1. Symbols and units (for additional symbols and units refer to (ISO-12213-2, 2006).

$$U^5 = \left(\sum_{i=1}^{N} y_i E_i^{5/2} \right)^2 + 2\sum_{i=1}^{N-1} \sum_{j=i+1}^{N} y_i y_j (U_{ij}^5 - 1)(E_i E_j)^{5/2}, \tag{19}$$

$$G = \sum_{i=1}^{N} y_i G_i + 2\sum_{i=1}^{N-1} \sum_{j=i+1}^{N} y_i y_j (G_{ij}^* - 1)(G_i + G_j), \tag{20}$$

$$Q = \sum_{i=1}^{N} y_i Q_i, \tag{21}$$

and

$$F = \sum_{i=1}^{N} y_i^2 F_i \,, \tag{22}$$

where, U_{ij} is the binary interaction parameter for mixture energy. The first and the second derivatives of the coefficients B and C_n^*, with respect to temperature are:

$$B' = -\sum_{n=1}^{18} a_n u_n T^{-u_n-1} \sum_{i=1}^{N}\sum_{j=1}^{N} y_i y_j B_{nij}^* E_{ij}^{u_n} (K_i K_j)^{3/2} \,, \tag{23}$$

$$B'' = \sum_{n=1}^{18} a_n u_n (u_n + 1) T^{-u_n-2} \sum_{i=1}^{N}\sum_{j=1}^{N} y_i y_j B_{nij}^* E_{ij}^{u_n} (K_i K_j)^{3/2} \,, \tag{24}$$

$$C_n^{*'} = -\frac{u_n}{T} C_n^* \,, \tag{25}$$

$$C_n^{*''} = -\frac{u_n + 1}{T} C_n^{*'} \,, \tag{26}$$

The ideal molar heat capacity c_{pI} is calculated by

$$c_{m,pI} = \sum_{j=1}^{N} y_j c_{m,pi}^j \,, \tag{27}$$

where y_j is the molar fraction of component j in the gas mixture and $c_{m,pi}^j$ is the molar heat capacity of the same component. The molar heat capacities of the ideal gas mixture components can be approximated by DIPPR/AIChE generic equations (DIPPR® Project 801, 2005), i.e.

$$c_{m,pi}^j = a_j + b_j \left(\frac{c_j / T}{\sinh(c_j / T)} \right)^2 + d_j \left(\frac{e_j / T}{\cosh(e_j / T)} \right)^2 , \tag{28}$$

where $c_{m,pi}^j$ is the molar heat capacity of the component j of the ideal gas mixture, a_j, b_j, c_j, d_j and e_j are the corresponding constants, and T is the temperature.

The partial derivative of pressure with respect to temperature at constant molar volume and the partial derivative of molar volume with respect to temperature at constant pressure are defined by the equations:

$$\left(\frac{\partial p}{\partial T}\right)_{v_m} = R\rho_m\left[Z + T\left(C_3 - \rho_r C_0\right)\right], \tag{29}$$

and

$$\left(\frac{\partial v_m}{\partial T}\right)_p = \frac{R}{p}\left[Z + \left(\frac{\partial Z}{\partial T}\right)_p T\right], \tag{30}$$

where,

$$C_3 = \sum_{n=13}^{58}\left(C_n^{*'} D_n^{*}\right), \tag{31}$$

$$D_n = \left(b_n - c_n k_n \rho_r^{k_n}\right)\rho_r^{b_n} e^{-c_n \rho_r^{k_n}}, \tag{32}$$

$$\left(\frac{\partial Z}{\partial T}\right)_p = \frac{R(TZ)^2 C_3 - pZ\left[TK^3 C_0 + C_4\right]}{R(TZ)^2 + pTC_4}, \tag{33}$$

$$C_4 = C_5 + \sum_{n=13}^{58} C_n^{*} D_{1n}, \tag{34}$$

$$C_5 = B - K^3 \sum_{n=13}^{18} C_n^{*} \tag{35}$$

and

$$D_{1n} = K^3\left[b_n^2 - c_n k_n\left(2b_n + k_n - c_n k_n \rho_r^{k_n}\right)\rho_r^{k_n}\right]\rho_r^{b_n - 1} e^{-c_n \rho_r^{k_n}} \tag{36}$$

The isentropic exponent is defined by the following relation

$$\kappa = -\frac{c_{m,p}}{c_{m,v}}\left(\frac{\partial p}{\partial v_m}\right)_T\left(\frac{v_m}{p}\right) = -\frac{c_{m,p}}{c_{m,v}\rho_m p}\left(\frac{\partial p}{\partial v_m}\right)_T \tag{37}$$

where

$$\left(\frac{\partial p}{\partial v_m}\right)_T = \left(\frac{\partial p}{\partial \rho_m}\right)_T \left(\frac{\partial \rho_m}{\partial v_m}\right)_T = -RT\rho_m^2\left(Z + \rho_m C_4\right) \tag{38}$$

The JT coefficient is defined by the following equation:

$$\mu_{JT} = \frac{RT^2}{pc_{m,p}}\left(\frac{\partial Z}{\partial T}\right)_p \tag{39}$$

The derivation of the Eq. (39) is elaborated in (Olander, 2007 & Maric, 2005).

3. Implementation in software

The procedure for the calculation of natural gas density, compression, molar heat capacity, isentropic exponent and the JT coefficient can be implemented in object oriented paradigm, which enables its easy integration into the software projects. The interface to the software object *S* is shown in Fig. 1. The input/output parameters and functions are accessible while the internal structure is hidden to the user. The function *"Calculate"* maps the input parameters (pressure, temperature and the molar fractions of natural gas components) into the output parameters (density, compression, molar heat capacity, isentropic exponent, JT coefficient, etc.).

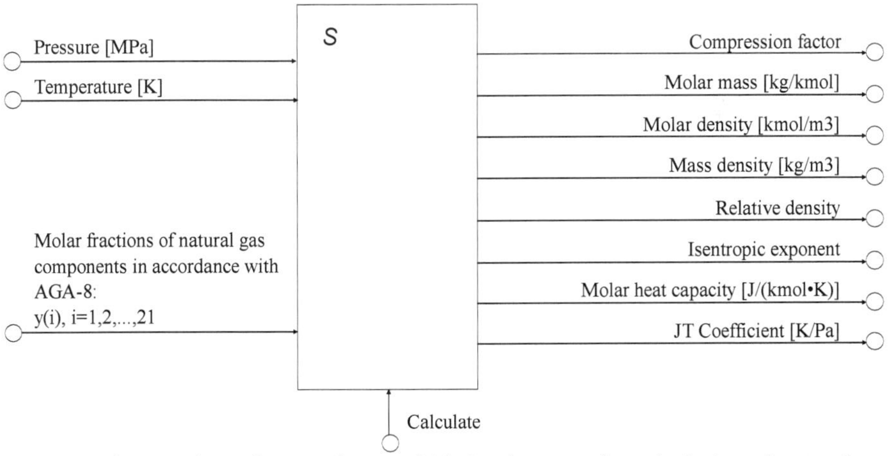

Fig. 1. Interface to the software object, which implements the calculation of natural gas properties.

Table 2 depicts the calculation procedure. Prior to the calculation of the molar heat capacities, isentropic exponent and JT coefficient, the density and the compression factor of a natural gas must be calculated. The false position method is combined with the successive bisection method to calculate the roots of the equation of state [Starling & Savidge, 1992].

Input parameters - constant:

- molar gas constant (R=8314.51 J/(kmol·K))
- natural gas equation of state parameters (a_n, b_n, c_n, k_n, u_n, g_n, q_n, f_n, s_n, w_n; n=1, 2,...,58), characterization parameters (M_i, E_i, K_i, G_i, Q_i, F_i, S_i, W_i; i=1,...,21) and binary interaction parameters ($E^*_{i,j}$, $U_{i,j}$, $K_{i,j}$, $G^*_{i,j}$) (see ISO 12213-2)
- DIPPR/AIChE gas heat capacity constants (a_j, b_j, c_j, d_j;, e_j; j=1,2,...,N)

Input parameters - time varying:

- absolute pressure: p [MPa]
- absolute temperature: T [K]
- molar fractions of the natural gas mixture: y_i; i=1,2,...,N

Calculation procedure:

1. mixture size parameter K (Eq. 13), second virial coefficient B (Eq. 14) and temperature dependent coefficients C_n^* (Eq. 18)
2. compression factor Z (Eq. 8) (see ISO-12213-2 for details of calculation)
3. molar density $\rho_m = p/RTZ$, density $\rho = M\rho_m$, reduced density $\rho_r = K^3\rho_m$ and molar volume $v_m = 1/\rho_m$.
4. coefficients D_n and D_{1n} (Eqs. 32 and 36)
5. 1st and 2nd derivative of the second virial coefficient B: B' (Eq. 23) and B'' (Eq. 24)
6. 1st and 2nd derivative of the coefficient C_n^*: $C_n^{*'}$ (Eq. 25) and $C_n^{*''}$ (Eq. 26)
7. 1st derivative of the compression factor Z: $(\partial Z/\partial T)_p$ (Eq. 33)
8. partial derivatives of pressure: $(\partial p/\partial T)_{v_m}$ (Eq. 29) and $(\partial p/\partial v_m)_T$ (Eq. 38)
9. ideal molar heat capacity of a gas mixture at constant pressure: $c_{m,pI}$ (Eq. 27)
10. molar heat capacity of a gas mixture at constant volume: $c_{m,v}$ (Eqs. 9)
11. molar heat capacity of a gas mixture at constant pressure: $c_{m,p}$ (Eqs. 4)
12. isentropic exponent κ (Eq. 37)
13. Joule-Thomson coefficient μ_{JT} (Eq. 39)

Table 2. The input/output parameters and the procedure for the computation of the natural gas properties.

4. Comparison with experimental results

In order to compare the calculation results, for the specific heat capacity c_p and the JT coefficient μ_{JT}, with the corresponding high accuracy measurement data (Ernst et al., 2001), we assume the identical artificial natural gas mixture with the following mole fractions: x_{CH4}=0.79942, x_{C2H6}=0.05029, x_{C3H8}=0.03000, x_{CO2}=0.02090 and x_{N2}=0.09939. The results of the measurements (Ernst et al., 2001) and the results of the calculation of the specific heat capacity c_p and the JT coefficient μ_{JT} of the natural gas mixture, for absolute pressure ranging from 0 MPa to 30 MPa in 0.5 MPa steps and for four upstream temperatures (250 K, 275 K, 300 K and 350 K), are shown in Fig. 2 and 3, respectively. The differences between the calculated values and the corresponding measurement results (Ernst et al., 2001), for the c_p and μ_{JT}, are shown in Table 3 and 4, respectively.

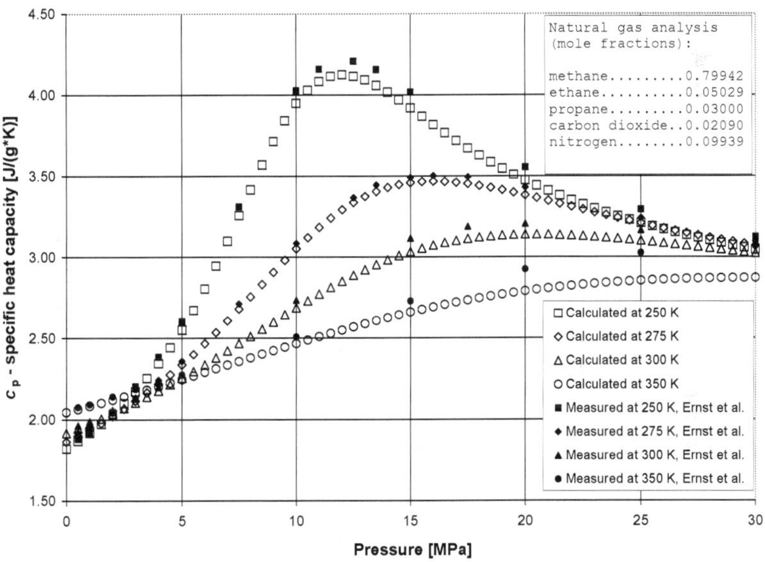

Fig. 2. Calculated and measured molar heat capacity at constant pressure of the natural gas mixture.

From Table 3 it can be seen that the calculated values of c_p are within ±0.08 J/(g*K) with the measurement results for the pressures up to 12 MPa. At higher pressures, up to 30 MPa, the difference increases but never exceeds ±0.2 J/(g*K). For pressures up to 12 MPa the relative difference between the calculated and experimentally obtained c_p never exceeds ±2.00%.

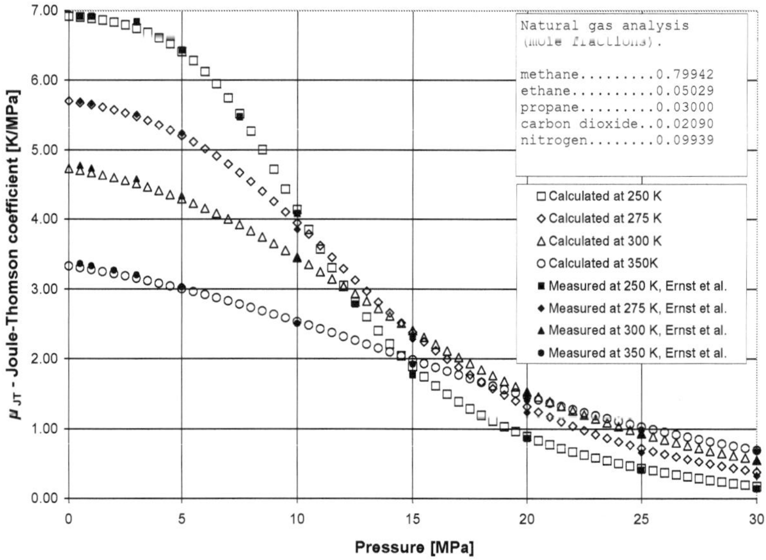

Fig. 3. Calculated and measured JT coefficient of the natural gas mixture.

P [MPa]: T [K]	250	275	300	350
	$(c_{p\_calculated} - c_{p\_measured})$ [J/(g*K)]			
0.5	-0.015	-0.018	-0.018	-0.012
1.0	-0.002	-0.014	-0.016	-0.011
2.0	-0.012	-0.019	-0.022	-0.020
3.0	-0.032	-0.020	-0.023	-0.026
4.0	-0.041	-0.023	-0.021	-0.027
5.0	-0.051	-0.022	-0.025	-0.029
7.5	-0.055	-0.032	-	-
10.0	-0.077	-0.033	-0.048	-0.042
11.0	-0.075	-	-	-
12.5	-0.092	-0.030	-	-
13.5	-0.097	-0.039	-	-
15.0	-0.098	-0.033	-0.082	-0.069
16.0	-	-0.036	-	-
17.5	-	-0.043	-0.075	-
20.0	-0.081	-0.048	-0.066	-0.134
25.0	-0.082	-0.033	-0.064	-0.171
30.0	-0.077	-0.025	-0.070	-0.194

Table 3. Difference between the calculated and measured specific heat capacity at constant pressure of a natural gas.

P [MPa]: T [K]	250	275	300	350
	$(\mu_{JT\_calculated} - \mu_{JT\_measured})$ [K/MPa]			
0.5	-0.014	-0.023	-0.075	-0.059
1.0	-0.032	-0.024	-0.068	-0.053
2.0	-	-	-	-0.051
3.0	-0.092	-0.032	-0.069	-0.049
5.0	-0.022	-0.036	-0.044	-0.026
7.5	0.043	-	-	-
10.0	0.060	0.096	0.019	0.030
12.5	0.034	-	-	-
15.0	0.113	0.093	0.050	0.061
20.0	0.029	0.084	0.009	0.047
25.0	0.025	0.059	0.002	0.043
30.0	0.031	0.052	0.005	0.012

Table 4. Difference between the calculated and measured JT coefficient of a natural gas.

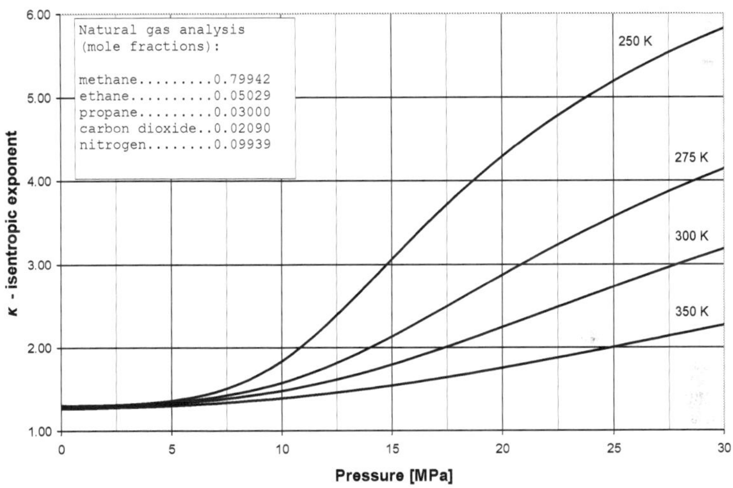

Fig. 4. Calculated isentropic exponent of the natural gas mixture.

From Table 4 it can be seen that the calculated values of μ_{JT} are within ±0.113 K/MPa with the experimental results for the pressures up to 30 MPa. The relative difference increases with the increase of pressure but never exceeds ±2.5% for the pressures up to 12 MPa. At higher pressures, when the values of μ_{JT} are close to zero, the relative difference may increase significantly. The calculation results obtained for pure methane and methane-ethane mixture are in considerably better agreement with the corresponding experimental data (Ernst et al., 2001) than for the natural gas mixture shown above. We estimate that the relative uncertainty of the calculated c_p and μ_{JT} of the AGA-8 natural gas mixtures in common industrial operating conditions (pressure range 0-12 MPa and temperature range

250-350 K) is unlikely to exceed ±3.00 % and ±4.00 %, respectively. Fig. 4 shows the results of the calculation of the isentropic exponent. Since the isentropic exponent is a theoretical parameter there exist no experimental data for its verification.

5. Flow rate measurement

Flow rate equations for differential pressure meters assume a constant fluid density of a fluid within the meter. This assumption applies only to incompressible flows. In the case of compressible flows, a correction must be made. This correction is known as adiabatic expansion factor, which depends on several parameters including differential pressure, absolute pressure, pipe inside diameter, differential device bore diameter and isentropic exponent. Isentropic exponent has a limited effect on the adiabatic correction factor but has to be calculated if accurate flow rate measurements are needed.

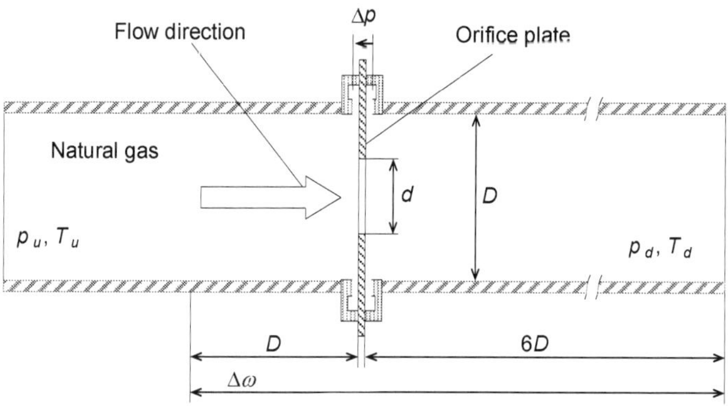

Fig. 5. The schematic diagram of the natural gas flow rate measurement using an orifice plate with corner taps.

When a gas expands through the restriction to a lower pressure it changes its temperature and density (Fig. 5). This process occurs under the conditions of constant enthalpy and is known as JT expansion (Shoemaker at al., 1996). It can also be considered as an adiabatic effect because the pressure change occurs too quickly for significant heat transfer to take place. The temperature change is related to pressure change and is characterized by the JT coefficient. The temperature change increases with the increase of the pressure drop and is proportional with the JT coefficient. According to (ISO5167, 2003) the upstream temperature is used for the calculation of flow rate but the temperature is preferably measured downstream of the differential device. The use of downstream instead of upstream temperature may cause a flow rate measurement error due to the difference in the gas density caused by the temperature change. Our objective is to derive the numerical procedure for the calculation of the natural gas specific heat capacity, isentropic exponent and JT coefficient that can be used for the compensation of flow rate error. In order to make the computationally intensive compensation procedure applicable to low computing power real-time measurement systems the low complexity surrogate models of original procedures will be derived using the computational intelligence methods: ANN and GMDH. The surrogate models have to be tailored to meet the constraints imposed on the approximation accuracy and the complexity of the model, i.e. the execution time (ET).

6. Compensation of flow rate error

We investigated the combined effect of the JT coefficient and the isentropic exponent of a natural gas on the accuracy of flow rate measurements based on differential devices. The measurement of a natural gas (ISO-12213-2, 2006) flowing in a pipeline through orifice plate with corner taps (Fig. 5) is assumed to be completely in accordance with the international standard (ISO-5167, 2003). The detailed description of the flow rate equation with the corresponding iterative computation scheme is given in (ISO-5167, 2003). The calculation of the natural gas flow rate depends on multiple parameters:

$$q_u = q(P_u, T_u, \Delta p, \rho_u, \gamma_u, \kappa_u, D, d), \tag{40}$$

where q_u, ρ_u, γ_u and κ_u represent the corresponding mass flowrate, density, viscosity and the isentropic exponent calculated at upstream pressure P_u and temperature T_u, while D and d denote the internal diameters of the pipe and the orifice, respectively. In case of the upstream pressure and the downstream temperature measurement, as suggested by (ISO-5167, 2003), the flow rate equation, Eq. (40), changes to:

$$q_d = q(P_u, T_d, \Delta p, \rho_d, \gamma_d, \kappa_d, D, d), \tag{41}$$

where q_d, ρ_d, γ_d and κ_d denote the corresponding mass flow rate, density, viscosity and the isentropic exponent calculated in *"downstream conditions"* i.e. at the upstream pressure p_u and the downstream temperature T_d. For certain natural gas compositions and operating conditions the flow rate q_d may differ significantly from q_u and the corresponding compensation for the temperature drop effects, due to JT expansion, may be necessary in order to preserve the requested measurement accuracy (Maric & Ivek, 2010).

The flow rate correction factor K can be obtained by dividing the true flow rate q_u calculated in the upstream conditions, Eq. (40), by the flow rate q_d calculated in the *"downstream conditions"*, Eq. (41):

$$K = \frac{q_u}{q_d} \tag{42}$$

For the given correction factor Eq. (42), the flow rate at the upstream pressure and temperature can be calculated directly from the flow rate computed in the *"downstream conditions"*, i.e. $q_u = K \cdot q_d$. Our objective is to derive the GMDH polynomial model of the flow rate correction factor. Given the surrogate model (K_{SM}) for the flow rate correction factor Eq. (42), the true flow rate q_u can be approximated by: $q_{SM} = K_{SM} \cdot q_d$, where q_{SM} denotes the corrected flow rate.

The flow rate through orifice is proportional to the expansibility factor ε, which is related to the isentropic exponent κ (ISO-5167, 2003):

$$\varepsilon = 1 - \left(0.351 + 0.256\beta^4 + 0.93\beta^8\right)\cdot\left[1 - \left(p_d/p_u\right)^{1/\kappa}\right], \tag{43}$$

where β denotes the ratio of the diameter of the orifice to the inside diameter of the pipe, while p_u and p_d are the absolute pressures upstream and downstream of the orifice plate, respectively. The corresponding temperature change (ΔT) of the gas for the orifice plate is defined by

$$\Delta T = T_u - T_d \approx \mu_{JT}(p_u, T_d)\Delta\omega\ , \tag{44}$$

where T_u and T_d indicate the corresponding temperatures upstream and downstream of the orifice plate, $\mu_{JT}(p_u, T_d)$ is the JT coefficient at upstream pressure p_u and downstream temperature T_d and $\Delta\omega$ is the pressure loss across the orifice plate (Urner, 1997)

$$\Delta\omega = \frac{\sqrt{1-\beta^4\left(1-C^2\right)}-C\beta^2}{\sqrt{1-\beta^4\left(1-C^2\right)}+C\beta^2}\Delta p\ , \tag{45}$$

where C denotes the coefficient of discharge for orifice plate with corner taps (ISO-5167, 2003) and ΔP is the pressure drop across the orifice plate. According to (ISO-5167, 2003), the temperature of the fluid shall preferably be measured downstream of the primary device but upstream temperature is to be used for the calculation of the flow rate. Within the limits of application of the international standard ISO-5167 it is generally assumed that the temperature drop across differential device can be neglected but it is also suggested to be taken into account if higher accuracies are required. It is also assumed that the isentropic exponent can be approximated by the ratio of the specific heat capacity at constant pressure to the specific heat capacity at constant volume of ideal gas. These approximations may produce a considerable measurement error. The relative flow measurement error E_r is estimated by comparing the approximate (q_d) and the corrected (q_u) mass flow rate i.e.

$$E_r = \left(q_d - q_u\right)/q_u \tag{46}$$

Step	Description
1	Calculate the natural gas properties (ρ_d , μ_{JT} and κ_d) at p_u, and T_d, (Table 2).
2	Calculate the dynamic viscosity γ_d at P_u, and T_d, using e.g. the residual viscosity equation (Poling, 2000).
3	Calculate the mass flow rate q_d and the discharge coefficient C at P_u, T_d and Δp (ISO-5167, 2003).
4	Calculate the pressure loss $\Delta\omega$, Eq. (45).
5	Calculate the upstream temperature T_u in accordance with Eq. (44).
6	Calculate the natural gas properties (ρ_u and κ_u) at p_u, and T_u, (Table 2).
7	Calculate the dynamic viscosity γ_u at p_u, and T_u, using e.g. the residual viscosity equation (Poling, 2000).
8	Calculate the mass flow rate q_u at p_u, T_u and Δp (ISO-5167, 2003).

Table 5. Precise correction of the flow rate based on downstream temperature measurement and on the computation of natural gas properties.

The individual and the combined relative errors due to the approximations of the temperature drop and the isentropic exponent can be estimated by using the Eq. (46). The precise correction of the natural gas flow rate, based on upstream pressure and downstream temperature measurement and on the computation of the corresponding natural gas properties, is summarized in Table 5.

The procedure in Table 5 requires a double calculation of both the flow rate and the properties of the natural gas. To reduce the computational burden we aim to derive a low-complexity flow rate correction factor model that will enable direct compensation of the flow rate error caused by the measurement of the downstream temperature. The correction factor model has to be simple enough in order to be executable in real-time and accurate enough to ensure the acceptable measurement accuracy.

7. Results of flow rate measurement simulations

In order to simulate a flow rate measurement error caused by the non-compensated temperature drop, a natural gas mixture (Gas 3) from Annex C of (ISO-12213-2, 2006) is assumed to flow through orifice plate with corner taps (ISO-5167, 2003) as illustrated in Fig. 5. Following the recommendations (ISO-5167, 2003), the absolute pressure is assumed to be measured upstream (p_u) and the temperature downstream (T_d) of the primary device. Fig. 6

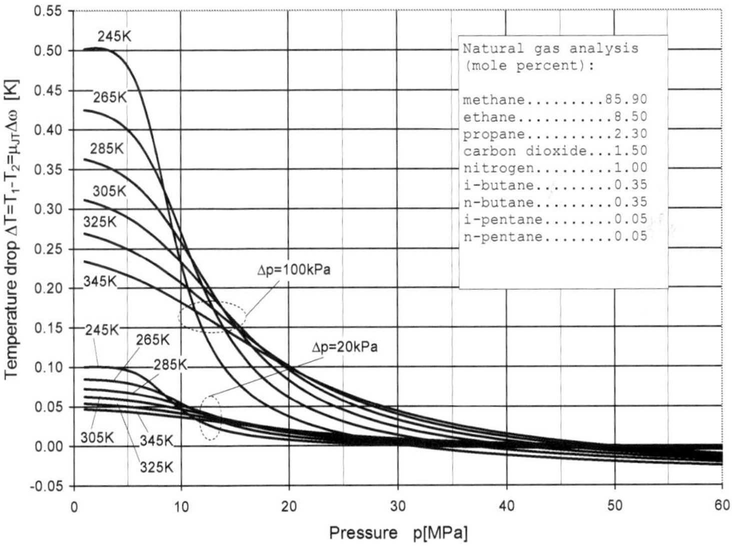

Fig. 6. Temperature drop due to JT effect $\Delta T = \mu_{JT} \Delta\omega$ when measuring flow rate of natural gas mixture through orifice plate with corner taps (ISO-5167, 2003). The upstream pressure varies from 1 MPa to 60 MPa in 1 MPa steps and upstream temperature from 245 K to 305 K in 20 K steps for each of the two differential pressures Δp (20 kPa and 100 kPa). The internal diameters of orifice and pipe are: d=120 mm and D=200 mm.

illustrates the temperature drop caused by the JT effect and calculated in accordance with the Eq. (44). The calculated results are given for two discrete differential pressures (Δp), 20kPa and 100kPa, for absolute pressure (p_u) ranging from 1 MPa to 60 MPa in 1 MPa steps and for six equidistant upstream temperatures (T_u) in the range from 245 to 345 K. From Fig. 6 it can be seen that for each temperature there exists the corresponding pressure where JT coefficient changes its sign and consequently alters the sign of the temperature change. A relative error in the flow rate measurements due to JT effect is shown in Fig. 7. The error is calculated in accordance with Eq. (46) by comparing the approximate mass flow rate (q_d) with the precisely calculated mass flow rate (q_u). The approximate flow rate and the corresponding natural gas properties (density, viscosity and isentropic exponent) are calculated at upstream pressure p_u, downstream temperature T_d and differential pressure Δp, by neglecting the temperature drop due to JT effect ($T_d = T_u$). The results are shown for two discrete differential pressures (Δp), 20kPa and 100kPa, for absolute upstream pressure (p_u) ranging from 1 MPa to 60 MPa in 1 MPa steps and for four equidistant downstream temperatures (T_d) in the range from 245 to 305 K.

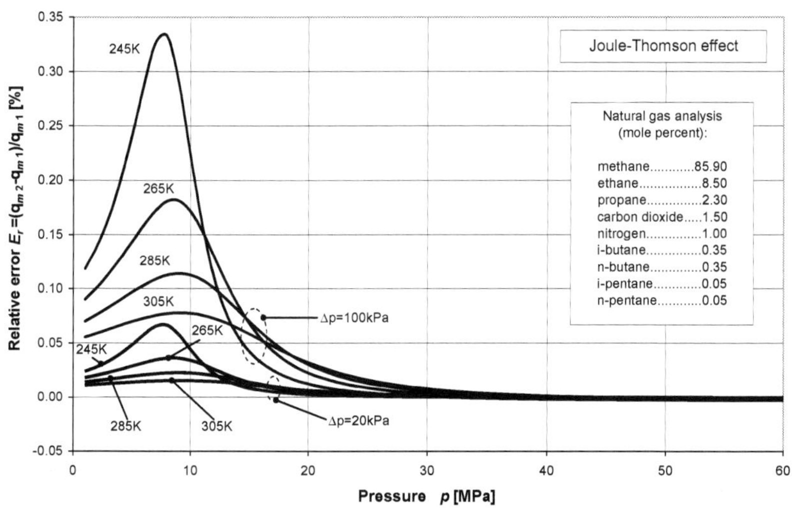

Fig. 7. Relative error $E_r = (q_d - q_u)/q_u$ in the flow rate of natural gas measured by orifice plate with corner taps (ISO-5167, 2003) when calculating flow rate using downstream temperature with no compensation of JT effect (q_d) instead of upstream temperature (q_u). The upstream pressure varies from 1 MPa to 60 MPa in 1 MPa steps and downstream temperature from 245 K to 305 K in 20 K steps for each of two differential pressures Δp (20 kPa and 100 kPa). The internal diameters of orifice and pipe are: d=120 mm and D=200 mm.

Fig. 8 illustrates the relative error in the flow rate measurements due to the approximation of the isentropic exponent by the ratio of the ideal molar heat capacities. The error is calculated by comparing the approximate mass flow rate (q_d) with the precisely calculated mass flow rate (q_u) in accordance with Eq. (46). The procedure for the precise correction of the mass flow rate is shown in Table 5. The approximate flow rate calculation is carried out in the same way with the exception of the isentropic exponent, which equals the ratio of the ideal molar heat capacities ($\kappa = c_{m,pI} / (c_{m,pI} - R)$). The results are shown for two discrete differential pressures Δp (20kPa and 100kPa), for absolute upstream pressure p_u ranging from 1 MPa to 60 MPa in 1 MPa steps and for four equidistant downstream temperatures T_d in the range from 245 to 305 K.

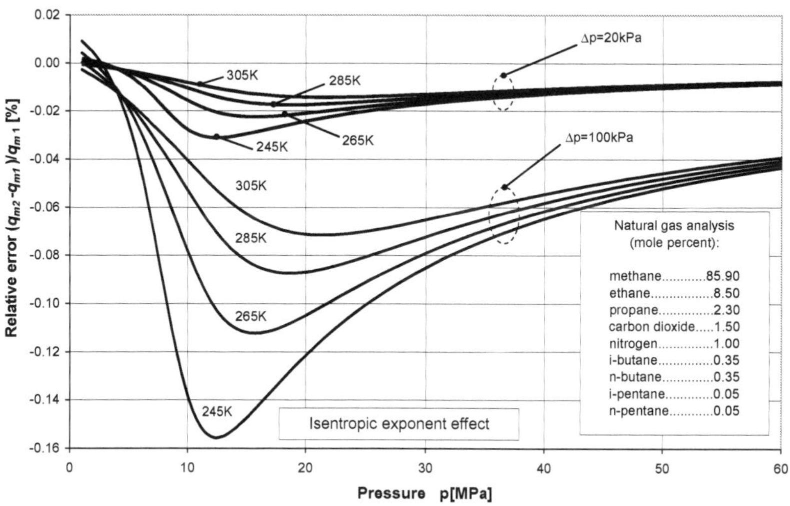

Fig. 8. Relative error $E_r = (q_d - q_u)/q_u$ in the flow rate of natural gas mixture measured by orifice plate with corner taps (ISO-5167, 2003) when using the isentropic exponent of ideal gas (q_d) instead of real gas (q_u). The upstream pressure varies from 1 MPa to 60 MPa in 1 MPa steps and downstream temperature from 245 K to 305 K in 20 K steps for each of two differential pressures Δp (20 kPa and 100 kPa). The internal diameters of orifice and pipe are: d=120 mm and D=200 mm.

Fig. 9 shows the flow rate measurement error produced by the combined effect of the JT and isentropic expansion. The error, Eq. (46), is calculated by comparing the approximate mass flow rate (q_d) with the mass flow rate (q_u) calculated precisely in accordance with the procedure depicted in Table 5. The approximate flow rate and the corresponding natural gas properties are calculated at upstream pressure p_u, downstream temperature T_d and differential pressure Δp, by neglecting the temperature drop due to JT effect ($T_d = T_u$) and by substituting the isentropic exponent by the ratio of the ideal molar heat capacities, $\kappa = c_{m,pI} / (c_{m,pI} - R)$. The results are shown for two discrete differential pressures Δp (20kPa and 100kPa), for absolute upstream pressure p_u ranging from 1 MPa to 60 MPa in 1

MPa steps and for four equidistant downstream temperatures T_d in the range from 245 to 305 K.

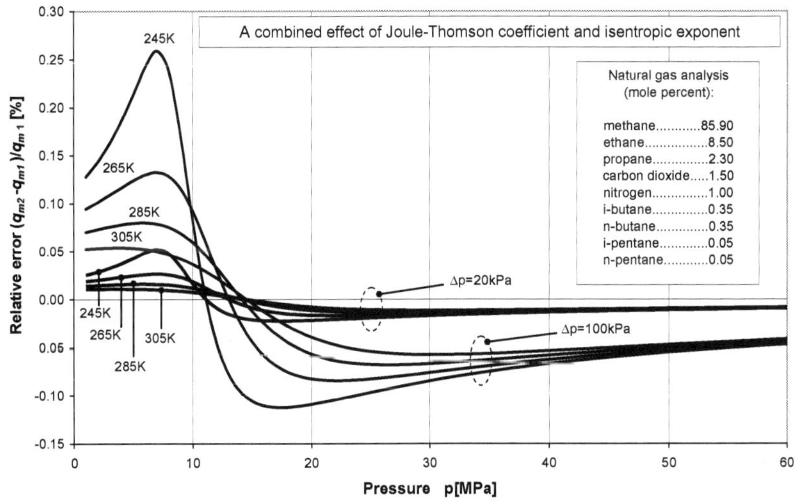

Fig. 9. Relative error $E_r = (q_d - q_u)/q_u$ in the flow rate of natural gas mixture measured by orifice plate with corner taps (ISO-5167, 2003) when using downstream temperature with no compensation of JT effect and the isentropic exponent of ideal gas at downstream temperature (q_d) instead of upstream temperature and the corresponding real gas isentropic exponent (q_u). The upstream pressure varies from 1 MPa to 60 MPa in 1 MPa steps and downstream temperature from 245 K to 305 K in 20 K steps for each of two differential pressures Δp (20 kPa and 100 kPa). The internal diameters of orifice and pipe are: d=120 mm and D=200 mm.

The results obtained for JT coefficient and isentropic exponent are in a complete agreement with the results obtained when using the procedures described in (Marić, 2005) and (Marić et al., 2005), which use a natural gas fugacity to derive the molar heat capacities. The calculation results are shown up to a pressure of 60 MPa, which lies within the wider ranges of application given in (ISO-12213-2, 2006), of 0 - 65 MPa. However, the lowest uncertainty for compressibility is for pressures up to 12 MPa and no uncertainty is quoted in reference (ISO-12213-2, 2006) for pressures above 30 MPa. Above this pressure, it would therefore seem sensible for the results of the JT and isentropic exponent calculations to be used with caution. From Fig. 9 it can be seen that the maximum combined error is lower than the maximum individual errors because the JT coefficient (Fig. 7) and the isentropic exponent (Fig. 8) show the counter effects on the flow rate error. The error always increases by decreasing the natural gas temperature. The total measurement error is still considerable especially at lower temperatures and higher differential pressures and can not be overlooked. The measurement error is also dependent on the natural gas mixture. For certain mixtures, like natural gas with high carbon dioxide content, the relative error in the flow rate may increase up to 0.5% at lower operating temperatures (245 K) and up to 1.0% at very low operating temperatures (225 K). Whilst modern flow computers have provision for applying a JT coefficient and isentropic exponent correction to measured temperatures, this usually takes the form of a fixed value supplied by the user. Our calculations show that any initial error in choosing this value, or subsequent operational changes in temperature, pressure or gas composition, could lead to significant systematic metering errors.

8. Flow rate correction factor meta-modeling

Precise compensation of the flow rate measurement error is numerically intensive and time-consuming procedure (Table 5) requesting double calculation of the flow rate and the properties of a natural gas. In the next section it will be demonstrated how the machine learning and the computational intelligence methods can help in reducing the complexity of the calculation procedures in order to make them applicable to real-time calculations. The machine learning and the computationally intelligence are widely used in modeling the complex systems. One possible application is meta-modeling, i.e. construction of a simplified surrogate of a complex model. For the detailed description of the procedure for meta-modeling the compensation of JT effect in natural gas flow rate measurements refer to (Marić & Ivek, IEEE, Marić & Ivek, 2010).

Approximation of complex multidimensional systems by self-organizing polynomials, also known as the Group Method of Data Handling (GMDH), was introduced by A.G. Ivakhnenko (Ivaknenkho, 1971). The GMDH models are constructed by combining the low-order polynomials into multi layered polynomial networks where the coefficients of the low-order polynomials (generally 2-dimensional 2nd-order polynomials) are obtained by polynomial regression. GMDH polynomials may achieve reasonable approximation accuracy at low complexity and are simple to implement in digital computers (Marić & Ivek, 2010). Also the ANNs can be efficiently used for the approximation of complex systems (Ferrari & Stengel, 2005). The main challenges of neural network applications regarding the architecture and the complexity are analyzed recently (Wilamowski, 2009).

The GMDH and the ANN are based on learning from examples. Therefore to derive a meta-model from the original high-complexity model it is necessary to (Marić & Ivek, 2010):

- generate sufficient training and validation examples from the original model
- learn the surrogate model on training data and verify it on validation data

We tailored GMDH and ANN models for a flow-computer (FC) prototype based on low-computing-power microcontroller (8-bit/16-MHz) with embedded FP subroutines for single precision addition and multiplication having the average ET approximately equal to 50 μs and 150 μs, respectively.

8.1 GMDH model of the flow rate correction factor

For the purpose of meta-modeling the procedure for the calculation of the correction factor was implemented in high speed digital computer. The training data set, validation data set and 10 test data sets, each consisting of 20000 samples of correction factor, were randomly sampled across the entire space of application. The maximum ET of the correction factor surrogate model in our FC prototype was limited to 35 ms ($T_{exe0} \leq 35$ ms) and the maximum root relative squared error (RRSE) was set to 4% ($E_{rrs0} \leq 4\%$). Fig. 10 illustrates a polynomial graph of the best discovered GMDH surrogate model of the flow rate correction factor obtained at layer 15 when using the compound error (CE) measure (Marić & Ivek, 2010). The

RRSE (E_{rrs}=3.967%) and the ET (T_{exe}=32 ms) of the model are both below the given thresholds (E_{rrs0}=≤4.0% and T_{exe0}≤35 ms) making the model suitable for implementation in the FC prototype.

GMDH polynomial model in recursive form

$$y=P_{31}(P_{30}(P_{28}(P_{27}(P_{26}(P_{25}(P_{20}(P_{19}(P_{18}(P_{17}(P_{11}(P_8(P_6(P_2(P_0(x_4,x_8),P_1(x_2,x_3)),P_5(P_3(x_3,x_4),P_4(x_6,x_7))),P_7(x_0,x_7)),P_{10}(P_9(x_0,x_4),P_1(x_2,x_3))),P_{16}(P_{15}(P_{13}(P_0(x_4,x_8),P_{12}(x_3,x_6)),P_{14}(x_2,x_5)),x_7)),P_{14}(x_2,x_5)),x_1),x_3),P_{24}(P_{22}(P_{21}(x_3,x_8),x_4),P_{23}(x_5,x_6)),x_6),P_{14}(x_2,x_5)),x_5),P_{29}(P_1(x_2,x_3),x_4)),x_3)$$

Basic regression polynomial

$$P_i(z_j,z_k)=a_0(i)+a_1(i)\,z_j+a_2(i)z_k+a_3(i)\,z_jz_j+a_4(i)\,z_kz_k+a_5(i)z_jz_k$$

Coefficients of the polynomials P_0 to P_{31}

i	a_0	a_1	a_2	a_3	a_4	a_5
0	1.0001E+0	-1.1357E-2	-6.8704E-4	2.5536E-4	8.0474E-4	8.4350E-3
1	9.8856E-1	-3.3090E-4	6.7325E-5	7.0360E-6	-1.0142E-7	7.3114E-7
2	-8.1858E+2	7.4253E+2	8.9596E+2	5.0943E+1	-2.5870E+1	-8.4398E+2
3	9.9012E-1	6.6260E-5	-4.1345E-2	-1.1050E-7	-2.7501E-5	1.1208E-4
4	1.0005E+0	5.2566E-3	-1.0140E-4	-5.5278E-3	9.3191E-7	4.1835E-6
5	-4.9380E+1	-3.2481E+1	1.3133E+2	-1.5787E+1	-9.7756E+1	6.5075E+1
6	-1.6081E+2	2.4385E+2	7.9023E+1	-1.7140E+1	6.5044E+1	-2.0896E+2
7	9.9774E-1	7.4210E-3	1.0690E-4	-6.6765E-3	-1.7098E-6	-2.4801E-4
8	-1.2395E+3	1.2696E+3	1.2113E+3	-3.1377E+1	-2.2670E+0	-1.2068E+3
9	9.9999E-1	8.5310E-4	-7.3055E-3	-7.3184E-3	4.8341E-4	-1.1245E-2
10	-4.3539E+2	1.2580E+3	-3.8654E+2	-1.0374E+2	7.1916E+2	-1.0505E+3
11	6.0579E+1	-8.4832E+1	-3.5432E+1	7.7879E+1	5.2456E+1	-6.9650E+1
12	9.8649E-1	6.4671E-5	5.4189E-3	-1.0113E-7	-7.4088E-3	1.0893E-5
13	-2.5121E+2	8.1232E+2	-3.0962E+2	4.1247E+1	6.0267E+2	-8.9441E+2
14	9.9954E-1	3.3668E-4	-5.4531E-5	-1.9968E-5	-2.5227E-9	3.5061E-6
15	-2.7176E+2	3.6409E+2	1.8065E+2	1.0868E+1	1.0229E+2	-3.8514E+2
16	-6.1959E+1	1.2610E+2	-2.8801E-2	-6.3142E+1	5.4548E-7	2.8761E-2
17	-3.0692E-1	1.6415E+1	-1.4806E+1	-1.8346E+1	-2.8921E+0	2.0936E+1
18	-1.8777E+2	1.1482E+2	2.6201E+2	6.4193E+1	-9.9645E+0	-2.4228E+2
19	-7.8929E+0	1.6780E+1	1.0244E+0	-7.8875E+0	5.9509E-3	-1.0252E+0
20	1.6250E+0	-2.4087E+0	5.0903E-4	1.7861E+0	2.4507E-8	-5.2458E-4
21	9.8493E-1	7.8212E-5	3.7369E-3	-1.0339E-7	8.8817E-4	-1.2276E-5
22	-8.8257E+1	1.7868E+2	-1.0451E+1	-8.9419E+1	-2.5096E-4	1.0451E+1
23	9.9690E-1	-3.3893E-6	8.3911E-3	-3.1845E-9	-6.8053E-3	-8.7023E-7
24	-8.0245E+2	6.4401E+2	9.6266E+2	2.1901E+1	-1.3782E+2	-6.8731E+2
25	2.0536E+1	1.4721E+2	-1.8732E+2	-1.2442E+2	4.2649E+1	1.0234E+2
26	-1.1994E+1	2.4927E+1	1.3707E-1	-1.1932E+1	7.7668E-4	-1.3829E-1
27	-3.3928E+1	-4.8502E+1	1.1742E+2	-2.0110E+1	-1.0364E+2	8.9758E+1
28	3.3045E+0	-5.6009E+0	-2.2026E-4	3.2964E+0	6.1967E-9	2.1961E-4
29	5.6656E+1	-1.1139E+2	-9.6569E+0	5.5730E+1	8.1188E-4	9.6565E+0
30	7.6042E+0	8.0651E+0	-2.2283E+1	1.6229E+0	1.6282E+1	-1.0291E+1
31	1.0721E+1	-2.0678E+1	7.3476E-4	1.0958E+1	8.2460E-9	-7.4024E-4

Table 6. GMDH polynomial model of the correction factor in recursive form with the corresponding coefficients of the second order two-dimensional polynomials.

The recursive equation of the flow rate correction factor model (Fig. 10) and the corresponding coefficients of the basic polynomials, rounded to 5 most significant decimal digits, are shown in Table 6, where $x_0,...,x_8$ denote the input parameters shown in Table 7. Table 7 also specifies the ranges of application of input parameters. The detailed description of the procedure for the selection of optimal input parameters is described in (Marić & Ivek, 2010). The layers in Fig. 10 are denoted by 'L00' to 'L15' and the polynomials by '$P_m(n)$', where 'm' indicates the order in which the polynomials are to be calculated recursively and 'n' denotes the total number of the basic polynomial calculations necessary to compute the mth polynomial by the corresponding recursive equation.

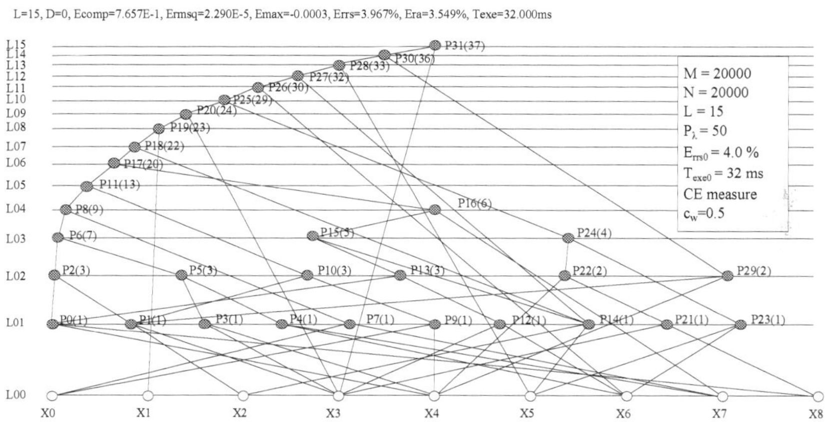

Fig. 10. Polynomial graph of the best GMDH surrogate model of the flow rate correction factor K (Marić & Ivek, 2010), obtained at layer 15 by using the CE measure with weighting coefficient c_w=0.5.

Index	Parameter description	Range of application
0	X_{CO2} - mole fraction of carbon dioxide	$0 \leq X_{CO2} \leq 0.20$
1	X_{H2} - mole fraction of hydrogen	$0 \leq X_{H2} \leq 0.10$
2	p - absolute pressure in MPa	$0 < p \leq 12$
3	T - temperature in K	$263 \leq T \leq 368$
4	Δp - differential pressure in MPa	$0 \leq \Delta p \leq 0.25p$
5	ρ - density in kg/m^3	unspecified
6	ρ_r - relative density	$9.55 \leq \rho_r \leq 0.80$
7	H_S - superior calorific value in MJ/m^3	$30 \leq H_S \leq 45$
8	β - orifice to pipe diameter ratio: d/D	$0.1 \leq \beta \leq 0.75$

Table 7. Input parameters for the natural gas flow rate correction factor modeling.

8.2 MLP model of the flow rate correction factor

Similarly, a simple feedforward ANN the multilayer perceptron (MLP), consisting of four nodes in a hidden layer and one output node (Fig. 11), with sigmoid activation function, $\sigma = \left(1 + e^{-\sum_{i=1}^{N}(w_i x_i + b)}\right)^{-1}$, has been trained to approximate the correction factor using the same data sets and the same constraints on the RRSE and ET as in GMDH example. The output (y) from MLP, can be written in the form:

$$y = \sigma\left(b_0 + \sum_{i=1}^{4}\left(w_i \cdot \sigma\left(b_i + \sum_{j=0}^{8}\left(w_{ij} \cdot x_j\right)\right)\right)\right), \tag{47}$$

where x_j represents the jth input parameter (Table 7), while b_i, w_i and w_{ij} denote the coefficients (Table 8), obtained after training the MLP by the Levenberg-Marquardt algorithm.

i	w_i	b_i	w_{0i}	w_{1i}	w_{2i}	w_{3i}
0	-	1.3996E+02	1.2540E-02	-1.9365E+00	1.6910E+00	2.8704E-01
1	-1.0130E+01	4.3451E+00	-1.7140E-02	1.1891E+00	-9.8349E-01	-2.2546E-01
2	-1.6963E+01	-3.9870E-01	8.7299E-04	3.1764E-02	-3.8641E-02	-1.3140E-02
3	-2.1044E+01	4.7731E-01	4.3873E-04	5.9977E-03	-6.8960E-03	-2.3687E-03
4	-1.0418E+02	4.0630E+00	8.1728E-02	-1.2010E+00	1.9725E+00	2.6340E+01
5	-	-	1.7424E-04	9.3751E-03	-4.0967E-03	7.4620E-04
6	-	-	-9.6008E-02	4.1284E-01	-1.7563E-01	9.0026E-02
7	-	-	-3.1605E-03	-2.5425E-02	2.6349E-02	8.6144E-03
8	-	-	-2.9468E+00	-2.1838E-02	5.6189E-02	3.6848E-02

Table 8. MLP coefficients truncated to 5 most significant digits.

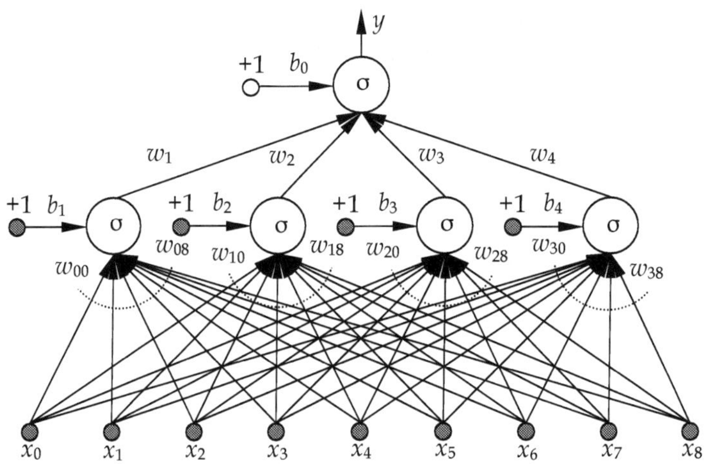

Fig. 11. MLP scheme for the flow rate correction factor modeling.

8.3 Flow rate correction error analysis

The execution times (complexities) of the MLP from Fig. 11 (T_{exe} =28 ms) and the GMDH model from Fig. 10 (T_{exe}=32 ms) are comparable but the embedding of MLP in FC software is slightly more complicated since it needs the implementation of the exponential function. The accuracy and the precision of the derived models were tested on 10 randomly generated data sets and the summary of the results is shown in Table 9. From Table 9 it can be seen that the standard deviation equals approximately 1% of the corresponding average value of RMSE and RRSE for both models and we may conclude that the derived correction factor approximates the correction procedure consistently in the entire range of application. In this particular application the MLP has significantly lower approximation error than the GMDH, both having approximately equal complexity. Note that RMSE and RRSE can be further decreased if increasing the number of layers (GMDH) or nodes (MLP) but this will also increase the corresponding execution time of the model. Fig. 12 illustrates the results of the simulation of a relative error, Eq. (46), in the measurement of a natural gas flow rate when ignoring the JT expansion effects (q_d), instead of its precise correction (q_u) in accordance with the procedure outlined in Table 5. The calculation of the flow rate is simulated by assuming the square-edged orifice plate with corner taps (ISO-5167, 2003), with orifice diameter of 20 mm, the pipe diameter of 200 mm, the differential pressure of 0.2 MPa, and with the downstream measurement of temperature. The error corresponds to the natural gas mixture 'Gas 3', given in Table G.1 of Annex G in (ISO-20765-1, 2005), which produces the largest temperature changes of all six mixtures given for validation purposes. The pressure varies from 1 MPa to 12 MPa in 0.5 MPa steps and the temperature from 263 K to 338 K in 10 K steps.

Validation set index	GMDH: E_{rms} x10^{-5}	GMDH: E_{rrs} [%]	MLP: E_{rms} x10^{-5}	MLP: E_{rrs} [%]
1	2.305	4.007	1.744	3.022
2	2.267	3.939	1.772	3.071
3	2.258	3.933	1.749	3.032
4	2.225	3.910	1.725	2.989
5	2.270	3.922	1.740	3.014
6	2.303	3.999	1.761	3.051
7	2.295	3.969	1.751	3.034
8	2.273	3.968	1.767	3.063
9	2.277	3.922	1.746	3.025
10	2.280	3.966	1.765	3.059
Mean value: $\bar{x} = \frac{1}{N}\sum_{1}^{N} x_i$	2.275	3.954	1.752	3.036
Standard deviation: $\sigma = \sqrt{\frac{N\sum_1^N x_i^2 - \left(\sum_1^N x_i\right)^2}{N(N-1)}}$	0.02356	0.03341	0.01437	0.02525

Table 9. Errors in the calculated correction coefficient when approximating the precise procedure (Table 5) by the best GMDH polynomial model (Fig. 10 and Table 6) and MLP (Fig. 11 and Table 8).

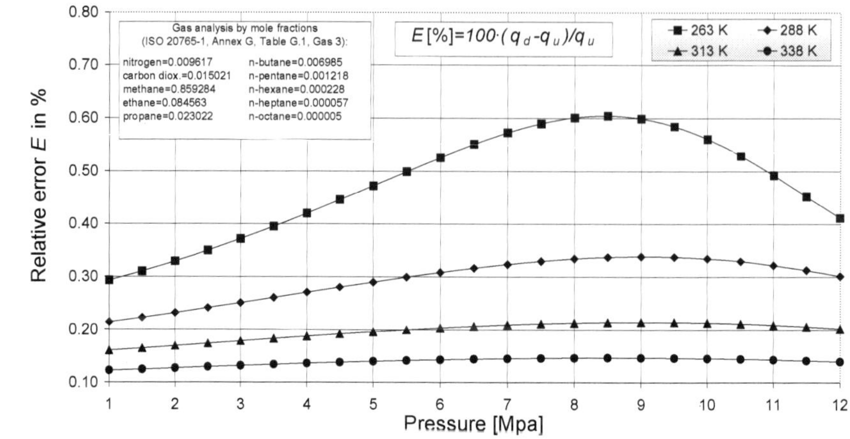

Fig. 12. Illustration of a relative error in the measurement of a natural gas flow rate by orifice plate with corner taps when ignoring the JT expansion effect.

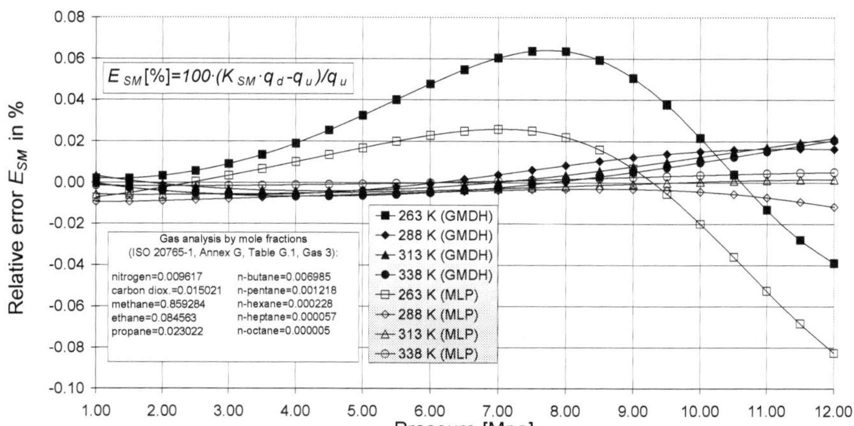

Fig. 13. Illustration of a relative error in the measurement of natural gas flow rate when using the GMDH (Fig. 10, Table 6) and MLP (Fig. 11, Table 8) surrogate models of the flow rate correction factor instead of the precise compensation procedure (Table 5).

From Fig. 12 it can be seen that the relative error slightly exceeds 0.6 % for the temperature of 263 K and for the pressures close to 8.5 MPa. The relative flow rate errors obtained for the remaining gas mixtures given in Table G.1 of (ISO-20765-1, 2005) are considerably lower.

Fig. 13 illustrates the relative flow rate error $E_{SM} = (K_{SM} \cdot q_d - q_u)/q_u$ when compensating the flow rate error by the GMDH (K_{SM}=K_{GMDH}) or by the corresponding MLP (K_{SM}=K_{MLP}), instead of its precise correction outlined in Table 5. The results in Fig. 13 are obtained by simulating the flow rate through the square-edged orifice plate with corner taps (ISO-5167,

2003), with orifice diameter of 20 mm, the pipe diameter of 200 mm, the differential pressure of 0.2 MPa, and with the downstream measurement of temperature. Again, the natural gas is taken from Table G.1 in (ISO-20765-1, 2005), and corresponds to the gas mixture denoted by 'Gas 3'. The pressure varies from 1 MPa to 12 MPa in 0.5 MPa increments and the temperature from 263 K to 338 K in 25 K increments.

From Fig. 13 it can be seen that the GMDH correction factor lowers the non-compensated relative error (Fig. 12) roughly by the order of magnitude in the entire pressure/temperature range. For the same complexity the MLP shows significantly better error performance characteristics than GMDH except at higher pressures close to 12 MPa. Both models have the error performance characteristics somewhat degraded at higher pressures and at lower temperatures but the absolute value of the relative error never exceeds 0.064% in case of GMDH and 0.083% in case of MLP. Similar results are obtained for the remaining gas mixtures from Table G.1 (ISO-20765-1, 2005) and for various randomly generated gas compositions. Almost identical error performance characteristics are obtained when applying the same GMDH model for the correction of the JT effect in the measurements using orifice plates with corner-, flange- or D&D/2-taps (ISO-5167, 2003).

The non-compensated flow rate error varies by varying the natural gas composition due to the corresponding variation of the JT coefficient. For a fixed natural gas mixture the absolute value of a JT coefficient (Marić, 2005 & 2007) is increasing by decreasing the temperature, thus increasing the temperature drop, Eq. (44), which increases the uncertainty of the calculated density of a natural gas and the uncertainty of the flow rate, as well. Also, the increase of the differential pressure and the decrease of the diameter ratio are increasing the pressure loss, Eq. (45), thus amplifying the temperature change, Eq. (44), and consequently the flow rate error.

The non-compensating flow rate error (Fig. 12) occurs when measuring the temperature downstream of the orifice plate and when assuming the same temperature upstream of the orifice plate. The procedure for the precise compensation of a temperature drop effect (Table 5) eliminates the corresponding flow rate error completely but it needs the calculation of both the flow rate and the properties of a natural gas to be executed twice and is therefore computationally intensive and time consuming and may be unacceptable for low-computing power measurement systems. The above described correction procedure performs a simple scaling of the flow rate, calculated using "downstream conditions", by the corresponding low-complexity surrogate of the correction coefficient (Eq. (42)). The correction procedure slightly increases the calculation time of a common procedure (ISO-5167, 2003) but it decreases the non-compensated flow rate error, due to the temperature drop, by one order of magnitude (Figs. 12 and 13). Most likely, the obtained surrogate models are not the best possible models. However, both derived models decrease the computational complexity of precise compensation (Table 5) significantly while preserving reasonable accuracy and are therefore applicable in low-computing-power systems. Hence, they make the error negligible with the acceptable degradation of the calculation time. For the same computational complexity the MLP surrogate of the correction procedure displays better approximation error characteristics than the GMDH model but it also exhibits slightly increased programming complexity when considering its implementation in low-computing-power microcomputer.

9. Conclusions

The above described procedure for the computation of thermodynamic properties of natural gas was originally published in the Journal Flow Measurement and Instrumentation (Marić, 2005 & 2007). The procedure is derived using fundamental thermodynamic equations (Olander, 2007), DIPPR AIChE (DIPPR® Project 801, 2005) generic ideal heat capacity equations, and AGA-8 (Starling & Savidge, 1992) extended virial-type equations of state. It specifies the calculation of specific heat capacities at a constant pressure c_p and at a constant volume c_v, the JT coefficient μ_{JT}, and the isentropic exponent κ of a natural gas. The thermodynamic properties calculated by this method are in very good agreement with the known experimental data (Ernst et al., 2001).
The effects of thermodynamic properties on the accuracy of natural gas flow rate measurements based on differential devices are analyzed. The computationally intensive procedure for the precise compensation of the flow rate error, caused by the JT expansion effects, is derived. In order to make the compensation for the flow rate error executable in real time on low-computing-power digital computers we propose the use of machine learning and the computational intelligence methods. The surrogate models of the flow rate correction procedure are derived by learning the GMDH polynomials (Marić & Ivek, 2010) and by training the MLP artificial neural network. The MLP and the GMDH surrogates significantly reduce the complexity of the compensation procedure while preserving high measurement accuracy, thus enabling the compensation of the flow rate error in real time by low-computing-power microcomputer. The same models can be equally applied for the compensation of the flow rate of natural gas measured by means of orifice plates with corner-, flange- or D and D/2-taps.

10. References

Baker, R.C. (2000). *Flow Measurement Handbook*, Cambridge University Press, ISBN: 0-521-48010-8, New York

DIPPR® Project 801, (2005). *Evaluated Process Design Data*, Design Institute for Physical Properties, Sponsored by AIChE, Electronic ISBN: 978-1-59124-881-1

Ernst, G., Keil, B., Wirbser, H. & Jaeschke, M. (2001). Flow calorimetric results for the massic heat capacity cp and Joule-Thomson coefficient of CH_4, of (0.85 CH_4 + 0.16 C_2H_6), and of a mixture similar to natural gas, *J. Chem. Thermodynamics*, Vol. 33, No. 6, June 2001, 601-613, ISSN: 0021-9614

Ferrari, S. & Stengel, R.F. (2005). Smooth Function Approximation Using Neural Networks, *IEEE Transactions on Neural Networks*, Vol. 16, No. 1, January 2005, 24-38, ISSN: 1045-9227

ISO-12213-2 (2006), *Natural gas -- Calculation of compression factor -- Part 1: Introduction and guidelines*, ISO, Ref. No. ISO-12213-2:2006(E), Geneva

ISO-20765-1, (2005), *Natural gas – Calculation of thermodynamic properties - Part1: Gas phase properties for transmission and distribution applications*, ISO, Ref. No. ISO-20765-1:2005, Geneva

ISO-5167 (2003). *Measurement of fluid flow by means of pressure differential devices inserted in circular-cross section conduits running full*, ISO, Ref. No. ISO-51671:2003(E), Geneva

Ivakhnenko, A. G. (1971). Polynomial Theory of Complex Systems, *IEEE Transactions on Systems Man, and Cybernetics*, Vol. SMC-1, No.4, Oct 1971, 364-378, ISSN: 0018-9472

Lemmon, E. W. & Starling, K. E. (2003). Speed of Sound and Related Thermodynamic Properties Calculated from the AGA Report No. 8 Detail Characterization Method Using a Helmholtz Energy Formulation, *AGA Operating Section Proceedings*, ISSN: 15535711, Phoenix, May 2004, American Gas Association.

Marić, I. & Ivek, I. (IEEE). Self-Organizing Polynomial Networks for Time-Constrained Applications, *IEEE Transactions on Industrial Electronics*, DOI: 10.1109/TIE.2010.2051934 , (accepted for publication)

Marić, I. & Ivek, I. (2010). Compensation for Joule-Thomson effect in flow rate measurements by GMDH polynomial, *Flow Measurement and Instrumentation*, Vol. 21, No. 2, June 2010, 134-142, ISSN: 0955-5986.

Marić, I. (2005). The Joule-Thomson effect in natural gas flow-rate measurements, *Flow Measurement and Instrumentation*, Vol. 16, No. 6, December 2005, 387-395, ISSN: 0955-5986

Marić, I. (2007). A procedure for the calculation of the natural gas molar heat capacity, the isentropic exponent, and the Joule-Thomson coefficient, *Flow Measurement and Instrumentation*, Vol. 18, No. 1, March 2007, 18-26, ISSN: 0955-5986

Marić, I., Galović, A. & Šmuc, T. (2005). Calculation of Natural Gas Isentropic Exponent, *Flow Measurement and Instrumentation*, Vol. 16, No. 1, March 2005, 13-20, ISSN: 0955-5986

Miller, E.W. (1996). *Flow Measurement Engineering Handbook*, McGraw-Hill, ISBN: 0-07-042366-0, New York

Nikolaev, N.Y. & Iba, H. (2003), "Learning Polynomial Feedforward Neural Networks by Genetic Programming and Backpropagation," *IEEE Transactions on Neural Networks*, Vol. 14, No. 2, March 2003, 337-350, ISSN: 1045-9227

Olander, D. R. (2007). *General Thermodynamics*, CRC Press, ISBN: 9780849374388, New York

Poling, B.E., Prausnitz, J.M. & O'Connell, J. (2000). *The Properties of Gases and Liquids*, McGraw-Hill Professional, ISBN: 0070116822, New York

Shoemaker, D.P., Garland, C.W. and Nibler, J.W. (1996). *Experiments in Physical Chemistry*, McGraw-Hill, ISBN: 9780072318210, New York

Span, R. & Wagner, W. (1996) A New Equation of State for Carbon Dioxide Covering the Fluid Region from the Triple-Point Temperature to 1100 K at Pressures up to 800 MPa, *J. Phys. Chem. Ref. Data*, Vol. 25, No. 6, November 1996, 1509-1596, ISSN 0047-2689

Span, R. & Wagner, W. (2003) Equations of State for Technical Applications. I. Simultaneously Optimized Functional Forms for Nonpolar and Polar Fluids, *Int. J. Thermophys.*, Vol. 24, No. 1, January 2003, 1-39, ISSN: 0195-928X

Starling, K. E. & Savidge, J. L. (1992). *Compressibility Factors for Natural Gas and Other Hydrocarbon Gases*, American Gas Association Transmission Measurement Committee Report No. 8, American Petroleum Institute (API) MPMS, chapter 14.2, Arlington

Urner, G. (1997), Pressure loss of orifice plates according to ISO 5167-1, *Flow Measurement and Instrumentation*, Vol. 8, No. 1, March 1997, 39-41, ISSN: 0955-5986

Wilamowski, B. M.; Cotton, N.J.; Kaynak, O. & Dündar, G. (2008). Computing Gradient Vector and Jacobian Matrix in Arbitrarily Connected Neural Networks, *IEEE Transactions on Industrial Electronics*, Vol. 55, No. 10, October 2008, 3784-3790, ISSN: 0278-0046

Wilamowski, B. M. (2009). Neural Network Architectures and Learning Algorithms, *IEEE Industrial Electronics Magazine*, Vol. 3, No. 4, December 2009, 56-63, ISSN: 1932-4529

22

Rarefied natural gas transport

Huei Chu Weng
Chung Yuan Christian University
Taiwan

1. Introduction

Natural gas (or simply gas), made up of around 82.0-89.6 mol% methane, 0.9-9.8 mol% nitrogen, 3.4-9.4 mol% ethane, 0.6-4.7 mol% propane, 0.1-1.7 mol% n-butane, and other gases (GPSA, 1998; Ivings et al., 2003; Schley et al., 2004), is a gaseous fossil fuel that has methane as its primary compound and is moved primarily by pipeline. It is desirable to understand the physical properties in rarefied natural gas transport and the corresponding behavior in pipelines.

Gas rarefaction was observed in many areas, such as gas bearings (Johnston & McCloskey, 1940; Carr, 1954; Burgdorfer, 1959; Hsia & Domoto, 1983), space vehicles (Ivanov & Gimelshein, 1998; Tsuboi & Matsumoto, 2005), microfluidic devices (Pfahler et al., 1991; Pong et al., 1994), etc. The effect of gas rarefaction is quantified by the Knudsen number Kn, which is defined as the ratio of the molecular mean free path to the characteristic length. Schaaf & Chambre (1961) have proposed the following ranges to determine the degree of gas rarefaction in a flow, as shown in Table 1. For $\mathrm{Kn} \le 0.01$, the flow is in the continuum regime, where the continuum hypothesis holds. The continuum field equations in conjunction with no-slip and no-jump boundary conditions describe the flow and thermal behavior. The rarefaction effect becomes noticeable when the value of Kn becomes greater than 0.01. The range 0.01-0.1 is referred to as the slip regime, where the conventional continuum field equations are still valid, but the conventional no-slip and no-jump boundary conditions may break down. It has been proven that the Maxwell-Smoluchowski (MS) slip and jump boundary conditions are valid (Arkilic et al., 1997; Beskok & Karniadakis, 1999; Ewart et al., 2007; Weng & Chen, 2008a). The range of $0.1 < \mathrm{Kn} \le 10$ represents the transition regime, where the molecular mean free path is comparable to the characteristic length, and the conventional continuum field equations may break down. For $10 < \mathrm{Kn}$, the flow is in the free-molecular regime, where the collisions among the molecules are negligible. Although conventional transport systems operate in the continuum regime, pipelines can operate in the slip regime due to the reduction of working pressure or the enhancement of working temperature (the molecular mean free path is related to the pressure and the temperature).

In this chapter, first we will provide the property formulas of gases in power-law form and present the physical properties of methane at the standard reference state, so as to simply predict the physical properties of natural gases. Then we will discuss in some detail the use of the mass, momentum, and energy conservation equations as well as the slip and jump boundary conditions as it is applied to fluid motion that is rarefied. Finally, we will develop the mathematical models of rarefied natural gas transport in basic driving mechanisms. We will obtain the analytical solutions of flow fields and characteristics, so as to realize the importance of gas rarefaction in natural gas transport.

After completing this chapter, you should be able to:

- use the property formulas of gases and the physical properties of methane at the standard reference state.
- use the mass, momentum, and energy conservation equations subject to the slip and jump boundary conditions.
- solve basic rarefied natural gas transport problems.
- use the analytical procedure shown in basic transport problems.
- apply the analytical solutions of basic transport problems.

Regime	Range
Continuum flow	$\mathrm{Kn} \le 0.01$
Slip flow	$0.01 < \mathrm{Kn} \le 0.1$
Transition flow	$0.1 < \mathrm{Kn} \le 10$
Free-molecular flow	$10 < \mathrm{Kn}$

Table 1. Flow regimes of rarefied gas dynamics (see Schaaf & Chambre, 1961).

2. Physical Properties

Any characteristic of a substance that can be observed or measured without changing the identity of the substance is called physical property. Some important physical properties in rarefied gas transport are shear viscosity μ, thermal conductivity k, constant-pressure specific heat c_p, density ρ, and molecular mean free path λ. In this part, we provide the formulas for these properties of gases in power-law form and present the physical properties of methane at the standard reference state. Then you can use the formulas and the properties to present further properties of other hydrocarbons, such as ethane, propane, butane, etc., to calculate further physical properties of natural gases in most common operating states, and to conduct further analyses of theoretical and experimental researches.

Various tests have been conducted to determine μ, k, c_p, and ρ of methane at different working pressures or temperatures (Mann & Dickins 1931; Johnston & McCloskey 1940; Clarke & Smith, 1969; Jansoone et al., 1970; Vennix et al., 1970; Younglove, 1974; Gammon & Douslin, 1976; Kerley, 1980; Kleinrahm & Wagner, 1986; Kleinrahm et al. 1986; Kurumov et al., 1988; Friend et al., 1989; Jin et al., 1992; Pátek & Klomfar, 2002; Hurly et al., 2003; Schley et al., 2004; Viswanathan, 2007). Except for density, temperature was found to be a relatively important factor in the gaseous state. It becomes necessary to have formulas for temperature-dependent physical properties of methane. The property formulas adopted in this chapter was originally proposed by Weng & Chen (2008b). According to the nature of gases, the power-law expressions for the shear viscosity, the thermal conductivity, the specific heat, and the density can be written, respectively, as ,

$$\mu = \mu_r \left(\frac{T}{T_r} \right)^{\omega} \tag{1}$$

$$k = k_r \left(\frac{T}{T_r} \right)^{\varpi} \tag{2}$$

$$c_p = c_{pr} \left(2 + s\left(\frac{T}{T_r} - 1 \right) - \left(\frac{T}{T_r} \right)^{\varpi - \omega} \right) \tag{3}$$

$$\rho = \rho_r \left(\frac{T}{T_r} \right)^{-1} \tag{4}$$

where the subscript r denotes the property values at a reference state, T is the temperature, ω and ϖ are the viscosity and conductivity indexes, respectively, and s is the specific-heat coefficient. Here, ideal gas behavior is assumed, which is quite reasonable in most common operating states. If the shear viscosity, thermal conductivity, and specific heat are known at two temperatures, one can determine the values of ω, ϖ and s. In Table 2, we use 300 K and 400 K for the constant determination of methane. In Fig. 1, we then verify the present formulas (1)-(4). The calculated results compare very well with the data listed in Friend et al. (1989) and Schley et al. (2004). The percentage absolute errors for μ, k, c_p, and ρ over $260\ \text{K} \le T \le 400\ \text{K}$ are less than 0.674, 2.01, 1.68, and 0.110, respectively. It should be noted that the data in the literature chosen for the comparison are modified to take into account the pressure effect.

Physical Property	Symbol	Unit	Value
Shear Viscosity	μ	$\text{N}\cdot\text{s/m}^2$	1.114×10^{-5}
Thermal conductivity	k	$\text{W/m}\cdot\text{K}$	0.0348
Constant-pressure specific heat	c_p	$\text{J/K}\cdot\text{kg}$	2223
Density	ρ	kg/m^3	0.648
Mean free path	λ	m	0.542×10^{-7}
Ratio of specific heats	γ	-	1.304
Prnadtl number	Pr	-	0. 712
Viscosity index	ω	-	0.835
Conductivity index	ϖ	-	1.225
Specific-heat coefficient	s	-	0.758

Table 2. Physical properties of methane at the standard reference state: 288.15 K and 101.325 kPa (see Johnston & McCloskey, 1940; Clarke & Smith, 1969; Haberman & John, 1980; Sonntag et al., 1998; GPSA, 1998; Ivings et al., 2003).

In addition, the average distance between molecular collisions, called the molecular mean free path, plays an important role in rarefied gas transport behavior, related to the temperature and pressure by

$$\lambda = \frac{\sqrt{\pi \hat{R} T / 2}\,\mu}{p} \tag{5}$$

where $\hat{R}$ is the specific gas constant. According to the power-law behavior considered here, the closed form is

$$\lambda = \lambda_r \left(\frac{T}{T_r} \right)^{\omega + 1/2} \tag{6}$$

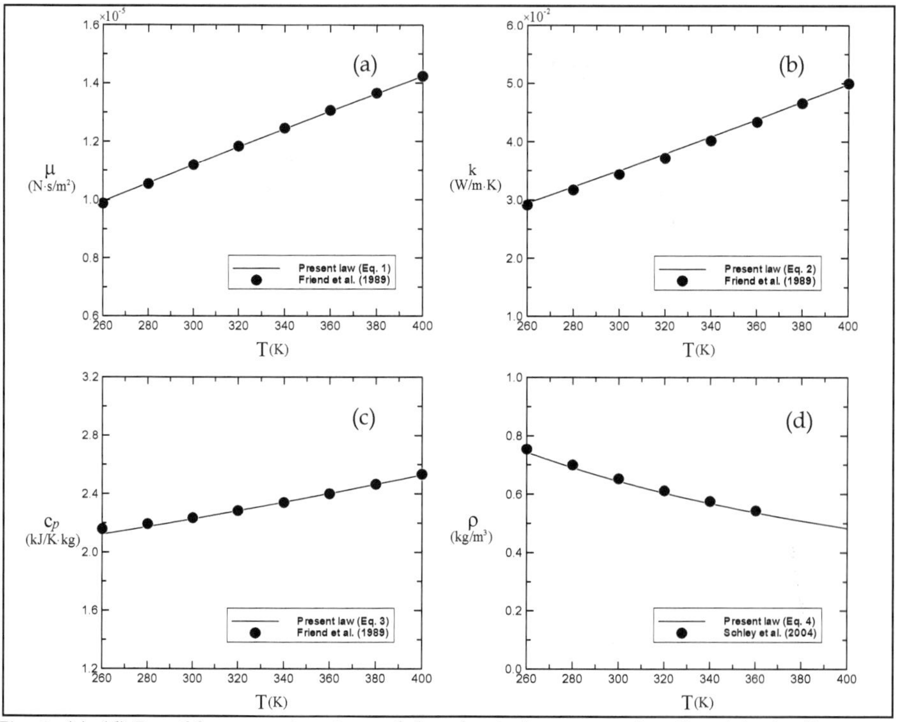

Fig. 1. (a)-(d) Possible approximations of the physical property data listed in the literature. $T_r = 298.15$ K

3. Fundamental Principles

In this part, we introduce the fundamental principles of rarefied natural gas transport. Then you can use the mass, momentum, and energy conservation equations subject to the slip and jump boundary conditions to solve further problems involving mass or volume flow rate, to solve further problems involving force related to momentum change, and to solve further problems involving losses due to friction and energy input by compressors or extraction by turbine.

3.1 Mass balance principle

Conservation of mass is the key to tracking flowing fluid. The conservation statement of mass for a control volume is

$$\frac{\partial M_{cv}}{\partial t} = \sum \dot{m}_{in} - \sum \dot{m}_{out} \tag{7}$$

where $\partial / \partial t$ is the time rate of change in a laboratory frame of reference, M is the instantaneous mass, $\dot{m}$ is the mass flow rate, the subscript cv denotes the control volume, and the subscripts in and out indicate the values for flow into and out of the control volume, respectively. The conservation statement requires that

$$\frac{d\rho}{dt} + \rho \nabla \cdot \mathbf{u} = 0 \tag{8}$$

where d / dt is the material derivative and $\mathbf{u}$ is the velocity vector.

Equation (8) is the mass conservation equation for compressible methane flow. In fact, the flow of a compressible fluid could be recognized as an incompressible flow while the flow speed is much less than the sound speed, i.e. Mach number<<1. For incompressible methane flow, the conservation equation is reduced to

$$\nabla \cdot \mathbf{u} = 0 \tag{9}$$

3.2 Momentum balance principle

Newton's second law of motion is the conservation statement of momentum. It leads to the conclusion that forces can result from or cause changes in a flowing fluid's velocity magnitude and/or direction. The conservation statement of momentum for a control volume is

$$\frac{\partial (Mu_n)_{cv}}{\partial t} = \sum \dot{m}u_n)_+ - \sum \dot{m}u_n)_- + \sum F_n)_+ - \sum F_n)_- \tag{10}$$

where u is the fluid velocity, Mu is the instantaneous momentum, $\dot{m}u$ is the momentum change rate associated with mass flow, F is the force acting on the control volume, the subscript n denotes the direction chosen for analysis, and the subscripts $+$ and $-$ indicate the values in the n or $-n$ direction on the control volume, respectively. The conservation statement requires that

$$\rho \frac{d\mathbf{u}}{dt} = \nabla \cdot \mathbf{T} + \rho \mathbf{f} \tag{11}$$

where $\mathbf{T}$ is the stress tensor and $\mathbf{f}$ is the body force vector per unit mass.

The stress tensor for compressible fluids can be related to the flow fields by the following constitutive relation:

$$\mathbf{T} = (-p + \eta \nabla \cdot \mathbf{u})\mathbf{I} + 2\mu \mathbf{D} \tag{12}$$

where p is the pressure, η is the bulk viscosity, related to the viscosity μ by $\eta = -2\mu/3$, $\mathbf{I}$ is the Kronecker delta tensor, and $\mathbf{D}$ is the deformation rate tensor, $\mathbf{D} = \left((\nabla\mathbf{u})^T + \nabla\mathbf{u}\right)/2$. Substituting Eq. (12) into Eq. (11) gives

$$\rho\frac{d\mathbf{u}}{dt} = -\nabla p + \nabla(\eta\nabla\cdot\mathbf{u}) + 2\nabla\cdot(\mu\mathbf{D}) + \rho\mathbf{f} \tag{13}$$

For incompressible methane flow with constant viscosity, the conservation equation is reduced to

$$\rho\frac{d\mathbf{u}}{dt} = -\nabla p + \mu\nabla^2\mathbf{u} + \rho\mathbf{f} \tag{14}$$

3.3 Energy balance principle

The first law of thermodynamics is the statement of conservation of energy. It leads to the conclusion that energies can result from or cause changes in a flowing fluid's velocity magnitude and/or temperature. The conservation statement of energy for a control volume is

$$\frac{\partial(Me)_{cv}}{\partial t} = \sum \dot{m}e)_{in} - \sum \dot{m}e)_{out} + \sum \dot{Q}_{in} - \sum \dot{Q}_{out} + \sum \dot{W} + \dot{O}, \tag{15}$$

where e is the specific internal energy, Me is the instantaneous energy, $\dot{m}e$ is the energy change rate associated with mass flow, $\dot{Q}$ is the energy change rate associated with heat conduction, and $\dot{W}$ is the energy change rate associated with force, and $\dot{O}$ is the internal heat generation rate. The conservation statement requires that

$$\rho\frac{de}{dt} = -\nabla\cdot\mathbf{q} + \mathbf{T}^S : \mathbf{D} + \rho o \tag{16}$$

where $\mathbf{q}$ is the heat flux vector, o is the internal heat generation rate per unit mass, and the superscript S indicates the symmetric tensor.
The heat flux vector can be related to the temperature field by the following constitutive relation (Fourier's law of conduction):

$$\mathbf{q} = -k\nabla T \tag{17}$$

Substituting Eqs. (12) & (17) into Eq. (16) and recognizing $\rho de/dt = \rho c_p dT/dt - \beta T\dot{p} + (p/\rho)d\rho/dt$ (Bejan, 2004) gives

$$\rho c_p\frac{dT}{dt} = \nabla\cdot(k\nabla T) + \beta T\frac{dp}{dt} + \mu\Phi + \rho o \tag{18}$$

where β is the thermal expansion coefficient and Φ is a function charactering the viscous dissipation, defined in index notation as

$$\Phi = \frac{\eta}{\mu}\left(u_{i,i}\right)^2 + \left(u_{i,j}u_{i,j} + u_{i,j}u_{j,i}\right) \tag{19}$$

For incompressible methane flow with constant thermal conductivity, the conservation equation is reduced to

$$\rho c_p \frac{dT}{dt} = k\nabla^2 T + \mu\Phi + \rho o \tag{20}$$

where

$$\Phi = u_{i,j}u_{i,j} + u_{i,j}u_{j,i} \tag{21}$$

3.4 Entropy balance principle

Any discussion of the fundamental principles must include the second law of thermodynamics because it is the basis for formulating and solving engineering problems. The second law of thermodynamics asserts that heat transfer processes are irreversible. To treat the second law quantitatively for processes, we use the conservation statement of entropy for a control volume:

$$\frac{\partial (Ms)_{cv}}{\partial t} = \sum \dot{m}s)_{in} - \sum \dot{m}s)_{out} + \sum \frac{\dot{Q}}{T}\Bigg)_{in} - \sum \frac{\dot{Q}}{T}\Bigg)_{out} + \dot{S}_{gen}, \tag{22}$$

were s is the specific entropy, Ms is the instantaneous entropy, $\dot{m}s$ is the entropy rate associated with mass flow, $\dot{Q}/T$ is the entropy rate associated with heat conduction, and $\dot{S}_{gen}$ is the entropy generation rate. The irreversibility of the process is measured by the value of $\dot{S}_{gen}$. The conservation statement requires that

$$\rho \frac{ds}{dt} = -\nabla \cdot \left(\mathbf{q}/T\right) + \rho s_{gen} \tag{23}$$

where s_{gen} is the entropy generation rate per unit mass. Substituting Eq. (17) into Eq. (23) gives

$$\rho \frac{ds}{dt} = \nabla \cdot \left(k\nabla T/T\right) + \rho s_{gen} \tag{24}$$

3.5 Maxwell's and Smoluchowski's principles of boundary conditions

The effect of gas rarefaction is quantified by the Knudsen number Kn, which is defined as the ratio of the molecular mean free path λ to the characteristic length l_c. When the value of Kn is greater than 0.01, the conventional no-slip and no-jump boundary conditions may break down. The slip boundary conditions are based on Maxwell's expression (Maxwell, 1879):

$$u_{sl} = u_{wall} \pm \frac{2-\sigma_m}{\sigma_m}\frac{1}{\rho\left(2\hat{R}T_{gw}/\pi\right)^{1/2}}\tau_{rt} + \frac{3}{4}\frac{\gamma-1}{\gamma}\frac{\Pr}{\rho\hat{R}T_{gw}}\left(-q_t\right) \tag{25}$$

The jump boundary conditions are based on Smoluchowski's expression (Kennard, 1938):

$$T_{ju} = T_{wall} \pm \frac{2-\sigma_e}{\sigma_e}\frac{2(\gamma-1)}{\gamma+1}\frac{1}{\hat{R}\rho\left(2\hat{R}T_{gw}/\pi\right)^{1/2}}(-q_r) \tag{26}$$

Here, u_{sl} and T_{ju} are the velocity slip and temperature jump, respectively, u_{wall} and T_{wall} are the wall velocity and wall temperature, respectively, τ_{ij} is the stress tensor expressed in index notation, q_i is the heat flux vector expressed in index notation, σ_m and σ_e are the tangential momentum and thermal accommodation coefficients, respectively, the subscript gw denotes the gas values near the wall surface, and the subscripts r and t denote the gas values near the wall surface in the normal and tangential directions, respectively. Note that σ_m and σ_e are the parameters that describe the gas-wall interaction and related to the gas composition, temperature, and velocity over the surface as well as the solid surface temperature, roughness, and chemical state. The two accommodation coefficients range from near 0 to 1, where the two values, respectively, represent specular accommodation and complete (or diffuse) accommodation.

Let r and t denote the two coordinates near the wall surface and let u_r and u_t denote the components of velocity field. According to Eqs. (12) & (17), the shear stress τ_{rt}, the tangential heat flux q_t and the normal heat flux q_r are

$$\tau_{rt} = \mu\left(\frac{\partial u_r}{\partial t} + \frac{\partial u_t}{\partial r}\right)_{gw}, \quad q_t = -k\left.\frac{\partial T}{\partial t}\right|_{gw}, \quad q_r = -k\left.\frac{\partial T}{\partial r}\right|_{gw} \tag{27}$$

Substituting Eq. (27) into Eqs. (25) & (26) gives

$$u_{sl} = u_{wall} \pm \frac{2-\sigma_m}{\sigma_m}\lambda\left.\frac{\partial u_t}{\partial r}\right|_{gw} + \frac{3}{2\pi}\frac{\gamma-1}{\gamma}\frac{c_p\rho}{\mu}\lambda^2\left.\frac{\partial T}{\partial t}\right|_{gw} \tag{28}$$

$$T_{ju} = T_{wall} \pm \frac{2-\sigma_e}{\sigma_e}\frac{2\gamma}{\gamma+1}\frac{1}{\Pr}\lambda\left.\frac{\partial T}{\partial r}\right|_{gw} \tag{29}$$

4. Basic Transport Problems

Weng & Chen (2008c) have indicated that, on the basis of pressure or temperature difference, the three basic driving mechanisms are (1) pressure (2) buoyancy and (3) thermocreep, as shown in Fig. 2. Pressure and buoyancy are two conventional mechanisms. Thermal creep phenomenon (or simply transpiration) is a tangential flow along the wall surface from a cooler region to a hotter region. Previous work focuses mainly on creep effect on pressure-driven flow or buoyancy-driven flow. The effect during an externally applied pressure gradient or internally induced density variation is to hasten the flow, resulting in a streaming potential, whereas, in the absence of an external pressure gradient or internal density variation, the creep induces fluid flow when wall surface temperature gradient exists. In this part, we develop the mathematical models of rarefied natural gas transport in the three basic driving mechanisms and obtain the analytical solutions of flow fields and characteristics. Then you can apply the analytical solutions of basic transport problems to determine further flow (or/and thermal) characteristics, predict and analyze further transport behavior of rarefied natural gas in pipelines, and understand why gas rarefaction in natural gas transport is so important; moreover, using the analytical procedure, you can conduct further analyses of theoretical researches.

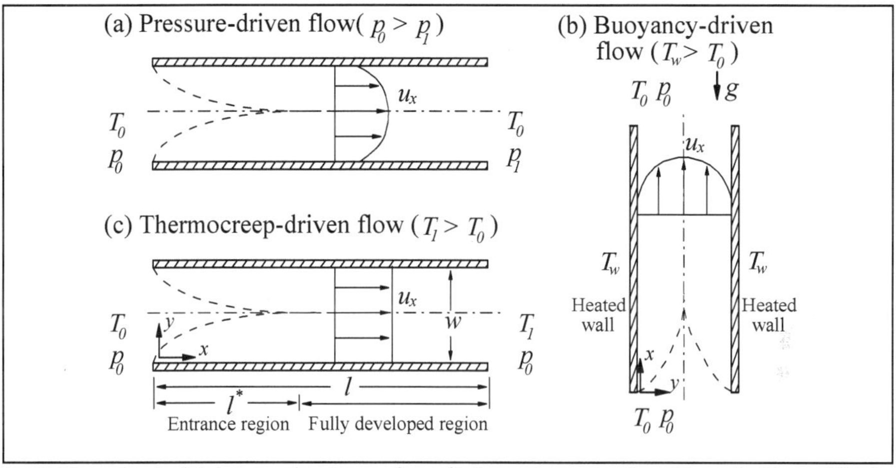

Fig. 2. Three basic driving mechanisms of rarefied gas dynamics (see Weng & Chen, 2008c).

Consider a sufficiently long open-ended parallel-plate channel of length l and width w. The flow section in the fully developed region is situated far from the entrance, and the flow is then fully developed. Assume that the flow originates from a reservoir in a reference state, enters the channel with a uniform velocity, and terminates in a discharge area. Let x and y denote the usual rectangular coordinates, let u_x and u_y denote the components of the velocity field, let the subscripts *0* and *1* denote the reservoir and discharge-area values, respectively, and let the subscript *i* denotes the inlet values.

4.1 Pressure-driven flow

If the driving mechanism is pressure, as shown in Fig. 2(a), then the reservoir pressure p_0 is greater than the discharge-area pressure p_1 and the discharge-area temperature T_1 is equal to the reservoir temperature T_0. Modeling the flow as a two-dimensional isothermal steady compressible flow, the field equations are given by (Eqs., 8 & 13)

$$\frac{\partial(\rho u_x)}{\partial x} + \frac{\partial(\rho u_y)}{\partial y} = 0 \tag{30}$$

$$\rho\left(u_x \frac{\partial u_x}{\partial x} + u_y \frac{\partial u_x}{\partial y}\right) = -\frac{\partial p}{\partial x} + \mu_r\left(\frac{\partial^2 u_x}{\partial x^2} + \frac{\partial^2 u_x}{\partial y^2} + \frac{1}{3}\left(\frac{\partial^2 u_x}{\partial x^2} + \frac{\partial^2 u_y}{\partial x \partial y}\right)\right) \tag{31}$$

$$\rho\left(u_x \frac{\partial u_y}{\partial x} + u_y \frac{\partial u_y}{\partial y}\right) = -\frac{\partial p}{\partial y} + \mu_r\left(\frac{\partial^2 u_y}{\partial x^2} + \frac{\partial^2 u_y}{\partial y^2} + \frac{1}{3}\left(\frac{\partial^2 u_x}{\partial x \partial y} + \frac{\partial^2 u_y}{\partial y^2}\right)\right) \tag{32}$$

The corresponding slip boundary conditions for a stationary unheated channel (Eq., 28) are

$$u_{sl} = \pm \frac{2-\sigma_m}{\sigma_m} \lambda \left.\frac{\partial u_x}{\partial y}\right|_{gw} \tag{33}$$

Here, we have neglected the body force for simplicity. By using a perturbation analysis (Weng & Chen, 2008a), the momentum conservation equations (31) & (32) for the flow through a sufficiently long channel can be reduced to the form:

$$\frac{dp}{dx} = \mu_r \frac{\partial^2 u_x}{\partial y^2} \tag{34}$$

Proceeding with the analysis, we introduce the following dimensionless parameters:

$$\left.\begin{aligned} &X = \frac{x}{l_c},\ Y = \frac{y}{l_c},\ U = \frac{u_x}{u_c},\ V = \frac{u_y}{u_c},\ P = \frac{p}{p_c}, \\ &\mathrm{Kn} = \frac{\lambda}{l_c},\ \mathrm{We} = \frac{\rho_r (p_i - p_o) l_c^2}{\mu_r^2},\ \Lambda = \frac{\rho_r \sqrt{\pi \hat{R} T / 2 l_c}}{\mu_r} \end{aligned}\right\} \tag{35}$$

where We is the pressure drop from the entrance to the exit, Λ is a material constant, and the subscript c denotes the characteristic values. Here, the characteristic length l_c, velocity u_c, and pressure p_c are, respectively, defined as

$$l_c = w,\ u_c = \frac{(p_i - p_o) l_c}{\mu_r},\ p_c = \frac{\mu_r^2}{\rho_r l_c^2} \tag{36}$$

Substituting Eq. (35) into Eqs. (30), (33), (34) gives

$$\frac{\partial (PU)}{\partial X} + \frac{\partial (PV)}{\partial Y} = 0 \tag{37}$$

$$\frac{dP}{dX} = \mathrm{We} \frac{\partial^2 U}{\partial Y^2} \tag{38}$$

$$\frac{u_{sl}}{u_c} = \pm \frac{2-\sigma_m}{\sigma_m} \mathrm{Kn} \left.\frac{\partial U}{\partial Y}\right|_{gw} \tag{39}$$

Solving momentum conservation equation (38) subject to the slip boundary conditions (39) gives

$$U(X,Y)=\frac{1}{2\mathrm{We}}\frac{dP(X)}{dX}\left(Y^2-Y-\frac{2-\sigma_m}{\sigma_m}\mathrm{Kn}(X)\right) \tag{40}$$

Substituting Eq. (40) into mass conservation equation (37) and integrating once in Y, we can derive an equation for the cross-flow velocity:

$$V(X,Y)=-\frac{1}{24\mathrm{We}}\left(\frac{1}{P}\frac{d^2P^2(X)}{dX^2}\left(2Y^3-3Y^2\right)-12\left(\frac{2-\sigma_m}{\sigma_m}\mathrm{Kn}(X)\frac{d^2P(X)}{dX^2}\right)Y\right) \tag{41}$$

Evaluating this at $Y=1$, where V must vanish, we can derive an equation for the pressure:

$$\frac{1}{12}P^2+\frac{2-\sigma_m}{\sigma_m}\Lambda P=C_1X+C_0 \tag{42}$$

where

$$\left.\begin{aligned}C_0&=\frac{1}{12}(\mathrm{We}+P_o)^2+\frac{2-\sigma_m}{\sigma_m}\Lambda(\mathrm{We}+P_o),\\ C_1&=-\frac{1}{12L}\left(\mathrm{We}^2+2P_o\mathrm{We}\right)-\frac{1}{L}\frac{2-\sigma_m}{\sigma_m}\Lambda\mathrm{We}\end{aligned}\right\} \tag{43}$$

The corresponding mass flow rate is

$$\dot{M}=\frac{p\int_0^W u_x dy}{p_c u_c l_c}=P\int_0^1 UdY=\frac{1}{24L}\left(\mathrm{We}+2\left(P_o+6\frac{2-\sigma_m}{\sigma_m}\Lambda\right)\right) \tag{44}$$

or

$$\dot{M}=\frac{\overline{P}}{12L}\left(1+6\frac{2-\sigma_m}{\sigma_m}\overline{\mathrm{Kn}}\right) \tag{45}$$

where L is the dimensionless channel length, $\overline{P}$ is the dimensionless average pressure, $\overline{P}=(P_i+P_o)/2$, and $\overline{\mathrm{Kn}}$ is the average Knudsen number calculated at $\overline{P}$.

4.2 Buoyancy-driven flow

If the driving mechanism is buoyancy, as shown in Fig. 2(b), then the wall temperature T_w is greater than the ambient temperature (that is, $T_1=T_0$ and $T_w>T_0$) and the discharge-area pressure p_1 is equal to the reservoir pressure p_0. Modeling the flow as a two-dimensional steady incompressible flow, the field equations under the Boussinesq approximation (Boussinesq, 1903) are given by (Eqs., 9, 14, 20)

$$\frac{\partial u_x}{\partial x}+\frac{\partial u_y}{\partial y}=0 \tag{46}$$

$$\rho_r\left(u_x\frac{\partial u_x}{\partial x}+u_y\frac{\partial u_x}{\partial y}\right)=-\frac{\partial \hat{p}}{\partial x}+\rho_r g\beta_r(T-T_r)+\mu_r\left(\frac{\partial^2 u_x}{\partial x^2}+\frac{\partial^2 u_x}{\partial y^2}\right) \tag{47}$$

$$\rho_r\left(u_x\frac{\partial u_y}{\partial x}+u_y\frac{\partial u_y}{\partial y}\right)=-\frac{\partial \hat{p}}{\partial y}+\mu_r\left(\frac{\partial^2 u_y}{\partial x^2}+\frac{\partial^2 u_y}{\partial y^2}\right) \tag{48}$$

$$\rho_r c_{pr}\left(u_x\frac{\partial T}{\partial x}+u_y\frac{\partial T}{\partial y}\right)=k_r\left(\frac{\partial^2 T}{\partial x^2}+\frac{\partial^2 T}{\partial y^2}\right) \tag{49}$$

where $\hat{p}$ is the pressure defect, related to p by $\hat{p}=p-p_h$, where p_h is the hydrostatic pressure. The corresponding slip and jump boundary conditions for a stationary isothermally heated channel (Eqs., 28 & 29) are

$$u_{sl}=\pm\frac{2-\sigma_m}{\sigma_m}\lambda_r\left.\frac{\partial u_x}{\partial y}\right|_{gw}+\frac{3}{2\pi}\frac{\gamma_r-1}{\gamma_r}\frac{c_{pr}\rho_r}{\mu_r}\lambda_r^2\left.\frac{\partial T}{\partial x}\right|_{gw} \tag{50}$$

$$T_{ju}=T_w\pm\frac{2-\sigma_e}{\sigma_e}\frac{2\gamma_r}{\gamma_r+1}\frac{1}{\Pr_r}\lambda_r\left.\frac{\partial T}{\partial y}\right|_{gw} \tag{51}$$

Here, we have neglected the internal heat generation for simplicity. It should be noted that assuming a small temperature difference between the wall and the ambient gas supports the constant-property assumption (Weng & Chen 2008b) and that considering the low-speed flow of a low-Prandtl-number fluid supports the neglect of viscous dissipation in the energy equation (Chen & Weng 2005; Weng & Chen 2008b).

We can think of the fully developed region as the flow section situated far from the entrance such that u_y is negligible. Based on this characterization, the mass conservation equation (46) requires that $\partial u_x/\partial x=0$. In most treatments, $u_y=0$ and $\partial u_x/\partial x=0$ are taken as a starting point in the analysis of fully developed flow. The momentum conservation equations (47) & (48) then become

$$0=\rho_r g\beta_r(T-T_r)+\mu_r\frac{d^2 u_x}{dy^2} \tag{52}$$

A solution of Eq. (52) in the form $u_x(Y)$ is only possible if T is a function of y position only, i.e., $\partial T/\partial X=0$. It implies that the assumption of a hydrodynamically fully developed flow necessarily means that the flow is also thermally fully developed. The energy conservation equations (49) and the slip boundary conditions (50) then be reduced to

$$0 = \frac{d^2T}{dy^2} \tag{53}$$

$$u_{sl} = \pm\frac{2-\sigma_m}{\sigma_m}\lambda_r\left.\frac{\partial u_x}{\partial y}\right|_{gw} \tag{54}$$

Proceeding with the analysis, we introduce the following dimensionless parameters:

$$X = \frac{x}{l_c},\ Y = \frac{y}{l_c},\ U = \frac{u_x}{u_c},\ \Theta = \frac{T-T_r}{T_w-T_r} \tag{55}$$

Here, the characteristic length l_c and velocity u_c are, respectively, defined as

$$l_c = w,\ u_c = \frac{\rho_r g\beta_r(T_w-T_r)l_c^2}{\mu_r} \tag{56}$$

Substituting Eq. (55) into Eqs. (51)-(54) gives

$$\frac{d^2U}{dY^2} = -\Theta \tag{57}$$

$$\frac{d^2\Theta}{dY^2} = 0 \tag{58}$$

$$\frac{u_{sl}}{u_c} = \pm\frac{2-\sigma_m}{\sigma_m}\mathrm{Kn}_r\left.\frac{dU}{dY}\right|_{gw} \tag{59}$$

$$\frac{T_{ju}-T_r}{T_w-T_r} = 1\pm\frac{2-\sigma_e}{\sigma_e}\frac{2\gamma_r}{\gamma_r+1}\frac{1}{\mathrm{Pr}_r}\mathrm{Kn}_r\left.\frac{d\Theta}{dY}\right|_{gw} \tag{60}$$

Equations (57) & (58) subject to (59) & (60) have the following velocity and temperature analytical solutions:

$$U(Y) = -\frac{1}{2}\left(Y^2 - Y - \frac{2-\sigma_m}{\sigma_m}\mathrm{Kn}_r\right) \tag{61}$$

$$\Theta(Y) = 1 \tag{62}$$

The corresponding mass flow rate is

$$\dot{M} = \frac{\int_0^w u_x dy}{u_c l_c} = \int_0^1 U dY = \frac{1}{2}\left(\frac{2-\sigma_m}{\sigma_m}\mathrm{Kn} + \frac{1}{6}\right), \tag{63}$$

4.3 Thermocreep-driven flow

If the driving mechanism is thermocreep, as shown in Fig. 2(c), then the reservoir temperature T_0 is less than the discharge-area pressure T_1 and the discharge-area pressure p_1 is equal to the reservoir pressure p_0. Modeling the flow as a two-dimensional steady incompressible constant-property flow, the momentum and energy equations under the fully developed flow limit ($u_y = 0$ and $\partial u_x / \partial x = 0$) are given by (Eqs., 9, 14, 20)

$$0 = -\frac{dp}{dx} + \mu_r \frac{d^2 u_x}{dy^2} \tag{64}$$

$$\rho_r c_{pr} u_x \frac{\partial T}{\partial x} = k_r \frac{\partial^2 T}{\partial y^2} + \mu_r \left(\frac{du_x}{dy} \right)^2 \tag{65}$$

Here, we have used the Prandtl boundary layer theory (Prandtl, 1904) to omit nonessential terms shown in field equations. It should be noted that assuming a small temperature difference between the reservoir and the discharge area supports the constant-property assumption.

Proceeding with the analysis, we introduce the following dimensionless parameters:

$$\left.\begin{array}{l} X = \dfrac{x}{l_c \mathrm{Gr}}, \; Y = \dfrac{y}{l_c}, \; U = \dfrac{u_x}{u_c}, \; \Theta = \dfrac{T - T_r}{T_1 - T_0}, \; P = \dfrac{p}{p_c}, \\ \mathrm{Gr} = \dfrac{\rho_r^2 g \beta_r (T_1 - T_0) l_c^3}{\mu_r^2}, \; \mathrm{Ec} = \dfrac{u_c^2}{c_{pr}(T_1 - T_0)} \end{array}\right\} \tag{66}$$

Here, the characteristic length l_c, velocity u_c, and pressure p_c are, respectively, defined as

$$l_c = w, \; u_c = \frac{\rho_r g \beta_r (T_1 - T_0) l_c^2}{\mu_r}, \; p_c = \rho_r u_c^2 \tag{67}$$

Substituting Eq. (66) into Eqs. (64), (65), and (50) gives

$$\frac{dP}{dX} - \frac{d^2 U}{dY^2} = 0 \tag{68}$$

$$\mathrm{Pr}_r U \frac{\partial \Theta}{\partial X} = \frac{\partial^2 \Theta}{\partial Y^2} + \mathrm{Pr}_r \mathrm{Ec} \left(\frac{dU}{dY} \right)^2 \tag{69}$$

$$\frac{u_{sl}}{u_c} = \pm \frac{2 - \sigma_m}{\sigma_m} \mathrm{Kn}_r \left. \frac{dU}{dY} \right|_{gw} + \frac{3}{2\pi} \frac{\gamma_r - 1}{\gamma_r} \frac{1}{\mathrm{Ec}} \mathrm{Kn}_r^2 \left. \frac{\partial \Theta}{\partial X} \right|_{gw} \tag{70}$$

From Eqs. (68) and (70), a solution of Eq. (68) in the form $U(Y)$ is only possible if dP/dX and $\partial \Theta / \partial X$ are constants (let C_0 and C_1, respectively). The constant value in temperature

gradient implies that the flow under the assumption of hydrodynamically fully developed flow is also thermally fully developed.
The momentum equation (68) can be integrated twice to obtain the streamwise velocity:

$$U(Y) = \frac{1}{2}C_0Y^2 + C_2Y + C_3 \tag{71}$$

Substituting Eq. (71) into the energy equation (69) and integrating the resultant with respect to Y twice and the thermally fully developed condition $\partial\Theta/\partial X = C_1$ with respect to X once, we obtain the temperature:

$$\begin{aligned}\Theta(X,Y) &= \mathrm{Pr}\,C_1\left(\frac{1}{24}C_0Y^4 + \frac{1}{6}C_2Y^3 + \frac{1}{2}C_3Y^2\right)\\ &- \mathrm{Pr}\,\mathrm{Ec}\left(\frac{1}{12}C_0^2Y^4 + \frac{1}{3}C_0C_2Y^3 + \frac{1}{2}C_2^2Y^2\right) + C_4Y + C_5 + C_1X.\end{aligned} \tag{72}$$

Integrating the pressure gradient $dP/dX = C_0$ with respect to X once, we obtain the pressure:

$$P(X) = P_0 + C_6 + C_0X \tag{73}$$

By applying the boundary conditions given in Eq. (70), the inlet condition $P(0) = P_0 - \dot{M}^2/2$ (Chen & Weng, 2006), the outlet conditions $P(L) = P_0$, the midline condition $\partial\Theta(X,1/2)/\partial Y = 0$, and the edge conditions $\Theta(L,0) = 1$ and $\Theta(0,0) = 0$, the six unknown constants C_0, C_1, C_2, C_3, C_4, C_5, and C_6 can be obtained as

$$\left.\begin{aligned}&C_0 = \frac{1}{2}\frac{\dot{M}^2}{L},\ C_1 = \frac{1}{L},\\ &C_2 = -\frac{1}{2}C_0,\ C_3 = -\frac{1}{2}\beta_mC_0 + \beta_{mc}C_1,\\ &C_4 = \frac{\mathrm{Pr}_r}{24}\left((1+6\beta_m)C_0C_1 - 12\beta_{mc}C_1^2 + \mathrm{Ec}C_0^2\right)\\ &C_5 = 0,\ C_6 = -\frac{1}{2}\dot{M}^2,\end{aligned}\right\} \tag{74}$$

where

$$\beta_m = \frac{2-\sigma_m}{\sigma_m}\mathrm{Kn},\ \beta_{mc} = \frac{3}{2\pi}\frac{\gamma-1}{\gamma}\frac{1}{\mathrm{Ec}}\mathrm{Kn}^2,\ L = \frac{l}{l_c\mathrm{Gr}},\ \dot{M} = \frac{u_i}{u_c} = \int_0^1 U dY \tag{75}$$

By using the flow-rate expression $\dot{M} = \int_0^1 U dY$, the channel length L can be obtained as

$$L = -\frac{1}{24}(1+6\beta_m)\dot{M} + \frac{\beta_{mc}}{\dot{M}} \tag{76}$$

5. Summary

In this chapter, the property formulas of natural gases are provided in power-law form. To simply predict the physical properties of natural gases, the physical properties of methane at the standard reference state are presented. The basic flows are analyzed by using important principles including conservation of mass, Newton's second law of motion, and the first and second laws of thermodynamics.

The following checklist provides a study guide for this chapter. When your study of the entire chapter has been completed, you should be able to

- use the property formulas of gases to present further properties of other hydrocarbons, such as ethane, propane, butane, etc.
- use the physical properties of methane, in conjunction with the properties of other gases as necessary, to calculate further physical properties of natural gases in most common operating states.
- use the property formulas of gases and the physical properties of methane at the standard reference state to conduct further analyses of theoretical and experimental researches.
- use the mass conservation equation to solve further problems involving mass or volume flow rate.
- use the momentum conservation equation subject to the slip boundary conditions to solve further problems involving force related to momentum change.
- use the energy conservation equation subject to the jump boundary conditions to solve further problems involving losses due to friction and energy input by compressors or extraction by turbine.
- use the analytical procedure shown in basic transport problems to conduct further analyses of theoretical researches.
- apply the analytical solutions of basic transport problems to determine further flow (or/and thermal) characteristics, predict and analyze further transport behavior of rarefied natural gas in pipelines, and understand why gas rarefaction in natural gas transport is so important.

6. Acknowledgment

The author would like to acknowledge financial support from the National Science Council in Taiwan as grant NSC 98-2218-E-033-003 and the CYCU Distinctive Research Area project as grant CYCU-98-CR-ME.

7. References

Arkilic, E. B.; Schmidt, M. A. & Breuer, K. S. (1997). Gaseous slip flow in long microchannels. *J. Microelectromech. Systems*, 6, 167–178.

Bejan, A., (2004). *Convection heat transfer*, John Wiley & Sons, 0471271500, 3rd edition.

Beskok, A. & Karniadakis, G. E. (1999). A model for flows in channels, pipes, and ducts at micro and nano scales. *Microscale Thermophys. Eng.*, 3, 43–77.

Boussinesq, J. (1903). *Theorie analytique de la chaleur heat dissipation of parallel plates by free convection*, Gauthier-Villars, Paris.

Burgdorfer, A. (1959). The influence of molecular mean free path on the performance of hydrodynamic gas lubricated bearings. *ASME J. Basic Eng.*, 81, 94–100.

Carr, N. L. (1954). Viscosity of hydrocarbon gases under pressure. *J. Pet. Technol.*, 6, 47–55.

Chen, C.-K. & Weng, H. C. (2005). Natural convection in a vertical microchannel. *ASME J. Heat Transfer*, 127, 1053–1056.

Chen, C.-K. & Weng, H. C. (2006). Developing natural convection with thermal creep in a vertical microchannel. *J. Phys. D*, 39, 3107–3118.

Clarke, A. G. & Smith, E. B. (1969). Low-temperature viscosities and intermolecular forces of simple gases. *J. Chem. Phys.*, 51, 4156–4161.

Ewart, T.; Perrier, P.; Graur, I. & Méolans, J. G. (2007). Mass flow rate measurements in a microchannel, from hydrodynamic to near free molecular regimes. *J. Fluid Mech.*, 584, 337–356.

Friend, D. G.; Ely, J. F. & Ingham, H. (1989). Thermophysical properties of methane. *J. Phys. Chem. Ref. Data*, 18, 583–638.

Gammon, B. E. & Douslin, D. R. (1976). The velocity of sound and heat capacity in methane from near-critical to subcritical conditions and equation-of-state implications. J. Chem. Phys., 64, 203–218.

GPSA (1998). *Engineering Data Book*, Section 23-Physical properties, Gas Processors Suppliers Association, Tulsa, Oklahoma, 11th edition.

Haberman, W. L. & John, J. E. A. (1980) *Engineering thermodynamics*, Allyn and Bacon, Inc., Boston.

Hurly, J. J.; Gillis, K. A.; Mehl, J. B. & Moldover, M. R. (2003). The viscosity of seven gases measured with a greenspan viscometer. *Int. J. Thermophys.*, 24, 1441–1474.

Hsia, Y. T. & Domoto, G. A. (1983). An experimental investigation of molecular rarefaction effects in gas lubricated bearings at ultra-low clearances. *ASME J. Lubrication Tech.*, 105, 120–130.

Ivanov, M. S. & Gimelshein, S. F. (1998). Computational hypersonic rarefied flows. *Ann. Rev. Fluid Mech.*, 30, 469–505.

Ivings, M. J.; Lea, C. J. & Ledin, H. S. (2003). Outstanding safety questions concerning the analysis of ventilation and gas dispersion in gas turbine enclosures: best practice guidelines for CFD, *Technical Report CM/03/12*, Health and Safety Laboratory.

Jansoone, V.; Gielen, H. & de Boelpaep, J. (1970). The pressure-temperature-volume relationship of methane near the critical point. *Physica.*, 46, 213–221.

Johnston, H. L. & McCloskey, K. E. (1940). Viscosities of Several Common Gases between 90°K. and Room Temperature. J. Phys. Chem., 44, 1038–1058.

Jin, G. X.; Tang, S. & Sengers, J. V. (1992). Thermodynamic properties of methane in the critical region. *Int. J. Thermophys.*, 13, 671–684.

Kennard, E. H. (1938) *Kinetic theory of gasses*, McGraw-Hill, New York.

Kerley, G. I. (1980). A theoretical equation of state for methane," *J. Appl. Phys.*, 51, 5368–5374.

Kleinrahm, R. & Wagner, W. (1986). Measurement and correlation of the equilibrium liquid and vapour densities and the vapour pressure along the coexistence curve of methane. *J. Chem. Thermodynam.*, 18, 739–760.

Kleinrahm, R.; Duschek, W. & Wagner, W. (1986) (Pressure, density, temperature) measurements in the critical region of methane. *J. Chem. Thermodynam.*, 18, 1103–1114.

Kurumov, D. S.; Olchowy, G. A. & Sengers, J. V. (1988). Thermodynamic properties of methane in the critical region. *Int. J. Thermophys.*, 9, 73–84.

Mann, W. B. & Dickins, B. G. (1931). The Thermal Conductivities of the Saturated Hydrocarbons in the Gaseous State. *Proc. R. Soc. Lond. A*, 134, 77–96.

Maxwell, J. C. (1879). On stress in rarefied gases from inequalities of temperature. *Philos. Trans. R. Soc. London*, 170, 231-256.

Pátek, J. & Klomfar, J. (2002). Measurement of the thermal conductivity of argon and methane: a test of a transient hot-wire apparatus. *Fluid Phase Equilib.*, 198, 147-163.

Pfahler, J.; Harley, J. C.; Bau, H. & Zemel, J. N. (1991). Gas and liquid flow in small channels, *Winter Annual Meeting of the American Society of Mechanical Engineers*, 32, 49–60, 0791808637, Atlanta, December 1991, ASME, New York.

Pong, K. C.; Ho, C. M.; Liu, J. & Tai, Y. C. (1994). Nonlinear pressure distribution in uniform microchannels, *Proceedings of the 1994 International Mechanical Engineering Congress and Exposition*, 197, pp. 51–56, Chicago, November 1994, ASME, New York.

Prandtl, L. (1904) On fluid motions with very small friction (in German), *Proceedings of the 3rd International Mathematical Congress*, pp. 484-491, Heidelberg, Germany.

Schaaf, S. A. & Chambre, P. L. (1961). *Flow of rarefied gases*, Princeton University Press, New Jersey.

Schley, P.; Jaeschke, M.; Kuchenmeister, C. & Vogel, E. (2004). Viscosity measurements and predictions for natural gas.

Sonntag, R. E.; Borgnakke, C. & Wylen, G. J., (1998). *Fundamentals of thermodynamics*, John Wiley & Sons, 047118361X, New York, 5th edition.

Tsuboi, N. & Matsumoto, Y. (2005). Experimental and numerical study of hypersonic rarefied gas flow over flat plates. *AIAA Journal*, 43, 1243–1255.

Vennix, A. J.; Leland Jr., T. W. & Kobayashi, R. (1970). Low-temperature volumetric properties of methane. *J. Chem. Eng. Data*, 15, 238–243.

Viswanathan, A. (2007). *Viscosities of natural gases at high pressures and high temperatures*, MS thesis, Texas A&M University, College. Station, Texas.

Weng, H. C. & Chen, C.-K. (2008a). A challenge in Navier–Stokes-based continuum modeling: Maxwell–Burnett slip law. *Phys. Fluids*, 20, 106101.

Weng, H. C. & Chen, C.-K. (2008b). Variable physical properties in natural convective gas microflow. *ASME J. Heat Transf.*, 130, 082401.

Weng, H. C. & Chen, C.-K. (2008c). Fully developed thermocreep-driven gas microflow. *Appl. Phys. Lett.*, 92, 094105.

Younglove, B. A. (1974). The specific heats cp and cv of compressed and liquified methane. *J. Res. Natl. Bur. Stand.*, 78A, 401-410.

23

Consequence analysis of large-scale liquefied natural gas spills on water

Hideyuki Oka
National Maritime Research Institute
Japan

1. Introduction

1.1 Background

As public concerns increase over global warming caused by the burning of fossil fuels, natural gas is gaining a lot of attention for the lowest emission of carbon dioxide among the fossil fuels. Thus, governments implementing national or regional plans to reduce greenhouse gas emissions may encourage its use to displace other fossil fuels. According to the Energy Information Administration, the worldwide natural gas consumption in 2030 will increase by about one and a half times as much as in 2006 (EIA, 2009), so that the number and frequency of seaborne transportation of liquefied natural gas (LNG) are expected to increase significantly around the world. In fact, there are a lot of projects to build new receiving terminals in the United States. Also, natural gas consumption is expected to rise rapidly in China and India. With such a growing global demand, recent LNG carriers (LNGCs) become larger up to a 266,000 m^3 cargo capacity, which are referred to as Q-Max vessels.

Due to the above change in the situation, there has recently been considerable interest concerning possible risks involved in the LNG carrier operations, though seaborne transportation of LNG has been conducted with a very good safety record since 1959. Hence, public authorities have raised their awareness of concern about the possibility of large-scale LNG spill hazards caused by accidental events or intentional attacks. As a result, a number of consequence analyses have been carried out in recent years in order to propose models and approaches or to assess hazards resulting from an unconfined LNG spill over water (Luketa-Hanlin, 2006). However, these studies showed a broad range of results due to their differences in models, approaches and assumptions, since the physics involved in such LNG spills and related phenomena is very complicated. In addition, because of the lack of experimental data for a large-scale LNG spill and subsequent combustion and/or dispersion events, there are many theoretical and experimental gaps related to understanding of the dynamics and limitations in predicting the associated hazards. Therefore, consequence assessment methods based on a combination of theoretical formulations and empirical relationships derived from laboratory and small-scale field experiments are the only practical measure to predict the hazards associated with large-scale LNG spills on water.

On the other hand, a broad range of results of these studies indicates how important it is to use appropriate assumptions, data, and models in trying to make an accurate assessment of hazards from an LNG spill. Although the results of recent consequence studies were compared in a few publications (Hightower et al., 2004), there was no comparison study on consequence models under the same scenarios in terms of LNG properties, release assumptions and weather conditions. Therefore, the current author compared and evaluated consequence models for pool fire hazards involving an LNG spill on water in order to clarify their characteristics (Oka & Ota, 2008).

In the above comparison study, attention was paid to thermal radiation hazards from pool fires, because there is a high possibility that an ignition source immediately after breaching a tank will be available (Hightower et al., 2004). Hence, the sensitivity analysis of a spill and the subsequent pool fire hazards to the hole size breached in a membrane-type tank of a conventional size LNGC (125,000 m^3 cargo capacity) were carried out using three major consequence assessment methods developed by the Federal Energy Regulatory Commission (FERC) (FERC, 2004), Sandia National Laboratories (SNL) (Hightower et al., 2004) and Fay (Fay, 2003). These methods were chosen based on an in-depth review of the recent literature available to the public. Through the sensitivity analysis, it was found that the FERC method was most appropriate for practical consequence analyses of incidents involving large-scale LNG spills on water from the practical viewpoint of applicability to any breach size.

Recent LNGCs are designed to have as much as a 266,000 m^3 cargo capacity, so that it is important to evaluate how much the extent of the hazard impact would increase due to the enlarged size and capacity of such carriers. Thus, thermal radiation hazards from pool fires involving spills from one of the latest and largest LNGCs (250,000m^3 cargo capacity) were assessed using the recommended FERC method, and the results were discussed in comparison with those for the conventional size LNGC. As a result, it was found that the maximum thermal hazard distance was only about 24 % longer than that for the conventional LNGC, while the spill volume was twice as much (Oka & Ota, 2008).

When the author focused on estimating LNG spill hazards from the latest LNGC, similar hazard assessments had not been covered at least in the publicly available literature. However, in almost the same period the U.S. Department of Energy requested that SNL conduct analyses of possible spill hazards from a breach of the latest LNGC (Luketa et al., 2008). The results of both studies were published at the same time. This updated SNL study presented somewhat different results in that the thermal hazard distances increased by approximately 7-8 % due to the increase in hydrostatic head and tank volume for the new, larger LNGC. In the scenarios used in the SNL studies (Luketa et al., 2008), the nominal breach size and the total spill volume from a single tank were determined as 5 m^2 and 41,000 m^3, respectively, so that a smaller breach size and a larger spill volume were used than those in the other study (Oka & Ota, 2008). Hence, for quantitative comparison, the current author carried out consequence analyses of pool fire hazards following an LNG spill from a breached tank of the conventional and latest LNGCs under the same scenarios as in the SNL studies (Oka, 2009). It was found that, as a whole, the thermal hazard consequences by the SNL method were in fairly good agreement with those by the FERC method.

1.2 Scope of the present study

The principal LNG hazards of interest for the present study are those posed by thermal radiation and flammable vapor dispersion resulting from unconfined LNG spills on water.

Cryogenic burns and asphyxiation are typically localized to LNG transport and storage areas, so that such secondary hazards are outside the scope of this study.

The two previous studies for the latest LNGC by the FERC method (Oka & Ota, 2008; Oka, 2009) were carried out under the following scenarios. In the first study (Oka & Ota, 2008), predicted consequences were compared only when the hole diameters were 1, 3 and 5 m as typical hole sizes, which were chosen from the recent literature on the assessment of the impacts of large-scale release from the conventional type LNGC. In the second study (Oka, 2009), two breach sizes of 5 and 12 m^2 were used as nominal tank breaches for near-shore and offshore LNG marine import operations, respectively, so as to compare the updated SNL study (Luketa et al., 2008). Therefore, no sensitivity analysis of pool fire hazards to the hole size has been carried out for the latest LNGC.

As for flammable vapor dispersion hazards, as far as the author knows, there is no study to assess consequences predicted by the FERC method for the latest LNGC. Though the sensitivity analysis of spills and the subsequent dispersion hazards to the breach size were conducted for the conventional size LNGC using the FERC method (Qiao et al., 2006), the averaging time used to estimate flammable gas concentrations was much larger than the recommended value in the FERC method. Thus, it is interesting to evaluate the sensitivity using the FERC method composed of all the recommended models and assumptions for the latest LNGC.

The present work considers the sensitivity of the flammable vapor and thermal radiation hazards to the hole diameter under release scenarios that a hole can develop just above the waterline level in the event of a breach of a single tank on the conventional and latest LNGCs. Under current circumstances, from the practical viewpoint of applicability to any breach size, the FERC method has been recommended in the previous studies (Oka & Ota, 2008; Oka, 2009), so that the present consequence analyses are carried out using the same method.

2. Overview of potential consequences

Currently, the potential for the dynamics and dispersion of a large spill and the associated hazards are not fully understood. As will be shown in Fig. 1 later, existing experimental data on LNG spill dynamics, dispersion, and burning over water cover only small amount of spill volumes that are two to three orders of magnitude less than those postulated in the recent literature (Luketa-Hanlin, 2006).

2.1 Brief description on major hazards of an LNG spill on water

The potential hazards associated with LNG spills include cryogenic damage caused by direct contact, pressure increase due to rapid phase transition (RPT), flash fires, pool fires, deflagrations and detonations. Because of its extremely low temperature, direct contact with LNG will result in brittle fracture of the ship's structure, which may cause cascading damage to additional LNG tanks. When LNG comes in contact with water at a significantly higher temperature than the boiling point of LNG, there is the possibility of RPT, which is a nearly instantaneous transition from the liquid to vapor phases and produces an associated rapid pressure increase. The impacts of RPT will be localized near the spill source and should not cause extensive structural damage.

LNG is comprised mostly of methane, so that LNG vapor is flammable in air approximately at 5 to 15 % by volume. At a 5 % concentration of gas in air, LNG vapor is at its lower flammability limit (LFL). Below the LFL, the cloud is too dilute for ignition. At a 15 % concentration of gas in air, LNG vapor is at its upper flammability limit (UFL), so that the cloud is too rich in LNG for ignition above the UFL.

The evaporating natural gas in the above range of combustible gas-air concentrations will burn above the LNG pool when it ignites immediately after LNG release. The resulting pool fire would spread as the LNG pool expands away from its source and continues evaporating. If released LNG does not ignite immediately, the LNG will form a vapor cloud that may drift some distance from the spill site at roughly the wind speed. Once it warms above approximately -108 °C, LNG vapor will become less dense than air and tend to rise and disperse more rapidly. However, LNG vapor at its normal boiling point -162 °C is 1.5 times denser than air at 25 °C. Typically, LNG vapor released into the atmosphere will remain negatively buoyant until after it disperses below its LFL. Therefore, the displacement of air by LNG vapor may cause asphyxiation as well as lung damage from breathing the cold vapor.

In the case of delayed ignition at downwind locations to which the spill vapor might spread, a flash fire will occur. This is a short duration fire that burns the vapor already mixed with air to flammable concentrations. The flame front may burn back through the vapor cloud to the spill site, resulting in a pool fire. A flash fire will burn slowly and is unlikely to generate damaging overpressures when it occurs in an unconfined space.

Explosions arising from combustion of flammable fuel-air mixtures are classified as either a detonation or a deflagration. Detonations generate very high overpressures, and hence are more damaging than deflagrations. It is pointed out that weak ignition of natural gas vapor in an unconfined and unobstructed environment is highly unlikely to result in deflagration-to-detonation transition (DDT). This transition is more likely in an environment with confinement such as with closely spaced obstacles, so that damaging overpressures could result from explosions in a confined space in cases that the flammable vapor leaks into a confined space inside LNGCs or other congested structures and then ignites.

2.2 Review of experiments on large-scale LNG spills

This subsection briefly reviews experiments on the vapor dispersion, pool fire and vapor cloud fire which are formed from unconfined LNG spills onto water. In reference to recent review papers (Luketa-Hanlin, 2006; Koopman & Ermak, 2007; Raj, 2007), only the largest spill volume tests are outlined chronologically in the following.

In 1973, the Esso Research and Engineering Company and the American Petroleum Institute carried out LNG dispersion tests in Matagorda Bay, Texas (Feldbauer et al., 1972). Volumes ranging from 0.73 to 10.2 m^3 were spilled. Pool radii ranging from 7 to 14 m were visually observed, and visible vapor clouds were very low in height compared to their lateral extent. In 1978, the U.S. Coast Guard China Lake tests (Schneider, 1980) were performed at the Naval Weapon Center (NWC) in China Lake, California in order to measure the thermal radiation output of pool fires as well as vapor cloud fires. The volumes of LNG ranged from 3 to 5.7 m^3 were released towards the middle of an unconfined water surface of a pond. The effective pool diameter was up to 15 m, and the flame lengths ranged from 25 to 55 m. In 1980, Shell Research conducted a series of experiments at Maplin Sands in England to obtain dispersion and thermal radiation data for 20 spills of 5 to 20 m^3 of LNG on the surface of the sea (Blackmore et al., 1982; Mizner & Eyre, 1983). An effective pool diameter of 30 m was calculated by approximating the flame base area as an ellipse. A pool fire was formed in one test, but it continued only for a few seconds before the fuel was consumed. Therefore, a fully developed pool fire was not achieved. At the same time as the Maplin Sands tests, the Burro series tests were conducted independently by the Lawrence Livermore National Laboratory (LLNL) at NWC (Koopman et al., 1982). The main objective of the Burro series was to obtain extensive data on LNG vapor dispersion under a variety of meteorological conditions. A total of eight LNG release onto water were performed with spill volumes ranging from 24 to 39 m^3. The pool radius measured in the tests was up to 5 m. The Coyote series tests (Rodean et al., 1984) followed the Burro series in 1981 so as to measure the characteristics of large vapor cloud fires and obtain more dispersion data from LNG spills ranging from 14.6 to 28 m^3 onto water. It was observed that the flame propagated toward the spill source and subsequently a pool fire occurred. However, measurements were not taken in the experiment of the flame propagation. After the Burro and Coyote series, the Falcon series tests (Brown et al., 1990) were conducted by LLNL in 1987. The main goal of the experiments were to provide a database on LNG vapor dispersion from spills in an environment with obstacles and to assess the effectiveness of vapor fences for mitigating dispersion hazards. The Falcon tests have been the largest spills so far, with release rates up to 30 m^3/min and spill volumes ranging from 21 to 66 m^3.

Figure 1 shows a comparison of the spill sizes tested to date with possible spill volume from a single LNG cargo tank through a hole just above the waterline level. It can be seen from this figure that the experimental tests were performed on considerably smaller scales compared with an LNGC tank size. In other words, there is a large disparity between the available experimental data and the scales of interest for consequence assessments, so that there are gaps and limitations in understanding and predicting the hazards associated with large-scale spills from a cargo tank. Therefore, a lot of consequence assessment methods for practical use can provide only rough estimates of the magnitude of effects for incidents involving large LNG release on water.

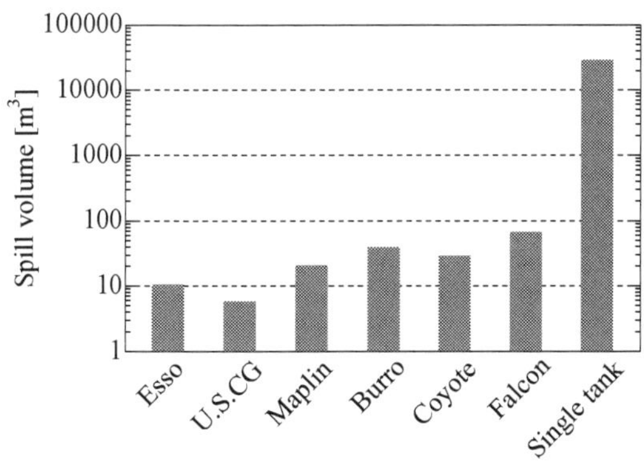

Fig. 1. Logarithmic scale comparison of spill volume level in field experiments with that of a possible cargo spill from a single tank of the latest LNGC.

3. Consequence assessment methods

In almost all of the studies on consequence modeling of LNG spill hazards, it is assumed that the reference LNGCs have membrane tanks. Qiao et al. investigated the influence of the geometric difference between membrane and Moss spherical tanks on the LNG release rate from a hole, but they did not carry out consequence analyses under the condition that LNG was released from a Moss spherical tank (Qiao et al., 2006). Hence, a membrane type LNGC is adopted as a reference vessel in accordance with the majority of studies. For the purpose of consequence assessment modeling, the geometry of a membrane tank is much simplified to a rectangular box, as shown in Fig. 2. Though an LNGC has a complete double hull in reality, a single hull structure is assumed on the side of the reference LNGC. The reason of this assumption will be described later.

The consequence analyses of LNG spill hazards are conducted in the following steps:

1. Calculate the LNG release rate from a non-pressurized tank with a single hole,
2. Calculate the diameter of the volatile liquid pool spreading on water,
3. In the scenario of immediate ignition, calculate the size of a pool fire and distances to specified radiative flux levels of concern. Otherwise, skip to the next step,
4. In the case of delayed or remote ignition, calculate downwind dispersion distances to specified concentration levels of concern.

Consequence models in each step, which constitute a consequence assessment method, are described in the following subsections.

3.1 LNG release from a cargo tank of a ship

In the absence of appropriate models that account for the complex structure of an LNGC and the physics of release of cryogenic LNG, a simple orifice model is employed in the FERC method on the assumption of a single hull structure of an LNGC. In spite of the complete double hull structure in reality, the orifice model is widely used even in the recent literature on consequence assessment (Luketa-Hanlin, 2006). Since this model assumes release from a single hole on the side of a ship with single hull structure, LNG flows directly from a tank onto the seawater without any leakage into the space between hulls.

The release rate from the tank to the seawater is expressed as a function of height through invoking Bernoulli's equation. Multiplied by a discharge coefficient, the mass flow rate is expressed as follows:

$$\dot{M} = C_d \rho_l \pi R^2 \sqrt{2gh} \,, \tag{1}$$

where $\dot{M}$ is the mass flow rate, C_d is the discharge coefficient to take account of the resistance given by the hole, ρ_l is the LNG density, R is the effective radius of the hull breach, h is the static head above the hull breach, g is the acceleration due to gravity.

Discharge coefficient C_d is often used to account for reduction below the theoretical exit velocity due to viscosity and secondary flow effects. In other words, it depends on the nozzle shape and the Reynolds number. In the case of an ideal frictionless discharge, it is reasonable that C_d is set to unity. In practice, however, a rough, irregular breach could occur in the wall of an LNG cargo tank, so that the friction would be expected to be larger than that in the case of a well-rounded, sharp-edged orifice. Thus, FERC recommended 0.65 as a reasonable estimate to account for the fact that friction retards the flow (FERC, 2004).

The orifice model does not attempt to account for the multi-hull construction of LNGCs, and therefore may overestimate the rate at which LNG would escape through a hole. Hence, the results should be interpreted as a rough guide to the rate of release for a given hole size.

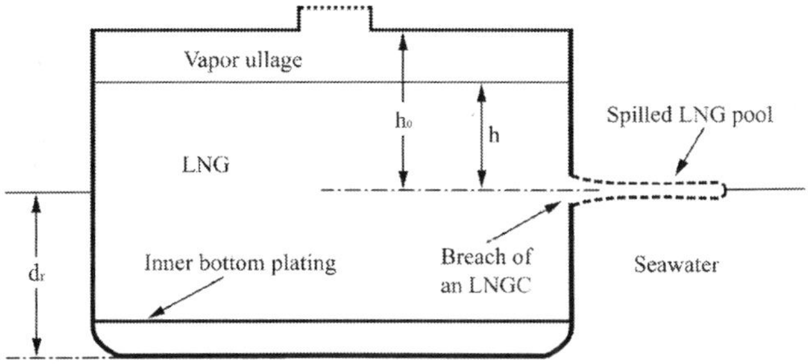

Fig. 2. Schematic view of a cross-section of an LNGC with a hole breached on the side. The amount of LNG just above the waterline is released through the hole over the seawater (Oka & Ota, 2008).

3.2 Spread of an unconfined, evaporating LNG pool on water

LNG spilled on water forms a floating pool because its density is roughly half that of water. This pool will spread over unconfined water, and will vaporize simultaneously due to the high heat transfer from the water and/or other sources. The ABSG study (ABSG, 2004) recommended the use of Webber's model (van den Bosch, 1997) since it has a sound theoretical basis and accounts for friction effects. This model is based on self-similar solutions of the shallow water equations and lubrication theory. In this formulation, resistance by turbulent or laminar friction effects is included in the pool spread equation as follows:

$$\frac{d^2 r}{dt^2} = \frac{4\Phi g_r \delta}{r} - C_F, \tag{2}$$

where r is the pool radius, t is the time, δ is the mean depth of the LNG pool, Φ is the dimensionless shape factor that describes the pool thickness profile, and C_F is the turbulent or viscous resistance term. The reduced acceleration due to gravity g_r is defined as follows:

$$g_r = \frac{\rho_w - \rho_l}{\rho_w} g, \tag{3}$$

where ρ_w is the seawater density. In order to close Eq. (2), Webber also provided theoretical and empirical models to determine δ, Φ and C_F (van den Bosch, 1997).

Next, film boiling effects on the above spreading model is briefly described. As an LNG pool spreads on a water surface, the heat transferred from the water and other sources will cause the liquid to vaporize. In the vapor dispersion scenario, vaporization is mainly controlled by heat transfer from the water to the LNG pool. The FERC recommended a film boiling heat flux of 85 [kW/m²] as a reasonable value, which was obtained in the Burro series tests (Koopman et al., 1982). In the pool fire scenario, vaporization is controlled by heat transfer from both water and fire to the LNG pool. The FERC recommended a mass burning rate per unit area $\dot{m}_b$ as 0.282 [kg/m²/s]. The film boiling and mass burning rates per unit area of the LNG pool are regarded as constant, but the total mass removal rate is dynamically linked to the spreading rate through the pool area.

In the present spread model of an evaporating pool, the physical effects of winds, waves, and currents are not taken into consideration. Several attempts to quantify some of these effects have been made in a few studies (Cornwell & Johnson, 2004; Spaulding et al., 2007), but it is difficult to validate them due to the lack of experimental data. On the other hand, Fay recently showed that the effects of ocean wave interaction on the spread of an evaporating LNG pool were only small or negligible in his classical and newly proposed models (Fay, 2007).

3.3 Thermal radiation from pool fires on water

LNG is known as a clean burning fuel, but significant smoke production is expected for large LNG pool fires (Luketa-Hanlin, 2006). This will tend to obscure the flame and reduce the thermal radiation emitted from the fire. Therefore, the FERC recommends the use of the two-zone solid flame model (Rew & Hulbert, 1996) for assessing the thermal hazards from pool fires. This model assumes that the flame is divided into lower and upper zones. Smoke does not obscure the flame in the lower zone, while it obscures the flame and reduces the amount of thermal radiation emitted from the upper zone. To determine the flame geometry, this model assumes that the flame is a solid, gray emitter having a regular well-defined shape such as an upright or tilted cylinder. The radiative heat flux upon an object can be determined by

$$q = \tau E F, \tag{4}$$

where τ is the atmospheric transmissivity, E is the surface emissive power, and F is the geometric view factor between the target and the cylindrical flame. The view factor F is determined from the dimension of flame area, which is characterized by the flame base diameter, visible flame height, and flame tilt. The flame base is equivalent to the pool size calculated by the pool spread model.

The flame height depends on the flame base diameter and the burning rate, and their correlation was developed by Thomas (Beyler, 2002) as follows:

$$\frac{H}{D} = 55\left(\frac{\dot{m}_b}{\rho_a\sqrt{gD}}\right)^{0.67} \cdot \left(u^*\right)^{-0.21}, \tag{5}$$

where H is the mean visible height of turbulent diffusion flames, D is the effective diameter of a pool, ρ_a is the ambient air density. The FERC method takes the effect of winds into consideration, so that the nondimensional wind speed u^* is determined by

$$u^* = \frac{u_w}{\left(g\dot{m}_b D/\rho_v\right)^{1/3}}, \tag{6}$$

where u_w is the wind speed measured at a height of 1.6 m, and ρ_v is the vapor density. However, u^* is assigned a value of unity if it is less than 1.

3.4 Vapor dispersion of LNG spills on water

When considering large release of LNG, dense-gas effects are important and must be taken into consideration in a dispersion model used for analysis. In the FERC method, the DEGADIS model (Spicer & Havens, 1987) was recommended for use in estimating the distances that flammable vapor might reach. DEGADIS accounts for dense-gas effects and was originally developed for the simulation of cryogenic flammable gas dispersion, particularly for LNG. The DEGADIS model are widely used in the public and private sectors due to the convenience of fast computational run time and ease of use. It has been validated against a wide range of laboratory and field test data. Furthermore, the federal siting requirements for onshore LNG facilities (CFR, 1980) specify the use of DEGADIS for the determination of dispersion distances.

DEGADIS is one of one-dimensional integral models which use similarity profiles that assume a specific shape for the crosswind profile of concentration and other properties. The similarity forms represent the plume as being composed of a horizontally homogeneous section with Gaussian concentration profile edges as follows:

$$c(x,y,z) = \begin{cases} c_c(x)\exp\left[-\left\{\frac{|y|-b(x)}{S_y(x)}\right\}^2 - \left\{\frac{z}{S_z(x)}\right\}^{1+\alpha}\right] & \text{for } |y| > b(x), \\ c_c(x)\exp\left[-\left\{\frac{z}{S_z(x)}\right\}^{1+\alpha}\right] & \text{for } |y| \le b(x), \end{cases} \tag{7}$$

where c is the concentration, c_c is the centerline, ground-level concentration, b is the half width of a horizontally homogeneous central section of gas plume, and S_y and S_z are the horizontal and vertical concentration scaling parameters, respectively. The downwind variations of spatially averaged, crosswind values are determined by using the conservation equations only in the downwind direction of x. Wind velocity u_x is assumed to be based on a power law profile as follows:

$$u_x = u_0\left(\frac{z}{z_0}\right)^{\alpha}, \tag{8}$$

where u_0 is the wind speed measured at $z = z_0$, and z_0 is the reference height in wind velocity profile specification. The power coefficient α in Eqs. (7) and (8) is a function of atmospheric stability and surface roughness. In DEGADIS, it is determined by a weighted least-squares fit of the logarithmic profile of wind speed.

Transient denser-than-air gas release cannot be represented as steady, continuous release, so that the spill is modelled as a series of pseudo-steady-state release in DEGADIS. It should be noted that the application of DEGADIS is limited to the description of atmospheric dispersion of denser-than-air gas release at ground level onto flat, unobstructed terrain or water. In other words, the weakness is that it cannot model the flow around obstacles or over complex terrain.

3.5 Summary of Consequence assessment methods

The consequence models for LNG release from a tank, volatile pool spread, thermal radiation from a pool fire, and denser-than-air gas dispersion have been briefly described in the previous subsections. These constitutive submodels in the FERC method are summarized in Table 1. In the pool spread process, its shape is assumed to be semi-circular because of the existence of a ship (Fay, 2003; FERC, 2004). The vaporization due to heat transfer from the fire and/or the water to the pool is taken into consideration, but environmental effects of waves, currents and winds are not incorporated into the spread model.

In general, since many of constitutive submodels for practical use, such as those in the FERC method, have limitations that can cause greater uncertainty in calculating release, spread, and subsequent hazards, these methods can provide only rough estimates of the magnitude of effects for incidents involving large LNG releases on water. The more detailed models based on computational fluid dynamics (CFD) techniques can be applied to improve analysis of site-specific hazards and consequences in higher hazard zones. In the vapor dispersion process, for example, it is important to appropriately represent the topography downwind of the release point so as to obtain precise estimates of effects in actual incident circumstances. However, CFD models have also their own limitations, and its further refinement is required to improve the degree of accuracy and reliability for consequence assessment modeling (Hightower et al., 2005). In addition, due to high computational costs, CFD models are not normally used for practical hazard assessment under the present circumstances.

LNG RELEASE	POOL SPREAD		VAPORI-ZATION	POOL FIRE		VAPOR DISPERSION	
Discharge coefficient	Time evolution	Friction effects included	Vaporization rate [kg/m²/s]	Flame model	Surface emissive power [kW/m²]	obstacles or terrain effects included	Averaging time
0.65	Unsteady	Yes	0.282 (Pool fire) 0.17 (Dispersion)	Two-zone solid cylinder that includes tilt for wind effects	265	No	Not more than a few seconds

Table 1. Summary of principal features of the FERC method

3.6 Consequence analysis conditions for LNG spill hazards

Large-scale LNG spill hazard scenarios (Oka & Ota, 2008) are shown in Table 2. These assumptions were originally employed in the ABSG study (ABSG, 2004; FERC, 2004) for the conventional size LNGC except for the total spill volume and the breach size. In the ABSG study, only two holes of 1 and 5 m in diameter were chosen to provide calculation examples of pool fire and vapor dispersion scenarios. In the present study, sensitivity to the breach size is analyzed in the range from 0.5 to 15 m in diameter. Unlike the ABSG study, the spill volume is determined based on Fay's study (Fay, 2003). He simplified the geometry of a membrane tank to a rectangular box and estimated the volume of the spilled LNG as follows. If d_r is the fully-loaded draft, the initial height h_0 (see Fig. 2) of the upper surface of LNG above the waterline level is about $1.1d_r$ for the conventional LNGC. The cargo surface area A_t is related to the cargo tank volume V_{ct} by the following equation,

$$A_t \simeq 0.52 \cdot \frac{V_{ct}}{d_r}. \tag{9}$$

For an LNGC of a 125,000 m^3 cargo capacity, with an 11.8 m draft and a 25,000 m^3 cargo tank volume, the initial height h_0 and the cargo surface area A_t are estimated to be 13 m and 1,100 m^2, respectively. Thus, the volume of the spilled LNG from the tank is given as $h_0 A_t$=14,300 m^3. Meanwhile, the height from the inner bottom plating to the load water line is easily derived from Eq. (9) as $0.82 d_r$, so that the depth of a double bottom is $0.18 d_r \simeq$ 2.1 m. This is a typical value for membrane type LNGCs. Therefore, Eq. (9) can be considered as a reasonable expression to easily estimate the typical dimensions of a membrane tank. Hence, the total spill volume for the latest LNGC is also determined in the same manner.

LNG carrier	Conventional	Latest
LNG properties:		
LNG composition	Methane	
LNG density	422.5 kg/m^3	
Release assumptions:		
Total cargo capacity	125,000 m^3	250,000 m^3
Volume of a cargo tank	25,000 m^3	50,000 m^3
Total spill volume	14,300 m^3	28,600 m^3
Initial LNG height above breach	13 m	13.2 m
Breach size	0.5 to 15 m in diameter	
Breach location	Just above the waterline	
Pool shape	Semi-circle	

Table 2. Release scenarios for an LNG spill from a tank of the conventional and latest LNGCs

Weather conditions at the time of the release have a major influence on the extent of dispersion. Thus, environmental conditions for the above spill hazard scenarios are provided in Table 3. These conditions were also used in the ABSG study (ABSG, 2004; FERC, 2004). In the vapor dispersion scenario, a wind speed of 2.0 m/s at 10 m above ground and an F stability class were used for an atmospheric stability condition. The F class is extremely stable and the atmospheric turbulence is very weak, so that it takes the greatest amount of time for the released gases to mix with the atmosphere. In other words, such low wind speed and stable atmospheric condition result in the greatest downwind distance to the LFL. In general, for a lot of one-dimensional integral models, topography is characterized by the surface roughness value. Since the surface roughness accounts for the effects of terrain on the vapor dispersion, a rougher surface will tend to cause more mixing with ambient air, which results in more rapid dispersion of a vapor cloud. As for the averaging time of gas concentration, the FERC method recommended that a short averaging time (not more than a few seconds) be used because a flammable cloud need only be within the flammable range for a very short time to be ignited. In the ABSG scenario (ABSG, 2004; FERC, 2004), its averaging time was set to 0 second, that is, a peak concentration was used.

Hazards	Pool fire	Vapor dispersion
Air temperature	300 K	295 K
Water temperature	294 K	294 K
Relative humidity	70 %	50 %
Wind speed	8.9 m/s	2.0 m/s
Pasquill stability class	-	F
Surface roughness	-	0.01 m

Table 3. Environmental conditions for the scenarios of pool fire and vapor dispersion hazards

4. Results and discussion

This work considers thermal radiation and flammable vapor hazards caused by unconfined LNG spills on water resulting from an LNG cargo release. The recommended FERC method is used to analyze the sensitivity of the LNG hazard consequences to the breach diameter in the following subsections. Based on the physical models and numerical algorithms of the FERC method, a computer program written in the Fortran 90 programming language was developed, except for the vapor dispersion model. The results calculated using this program was carefully checked against those of consequence assessment examples in the ABSG study (FERC, 2004) to verify and validate the program. Unlike this study, the computations presented in the ABSG study were performed with the assistance of the Mathcad computer software.

4.1 LNG release process

Figures 3 and 4 show the influence of the breach diameter on the time taken to empty a tank above the waterline level and the time to vaporize all of the LNG released on water under the pool fire scenario and under the vapor dispersion scenario, respectively. In other words, the former time corresponds to total spill duration in both scenarios. The latter can be referred to as total fire duration in the pool fire scenario and as total evaporation duration in the vapor dispersion scenario.

The orifice model is used to calculate LNG release rate from a tank. Integrating Eq. (1) with the initial condition, $h = h_0$ at $t = 0$, one can easily obtain the analytical expression of the spill duration t_s as follows:

$$t_s = \sqrt{\frac{32h_0}{g}} \frac{A_t}{C_d \pi} d^{-2}, \tag{10}$$

where d is the hole diameter, and h_0 and A_t depend on the size and capacity of an LNGC. Hence, as shown in Figs. 3 and 4, the total spill duration is depicted as a linear function of the breach diameter with a slope of -2 on a double logarithmic graph. On the other hand, the total duration of fire and that of evaporation can be obtained as a solution of the pool spread model. As for the total fire duration, it is equal to the total spill duration when the breach diameters are less than 2 and 3 m in Figs. 3(a) and 3(b), respectively. With the increase in the breach diameter, however, the curve representing the fire duration begins to deviate from the straight line representing the spill duration. The total spill duration is much shorter than the total fire duration when the breach diameters are larger than about 5 and 6 m for the conventional and latest LNGCs, respectively.

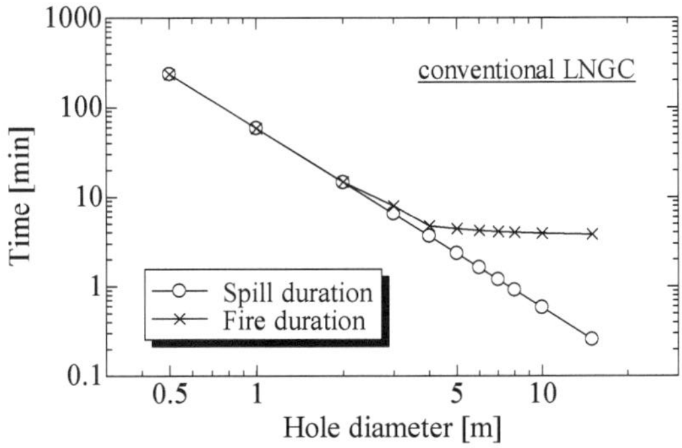

(a)

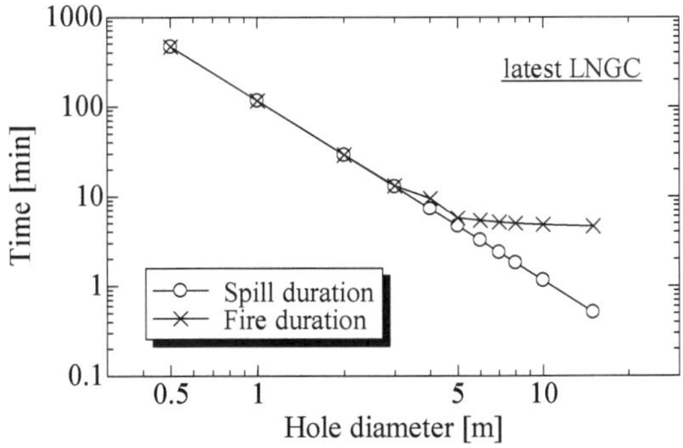

(b)

Fig. 3. Effect of breach diameter on the total duration of spill and that of fire under the pool fire scenario: (a) the conventional LNGC; (b) the latest LNGC.

From these findings, an LNG spill can be characterized as either a long duration release or a large-scale release of short duration, depending upon the breach diameter. In general, the former is referred to as a continuous spill, and the latter as an instantaneous spill. The instantaneous spill in the literal sense is unlikely to occur, and represents an ideal limiting case. In reality, it represents a large-scale spill for a short time. Under the present pool fire scenario, a release can be classified into the instantaneous spill type when the breach diameters are greater than about 5 and 6 m for the conventional and latest LNGCs, respectively. In the same manner, when the breach diameters are less than 2 and 3 m, respectively, it can be classified into the continuous spill type. Any release from a breach whose diameter lies between these two ranges is considered to be in transition from the continuous spill type to the instantaneous spill type.

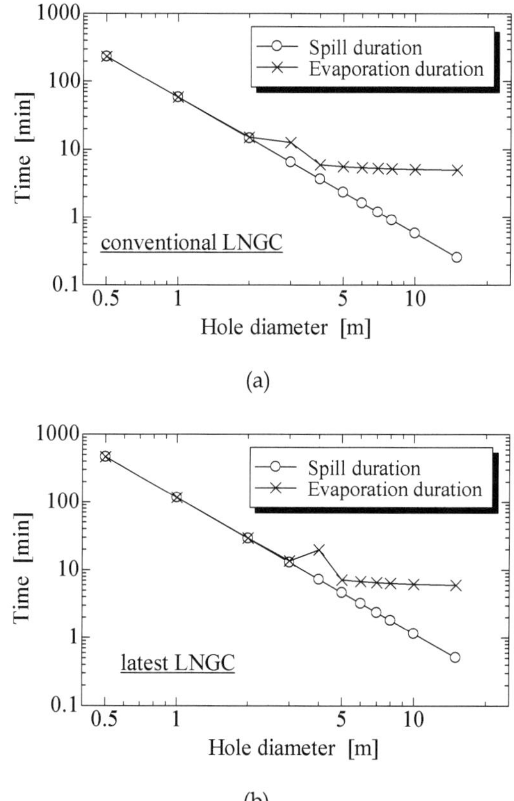

Fig. 4. Effect of breach diameter on the total duration of spill and that of evaporation under the vapor dispersion scenario: (a) the conventional size LNGC; (b) the latest LNGC.

On the Whole, the above discussion holds true for the total evaporation duration under the vapor dispersion scenario in Fig. 4. Unlike in Fig. 3, however, the curve representing the evaporation duration is markedly out of alignment in the transitional range between the continuous and instantaneous spill types. This is attributed to the difference of the vaporization rates of an LNG pool, i.e., the difference between the film boiling rate and the mass burning rate. The pool spread model recommended in the FERC method is based on an integral approach that can avoid the need to characterize a spill type as either instantaneous or continuous. However, the lack of smoothness of these duration data suggests that it is necessary to improve the pool spread model in view of the transitional spill type range.

4.2 Pool spread process

Figures 5(a) and 5(b) show the sensitivity of the maximum pool radius to the breach size under the pool fire and vapor dispersion scenarios, respectively. In both scenarios the LNG pool radius increases with the increase in the breach diameter. Then, it reaches an asymptotic value when the breach diameters are greater than about 5 and 6 m for the conventional and latest LNGCs, respectively.

In the pool fire scenario, it can be seen from Figs. 3 and 5(a) that the maximum pool size is independent of the hole size in the instantaneous spill range. The asymptotic value for the latest LNGC is approximately 430 m, and is about 30 % longer than that for the conventional size. On the other hand, the maximum pool radius increases almost linearly in the continuous spill range. In particular, when the breach diameter is less than about 2 m, there is no significant difference of the maximum pool radius between the conventional and latest LNGCs. In this hole diameter range, the maximum pool radius can be approximately estimated on the assumption that the vaporization rate matches the release rate from a tank for most of the total spill duration.

The above discussion holds true for the results in the vapor dispersion scenario, except for the asymptotic value of a maximum pool radius. As shown in Fig. 5(b), it is approximately 480 m for the latest LNGC, and is longer than that in the case of the pool fire scenario because the vaporization rate is lower than the mass burning rate. Similarly to the pool fire case, the asymptotic value of the maximum pool radius for the latest type LNGC increases by 30 % as compared to the conventional type, whereas the spill volume doubles. The reason for this is as follows: The LNG release rates calculated by the orifice model are proportional to $\sqrt{h}$ as shown in Eq. (1) and the initial height h_0 has almost the same value for the two size LNGCs, so that the release rates from the latest type LNGC are almost equal to those from the conventional type at the initial stage of a spill. Therefore, the maximum pool radius does not expand significantly even if the cargo capacity becomes twice as large (Oka & Ota, 2008).

In the present pool spread model, it is assumed that a single, semi-circular pool can be formed on water. In fact, however, the shape and size of the pool could be affected by environmental conditions, such as wind, waves and currents. Therefore, it may be more likely that the waves would break up a single pool into multiple irregular-shaped pools.

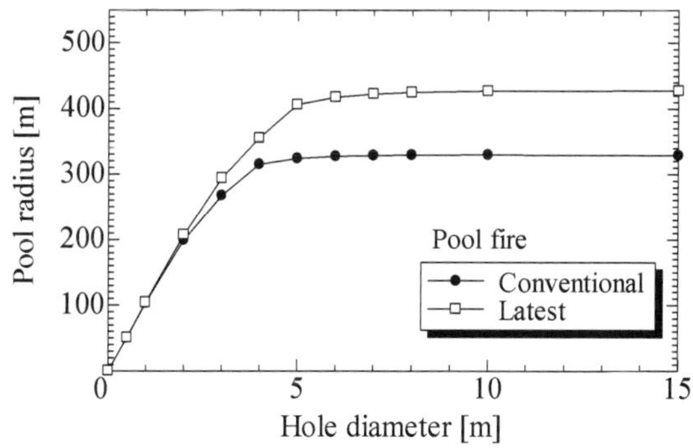

(a)

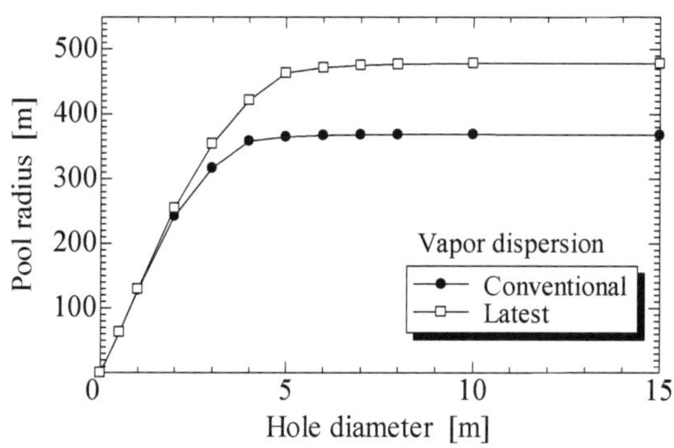

(b)

Fig. 5. Sensitivity of the maximum pool radius to the breach diameter under (a) the pool fire scenario and (b) the vapor dispersion scenario. The results for the conventional and latest LNGCs are compared in each scenario.

4.3 Pool fire process

For the conventional size LNGC, the sensitivity of the thermal radiation hazard distance to the breach diameter has already been investigated (Oka & Ota, 2008), but not for the latest one. Thus, the distances to 5 kW/m^2 are compared in Fig. 6 as a function of the hole diameter.

This intensity level is specified as a level of concern by the United States Federal Safety Standards for Liquefied Natural Gas Facilities (CFR, 1980). According to the Federal Safety Standards, the heat flux of 5 kW/m^2 is an acceptable level of concern for direct exposure of human beings. For bare skin exposure, a heat flux at this intensity level will result in unbearable pain after an exposure of 13 seconds and second degree burns after an exposure of 40 seconds (Mudan, 1984). In general, the intensity level of 5 kW/m^2 is used as a criterion for injury in a thermal radiation hazard assessment.

The downwind distance profiles shown in Fig. 6 are each calculated based on the maximum pool radii for its corresponding LNGC size, so that they give profiles similar to those of the LNG pool radii shown in Fig. 5(a). In the cases where breaches are less than 2 m in diameter, there is not much difference in the downwind distance between the two LNGC sizes, though the total volume spilled from the latest LNGC is twice that from the conventional size. The reason for this is that the LNG released from the tank is in the continuous spill range. When the breach diameters are greater than approximately 5 and 6 m for the conventional and latest LNGCs, respectively, the effect of the breach diameter on the thermal hazard distance is negligible. The asymptotic value of the downwind distance to 5 kW/m^2 extends approximately from 1,600 m up to 2,000 m due to the enlarged capacity of the latest LNGC. Consequently, while the spill volume doubles, the maximum thermal hazard distance for the latest LNGC increases by only 25 % than that for the conventional size because of the same reason as discussed in the previous section on the pool spread process.

The present study assumes that a single, coherent pool fire can be maintained for a very large pool diameter. However, this assumption may not be appropriate due to the inability of air to reach the interior of a fire and maintain combustion over such a large LNG pool (Luketa-Hanlin, 2006). Instead, the flame envelope would break up into several smaller, shorter flames at some very large size due to the environmental conditions, such as wind, waves and currents. The SNL study (Hightower et al., 2004) noted that these factors could reduce the thermal hazard distance by a factor of two to three. However, it is not yet known how to determine the limiting breakup diameter for a given LNG pool fire on water. The pool diameters presented here are speculative because experiments for large pool fires have never been performed (Luketa-Hanlin, 2006; Raj, 2007). Therefore, due to the assumption of a single, coherent pool fire, the hazard distances obtained in the present analyses should rather be considered as conservative estimates.

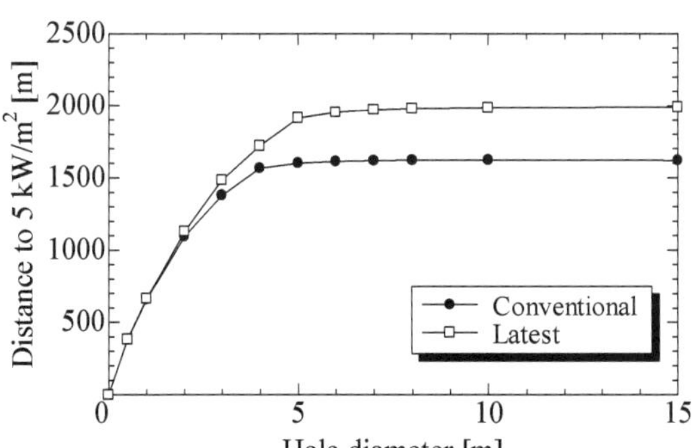

Fig. 6. Sensitivity of the downwind distance to 5 kW/m^2 to the hole diameter of a single tank for the conventional and latest LNGCs.

4.4 Vapor cloud dispersion process

For flammable vapor dispersion distance calculation, the level of concern is generally taken as the LFL for the substance in the case of a flash fire. In addition, the level of concern is often defined as half the LFL to account for the localized pockets of higher gas concentrations that may occur in an actual release. The use of half the LFL for LNG is also supported by the Federal Safety Standards (CFR, 1980), which specifies the use of an average gas concentration in air of 2.5 % for onshore exclusion zones. For the present calculations, hazard distances are provided for the LFL.

Figure 7 shows the effect of the hole diameter on the maximum distance to the LFL. The dispersion calculations were conducted under atmospheric stability class F as the worst-case scenario. Similarly to the calculation of the pool radius and the thermal hazard distance, profiles of the distance to reach the LFL are given as a function of the hole diameter, and it reaches an asymptotic value with the increase in the breach diameter. However, unlike the pool spread and pool fire processes, the vapor dispersion distance approaches asymptotically to an averagely constant level when the breach diameters are greater than about 3 and 4 m for the conventional and latest LNGCs, respectively. This inconsistency is attributed to the total evaporation duration which is singularly longer in the transitional spill range, as shown in Fig. 4. The asymptotic value of the distance to the LFL for the latest LNGC is only about 30 % longer than that for the conventional size, while the total spill volume from the latest LNGC is twice as much. This reason is the same as elaborated in the pool spread process.

From the above discussion, it has been found that the evolution of an LNG vapor cloud is strongly influenced by the characteristics of the LNG pool spread process, i.e., the source conditions. This fact is consistent with the dispersion behavior of denser-than-air gas observed in field experiments (Blackmore et al., 1982). Therefore, it is necessary to improve the present pool spread model so as to provide more accurate source conditions for vapor dispersion calculation.

As mentioned earlier in the first section, Qiao et al. investigated the sensitivity of vapor dispersion consequences to the breach diameter for the conventional size LNGC using the FERC method (Qiao et al., 2006). In their study, the averaging time to estimate flammable gas concentrations was set to 1 minute, though the use of a much shorter period of time was recommended in the FERC method (see Table 1). In the vapor dispersion scenarios of the ABSG study (ABSG, 2004; FERC, 2004), the averaging time was set to 0 second, so that the same averaging time is also used in this study. Qiao et al. employed completely the same scenarios as those in the ABSG study, which provided results of two example dispersion calculations (FERC, 2004). The downwind distances to the LFL shown in the ABSG study were about 3,400 and 4,100 m for 1 m and 5 m hole diameters, respectively, whereas the corresponding results by Qiao et al. were about 3,400 and 3,300 m. Their results are in contradiction to the previous experimental observation that the higher the vaporization rate is, the greater the distance to the LFL becomes. In addition, they performed curve fitting for their calculated values without taking account of the results in the cases where the hole diameters were 1, 4 and 6 m. Nevertheless, they drew a questionable conclusion that the distance to the LFL approached asymptotically to an almost constant level when the breach diameter was greater than 5 m for the conventional size LNGC.

Finally, the present results are briefly compared to those in the ABSG study (FERC, 2004). When the hole diameters are 1 and 5 m, the distances to the LFL in this study are about 3,340 and 4,940 m for the conventional size LNGC, respectively. Since the LNG spill volume in the present calculation is greater than that in the ABSG study, both results are not compared quantitatively in a strict sense. In addition, for the purpose of conservative estimation, the maximum distances are determined by the LFL concentrations at the ground level, in contrast to a default height of 0.5 m to calculate the flammability contours in DEGADIS. In the case with 1 m hole diameter, however, almost the same results are obtained in both of the studies, because the release is in the continuous spill range. On the other hand, since the LNG release from a hole with a diameter of 5 m is classified into the instantaneous spill type, the dispersion distance in the present study is longer than that in the ABSG study due to the effects of the larger spill volume and the difference of the elevation level to measure the LFL concentration.

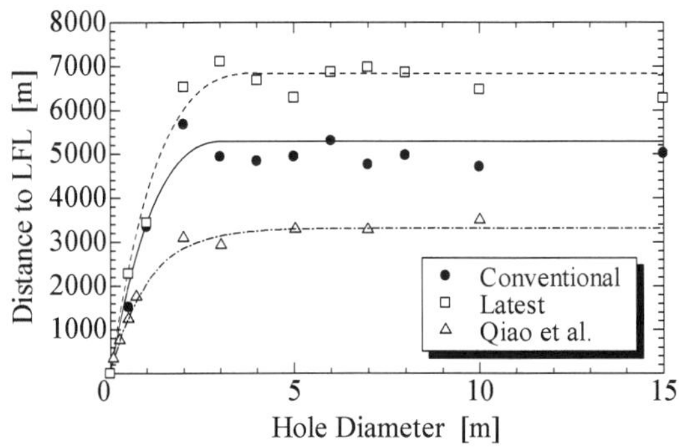

Fig. 7. Sensitivity of the downwind distance to the LFL to the hole diameter of a single tank for the conventional and latest LNGCs. For comparison, results from the previous study (Qiao et al., 2006) are also shown in the figure.

5. Concluding remarks

Consequence analyses of large-scale liquefied natural gas spills on water have been carried out using the method proposed by the Federal Energy Regulatory Commission (FERC, 2004). The principal LNG hazards of interest for the present study are those posed by thermal radiation and flammable vapor dispersion following unconfined LNG spills on water. In particular, this study has focused on the sensitivity of the LNG release duration, of the volatile pool spread, and of the pool fire and vapor dispersion hazards to the size of a hole breached in a membrane-type tank of the conventional and latest LNGCs. From the practical viewpoint of applicability to any breach size, the use of the FERC models has been recommended as the most appropriate, practical method at the present time (Oka & Ota, 2008; Oka, 2009).

The present sensitivity analyses have shown that the consequences are strongly dependent upon the breach size in the ranges associated with the continuous and transitional spill types under the present release assumption. On the other hand, when the breach diameter is larger than a certain critical value, there is little influence on the consequences regardless of the scenarios of pool fire and vapor dispersion hazards.

In the pool fire scenario, the critical values of the hole diameter are about 5 and 6 m for the conventional and latest LNGCs, respectively. In the vapor dispersion scenario, on the other hand, the critical diameters for the distance to the LFL are approximately 3 and 4 m for the two size LNGCs, respectively. This inconsistency is attributed to the singularly long evaporation duration of an LNG pool in the transitional spill range. These singular solutions obtained from the pool spread model indicate the lack of appropriate dynamic nature for transitional spills in the present integral approach.

Therefore, it is important to develop a simple, but accurate pool spread model without dependence on spill types. On the other hand, it should be noted that practical consequence assessment methods can generally provide only rough estimates of the magnitude of effects for incidents involving large LNG release on water because of the variability in actual incident circumstances as well as the uncertainty inherent in the methods used.

6. References

ABSG (2004). Consequence Assessment Methods for Incidents Involving Releases from Liquefied Natural Gas Carriers, *Contract report for the Federal Energy Regulatory Commission*, ABSG Consulting Inc., FERC04C40196, http://www.ferc.gov/industries/lng/safety/reports/cons-model-comments.pdf

Beyler, C. L. (2002). Fire Hazard Calculations for Large, Open Hydrocarbon Fires, In: *The SFPE Handbook of Fire Protection Engineering, Third Edition*, Chapter 11, 3-268-3-314, National Fire Protection Association, ISBN: 0877654514, Massachusetts

Blackmore, D. R.; Eyre, J. A. & Summers, G. G. (1982). Dispersion and Combustion Behavior of Gas Clouds Resulting from Large Spillages of LNG and LPG onto the Sea, *Transactions of the Institute of Marine Engineers (TM) 94*, Paper 29, 1-18 , ISSN: 0268-4152

Brown, T. C.; Cederwall, R. T., & Chan, S. T., et al. (1990). Falcon Series Data Report: 1987 LNG Vapor Barrier Verification Field Trials, *Final Report*, Gas Research Institute, GRI-89/0138

CFR (1980). Code of Federal Regulations, Title 49: Transportation, Part 193 Liquefied Natural Gas Facilities: *Federal Safety Standards*. U.S. Government Printing Office, Washington, DC

Cornwell, J. B. & Johnson, D. W. (2004). Modeling LNG Spills on Water, *AIChE 2004 Spring National Meeting*, New Orleans, Louisiana, April 25-29, 2004

EIA (2009). *International Energy Outlook 2009*, Energy Information Administration, Office of Integrated Analysis and Forecasting, U.S. Department of Energy, DOE/EIA-0484, Washington, DC 20585

Fay, J. A. (2003). Model of Spills and Fires from LNG and Oil Tankers, *Journal of Hazardous Materials*, Vol. B96, 171-188, ISSN: 0304-3894

Fay, J. A. (2007). Spread of Large LNG Pools on the Sea, *Journal of Hazardous Materials*, Vol. 140, 541-551, ISSN: 0304-3894

Feldbauer, G. F.; Heigl, J. J. & McQueen, W., et al. (1972). Spills of LNG on Water-Vaporization and Downwind Drift of Combustible Mixtures, Report No. EE61E-72, Esso Research & Engineering Company

FERC (2004). *Staff's Responses to Comments on Consequence Assessment Methods for Incidents Involving Releases from Liquefied Natural Gas Carriers*, Federal Energy Regulatory Commission, Docket No. AD04-6-000, http://www.ferc.gov/industries/lng/safety/reports/cons-model-comments.pdf

Hightower, M.; Gritzo, L. & Luketa-Hanlin, A., et al. (2004). Guidance on Risk Analysis and Safety Implications of a Large Liquefied Natural Gas (LNG) Spill Over Water, *SANDIA REPORT*, SAND2004-6258, Sandia National Laboratories, Albuquerque, NM

Hightower, M.; Gritzo, L. & Luketa-Hanlin, A. (2005). Safety Implications of a Large LNG Tanker Spill Over Water, *Process Safety Progress*, Vol. 24, 168-174, ISSN: 1066-8527

Koopman, R. P.; Cederwall, R. T. & Ermak, D. L., et al. (1982). Analysis of Burro Series 40-m^3 LNG Spill Experiments, *Journal of Hazardous Materials*, Vol. 6, No. 1-2, 43-83, ISSN: 0304-3894

Koopman, R. P. & Ermak, D. L. (2007). Lessons Learned from LNG Safety Research, *Journal of Hazardous Materials*, Vol. 140, Issue 3, 412-428, ISSN: 0304-3894

Luketa-Hanlin, A. (2006). A Review of Large-Scale LNG Spills: Experiments and Modeling, *Journal of Hazardous Materials*, Vol. A132, 119-140, ISSN: 0304-3894

Luketa, A.; Hightower, M. M. & Attaway, S. (2008). Breach and Safety Analysis of Spills Over Water from Large Liquefied Natural Gas Carriers, *SANDIA REPORT*, SAND2008-3153, Sandia National Laboratories, Albuquerque, NM

Mizner, G. A. & Eyre, J. A. (1983). Radiation from Liquefied Gas Fires on Water, *Combustion Science and Technology*, Vol. 35, Issue 1-4, 33-57, ISSN: 0010-2202

Mudan, K. S. (1984). Thermal Radiation Hazards from Hydrocarbon Pool Fires, *Progress in Energy and Combustion Science*, Vol. 10, 59-80, ISSN: 0360-1285

Oka, H. & Ota, S. (2008). Evaluation of Consequence Assessment Methods for Pool Fires on Water Involving Large Spills from Liquefied Natural Gas Carriers, *Journal of Marine Science and Technology*, Vol. 13, No. 2, 178-188, ISSN: 0948-4280

Oka, H. (2009). Consequence Analysis of Pool Fire Hazards from Large-Scale Liquefied Natural Gas Spills Over Water, *Hydrocarbon World*, Vol. 4, Issue 1, 90-93, ISSN: 1753-3899

Qiao, Y.; West, H. H. & Sam Mannan, M., et al. (2006). Assessment of the Effects of Release Variables on the Consequences of LNG Spillage onto Water Using FERC Models, *Journal of Hazardous Materials*, Vol. 130, 155-162, ISSN: 0304-3894

Raj, P. K. (2007). LNG Fires: A Review of Experimental Results, Models and Hazard Prediction Challenges, *Journal of Hazardous Materials*, Vol. 140, Issue 3, 444-464, ISSN: 0304-3894

Rew, P. J. & Hulbert, W. G. (1996). Development of Pool Fire Thermal Radiation Model, *Health and Safety Executive Contract Research Report*, No. 96/1996, ISBN: 0717610845

Rodean, H. C.; Hogan, W. J. & Urtiew, H. C., et al. (1984). Vapor Burn Analysis for the Coyote Series LNG Spill Experiments, *UCRL-53530*, Lawrence Livermore National Laboratory, Livermore, California

Schneider, A. L. (1980). Liquefied Natural Gas Spills on Water: Fire Modeling, *Journal of Fire and Flammability*, Vol. 12, No. 4, 302-313, ISSN: 0022-1104

Spaulding, M. L.; Craig Swanson, J. & Jayko, K., et al. (2007). An LNG Release, Transport, and Fate Model System for Marine Spills, *Journal of Hazardous Materials*, Vol. 140, 488-503, ISSN: 0304-3894

Spicer, T. O. & Havens, J. A. (1987). Field Test Validation of the DEGADIS Model, *Journal of Hazardous Materials*, Vol. 16, 231-245, ISSN: 0304-3894 van den Bosch, C. J. H. (1997). Pool Evaporation, *Methods for the Calculation of Physical Effects (TNO Yellow Book, CPR14E(Part 1), 3rd edn),* van den Bosch, C. J. H. & Weterings, R. A. P. M., (Ed.), 3.1-3.126, Sdu Uitgevers, The Netherlands, ISBN: 9012084970

24

Risk assessment of marine LNG operations

Tarek Elsayed, Associate Professor
Arab Academy for Science & Technology & Maritime Transport
Alexandria, Egypt

1. Introduction

The safety and reliability of LNG transfer operations is a major concern for LNG operating companies. LNG hazards have a high potential financial impact in addition to shutdown and failure of delivery. The LNG industry has developed and refined its practices gradually over the past 30 years, achieving very good results. Risk assessment/management of LNG marine terminal operations is an essential tool, however, for maintaining the industry's record for safe operations at LNG terminals.

This chapter presents various methodologies for the risk assessment of LNG transfer operations at the ship-shore interface of gas terminals. Hazards are identified as well as potential accident consequences. Various risk assessment approaches for modeling LNG accident scenarios at gas terminals are presented. These include qualitative risk matrices, multiple attribute utility models and a fuzzy inference system. LNG accident consequences (SIGTTO, 1999) involve multiple consequence classes such as personnel injuries, environmental pollution and loss of material assets. These consequences have different measurement scales and need to be combined in order to assess/rank risks arising from various hazardous scenarios. Fig. 1 shows the different consequence classes for an LNG ship accident while loading/offloading at the terminal (Elsayed et al., 2009).

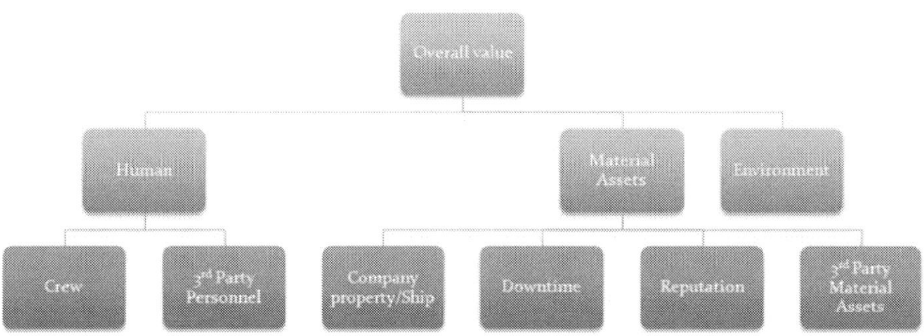

Fig. 1. Consequence classes resulting from hazardous sceanrios

2. Qualitative Risk Assessment

A standard qualitative risk assessment approach involves the evaluation of likelihood or probability of different accident scenarios (American Bureau of Shipping, 2000; HSE, 2002). Next an evaluation of the impact of the different accident scenarios with respect to the different consequence attributes is carried out. A risk matrix is often used in this approach. A risk matrix combines the likelihood of an event with its consequence severity into a risk level. Table 1 shows a sample qualitative assessment of the probabilities of occurrence of an LNG ship accident while loading/unloading at the terminal. Probabilities are assigned letters such as A, B, C, D and E corresponding to a linguistic scale: `frequent', `probable', `occasional', `remote' and `improbable' likelihoods. An indicative quantitative frequency range is associated with each probability level.

Level	Description	Indicative Frequency (per vessel year)	Definition
A	Frequent	>0.5	Will occur frequently
B	Probable	0.05-0.5	May occur several times
C	Occasional	0.005-0.05	Likely to occur during lifetime
D	Remote	0.0005-0.005	Unlikely to occur during lifetime
E	Improbable	< 0.0005	So unlikely event it may never be experienced

Table 1. Definition of likelihood levels

Table 2 shows a similar qualitative assessment of LNG accident consequence categories. Consequences are assigned numbers such as 1, 2, 3 and 4 corresponding to `minor', `major', 'critical' and `catastrophic' severity level. An indicative descriptive linguistic range is associated with each consequence attribute and severity level.

Consequence Class	1	2	3	4
	Minor	**Major**	**Critical**	**Catastrophic**
Crew	Minor injury	Serious injury	One fatality	Several fatalities
3rd Party personnel	No injury	Minor injury	Serious injury	Fatalities
Environmental	Negligible pollution	Pollution reportable to regulatory authorities Minor release	Pollution reportable to regulatory authorities Major release	Pollution reportable to regulatory authorities Uncontrolled pollution
Ship damage	Minor damage	Moderate damage	Major damage	Loss of ship
Downtime	Negligible	One day	One week	More than one week
Reputation	Negligible	Affected localy	Affected nationaly	Loss of reputation
3rd party assets	No effect	Minor damage	Major damage	Extensive damage

Table 2. Definition of severity levels of accident consequences

Table 3 shows an example of the risk matrix used by ship classification societies. The risk matrix combines likelihood and severity into an output linguistic risk level for each scenario and consequence attribute. These linguistic risk values are then combined to give an overall linguistic risk value for each accident scenario being evaluated. Output risk level are denoted linguistically as `low', `medium' or `high' (Skramstad & Musaeus, 2000).

		SEVERITY			
		1	2	3	4
LIKELIHOOD	A	M	H	H	H
	B	M	M	H	H
	C	L	M	M	H
	D	L	L	M	M
	E	L	L	L	L

Table 3. Example of a qualitative risk matrix

3. Multiple Attribute Utility Risk Model

A multiple attribute risk assessment approach using utility theory is presented in this section (Elsayed et al., 2009). Multiple attribute risk assessment based on utility theory has many advantages. Most importantly, it allows LNG operating companies to identify/rank operational risks and to express their expectations about the consequences of various hazardous scenarios. It also provides insights into how the uncertainty of their expectations affects the ranking of risk scenarios. In addition, multiple attribute risk assessment provides a systematic method for evaluating an organization's risks using the best available hazard information. As operating companies gain better hazard/consequence information, the risk models can be easily updated with new input data and the marginal effect on risk assessment can be measured. The value of a multiple attribute risk assessment is not only in the numbers produced, but also in the insights that operating companies gain during sensitivity analyses and each refinement step of the assessment.

3.1 Modelling Consequences Using Utility Functions

Utility is a number measuring the attractiveness of a consequence, the higher the utility, the more desirable the consequence, the measurement sometimes being made on a probability scale (Clemen, 1997; Lindley, 1992;). Different people and/or organizations have different risk attitudes and thus are willing to accept different levels of risk (Oliver & Marshall, 1997). Some are prone to taking risks while others are more conservative and tend to avoid risk. Individuals who are unwilling to risk a substantial part of their assets even for positive expected return are said to be risk averse. Those willing to take a risky venture for a negative expected return are said to be risk seeking. Finally, an individual can be risk neutral. Risk neutrality is reflected by a utility curve that is a simple straight line. A decision maker that has a constant aversion to risk, is referred to as constantly risk-averse decision maker. In this work, the constant risk aversion utility model is adopted. This is to reflect the fact that LNG accident consequences are acute in nature with very severe consequences and LNG ship operators are constantly averse in taking accident risks. An example of a utility function is the exponential utility function with constant risk aversion and can be expressed as:

$$u(x) = \frac{e^{-b(x - x_{min})} - 1}{e^{-b(x_{max} - x_{min})} - 1} \quad (1)$$

Where x_{max} and x_{min} are best (most preferred) and worst (least preferred) values of the consequence attribute and b is a coefficient of risk aversion. In order to model the consequence classes shown in Fig.1, seven utility functions are needed corresponding to the seven consequence attributes. Each utility function is constructed such that the most preferred value x_{max} for the consequence of interest would be 'minor' or 'negligible' consequence on a qualitative scale and would correspond to a utility value of 1. Whereas the least preferred value would correspond to 'catastrophic', corresponding to a utility value of 0.

3.2 Probabilistic Multiple Attribute Utility Risk Model

A probabilistic multiple attribute risk model can be used for modeling situations during LNG ships loading/offloading at the LNG ship/terminal interface where risks needs to be assessed and ranked in terms of severity. Various resulting hazard consequences are taken into account, and a systematic and consistent evaluation of various risk alternatives is carried out to determine most/least severe risk alternative. These include environmental pollution, injuries/fatalities to crew or 3rd party personnel and material assets such as ship damage, down time, reputation and third party material assets.

Multiple attribute utility theory is then used to combine the effects of different consequences into a unified utility measure. According to the maximum expected utility (MEU) concept (Chen & Hwang, 1992), a maximum risk alternative is selected such that:

$$R_{max} = \min_{1 \leq i \leq M} \sum_{j=1}^{N} k_{ij} u_j \quad (2)$$

R_{max} = maximum risk alternative.
M = number of risk alternatives or hazards.
N = number of consequences.
k_{ij} = weight of importance of the jth consequence.
u_j = measure of consequence, utility, of the ith consequence in terms of jth risk alternative

This semi-quantitative approach assigns a numeric expected utility value for each risk scenario thus allowing the ranking of various hazardous scenarios. Software tools can be used to implement the abovementioned risk model.

4. Fuzzy Risk Assessment

4.1 Modelling of Probabilities and Consequences as Fuzzy Sets

In many engineering situations there is pervasive fuzzy information, *i.e.* information that is vague/qualitative, linguistic and/or imprecise (Bellman & Zadeh, 1970; Chen & Hwang, 1992; Zadeh, 1965; Zadeh, 1975; Zimmerman, 1976; Zimmerman, 1987). This is often the case when trying to assess accident probabilities/consequences that are not known a priori and/or difficult to quantify mathematically (Elsayed et al., 2008). The assignment of accident probabilities is usually based on reliability methods and/or historical failure data. Reliability methods require knowledge of the relevant physical process and the specification of a limit state function (Elsayed & Mansour, 2003). In many cases, historical failure data can be lacking and/or unreliable. When historical failure data is available, it can be supplemented with expert judgment (Cooke, 1996). These approaches however are not sufficient to predict accident probabilities under all relevant circumstances. This is due to lack of knowledge of physical conditions and processes, change of industry practice over the years, and lack/unreliability of data. Hence, predictions of accident probabilities are often associated with significant uncertainties. In fact it is because of these uncertainties that many risk assessment tools avoid absolute probability values all together and stick to relative probabilities (American Gas Association, 1990).

LNG accident consequences (Elsayed et al., 2009; Gyles, 1992; Skramstad & Musaeus, 2000; McGuire & White, 1999) vary from personnel injuries to environmental pollution and loss of material assets. These consequences are imprecise in nature, each with its own measurement scale, and cannot be added mathematically. They may however be defined linguistically or on a qualitative scale. In this section, a new approach for the risk assessment of LNG carriers using a fuzzy inference system FIS is adopted. The main advantage of the use of the fuzzy inference system is its ability to handle imprecise data. The approach uses the concept of a pure fuzzy logic system. A fuzzy rule base is constructed to follow the logic used by the risk assessor when using the traditional qualitative risk matrix approach. The fuzzy inference engine uses these rules to determine a mapping from probability and consequences, modeled as fuzzy sets, to a fuzzy output set of risk values. In doing so, it is implied that probabilities/consequences used in the risk assessment process have an inherent degree of uncertainty.

4.2 Fuzzy inference System

Fuzzy inference is the process of mapping from a given input set to an output set using fuzzy logic. Membership functions, fuzzy logic operators and *if-then* rules are used in this process. The fuzzy inference system FIS is known in the literature by a number of names, such as fuzzy-rule-based system, fuzzy expert system or simply a fuzzy system (Kandel, 1992). The basic advantage of such system is its tolerability to linguistic/imprecise data. In this work, the Mamdani and the Sugeno fuzzy inference methods are adopted (Mathworks, Inc., 2006). In the Mamdani type of inference, the output membership functions are fuzzy sets. These are in turn defuzzified to obtain a crisp output risk value for each consequence alternative.

In the Sugeno method of fuzzy inference, output membership functions are either linear or constant. A typical rule in a Sugeno fuzzy model has the form:

$$If\ Input\ 1 = x\ and\ Input\ 2 = y, then\ Output\ is\ z - ax + by + c \quad (3)$$

Where a, b and c are the consequence parameters of the rule. The output level zi of each rule is weighted by the firing strength wi of the rule. For example, for an AND rule with Input 1 = x and Input 2 = y, the firing strength is

$$w_i = AndMethod(F_1(x)F_2(y)) \quad (4)$$

where $F1,2$ (.) are the membership functions for inputs 1 and 2.
The final output of the system is the weighted average of all rule outputs, computed as

$$Final\ Output = \frac{\sum_{i=1}^{N} w_i z_i}{\sum_{i=1}^{N} w_i} \quad (5)$$

where N is the number of rules.
Fig 2 shows the structure of the Mamdani fuzzy inference system FIS used for the assessment of a risk value for each consequence class or attribute.

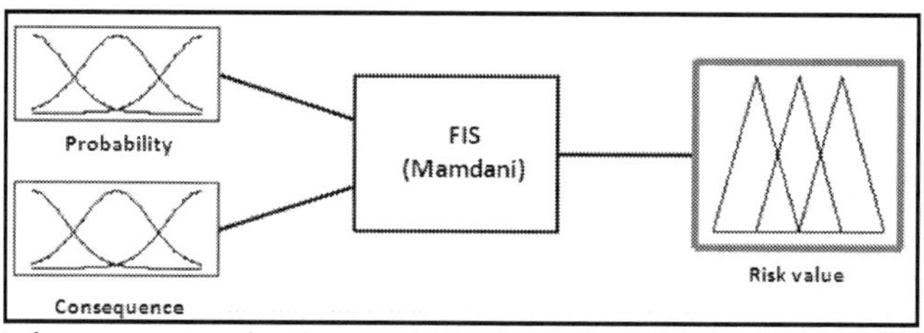

Fig. 2. Fuzzy inference system for the assessment of risk values

Fig. 3 shows a Sugeno FIS including two input variables x, y, and one output variable z.

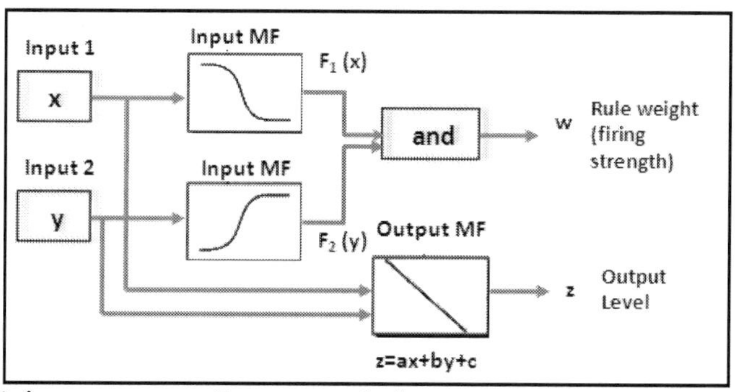

Fig. 3. Fuzzy inference process - Sugeno s method

5. Case Study: Assessment of LNG Risks during Loading/offloading at Terminals

A case study is used to demonstrate the above mentioned approaches for an LNG carrier loading at the terminal. Six hazardous scenarios are evaluated during LNG loading/offloading at the terminal (Elsayed, 2009). These are summarized in Table 4. As seen from Table 4, each consequence is denoted by a letter followed by a number. The letters (C,P,E,S,D,R,M) correspond to the consequence class for e.g. (crew, 3rd party personnel, environment, ship, downtime, reputation and 3rd party material assets). The numbers (1, 2, 3, and 4) correspond to the degree of severity of the consequence, for e.g. 'minor', 'major', 'critical' and 'catastrophic') on a qualitative scale.

Haz id	Hazard Description	Likelihood	Consequences
1	**Leak on the cargo system:** unignited release - continuos flow. This comprises all leak sizes that cannot easily be stopped by operatioal routines to a rupture in apipe. Potential consequences is brittle fracture of hull or secondary structure. Frost burns for personnel. No consequences to 3rd party anticipated.	Occasional (C)	C2, P1, E1, S2, D3, R3, M1
2	**Release of liquid nitrogen**: can give local effects to steel due to low temperature. Possible frost burn for personnel.	Occasional (C)	C2, P1, E1, S2, D2, R1, M1
3	**Release of bunker oil during loading operation.** Very low risk of fire and personnel injuries. The oil may mess up nearby quays affecting 3rd party assets.	Occasional (C)	C1, P1, E3, S1, D2, R3, M2
4	**Fire in the engine room**. Since always manned during this operational mode, the escalation potential is considered low. The event is considered not to affect 3rd party. All fires will have to be reported terminal, thus local reputation is affected.	Remote (D)	C3, P1, E1, S2, D3, R3, M1
5	**Accomodation fires**. The crew present in the accomodation most likely quickly extinguishes these fires. It is considered to be less likely to occur than fire in the engine room, but still a remote probability. If developing to a large fire more crew members may be affected by the accident, than for an engine room fire. 3rd party not likely to be affected.	Remote (D)	C3, P1, E1, S2, D3, R3, M1
6	**Fires on open deck.** Ignited cargo release. The consequences depend on the release size and the development of the event, including shutdown. Early ignition gives smaller consequences than late ignition. Most likely there is a flash fire which burn back to a smaller fire at the release location (jet or diffusive, depending on pressure in the system and if the release hits obstructions or not). Wether the fire may escalate to the LNG tanks depend on the possibility to shut down fuel to the fire.	Remote (D)	C4, P3, E1, S3, D4, R3, M2

Table 4. Hazards considered during LNG loading/offloading at terminal

5.1 Qualitative Risk Assessment Results

Table 5 provides a summary of the calculated qualitative risk values for the six hazardous scenarios and seven consequence attributes. As can be as seen the various consequence attributes are assigned linguistic risk values (low, medium, high) using the qualitative risk matrix outlined above. These consequences are then combined to provide an overall linguistic risk value for each accident scenario.

	Leak on the cargo system	Release of liquid nitrogen	Release of bunker oil	Fire in engine room	Accommodation Fires	Fires on open deck
Crew	Medium	Medium	Low	Medium	Medium	Medium
3rd party personnel	Low	Low	Low	Low	Low	Medium
Environment	Low	Low	Medium	Low	Low	Low
Ship	Medium	Medium	Low	Low	Low	Medium
Downtime	Medium	Medium	Medium	Medium	Medium	Medium
Reputation	Medium	Low	Medium	Medium	Medium	Medium
3rd party material assets	Low	Low	Medium	Low	Low	Low
Final rating	Medium	Low	Medium	Low	Low	Medium

Table 5. Summary of calculated qualitative risk values for six hazardous scenarios and seven consequence attributes.

5.2 Multiple Attribute Utility Risk Assessment Results

A software tool has been written using a decision analysis software suite (Treeage, 2006) to implement the abovementioned risk model. Fig. 4 shows the risk model used for modeling of the LNG carrier loading/offloading at the terminal. Emanating from the hazard node are the six accident scenarios. These are 'leak on the cargo system', 'release of liquid nitrogen', 'release of bunker oil', 'fire in the engine room', 'accommodation fires' and 'fires on open deck'. Each hazard is associated with a probability level. An overall consequence for each scenario is measured by seven consequences (crew, third-party personnel, environment, ship, down time, reputation and third-party material assets). Each consequence is modeled using a utility function. Formulated in this way, the optimum or minimum risk alternative corresponds to the highest maximum expected utility (MEU). The decision analysis software DATA (Decision Analysis by Treeage) was used for the modeling of the risk model. Sensitivity of hazardous scenarios to various model variables can also be carried out. Fig. 5 shows the output of the risk model for the six hazardous scenarios. The risk model shows the most severe scenario, 'Fire on Open Deck', in this case corresponding to minimum total expected utility (MTEU) value of 0.00406. The program windows interface shows a red color, which indicates the most severe scenario.

Fig. 6 shows a snapshot of a one-way sensitivity analysis with respect to 'occasional probability'. As can be seen the recommendations of the risk model are not affected by the change in the probability level within the range considered (0.005–0.05)

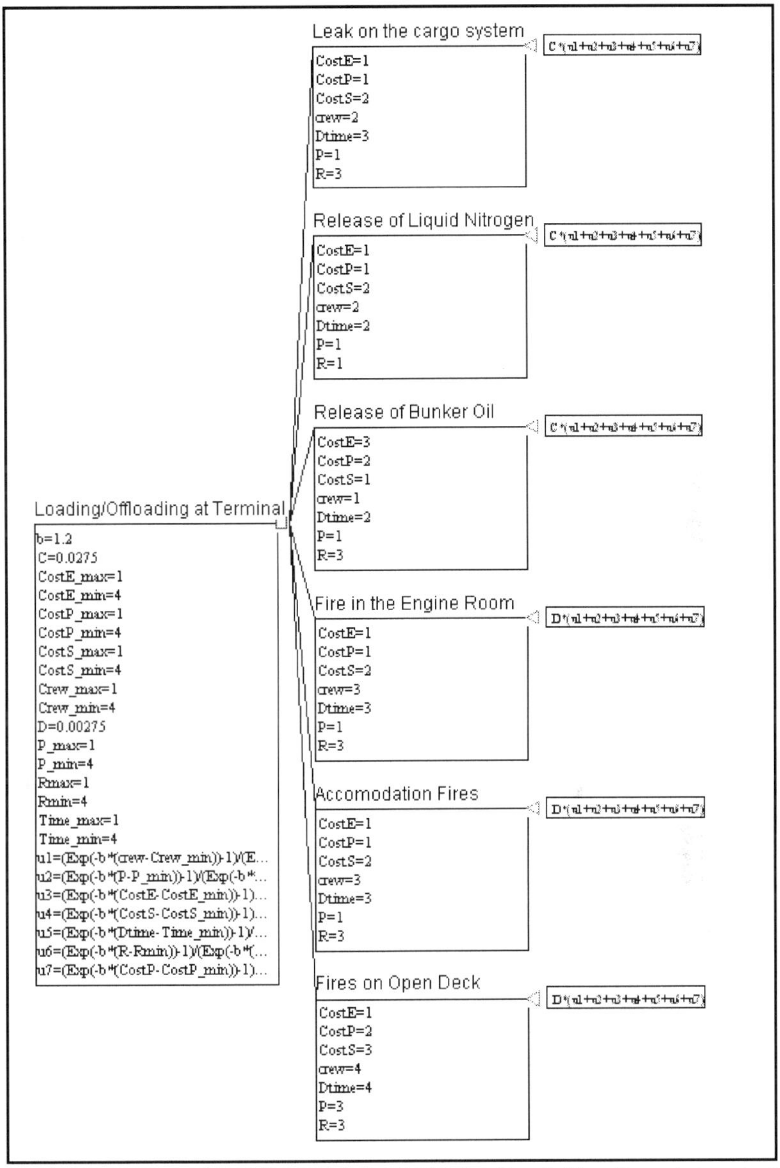

Fig. 4. Basic framework for risk model for LNG loading/offloading at terminal using multiple attribute utility theory

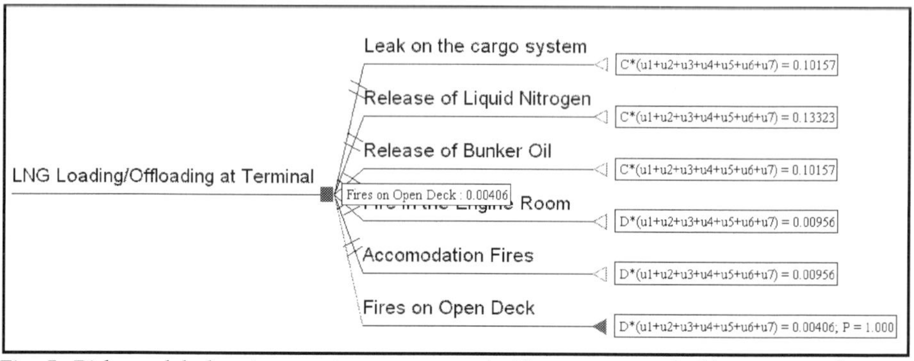

Fig. 5. Risk model showing most severe scenario 'fires on open deck', corresponding to minimum total expected utility

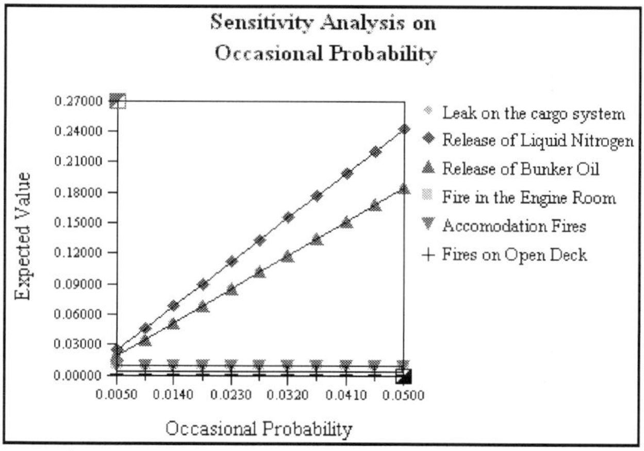

Fig. 6. Sensitivity analysis on 'occasional probability'.

5.3 Fuzzy Risk Assessment Results

5.3.1 Determination of Membership Functions and Rule Base

In order to adopt this approach for risk assessment, probabilities of accident scenarios as well as accident consequences are modeled as fuzzy sets. In doing so, it is implied that probabilities/consequences are by themselves uncertain or at least a degree of uncertainty is associated with their values. Several approaches for building and adapting membership functions exist (Zhou et al., 1997). In this work, a fixed center-based membership function approach using the symmetric Gaussian membership function was adopted. One membership function is assigned to each value of the fuzzy variable. The Gaussian membership function depends on two parameters and is given by:

$$f(x,\sigma,c) = e^{\frac{-(x-c)^2}{2\sigma^2}} \tag{6}$$

Where c is the mean value and σ is the standard deviation. Membership functions are centered, each at the mid-value of the numeric interval associated with each fuzzy variable. For example for 'remote probability' the Gaussian membership functions is centered on the average value of the interval (0.005-0.0005) which would correspond to a mean value c=0.00275, as shown in Table 6. The standard deviation parameters σ for the Gaussian functions were chosen such that membership function curves are completed with the minimum and maximum points of the interval associated with each of the fuzzy variables (Zhou et al., 1997). With these membership functions each input value will belong to no more than two fuzzy sets. Fig. 7 shows the membership functions for the probability of occurrence and consequences respectively modeled as fuzzy sets. Probabilities of occurrence are represented by fuzzy sets whose ranges are chosen to coincide with the indicative frequency ranges shown in Table 1. Similarly, accident consequences are represented by fuzzy sets whose ranges are chosen to coincide with the indicative severity levels shown in Table 2. Figure 8 shows the membership functions for the resulting risk value. A scale of 1 to 10 was adopted to represent output risk values. As can be seen, the use of fuzzy sets allows representation of linguistic terms, such as 'frequent', 'probable', 'occasional', 'remote' and 'improbable' for likelihoods. Consequences are also represented linguistically as 'minor', 'major', 'critical' and 'catastrophic'. Finally output risk values are denoted as 'low', 'medium' and 'high'. Table 6 summarizes the membership type and parameters adopted in the developed fuzzy inference system FIS.

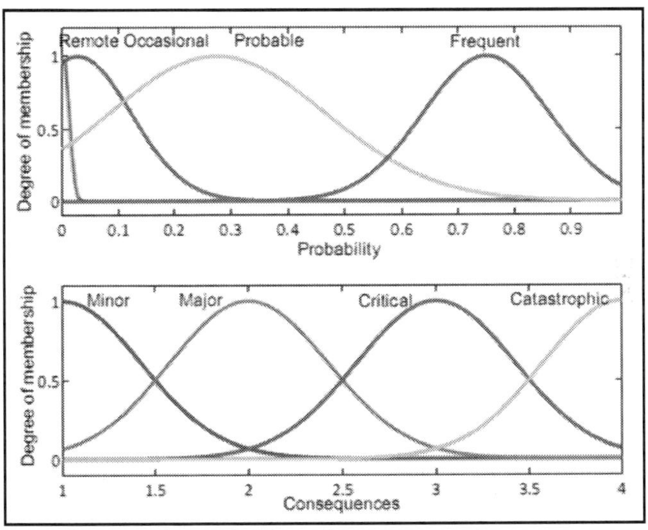

Fig. 7. Membership functions for probability of occurrence and consequence severity levels

Table 6. Membership type and parameters in the fuzzy inference system.

Variable	Value/Range	Membership Type	Membership parameters (σ,c)
Probability -' frequent'	> 0.5	Gauss	(0.11,0.75)
Probability - 'probable'	0.5-0.05	Gauss	(0.19,0.275)
Probability - 'occasional'	0.05-0.005	Gauss	(0.09,0.0275)
Probability- 'remote'	0.005 -0.0005	Gauss	(0.009,0.00275)
Probability- 'improbable'	< 0.0005	Gauss	(0.0009,0.000275)
Consequence- 'catastrophic'	4	Gauss	(0.4247,4)
Consequence- 'critical'	3	Gauss	(0.4247,3)
Consequence- 'major'	2	Gauss	(0.4247,2)
Consequence- 'minor'	1	Gauss	(0.4247,1)
Risk - 'high'	10	Gauss	(1.911,10)
Risk - 'medium'	5.5	Gauss	(1.911,5.5)
Risk - 'low'	1	Gauss	(1.911,1)

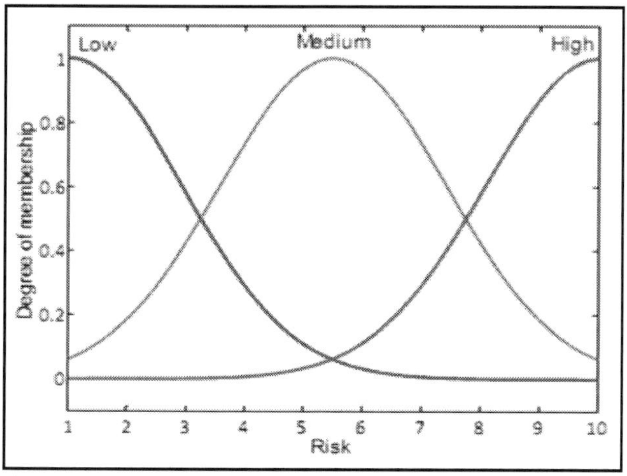

Fig. 8. Membership functions for risk values

The mapping between probability, consequences and final risk value is accomplished by the use of fuzzy *if-then* rules. For a single attribute risk problem, a total of twenty *if-then* rules can be used in the fuzzy inference system FIS to provide the mapping between probability, consequence and the computed risk value. The rules are designed to follow the logic of the risk assessor when using the qualitative risk matrix approach outlined earlier. First twelve rules of the developed system are listed below.

Rule 1: if (Probability is Frequent) and (Consequence is Minor) then Risk is Medium.
Rule 2: if (Probability is Frequent) and (Consequence is Major) then Risk is High.
Rule 3: if (Probability is Frequent) and (Consequence is Critical) then Risk is High.
Rule 4: if (Probability is Frequent) and (Consequence is Catastrophic) then Risk is High.
Rule 5: if (Probability is Probable) and (Consequence is Minor) then Risk is Medium.
Rule 6: if (Probability is Probable) and (Consequence is Major) then Risk is Medium.
Rule 7: if (Probability is Probable) and (Consequence is Critical) then Risk is High.

Rule 8: if (Probability is Probable) and (Consequence is Catastrophic) then Risk is High.
Rule 9: if (Probability is Occasional) and (Consequence is Minor) then Risk is Low.
Rule 10: if (Probability is Occasional) and (Consequence is Major) then Risk is Medium.
Rule 11: if (Probability is Occasional) and (Consequence is Critical) then Risk is Medium
Rule 12: if (Probability is Occasional) and (Consequence is Catastrophic) then Risk is High.

As can be seen, the first four rules represent the first row in the qualitative risk matrix given in Table 3. The second row in the matrix is represented by the next four rules *i.e.* rules 5-8 and so on. For the qualitative risk matrix given in Table 3, (Skramstad & Musaeus, 2000), the numbers of rows is five and the number of columns is four, *i.e.* a total of twenty rules are needed for modeling the logic embedded in this matrix. As such, the total number of rules needed to construct the fuzzy inference engine can be expressed as:

$$N = m \times n \tag{7}$$

where N = number of fuzzy *if-then* rules
m = number of rows in qualitative risk matrix.
n = number of columns in qualitative risk matrix.

These rules provide the mapping for each hazardous scenario for only one consequence attribute.

5.3.2 Application to Hazardous Scenarios

Often LNG risk assessment problems involve multiple consequence attributes for each hazardous scenario, such as material assets, human life and/or environmental pollution. These consequences are combined to provide an overall risk value for each accident scenario. In this work, a fuzzy risk index FRI is used to combine the various consequence attribute risks into a unified risk measure. The fuzzy risk index FRI value is an average aggregation operator for each accident scenario can be calculated by (Yager, 1988)

$$FRI = \frac{(\sum_{i=1}^{N} k_i Risk_i)/N}{\sum_{i=1}^{N} k_i} \tag{8}$$

where N = number of consequences.
k_i = weight factor for each consequence.
$Risk_i$ = calculated fuzzy risk value for each consequence attribute.

The weighting factors k_i reflects the attribute's relative importance. Fig. 9 shows the structural hierarchy and information storage for the fuzzy inference system used. The FIS structure contains various substructures which in turn contain variable names, membership function definitions and computation method.

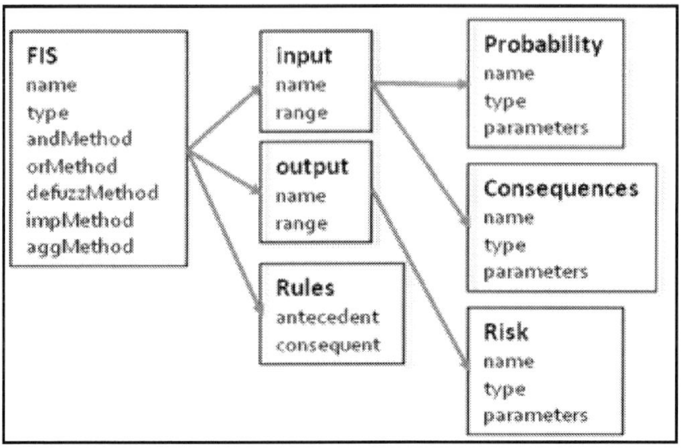

Fig. 9. Fuzzy inference system structural hierarchy

Figures 10 and 11 show the resulting output surface envelopes for both the Mamdani/Sugeno methods for two fuzzy inputs, probability and consequence as well as the fuzzy output risk. A zero-order Sugeno model was adopted for computation of final risk values.

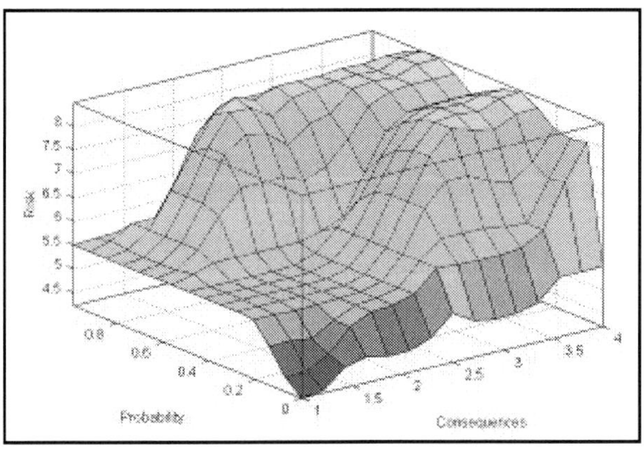

Fig. 10. Output risk surface envelope (Mamdani) for two fuzzy inputs: probability and consequence.

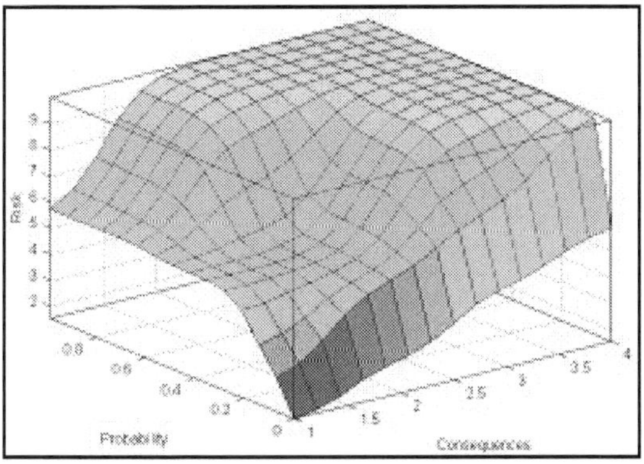

Fig. 11. Output risk surface envelope (Sugeno) for two fuzzy inputs: probability and consequence

Table 7 provides a summary of the calculated fuzzy risk values for the six scenarios and seven consequence attributes. Both the Mamdani/Sugeno methods of inference were used in the calculation of final risk output values. Table 8 shows a comparison between qualitative and fuzzy risk assessment results for the six scenarios considered. Figures 12 and 13 show the Sugeno fuzzy risk values for material assets and crew respectively. Figures 14 and 15 show the Mamdani fuzzy risk values for material assets and crew respectively.

	Leak on the cargo system		Release of liquid nitrogen		Release of bunker oil		Fire in engine room		Accommodation Fires		Fires on open deck	
	M	S	M	S	M	S	M	S	M	S	M	S
Crew	5.50	5.30	5.50	5.30	4.35	2.51	5.76	6.06	5.76	6.06	6.31	7.83
3rd party personnel	4.35	2.51	4.35	2.51	4.35	2.51	4.23	1.80	4.23	1.80	5.76	6.06
Environment	4.35	2.51	4.35	2.51	5.84	6.92	4.23	1.80	4.23	1.80	4.23	1.80
Ship	5.50	5.30	5.50	5.30	4.35	2.51	4.67	3.56	4.67	3.56	5.76	6.06
Downtime	5.84	6.92	5.50	5.30	5.50	5.30	5.76	6.06	5.76	6.06	6.31	7.83
Reputation	5.84	6.92	4.35	2.51	5.84	6.92	5.76	6.06	5.76	6.06	5.76	6.06
3rd party material assets	4.35	2.51	4.35	2.51	5.50	5.30	4.23	1.80	4.23	1.80	4.67	3.56
Final rating	5.10	4.57	4.83	3.70	5.10	4.57	4.95	3.88	4.95	3.88	5.54	5.60

Table 7. Summary of calculated fuzzy risk values for six hazardous scenarios and seven consequence attributes. (M=Mamdani method, S=Sugeno method).

Id	Hazard	Qualitative risk value	*Mamdani* fuzzy risk value	*Sugeno* fuzzy risk value
1	Leak on the cargo system	Medium	5.1073	4.5706
2	Release of liquid nitrogen	Low	4.8437	3.7091
3	Release of bunker oil	Medium	5.1073	4.5706
4	Fire in engine room	Low	4.9524	3.8820
5	Accommodation fires	Low	4.9524	3.8820
6	Fires on open deck	Medium	5.5461	5.6054

Table 8. Comparison between qualitative and fuzzy risk values for six hazardous scenarios.

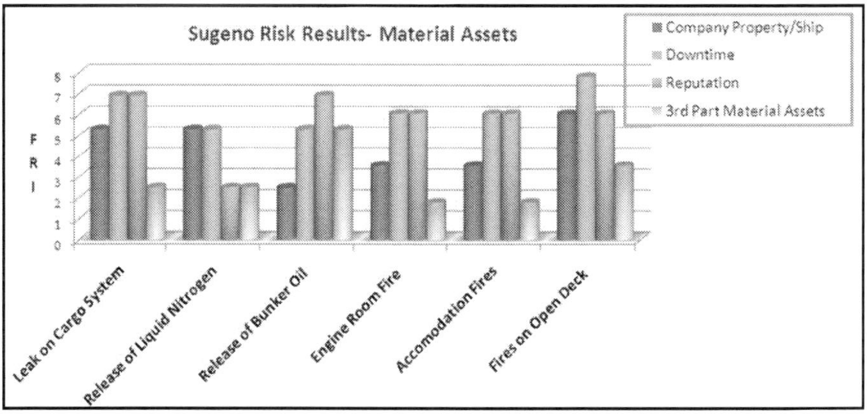

Fig. 12. Sugeno fuzzy risk values for material assets

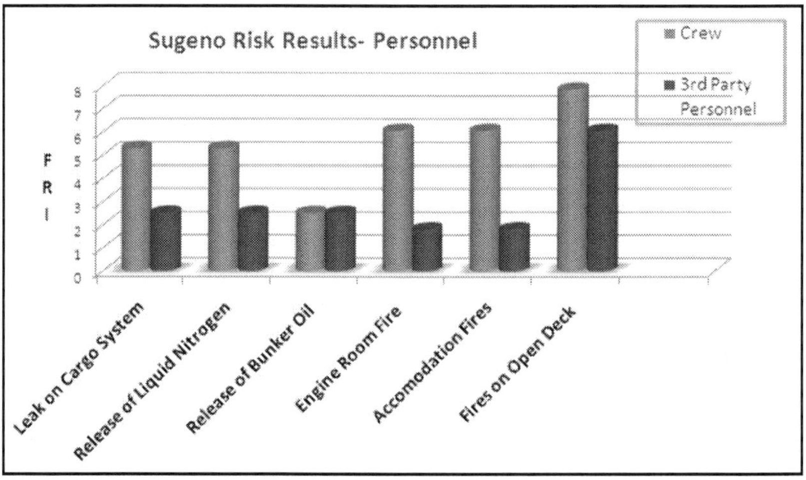

Fig. 13. Sugeno fuzzy risk values for personnel

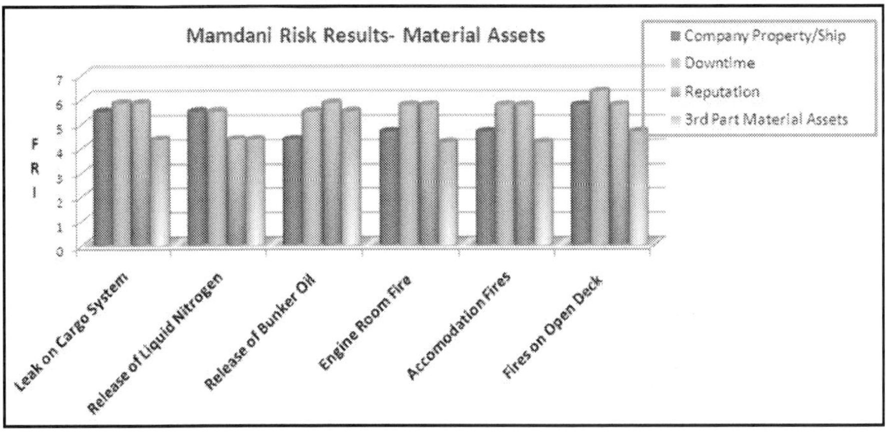

Fig. 14. Mamdani fuzzy risk values for material assets

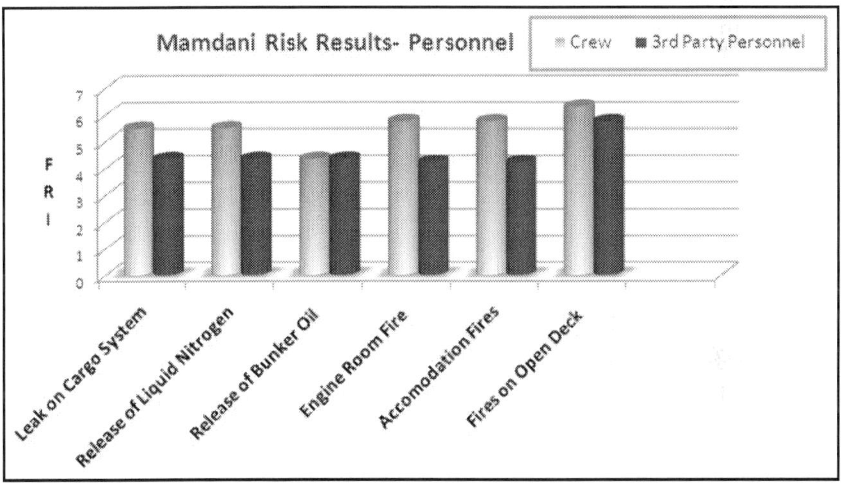

Fig. 15. Mamdani fuzzy risk values for personnel

As can be seen in Table 4, scenarios 1 and 3 are expected to have equal risk values. Both scenarios have a probability level of 'occasional' and the same combined overall consequence level of (3 'minor', 2 'major' and 2 'critical'). As can be seen in Table 7, the computed fuzzy risk values for these two scenarios are indeed equal. Same situation applies to scenarios 4 and 5. Their corresponding computed fuzzy risk values provided in

Table 7 are also equal. Table 9 provides a comparison between risk results for crew obtained using a qualitative risk matrix approach (Skramstad & Musaeus, 2000) and those using a fuzzy risk index measure. Scenarios are ranked from least severe to most severe with respect to risks to crew members. As can be seen, the same ranking is obtained using both methods for the six hazardous scenarios under consideration.

Hazard	Qualitative risk values	*Mamdani* fuzzy risk value	*Sugeno* fuzzy risk value
Release of bunker oil	Low	4.35	2.51
Leak on the cargo system	Medium	5.50	5.30
Release of liquid nitrogen	Medium	5.50	5.30
Fire in engine room	Medium	5.76	6.06
Accommodation fires	Medium	5.76	6.06
Fires on open deck	Medium	6.31	7.83

Table 9. Comparison between fuzzy risk results and qualitative risk values for crew

6. Conclusion

Various methodologies for the risk assessment of LNG transfer operations at the ship-shore interface of gas terminals were presented. These include a qualitative risk matrix approach, a multiple attribute utility model and a fuzzy inference system. The use of multiple attribute utility theory in risk assessment of LNG operations allows the ranking of risk alternatives based on a unified utility measure. A maximum risk alternative is selected to minimize the overall expected utility. This methodology allows modeling of the decision maker's attitude towards risk, i.e., risk aversion/neutral and/or risk taker. Available software tools allow ranking of risk alternatives and sensitivity analyses to be carried out to assess the sensitivities of the risk model's recommendations to various modeling variables.

An approach for the assessment of multiple attribute risk using fuzzy set theory was also presented. The developed methodology is an alternative to qualitative risk assessment matrices currently used in many industries and by ship classification societies. A three dimensional risk envelope or surface is generated and used for computation of risk values as replacement to the traditional risk matrix.The use of fuzzy sets and a fuzzy inference engine is suited for handling imprecision often associated with accident likelihood and consequence data. The total number of rules needed to construct the fuzzy inference engine is the product of the number of rows and the number of columns for the corresponding qualitative risk matrix. The proposed approach improves upon existing qualitative methods and allows the ranking of risk alternatives based on a unified measure. A fuzzy risk index was adopted for aggregation of multiple consequences into a unified measure. Both the Mamdani and Sugeno type inference methods were adopted. Results show that while the Mamdani method is intuitive and well suited to human input, the Sugeno method is computationally more efficient and guarantees continuity of the final risk output surface. It was also found that computed risk results using a fuzzy risk index measure are consistent with those obtained using a qualitative risk matrix approach. The use of a fuzzy inferene system provides more output information than the traditional risk matrix approach. Such approach is applicable to other ship operating modes such as transit in open sea and/or entering/leaving port

7. References

American Bureau of Shipping, ABS, (2000) "Guidance Notes: Risk Assessment for the Marine and Offshore Oil and Gas Industries", Houston, TX 77060.

American Gas Association, (1990) "Methods for Prioritizing Pipeline Maintenance and Rehabilitation", Battelle Corporation.

Bellman, R.A and Zadeh, L.A., (1970) "Decision making in a Fuzzy Environment", Management Sciences, Ser. B 17, pp. 141-164.

Chen, S. and Hwang, C. (1992), Fuzzy Multiple Attribute Decision Making Methods and Applications, Lecture Notes in Economics and Mathematical Systems, New York, Spinger-Verlag.

Clemen, R. (1997). Making hard decisions: an introduction to decision analysis (business statistics). 2nd ed. Belmont, CA: Duxbury Press.

Cooke, R. & Jager, E., (1996) "Methods for Assessing the Failure Frequency of Underground Gas Pipelines with Historical Data and Structured Expert Judgment", Report of the Faculty of Technical Mathematics and Informatics, 96-75, Delft University of Technology.

Elsayed, T. & Mansour, A.E., (2003), Reliability-Based Unfairness Tolerance Limits for Stiffened Plates, Journal of Ship Research.

Elsayed, T. et al,(2008), "Assessment of Operational Risks to Personnel on board Fixed Offshore Structures", Proc. IMechE Vol. 222 Part M: Journal of Engineering for the Maritime Environment.

Elsayed, T. et al, (2009) "Multi-attribute Risk Assessment of LNG Carriers during Loading/Offloading at Terminals", Ships and Offshore Structures, Volume 4, Issue 2.

Elsayed, T., (2009) "Fuzzy Inference System for the Risk Assessment of Liquefied Natural Gas Carriers during Loading/Offloading at Terminals", Applied Ocean Research, Volume 31, Issue 3.

Gyles, J., (1992) "LNG Terminal Safety", Tenth International Conference on Liquefied Natural Gas, (LNG 10). Kuala Lumpur, Malaysia.

HSE, (2002), "Marine Risk assessment", Offshore Technology Report 2001/063.

Kandel, A. (1992) Fuzzy Expert Systems, Boca Raton, FL: CRC Press.

Lindley, D.V. (1992). Making decisions. 2nd Ed. New York: John Wiley & Sons

Mathworks, Inc. (2006), Fuzzy Logic Toolbox; User s Guide, Natick, MA.

McGuire, J. & White, B. (1999) Liquefied Gas Handling Principles on Ships & Terminals, Witherby & Co Ltd; 3Rev Edition.

Oliver, R. & Marshall K. (1997). Decision-making and forecasting: with emphasis on model building and policy analysis. New York: McGraw-Hill.

Skramstad, E. & Musaeus, S., (2000) "Use of Risk Analysis for Emergency Planning of LNG Carriers", Gas Technology Conference, Houston, TX, USA.

Society of International Gas Tanker and Terminal Operators, SIGTTO, (1999) LNG Shipping Incident Report.

Treeage Software, Inc. (2006). Manual for decision analysis by Treeage, DATA. Williamstown, MA, USA.

Yager, R.R. (1988), "Ordered weighted averaging aggregation operators in multi-criteria decision making, IEEE Trans. On Systems, Man and Cybernetics, 18, pp. 183-190.

Zadeh, L.A. (1965) "Fuzzy Sets", Information and Control, Volume 8, pp. 338-353.

Zadeh, L.A. (1975) "The concept of a linguistic variable and its application to approximate reasoning-I", Information Sciences, Volume 8, pp. 199-249.

Zhou, Q., Purvis, M., & Kasabov, N. (1997), "A membership function selection method for fuzzy neural networks", Proceedings of the International Conference on Neural Information Processing and Intelligent Systems, Springer, Singapore, pp. 785-788.

Zimmerman, H.J. (1987), Fuzzy Sets, Decision Making and Expert Systems, Kluwer Academic Publishers.

Zimmerman, H.J. (1976), "Description and Optimization of Fuzzy Systems, Int. J. General Systems, Vol.2, pp. 209-215.

Reliability measures for liquefied natural gas receiving terminal based on the failure information of emergency shutdown system

Bi-Min Hsu[a], Ming-Hung Shu[b]*, and Min Tsao[c]
[a]Department of Industrial Engineering & Management
Cheng Shiu University, Kaohsiung County 833, Taiwan
[b]Department of Industrial Engineering & Management
National Kaohsiung University of Applied Sciences, Kaohsiung 807, Taiwan
[c]Department of Mathematics & Statistics
University of Victoria, British Columbia, Canada

Abstract
Natural Gas (NG), one of the cleanest, most efficient and useful of all energy sources, is a vital component of the world's energy supply. To make the NG more convenient for storage and transportation, it is refined and condensed into a liquid called liquefied natural gas (LNG). In an LNG site, safety is a long-team critical issue. The emergency shutdown system (ESS) in the LNG receiving terminal is used to automatically stop the pumps and isolate the leakage section. Fault tree analysis (FTA) has been widely used to characterize the logical functional relationships among components and subsystems of a system, and to identify the root causes of failures in a system. In the conventional FTA for the ESS, we assume that exact failure probabilities of events are available. However, in real applications the FTA for the ESS needs to be done at an early design or manufacturing stage at which certain new components may have to be used without failure data. Also, sometimes due to environmental changes in the ESS during the operation periods, it is difficult to gather past exact failures data for the FTA. Hence there may not be sufficient information for a conventional FTA. In this research, we propose an intuitionistic fuzzy (IF) sets theory based approach for the FTA which can be used when the conventional FTA cannot. We generate the IF fault-tree interval and the IF reliability interval for the ESS. Based on IF-FTA, we also present an algorithm for finding the critical components and determining weak paths in the ESS for which the key improvement event must be made.

Keywords: Liquefied natural gas, Emergency shutdown system, Fault-tree analysis, Intuitionistic fuzzy sets

*corresponding author with e-mail: workman@cc.kuas.edu.tw

1. Introduction

The natural gas (NG), one of the cleanest, most efficient and useful of all energy sources for residential and industrial customers, is a vital element of the world's energy supply. It is a combustible mixture of hydrocarbon gases and its composition can vary a great deal. Table 1 shows the main ingredients and their percentages; the primary ingredient is the methane (CH_4) but heavier gaseous hydrocarbons such as ethane (C_2H_6), propane (C_3H_8) and butane (C_4H_{10}) and trace gases are also present.

Component	Typical Weight %
Methane CH_4	70-90
Ethane C_2H_6	5-15
Propane (C_3H_8) and Butane (C_4H_{10})	< 5
CO_2, N_2, H_2S, etc.	Balance

Table 1. Typical composition of natural gas

To make the NG more convenient in further storage and transportation, it is refined to remove impurities such as water, hydrogen sulfide and other compounds which could cause problems for downstream conveyance or environmental pollution. After refining, the clean NG at nearly atmospheric pressure is condensed by cooling it to approximately -162 degrees Celsius into a liquid form, resulting in the liquefied natural gas (LNG). The LNG is about 1/600th the volume of that of the NG at standard temperature and pressure. It can be delivered by specially designed cryogenic vessels and cryogenic tankers over long distances. It is returned to the gas form through gasification at end-use facilities.

Generally, mass volumes of the LNG are conveyed and stored often in the proximity of densely populated area. Due to its highly flammable and explosive nature, accidents involving LNG can lead to loss of human lives and serious damages to industrial facilities and the natural environment. Because of these, high reliability and safety is a long-term crucial issue for the LNG industry. The reliability of a huge quantity of the LNG stockpiled in a conveying system (which mostly consists of pipes and storage tanks) is a major issue affecting the LNG receiving terminal safety. During the LNG processing process, even a small amount of the LNG leakage may cause considerable contamination, fire accidents or explosions. Consequently, to prevent leakage, an emergency shutdown system (ESS) in the LNG receiving terminal is implemented to automatically stop the LNG pumping and isolate the leakage condition.

For the reliability of equipments and operational procedures at the LNG receiving terminals, the failure information provided by the ESS is considered to be the most vital resources for the safety and thus deserves particular attention. A typical LNG plant devotes a substantial amount of manpower and capital towards the monitoring and investigation of failure events which trigger off the ESS in order to learn the underlying causes of these failure events. In order to understand the LNG receiving terminal reliability, an effective analysis and performance measure based on the failure information gathered by the ESS is required. The fault tree analysis (FTA) has been widely employed in variety of systems for providing logical functional relationships among components and subsystems of a system, and identifying root causes of the undesired system failures (9; 12).

In this research, we first describe the detailed LNG receiving procedure and then its FTA on the basis of the failure information from the ESS. For this description of the FTA, we assume that all the malfunction events provided by the ESS are fully understood; that is, exact data of their failure probability collected from normal operations of the LNG receiving terminal are available. We then present the traditional reliability measure of the FTA for the LNG receiving terminal based on the failure information of the ESS.

However, collecting precise failures data for the FTA requires substantial amount of time and knowledge of operations and maintenance on the LNG receiving terminal. In real operations, the following scenarios often occur:

- FTA for the ESS needs to be done at an early design or manufacturing stage at which certain new components may have to be used without prior failure data, and
- due to environmental changes in the ESS during the operation periods, it may be difficult to gather past exact failures data for the FTA.

Under these uncertain situations, traditionally system engineers usually omit ambiguous failure events of the ESS when they construct or analyze the fault tree. But such omitted events may actually be critical, and the measure of reliability of the LNG receiving terminal that does not take into consideration such events may be unreliable.

In order to handle inevitable imprecise failure information in diversified real applications, many research works have taken the uncertain situations into consideration. Chen (7) and Mon et al. (15; 16) carried out system reliability analysis by using the fuzzy set theory. Suresh et al. (17), Antonio et al. (1), Tanaka et al. (20), and Huang et al. (11) proposed the fuzzy FTA for certain systems applications. The concept of an intuitionistic fuzzy (IF) sets can be viewed as an alternative approach to define a fuzzy set in cases where available information is not sufficient for the definition of an imprecise concept by means of a conventional fuzzy set (2; 3). Bustince and Burillo (6) showed that the notion of vague sets coincides with that of IF sets; that is, fuzzy sets are IF sets, but the converse is not necessarily true (2; 3). IF sets theory has been widely applied in different areas such as logic programming (4; 5), decision making problems (13; 18; 19) in medical diagnosis (8), and pattern recognition (14).

In this research, with imprecise failure information from the ESS, we apply fuzzy fault tree (20) and Posbist fault tree (11) methods to construct fuzzy reliability measures for the LNG receiving terminal and provide the corresponding IF fault-tree interval and the IF reliability interval. We also compare the results of these proposed reliability measures for the FTA methods. Further, we will discuss identification of the most critical component of the LNG receiving terminal which is essential for determining weak paths and areas where the key improvements must be made.

2. LNG-ESS Fault Diagnosis

2.1 The Operation Process of the LNG Receiving Terminal

Most LNG is imported from exporters such as Indonesia, Malaysia and Qatar by long-term contract carriers. In this paper, we investigate an LNG receiving terminal located in Asia, Taiwan. When the LNG vessels arrive at the LNG terminal, the LNG they carry is discharged and stored at about -160^0C and $0.2kg/cm^2$ in storage tanks. Through an open rack vaporizer, the stored LNG is reheated and gasified into natural gas. The open rack vaporizer is connected to a storage and trunk-line distribution network through which the natural gas is transported to local distribution companies, independent power plants and households. A typical process diagram of the LNG receiving terminal is given in Figure 1 which shows the receiving, storage, vaporization and distribution components of a receiving terminal and how these components are connected.

Normally, the LNG must be kept cold in order to remain in liquid form. However, because of heat coming from the outside ambient atmosphere, there is inevitably a certain amount of boil-off gas (BOG). The BOG can be re-liquefied through a BOG compressor and a recondenser. The recondenser has an emergency isolation valve to keep the liquid lever from falling too low or raising too high to prevent the internal pressure from rising abnormally. It has two primary functions. First, it recycles BOG when the LNG is stored and transported through pipelines. Second, through secondary stage pumps which are submerged high-pressure centrifugal pumps, it provides buffer control to LNG which is flammable even at ultra-low temperatures. The secondary stage pumps are used to collect the LNG from the recondenser, and then pressurize and pump the LNG to the open rack vaporizer. The open rack vaporizer consists of finned tubes submerged in seawater. When the LNG flows through the tubes, heat exchange between the seawater outside of the tubes and the LNG inside takes place, and the LNG is re-gasifies and return to its original gaseous state. Before leaving the receiving terminal, the natural gas is measured for its quantities through a measure station. Other related systems such as the cold power generator (CPG), pressure power generator (PPG) and air separation plant (ASP) are set up for the purposes that achieve the goals of energy conservation and energy recycling.

In case of a LNG leakage, the emergency shutdown system (ESS) in the LNG receiving terminal can be automatically invoked to isolate the leakage pipe section in the unloaded dock district and the tank district and to stop the primary pumps.

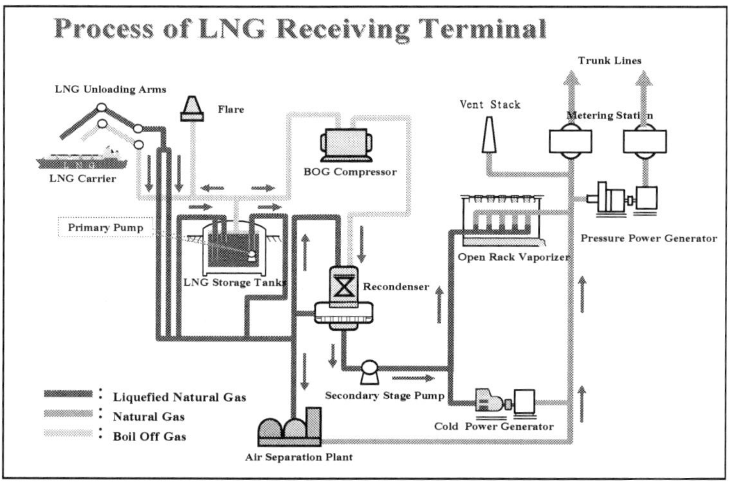

Fig. 1. The operation process of the LNG receiving terminal.

2.2 Fault-Tree Analysis of the ESS

Prior to the actual construction of the ESS fault tree, it is essential to have an in-depth understanding about related equipments involved in the ESS. Incidents related to the LNG facilities are generally classified into two classes, namely internal events and external events. The former include equipment failures, miss-operation and other incidents resulted from internal causes within a site. The latter include the device breakdown and the pipe leakage due to typhoon or earthquake. In this paper, we make the following assumptions which are necessary for the construction of the fault-tree analysis (FTA) of the ESS.

- Our primary concern is focused on internal events with the ESS.
- We consider only the isolating valve closest to the point of leakage; in other word, only the first level of isolating mechanism was taken into account.
- The entire isolation procedure is considered to have failed if the isolating device did not function correctly.
- All failures are independent events.

Based on the descriptions in Sections 2.1 and 2.2, the fault tree of the ESS is developed and shown in Figure 2, whose subevents and bottom events are listed in Tables 2 and 3.

Code	Fault
I	Emergency process isolation of the ESS fails
II	Primary pump shut-down of the ESS fails
A	Isolation valve of tank inlet fails to close
B	Isolation valve of tank outlet fails to close
C	Isolation valve of BOG pipe fails to close
D	Isolation valve of ICD (Initial Cooling Down) pipe fails to close
E	Circuit breaker of pump fails to open
F	Pump S/D control logic failure
G	Loss of pump stopping signal

Table 2. Descriptions of sub-events of the ESS fault

2.3 Traditional Reliability Measure of FTA

Traditionally, the reliability measure of the FTA of the "ESS Fault" can be obtained as follows:

$$\begin{aligned} \text{ESS Fault} &= \text{I} \cup \text{II} \\ &= (A \cup B \cup C \cup D) \cup (E \cup F \cup G) \\ &= (A_1 \cup A_2 \cup A_3 \cup A_4 \cup A_5 \cup A_6) \cup (B_1 \cup B_2 \cup B_3 \cup B_4 \cup B_5 \cup B_6) \cup \\ &\quad (C_1 \cup C_2 \cup C_3 \cup C_4 \cup C_5 \cup C_6) \cup [D_1 \cup (C_{21} \cup D_{22}] \cup \\ &\quad \{E \cup (F_1 \cup F_2 \cup F_3 \cup F_4) \cup [(G_{11} \cup G_{12}) \cap (G_{11} \cap G_{12}) \cap G_3]\}, \end{aligned} \tag{1}$$

where ∩ means relation of parallel "and" operation) and ∪ means series ("or" operation). Let f_i represent the crisp (precise) failure rate of event i. Then the crisp failure probability of the "ESS Fault", denoted by f_T, can be computed as follows

$$\begin{aligned} f_T &= 1 - [(1-f_{A_1})(1-f_{A_2})(1-f_{A_3})(1-f_{A_4})(1-f_{A_5})(1-f_{A_6})] \\ &\quad [(1-f_{B_1})(1-f_{B_2})(1-f_{B_3})(1-f_{B_4})(1-f_{B_5})(1-f_{B_6})] \\ &\quad [(1-f_{C_1})(1-f_{C_2})(1-f_{C_3})(1-f_{C_4})(1-f_{C_5})(1-f_{C_6})] \\ &\quad [(1-f_{D_1})(1-f_{D_{21}})(1-f_{D_{22}})][(1-f_E)] \\ &\quad [(1-f_{F_1})(1-f_{F_2})(1-f_{F_3})(1-f_{F_4})] \\ &\quad \{[1-(1-f_{G_{11}})(1-f_{G_{12}})](f_{G_{21}} f_{G_{22}} f_{G_3})\}. \end{aligned} \tag{2}$$

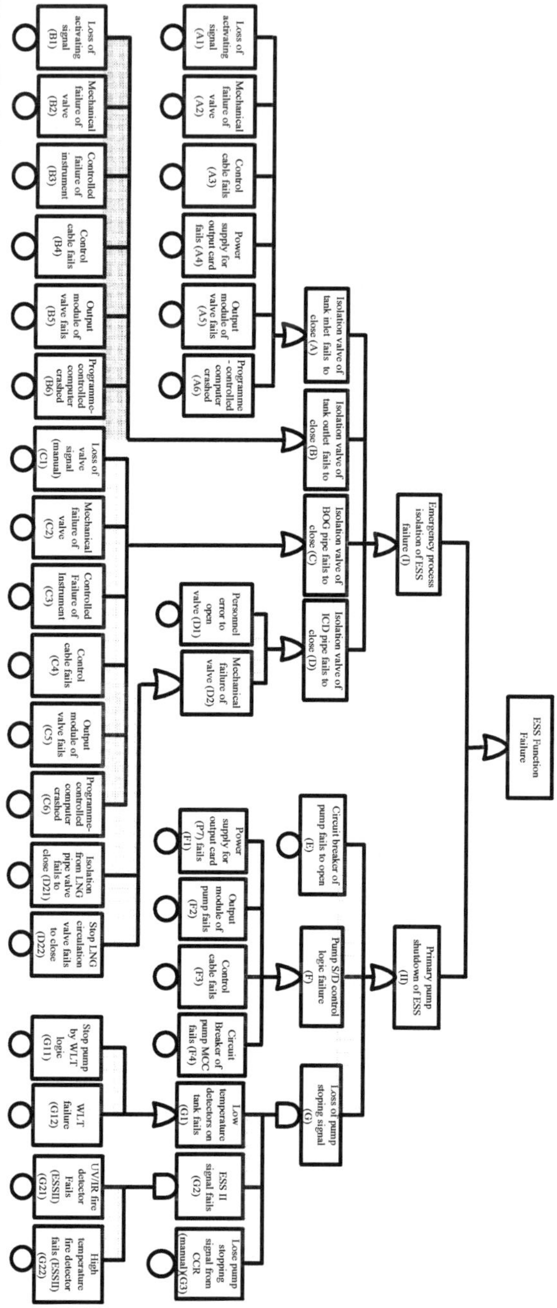

Fig. 2. The fault tree of the ESS.

Code	Fault	Code	Fault
A_1	Loss of activating signal	C_5	Output module of valve fails
A_2	Mechanical failure of valve	C_6	Program-controlled computer crashed
A_3	Control cable fails	D_1	Personnel error to open valve
A_4	Power supply for output card fails	D_{21}	Isolation from shore LNG pipe valve fails to close
A_5	Output module of valve fails	D_{22}	Stop LNG circulation valve fails to close
A_6	Program-controlled computer crashed	E	Circuit breaker of pump fails to open
B_1	Loss of activating signal	F_1	Power supply for output card fails
B_2	Mechanical failure of valve	F_2	Output module of pump fails
B_3	Controlled failure of instrument	F_3	Control cable fails
B_4	Control cable fails	F_4	Circuit breaker of pump MCC fails
B_5	Output module of valve fails	G_{11}	Stop pump by WLT logic
B_6	Program-controlled computer crashed	G_{12}	WLT fails
C_1	Loss of activating signal(manual)	G_{21}	UV/IR fire detector fails (ESSII)
C_2	Mechanical failure of valve	G_{22}	High temperature fire detector fails (ESSII)
C_3	Controlled failure of instrument	G_3	Lose pump stopping signal from CCR (manual)
C_4	Control cable fails		

Table 3. Descriptions of the bottom events of the ESS fault

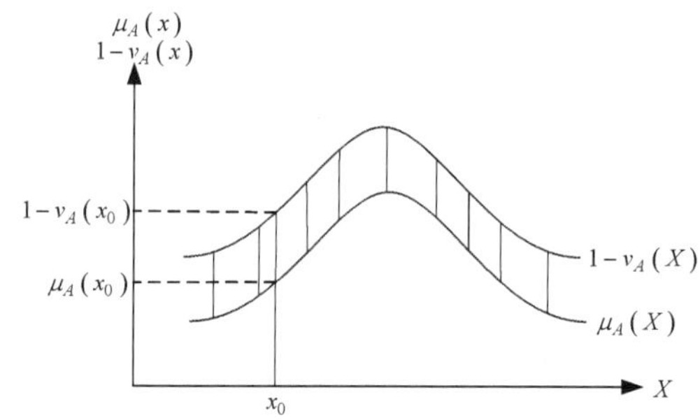

Fig. 3. IF set of a real number R.

3. Intuitionistic Fuzzy Reliability Measure of FTA

In the conventional FTA for the ESS of the LNG terminal, we must fully understand the ESS. Usually, we assume that exact failure probabilities of events are available. However, collecting failures data for the FTA is a challenging task requiring extensive human expertise and knowledge of operations and maintenance on the system. In real operations, this may not even be possible as the FTA for the ESS of the LNG receiving terminal needs to be made at an early design or manufacturing stage at which we have no failure data on new components. Furthermore, sometimes the environmental change in the system during the operation periods can also make it more difficult to gather past exact failures data for the FTA. In such uncertain situations, traditionally system engineers usually omit some ambiguous failure events of the ESS when measuring the reliability of the LNG receiving terminal. But the missing events or probability information might be critical and thus omitting these may lead to unreliable decision results. In order to handle inevitable imprecise failure information of the ESS, which has been recognized as one of the uncertainties in the real world, a possible solution is to use intuitionistic fuzzy (IF) sets, defined by Atanassov (2; 3).

3.1 IF-FTA on the ESS

Definition 3.1. Let a set U be fixed. An intuitionistic fuzzy (IF) set $\tilde{a}$ of U is an object having the form, $\tilde{a} = \{x, u(x), v(x) | x \in U\}$ where the function $u_{\tilde{a}} : U \to [0,1]$ and $v_{\tilde{a}} : U \to [0,1]$ measure the degree of membership and the degree of non-membership, respectively, of an $x \in U$ as a potential member of set $\tilde{a} \subset U$, and $0 \leq u(x) + v(x) \leq 1$ for $x \in U$.

Clearly, the IF set uses a degree of truth membership function $\mu_{\tilde{a}}(x)$ and a degree of falsity membership function $v_{\tilde{a}}(x)$ to represent lower bound $\mu_{\tilde{a}}(x)$ and upper bound $1 - v_{\tilde{a}}(x)$ such that $\mu_{\tilde{a}}(x) + v_{\tilde{a}}(x) \leq 1$. By complementing the membership degree with a non-membership degree that expresses to what extent the element does not belong to the IF set, the interval $[\mu_{\tilde{a}}(x), 1 - v_{\tilde{a}}(x)]$ can extend the fuzzy set of membership function. The uncertainty or hesitation can be quantified for each x in $\tilde{a}$ by the length of the interval $\pi_{\tilde{a}}(x) = 1 - v_{\tilde{a}}(x) - \mu_{\tilde{a}}(x)$. A small $\pi_{\tilde{a}}(x)$ represents that we are more decisive about x, and a large $\pi_{\tilde{a}}(x)$ represents that we are more uncertain about x. Obviously, when $\mu_{\tilde{a}}(x) = 1 - v_{\tilde{a}}(x)$ for all elements of the universe, the traditional fuzzy set concept is recovered. As an example, Figure 3 shows an IF set of a real number R.

Note that when $a_1 = a_1'$, $c_1 = c_1'$ and $a_2 = a_2'$, $c_2 = c_2'$, the IF set is changed from Figure 4 to Figure 5, and its four arithmetic operations become much more easy.

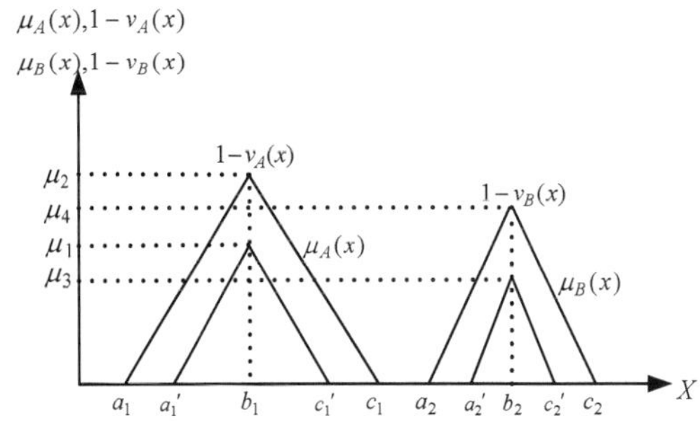

Fig. 4. A triangle IF set.

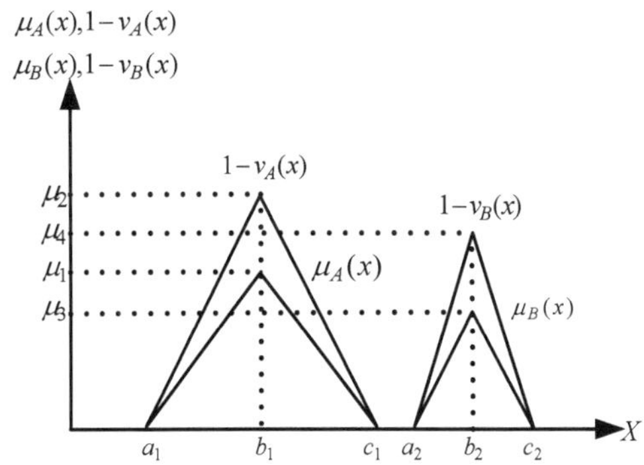

Fig. 5. A triangle IF set.

Based on definition of a triangle IF set shown in Figure 4, we propose failure possibility operations for the FTA on the ESS as follows. Let $\tilde{f}_A$ and $\tilde{f}_B$ be failure possibilities of two triangular IF sets, truly, $\tilde{f}_A > 0$ and $\tilde{f}_B > 0$:

$$\begin{aligned} \tilde{f}_A &= \{(a_1', b_1, c_1'); \mu_A, (a_1, b_1, c_1); 1 - v_A\}, \\ \tilde{f}_B &= \{(a_2', b_1, c_2'); \mu_A, (a_2, b_2, c_2); 1 - v_B\}. \end{aligned}$$

Let $\oplus$, $\ominus$ and $\otimes$ be binary operations between two IF sets $\tilde{f}_A$ and $\tilde{f}_B$ corresponding to the operations $\circ = +, -,$ and $\times$, respectively. Then we have the following useful results of operations on the IF set (2).

Proposition 3.1. Let $\tilde{f}_A$ and $\tilde{f}_B$ be two triangular IF set numbers. Then $\tilde{f}_A \oplus \tilde{f}_B$, $\tilde{f}_A \ominus \tilde{f}_B$ $\tilde{f}_A \otimes_{\text{product}} \tilde{f}_B$, and $\tilde{f}_A \otimes_{\min} \tilde{f}_B$ are also triangular IF set numbers. They have the following operations.

$$
\begin{aligned}
\tilde{f}_A \oplus \tilde{f}_B =& \{(a_1' + a_2', b_1 + b_2, c_1' + c_2'); \min(\mu_A, \mu_B), \\
& (a_1 + a_2, b_1 + b_2, c_1 + c_2); \min(1 - v_A, 1 - v_B)\} \\
\tilde{f}_A \ominus \tilde{f}_B =& \{(a_1' - c_2', b_1 - b_2, c_1' + a_2'); \min(\mu_A, \mu_B), \\
& (a_1 - c_2, b_1 - b_2, c_1 - a_2); \min(1 - v_A, 1 - v_B)\} \\
\tilde{f}_A \otimes_{\text{product}} \tilde{f}_B =& \{(a_1'a_2', b_1b_2, c_1'c_2'); \min(\mu_A, \mu_B), \\
& (a_1a_2, b_1b_2, c_1c_2); \min(1 - v_A, 1 - v_B)\} \\
\tilde{f}_A \otimes_{\min} \tilde{f}_B =& \{(\min(a_1', a_2'), \min(b_1, b_2), \min(c_1', c_2')); \min(\mu_A, \mu_B), \\
& (\min(a_1', a_2'), \min(b_1, b_2), \min(c_1', c_2')); \min(1 - v_A, 1 - v_B)\}
\end{aligned}
$$

$\tilde{a}$ is a crisp number with value m if its membership function is defined by

$$
u_{\tilde{a}}(x) = \begin{cases} 1 & \text{if } x = m \\ 0 & \text{if } x \neq m, \end{cases}
$$

which is also denoted by $\tilde{1}_{\{m\}}$.

According to equation (2), the IF set failure possibility of the "ESS Fault", denoted by $\tilde{f}_T$, can be computed by

$$
\begin{aligned}
\tilde{f}_T =& \tilde{1}_{\{m\}} \ominus [(\tilde{1}_{\{m\}} \ominus \tilde{f}_{A_1}) \otimes (\tilde{1}_{\{m\}} \ominus \tilde{f}_{A_2}) \otimes (\tilde{1}_{\{m\}} \ominus \tilde{f}_{A_3}) \otimes (\tilde{1}_{\{m\}} \ominus \tilde{f}_{A_4}) \otimes \\
& (\tilde{1}_{\{m\}} \ominus \tilde{f}_{A_5}) \otimes (\tilde{1}_{\{m\}} \ominus \tilde{f}_{A_6})] \otimes \\
& [(\tilde{1}_{\{m\}} \ominus \tilde{f}_{B_1}) \otimes (\tilde{1}_{\{m\}} \ominus \tilde{f}_{B_2}) \otimes (\tilde{1}_{\{m\}} \ominus \tilde{f}_{B_3}) \otimes (\tilde{1}_{\{m\}} \ominus \tilde{f}_{B_4}) \otimes (\tilde{1}_{\{m\}} \ominus \tilde{f}_{B_5}) \otimes \\
& (\tilde{1}_{\{m\}} \ominus \tilde{f}_{B_6})] \otimes \\
& [(\tilde{1}_{\{m\}} \ominus \tilde{f}_{C_1}) \otimes (\tilde{1}_{\{m\}} \ominus \tilde{f}_{C_2}) \otimes (\tilde{1}_{\{m\}} \ominus \tilde{f}_{C_3}) \otimes (\tilde{1}_{\{m\}} \ominus \tilde{f}_{C_4}) \otimes (\tilde{1}_{\{m\}} \ominus \tilde{f}_{C_5}) \otimes \\
& (\tilde{1}_{\{m\}} \ominus \tilde{f}_{C_6})] \otimes \\
& [(\tilde{1}_{\{m\}} \ominus \tilde{f}_{D_1}) \otimes (\tilde{1}_{\{m\}} \ominus \tilde{f}_{D_{21}}) \otimes (\tilde{1}_{\{m\}} \ominus \tilde{f}_{D_{22}})] \times [(\tilde{1}_{\{m\}} \ominus \tilde{f}_E)] \otimes \\
& [(\tilde{1}_{\{m\}} \ominus \tilde{f}_{F_1}) \otimes (\tilde{1}_{\{m\}} \ominus \tilde{f}_{F_2}) \otimes (\tilde{1}_{\{m\}} \ominus \tilde{f}_{F_3}) \otimes (\tilde{1}_{\{m\}} \ominus \tilde{f}_{F_4})] \otimes \\
& \{[\tilde{1}_{\{m\}} \ominus (\tilde{1}_{\{m\}} \ominus \tilde{f}_{G_{11}}) \otimes (\tilde{1}_{\{m\}} \ominus \tilde{f}_{G_{12}})](\tilde{f}_{G_{21}} \otimes \tilde{f}_{G_{22}} \otimes \tilde{f}_{G_3})\}.
\end{aligned} \tag{3}
$$

It should be noted that $\tilde{f}_A \otimes \tilde{f}_B$ is represented by either $\tilde{f}_A \otimes_{\text{product}} \tilde{f}_B$ or $\tilde{f}_A \otimes_{\min} \tilde{f}_B$, whose operations are described in Proposition 3.1. The collected data of IF failure interval are listed in Table 4, which is based on the representation of the triangle IF set. The IF reliability interval for the ESS results are

$$
\tilde{f}_{\text{ESS Fault}} = \tilde{f}_{T\text{product}} = \{(0.0619, 0.0746, 0.0816); 0.6, (0.0440, 0.0746, 0.0966); 0.7\} \tag{4}
$$

$$
\tilde{f}_{\text{ESS Fault}} = \tilde{f}_{T\min} = \{(0.0650, 0.0772, 0.0836); 0.6, (0.0478, 0.0772, 0.0980); 0.7\} \tag{5}
$$

3.2 The Critical Components on the ESS

In order to find the critical components in the system based on IF-FTA and determine weak paths in the ESS where key improvement event must be made, we expand Tanaka et al's (20) fuzzy-FTA definition and redefine the influence degree of every bottom event through implementing four arithmetic operations of the triangle IF set as shown in Proposition 3.1.

Definition 3.2. Denote by $\tilde{f}_{T_i}$ the computation result that the ith bottom event of failure interval (delete the ith bottom event) is not included in the $\tilde{f}_T$ shown in equation (3), and denote by $\mathcal{V}(\tilde{f}_T, \tilde{f}_{T_i})$ the difference between $\tilde{f}_T$ and $\tilde{f}_{T_i}$; that is,

$$\mathcal{V}(\tilde{f}_T, \tilde{f}_{T_i}) = (a'_T - a'_{T_i}) + (a_T - a_{T_i}) + (b_T - b_{T_i}) + (c'_T - c'_{T_i}) + (c_T - c_{T_i}). \quad (6)$$

A larger value of $\mathcal{V}(\tilde{f}_T, \tilde{f}_{T_i})$ represents the ith bottom event has a greater influence on $\tilde{f}_T$.

Therefore, according to Definition 3.2, we can calculate $\mathcal{V}(\tilde{f}_T, \tilde{f}_{T_i})$ for $i = A, B, \cdots, G$, the IF failure difference between overall and partial (with second level nodes deleted) fault-tree, for obtaining the most critical system event of the "ESS Fault". Table 5 shows the ranks of such differences. Based on these results, the failure of BOG (Boil Off Gas) pipes and isolation valve of BOG pipe failing to close (event "C") and ICD pipes and isolation valve of ICD pipe failing to close (event "D") are the first and second significant events leading to ESD failure. Because of this, the components involved in these events require particular attention in daily maintenance. From the well known 80/20 rule, we can effectively reduce 80% of risk if we can have 20% of critical equipments under our control. Daily monitoring of such critical components will help to significantly reduce the change of failure.

Finally, for ease of implementation in real applications, we provide a step-by-step procedure of the IF-FTA on the ESS as follows:

Step 1. Construct fault-tree logic diagram, fault-tree logical symbols such as "AND" gate and "OR" gate, for all the faults under the top level event shown in Figure 2. Use these to represent the sequence of faults and causes and trace back whole process from top to bottom events.

Step 2. Obtain the possible failure intervals of bottom events shown in Table 4 based on the aggregation of the ESS information and expert's knowledge and experience.

Step 3. calculate the "ESS Fault" reliability result by using equation (3).

Step 4. Find the influential bottom events of the system reliability by using equation (6).

Step 5. Discuss the results and make suggestions.

4. Reliability Measures Methods for FTA

In this section, we briefly review existing reliability measures for the FTA within reliability theory and compare the results of the existing approaches and our proposed methods.

Traditionally, probability method is the method for dealing with the heterogeneous problems, and probability can only show the randomness of success or failure events. The usage of this method depends on the availability of a large amount of sample data and complete knowledge of all event outcomes. We calculated the failure possibility of the top event "ESS Fault" based on equation (2) using the crisp failure probabilities, b_i, in Table 4 and obtained $f_T = 7.4631 \times 10^{-2}$.

Posbist reliability theory, developed by Cai (10), is one of the forms of fuzzy reliability theories that have been proposed. It uses the possibility assumption and the fuzzy state assumption in place of the probability assumption or the binary state assumption. For systems with extremely small failure probabilities or when necessary statistical data are scarce, the posbist reliability theory has certain advantages for evaluating system reliability and safety (11). Based on the fault tree of failure of the ESS shown in Figure 3 and the corresponding failure possibility of every basic event listed in Table 4, we have identified the failure possibility of the top event by using the Posbist reliability theory and this value is $f_T = 3.95 \times 10^{-2}$.
The results of the existing approaches and our proposed methods are shown in Figure 6 and Table 6. For comparisons, we find that the failure possibilities of Posbist under different α-level are the same with crisp failure possibility of event "C_4". This is because the Posbist select the maximal failure possibility of bottom events. In this case, this simple assessment could possibly overlook the potential risk on the ESS reliability and safety. The results of the proposed "and-by-product" or "and-by-min" method could cover both the results of the crisp failure possibility and the fuzzy FTA approach (20). The differences between proposed two "and" operations are that "and-by-product" gives more consistent results because the center of its estimated result is the same with the crisp failure possibility and the fuzzy FTA approach. Also, "and-by-min" can give the more safety results because it selects the least possibility in each event connected by the "and" gate without underestimating the failure interval.

5. Conclusions

Traditionally, we assume that exact probabilities of failure events are available and fully understood. However, in real applications for various reasons it is often difficult to obtain the past exact failures data. In this paper, to handle uncertain situations and inevitable imprecise information occurring in the liquefied natural gas (LNG) terminal emergency shut-down system (ESS), we propose a new approach which integrates intuitionistic fuzzy (IF) set operations on fault-tree analysis (FTA) to compute the IF fault-tree interval, traditional reliability and IF reliability interval based on the failure information gathered by the ESS. Moreover, based on IF-FTA, we present an algorithm to find the critical components and determine weak paths in the ESS where key improvement events must be made. The failures of BOG (Boil Off Gas) pipes, isolation valve of BOG pipe failing to close (event "C") and ICD pipes, isolation valve of ICD pipe failing to close (event "D"), are the first and second significant events leading to ESD failure. As such, particular attention must be paid to the related components in the daily maintenance to effectively reduce risk; 80% of the risk can be removed when 20% of critical equipments are under our control. A step-by-step procedure of the IF-FTA on the ESS is also presented for easy implementation in real applications. Finally, the result of this proposed methodology is briefly compared with the existing FTA approaches.

Bottom Event	a_i	a'_i	b_i	c'_i	c_i	u_i	$1-v_i$
A_1	2.26E-05	2.59E-05	3.37E-05	3.67E-05	4.11E-05	0.60	0.70
A_2	4.86E-04	6.23E-04	7.16E-04	7.88E-04	9.45E-04	0.70	0.90
A_3	3.58E-05	4.36E-05	5.52E-05	6.29E-05	6.99E-05	0.80	0.90
A_4	2.43E-05	2.76E-05	3.10E-05	3.35E-05	3.62E-05	0.90	0.95
A_5	1.97E-05	2.85E-05	3.10E-05	3.63E-05	4.28E-05	0.65	0.85
A_6	1.71E-05	2.01E-05	2.28E-05	2.74E-05	2.98E-05	0.80	0.85
B_1	1.78E-05	2.34E-05	3.39E-05	3.90E-05	4.37E-05	0.85	0.90
B_2	4.11E-04	6.23E-04	7.16E-04	8.09E-04	9.87E-04	0.70	0.90
B_3	1.17E-03	1.41E-03	1.70E-03	1.85E-03	2.37E-03	0.60	0.75
B_4	3.27E-05	4.36E-05	5.52E-05	6.29E-05	7.80E-05	0.75	0.85
B_5	1.88E-05	2.73E-05	3.10E-05	3.63E-05	4.79E-05	0.80	0.90
B_6	1.74E-05	1.92E-05	2.28E-05	2.76E-05	3.70E-05	0.85	0.90
C_1	2.00E-03	2.28E-03	2.68E-03	3.11E-03	3.79E-03	0.70	0.80
C_2	4.41E-04	5.44E-04	7.16E-04	8.02E-04	9.86E-04	0.80	0.95
C_3	1.12E-03	1.51E-03	1.70E-03	1.84E-03	2.39E-03	0.60	0.75
C_4	2.22E-02	3.12E-02	3.95E-02	4.31E-02	5.30E-02	0.75	0.90
C_5	1.24E-02	1.94E-02	2.23E-02	2.36E-02	2.58E-02	0.60	0.80
C_6	1.30E-05	1.92E-05	2.28E-05	2.67E-05	3.15E-05	0.65	0.90
D_1	7.70E-05	8.55E-05	1.14E-05	1.38E-04	1.53E-04	0.90	1.00
D_{21}	1.98E-03	2.39E-03	2.57E-03	3.16E-03	3.95E-03	0.80	1.00
D_{22}	1.60E-03	2.10E-03	2.53E-03	3.14E-03	3.76E-03	0.90	0.95
E	2.33E-04	3.38E-04	3.98E-04	4.86E-04	5.68E-04	0.85	1.00
F_1	1.62E-05	2.45E-05	3.98E-04	3.57E-05	3.89E-05	0.75	0.90
F_2	2.10E-05	2.39E-05	3.10E-05	3.35E-05	3.95E-05	0.90	1.00
F_3	4.32E-05	5.02E-05	5.52E-05	6.57E-05	8.21E-05	0.70	0.85
F_4	2.76E-04	3.54E-04	3.98E-04	4.86E-04	5.63E-04	0.65	0.80
G_{11}	1.85E-02	2.13E-02	3.09E-02	3.83E-02	4.52E-02	0.80	0.90
G_{12}	8.42E-04	9.90E-04	1.32E-03	1.53E-03	1.67E-03	0.70	0.90
G_{21}	2.54E-02	3.02E-02	3.64E-02	4.33E-02	5.24E-02	0.85	0.90
G_{22}	1.46E-03	2.24E-03	2.73E-03	3.25E-03	4.00E-03	0.85	1.00
G_3	1.00E-02	1.28E-02	1.44E-02	1.74E-02	1.90E-02	0.80	0.90

Table 4. The possible ranges of bottom events of ESS failures

	"and" operation type	
	and-by-product	and-by-min
$V(\tilde{f}_T,\tilde{f}_{T_A})$	**4.0824E-03(4)**	**4.0720E-03(4)**
$V(\tilde{f}_T,\tilde{f}_{T_B})$	**1.1820e-02(3)**	**3.2951E-03(5)**
$V(\tilde{f}_T,\tilde{f}_{T_C})$	**3.0936E-01(1)**	**3.0858E-01(1)**
$V(\tilde{f}_T,\tilde{f}_{T_D})$	**2.5745E-02(2)**	**2.5680E-02(2)**
$V(\tilde{f}_T,\tilde{f}_{T_E})$	**1.8688E-03(6)**	**1.8641E-03(7)**
$V(\tilde{f}_T,\tilde{f}_{T_F})$	**2.4672E-03(5)**	**2.4609E-03(6)**
$V(\tilde{f}_T,\tilde{f}_{T_G})$	**2.1484E-07(7)**	**1.2764E-02(3)**

Table 5. The failure difference between deleting any fault-tree node in evel 2

α-level	Fuzzy fault tree			Crisp	Posbist	Proposed intuitionstic fuzzy fault tree analysis									
	Left point	Middle point	Right end point			and-by-product					and-by-min				
						a	a'	b	c'	c	a	a'	b	c'	c
1.0	7.46E-02	7.46E-02	7.46E-02	7.46E-02	3.95E-02										
0.9	7.34E-02	7.46E-02	7.53E-02	7.46E-02	3.95E-02										
0.8	7.21E-02	7.46E-02	7.60E-02	7.46E-02	3.95E-02										
0.7	7.08E-02	7.46E-02	7.67E-02	7.46E-02	3.95E-02	7.46E-02		7.46E-02		7.46E-02	7.72E-02		7.72E-02		7.72E-02
0.6	6.96E-02	7.46E-02	7.74E-02	7.46E-02	3.95E-02	7.03E-02	7.46E-02	7.46E-02	7.46E-02	7.78E-02	7.30E-02	7.72E-02	7.72E-02	7.72E-02	8.01E-02
0.5	6.83E-02	7.46E-02	7.81E-02	7.46E-02	3.95E-02	6.59E-02	7.25E-02	7.46E-02	7.58E-02	8.09E-02	6.88E-02	7.51E-02	7.72E-02	7.82E-02	8.31E-02
0.4	6.70E-02	7.46E-02	7.88E-02	7.46E-02	3.95E-02	6.15E-02	7.04E-02	7.46E-02	7.69E-02	8.41E-02	6.46E-02	7.31E-02	7.72E-02	7.93E-02	8.61E-02
0.3	6.57E-02	7.46E-02	7.95E-02	7.46E-02	3.95E-02	5.71E-02	6.83E-02	7.46E-02	7.81E-02	8.72E-02	6.04E-02	7.11E-02	7.72E-02	8.04E-02	8.90E-02
0.2	6.45E-02	7.46E-02	8.02E-02	7.46E-02	3.95E-02	5.28E-02	6.62E-02	7.46E-02	7.93E-02	9.04E-02	5.62E-02	6.90E-02	7.72E-02	8.15E-02	9.20E-02
0.1	6.32E-02	7.46E-02	8.09E-02	7.46E-02	3.95E-02	4.84E-02	6.41E-02	7.46E-02	8.04E-02	9.35E-02	5.20E-02	6.70E-02	7.72E-02	8.25E-02	9.50E-02
0.0	6.19E-02	7.46E-02	8.16E-02	7.46E-02	3.95E-02	4.40E-02	6.19E-02	7.46E-02	8.16E-02	9.66E-02	4.78E-02	6.50E-02	7.72E-02	8.36E-02	9.80E-02

Table 6. Comparisions with other fault analysis methods

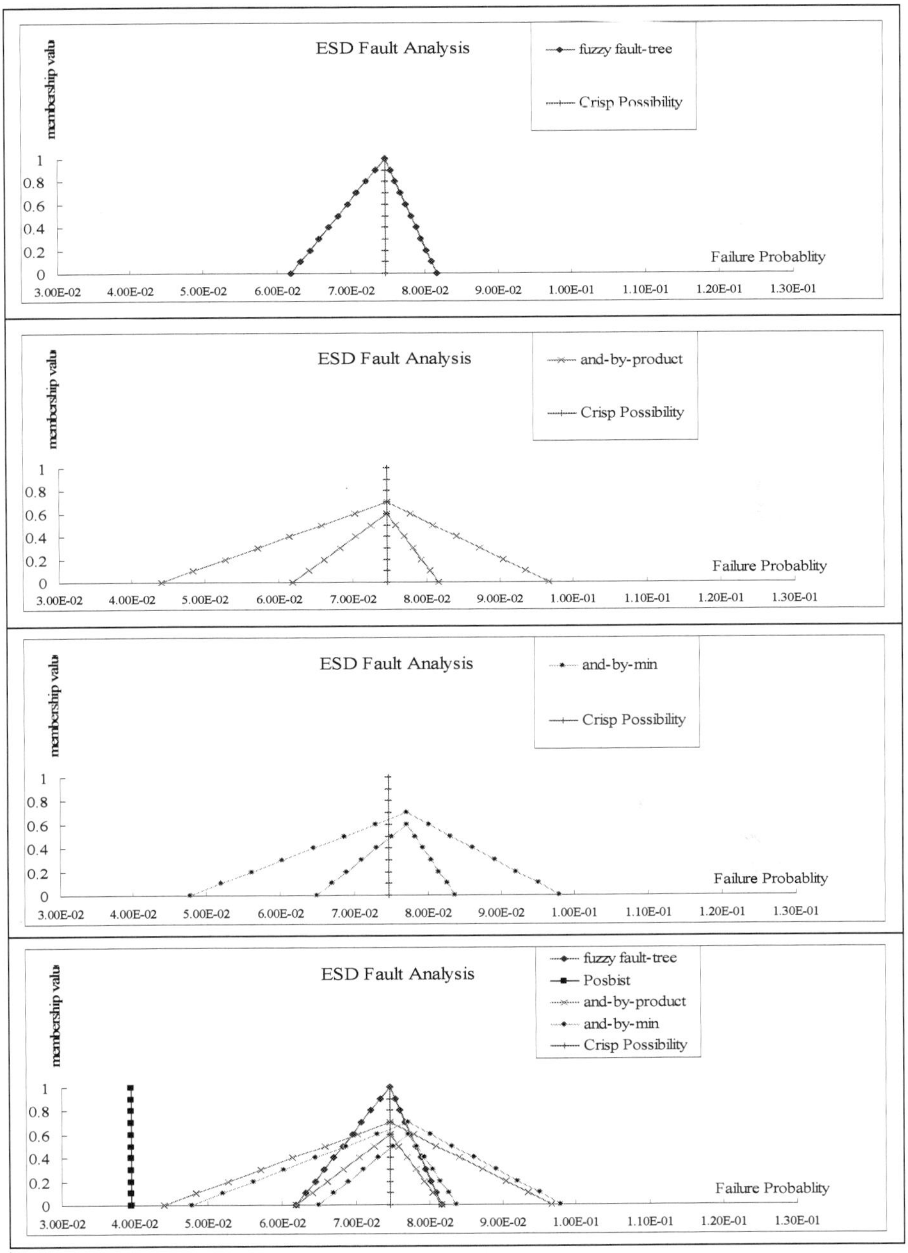

Fig. 6. Membership function for top event of ESS fault.

6. References

[1] Antonio, C.F.G. and Nelson, F.F.E., 1999. Fuzzy FTA: a fuzzy fault tree system for uncertainty analysis. *Annals of Nuclear Energy*, 26, 523-532.

[2] Atanassov, K.T. *Intuitionistic Fuzzy Sets*. Central Tech. Library, Bulgarian Academy Science, Sofia, Bulgaria, Rep. No. 1697/84, 1983.

[3] Atanassov, K.T., 1986. Intuitionistic fuzzy sets. *Fuzzy Sets and Systems*, 20, 87-96.

[4] Atanassov, K. and Gargov, G., 1990. Intuitionistic fuzzy logic. *C.R. Acadamy Bulgare Science*, 43(3), 9-12.

[5] Atanassov, K. and Georgeiv, C., 1993. Intuitionistic fuzzy prolog. *Fuzzy Sets and Systems*, 53, 121-128.

[6] Bustince, H. and Burillo, P., 1996. Vague sets are intuitionistic fuzzy sets. *Fuzzy Sets and Systems*, 79, 403-405.

[7] Chen, S.M., 1994. Fuzzy system reliability analysis using fuzzy number arithmetic operations. *Fuzzy Sets and systems*, 64, 31-38.

[8] De, S.K., Biswas, R., Roy, A.R., 2001. An application of intuitionistic fuzzy sets in medical diagnosis, *Fuzzy Sets and Systems*, 117, 209-213.

[9] Ejlali, A. and Miremadi, S.G., 2004. FPGA-based Monte Carlo Simulation for Fault Tree Analysis. *Microelectronics Reliability*, 44(6), 1017-1028.

[10] Cai, K.Y., Wen, C.Y., Zhang, M.L., 1991. Fuzzy variables as a basis for a theory of fuzzy reliability in the possibility context. *Fuzzy Sets and Systems*, 42, 145-72.

[11] Huang, H.Z., Tong, X. and Zuo, M.J., 2004. Posbist fault tree analysis of coherent systems. *Reliability Engineering and system safety*, 84, 141-148.

[12] Kales P., *Reliability for Technology, Engineering, and Management*, Prentice-Hall, 1998.

[13] Li, D.F., 2005. Multiattribute decision making models and methods using intuitionistic fuzzy sets. *Journal of Computer and System Sciences*, 70, 73-85.

[14] Li, D.F. and Cheng, C., 2002. Newsimilarity measures of intuitionistic fuzzy sets and application to pattern recognitions. *Pattern Recognition Letter*, 23(1-3), 221-225.

[15] Mon, D.L. and Cheng, C.H., 1993. Fuzzy system reliability analysis by interval of confidence. *Fuzzy Sets and Systems*, 56, 29-35.

[16] Mon, D.L. and Cheng, C.H., 1994. Fuzzy system reliability analysis for components with different membership functions. *Fuzzy Sets and Systems*, 64, 145-157.

[17] Suresh, P.V. and Babar, A.K., Raj, V.V., 1996. Uncertainty in fault tree analysis: A fuzzy approach. *Fuzzy Sets and Systems*, 83, 135-141.

[18] Szmidt, E. and Kacprzyk, J., 1996. Intuitionistic fuzzy sets in group decision making. *NIFS*, 2(1), 11-14.

[19] Szmidt, E. and Kacprzyk, J., 1996. Remarks on some applications of intuitionistic fuzzy sets in decision making. *NIFS*, 2(3), 22-31.

[20] Tanaka, H., Fan, L.T., Lai, F.S., Toguchi, K., 1983. Fault-tree analysis by fuzzy probability. *IEEE Transactions on Reliability*, 32, 150-163.